Preface

The book is primarily a collection of papers, many unpublished, solicited from U. S. Universities or Institutions with a history of programs in Aerospace/Aeronautical engineering. The response from the Institutions was almost unanimous. This collection of papers represents an authoritative story of the development of educational programs in the nation that were devoted to human flight. Most of these programs are still in existence but a few papers cover the history of programs that are no longer in operation. This collection, undertaken in 2003, is only one of many activities throughout the year that will commemorate the 100th anniversary of the Wright brothers' accomplishment. On December 17, 1903 at Kitty Hawk, NC. The brothers, Wilbur and Orville, succeeded in achieving controlled flight in a heavier-than-air machine. This feat was accomplished by them only after meticulous experiments and a study of the work of others before them like Sir George Cayley, Otto Lilienthal, and Samuel Langley.

The first evidence of the academic community becoming interested in human flight is found in 1883 when Professor J. J. Montgomery of Santa Clara College conducted a series of glider tests. Seven years later, in 1890, Octave Chanute presented a number of lectures to students of Sibley College, Cornell University titled "Aerial Navigation." A comprehensive treatment of these early beginnings of aerospace education are documented in Part I, Chapter 1 up until the early 1940s. At that time a rapid expansion of educational programs relating to aeronautical engineering began. These later developments are covered collectively in Chapter 2. Part II, Chapters 3 through 6, are devoted to those four schools that were approximately four or five years ahead of everyone else in establishing formal programs. Part III, Chapter 7 describes the activities of the Guggenheim Foundation that spurred much of the development of programs in aeronautical engineering. In Part IV the colleges and universities are arranged approximately in the chronological order in which the programs were formally established in the mid-1930s to the present. However, the last two parts contain schools with specific characteristics in their development. The military institutions are grouped together in Part V, while Part VI presents the histories of those programs that evolved from proprietary institutions. In general each chapter in this book is a paper or article submitted by current or emeritus faculty from the many universities and colleges.

Table of Contents

Part V
Military School Programs in Aerospace Engineering

Part VI
Proprietary School Programs in Aerospace Engineering

Part I

A Comprehensive Look at the History of Education
in Aerospace Engineering

Chapter 1

The Evolution of Technology and Education in Applied Aerodynamics

Barnes W. McCormick
The Pennsylvania State University

Abstract

The history of education in applied aerodynamics is traced from its earliest beginnings in the United States and correlated with the development of the technology. The early programs, courses and texts are examined in chronological order with some interesting facets presented which relate to the early developers of both the technology and education.

Introduction

The brief chronology which follows presents some milestones in the evolution of aeronautical technology and in the corresponding educational system developed to train aeronautical engineers. Only a sampling of the important developments in aeronautics is presented but, interspersed with educational activities, it may give the reader a feeling for how the history of the academic programs relates to the technical developments. In the beginning we see persons striving to understand the principles involved with flight. The problem of estimating the lift (called pressure) from a flat plate wing (called an aerofoil) was a major one and many conflicting methods on how to do this can be found in the literature prior to 1900. Slowly, data bases were formed by researchers like Cayley, Lilienthal, Dines and Langley from which one could make a fairly reasonable estimate of lift and drag. Prior to 1900, there were only two activities noted with some link to academia; Professor Montgomery's gliding experiments at Santa Clara College and Octave Chanute's lecture at Cornell. However neither of these activities appear to have contributed to furthering the development of educational programs in aeronautics.

<u>Chronology</u>
(Educational activities in bold font)

1687 Sir Isaac Newton publishes *Philosophiae Naturalis Principia Mathematica*
1735 M. Clare, <u>The Motion of Air and Water</u>
1738 Daniel Bernoulli published his book, *Hydrodynamics*, and thus coined the word.
1752 Leonhard Euler properly formulated equations of motion.
1799 Sir George Cayley glider including vertical and horizontal tails.
1804 Cayley's whirling arm apparatus for testing
airfoils. Flew first model glider.
1852 Cayley's paper on man-carrying glider.
1853 Cayley built first man-carrying glider. Flown by his Coachman.
1865 W. J. M. Rankine's momentum theory of propellers.
1873 Otto Lilienthal begins lift and drag measurements on uncambered and cambered thin wings using whirling arm apparatus.

1883 Prof. J.J. Montgomery, Santa Clara College, CA, begins glider tests.
1883 Reynolds, *An Experimental Investigation of the Circumstances Which Determine*
Whether the Motion of Water Shall Be Direct or Sinuous and of the Law of Resistance
in Parallel Channels
1877 E. J. Routh, "Essay on Stability of a Given State of Motion", London
1890 Dines research concerning forces on inclined surfaces published.
1890 Octave Chanute's lecture to students of Sibley College, Cornell University titled "Aerial Navigation"
1891 Samuel P. Langley's book, *Experiments in Aerodynamics,* published by Smithsonian.
1894 Lanchester began experiments on wings
1895 Lamb's, *Hydrodynamics, 2nd Ed.*
1895 Boston Aeronautical Society formed including MIT Profs. Lanza, Means and Howard.
1896 Otto lilienthal had made over 2000 glider flights but was killed on August 9 when he stalled due to a gust and fell from about 50 ft. His last words [1] were "Opfer müssen gebracht werden". (Sacrifices must be made)
1896 A. J. Wells, MIT thesis in ME Dept *An Investigation of Wind Pressures On Surfaces*
1897-1912 Six more theses published at MIT on aeronautics
1902 Kutta-Joukowsky theorem relating to circulation.
1903 Wright Brother's first flight (Dec. 17)
1904 C. Felix Klein brings Prandtl and Runge to Göttingen
1904 Prandtl's paper on boundary layers
1904 G. H. Bryan and Williams paper on Stability in Aviation
1907 Lanchester's publication, *Aerodynamics* (3-D
wing theory - subjective in nature)
1909 Lanchester;s publication, *Aerodonetics*
1909 MIT Aero club formed
1910 Prof. Lucien Marchis taught first course ever on aerodynamics, University of Paris
1911 Students at U. of Michigan built small wind tunnel, experimented with gliders as kites.
1911 Prof. Montgomery killed in gliding accident (see 1883)
1911 von Mises published general equations of motion including propeller gyroscopic effects in Encyclopaedie der mathematischen Wisssenschaften, Vol IV
1911 Grover C. Loening, <u>Monoplanes and Biplanes</u>
1912 Full-scale testing resulted in two MIT theses. One done using whirling arm at WorcesterPolytech.
1912 Charles B. Hayward, <u>Practical Aeronautics</u>, 1912, American School of Correspondence, Chicago
1913 A.A. Merrill presented seven lectures at MIT covering early experiments with gliders, research of Langley, Lillienthal, Maxim, Hargraves and Eiffel (application of resistantance coefficients), methods of control. Students given problems based on data from blueprints of "modern" biplane.
1913 von Mises taught course for first time on mechanics of flight. Published elementary textbook, <u>Fluglehre</u>.
1913 First course in Aeronautics taught at University of Michigan by Prof. Felix Pawlowski (student of Marchis-see 1910)
1914 Wind tunnel installed at MIT under direction of Hunsaker. Hunsaker teaches first course in aeronautics.
1915 NACA founded
1915 Two new courses in Aeronautics at U. of Michigan,*Propulsion of Airplanes* and *Airplane Design.* **16 students enrolled.**
1915 Hou-Kun Chow awarded MS degree in Aeronautical Engineering at MIT. Thesis entitled "Experimental Determination of Damping Coefficients in the Stability of Aeroplanes"

1916 Boeing Company founded, donated wind tunnel and established chair in Aeronautics at U. of Washington.

1916 William S. Durand began research at Stanford University.

1916 Aeronautics Department established at U. of Michigan. 4-year BS program in Aeronautical Engineering.

1916 Jerome C. Hunsaker received Eng. D. from MIT and returned to the Navy. Thesis entitled "Dynamical Stability of Aeroplanes". Alexander Klemin succeeded Hunsaker at MIT

1917 E.P. Warner succeeded Klemin at MIT until 1926 when he was named Asst. Sec. of Navy for Aeronautics

1917 William Frederick Gerhardt first MS in Aero from U. Of Michigan. Flavius E Loudy first BS

1918 Prandtl presents three-dimensional wing theory

1918 Karman-Trefftz paper on Kármán-Trefftz profile

1920 Bairstow's *Applied Aerodynamics* published

1920 U. Of Michigan had graduated over 100 students by this time in aeronautical engineering.

1920 Langley Memorial Aeronautical Laboratory opened directed by Max Munk (pupil of Prandtl)

1920 NACA first published report on airfoils contained much data from Göttingen

1921 von Kárman developed idea of turbulent boundary layer in Z. Angew. Mathem. Mechanik

1921 Louis Bréguet paper on range and endurance

1921 Guggenheim funded Cal Tech

1922 Practical course in aeronautical engineering first taught at NYU by Alexander Klemin

1925 Guggenheim funded NYU

1926 Daniel Guggenheim School of Aeronautics established at CalTech directed by von Karman

1926 Guggenheim funded Stanford

1926 Guggenheim funded U. of Michigan

1926 "Department" of Aeronautical Engineering (under control of Physics Department) formed at MIT and undergraduate courses offered first time.

1927 Guggenheim funded MIT

1927 E. P. Warner's Airplane **Design Aerodynamics**

1928 Guggenheim funded U. of Washington

1928 H. Glauert formulates transformation for equivalent for compressible flow.

1928 Diehl, W. S., **Engineering Aerodynamics**

1929 Guggenheim funded Cal Tech and U. of Dayton (with airship grants)

1930 Guggenheim funded Ga. Tech

1931 Department of Aeronautical Engineering formed at Ga. Tech.

1934 Aerodynamic Theory, Durand

1935 K. D. Wood's **Technical Aerodynamics**

1936 First accreditation (ECPD) of Aeronautical Engineering (two).

1939 Aeronautical Engineering became a separate department at MIT with Hunsaker as first Head.

1940 Twelve programs

1941 Thirty seven programs

1943 40 programs. Aero option in ME offered at Penn State.

1945 Department of Aeronautical Engineering formed at Penn State, David J. Peery (author of *Aircraft Structures*), Head

1949 49 programs

1998 currently about 59 programs

Development of the Technology

A recent book by Anderson[2] on the history of aerodynamics is a "must" for anyone interested in the history of theoretical and experimental developments of flight. Some of the material found in this paper is repetitive of Anderson's book. Indeed, a certain degree of success was felt in researching this paper when a historical tidbit not mentioned by Anderson was uncovered.. There are a few of these contained herein including the work by Clare mentioned below and the history of some of the academic programs in aeronautical engineering.

In 1687 Sir Isaac Newton published his *Philosophiae Naturalis Principia Mathematica* which became the foundation for all ensuing developments in solid and fluid mechanics. A book by M. Clare[3], found in the rare books division of the Pattee Library of Penn State, dates back to 1735. This book, entitled The Motion of Air and Water. was published in 1735 by Edward Symon, over-against the Royal-Exchange in Cornhill and is printed in old English font. Any reader who is a Free and Accepted Mason will find it interesting to note that this book is dedicated to "to the Right Honourable Thomas Thynne, Lord Viscount, Baron of Warminster, Grand Master of the Ancient and Honourable Society of Free and Accepted Masons." The book contains 67 very short chapters on various topics: Chapter titles include "Fluidity in General", "Hydrostatic Principles", "Attraction of Cohesion", "Syphon", "Pumps", "Friction in Water-Works", "Properties of Air", "On the Heart and Circulation of Blood", "Resistance of Fluids to Bodies Moving Therein". "Art of Diving",and "Barometer". The importance of this publication to applied aerodynamics lies with the reference to what we now call skin-friction and pressure drag. To quote Clare, *"There are two kinds of resistance in Fluids; the one is from the viscidity, tenacity or cohesion of their parts, the other from the inactivity, in what Sir Isaac Newton calls, the Vis Inertia of Matter."* Many who followed this time period, including Langley, did not recognize viscoscity as a source of drag. One wonders where Clare got his information and why others to follow did not have the benefit of his knowledge.

Daniel Bernoulli's (1700-82) coined the word, *Hydrodynamics* with his book of the same name published in 1738[4]. This, in a sense began the era of the hydrodynamicists. In 1752 Leonhard Euler (1707-83) properly formulated equations based on his own work, that of Daniel and John Bernoulli (1667-1748) and Jean le Rond d'Alembert (1717-83). Euler was followed, chronologically through the 19th century, by such persons as George Green (1793-1841), Sir George Stokes (1819-1903), William J. M. Rankine (1820-72), Hermann L.F. von Helmholtz (1821-94), Gustav Kirchoff (1824-97), Osborne Reynolds (1842-1912) and Sir Horace Lamb (1849-1934). We then see, with some overlapping in time, the emergence of theoretical aerodynamicists including Nikolai Joukowsky (1847-1921), Frederick Lanchester (1868-1946), and Ludwig Prandtl (1875-1953).

Development of Educational Programs

The first formal course of record appears to be one on aerodynamic principles offered by Professor Lucien Marchis in 1910 at the University of Paris.[5] Even before then, in 1896, the Boston Aeronautical Society was formed which included several professors from MIT and, that same year, a thesis was prepared at MIT by A. J. Wells in the ME Department entitled *An Investigation of Wind Pressures on Surfaces.* In today's terminology, *force* would be substituted for *pressure.* Wells' experimental investigation was done in a small wind tunnel which he built specifically for his thesis. In 1912, full-scale testing resulted in two additional MIT theses. One of these was done using a whirling arm facility at Worcester Polytechnical Institute.

In 1913 at MIT[6] airplane designer A.A. Merrill presented an informal course consisting of seven lectures covering early experiments with gliders, research done by Langley, Lilienthal, Maxim, Hargraves and Eiffel (application of resistance coefficients), and methods of control. Students were given problems based on data from blueprints of "modern" biplanes. Note that the work of the

Wright Brothers is not mentioned in the course description. This may be because the Wrights zealously guarded their results to protect patents or because the academic community regarded them as a "couple of bicycle mechanics." Also in 1913, von Mises taught a course for the first time on mechanics of flight and published an elementary textbook, Fluglehre, which became the basis for his later book, Theory of Flight. [7] That same year, the first course in Aeronautics was taught at the University of Michigan by Professor Felix Pawlowski, a student of Marchis [11, 12,13].

Again in 1913, a wind tunnel was installed at MIT under the direction of Jerome C. Hunsaker. Hunsaker, who was to receive the first doctorate in Aeronautical Engineering from MIT in 1916, also taught in 1913 the first formal course in aeronautics at MIT. In 1915, Hou-Kun Chow was awarded the first MS degree specifically in Aeronautical Engineering at MIT. His thesis was entitled "Experimental Determination of Damping Coefficients in the Stability of Aeroplanes" All of this early work at MIT was at the graduate level.

Thirteen years after the Wright Brothers "lifted off", in 1916, the Aeronautics Department was established at the University of Michigan. [8,9,10] It is not too surprising that it took this long to create the first department in a new discipline. Probably, academic administrators had to be assured that aeronautical engineering was, indeed, a new discipline and of sufficient importance to warrant a separate department and program.

In 1920, at MIT, E. P. Warner was put in charge of an aeronautical engineering program under the control of the Physics Department but it was not until 1939 that a formal Department of Aeronautical Engineering was formed at MIT offering undergraduate courses with Hunsaker as the Head.

It is difficult to say that MIT or the University of Michigan was first in instituting educational programs in aeronautics. Developments at the two institutions appeared to be nearly concurrent. At the graduate level, MIT was clearly the leader whereas UM established a formal undergraduate program well before MIT.

There were other universities that appeared early in the game. In 1916, the Boeing Company was founded. It immediately donated a wind tunnel to the University of Washington and established a chair in Aeronautics. Also, in 1916, Alexander Klemin succeeded Hunsaker at MIT but left shortly thereafter. In 1922, Klemin taught the first practical course in aeronautical engineering at NYU, the first school to receive Guggenheim funds in 1925. From then through 1930, the Guggenheim Fund supported seven additional schools; Stanford, U. Of Michigan, MIT, U. Of Washington, Caltech, Ga. Tech and the U. Of Dayton.

The Daniel Guggenheim Graduate School of Aeronautics was founded at Caltech in 1926 with Theodore von Kármán as its head. Von Karman brought the personality and organization of Göttingen-style applied science to Caltech. To quote Hanle [11], "He brought a kind of academic entrepreneurship of aviation-pursuing basic scientific research in aerodynamics while teaching, giving popular talks, and helping to build airplanes". In the beginning, Caltech leaned more toward scientific research supported by government programs and foundations while, in contrast, MIT and NYU had especially strong ties to industry.

The first classes at Georgia Tech started in the Guggenheim Building in 1931 [12] with 18 students, 2 faculty and budget of $10,000. Before that time, in 1917 a military aeronautics school had been established there by the U.S. Army. It was essentially a ground school for *aviators* and supply officers. By 1950, the faculty size in Aeronautical Engineering had increased to 11. Somewhat surprising is the fact that the first Ph.D. degree at Georgia Tech in Aerospace Engineering was not awarded until 1966.

Around 1940 there was a quantum increase in the number of schools in the U. S. offering programs in aeronautical engineering, probably driven by the developing World War II. Typical of schools that came along around this time is Penn State. Like many aeronautical engineering programs the one at Penn State was started as an option in ME in 1943. In 1945, the program was made into a separate department with David J. Peery, author of *Aircraft Structures*, as Department Head. Initially, the Department had only 2 professors, increasing to three in 1946 supplemented by 2 instructors.

Early Texts

When organized educational activities began in aeronautical engineering, a fairly large base of data and theories existed which could be utilized. It was noted earlier that formal lectures at universities began around 1913 but it is interesting to note that the first formal instruction was apparently given by correspondence through the American School of Correspondence in Chicago. A text was published by them in 1912 authored by Charles B. Hayward. However before Hayward was a book by Loening which was a text of sorts since it included an example of how to design an airplane.

Loening (1911) [13]

Most of the writing of this book, *Monoplanes and Biplanes*, was done to fulfill the requirements for a Master of Arts from Columbia University and the author acknowledges the help of a physics professor, Dr. Charles C. Trowbridge. Loening, later to become known for the design of large seaplanes, covers the work of Langley, Lilienthal, Chanute and others. He credits Chanute with over 700 glides without an accident. He also presents Duchemin's formula for the resistance of a flat plate (see Hayward below) and uses *drift* to mean *drag*. He discusses Lilienthal's results for curved surfaces and presents the effect of *curvature* (camber) and aspect ratio on lift and *drift*.

In his numerical example of the design of an airplane, it is interesting to note that he starts by assuming a weight, speed and angle-of-attack. He then uses Lilienthal's and Eiffel's tables to determine drag and power. On the propeller, he states that "the propeller industry is well established" and thus does little in this regard.

Loening devotes a whole chapter to accidents. He states that the press plays them up and *"aviation is as safe if not safer than automobiling."* Nothing has changed in 80 years! He goes on to say that *"Negligence on part of aviator"* is the main cause of accidents.

Loening presents some interesting tables which summarize airplane parameters at the time. Regarding speed, he list 18 monoplanes with speeds between 39 and 69 mph. The listed speeds of 23 biplanes lie between 33 and 68 mph. The pounds per horsepower for all of the airplanes range between 7.5 and 41. The wing loadings in psf vary from 2 to 5.8 for monoplanes and from 2.05 to 5.92 for biplanes. The average wing loading for the biplanes is lower at 3.2 than the monoplanes at 3.89.

Hayward (1912) [14]

The one-page introduction in this text, Practical Aeronautics, was written by none other than Orville Wright and is worth repeating here since it provides a little insight into the character of the co-inventor of the airplane. His writing takes a few unnamed persons to task but is also gracious to Hayward. The signature is duplicated exactly from the reference.

INTRODUCTION

The achievement of flight by man after ages of disappointment has so aroused the imagination and the interest of the public that a great demand has been created for works treating on this subject. A number of authors have attempted to supply this demand. Some, having no real historical or scientific knowledge of the subject, have been compelled to draw their materials from the imaginative stories of newspaper writers. Others, with some knowledge of engineering and physics but with no practical experience in aeronautics, have fallen into serious errors in their attempts to explain the principles of flight. This has resulted in the publication of a great many works that might better have been left unprinted.

At the request of the author I have looked over some of the proofs of the the present work. On account of lack of time I have not been able to read all the chapters as I should like, but those I have examined, such as the chapters treating of the work of the early experimenters and the present status of the patent litigation, are remarkably free from the errors usually found in aeronautical works of this character. The story of the early work of my brother and myself is also correct, and is taken almost verbatim from an article written by us several years ago for the Century Magazine. The chapter on the patent litigation is the best and clearest presentation of the legal aspect of the subject that has come to my notice. If the portions of the book which I have not examined have been prepared with the same care and accuracy as those I have read, I am sure the work will be a valuable addition to the literature of Aeronautics."

Orville Wright

Hayward begins with material on dirigibles and balloons followed by what he calls *Theory of Aviation* beginning with the experimental work of Cayley. He describes the Eiffel laboratory with an open jet wind tunnel equipped with a balance. On the lift of a wing, he presents Duchemin's formula

$$P_1 = P\frac{2\sin\alpha}{1 + \sin^2\alpha}$$

where P equals the "pressure" (normal force) on the plate when the plate is normal to the flow and P1 equals the "pressure" on the plate at an angle of attack. How good is this equation? Anderson notes that this equation was used by Octave Chanute in his book published in 1894, *Progress in Flying Machines* where Chanute says it agrees well with Langley's data. However, at the time the of the book, there was still a lot of uncertainty (at least for applied aerodynamicists) concerning the calculation of aerodynamic forces. Hayward calculates a "pressure" (force) on a flat plate normal to the flow as equal to the momentum flux of the free stream over the area of the plate and obtains a pressure for a 1 square foot plate as $P = 0.0054\ V^2$ using a weight of air of 0.0807 lb/cu. ft..(equivalent to a density of 0.002506 slugs/cu. ft.). With a drag coefficient of 1.18 for a square flat plate (using his density), the equation should be $0.00318\ V^2$.

Hayward tabulates results for a number of investigators in the form of $P=KV^2$; Col. Renard 1887(K=.00348), Langley 1888(K=.00389 to .00320), Lilienthal 1889(K=.005), Voisin 1900 (K=.0025), Wright Brothers 1901 (K=.0033), and Eiffel 1903 (K=.0031).

He says the "theoretical" value of .005 is too high and settles on .0031 which is close to the value we would use today. Examining Duchemin's equation, since P includes the area, the slope of the lift coefficient curve is calculated as 2.13 /degree which is close to what one would estimate for a wing with an aspect ratio of 1.0. (Approximately 2.44). Of course, since the data was obtained with 1-foot square flat plates, the application of Duchemin's formula is very limited.

Hayward states that "*The planform and its aspect or direction of presentation are items of the greatest importance in determining the lifting power per unit of area.*" He then presents a simpilified explanation of aspect ratio effects. Showing three rectangular planforms each with the same planform area, Hayward states that the lift of each will be in direct proportion to the aspect ratio. But then he backs off from this statement by saying that "*Of course, these figures are modified considerably by the heights and depths to which the air is acted upon by the passage of the surfaces, and, since these values vary directly but in reduced ratios, as the axial lengths of the compared surfaces, it will be apparent that C (AR=4) will not actually lift eight times as much weight as B (AR=0.5). However, the rate of gain is very much higher thabn the rate of loss, as the width or spread is extended and the depth corrrespondingly decreased, as strikingly shown by the simple comparison just given.*"

A more rational explanation of aspect ratio effects was given in 1907 by Lanchester but with the communications at that time, Hayward apparently did not know of Lanchester's theories. Prandtl's quantative solution of aspect ratio effects was produced later in 1918. Of course, by 1918 a lot of airplanes had been designed, built and tested. The Wright Brothers were aware of the effect of camber and aspect ratio on lift, center of pressure, profile drag and induced drag Unfortunately, even though Orville provided an introduction, he did not share with Hayward the reams of data on these effects which the Brothers has collected.

On this topic, it is interesting to note an exchange [15] between Wilbur Wright and Octave Chanute where Langley is taken to task for the interpretation of some of his lift data. After stalling, the lift of a wing drops but then will increase if the angle of attack is increased well above the stall angle. This produces a bump in the lift curve which Langley refused to believe and simply faired a curve through his data. This curve, which he provided to the Wrights showed the lift continuously increasing with angle of attack in the stall region. When the Wrights kept getting a bump from their wind tunnel data, they finally wrote to the Smithsonian in frustation asking for a copy of Langley's original data. When they received the data, they found it had the same bump in the lift curve as given by their own measurement. On December 1, 1901, regarding this incident, Wilbur wrote to Octave Chanute,

"*If he (Langley) had followed his observations, his line would probably have been nearer the truth. I have myself sometimes found it difficult to let the lines run where they will, instead of running them where I think they ought to go. My conclusion is that it is safest to follow the observations exactly and let others do their own correcting if they wish.*"

Apparently this trait was one of Langley's failings. Lanchester [16] takes him to task for neglecting, or in not recognizing, viscosity as a source of drag. On the drag of flat plate plates, Langley faired a curve to zero at an angle of attack of zero claiming the finite value of C_D at an alpha of zero was attributable to experimental error. Anderson shows that Langleys data was close to what one would predict based on today's knowledge. To quote Lanchester,

"*It is from no wish to belittle the work of the late Prof. Langley that attention has so frequently been drawn to the point at issue. Langley's name will always stand as one of the most distinguished pioneers of experimental aerodynamics. The whole of the mis-statements to which attention has been directed hinge upon the one fundamental error, that of the assumption of the negligibility of skin friction; and if the whole Memoir be prefaced by the words, "neglecting the influence of skin-firiction," Langley's position would be substantially regularised.*"

Professor Langley's work [17] *has, however, been widely read, and his statements, unqualified as they stand, have been commonly accepted, and it is therefore impossible in a work of this type to be too emphatic in denouncing the errors in question.*

So far as the experimental work itself is concerned, apart from inference, it is undoubtedly the most valuable contribution to our knowledge that has so far appeared, with the exception perhaps of the work of Dines already discussed.

This early reference by Hayward contains some rudimentary material on propellers and on stability of airplanes. It is somewhat surprising to find at this early date a presentation on the employment of gyroscopic stabilization of airplanes. Apparently there was a widespread interest in this topic at the time including several experimental attempts to achieve this goal. Hayward has no equations relating to longitudinal or lateral open-loop stability. However, his qualitative statements on these subjects are quite accurate. He says to achieve longitudinal stability, simply place the cg in front of the *normal* center of *air* pressure and incline the *forward planes* at a greater angle than the *following planes*. Today, we would say that this assures an airplane which is trimmed at a positive lift coefficient with a negative slope of moment coefficient vs. lift coefficient. On lateral stability, Hayward says to employ dihedral or use vertical surfaces. There is no mention of parameters such as moment coefficients, aerodynamic center, or neutral point in Hayward's book but there is a discussion and graphs relating to the center-of-pressure.

Over 60 years after Clare's book, Sir George Cayley began his experiments relating to flight. Anderson refers to Cayley in a chapter title as *"the true inventor of the airplane"* but tempers that statement later calling him the *"first true aeronautical engineer"*. The latter designation is certainly appropriate since Cayley methodically tested wings (referred to as *airfoils)* on a whirling arm apparatus and created a data base available for others to use. Cayley was also concerned with control and flew a man-carrying glider that incorporated elevator and rudder control. However, he did not succeed in achieving fully-controlled, powered flight so the reference by Anderson to Cayley as the inventor of the airplane might be questioned.

Bairstow (1920) [18]

Bairstow's *Applied Aerodynamics,* published in 1920, eight years after Hayward, was, technically, a notch above Hayward's book. Bairstow was a member of the (British) Advisory Committee for Aeronautics and past superintendent of the Aerodynamics Department of the National Physical Laboratory. In his introduction, he discusses the work of Maxim. *"In England, Maxim attempted the design of a large aeroplane and engine, and achieved a notable result when he built an engine, exclusive of boilers and water, which weighed 180 lbs. and developed 360 horsepower. The aeroplane was made captive by fixing wheels between upper and lower rails. The experiments carried out were very few in number, but a lift of 10,000 lbs. was obtained before one of the wheels carried away after contact with the upper rail."*

Bairstow also gives due credit to the Wright Brothers in his introduction. *[" The lack of control existed chiefly in the lateral balance of the aeroplanes, it being difficult to keep the wings horizontal by use of the rudder alone. The revolutionary step came from the Brothers Wright, in America as the result of a patient study of the problems of gliding. A lateral control was developed which depended on the twisting or warping of the aeroplane wings so that the lift on the depressed wing could be increased in order to raise it, with a corresponding decrease of lift on the other wing.]*

[From the time of the Wrights' first public flight in Europe in 1908 the aviators of the world began to increase the duration of their flights from minutes to hours.]

Bairstow's treatment of stability is fairly complete and, in his introduction he discusses the history of that technical area. "*As soon as the problems of sustaining the weight of an aeroplane and of controlling the motion through the air had been solved, many investigations were attempted of stability so as to elucidate the requirements in an aeroplane which would render it able to control itself. Partial attempts, were made in France for the aeroplane by Ferber, Seé and others, but the most satisfactory treatment is due to Bryan. Starting in 1903 in collaboration with Williams, Bryan applied the standard mathematical equations of motion of a rigid body to the disturbed motions of an aeroplane, and the culmination of this work appeared in 1911.*

Except for considerations of compressible flow, the material contained in Bairstow's book is not too far removed from that found in a modern book on aeronautics. There is material on lift, drag, scale effects, propulsion and airplane performance. The material on stability and control develops six degree-of-freedom equations of motion, linearizes them and then discusses short period and phugoid motion. An appendix even presents a method for extracting complex roots from polynomials. The only information given in Bairstow's book regarding aspect ratio effects are a few graphs showing lift coefficient and L/D vs. angle of attack for aspect ratio values from 3 to 7. Glauert's explanation and lifting line method is not included in the material.

Wilson (1920) [19]

The text by Edwin B. Wilson, *Aeronautics*, was published the same year as Bairstow's book. The material is somewhat similar to Bairstow's but Wilson's book is less applied in nature than Bairstow's and considerable smaller. (260 pages compared to 560). In the preface to his book, Wilson states,"*For several years I have been giving, at the Massachusetts Institute of Technology, courses of lectures on those portions of dnamics, both rigid and fluid, which are fundamental in aeronautical engineering.*
Wilson (and Bairstow) continue to use the term "pressure" to denote total force on a plate and his discussion of the effect of aspect ratio on the "pressure" is still rather elementary. He says "*The two tables show that as a normal plane is smaller the pressure is relatively less, and as the plane is more elongated across the wind the pressure is relatively greater.* (i.e. - lift curve slope increases with aspect ratio) As with Bairstow, there is no analytical explanation given for aspect ratio effects.

As a text, Wilson presents questions at the end of each chapter. A few examples follow.

*   Given $W \, du/dt = 32.2F$ in the English system. Make the change of units to find the numerical multiplier if TV is in kg, u in m/sec, F in kg. The answer should be about 9.81.

*   If an automobile with top up exposes 6 ft^2 more of effective normal surface to the wind and if the automobile is driving 25 mil/hr into a 25-mile wind, how much is the additional force due to the top and how much additional power is required?

*   A plane 50 by 8 ft moves in a circle of 200-ft radius with a velocity of 100 ft/sec and an angle of $i = 8^o$. Find the pressure if the surface is assumed centered, find the actual excess pressure, the moment about the central line, and the distance of the center of pressure from the line.

In his treatment of stability and control, Wilson also develops equations for motion in a vertical plane relative to moving axes. He then expands aerodynamic forces in a Taylor series and examines derivatives. Two modes are obtained and the periods and time to damp to half amplitude are presented. His calculations show that at low speed both the "Clark" and the "JN-2" become unstable. Some of his calculations for the JN-2 follow:

Data for Curtiss JN-2. V = mi/hr, W =1800, k^2_B=34. Note: I =angle of attack, degrees

V	79	51.8	45.2	43.7
	stable	stable	unstable	unstable
period T	34.3	16.7	12	11.6
time t (1/2)	10.8	17.7	16	19.3

Wilson also covers lateral stability and formulates three equations independent of longitudinal motion to describe lateral-directional motion including what he calls the "Dutch roll" and "spiral" motion. (Identical to modern practice)

Warner (1927)[20]

It is interesting to note that the text by Edward P. Warner, <u>Airplane Design Aerodynamics</u>, is dedicated to his Mother-in-Law, Mr. Lucy S. Pearson *"Without whose early guidance it would never have been writen, .."* There are probably not too many books written which are dedicated to the author's Mother-in-Law. The preface to this book is rather philosophical and profound viewed in light of today's burgeoning explosion of technology and its transfer. Selected parts of it follow..

PREFACE

[The student who embarks upon aeronautical studies in 1927 is subject, in some degree, to a contrary embarrassment. Experimental data inundate him.]

[The shelves upon shelves of books and pamphlets that house the primary expression of the studies that have, during the past two decades, extended over the whole world, have piled upon him gradually. To the novice, confronted with the whole mass of this material all at once, it brings bewilderment. The need is for systematization, correlation, coordination, that the prior art and the existing state of knowledge may be presented as a whole and in some harmonious scheme.]

[Bare formulae without explanation, to be applied by rote, have been avoided where possible, but even a crude picture of the physical phenomena which can be tied up to the formula, and from the adoption of which by assumption the formula could be reproduced if forgotten, has been accepted in lieu of complete analytic treatment. The charge that rigor is lacking must then leave me unmoved. For rigor qua rigor I have not gone out of my way to seek.]

Warner's text mentions *helicopter* and *ornithopter* but says neither is yet successful. On terminology, he notes the use of *airplane* and *propeller* in the U.S. and *aeroplane and* airscrew by the British. Warner provides a lengthy description of struts, wires and their purpose.
Reynolds number is defined the same as today but Mach number is not defined but simply referred to as the *ratio of speed of motion to speed of sound*. Warner says that the ratio is not important except at propeller tips.

Warner defines the system of coefficients in use at the time; namely,

$L=L_C SV^2$ where L_C is engineering coefficient used by army and navy and private laboratories

$L= L_c V^2S$ where L_c is absolute coefficient and adopted by British.
$L=1/2 \_ V^2 S C_L$ where C_L is absolute coefficient and adopted by NACA and Germany.

As with Bairstow, Warner discusses the motion of center-of-pressure in some detail but there is no mention of aerodynamic center. In that period, camber apparently referred to the shape of the airfoil surfaces and Warner shows the effect of <u>upper and lower</u> camber on maximum C_L, L/D and minimum D. The results of Glauert's wing theory are reflected in Warner's text. We see for the first time a reference to induced drag.

"Throwing the coefficients into the form used by the National Advisory Committee for Aeronautics..."

$$C_{D_i} = \frac{C_L^2}{\pi R}$$

Warner states that *"Variable-lift devices can be classified under three general headings, dependent, respectively, on variation of angle, of area, and of airfoil section."*

He continues, *"...a few experimental machines of variable area have been constructed,....Most of them provide for the simultaneous variation of area and change of sectional form of the airfoil..."*

Following this, he discusses the use of variable camber, flaps, leading edge flaps and slots for increasing maximum lift. He also describes a "geared" flaps which we now call "drooped ailerons". He describes a turbulent boundary layer with a laminar sub-layer and presents a graph of _ up to 40,000 ft. which is close to today's standard atmosphere. On stability and control, Warner notes that a left-handed system is used in America but a right-handed one in British publications. His material includes many effects considered today like downwash at the tail and slipstream effects but there is no mention of neutral points. On dynamic longitudinal stability, he references Stability in Aviation, G. H. Bryan, London, 1911, develops the quadratic characteristic equation and discusses two modes. He references Bairstow for an approximate solution and states that the quadratic factors into two solutions where the first is an oscillation of very short period, heavily damped and of no practical importance. The other is a longer period oscillation. He then presents an example (JN2 training airplane) and at _ of 1 degree, obtains a stable mode with a period of 34 secs. And a time to halve of 11 secs. For _ of 14 degs., the mode becomes unstable with a period is 11.5 secs and a time to double of 24.7 secs. These numbers agree fairly well with Wilson's results tabulated above where I in the table refers to angle of attack.

Diehl (1928) [21]

The original version of *Engineering Aerodynamics* was published in 1928 with a second edition in 1936. An interesting stamp was found inside the cover of the second edition which read "This book is supplied by the publisher on a school order for textbook use and is to be sold to students only at a special price of $5.00 plus transportation charges." This sounds as if it is inexpensive but in 1936 that amount was probably appreciable. Diehl was a Lieutenant in the US Navy, Scientific Section, Bureau of Aeronautics and as noted in his book in 1936, was a Fellow of the IAS, fore-runner of the AIAA. The contents of the 1928 book are rather comprehensive as noted by the chapter titles: Wing section data, Wing theory, Parasite drag data, Control surface design, Engine and propeller considerations, Performance calculation, Variation of rate of climb with altitude, Aspect ratio and parasite drag, Reduction of observed performance to standard conditions, notes on flight testing, Range and endurance, Special flight problems, Performance estimation, and Seaplane floats.

Diehl states that practically all of the early work in this country was given in the form of coefficients having the dimensions of lbs./sq.ft./(mi/hr)^2 , using the symbols Ky for lift and Kx for drag."

$$L = K_y \frac{\rho}{\rho} \frac{\rho}{\rho_o} \frac{\rho}{\rho} S V^2$$

and that about 1919, NACA recommended *absolute coefficients*, Lc and Dc, in consistent units. In 1921, copying Prandtl, NACA adopted the modern terminology for coefficients.

He presents data on Cd and CL from Washington Navy Yard tests using airfoils labled RAF-15, USA-27, Clark Y, N-22 and G-398. The data was obtained on 5" x 30" models at 40 mph in an 8'x8' Washington Navy Yard Wind Tunnel and corrected for wall interference.

Diehl defines stream function, velocity potential, circulation and uses the Kutta-Joukowski equation. He utilizes Prandtl (1911) and derives relation between circulation and downwash including induced drag for the elliptical lift distribution. He also presents a lot of material on biplanes and states, *"Munk has shown in a remarkable paper that the classical treatment of wing theory by means of vortices may be replaced by kinetic energy considerations. In the paper he shows that the lift due to the curvature of a wing acts at 50% of the chord while lift due to angle of attack acts at 25% of the chord. Consequently the moment about a point on the chord 25% aft of the leading edge is constant."* (Max M. Munk, "Elements of the Wing Section Theory and of the Wing Theory," NACA TR 191, (1924)

Hence, in 1928 the student is made aware of the aerodynamic center but it is not defined as such until the 1936 edition.

Books of the Thirties and Forties

Following Diehl, a number of aerodynamic texts appeared in the late thirties and early thirties including K. D. Wood (1935)[22], Jones (1936)[23], and Hemke[24] (1946). By this time period, subsonic aerodynamics including wing and airfoil theory was well in hand. Wood was a very practical text whereas Hemke presented a challenge to the student with his presentation of potential flow.

Development of Programs

The first Department of Aeronautical Engineering was established at the University of Michigan in Aeronautical Engineering in 1916 but the program was in effect before that time and appeared in print[25] in 1915. The program required 132 total semester hours including 16 hours of language, 9 of chemistry, 8 of algebra and analytical geometry, 12 of shop and descriptive geometry, 10 of calculus, and 10 of physics. Courses were also required in machine drawing, kinematics, surveying, dynamics, thermodynamics, machine design, heat engines, hydromechanics, theory of structures, mechanical lab, internal cimbustion engines, English, mechanical lab, and engineering materials. Courses specific to the major (listed as aero courses) included general aeronautics, theory of aviation, theory and design of propellers, aeroplane design, aerodynamics lab and design of aeronautical models. There was also room in the program for 10 credits of electives. The preface to the program states, *"It is expected that the students will gain much information and also practical experience in connection with the work done at the Packard Motor Car Company of Detroit. The aim of the course is to teach the theory of aeroplanes and to enable students to secure positions in manufacturing plants.* A description of the courses specific to applied aerodynamics follow.

<u>General Aeronautics</u> Lectures and recitations. Two hours. First semester.

An introductory course giving the essential principles of aeronautics. (Balloons, dirigibles, ornithopters, helicopters, aeroplanes, helicoplanes and kites), history f flight and description of modern aircraft.

<u>Theory of Aviation</u> Lectures and recitations. Two hours. Second semester.

The course deals with the following questions: properties of the air, general discussion of aerodynamics, aerodynamical properties of planes and various constructive elements of an aeroplane, power necessary for flight, equilibrium of aeroplanes, stability of aeroplanes, air currents

<u>Aerodynamic Design</u> Lectures and drawing. Three hours. First semester

This course includes the investigation of the design of the aeroplane from the aeronautical and strength standpoints. The strength and design of all the detail are discussed and a completed design prepared.

<u>Aeronautical Laboratory.</u> One hour. Second semester An elementary course covring use of instruments, investigation of aerodynamical properties of the various bodies used in aeroplanes and airships, test of propellers.

<u>Advanced Stability.</u> Lectures and recitations.

Advanced study of more complicated phenomena of stability according to Ferber, Bothesat, Bryan and Bairstow.

One wonders what texts, if any, were used for these first courses. Loening's book was available but it was limited in its coverage of the technology available in 1915. Probably, notes were used extensively by Professors Sadler and Pawlowski. Although early in the development of aerodynamics, there was still a lot that had been learned by the theoreticians such as Prandtl, Joukowski, Rankine, Bryan, Lanchester and the experimenters such as the Wright Brothers, Dines, Langley, Eiffel, Cayley and Lilienthal. With the technology available to design and build the airplanes of World War I, the material in aerodynamics should have contained fairly good explanations of airfoil and wing behavior. One major item missing at that time would have been Prandtl's wing theory and the application of potential flow methods using elementary flow functions. Of course, in 1915, the systematic collection of airfoil data done by the NACA (founded in that year) did not exist. Also, of course, material on compressible flows did not evolve in educational programs until the early 1940's.

Following Michigan's lead, many other undergraduate programs were established. By 1922, programs existed at MIT, Cal Tech, Michigan, Washington and Stanford. In 1936, the Engineer's Council for Professional Development (ECPD), now Accrediting Board for Engineering and Technology (ABET), began accrediting programs in Aeronautical Engineering at the undergraduate level. Two programs were initially accredited, MIT and NYU. In 1937, the list grew to ten and included Alabama, Michigan, Cincinnati, Detroit, Georgia Tech, Minnesota and Washington. In 1938, RPI and Catholic University were added. The list remained the same until 1940 when Aeronautical Engineering was listed as option as part of another accredited curricula. Twenty-five additional schools were added in this manner. All of these were under M.E. In 1942, Notre Dame and Texas A&M were added directly under Aeronautical Engineering. By 1950, fifty programs were accredited. Today that number has grown to about 57.

Conclusions and Observations

The history of aeronautics, and the development of educational programs in the field, is an interesting one. Serious attempts at flight began around 1800 with Cayley but success was not achieved until around a hundred years later by the Wright Brothers. A painful and slow process was undertaken to understand the principles of flight and even, when it was finally achieved, there was much that was unknown to the aviation pioneers. Flight and its perfection might have come much quicker if more communication had taken place between the practitioners and the theoreticians. This kind of technology transfer began to occur with the development of individual courses in aeronautics around 1910 and in the development of formal programs in 1915.

References

1. *History of Aviation*, John W. R. Taylor and Kenneth Munson, 1977, Crown Publishers, Inc., NY
2. Anderson, John D., *A History of Aerodynamics and Its Impact on Flying Machines*, Cambridge University Press, 1997
3. M. Clare, *The Motion of Air and Water*, 1735, published by Edward Symon, over-against the Royal-Exchange in Cornhill (font is old English)
4. *Robertson, James S., Hydronamics in Theory and Application*, Prentice-Hall, 1965
5. Dutton, Donnell W., "A Brief History of Aerospace Engineering Education and Curriculum Changes", ASEE 86th Annual Conference, U. Of British Columbia, June, 1978
6. Correspondence with Prof. Eric Feron, MIT, July, 1998
7. Von Mises, Richard, *Theory of Flight*, Dover Publications, 1945
8. anonymous, *The University of Michigan, A Century of Engineering Education*, U. Of Michigan Press, 1954
9. Correspondence with Prof. Joseph Eisley, University of Michigan, April, 1998
10. Robert P. Weeks, *The First Fifty Years, (A Fragmentary, Anecdotal History) Department of Aeronautical and Astronautical Engineering*, The University of Michigan, Aero Golden Anniversary, October 1954, supplied by Department
11. Hanle, Paul A., *Bringing Aerodynamics to America*, MIT Press, 1982
12. Correspondence with Prof. Robert Loewy, Ga. Tech., April 1998
13. Loening, Grover C., *Monoplanes and Biplanes*, Munn & Co., NY 1912
14. Charles B. Hayward, *Practical Aeronautics*, 1912, American School of Correspondence, Chicago Constable & Company Ltd., London, 1909
15. McFarland, Marvin W., Edit., *The Papers of Wilbur and Orville Wright*, McGraw-Hill, 1953
16. Lanchester, F. W., *Aerodynamics* (Constituting the First Volume of a Complete Work on Aerial Flight),
17. Langley, Samuel P., "Experiments in Aerodynamics", Smithsonian Institution, 1891
18. Bairstow, Leonard, F.R.S., C.B.E., *Applied Aerodynamics*, Longmans, Green and Co., London, 1920
19. Wilson, E. B., *Aeronautics*, John Wiley & Sons, 1920
20. Warner, Edward P., *Airplane Design Aerodynamics*, McGraw-Hill, 1927
21. Diehl, W. S., *Engineering Aerodynamics*, Ronald Press, New York, 1928, (revised 1936)
22 Karl D. Wood, *Technical Aerodynamics*, 1935, McGraw-Hill
23 . Bradley Jones, *Elements of Practical Aerodynamics*, 1936, 1939, 1942
24. Paul E. Hemke, *Elementary Applied Aerodynamics*, 1946, Prentice-Hall
25 anonymous, "The University of Michigan Course in Aeronautics", reprinted from Aerial Age, 1915

Chapter 2

The Growth of Aerospace Education following its Beginning

Barnes W. McCormick1
The Pennsylvania State University

Introduction

Chapter 1 traces the early beginnings of aeronautics and notes the concurrent development of. educational activity relating to the developing technology. The first educational activity in aeronautics appears to have taken place on the campus of Santa Clara College in California where, in 1883, Professor J. J. Montgomery began his testing of gliders. In 1888, Albert Zahm launched man-carrying gliders from the roof of Science Hall at Notre Dame. Then, Octave Chanute presented lectures in 1890 to Cornell students on "aerial navigation" . In 1895, a Boston Aeronautical Society was formed by three MIT faculty; Professors Lanza, Means, and Howard. It was not until 1910 that the first formal course in aerodynamics was taught by Professor Lucien Marchis at the University of Paris.

Subsequently, many educational activities followed including the writing of texts, presentation of new courses and finally, in 1916, the establishment of a 4-year program and Department of Aeronautical Engineering at the University of Michigan. In 1926 a department was formed at MIT. However, before that time, MIT offered several aeronautics courses and awarded graduate degrees relating to aeronautical engineering. In 1936, the Engineers Council for Professional Development (ECPD) accredited the first programs at MIT and NYU in Aeronautical Engineering. By 1940 there were 12 programs and this number increased to 37 in 1941.

This Chapter begins around the period when courses and programs were first being developed, i.e., in the late 20's and early 30's. Air transportation was just about to burgeon with the introduction of the DC-3. Also, World War I and subsequent flight testing, particularly by the Navy, had proven the value of military aircraft.

Development of Programs and Departments

There appear to be four happenings that greatly affected the development of education in aeronautics and astronautics. First, of course, was the Wright Brother's success on December 17, 1903. They proved that one could fly. Then there was World War I. This tragic era proved that, not only could one get off the ground but that the airplane was capable of performing many useful tasks. The next real impetus was World War II. The money that was poured into this gargantuan effort resulted in several results. First the technology was rapidly advanced because of the need for airplanes that could fly faster, further, higher and with more payload. The turbojet engine was developed toward the end of this saga but did not see many applications except for, fortunately, only a few German fighter aircraft.

---

1 Professor Emeritus of Aerospace Engineering, The Pennsylvania State University

The military also provided impetus to the development of academic programs by establishing training programs on many campuses. The Navy operated V-5 and V-12 programs intended to give some college-level education to future flight and deck officers. The Army operated ASTP programs that paralleled the Navy's effort. Even Industry established programs on the campuses. Many alumni of that era remember (with a smile) the Curtiss-Wright Cadettes. Purdue, Penn State and Iowa State, to name a few, were benefactors of these three programs. Another important effect, at least in the U. S., was the GI bill that followed. Thousands of GI's with their costs covered returned in the middle 40's to pursue or complete their college degree. Many of these had been in the Army Air Corps or the Navy and Marines and had developed an interest in aeronautics.

The final major influence on education in aeronautics and astronautics came in the 50's with the launching of the space age. After Russia shocked the world on October 4, 1957 with the launching of Sputnik, the West hurried to catch up and duplicate their feat. Many of the engineers attempting to do this were aeronautical engineers who knew little about rocket propulsion. orbital mechanics and such. The academic community hurried to fill this void and most of this effort was done within the existing departments of aeronautical engineering. It was a natural transition then to find that these departments, beginning around 1957 to about 1965, changed their names to reflect the activities in astronautics. Those that adopted something like "Aeronautics and Astronautics" were realistic. The term "Aerospace Engineer" immediately brings to mind, to the man on the street, a "rocket scientist". There are still many of us who would rather be called "aeronautical engineers" but that went out of fashion in the first half of the 60's.

In preparing this paper the help of those Departments whose Heads belong to the Aerospace Department Chairman's Association, ADCA, was solicited. Not all of them responded but a sufficient number did so giving results that should typify the academic community. Using the same format as Reference 1, a chronology of events will first be given followed by a more detailed discussion of events along the time path. I apologize for using first-person but, considering my age, I lived through much of the period of interest and interacted with many of the wonderful people to whom I will refer.

Chronology

NOTE: Bold-faced items are technology developments.
1903 WRIGHT brothers first flight
1910 First formal course in aerodynamics taught at U. of Paris by Prof. Lucien Marchis
1916 Department and 4-year program established at U. of Michigan
1916 Boeing Airplane Co. formed and recruited U. of Washington Engineering faculty
1918 WORLD WAR I
1921-1922 elective courses in Mechanical Engineering offered on engines, meteorology and aircraft design at Purdue
1923 Aero activity began at North Carolina State but no graduates until 1950 as an option in ME
1926 Alexander Vallance teaches first course for "budding" Aeronautical Engineering Dept at Texas. However BS in Aero suspended and not revived until 1942 as result of WW II
1926 Department formed at MIT but courses taught prior to that time.
1926 Dept of Math and Materials offered courses in aerodynamics at Minnesota
1927 Program and Aeronautics Dept formed at Cal Poly San Luis Obispo
1927 August 1, Oliver Lafayette Parks, a Chevy salesman. opened Parks
Air College
1928 First course, Airplanes (3 credits) offered at Penn State in ME.

1928   4-year course leading to BS in Aeronautical Engineering at university of Wichita. Alex Petroff named Director of the School of Aeronautics. Tau Omega formed in 1927. Merged in 1953 to form Sigma Gamma Tau

1928   At Iowa State, 3 grad courses and 1 UG course in aero offered by ME and expanded to 7 course offerings in 1929

1929 FORD TRI-MOTOR

1929 Oklahoma established aero option in ME. Aero honor society Tau Omega formed in 1928. Later merged to for Sigma Gamma Tau

1929 Department formed at Minnesota (accredited in 1936)

1929   Department formed at University of Cincinatti with Bradley Jones as first   Head. Remained head for 28 years until his death. Program required several years of German since Jones felt you needed german to study aeronautics

1930   Guggenheim Hall of Aeronautics opened at University of Washington.   and the first BS awarded in aero by them in 1931.

1931 University of Colorado offers option in ME

1932 Department began at Mississippi State with two faculty. (Accredited in 1949)

1934 Parks College establishes 4 year program in aeronautical engineering.

1935 Notre Dame forms Department of Aeronautical Engineering (Frank N. M. Brown as head)

1936 DC-3

1938   Kirsten wind tunnel completed at U. of Washington

1938 Aero option in ME at Virginia Tech.

1939 FIRST JET FLOWN, HEINKEL He 178

1939 Aero curriculum started by Purdue under ME. First grads in 1943

1940 Prof. John Younger begins aeronautical experiments at U. of Maryland

1940 Texas A& M forms Dept. of Aeronautical Engineering with Howard W. Barlow as Head.

1941 PEARL HARBOR

1941 Prof. Younger at Maryland receives "Spirit of St. Louis" Gold Medal for service to aviation.

1941 Iowa State curriculum in Aero approved to be offered through ME

1942 BELL AIRCOMET, 1ST U. S. JET

1942 Iowa State forms separate Aero Dept with Wilbur C. Nelson as head.

1942 Texas re-establishes Aero Dept with Milton J. Thompson as Head.

1943 V-5, V-12, ASTP and Curtiss-Wright Cadettes at Iowa State

1943   First graduates of Aeronautical Option in ME at Penn State.

1943 Separate Aero Dept. formed at Virginia with Leon Seltzer as Head

1943 BS in aero approved at Colorado

1944 Glenn L.. Martin donates $1.7 million to U. of Maryland to establish research and instructional program.

1944   Parker named head of Dept. of ME and Aeronautical Engineering and K.D. Wood appointed at Colorado.

1944 Department of Aeronautical Engineering formed at Penn State with 3 faculty and David J. Peery as head.

1944 Parks College completes training of WW II pilot cadets. Over 27,000 received commissions. 10% of all air force pilots received primary flight training at Parks.

1945 Purdue forms School of Aeronautics separate from ME.

1946 Parks becomes Parks College of Aeronautical Technology of Saint Louis University. Accredited by North Central Association in 1949. Leon Selzer becomes head in 1963 and pushes for ECPD accreditation but unsuccessfully..

1946 Colorado forms separate Department with K. D. Wood as Head.

1946 AFIT opens 2-year programs comparable to last two years of UG degree. Programs evolve into BS programs in EE and Aero

1946 MS program started at University of Washington

1946 Iowa State has four faculty including T. A. Wilson. Nelson goes to Michigan and Carl Sanford named head.

1947 First degree given by Naval Post-Graduate School but courses offered in early 1930's. Prof. Coates becomes Chairman of Aeronautics Department

1948 First BS awarded at U. of Florida by Dept. of Aeronautical Engineering

1949 DE HAVILLAND COMET FLOWN. FIRST COMMERCIAL JET

1949 Department of Aeronautical Engineering established at U. of Maryland with A. Wiley Sherwood as first chairman.

1950 University of Wichita offers evening classes in attempt to save program.

1950 First MS awarded at U. of Florida.

1951 Oklahoma forms School of Aeronautical Engineering and 8 BSAE degrees awarded in 1953

1951 MS and PhD programs started at AFIT

1951 NPS moved from Annapolis to Monterey to occupy site and grounds of world-renowned Hotel Del Monte

1953 First PhD awarded at Penn State

1954 FIRST FLIGHT OF BOEING 707 (on my birthday)

1957 SPUTNIK LAUNCHED

1958 At Minnesota, Department combined with Mechanics and Materials.

1959 PhD program started at U. of Washington

1960 First PhD (to Joe Cornish) at Mississippi State

1960 Purdue merges Aero with Engineering Sciences to form School of Aeronautical and Engineering Sciences (renamed Aeronautics, Astronautics and Engineering Science in 1965).

1961 Name of Department changed to Aerospace Engineering at Penn State

1963 Oklahoma- Aero and ME merges

1963 U. of Maryland awards first PhD in Aero

1965 U. of Florida awards first PhD.

1965 Colorado awards first PhD to Colonel in South Vietnamese Air Force

1967 Aero at Notre Dame combines with ME

1967 MS program initiated at Howard. First degree in1969

1968 Department at U. of Texas merged with Engineering Mechanics

1969 First BS in Aero granted by U. of Missouri-Rolla

1970-73 70% drop in UG enrollment at U. of Maryland (typical of all)

1973-80 At the U. of Maryland: Prof. Gerald Corning dies, Bob Rivello and Wiley Sherwood retire.

1973 Iowa State starts PhD program

1976 PhD program initiated at Howard

1977 Howard University initiates PhD program.

1980 U. of Maryland names Alfred Gessow as Head

1981 U. of Maryland establishes Center for Rotorcraft Education.

1986 U. of Central Florida opened and Dept of ME and Aerospace Engineering formed with three Aero faculty.

Over the Years

Most programs seemed to develop from existing engineering departments, particularly, mechanical engineering. There were a few notable exceptions. Parks College and Embry-Riddle both began as flight schools and Wichita first offered an aeronautical engineering program before expanding into CE and ME. Indeed, if I remember correctly, ME was first offered at Wichita as an option in Aero!

It is also interesting to reflect on how small and close-knit the Aerospace Academic community was, particularly back in the 40's and 50's when it was all focused toward aeronautics. As I read the literature from various schools, several names kept popping up in different places. K. D. Wood was one of these.

After serving a while in industry at Consolidated-Vultee (later Convair) in San Diego, "KD" started his academic career, sometime before 1935, teaching aerodynamics at Cornell. He joined Purdue around 1937 leaving in 1944 to become Head at Colorado. While at Cornell he authored one of the first comprehensive texts in aerodynamics entitled *Technical Aerodynamics*. Many schools used it in the 40's. I still have my own well-worn copy that I used as a student at Penn State. Also in 1935, he published a practical manual on airplane design. I also have a well-worn copy of the 7th edition published while he was at Colorado. Later, I had the privilege of getting to know KD when I attended a summer course at Colorado in the late 60's. He simply coordinated the course on Operations Analysis with his only lecture being on his coin collection and the value of such a pastime. A fine gentleman, he invited us all out to his house for an entertaining and relaxing evening. KD had a daughter, Cecibel Maxine. He called her CL-max for short. What a dedicated aerodynamicist!!

Reflecting about KD brings out the fact that we know so many of the early pioneers in aerospace education through their books. In structures, the names of E. F. Bruhn at Purdue, D. J. Peery at Penn State or A.S. Niles and J. S. Newell will always be remembered for their classic texts. In the 50's, it was said at Wichita "If Prof. Everett Cook ever had twins, he would name them Niles and Newell".

Cal Poly, San Luis Obispo formed an Aeronautical Engineering Department in 1927 in conjunction with the Automotive Department. It appears to be a misnomer for the program was more that of an A&E school. Indeed, the students built and flew several airplanes in the late 20's and Amelia Earhart landed there in 1935 for repairs to her airplane. Like Parks and Embry Riddle, the program at Cal Poly, San Luis Obispo evolved into a more traditional one and was accredited by ECPD in 1960. However, as one of their recent newsletters states, "students have maintained their 'hands on' approach to education". Prominent alumni from there include Robert "Hoot" Gibson, the astronaut and Elburt "Burt" Rutan, the well-known airplane designer who believes that horizontal tails of airplanes should be in the front. This feeling may have come from the fact that he won an AIAA Student Paper Competition while an undergraduate for his paper extolling the virtues of the canard configuration.

The Aero Department at the University of Cincinnati was created in 1929 with Bradley Jones as the first Head. Of interest is the fact that their students back then had to learn the German language. Jones believed it to be necessary in order to read the literature in aeronautics. Jones may have set a tenure record for he remained as Head until his death around 1956. Over the years they have had seven Heads, two of whom I have known fairly well, Paul Harrington and Gary Slater. In 1958, like many other Departments, they changed their name to Aerospace Engineering. One of their most prominent alumni is Raymond (Ray) L. Bisplinghoff who certainly deserves a place in the history of Aerospace Education. In 1955, Bisplinghoff, together with Holt Ashley and Robert L. Halfman published the classic reference *Aeroelasticity*, still one of the most valuable references around on the subject.

Aeronautical engineering began at Iowa State University in 1928 with three graduate courses and one undergraduate course offered through ME. This was expanded to seven UG courses in 1929 and an aeronautical laboratory was built in 1935. However, a separate department was not formed until 1942 with Wilbur C. Nelson as Head. Iowa State is rich in the history of Aerospace Education and I will therefore refer the reader to Reference 8 that is being presented at the same AIAA session as this one. However, I would like to note that one of their most renowned alumni was also a faculty member in 1946; namely, T. A. Wilson who became CEO of the Boeing Airplane Co. and played a large part in making Boeing what it is today. I met "TA" for the first time at an AIAA reception. My wife, Emily, was with me at the time and I was a little taken back when she told the CEO of Boeing she was teaching airplanes to her first grade class and maybe he could send her some pictures. Would you believe a large box came to the house shortly after that meeting!

Figure 1 shows the number of BS graduates in Aerospace Engineering from Iowa State from 1943 to 2000. I show it because it is representative of what occurred during this same period at other schools. Notice particularly the sharp drop after 1970, which meant a similar drop in enrollment four years earlier. Similarly, between 1970 and 1973, the University of Maryland experienced a 70% drop in enrollment This was a precipitous time for Aero Departments and many, like LSU, disappeared while others were combined with other Engineering Departments. Aero Departments were concerned and also felt that Aero programs were getting a bad rap for the problems of industry.

The headlines read "Thousands of Aerospace Engineers Laid Off. The TV show, *60 Minutes*, had a segment where they interviewed a PhD identified as an "Aerospace Engineer" who had just opened up a hamburger stand. What parent, in their right mind was going to allow their offspring to study such a subject? At that time, a meeting at the University of Florida was held of Aero Department Heads at the urging of Knox Milsaps. I was there and we formed ADCA, the Aerospace Department Chairman's Association. ADCA's first act was to survey industry to learn more about the layoffs. What they found was that 99% of those laid off did not hold degrees in Aerospace Engineering; they were EE's, CE's and ME's. Bob Brodsky, Head of Iowa state, Jan Roskam of Kansas and myself gave interviews to Aviation Week Magazine to try to remedy the situation but I do not believe they did any good. It was only with the improvement in industry that the enrollment leveled out and then began to increase. The enrollment went up from around 1973 to 1983 as fast as it had gone done.

During the 70's there was a lot of bitterness expressed by ADCA toward the AIAA. The members felt that Aerospace Engineering did not have a professional society that truly represented them. ME's had ASME, CE's had ASCE, EE's had IEEE but AIAA was, and still is, an organization that represents many facets of engineering with no particular homage paid to Aerospace Engineering. However, the rift between ADCA and AIAA has disappeared and, indeed, ADCA is now officially a part of AIAA.

The Department of Aeronautical Engineering at the **University of Minnesota** was established in 1929, three years after their Department of Mathematics and Materials first offered a course in aerodynamics. John D. Ackerman, the first Head and Joseph A. Wise taught all of the aero courses. Ackerman was Latvian-born and had studied under Nickolaia Joukowski (Zhoukowski). In 1930 they graduated their first class of six persons in aeronautical engineering. In 1935 they finally got their own quarters, adapting the Atheletic Department's facilities including a swimming pool. In 1958 the Department was merged with Mechanics and Materials after a bitter debate between Ackerman and the new dean, Athelstan Spilhaus. It was predictable that the dean won the debate. Akerman was named head of the Rosemount Aeronautical Laboratory and the two departments were merged with the head of the combined department being the former head of Mechanics and Materials.

Aeronautical Engineering had its start at **North Carolina State University** sometime around 1923 with a few courses being taught in the subject. It is not clear whether a department was formed at that time but, in any case, the activity ceased until the upsurge of enrollment in 1946. There were 14 graduates in 1950 but the degrees were in ME with an option in Aeronautical Engineering. It was not until the late 70's that the degrees were changed to read "Aerospace Engineering". Sometime in the early 1980's, the faculty attempted to establish a separate department but it never happened. Today, the department is now the Department of Mechanical and Aerospace Engineering.

Notre Dame's beginnings were mentioned earlier with the work of Albert Zahm. In 1888 he launched man-carrying gliders from the roof of Science Hall. Later Zahm suspended a foot-powered flying machine on a 50-foot rope from the top of Science Museum. The purpose was to study propellers. To keep from hitting a wall, the operator had to put his feet out in front of machine and onto the wall. The next morning, Brother Benedict, the custodian, was horrified to see footprints high up on the walls and was convinced only the devil could be responsible. He lost no time in sprinkling holy water all about the museum. In 1893 Zahm left and joined Catholic University in D.C. where he built the first U.S. wind tunnel, 50 feet long, 6 feet square and with 12 hp it had winds of 25 mph.

In a P.S. to his letter to me, Notre Dame's Bob Nelson writes. "The notoriety of the crash that killed Knute Rockne was credited with TWA asking Douglas Aircraft to build an all metal aircraft. So even our football program had an effect on aviation." Only the Irish would stretch things to this extent.

The Department of Aeronautical Engineering was founded **at Mississippi State University** in 1933 with two faculty; Kenneth Withington (Head) and Mason Summer Camp. The first BS was awarded in 1935. In 1962 it was renamed Aerospace Engineering and then merged in 1967 with the Department of Aerophysics. However, in 1977, the combined department name reverted to Aerospace Engineering.

The program was accredited in 1949. It was that same year that August Raspet and his team pioneered the application of suction BLC for high lift systems. One of his team members was Joseph Cornish who was the first person to receive the PhD from MSU. Joe later became Vice-President for Lockheed in Marietta, GA.

The research back then at MSU was done in a frugal manner. I learned during a visit to MSU in the middle 50's that Gus Raspet had recruited his office staff to poke holes in the fabric of a wing for purposes of BLC. Later, he modified his wife's Mixmaster to run along a rack placed on the wing's surface and thus, automated the hole punching. I had the pleasure of taking a ride with him at that time in a sailplane with BLC. Suffice it to say that the comparison of the stall performance with and without the BLC was impressive. I was also impressed when I climbed in the back. The seat was a formed-metal tractor seat nailed on a board. We closed the canopy and secured it with a screen-door latch. I think the results of his research showed rather conclusively that one does not need "megabucks" to accomplish meaningful research.

As early as 1921, **Purdue University** offered elective courses in ME aircraft design. Beginning in 1937, the offerings were expanded in ME to include rotary and fixed-wing aircraft, propellers, performance, stability and control, power plants and airfoils. In 1939, an Aeronautical Engineering Curriculum was offered by ME . The first graduates finished in 1943 under a wartime accelerated program of three semesters per year.

In 1945 the School of Aeronautics was formed and AE separated from ME. Elmer F. Bruhn was named head. At a salary of around $5700 a year, he probably had a greater income from his popular book on aircraft structures. Bruhn had one associate professor and two instructors to help him. By 1950, the number of faculty had grown to 15. The program required 152 semester credit hours for the BS. Purdue's program, up to that point, had a practical bent. The university owned the airport, had an air transportation program and a glider club. It is worth noting that, starting in 1952, Neil Armstrong was chairman of the club.

When Milton Clauser became head in 1950, he felt that the curriculum was not rigorous enough and too slanted toward industry. Therefore, a Theoretical Aeronautics Option was introduced and entry to the Air Transportation Program stopped in 1956. Clauser went on leave in 1954 and did not return. In 1956 the name was changed to School of Aeronautical Engineering and then in 1965 to School of Aeronautics, Astronautics and Engineering Sciences. However the science option did not attract many students and the name was changed sometime around 1980 to the School of Aeronautic and Astronautics. It's interesting to note that twenty astronauts have graduated from Purdue, 13 of them with Aero degrees. Names like Kincheloe, Armstrong, Chafee and Grissom are well known. . The first "space pioneer" to graduate in 1949 was Ivan Carl Kincheloe who set an altitude record in the rocket-powered X-2 in 1956. He was killed in 1958 in a Jet crash at Edwards AFB. Neil Armstrong entered as a freshman in 1947 but dropped out and flew in the Korean war. He came back and finished his BSAE in 1955. He was, of course, the Apollo 11 Commander and the first man to step on the moon July 20, 1969. Roger Chafee, BSAE 1957 and Virgil (Gus) Grissom, BSME died in a tragic fire on a Cape Kennedy launch pad. There is now a Grissom Hall and a Chafee Hall on Purdue's campus.

The ME Department at **Penn State University** offered it's first course in "Airplanes" in 1928. This offering was subsequently expanded and in 1941, a 3'x4' subsonic wind tunnel was built. The balance for the tunnel was built by William Brown, Jr. and that name may ring a bell if you were a model airplane enthusiast in the 40's. Bill Brown invented the first gasoline model airplane engine. The balance was designed with the help of W. E. Diefenderker who later became president of Hamilton-Standard.

ME established a four-year program in AE as an option in ME. The first graduates of that option were in 1943. That same year, the Department of Aeronautical Engineering was formed at Penn State. There were only three faculty including the head, David J. Peery whose text, "Aircraft Structures" is still found today on many bookshelves. 150 semester credits were required for the BS including foundry, machine shop and sheet metal working.

In 1961 the name of the Department was changed to Aerospace Engineering and courses added to cover topics in astronautics as well as aeronautics. Graduate programs were begun shortly after the Department was formed. The first MS degrees were awarded around 1946 and the first PhD was given in 1953.

Many alumni from Penn State have distinguished themselves. To name a few, Paul Weitz spent 30 days in Skylab and commanded the first flight of the Challenger. Guion (Guy) Bluford was the nation's first black astronaut and has flown on several Shuttle missions. Mary Ilgen was the Chief of Commercial Flight Testing for McDonnell-Douglas at the time the DC-9 and DC-10 were certified. Glenn Spacht was the manager of the X-29 program and, later, a Vice=President for Grumman.

The Department of Aeronautical Engineering was formed at the **University of Illinois** in 1944 and was headed by Joseph (Shel) Stillwell until 1976. Beginning with only two srudents, it is presently one of the largest programs as measured by student numbers in the country. In 1969 the name was changed to Aeronautical and Astronautical Engineering. Illinois claims many renown alumni including an astronaut, Steven R. Nagel, an educator, Earl H. Dowell and Robert H. Liebeck who has contributed much to low-speed aerodynamics.

It may be of interst to those who know him to repeat a historical note found in reference 10 which states that "in 1947 at the tender age of 21, Allen I. Ormsbee was the youngest person in a class of World War II veterans (14 persons). He was the instructor."

Howard University has a small program in Aerospace Engineering with options at all three degree levels within the Mechanical Engineering Department.
The B.S. level programs date back to the early 1900's, although it is uncertain when the Aero option was first introduced. MS programs were initiated in **1967** with the first degrees awarded in 1969. One of them was in Aerospace.

Ph. D. level programs were initiated in the 1976-77 academic year and on the average, have produced about 1 Ph.D per year. The majority have been in Aerospace Dynamics, Fluid-Thermo, or Aerospace Structures and Materials. This is for a Department which has averaged only about 3 FTE faculty devoted to Aerospace Engineering.

The program at the **University of Central Florida** is relatively new compared to other programs The University did not open until 1986 at which time the Department of Mechanical Engineering and Aerospace Sciences was formed. The program was accredited by ECPD in 1987. In 1991 it was changed to Mechanical and Aerospace Engineering and then in 1996 to Mechanical, Materials and Aerospace Engineering. In 1986 there were three aero faculty with the program requiring 132 hours. It now requires 128 hours by State mandate. Presently the program has eight faculty devoted primarily to aerospace engineering.

Summary

This paper has barely scratched the surface of the information available on the 50 plus programs of Aerospace Engineering found in the U.S. Because of time and space limitations, many notable programs have not been mentioned and even those that are referred to in the paper are deserving of better coverage. The Departments of Aerospace Engineering can be proud of providing the base for the rapid development of the technology that has produced the modern aircraft and spacecraft. The jet transport is a beautiful machine meticulously designed to carry hundreds of passengers with utmost safety at speeds of 500 mph or faster above the weather to destinations all over the world. In two years from this December we will celebrate the 100th anniversary of the Wright Brother's achievement. I am proud to call myself an Aeronautical Engineer but extremely disturbed and saddened by the despicable and cowardly act of terrorism on Tuesday, September 12, 2001, that turned beautiful machines into weapons of horror.

References

1.   The Evolution of Technology and Education in Applied Aerodynamics,   Barnes W. McCormick, AIAA Annual Conference, Reno, NV, January 2000

2. History of the Aeronautical Engineering Department at the University of Minnesota 1929 to1962
Amy E. Foster, Master's Paper, U. of MN, 1999 (on the web)

3.   One Small step: The History of Aerospace Engineering at Purdue University, A. F. Grandt, W. A. Gustafsib, L. T. Cargnino, School of Aeronautics & Astronautics, Purdue University, 1995
4.   Parks College: Legacy of an Aviation Pioneer, William B. Faherty, S. J., Harris & Friedrich, 1990

5.   Proud Past…Bright Future: A History of the College of Engineering at the university of Colorado, 1893-1966 , Siegfried Mandel and Margaret Shipley, University of Colorado/College of Engineering

6. History of the College of Engineering, Wichita State Universtiy, Melvin H. Snyder, published by author

7.   Aeronautical/Aerospace Engineering at Iowa State; One Person's Story, Paul University

8.   Three Quarters of a Century of Aero at Iowa State University, Paul J. man, AIAA Paper No. 2002-0565, Jan. 2002

9.   Who Was Albert F. Zahm?, H. Dethloff and L. Snaples, AIAA Paper January 2000

10. Department of Aeronautical and Astronautical Engineering, University of Illinois at Urbana-Champaign, 1944-1994, booklet by Department celebrating 50th anniversary.

Figure 1

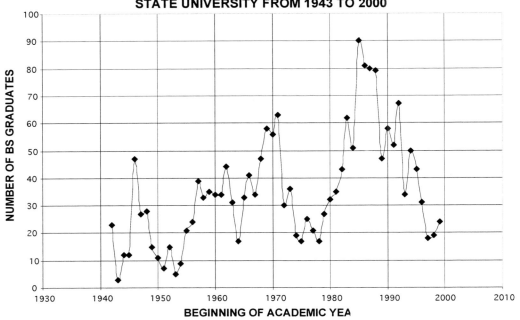

Part II

The Pioneering Universities

Chapter 3

A Century of Aerospace Education at MIT

by Lauren Clark[*] and Eric Feron[**]

Introduction

The aerospace industry and aerospace education at the Massachusetts Institute of Technology (MIT) came of age together. They shaped one another and were shaped by the same key figures, many of whom moved back and forth between industry, government, the military, and academia throughout their careers.

MIT's Department of Aeronautics and Astronautics is renowned for its key contributions to air- and spacecraft design, the internal combustion engine, aeroelasticity, and guidance and control, not to mention aerospace policy and administration. From the very start, the Department established a tradition of providing leadership to the field.

The link between the growth of the aerospace industry and the evolution of the aeronautics and astronautics curriculum at MIT is illustrated by the careers of two men who had enormous influence on both: Jerome Clarke Hunsaker and Charles Stark Draper.

In the early days of aviation, Hunsaker did more than anyone in the U.S. to "foster engineering science, transferring European theories of aerodynamics to America and bridging the gap between theory and practice," writes aviation historian William F. Trimble.[1] Part of this accomplishment was establishing an aeronautics course—the nation's first—at MIT.

Draper, in addition to his achievements in guidance and control, helped define the modern aerospace curriculum, one that produced many of the field's leaders and innovators. Together, Hunsaker's and Draper's careers spanned most of the first 100 years of aerospace education.

This chapter recognizes the contributions of other MIT faculty past and present, including Edward P. Warner, Charles F. Taylor, Raymond L. Bisplinghoff, Robert C. Seamans, Rene H. Miller, Jack L. Kerrebrock, and Sheila E. Widnall. We acknowledge the many contributions to the aerospace industry by alumni who did not return to campus after earning their degrees. We do not detail them, however, since our focus is on aerospace education at MIT.

The authors acknowledge the archives of MIT's Department of Aeronautics and Astronautics, the Institute Archives, the MIT Libraries, and the MIT Museum for information presented in this chapter, except where other references are noted. Also, they thank Deborah Douglas, the MIT Museum's curator of science and technology, and William Litant,

---

Massachusetts Institute of Technology, Cambridge, MA 02139
*School of Engineering Communications Office
** Laboratory for Information and Decision Systems, Department of Aeronautics and Astronautics.
Ph: 617-253-1991. Email: feron@mit.edu.

communications director of the Department of Aeronautics and Astronautics, for contributing their knowledge about the Department.

I. The Beginnings of Aerospace Education at MIT

The study of aeronautics at MIT dates back to 1896, when mechanical engineering student Albert J. Wells designed and built a 30-square-inch wind tunnel as part of his thesis, "An Investigation of Wind Pressure upon Surfaces." Interest in aeronautics grew in the decade that followed. The Tech Aero Club was founded in 1909 by a group of students that included Frank W. Caldwell, who later introduced a practical variable-pitch propellor. The same year, Naval Academy graduate Jerome C. Hunsaker enrolled in MIT's graduate program in naval construction, concurrent with an appointment at the Boston Navy Yard.

Class in shop working on alignment, 1918. In the early years of aerospace education at MIT, students learned using actual aircraft. *Courtesy MIT Museum*

MIT's educational partnership with the U.S. Navy, which began in 1893 with the creation of the Department of Naval Architecture, would be crucial to the development of its aeronautical engineering program. The Navy was one of the first "markets" for aeronautical engineers, and it continues to support much research in the Department today.

As student and faculty enthusiasm for aeronautics at MIT grew, MIT President Richard C. Maclaurin worked to establish the subject as part of the curriculum. He "saw aerodynamics, a field in which Europeans had made rapid advances in recent years, as one of the priorities for the 'New Technology,'" writes Trimble.[2] In 1913, following a series of well-received lectures by Albert A. Merrill, an aviator and founding member of the Boston aeronautical society, the Executive Committee requested that a graduate course in aeronautics be established. Maclaurin asked the Navy to detail Hunsaker to MIT to develop the new curriculum.

While studying ship propeller design, the young naval engineer had developed a fascination with the aeronautical literature in MIT's library and become an aviation enthusiast. He published the first English translations of two important French works in aerodynamics: E. LaPointe's 1910 article, "Aviation in the Navy," on the mathematical foundations of the study of aeronautics; and *The Resistance of the Air and Aviation*, the classic book on aerodynamics by Gustave Eiffel (best known for his eponymous tower in Paris).

Hunsaker spent the summer and fall of 1913 surveying aeronautical laboratories in England, France, and Germany. He met eminent aerodynamicists Eiffel and Ludwig Prandtl. His European survey, and the resulting implemention of an aeronautical engineering program at MIT, initiated a shift in emphasis in U.S. aeronautics—from the trial-and-error experimentation of the Wright brothers to research firmly grounded in theory and mathematical principles.

"We are at the point where the inventor can lead us but little farther, and it is to the physicist and the engineer that we must look for the perfection of air craft and the development of a new industry growing out of their manufacture and operation," wrote Hunsaker.[3]

In 1914, MIT offered the nation's first course in aeronautical engineering: 13.72, Aeronautics for Naval Constructors, in the Department of Naval Architecture. (The nation's first full-fledged department of aeronautics was established at the University of Michigan-Ann Arbor in 1916-1917.[4] At MIT, Aeronautics moved to Physics, then to Mechanical Engineering, before becoming its own department in 1939.) The head of the Department of Naval Architecture, Cecil H. Peabody, and professor of mathematics Edwin B. Wilson helped Hunsaker design and administer the curriculum. Subjects included theoretical hydrodynamics, applied hydrodynamics, wind tunnel laboratory, airship theory, aerial propellers, and theory and practice of airplane design.

As part of the new course, Hunsaker built a wind tunnel modeled after the one at Great Britain's National Physical Laboratory (NPL). The first structure on MIT's new campus (recently relocated from Boston to Cambridge), the open-circuit tunnel was four feet square (1.2 by 1.2 meters) with a speed of 30 miles per hour. Hunsaker's assistant on the wind tunnel project was Donald W. Douglas (SB '14), who later established the aircraft company bearing his name.[5]

The first person to complete the course in aeronautics, thus earning the nation's first master of science (SM) in aeronautical engineering, was Hou-Kun Chow. His 1915 thesis, *Experimental Determination of Damping Coefficients in the Stability of Aeroplanes*, describes a problem that remains difficult to solve even today.

The first American doctorate in aeronautical engineering was conferred on Hunsaker a year later. His research, culminating in a thesis titled *Dynamical Stability of Aeroplanes*, was the subject of a technical report published by the National Advisory Committee for Aeronautics (NACA) in its first annual report in 1915. MIT President Maclaurin was one of the first 12 members appointed to the NACA, the predecessor of the National Aeronautics and Space Administration (NASA).

Hunsaker's first association with MIT ended in 1916, during the United States' effort to bolster its defenses prior to its involvement in World War I. He went to Washington, D.C. to head the Aircraft Division in the Bureau of Construction and Repair. He would spend the next 17 years organizing the U.S. Navy's aviation design and procurement program, then working on aerial navigation systems for Bell Labs and commercial airships for Goodyear-Zeppelin, before returning to MIT in 1933.

During World War I, MIT leased its wind tunnel to the U.S. Army and gave Army and Navy officers non-degree training in aeronautics. The war, which marked the beginning of air power as a weapon, transformed aeronautical technology. By 1918, MIT had conferred 17 graduate degrees in aeronautics, and 81 students were enrolled in Course 13.72. After the war, the wind tunnel was kept busy with Army contracts.[6]

In 1920, Aeronautics moved to the Department of Physics, and Edward P. Warner became the new head. Warner, having earned his SM in 1918, went to the NACA Laboratory at Langley Field, where he started a program of flight research and the design and construction of the NACA's first wind tunnel.[7] On his return to MIT, he revised and expanded the graduate program in aeronautics and built new wind tunnels. Warner left MIT in 1926 to be Assistant Secretary of the Navy for Aeronautics. He went on to make significant contributions to aeronautics as editor of *Aviation*; a member of the NACA, the Federal Aviation Commission, and the Civil Air Board; and president of the Council of the International Civil Aviation Organization.

MIT's first undergraduate course in aeronautical engineering—Course 16—began in 1926-1927 and quickly filled up. "The influx of students became so large that starting in the fall of 1928 the sophomore class was limited to 30 men," wrote longtime MIT aeronautics professor Shatswell Ober.[8] By 1929, 29 students had earned the SB.

Charles Fay Taylor left his position as chief engineer of the Wright Aeronautical Corp. in 1926 to join the MIT aeronautics faculty and head the department. Taylor had a long and distinguished career at the Institute and is best known as a pioneer in the development of the internal combustion engine, especially those used in aircraft. While at Wright, he was involved in developing the air-cooled, "Whirlwind" engine used on the historic flights of Charles Lindbergh and Robert Byrd. His research and teaching at MIT formed the basis of the scientific framework for engine design and operation still in use today. It established MIT as an internationally renowned center in this field.[9]

MIT's Daniel Guggenheim Aeronautical Laboratory opened in 1928, thanks to a $230,000 grant from the Daniel Guggenheim Fund for the Promotion of Aeronautics. The grant was followed by more support from private philanthropy and industry, including a gift in 1929 from General Motors CEO Alfred P. Sloan, Jr., for an internal combustion engine laboratory—the Sloan Automotive Laboratory—which remains a vital research laboratory to this day.

In 1933, Jerome Hunsaker returned to MIT as head of the Department of Mechanical Engineering, where the Aeronautical Engineering course then resided. He immediately set about updating the undergraduate curriculum, placing emphasis on fluid dynamics, thermodynamics, and electrical engineering, and making room for more electives. At the graduate level, Otto C. Koppen's class on airplane design "reflected some of the new thinking that emphasized the airplane as part of a larger technological system," writes Trimble.[10]

Hunsaker also led MIT's effort to acquire a state-of-the-art wind tunnel. The existing 4-foot, 5-foot, and 7.5-foot diameter wind tunnels had become virtually obsolete due to advances in the speed and size of aircraft. Hunsaker raised funds from aviation companies and industry executives—many of them MIT alumni—which led to the construction of the Wright Brothers Memorial Wind Tunnel.[11] Dedicated in 1938, the variable-pressure tunnel was able to simulate altitudes of up to 37,000 feet. It was the first of MIT's large-scale facilities for advanced research in aerodynamics and became a center for aeronautics research and testing during World War II.[12] Still active on campus today, the tunnel is undergoing modernization for many more years of profitable use.

Jerome C. Hunsaker (left), who established MIT's aeronautics program, dedicates the Naval Supersonic Wind Tunnel (1949) as lab director John Markham (seated left) and MIT President James Killian (seated right) observe.   *Courtesy MIT Museum*

When Aeronautics became its own department, in 1939, Hunsaker remarked, "The total effect of our graduates on the airplane industry cannot be estimated. But it is of interest to note that MIT graduates include the chief engineers or engineering directors of Curtiss Wright, Glenn L. Martin, Pratt & Whitney, Vought, Hamilton Standard, Lockheed, Stearman, and Douglas, as well as the engineer officers of the Naval Aircraft Factory and of Wright Field."[13] The early, "try-and-fly" days of aviation were over. The era of the engineered aircraft had fully emerged.

II. The Boom Years – World War II to the Late 1960s

Charles Stark Draper, who had joined the aeronautics staff as a research assistant in 1929, was an early addition to the faculty under Hunsaker's stewardship. In the 1930s, he single-handedly established a course of study in instrumentation and in 1935 founded the Instrumentation Laboratory. The Lab started as a small group of graduate students doing research on discarded aircraft instruments: altimeters, airspeed meters, magnetic compasses. It would eventually become the world's foremost academic center for inertial guidance research and development, "an academic empire rivaling MIT itself," writes science and technology historian Stuart W. Leslie.

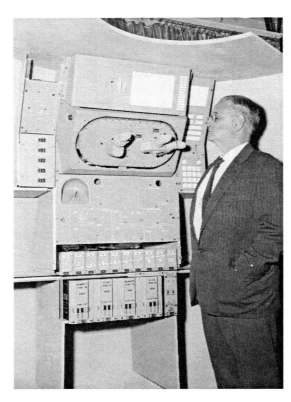

Charles Stark Draper, founder of the MIT Instrumentation Laboratory, inspects a mock-up of the Apollo guidance, navigation, and control system.

– Courtesy MIT Museum

During World War II, MIT's Department of Aeronautics expanded rapidly to meet the needs of the U.S. military. The caps on enrollment that had been instituted in the 1920s were removed in response to the urgent demand for aeronautical engineers—a demand that would continue during the cold war. While there were only 10 ScDs between 1933 and 1946, there were 47 doctoral degrees between 1947 and 1959.

As in World War I, MIT gave special training to Army and Navy officers. Approximately 600 officers received aviation engineering training with specialization in structures or engines.

The size and importance of the Instrumentation Lab increased dramatically during the war. Through contracts with the Sperry Gyroscopic Company, the Lab developed the Mark 14 gunsight for the Navy. This led to a new contract with the Navy to further develop gyroscopic gunsight technology. It also led to a new name that reflected the lab's focus on classified research—the Confidential Instrument Development Laboratory, or CIDL. The CIDL collaborated with Gordon S. Brown's Servomechanisms Laboratory, which was also conducting research in gunfire control.

War-related research took place at the Department of Aeronautics' other laboratories as well. The Wright Brothers Wind Tunnel operated 24 hours a day, testing models of new aircraft designs for Martin, Grumman, Lockheed, and other manufacturers. Charles F. Taylor and his brother, Edward S. Taylor, worked on aviation fuel studies at the Sloan Laboratory. The Vibrations Measurements, Flutter Research, and Structures laboratories studied aircraft vibrations at high speed.

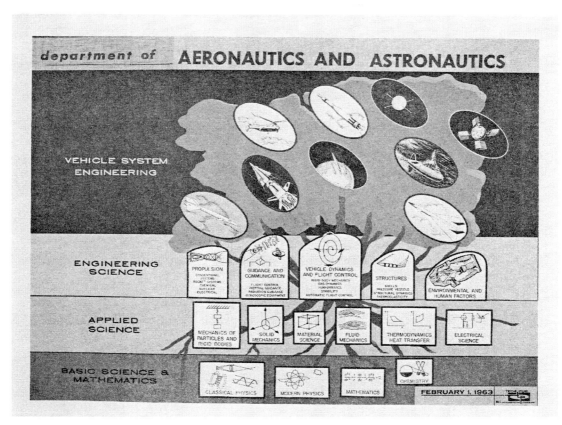

A 1963 illustration by Draper depicts the evolution of the aerospace industry as a tree, with Basic Science & Mathematics at the roots and Vehicle System Engineering at the leaves.

Courtesy MIT Museum

MIT was recognized for having played an important role in winning the war through many technological advancements. In addition to the Department of Aeronautics' contributions, the Radiation Laboratory transformed British radar technology into one of the key instruments of air war. (Interestingly, while MIT advanced radar, a team of Harvard physicists led by Fred R. Whipple developed radar countermeasures.) Faculty and administrators served as leaders of government research and development. The experience helped MIT become one of the world's leading science- and technology-based universities.

After World War II, almost all major research in MIT's Department of Aeronautics was performed for the U.S. military. MIT, in fact, emerged from the war as the nation's largest nonindustrial defense contractor.[14] The Instrumentation Lab, in particular, was transformed by its war-related activities. Its staff had grown to 100, its annual budget to six figures.

Other laboratories in the Department also became prominent after the war. The Gas Turbine Laboratory advanced the new technology of the gas turbine engine and quickly grew into a top academic center. The Aeroelastic and Structures Laboratory, under the guidance of Raymond Bisplinghoff, led the design of aircraft structures for high-speed flight, virtually creating the modern specialty of aeroelasticity.[15] The Naval Supersonic Laboratory achieved supersonic flow at Mach 2.0 and, through Project Meteor, helped develop a supersonic air-to-air missile.

Meanwhile, the Instrumentation Laboratory began its pioneering research on inertial guidance. In 1953, Draper, who replaced Hunsaker as head of the Department of Aeronautics two years earlier, flew from Massachusetts to Los Angeles using his SPace Inertial Reference Equipment (SPIRE). This was the first long-distance inertially navigated aircraft flight. The Instrumentation Lab later designed the Polaris missile's inertial guidance system.

From just after World War II until 1959, the aeronautical engineering program grew and branched into specializations as did the industry at large. Draper and his colleagues on the faculty worked to compress the myriad areas of the field into a managable, four-year course. In the process, they helped define the modern aerospace curriculum. "Ever-increasing performance requirements for today's and tomorrow's aircraft force the engineers who devise, construct and prove them to become familiar with a broader range of professional fields," Draper wrote in 1952. "In an industry that is now acutely feeling the shortage of professionally trained men, the greatest demand is for engineers educated with the systems viewpoint."[16]

One of the new areas of emphasis was Draper's specialty, automatic control. That emphasis grew in tandem with the increased interest in space research as the decade progressed.

"The airplane is here for a long time, and we will continue to regard aeronautics as fundamental. But the sky, or speaking more precisely, the air, is no longer the limit. Interplanetary travel is yet to be accomplished but clearly it will be feasible," said Draper when MIT's Aeronautics Department became the Department of Aeronautics and Astronautics in 1959.[17]

The change coincided with a national concern over a "missile gap" between the U.S. and the Soviet Union following the Sputnik launch in 1957. The cold war, the space race, and massive government investment in missile technology led the Department to expand instruction and research in astronautics, instrumentation, and guidance. Enrollment rose sharply at this time and again in the mid-1960s, when NASA's Apollo program was in full swing.

In 1961, President John F. Kennedy gave a speech urging the U.S. to commit itself to sending astronauts to the moon by the end of the decade. The speech was based on a plan prepared by Robert C. Seamans, Jr., Deputy Administrator of NASA. (The NACA's transformation to NASA in 1958 came at the recommendation of former MIT President James R. Killian, Jr.) Seamans earned his ScD in Instrumentation under Draper in 1951, taught at MIT until 1955, and returned to Cambridge in 1968 when appointed to the Jerome Clarke Hunsaker Professorship. The professorship was established in part by Maj. Lester D. Gardner (SB 1898), whose bequest also made possible an annual lecture on the history of aeronautics. Starting in 1959, the list of Gardner Lecturers includes many luminaries in the field.

Astronautics research at MIT accelerated in 1963-4 with the opening of the Center for Space Research (funded by NASA), the Experimental Astronomy Laboratory (later the Measurements Systems Laboratory), and the Space Propulsion and Man-Vehicle laboratories. In 1965, MIT's Flight Transportation Lab (FTL) was founded. With the participation of the Electronic Systems Laboratory (formerly the Servomechanisms Laboratory and now the Laboratory for Information and Decision Systems), the FTL pioneered research in commercial aviation.

The late 1960s are distinguished by MIT's contribution to the Apollo program. In 1968, Apollo 8 circumnavigated the moon with a guidance, navigation, and control system developed by the Instrumentation Lab. The system performed flawlessly on this mission and on all subsequent Apollo missions, notably through the contributions of Richard H. Battin to computer-assisted spacecraft guidance and control. Apollo 11 astronaut Buzz Aldrin (PhD '63) was the second man to set foot on the moon. Including Aldrin, four of the 12 Apollo astronauts who walked on the moon were Course 16 graduates. The others were Edgar D. Mitchell (ScD '64), David R. Scott (SM '62), and Charles M. Duke (SM '64).

Draper stepped down as Department head in 1968 and was replaced by Rene H. Miller, faculty member since 1944 and an expert in helicopters and supersonic transport. Miller filled the post until 1978.

In the late 1960s and early 1970s, MIT was criticized by anti-war protesters for its support of defense research. The Instrumentation Lab, the center of much of MIT's classified research for the military, was a particular target.[18] In response, MIT renamed the lab the Charles Stark Draper Laboratory and made it a separate division within the Institute. By the time Draper retired in 1969, his lab boasted 800 alumni and a $54 million budget—a budget equal to the rest of MIT's laboratory budgets (excluding Lincoln Laboratory's) combined. Draper Laboratory was finally divested from MIT in 1973.

III. Contributions to Late 20th Century Aerospace Systems and Education

The anti-military protests, the maturation of the aerospace industry, and the emergence of computer science as the most alluring field for engineers all contributed to a drop in enrollment in aerospace at MIT. In 1974, the number of incoming sophomores (MIT students generally declare their majors in their sophomore year) fell to a pre-World War II low of 12. The Department developed Unified Engineering "as a way to teach a small, informal class which clearly showed the interrelationship of different disciplines when applied to aerospace engineering."[19]

One of the new curriculum's authors was Jack Kerrebrock, who introduced computational fluid dynamics and blowdown compressor testing as new research tools in the Gas Turbine Laboratory. He would replace Rene Miller as Department head in 1978.

The challenging curriculum, which students took as sophomores, crammed into two semesters the disciplines of statics, solid mechanics, and materials; dynamics; fluid mechanics; thermodynamics and propulsion; and linear systems. It pointed out the connections between the disciplines and emphasized aerospace engineering as a systems business. It also put forth the idea that leaders in the field consider the interaction of technical solutions with economic, political, social, and environmental needs and constraints of society.[20]

Although enrollment in the early 1970s had reached a low point, the quality of the students in the Department remained high. The members of the 1974 sophomore class included future MIT faculty Edward F. Crawley, head of the Department from 1996 to 2003 (Wesley L. Harris took over the position in June 2003), and Paul Lagace, an international authority on the response and failure of composite structures.

MIT Aeronautics and Astronautics Sophomore Enrollment: 1943-2000

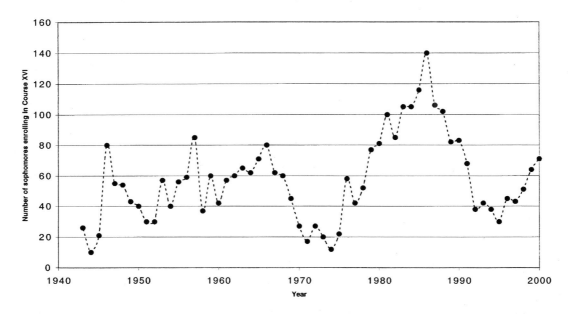

Data sources: Hollister, W.M., Crawley E.F. and AmirA.R. "Unified Engineering: A Twenty Year Experiment in Sophomore Aerospace Education at MIT", 32nd Aerospace Sciences Meeting & Exhibit, January 10-13, 1994, Reno, NV, AIAA Paper no.94-0851; MIT Aeronautics and Astronautics Visiting Committee Presentations, 1977 and 1997; The MIT Aeronautics and Astronautics Undergraduate Office.

— Graph by Annalisa L. Weigel

Two aerospace firsts in the late 1970s and early 1980s involved MIT alumni. A. Thomas Young (ScD '72) was director of NASA's Viking I and II missions to Mars in 1976. The space shuttle program, which began in 1981 when *Columbia* was launched, was led by James A. Abrahamson (SB '55) from 1981 to 1984. In 1983, experiments developed by the Man Vehicle Laboratory (MVL) flew aboard the Shuttle for the Spacelab 1 mission. It would be the first of a series of MVL experiments to accompany Shuttle missions.

After Ronald E. McNair (PhD '77) died in the *Challenger* accident in 1986, Department Head Eugene E. Covert served on the presidential commission that investigated the accident, recommending changes in the Shuttle's design and management. That incident led to a spike in enrollment in aeronautics and astronautics, which had been rising steadily since the mid-1970s.

In 1988, a team of 40 MIT students and alumni set a world record for human-powered flight. Their craft, *Daedalus*, flew 74 miles from Crete to Santorini in a recreation of the mythological flight of the craft's namesake. In 1989, the Space Shuttle *Atlantis* launched the Magellan spacecraft to conduct a radar mapping of Venus. Building on MIT's extensive role in developing radar, an Institute team headed by Gordon H. Pettengill designed the radar instrument, which mapped more than 98 percent of Venus's surface.

The Department went through a period of change in the 1990s, as several senior professors who had made their careers during the cold war retired and enrollment declined in the first part of the decade (though not as much as it did in the 1970s). Since the mid-1990s, undergraduate enrollment has steadied to an average of 50 to 60 in the sophomore class.

In 1993, Sheila Widnall, the first female faculty in MIT's Department of Aeronautics and Astronautics, was appointed the first female Secretary of the Air Force. (More recently, she was appointed to the panel investigating the 2003 *Columbia* shuttle accident.) In 1997, Pathfinder landed on Mars, and its sister ship, Mars Global Surveyor, conducted the most extensive mapping of the planet to date. MIT scientists developed the Surveyor's laser altimeter, which provided high-resolution elevation maps of Mars' surface.

MIT's Center for Space Research was responsible for the CCD Imaging Spectrometer for the Chandra X-Ray Observatory, which was deployed in 1999. Chandra is an advanced astrophysics facility whose control center in Cambridge, MA, collects data for the study of exploding stars, quasars and black holes.

Overall, MIT has graduated more astronauts than any other university other than the service academies. Its alumni have taken part in more than one-third of U.S. space flights and have collectively logged more than 10,000 hours in space. Five MIT faculty have served as chief scientist for the Air Force: Eugene E. Covert, Winston R. Markey, James W. Mar, H. Guyford Stever, and Daniel Hastings. Moreover, MIT has widely influenced academia. More than 25 percent of professors in the nation's leading aerospace programs are MIT alumni. In 2000, several MIT alumni and former faculty were heading aerospace programs at other major institutions: George S. Springer at Stanford, Robert G. Loewy (MS '48) at Georgia Tech, David C. Hyland (PhD '69) at Michigan, and Dennis K. McLaughlin (PhD '70) at Penn State.

IV. MIT's Place in Aerospace Engineering – Present and Future

With the end of the emphasis, during the cold war, on technology driven by defense, both the aerospace profession and aerospace education reached a crossroads. The American aerospace industry adjusted to the new reality through massive consolidation and an emphasis on cost over performance. The late 1990s and early 2000s introduced new opportunities for aerospace in the areas of transportation, commerce, communications, and security.

At the same time, the research and development priorities of aerospace education shifted, as did its leaders' areas of expertise. The MIT Department of Aeronautics and Astronautics, therefore, is repositioning itself in order to continue leading the field, particularly in the areas of information engineering, vehicle engineering, and systems architecture and engineering.

Under the leadership of Edward Crawley, the year 2000 marked the beginning of a new focus for the Department. Realizing that engineering education had in recent years emphasized engineering science at the expense of engineering practice, the Department embarked on an overhaul of its curriculum. After two years of discussion with industry, benchmarking of other universities, reviewing literature, and surveying alumni, the Department proposed an educational initiative, called CDIO, that will have fundamental, long-lasting institutional impact at MIT, and more broadly, in university engineering education.

View of the new hangar space inside the recently renovated Guggenheim Aeronautical Laboratory, which originally opened in 1928. – *Photo by Nick Wheeler, courtesy Cambridge Seven Associates, Inc.*

The Department is committed to making the Conception, Design, Implementation, and Operation (CDIO) of aerospace and related complex high-performance systems the engineering context of the education it provides. With CDIO, the Department is creating a new, integrated education whose overall aim is to produce technically expert, socially aware, and entrepreneurially astute aerospace professionals who will become—like their accomplished predecessors—the next engineering leaders.

ENDNOTES

[1] William F. Trimble, *Jerome C. Hunsaker and the Rise of American Aeronautics* (Smithsonian Institution Press, 2002), p. 2

[2] Trimble, p. 23

[3] Jerome C. Hunsaker, "Aeronautical Research," December 20, 1913. Folder 29, box 2, JCHP/MIT Archives

[4] Emerson W. Conlon, "The Department of Aeronautical Engineering," University of Michigan College of Engineering (early 1950s)

[5] Stuart W. Leslie, *The Cold War and American Science: The Military-Industrial-Academic Complex at MIT and Stanford* (Columbia University Press, 1993), p. 12

[6] Leslie, p. 78

[7] Shatswell Ober, "The Story of Aeronautics at MIT 1895-1960" (Department history, 1965), p. 12

[8] Ober, p. 18

[9] *Tech Talk*, "C.F. Taylor Sr. dies at 101," July 24, 1996

[10] Trimble, p. 146

[11] Leslie, p. 79

[12] Leslie, p. 146

[13] Leslie, p. 79

[14] Leslie, p. 14

[15] Leslie, p. 88

[16] Charles S. Draper, *Course XVI Takes a New Slant* (MIT course booklet, 1952), p. 2

[17] *The Tech*, "Course XVI Hints at Future in Added Astronautics Title," January 9, 1959

[18] Leslie, 235-8

[19] Walter M. Hollister, Edward F. Crawley, and Amir R. Amir, "Unified Engineering: A Twenty-Year Experiment in Sophomore Aerospace Education at MIT," (AIAA 94-0851), p. 1

[20] Hollister, Crawley, Amir, p. 2

Chapter 4

Aeronautical and Aerospace Engineering
At The University of Michigan

Thomas C. Adamson, Jr. *
University of Michigan
Ann Arbor, Michigan

ABSTRACT

A brief history is given of the Department of Aerospace Engineering at the University of Michigan and its contributions to the Aerospace field.

The University of Michigan has been involved in aerospace education since the very inception of the field. Indeed, our founder, Felix Pawlowski, is shown with Orville Wright and other early aviation pioneers in Figure 1. To the best of our knowledge, ours

FIGURE 1   Aviation pioneers, including Orville Wright (4th from right, front row) and University of Michigan Aeronautical Engineering Program Founder Felix Pawlowski (3rd from right, front row)

was the first undergraduate program for aeronautical engineering offered anywhere. Because of this, we have a large group of graduates, a total of 6403, with 5094 still living.

In the following, I concentrate on an anecdotal history of the department and its graduates and their contributions, rather then repeating only facts and figures. My material comes, in the main, from three sources: A University Publication, "A Century of Engineering Education", (1) published in 1954; "The First Fifty Years (a Fragmentary, Anecdotal History)" (2), written in 1964 on the occasion of our 50th Anniversary; and "The Third Quarter Century, More Fragmentary, Anecdotal History" (3) written in 1989 on our 75th Anniversary. I cheerfully admit to using portions of these documents for this manuscript.

Our Department owes its inception to the intense interest in aviation of Professor Herbert C. Sadler, Chairman of the Department of Naval Architecture and Marine Engineering. This interest was evidently traditional in his family – his great-grand-uncle James Sadler was the first English balloonist (late 1700's early 1800's) and two of James's sons became balloonists. Also, Sadler worked at the University of Glasgow (before the U of M) with a man who was a British pioneer in glider flying and a follower of Lilienthal.

In 1911, the U of M Aero Club was started and students built a small wind tunnel for experiments. They also built a glider and flew it as a kite around Ann Arbor, with a student "flying" in it. For lateral control, there were two helpers with lines to the wing tips; the ground helpers were often lifted off the ground by sudden wind gusts or the pilot's too enthusiastic use of the elevator. Sadler was the advisor and repeated the warning given him by Wilbur Wright; "If you will advise them (the students) to build a glider and to fly it, do not let them build it too light!".

Felix Wladislaw Pawlowski, who had taken the first course in aeronautical engineering ever given, by Lucien Marchis at the University of Paris, came to this country in 1910 and spent two years in Chicago as a designer in the automobile industry. In 1911-1912 he wrote to a number of engineering colleges and technological institutes wanting to develop courses in aeronautics. Most gave negative replies ("e.g. aviation very likely will never amount to anything!"). However, he did receive two encouraging replies, one from MIT which had to decline "for the present" due to lack of funds, and one from the Dean of Engineering at the University of Michigan, Mortimer Cooley, which resulted in his being appointed in 1913 as a teaching assistant in Mechanical Engineering. He was appointed with the understanding that he would be permitted to introduce courses in aeronautical engineering, and became an Instructor in 1914. His job as a teaching assistant paid $800 per year! Pawlowski, was a very talented, charismatic person, and also was evidently quite a character. At the age of 32, he had seen the Wright Brothers fly, and was so stirred that he decided to become an aeronautical engineer. He could not afford to learn to fly at the school set up by the Wrights in Paris, France, so he returned to the University of Paris, where he was doing graduate work in Mechanical Engineering. Somehow, he taught himself to fly in the fields outside the city in a monoplane similar to the one Bleriot had used to cross the channel the previous summer. His plan to become an aeronautical engineer was reinforced by the aforementioned course by Marchis, at the University of Paris, where he also received the <u>certificat d'étude.</u>

THE UNIVERSITY OF MICHIGAN COURSE
IN AERONAUTICS

THE faculty of the College of Engineering of the University of Michigan is developing the course in aeronautics which they offer and it is to be their endeavor to make it as comprehensive as possible. It is expected that the students will gain much information and also practical experience in connection with the work done at the Packard Motor Car Company of Detroit. The aim of the course is to teach the theory of aeroplanes and to enable students to secure positions in manufacturing plants.

The course is under the direction of Professor H. C. Sadler and Assistant Professor Felix Pawlowski, one of our contributing technical editors. The summary of the course is as follows:

1. GENERAL AERONAUTICS. Lectures and recitations. *Two hours.* First semester.

    An introductory course giving the essential principles of aeronautics (balloons, dirigibles, ornithopters, helicopthers, aeroplanes, helicoplanes and kites), history of flight and description of modern aircraft.

    Open to junior students. Must be preceded by E. M. 2 and 3.

2. THEORY OF AVIATION. Lectures and recitations. *Two hours.* Second semester.

    The course deals with the following questions: properties of the air, general discussion of aerodynamics, aerodynamical properties of planes and various constructive elements of an aeroplane, power necessary for flight, equilibrium of aeroplanes, stability of aeroplanes, air currents.

    Must be preceded by Course 1.

3. THEORY AND DESIGN OF PROPELLERS. Lectures, recitations and drawing. *Two hours.* First semester.

    Theory of propellers on the Drzewiecki system; Eiffel's method of propeller analysis and graphical method of determining propellers for specified conditions; strength of propellers and influence of gyrostatic moments in quick turns. The student will design a propeller and analyze the distribution of stresses in the blades. Must be preceded by Course 2.

4. AEROPLANE DESIGN. Lectures and drawing. *Three hours.* First semester.

    This course includes the investigation of the design of the aeroplane from the aeronautical and strength standpoints. The strength and design of all the detail are discussed and a completed design prepared.

    Must be preceded or accompanied by Course 3 and preceded by M. E. 6.

5. AERODYNAMIC LABORATORY. *One hour.* Second semester.

    An elementary course covering use of instruments, investigation of aerodynamical properties of the various bodies used in aeroplanes and airships, test of propellers.

    Must be preceded or accompanied by Courses 2 and 3, and preceded by M. E. 7.

6. DESIGN OF AERONAUTICAL MOTORS. Lectures and drawing. *Two hours.* Second semester.

    Complementary course to M. E. 15, dealing with special features of the aeronautical motors, critical study of various types of motors and design of a complete motor of certain type.

    Must be preceded by M. E. 15.

7. THEORY OF BALLOONS AND DIRIGIBLES. Lectures and recitations. *Two hours.*

    Study of equilibrium and stability of spherical balloons and dirigibles; description of French, German and Italian types; resistance and propulsion, dynamical stability of dirigibles; operation and maintenance of balloons and dirigibles.

    Must be preceded by Courses 1, 2, and 3.

8. DESIGN OF BALLOONS AND DIRIGIBLES. Lectures and drawing. *Two hours.*

    Investigation of the design of a balloon and a dirigible from the aeronautical and strength standpoints. Questions of strength and design of all the details of the non-rigid, semi-rigid, and rigid types are discussed and a completed design of one type prepared.

    Must be preceded by Course 7.

9. THEORY AND DESIGN OF KITES. Lectures, recitations and drawing. *Two hours.*

    Critical study of various types of man-carrying kites and the launching devices. Investigation of the design from the aeronautical and strength standpoints. Completed design of a kite train of one type is prepared.

    Must be preceded by Courses 1, 2, and 7.

10. DESIGN OF AERODROMES AND HANGARS. Lectures, recitations and drawing. *Two hours.*

    Planning and equipment of aerodromes and aero-ports; construction of transportable, stationary, revolving and floating hangars. Completed design of one type is prepared.

    Must be preceded by Courses 2 and 7.

11. ADVANCED STABILITY. Lectures and recitations. Advanced study of more complicated phenomena of stability according to Ferber, Botheast, Bryan, and Bairstow.

    Must be preceded by Course 2 and Math. 9 (Differential Equations).

12. AERONAUTICS. Advanced Reading and Seminary.

13. AERONAUTICS. Advanced Design.

14. AERONAUTICS. Advanced Research.

The program which students taking the complete course have to take is as follows:

FIRST YEAR

FIRST SEMESTER		SECOND SEMESTER	
*Modern Language	4	*Modern Language	4
Gen. Chem. (2E), or		Engl. or Gen. Chem. (2E)	
Engl. 1	5 or 4		4 or 5
Alg. and Anal. Geom. (Math. 1)	4	Alg. and Anal. Geom. (Math. 2)	4
Shop 1 or 2 and Des. Geom. 4	4	Des. Geom. 5 and Shop 1 or 2	4
Total hours	17 or 16	Total hours	16 or 17

SECOND YEAR

*Language	4	*Language	4
Calculus I (Math. 3E)	5	Calculus II (Math. 4E)	5
Mech., Sound, Heat (Phys. 1E)	5	Magn., Elec., Lt. (Phys. 2E)	5
Surveying 4	2	Kinematics, etc. (E. M. 1)	4
Machine Draw. (M. E. 1)	2		
Total hours	18	Total hours	18

SUMMER SESSION

Shop 3	4
Elect. App. I (E.E. 2)	4
Total hours	8

THIRD YEAR

Shop 4	4	Hydromechanics (E. M. 4)	2
Strength, Elec. (E. M. 2)	3	Thermodynamics (M.E.) 5)	3
Dynamics (E. M. 3)	3	Machine Design (M. E. 6)	4
El. Mach. Des. (M. E. 2)	3	Eng. Materials (Ch. E. 1)	3
Heat Engines (M. E. 3)	4	Theory of Struct. (C. E. 2)	3
Gen. Aeronautics (Aero. 1)	2	Theory of Avia. (Aero. 2)	2
Total hours	19	Total hours	17

FOURTH YEAR

Mech. Lab. (M. E. 7)	2	English 5, 6, 9 or 10	2
Internal Com. Eng. (M. E. 15)	3	Mech. Lab. (M. E. 32)	2
Theory and Design of Propell. (Aero. 3)	2	Aerodynam. Lab. (Aero. 5)	1
Aeropl. Design (Aero. 4)	3	Design of Aeronaut. Mod. (Aero. 6)	2
Elective	5	Elective	5
Total hours	15	Total hours	12

Reprinted from *Aerial Age*, 1915.

FIGURE 2   First curriculum for Aeronautics at the University of Michigan

Interest in aeronautics was stimulated in 1913 by a series of lectures by Pawlowski and Professor Marchis, who came to Ann Arbor from Paris to deliver his talks. His principal subject was the practical application of physics, so that his lectures were not limited to aeronautical ideas, but the appearance of a world-famous authority on an American campus strengthened the increasing academic respectability of aeronautical engineering in this country.

The first course, Theory of Aviation, was introduced in 1914 for two hours of credit; it dealt with the principles of aerodynamics and mechanics of flight. In his autobiography Dean Cooley said, "I hid this course in the Department of Marine Engineering and Naval Architecture for a time, for aeronautical engineering was not considered important enough to make it conspicuous... !" Previous to this, courses were offered without credit to members of the aero club, which built another ("not better") glider, a biplane. It was again flown as a kite; in addition, probably for the first time in the history of aviation, an automobile was used to tow it. In 1915-1916 two new courses, Propulsion of Aeroplanes and Aeroplane Design were added.

The first regular courses in aeronautics and the first curriculum were established at this time. A reproduction of the coursework covered is shown in Figure 2; it was printed in a 1915 issue of "Aerial Age". Of the 14 courses listed, only the first six were required as a minimum to qualify for a degree in aeronautical engineering. The remainder were offered as electives. In 1916-1917 a four-year program leading to a bachelor's degree in aeronautical engineering was arranged and included in the Department of Naval Architecture Marine Engineering and Aeronautics! In June 1917, William Frederick Gerhardt was the first student to receive this degree of Bachelor of Science in Engineering (Aeronautical Engineering), although some controversy was evidently raised when in 1929 one of the first group of students somehow arranged for the Regents to give him a degree predated to 1916. Mr. Gerhardt also became the first student in the department to receive an M.S. in aeronautical engineering, in June, 1918. In later years he became known as the designer of the "venetian blind" multiplane aircraft built at McCoole field and evidently seen regularly in the film made to show spectacularly unsuccessful designs. It should be noted that during this time, and indeed until 1930, there was no separate Department of Aeronautical Engineering.

It is no coincidence that the men who did the most to establish the idea, in Ann Arbor, that aeronautical engineering was a suitable field for university instruction and research – Sadler, Pawlowski and Marchis - were from Great Britain and the Continent. For the French, Russians, Italians, British and Germans had by 1910 long recognized the value of applying science to the problems of aeronautics and were engaged in aeronautical research at universities; such as in Prandtl's laboratory at Göttingen; at military installations, such as the French army aero lab at Chalais-Meudon, at government installations, such as in Alderhof in Germany; and at numerous private laboratories, such as Riabduchinski's in Koutchino, Russia, and those of the English scientists Cayley, Wenham, and Phillips. During this same period the U.S. was dependent largely upon the efforts of a host of amateur inventors who approached the problem empirically and with limited means but great ingenuity. This situation is in part reflected in the number of military aircraft possessed by each of the leading powers at the outbreak of World War I in 1914: France - 1,400; Germany - 1,000; Russia - 800; Great Britain - 400; and the U.S. - 23. But perhaps it is more accurately reflected in the fact that the stimulus and model for university instruction and research not merely in aeronautical engineering but through the entire range of the physical and medical sciences came, to a large extent, from abroad.

During World War 1, early in 1917, Professor Pawlowski was granted a leave to accept the position of aeronautical engineer for the U.S. Army. However, the War Department accepted the advice of experts of our Allies and abandoned attempts to develop original designs, to concentrate upon using the country's enormous manufacturing potential. Thus, Pawlowski returned to the University in the fall of 1917 to assist in conducting a special course, Principles of Aviation, which permitted students drafted into the Army to qualify for or to claim preference for Air Corps service. He took another leave of absence in 1919 to organize aeronautical research for the Polish Army and returned in 1920 to teach nearly all of the courses in aeronautical engineering at the University.

As pressure for research capabilities grew, it was decided to build a wind tunnel and so one was included in the plans for the East Engineering building; the wind tunnel was built into the foundation of the building. Started in 1924, it was completed in 1926 with the aid of a gift of $28,000 from the Guggenheim Fund. In addition, the Guggenheim fund provided $50,000 for a professorship of applied aeronautics for ten years. Mr. Laurence Kerber, Class of 1918, was first appointed to this position. Professor Kerber, with Mr. Gerhardt, wrote the Manual of Flight Test Procedure; his interest in this area led to his association with the CAA where he was instrumental in establishing the first set of procedures for obtaining the "approved type certificate". Later, Professor Pawlowski was appointed to this professorship.

Finally, in 1930, nearly 20 years after Sadler and Pawlowski had aroused interest in aviation at the University of Michigan, the Department of Aeronautical Engineering was established as a separate department. The first chairman was Edward A. Stalker (AeE' 19, MSE, '23).

There is no more colorful, adventurous chapter in the history of aeronautical engineering at the University of Michigan, than the one recounting student efforts to fly – in gliders, balloons, and primitive airplanes. In one example a model "B" hydroplane built by the Wrights in 1912 was donated to the Aero Club in 1915 by two wealthy Detroiters. During a trial flight from Barton Pond, shortly after the airplane arrived, the hydroplane crashed and was ruined. Happily, the untrained student pilot survived. During this time also, gliders were built and used to train student fliers, both as kites and as free gliders pulled up behind automobiles as mentioned previously. Many of the glider enthusiasts went on to become distinguished pilots, both as test pilots and in the armed services. The most adventurous of the activities however, were those connected with free ballooning, begun in 1926. Indeed, although Lindbergh's flight in 1927 electrified the world, the balloonists were not overly impressed. After all, they had persisted in spite of the stench of coal gas, the complications of rotting fabric, inadequate funds, being shot at by farmers, struck by lighting, caught in trees, nearly drowned or frozen to death, or lost in the wilds of Ontario! They really wanted to fly! The wilds of Ontario are mentioned because one of the students decided to take a balloon trip from Cleveland to Ann Arbor to attend the Michigan-Minnesota football game. He left Cleveland at 11:00 PM on the Friday before Thanksgiving in 1931, was caught in a violent snow and sleet storm over Lake Erie, spent all night going down and up as ice formed and then melted, sighted a shoreline after 18 hours, landed in a fire-charred desolate woods, and after three days came stumbling out of the woods in northwest Ontario, 70 miles north of Michigan. The balloon was never found; he missed the game.

The aforementioned wind tunnel supported by the Guggenheim fund had an openthroat test section with a maximum size of eight feet across the flat sides of its octagonal cross section. It had curved guide vanes at corners or bends and short lengths of stove pipes used as straighteners at several points. Initially the model was supported primarily by three vertical wires, two at the leading edge and one near the tail, also used to change the angle of attack. Forces were measured using the wire balance systems first developed by Ludwig Prandtl, whose laboratory, Pawlowski had visited. A two-bladed propeller was powered by two electric motors, one 200 horsepower, the other 50 horsepower, with fairly rough control systems. Maximum airspeed was 80 MPH; this tunnel probably had a high turbulence and noise level, but was very useful, nonetheless.

Indeed, work in this wind tunnel led to the start of the career of our department's, and the nation's, most famed designer, Clarence "Kelly" Johnson. He was hired in 1930 as a student assistant by the department Chairman, Edward Stalker, and was put to work in the wind tunnel. Seeing its potential, Johnson asked if he could rent the tunnel (for $35 per day plus power charges) and so he and his college friend, Don Palmer, became part-time proprietors of the University of Michigan wind tunnel (4). They immediately approached the Studebaker Motor Company and were hired to test the Pierce Silver Arrow; they found that the big, ugly headlamps on Studebakers were eating up 16% of the engine power at 65 MPH and had them shaped into the fenders. This and many other jobs kept them relatively well to do, but they stopped this consulting when the faculty noticed how lucrative it became!

After graduating in 1932, Kelly and his friend traveled to the west coast in a car borrowed from a professor and found that there was no work. However, The Lockheed Company had just been purchased from receivers for $40,000 (!) and was being reorganized, and the chief engineer suggested that Kelly return to the University of Michigan for a Masters degree and then come to work for them. He did this, and while working for the University in the wind tunnel once again, performed some tests on the new airplane being proposed by Lockheed, the Electra; (see Figure (3)). At the end of this year, Kelly returned to Lockheed. The first thing he did was to inform the secretary of the company and the chief engineer was that he didn't agree with the official report of the University and that their airplane was unstable! To the credit of the Lockheed Company, Johnson was sent back to the University of Michigan with the Electra Model (he drove). After 72 tunnel runs, he found that removing the wing-body fillet and adding end-plates on the horizontal tail made this tail more effective, and that more rudder area was needed for directional control for one-engine operation. And so a double vertical tail was the answer to the problems, a feature of several Lockheed airplanes; this was extended to three rudders in the Constellation. The Company was very impressed. Kelly went on to be responsible for the designs of the Hudson Bomber, the Constellation and Super constellation, as well as the P-38, C-130, T-33 trainer, F-80, F-104, U-2, YF12-A and the SR-71 Blackbird, among others. He was, arguably, the top airplane designer in this country and, indeed, in the world.

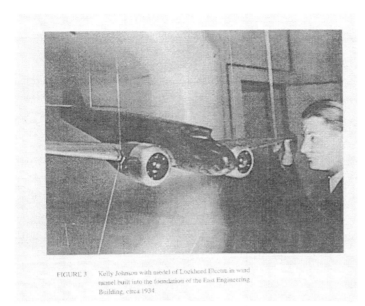

FIGURE 3   Kelly Johnson with model of Lockheed Electra in wind tunnel built into the foundation of the East Engineering Building, circa 1934.

In the 30's and 40's research became increasingly important in the development of the department. Improvements to the existing wind tunnel, the addition of another smaller subsonic tunnel, supersonic tunnels and structures laboratories and testing equipment allowed broader research interests to flourish and this in turn led to the introduction of more sophisticated course work. With the advent of World War II, the Army and Navy sent graduate officers to the University of Michigan for education in aeronautics. In addition, faculty positions and the number of students enrolled in aeronautical engineering increased enormously during this 20 year period except, of course, during the actual war years from 1941 to 1945. In 1946, more than thirty years after he had first kindled interest in aviation at the University of Michigan, Professor Pawlowski retired to live in Paris, France. He died in 1951; all who knew him experienced a great feeling of loss.

It was at the end of this 20 year period, in 1950, that a major impact on our field was made with the introduction of the book *Foundations of Aerodynamics* by Professors Arnold Kuethe and Jay Schetzer. This text became an instant success and was used by nearly every major department in this country and throughout the world. Indeed, the latest edition, with Professor Chuen-Yen Chow replacing Jay Schetzer as the second author, is still in use; it ranks with the important texts in aerodynamics.

A significant far-reaching event in the research program occurred in 1946 when Professor Myron Nichols brought to the Aero Department several engineers and physicists from the Palmer Physical Laboratory at Princeton. They, along with several other people already in Ann Arbor, formed a group identified only as "Research Techniques" which was installed in a laboratory at Willow Run Airport. They concentrated on two areas, the development and use of analog computers and differential analyzers, and the structure of the upper atmosphere. Each of these studies resulted in the development of large research and educational programs.

The first of these activities led to formation of the well-known graduate program in Instrumentation Engineering, which evolved into the graduate program in Information and Control Engineering. Fundamental research involving many PhD students resulted from this foray into the new area of control theory. As the research broadened the program was enlarged to become a College of Engineering graduate program called Computer, Information and Control Engineering (CICE). It involved faculty from Electrical Engineering and Mechanical Engineering as well as those from the Aero Department. Finally, in the 1980's CICE was discontinued as a program, with continued work being carried out in each faculty member's home department.

The second activity was the beginning of a long and successful program of upper atmosphere research and led to the formation of a High Altitude Research Laboratory. The research focused on the structure of the atmosphere, now extended to the limit of the terrestrial atmosphere, and included phenomena of meteorological significance. Some of the early experiments involved scientific payloads placed on V-2 Rockets captured from Germany and fired at White Sands Proving Ground. The Army invited laboratories at Johns Hopkins, Princeton and Michigan, all of whom had upper-air research programs, to install payloads on their flights. This group formed a committee, the V-2 Rocket Panel, which served until the formation of NASA as a quasi-official commission guiding upper air rocket research in the United States. This panel was very influential, and as its crowning achievement, set up for the National Academy of Science, the U.S. program in rocketry for the International Geophysical Year (IGY). It also published the first standard atmospheric table based on in situ measurements by rockets. Later this included several solid propellant rockets. In this regard, the Nike – Cajun rocket used throughout the world as a sounding rocket was developed at the University of Michigan High Altitude Research Laboratory, as were the 3-stage Exos and 5-stage Strongarm sounding rockets. In ongoing efforts to refine the measurement techniques, the use of very delicate instruments, including mass spectrometers, in the high g-load environment of a rocket payload, was pursued with eventual great success. This research in aeronomy was carried out until the 1980's, when support decreased.

Although Air Force Officers had been sent to the University of Michigan in the late 1940's for training in the field of Pilotless Aircraft, a new Guided Missiles program for Air Force personnel was begun in the early 1950's, at the request of the Air Force Institute of Technology. The curriculum was set up such that in two calendar years and one summer session, the enrollees received two master's degrees, one in aeronautical engineering and one in instrumentation engineering. Many of the Air Force Officers who went on to work in space-related activities, including astronaut training, were graduates of this program. Also officers from several foreign countries attended these classes.

Several of the Officers trained at the University of Michigan went on to become astronauts. Indeed, several space flights were all Michigan. The first of these was the Gemini GT-4 mission flown in 1965. It was an earth-orbital flight with the first attempt at rendezvous and the first space walk by Ed White with Jim McDivitt as pilot. These two astronauts were granted honorary Doctor of Aerospace Engineering Degrees in a convocation in which an amusing but far reaching gaffe occurred. In 1964 we had changed our name to the Department of Aeronautical and Astronautical Engineering. At the convocation our University President introduced our chairman, as the Chair of Aeronautical and Astronomical Engineering. With great fear of the time when astrological might replace astronomical, we changed our name once again, to Aerospace Engineering; in 1966, this became the Department's name and remains thus to this day. The second all Michigan flight was Apollo 15, with Astronauts Al Worden, Jim Irwin, and Dave Scott. They left on the moon a document representing the Alumni Association of the University of Michigan, Charter Number 1, certifying that the U of M Club of the Moon is a duly constituted unit of the Alumni Association, Figure (4). Finally, one of our Astronauts, Jack Lousma, was on Skylab for 57 days, and later piloted the shuttle on one of its flights. He has been an adjunct Professor in the Aerospace Engineering Department. It is also important to note that one of our faculty, Professor Harm Buning, spent considerable time at the NASA Johnson Space Center in Houston, giving courses in orbital mechanics to the first two groups of astronauts.

**The Alumni Association
of
The University of Michigan**

Charter Number 1

This is to certify that

The University of Michigan Club of

The Moon

is a duly constituted unit of the Alumni Association and

entitled to all the rights and privileges under the Association's Constitution

President of the University   President, Alumni Assn.

Vice-President, Univ. Relations   Past President, Alumni Assn.

Executive Director, Alumni Assn.   Chairman, Clubs Council

Director of Field Activities   Past Chairman, Clubs Council

FIGURE 4   Alumni Seal for the University of Michigan Club of the Moon, Charter Number 1, left on the moon by astronauts John Irwin, Al Warden, and Dave Scott, Apollo 15.

Although exploits in space were headline news in the 1960's, one of our faculty members did his bit to advance the art of aircraft design. Professor Ed Lesher designed, built, and flew two of the pusher-prop designs for which he became internationally known. The first was built to prove the design feature, in particular the long shaft between the engine mounted immediately behind the pilot and the pusher propeller mounted at the rear of the fuselage. The dynamic problems conquered, Ed then built a smaller, lighter version conforming to the FAI class for aircraft with a total maximum weight of 500 kg Figure (5). The Teal first flew in 1965 and by 1967 was beginning its series of record breaking flights, the first three for speed in a closed circuit. Then, in 1970 Ed, broke the previous record for distance in a closed circuit by 311 miles, roughly 25% longer than the record. Next, he broke two records for speed over a measured course. Finally, for his seventh record flight Ed set the record for distance in a straight line (1835.4 miles from St. Augustine, Florida to Goodyear Arizona). One of the greatest demands for this flight was his diet; each pound he lost was a pound of fuel added! All in all, Professor Lesher was awarded four Bleriot medals by the FAI. It should be noted that the students in Ed's design class checked all numbers. They were aiding in the design of a real airplane and were very involved and interested in it. Not many of them realized how fortunate they were to be taught by a man who could

FIGURE 5   Professor Ed Lesher flying over Ann Arbor in his Teal Airplane, holder of seven world records for the FAI class of aircraft with a total weight of 500 kg.

design, build, and then fly an airplane, let alone one with such innovative and creative ideas in its design.

Faculty at the University of Michigan have been fortunate throughout the years in having excellent facilities. After World War Two, new subsonic and supersonic wind tunnels were added, as were the latest strength testing machines, and analog computer equipment since replaced by digital computers. This led to the formation of several laboratories and research groups, each involving several faculty members and many graduate students, which have evolved to the present state of the curriculum and research.

In the aerodynamics laboratory group, founded by Professor Arnold Kuethe, fundamental work in turbulent flows, supersonic and subsonic mixing, and flow separation has been carried out. Many papers were published, also, on theoretical works in transonic flow and wing theory, for both stationary and rotating wings. More recently, a research group in computational fluid dynamics was begun by Professor Bram van Leer, this being the genesis of the W.M. Keck Laboratory for Computational Fluid Dynamics. From this laboratory have come advances in the accuracy robustness and efficiency of numerical methods for compressible flow, including work on genuinely multi-dimensional methods, solution-adaptive methods, optimal preconditioning, and multigrid acceleration techniques. Additionally, contributions have been made to numerical methods for electromagnetics, aeroacoustics, magnetohydrodynamics, non-continuum flows, modeling of advanced propulsion devices, and meteorological flows.

The analog computer laboratory begun by Professor Myron Nichols grew to include work in controls, as mentioned previously, and indeed spawned a departmental instrumentation program which evolved into a college wide program in computer information and control engineering, lasting until 1983. In the Department of Aerospace Engineering this work was and has been carried out by the faculty in the flight dynamics and control section. Important contributions have been made in computer simulation, the foundation of control system theory, optimal aircraft and spacecraft maneuvers, control of flexible space structures, and aircraft and spacecraft dynamics and control. A company named Applied Dynamics, specializing in very fast computers for real-time simulation was started by four faculty members and still flourishes. Finally two renowned text books, *Principles of Dynamics and Classical Dynamics* written by Professor D.T. Greenwood, are used throughout the world.

The discipline of structural mechanics has undergone remarkable changes since the early days of aircraft design. The ubiquitous concern to minimize weight first led our faculty and others to do research in efficient structural design and methods for accurate structural analysis. Considerable work also was done, on optimal structures. More recently, a dramatic modernization of laboratory facilities, including state of the art testing machines, materials processing and characterization instrumentation, and intensive computing facilities has taken place. This has allowed more fundamental research in the interaction of structures with electro-magnetic fields and new interactive materials as well as with the more traditional interactions with fluids, thermal fields, and in failure processes. More specifically, research is being conducted in the areas of aeroservoelasticity of helicopter rotors, advanced composites, modeling of composite rotor blades, life cycle durability of components and the use and fundamental behavior of adaptive materials, such as shape memory alloys, piezoelectric material, magnetorheological solids, and eletrodynamic membranes. In addition, the modeling of fracture and the effect of microstructure is being studied. Both experimental and theoretical work involving many graduate students is being carried out in these areas.

A propulsion laboratory begun by Professor Richard Morrison at facilities at the Willow Run Airport became the Gas Dynamics Laboratory back on North Campus and grew to include more general research in reacting gas flows. A great deal of work was and has been done on detonation phenomena in this laboratory. Indeed, Professor Arthur Nicholls, who became its director, was the first to establish a standing detonation wave. Considerable work was also done on combustion instabilities, underexpanded nozzle flows, and steady and unsteady transonic channel flows. Now headed by Professor G. Faeth, typical research areas are supersonic mixing and combustion, microgravity combustion, micro fluids and combustion systems for propulsion, turbulent flow drag reduction and turbulent flames. Fundamental work is carried out also in determining the formation, structure, and combustion of sprays, as well as in the formation growth, and radiative properties of soot. In addition, new diagnostic techniques are being employed to aid in measuring combustion products in engines; these studies aid in developing realistic models of engine combustion. This work, experimental and theoretical, has resulted in a multitude of PhD theses over the years.

A relatively recent facility, one of the largest vacuum chambers in the country, was obtained from the Bendix Corporation when it left Ann Arbor; testing for the Apollo program was carried out in the chamber. In this laboratory, supervised by Professor Alec Gallimore, experiments in electric space propulsion are carried out. The chamber is large enough that propulsion units can be run for significant test times in near vacuum conditions found in space. Figure (6) shows a test run of a P5

FIGURE 6   Hall Thruster running at full power (9.2 kw, 2800 s, 400 mn) in vacuum chamber at 10^{-5} torr.

Hall thruster designed by the people in this lab and built by the USAF; it is operating at full power (9.2 kw, 2800s, 400 mn) with a chamber pressure maintained at roughly 10^{-5} torr.

As usual, many of the subjects in the above mentioned research areas are later found in graduate and undergraduate courses. A comparison of these subjects with those covered in our first curriculum, Figure 2, illustrates the incredible increase in sophistication and breadth of study that has occurred in the past eighty-eight years. Also apparent, in 1914 and in the intervening years to the present, is the fact that the aerospace field truly exemplifies cutting-edge technology

It is often of interest to know where faculty members of a given department received their graduate training. Of the twenty-one members of the Department of Aerospace Engineering at the U of M, four come from Caltech, three each from MIT and The University of Michigan, two each from Princeton and the University of Texas at Austin, and one each from Columbia, Penn State, Brown, Georgia Tech, Cambridge, (U.K), Leiden, (Holland) and the University of South Hampton (U.K.). Seven work in the areas of propulsion, aerodynamics, and combustion; three in computational fluid dynamics, six in structural mechanics, and five in flight dynamics and control.

Finally, a note about the superlative support given to the Aerospace Department and aerospace engineering in general by the François-Xavier Bagnoud Association. In 1982, a young Swiss man, François-Xavier Bagnoud, graduated with a degree in Aerospace Engineering, and joined his father at Air Glaciers, the largest private Alpine rescue and mountain flying company in Switzerland. He became, at age 23, the youngest professional IFR pilot in Europe, of both airplanes and helicopters. Within three years, in addition to his regular piloting responsibilities, he completed 300 successful rescue and flight operations in the Alps and two Paris – Dakar races. Tragically, François-Xavier flew a fatal helicopter mission in 1986 in the desert in Mali, West Africa, while flying for another Paris-Dakar road race.The François-Xavier Bagnoud Foundation was born out of the desire of the parents of Francois-Xavier, Countess Albina du Boisrouvroy and Bruno Bagnoud, his stepfather, and close friends to commemorate his caring for others and his passion for all things aerospace. This Foundation has been instrumental in philanthropic aid for worthy causes, especially those involving sick children all over the globe. Also, in memory of his great interest in and fond memories of the Department of Aerospace Engineering at the University of Michigan, the Association provided major funding for our new building, four graduate fellowships, a chaired Professorship, and a center for Rotary and Fixed Wing Aircraft Design. In addition the François-Xavier Bagnoud Aerospace prize was created and is administered by the University of Michigan.

The François-Xavier Bagnoud Building is an outstanding facility Figure (7). Containing 94000 square feet of area, with a large atrium, it houses 19 laboratories, several with 18 foot ceilings, three which are blast resistant, and one a CFD lab, three classrooms, a design lab, a large lecture hall, 32 offices for faculty and staff, 25 offices for graduate students, a library, three conference rooms, and two student organization offices; it is truly one of the outstanding educational facilities in the country. Figure (8) shows the setup for an experiment on vibration control of a large-aperture spaceborne telescope for astronomy, illustrating the space available for relatively large scale equipment.

FIGURE 7   View of the François-Xavier Bagnoud (FXB) Building for Aerospace Engineering at the University of Michigan.

The FXB Fellowships provide for up to five years of graduate study leading to a Ph.D. including tuition, fees, excellent stipends, and one trip home each year. These are our most prestigious fellowships.

The FXB Center for Rotary and Fixed Wing Air Vehicle Design is headed by Professor Peretz Friedmann, who also holds the FXB Professorship. This center focuses on multidisciplinary analysis that plays a key role in the design of manned and unmanned air vehicles. The areas emphasized are interactions between computational aeroelasticity and aerodynamics, controls, flight mechanics, active materials and composite structures, including innovative lightweight nanotube based composites, and high temperature aerospace vehicle structures. The goal is the development of lightweight, highly efficient vehicles with low vibration and noise levels, good damage tolerance characteristics, and low cost.

Lastly, the FXB Aerospace Prize consists of a $250,000 prize for outstanding accomplishments in the aerospace field. It is awarded biannually and is international in scope. The awardee is chosen by an international selection committee, representing the aerospace community; this committee is nominated by the FXB Prize Board.

In summary, the Department of Aerospace Engineering at the University of Michigan has a long and proud history and tradition. Its graduates have attained many positions of great responsibility as designers, engineers, pilots, top executives in industry and government, teachers, and researchers. As we enter our eighty-eighth year of operation, we look forward to continuing contributions to the aerospace community. Further information on our programs and faculty can be found on the web at www.engin.umich.edu/dept/aero.

FIGURE 8   Experiment on vibration control of a large aperture
spaceborne telescope for astronomy, illustrating high bay
laboratories available for large scale experiments in the
FXB Aerospace Engineering Building.

Acknowledgements

The author wishes to express his gratitude to his colleagues, Professors H. Buning, E. Gilbert, M. Sichel, R. Howe, and K. Powell, for their advice and suggestions in preparing this manuscript and Debbie Laird for typing it.

References

1)   A Century of Engineering Education, University of Michigan Press, Ann Arbor; 1954, pp 1181-1189.

2)   The First Fifty Years (A Fragmentary, Anecdotal History), Robert P. Weeks, Professor of English, College of Engineering, October, 1964.

3)   The Third Quarter Century (More Fragmentary, Anecdotal History of the Department of Aerospace Engineering), Helena S. Buning, October, 1989.

4)   Kelly, More Than My Share of It All, Clarence L. "Kelly" Johnson with Maggie Smith, Smithsonian Institution Press, Washington, D.C., 1985.

Chapter 5

Aeronautical Engineering at the University of Detroit

Herman S. Muller, Professor of Economics
Leo E. Hanifin, Dean and Chrysler Professor of Engineering

The aeronautical engineering program was established at the University of Detroit (now University of Detroit Mercy) in 1921. The *Detroit Free Press* broke the news: "Looking into the future and seeing there its home city the center of the aeronautic industry as she has been the center of the automotive development of the country, the University of Detroit is developing a five-year course in aeronautical engineering, the first of its kind in the country." It was to teach its citizenry the science of designing, building and flying aeroplanes that the University of Detroit launched its new five-year program.

Specialized instruction in aeronautics fell to Professor Clarence H. Powell who had been associated in England with the National Physical Laboratories and with the Sopwith Aeroplane Company. Before coming to Detroit, he had been at the Massachusetts Institute of Technology.

When the university hosted an exposition on aeronautical engineering, President Warren Harding showed his hearty approval in a personal letter to Dean Dunn. "I cannot but regard it as particularly fortunate," he wrote, "that an institution such as the University of Detroit should interest itself in an effort to coordinate the purely scientific with the practical and mechanical phases. The field is a great and important one, and I hope your exposition will produce a real contribution to the interest of American aviation."

Prototype Aircraft in 1939

Glider Designs at U of D in the '50's

Student Competitors and Their Trophy

A few years later, the U of D Flying Club designed and built the Powell Racer. At the International Air Races in New York in 1925 it was awarded the *Aero Digest* Trophy, *Dayton Daily News* Trophy and the *Scientific American* Trophy, plus some $2,000 in cash prizes. Three years later the U of D Glider Club won several prizes with a glider it had designed and built. In 1930, at the Cleveland Air Races, the Club received the *Air Digest* Glider Trophy.

A prototype jet engine at U of D in the 1950's

Activities in U of D's High Bay

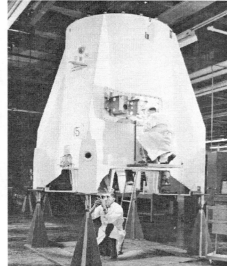

U of D Wind Tunnel

On Tuesday, October 26, 1937, the College of Engineering of the University of Detroit received the highest accreditation possible for an engineering school, that of the Engineers' Council for Professional Development. At the time it was the only school in the Detroit Metropolitan Area so to be recognized in **aeronautical**, architectural, civil, electrical and mechanical engineering.

Prior to World War II, the University of Detroit conducted primary and secondary Civil Aeronautics Authority (CAA) flight training programs.

In June of 1965 the University of Detroit phased out of the Aeronautical Department. While, according to the *Detroit Free Press*, U of D was the first to have such a program . . . and for many years its graduates had been achieving prominence in the field . . . the university's administration recognized that it was the automobile, not the airplane that would dominate industry in Detroit.

Many graduates from University of Detroit's Aeronautical Engineering Program have gone on to important leadership positions in the nation's aircraft and defense industries. The following is a list of a few of these leaders:

- Thomas Basacchi – Vice President – Boeing
- George Graff – President – McDonnell Aircraft
- John Griffin – Director of Engineering – B2 Program Office,
     Chief Engineering – F15 Fighter
     Chief Airframe Engineer – F16 Fighter and Air
     Launched Cruise Missile
     Wright Patterson Air Force Base
- Glynn Lunney – Vice President – Rockwell Space Operations and
     Vice President –Rockwell Satellite Systems Division

Chapter 6

A Brief History of
The Daniel Guggenheim School of Aeronautics
College of Engineering, New York University

Victor Zakkay, Retired Vincent Astor Professor

On December 17,1903, a bird took wing over the lonely sand dunes near Kitty Hawk, North Carolina--a mechanical bird built by men from wire, wood and linen. The men who built it called it the Flyer. One of them, Orville Wright, lay prone in the skeleton as it rose and pushed through the air at 30 miles per hour. The flyer flew for 12 seconds before it plunged down into the sand. The air age was born.

At the beginning of the 20th century, there existed at New York University a school of Applied Science. The school of Applied Science was one year short of its fiftieth birthday, when subsequently it became known as the College of Engineering of New York University. It had been organized as the school of Civil Engineering & Architecture, but by the turn of the century had enlarged its offerings to include mechanical and chemical engineering.

In 1921 the growing importance of aviation led to an investigation of its educational aspects by the College of Engineering and to a course of lectures given by Alexander Klemin to senior students during the academic year 1921-1922. The Council of New York University later authorized an aeronautical option for an experimental period of three years. These courses were made possible by a very modest budget underwritten by one 100 men prominent in aeronautics banking and business.

The first courses in aeronautical engineering were started in September 1922 by Dr. Alexander Klemin. Only one or two courses were offered that year as permissible electives in mechanical engineering. Dr. Klemin had been chief engineer and partner in an aircraft concern that had up to 1921 been unusually successful in design competitions for the Army and Navy. He is credited with having built the first amphibious gear in this country, and the first ambulance plane.

The success of these courses, the enthusiastic response of the student body, the immediate placement of graduating students and the growing interest in aviation indicated that this aeronautical instruction should be placed on a permanent footing. An organizing committee formed in March, 1925, formulated plans for a permanent endowment. The chairman of the committee was the late General John J. Carty; Vice-Chairmen were F. Trubee Davison, Artemus

L. Gates and Harry F. Guggenheim, with the former Dean Collin P. Bliss and Professor Klemin as University representatives.

At a historic luncheon at the Princeton Club, Harry F. Guggenheim, who had served with high distinction as a Naval aviator during the war, requested the late Dr. Elmer A. Brown, then Chancellor of the University, to prepare a letter setting forth the desirability of a permanent educational aeronautical center in New York City. Mr. Guggenheim undertook to show this letter to two of his uncles as possible patrons of the undertaking, after first securing the criticisms of his father, Mr. Daniel Guggenheim. When Daniel Guggenheim read the Chancellor's letter with the rapidity and concentration which were habitual to him, he said immediately: "Don't show this letter to your uncles, Harry, I will do it myself. I have given all my life to work underground; now let me see what I can do to help above ground". It was highly fortunate for American aeronautics that Mr. Guggenheim, always of an imaginative and pioneering mind, had been fired with the potentialities of aviation. The Daniel Guggenheim Fund for the Promotion of Aeronautics with its manifold consequences undoubtedly had its indirect inception on that day. Moreover, the organizing Committee, which had expected a long and arduous campaign to secure its objectives, found its plan immediately realized. Just three days after reading the Chancellor's letter, Mr. Guggenheim signified his intention of endowing a school of aeronautics with a gift of $500,000, and in a formal letter to Dr. Brown defined the motives that actuated his action as follows:

"For some time I have been impressed with the need of placing aeronautics on the same educational plane that other branches of engineering enjoy. It has seemed to me that aviation is capable of rendering much service to our nation's business and economic welfare as well as to its defense, that our universities should concern themselves with the education of highly trained engineers capable of building better and safer commercial aircraft, and industrial engineers capable of making the operation of aircraft as a business proposition comparable to the operation of railroads. In this way we shall give America the plane in the air to which her inventive genius entitles her".

A Permanent Advisory Committee

A permanent advisory committee was formed now including Messrs. Carty, Davison and Gates. Orville Wright consented to act as Chairman, and Harry F. Guggenheim as Vice-Chairman. Other early members included such distinguished aeronautical personalities as Charles L. Lawrence, Sherman M. Fairchild, Graham Grosvenor, Maurice Holland, Grover Loening, William P. MacCracken, Jr., Earl D. Osborn, and Juan T. Trippe. This Advisory Committee with Mr. Guggenheim as the active chairman continued to give keen interest and sage counsel to the Guggenheim School since its inception, with great benefit to all its activities.

Construction and Equipment of Building

In the administration of the $500,000 fund, it was soon decided that $225,000 should be expended in the construction of a building and equipment and $275,000 be allocated to an income producing endowment.

The design of the building for the new school was rapidly pushed to completion. Ground for it was broken in the fall of 1925 and classes met for the first time in the new quarters in the fall of 1926. The Guggenheim Building is a two-story structure, 120 x 85 feet with a first story projection of the main wind tunnel. Approximately $250,000 was expended on the building and the housing and structure of the wind tunnel, leaving $75,000 for the scientific equipment.

A third story was added to the aeronautical building. It housed a drafting room, a lecture room, a model radio, and a dispatching room, as well as the instrument and vibration laboratory. A student seminar room and offices were also added

Experimental facilities were organized to cover almost the entire field of aeronautical research (aerodynamics, rotary aircraft, structures, engines, instruments, vibration and seaplanes). In connection with the main wind tunnel, the following auxiliary equipment was outstanding:

1. The first Moving Ground Belt for the investigation of aerodynamic ground effects ever successfully constructed
2. The first mirror pitot.
3. The only balance in the world for the investigation of the spinning airplane in a horizontal tunnel.
4. Only apparatus available for testing autogiro models with elimination of scale effects.
5. The engine laboratory, as far is known, is the only one of university character approved by the Bureau of Air Commerce and its successor, the Civil Aeronautics Authority, for the certification of aircraft engines.
6. The towing basin, with a length of 150 feet and a width of 8 feet, is available for the seaplane investigation. It has proved very useful for graduate research particularly in regard to the study of seaplane porpoising.

The building of the Guggenheim School of Aeronautics was completed in the fall of 1926. Within the structure, work had started on what was then the largest nongovernmental wind tunnel in the United States. Renato Contini, a member of the first class of mechanical engineering graduates who had studied under the aeronautical engineering program, drew the plans for the tunnel. The tunnel was 94 feet long, and from 18 feet high at the entrance end to 14 feet at the exit. The working section was octagonal, in order to approximate a circular nine foot crosssection aerodynamically. To produce an air flow the working section that was symmetrical about a longitudinal axis, the tunnel had fully enclosed double returns. Flow was generated by an eight-bladed cast aluminum alloy propeller 13' 8" in diameter, powered by a 250 hp d.c. motor. Tunnel wind speed went to 90 mph.

In the twenties and thirties the airplane was streamlined,the biplane giving way to the monoplane. All-metal planes were built with stressed skins, engineering advances produced retractable landing gear, higher wing loadings, trailing edge flaps, and improved wing section design.

Evolution of educational programs in the Guggenheim school of Aeronautics at New York University mirrored these and other developments. Industrial employers expected the early graduates in Aeronautical Engineering to be jacks of all trades. But with growing specialization, the demand was not for the all-embracing "Aeronautical Engineer" but for aerodynamicists, stress analysts, detail designers , tool engineers, and production men. Air transport positions called for, specialists in meteorology, navigation, piloting, and flight research.

In the depth of the depression, aircraft manufacturers and airlines might well have been heard echoing the quip of barnstormers of a decade before, that "the greatest hazard in flying is the risk of starving to death." Nevertheless, the Guggenheim School of the New York University College of Engineering enlarged its educational offerings to include the civil transportation aspects of aviation. About 1930 an air transport option to the technical program was established, the first such curriculum in the United States.

The Undergraduate Curriculum in
Aeronautical Engineering

From the very beginning of the aeronautical work at New York University, definite educational policies have been pursued, which have been fully justified by experience. In spite of some pressure by students and parents, there has been strict adherence to the contention that early specialization in aeronautical work for a college student is undesirable, and that aeronautical engineering should achieve a solid foundation in physics, mathematics, applied mechanics and mechanical engineering before undertaking specialized work in the second term of the junior year. In the senior year there is almost exclusive specialization in aeronautical subjects; the purpose is to train engineers rather than pure aerodynamicists -- men who can take a full part in the important work of developing, designing and constructing aircraft and aircraft engines. To the now almost classical curriculum of aeronautical engineering (aerodynamics, stress analysis, propeller theory, etc) the aeronautical engineering school made two pioneer contributions: first, a course in airplane design, where students work under engineering and drafting office conditions, and where excellence is rewarded by the Chance Vought Memorial Prize established by Mrs. Anni Vought in memory of her brilliant son. Second a course in airplane detail design which tried to remove the criticism of industry that aeronautical graduates acquainted with advanced aerodynamics theory were incapable of designing even a simple fitting. The drawings or designs which were the result of these two courses are continually receiving the commendation of practicing engineers.

Thus in the first academic year 1925-1926 of the Guggenheim School, the seniors enrolled numbered eight or nine men. Each subsequent year has invariably seen a rise in enrollment and in the academic year 1938-1939 the number of seniors alone taking the technical option of Aeronautical Engineering had risen to forty-four and those taking the Air Transport senior year number fifteen. Such increase in senior enrollment had occurred in spite of a selective policy of admission to the senior year in Aeronautical Engineering

Graduate Studies

Beginning with the academic year 1938-1939, New York University was one of the first American institutions to offer opportunities for advanced graduate study and research in aeronautical engineering. The degree offered for such work provided a Master of Aeronautical Engineering. Initially there were four to five graduate students in 1928-1929, thereafter the graduate enrollment rose to twenty- two, four of whom proceeded to the Doctorate in Engineering Science. With the increasing complexity of experimental research work in aeronautics, it became increasingly necessary for able and ambitious students to undertake the most advanced studies in aerodynamics, dynamics of aircraft, structures and engine flutter and vibration, engine design, etc.

Seaplane Research and Fellowship

With New York City surrounded by many waters, and the terminal for transatlantic flying, and with a towing basin available at the University, it is natural that advanced study and research in Seaplane Engineering should be undertaken. Such work was given additional impetus by the establishment early in 1938 of the Richard Hoyt Memorial Seaplane Fellowship. Guy W. Vaughan, and others. The first incumbent, John D. Pierson, undertook the investigation of the difficult subject of seaplane porpoising, and was co-author of a paper, which was published in the Journal , and presented in 1939 at a meeting of the Institute of the Aeronautical Sciences.

Rotary Aircraft Curriculum and Fellowship

Beginning with a paper, "Introduction to the Helicopter", which was probably the first serious study of this subject published in the United States, presented before the American Society of Mechanical Engineers in November 1924, The Guggenheim School of Aeronautics maintained an unvarying attitude of support for rotary aircraft work, sometimes in the face of discouragement from the aviation world. There has been research by the staff, research in theses of graduate students, development of special apparatus for rotary research, and constant research aids to all the chief workers in the rotary fields such a Pitcairn, Kellett, Wilford, Platt Herrick, etc.

For the academic year 1939-1940, the school took another pioneer step in the provision of a new graduate curriculum in Rotary Aircraft - the first to be established anywhere in the world. The curriculum included the "aerodynamics and structural theory to Rotary Aircraft," and a series of closely related courses. The chief of Army Air Corps detailed an officer with autogiro flight experience to take advantage of this offering.

Air Transport

When Daniel Guggenheim endowed the school at New York University, he stated that one of his objectives was the training of engineers capable of advancing the actual operation of aircraft. Faithful in this, as in other respects of the vision of its founder, the Guggenheim School took a pioneer step by providing an option in Air Transport during the academic year of 1930-1932. As far as is known, this was the first curriculum in Air Transport of university character to be presented anywhere, and recent history has shown the step to have been fully justified. Students in Air Transport took the basic engineering training common to all aeronautical engineering students, and a number of basic aeronautical subjects.

In the early days of American air transport, operations were carried out and directed mainly by experienced and skillful pilots, somewhat distrustful of technicians and hence of college graduates. But as the technique of operation, and the aids to navigation increased in complexity, a more technical attitude prevailed and the modern air line operator fully recognize the necessity of specialized university training when selecting personnel. Accordingly, the number of men enrolled in the Air Transport option grew and graduates of this option found increasing opportunities in the meteorological engineering and operational aspects of air transport.

By the beginning of World War ll, hundreds of graduates of the Guggenheim School of Aeronautics at New York University had fanned out into the plants and research laboratories of the aircraft industry and of the government, into the offices of airlines, into classrooms at other colleges and universities, to teach growing numbers of students in aeronautical science and technology.

With New York University as a model, seven additional Daniel Guggenheim aeronautical schools and laboratories were established at other universities from 1926 to 1930.

The Period 1964-1973

All the work described above has dealt with low speed aerodynamics. With the advent of high speed aerodynamics which had been adopted by several universities in the country for several years, the New York School of Aeronautical Engineering, headed by Dr. Lee Arnold, desired to expand its curriculum to include both theoretical and experimental high speed aerodynamics.

In 1964 Dr. Lee Arnold, chairman of the department of aeronautical engineering made an offer to Professor Antonio Ferri, a prominent high speed aerodynamist, to join New York University and to create a high speed aerodynamics laboratory coupled with a curriculum in the same subject. Professor Ferri was the director and chairman at the Polytechnic Institute of Brooklyn (presently called, the Polytechnic University), and had organized an advanced high speed experimental facility and a graduate program in high speed aerodynamics. At the same time an offer was made to some of the key faculty of the Polytechnic University. The offer was also made to Professor's Lu Ting , Victor Zakkay, Vaglio Laurin, and Herbert Fox. Several students also transferred to NYU.

In 1964, the National Aeronautics and Space Administration (NASA) was offering several universities funding to establish new advanced facilities in high speed aerodynamics. Drs. Arnold and Ferri applied for a grant from NASA to establish an advanced aerodynamics facility. The proposal was accepted by NASA, and a plan for the laboratory was initiated in 1964. A location for this facility had to be established as well the selection of a staff that would have a strong background in theoretical and experimental hypersonic aerodynamics.

In the latter part of 1964, a site was designated to house the new NYU Aerospace Facility. The location was chosen adjacent to the Harlem River, which was in close proximity to the NYU Bronx campus. A substantial amount of preparation of the land had to be undertaken prior to the installation of the building. Since the construction of the facility would take two years, it was decided to build a temporary hypersonic facility in the NYU Guggenheim Building at the Bronx campus.

Dr. Ferri decided to concentrate on research in experimental and theoretical high Reynolds and high temperature. The first facility had a capacity of Mach 6, capable of pressures of up to 130 atmospheres, and a temperature of 1,400 degrees Fahrenheit. This facility was operational in 1964 . Simultaneously, a facility for simulating reentry flow fields over blunt-nosed bodies was designed and built. Utilizing an H20-O2 mixture, it was shown that simulation of reentry flow field at the stagnation point might be achieved at a lower stagnation temperature

In the latter part of 1966, the NYU Aerodynamics Facility on the Harlem River was completed, and a dedication by the city of New York was made with Mayor John Lindsey presiding.

Because of the great interest generated at that time in the aerodynamics of high mach number flight connected with the possibility of designing recoverable vehicle space planes and, in general high speed airplanes flying in the atmosphere, the possibility of using air-breathing engines or lifting devices of various types were considered. All such vehicles required the knowledge of boundary layer phenomena in the presence of pressure gradients. It was then decided to build a facility that would be capable of performing research at high mach number and Reynolds number.

The Aerospace Laboratory consisted of the following components: a 40,000 cubic feet vacuum sphere, and a high-pressure air supply capable of pressures of up to 150 atmospheres. Two high mach number high Reynolds number facilities were also designed and built. The first facility was unique in the United States. It was capable of operating at a mach number of 14, and a Reynolds number as high as 20 million per foot, operating up to 2,000 atmosphere, with a stagnation temperature of 2,500 degrees Rankine. This work was supported by the United States Wright Patterson Air Force Base (WPAFB).

The research performed during the period 1967 -1970 covered a broad range of technical areas ranging from the utilization of interference effects on different parts of air breathing vehicle to heat transfer to hypersonic bodies. This work was primarily performed by Drs. Ferri and Ting. Other research performed for the Aerospace Research Laboratories (ARL) consisted of investigating problems at hypersonic speeds. Flight at hypersonic speed introduced several problems connected with aerodynamic heating, thereby imposing severe requirements for cooling the structure of the vehicle. The purpose of the research was to obtain basic information on practical problems related to viscous flow phenomena in the hypersonic range.

The possibility of scram jet propulsion technology and the possibility of using supersonic combustion air-breathing engines as propulsion systems for a hypersonic vehicle was investigated both theoretically and experimentally, and in January 1975, Dr. Ferri delivered the Dryden Lecture on "Possibilities and goals for future SSTs" at the American Institute of Aeronautics and Astronautics (AIAA)

A significant research program at the NYU Aerospace Laboratory was its research contribution to the space shuttle program, which was initiated in 1973 by Drs. Ferri and Zakkay. The research concentrated on determining the complicated flow field on the leeward side of the space shuttle and the shock boundary layer interaction present in that region at various angles of attack. The research indicated that there are two distinct types of high heating rates on the leeward side of a space shuttle configuration: peak heating owing to to boundary layer transition and peak heating associated with vortex interaction within a separated flow region. The above research resulted in a NASA award being given to the Laboratory in 1982. The award consisted of a plaque containing a United States flag and a space shuttle emblem flown on the space shuttle Columbia on April 12, 1981. The award was signed by astronauts John Young and Bob Crippen.

In February 1972, Dr Ferri became the last Chairman of the Department of Aeronautics and Astronautics at NYU. In 1973, NYU decided to close its School of Engineering and Science at the Heights campus, and an agreement was reached with New York State that the entire NYU Engineering faculty would be transferred to the Polytechnic Institute of Brooklyn. Realizing his invaluable contribution, NYU retained Dr. Ferri together with Drs. Ting , Zakkay, and Vaglio Laurin. A new Division of Applied Science was then founded and the Aerospace Laboratory was relocated to Westbury, Long Island. In 1975 Dr. Ferri passed away. Thus ended an era that spanned a period of sixty years.

It is strange that the engineering program at NYU started as an Applied Science school at the beginning of the 20th century and became "The Department of Applied Science " in December 1978, with Dr. Victor Zakkay as its first chairman.

References

1. NYU archives Bobst library.
2. Klemin, Alexander: The Guggenheim school of Aeronautics, History New York University. January 1931
3. Klemin, Alexander: Pioneer education in the air age. 1935
4. Teichmann, Fredrick. K, A brief history of the NYU school of

Aeronautics, 1945.

5. Zakkay, Victor (Editor). Antonio Ferri , Vincent Astor Professor of Aerospace Sciences. Memorial book dedicated to Professor Ferri June , 2,000

Part III

The Guggenheim Support

Chapter 7

The 2 Men Behind the 7 Guggenheim Schools of Aeronautics and their further contributions to aerospace engineering in the United States

Robert G. Loewy
William R.T. Oakes Professor and Chair
School of Aerospace Engineering, Georgia Institute of Technology

Who were the Guggenheims? And why should that name – so relatively unfamiliar to the general public – be linked to those of such giants of aviation – familiar to almost everyone – as Orville Wright, Glenn Curtis, Charles Lindbergh, Jimmy Doolittle, Theodore Von Karman and others? Because Daniel and Harry Guggenheim, father and son, influenced the advent of flight in the United State in so profound, far reaching and long lasting ways as to fully justify the respect, admiration, and gratitude that was, in fact, freely given them by all these more well-known heroes and many others, who followed.

Background

Born in 1856 in Philadelphia, Daniel Guggenheim had lived about 60 of his 74 years before aviation became of any real consequence to him. He was the second of eleven children raised by a successful businessman, Meyer Guggenheim. This financier and patriarch sent Daniel to Europe to manage the Swiss branch of a family firm at age 17, ending his formal education. On returning to America eleven years later, he entered another part of the family business, mining and smelting, and by the turn of the century had become a very wealthy – and highly philanthropic - man.

By the time the United States entered World War I, Daniel Guggenheim had two grown sons, Robert and Harry. Both had enlisted in the armed forces, and Harry, a navy lieutenant, became a pilot. He served in France, England and Italy until the armistice. As an American flying during the war, Daniel's son Harry had seen at first hand– like all World War I flyers <u>and</u> the entire earth-bound European population – the power and promise of aviation. It was a great disappointment to him that by contrast, his countrymen - the nation of the Wright Brothers and Glenn Curtiss - were unable to provide a single combat aircraft designed and built in America to the conflict in which he and his fellow servicemen flew. U.S. fighter pilots, for the most part flew French Nieuports and Spads, and our reconnaissance and bomber pilots the British DeHaviland DH-4's.

Harry's understanding and dismay regarding the differences between American and European attitudes toward aviation continued following the war years. He saw England and France begin scheduled air flights between London and Paris as early as 1919. By 1923 there were 30 scheduled British flights across the Channel daily. One needed little more evidence that, in the United States, aviation was just a kind of exciting amusement, than the fact that, in this same era, the U.S. air mail fleet was mostly the same DH-4's used during WWI, converted to mail carriers.

Further, of the first 49 pilots hired by the U.S. Post Office Department in 1919, 31 had been killed in crashes by 1925.

Involvement

That spring, Harry Guggenheim was invited, because of his widely known interest in aviation, to a meeting at New York University, called by its Chancellor. Prof Alexander Klemin had been teaching some aeronautical courses there and proposed that the College of Engineering establish a School of Aeronautics. Prof. Klemin was at the Chancellor's meeting, along with several flyers and public relations experts. The university's plan, a public campaign to raise $500,000, was presented and discussion invited. Harry [2]Guggenheim was negative. "A campaign would be futile," he argued, "At present the American people aren't much interested in aviation. They're likely to fear their money would be wasted."[1] This comment, quite naturally, was followed by a "what would you do " reaction. "Guggenheim, articulate and forceful, suggested that the best way to raise money was to seek out a single individual with sufficient vision to see the potential of an aviation school, and the financial resources to endow one. He went a step further; he offered to draft a letter for submission to a selected group of wealthy men, adding the condition that no members of his own family would be solicited."[1] The NYU people readily agreed.

Although he had his own nearby place on Long Island, called Falaise, where in just a few years he would routinely provide quarters for Charles Lindbergh, when the famous flyer would feel the need to escape the unwelcome attention of adoring throngs,[2] that night Harry Guggenheim was staying at his father's mansion, Hempstead House. During the evening, Harry drafted the letter soliciting support for NYU, told his father the agreement he had made with the Chancellor, and asked Daniel to critique the letter, since many of Daniel's friends and acquaintances were likely to be approached. "Daniel read what his son had written, then put the letter into the pocket of his dressing gown. Let me think about this overnight, he said. Both father and son breakfasted together the next morning and Daniel wasted little time in delivering his verdict. 'Well, Harry, I've thought about your letter and I've decided to endow the school myself.'"[1]

The Daniel Guggenheim Fund for the Promotion of Aeronautics

Daniel Guggenheim wasted no time in acting on his decision. Only a few months after the Chancellor's planning meeting on June 12, 1925 he made a grant of $500,000 to NYU and, on October 23, 1925, at a ground-breaking ceremony, he turned up a symbolic spadeful of earth, personally, for the first school of aeronautics at a major U.S. university. To the more than 400 onlookers, students, faculty, staff and aviation pioneers in attendance – he remarked, "As I am an old man whose active days are past, I shall dedicate the rest of my life, with the active aid of my son, Harry F. Guggenheim, to the study and promotion of the science of aeronautics. I shall do this as a part of my duty to my country, whose ample opportunities have ever been at my hand and whose bountiful blessings I have had the good fortune to enjoy."[1]

Daniel Guggenheim's statement at the NYU groundbreaking was more than an appropriate pronouncement crafted for the occasion. To give his dedication flesh and bones, he and his son, Harry, formulated an ambitious over-all plan, and Harry met with President of the United States, Calvin Coolidge, and Secretary of Commerce, Herbert Hoover, in late 1925, to present it. Shortly thereafter, on January 16, 1926, Daniel Guggenheim sent a letter to Herbert Hoover, in which he announced establishing the Daniel Guggenheim Fund for the Promotion of Aeronautics. He endowed the Fund with an initial grant of $500,000, followed by a second of $2,000,000 and a third of another $500,000. The announcement set four goals for the fund:

"1. To promote aeronautical education both in institutions of learning and among the general public.
2. To assist in the extension of fundamental aeronautical science.
3. To assist in the development of commercial aircraft and aircraft equipment.
4. To further the application of aircraft and business, industry and other economic and social activities of the nation."

The Fund's original Board of Trustees had Harry as President; Rear Admiral Hutchinson I. Cone as Vice President, Assistant Secretary of War for Air, F. Trubee Davidson; Senator Dwight W. Morrow (soon to be Lindbergh's father-in-law); leading attorney, Elihu Root, Jr.; and financier John D. Ryan. In later years, others – equally illustrious – were added; including Orville Wright, two Nobel Laureats in physics – Robert A. Millikan and Albert A. Michelson, and General George Goethals, famous for his role in building the Panama Canal.

In an action that would be repeated by the federal government after WWII, the Fund's first operation was a tacit concession that the United States had much to learn in aeronautics from Europeans. It sent Harry and Admiral Cone in February of the same year, 1926, to England, France, Germany, the Netherlands, and Spain to visit the major research and education centers for aeronautics and to speak with their leading authorities. At the trip's outset, they met in London with Jerome C. Hunsaker, who had just resigned after over 20 year's active service in the Navy, and who – in a distinguished career - would found the aeronautics department at MIT and chair the National Advisory Committee for Aeronautics (NACA, the predecessor to NASA) for fifteen years. He was invited to join Guggenheim and Cone in at least some of their trips and did[3]. The report which resulted confirmed Harry's impression as an informed layman. It's concluding sentence stated. "Europe today, is far ahead of the United States in the number of able men, both in and out of government service, who are devoting their lives to the solution of the problems confronting civil aviation."[1] However, the Fund's representatives were not fully approving of everything European they saw and heard in making their aeronautical assessments. They noted that aeronautical engineering education was no more advanced in Europe than at home, and only a handful of laboratories were equipped for comprehensive aeronautical research and development. Of considerable interest to them was the discovery that an impressive fraction of travelers of passenger-carrying air operations in Europe was constituted by American tourists. This meant to them that airlines in the U.S. would have a reasonable passenger clientele, if they offered safe, reliable options.

Armed with the report's findings, the Fund moved in a number and a variety of directions, all seen as important to achieving its goals. While prepared to implement a plan to lay a firm foundation for aeronautical engineering education in the United States, the fund didn't do so simply by making money available. In the summer of 1926, for example, when Dr. Robert Millikan, head of the California Institute of Technology, went to New York to tell Daniel Guggenheim that Cal Tech should have a Guggenheim endowment, because "Southern California was destined by climate and land availability to become a national center of airplane production, Guggenheim listened attentively. 'All right,' he finally told Millikan, 'I will give Cal Tech money, but on one condition. Find me somebody from Europe who is familiar with the scientific side of aviation to build the laboratory." Millikan said he would, and from that came the emigration of Dr. Theodore von Karman, a Hungarian engineer, from Germany to the United States. Von Karman first gave seminars and advice on the establishment of the Guggenheim Aeronautical Laboratory at the California Institute of Technology (GALCIT), then became a Research Associate in 1928, and finally Director in 1930. From those early days until his death in May of 1963, Von Karman was regarded by many aerospace executives – including generations of Air Force leadership – as the Dean of U.S. Aerospace engineers, "a towering figure in the flight sciences."[1]

Of his first meeting with the philanthropist, Von Karman wrote, "Daniel Guggenheim had a real appreciation of science and a good sense of humor. I recall that during one conversation he asked me what the Guggenheim Foundation should do – beyond building aeronautical laboratories and subsidizing research – to advance aeronautical science in the United States. I suggested the financing of a series of handbooks for young scientists. I also casually mentioned that one thing that promoted science on the continent was that scientists could meet informally in cafes to iron out scientific problems. America didn't have this tradition and I thought she suffered accordingly. Mr. Guggenheim said he would be willing to subsidize a series of handbooks, 'But,' he added with a smile, 'I will not enter the café business, even for Science's sake.' Incidentally, this conversation led in 1934 to the creation of the six-volume "Aerodynamic Theory", edited by William F. Durand, the first encyclopedia of aeronautics, and one that is still used as a basic reference." [4] While Von Karman mentions the Durand series by starting with "Incidentally", in fact, his suggestion induced the Fund to contribute $60,000 to the preparation of the classic texts.[6]

Following Millikan's recommendation, the fund made grants of $350,000 and $195,000 to Cal Tech and Stanford, respectively, on the same date, August 24, 1926. The latter grant, reflecting the perceived needs of the era, specifically emphasized propeller research. Continuing its populating areas of the nation with schools of aeronautics, the Fund made a grant of $78,000 to the University of Michigan in December 1926 to establish the Daniel Guggenheim Professorship of Applied Aeronautics, providing also for the completion of that University's "aviation laboratory" [1]. MIT received $230,000 in January 1927 for a Daniel Guggenheim Aeronautical Laboratory. The latter grant was followed by another of $34,000 for a graduate course in meteorology and "a new meteorological research group."

Influence on Air Operations

Meteorology today might seem somewhat removed from mainstream aerospace engineering, but not so in 1927. Convinced that the national Weather Bureau's services wouldn't serve the purposes of an air transportation system, in August 1927 the fund created The Committee on Aeronautical Meteorology to develop systems allowing bad weather operations. Chaired by a Swedish-born meteorologist, Carl-Gustaf Rossby, two major projects were generated by the Committee. The first was to establish what they called the Full Flight Laboratory. To head it they chose Dr. (soon to be Air Force General) Jimmy Doolittle, who took charge in the fall of 1928. As part of the Full Flight Laboratory's equipment, two aircraft were purchased at the total cost of $26,000. One was a consolidate NY-2 "Husky", the other a Navy Vought Corsair 02U-1. The latter was accepted by the lab on Nov. 21, 1928, and Harry Guggenheim was its first passenger. Jimmy Doolittle's famous demonstration of control on take-off and landing from within a hooded cockpit on September 24, 1929, was a culmination of the labs' work, providing important evidence of the effectiveness of radio implemented blind flying devices. Harry Guggenheim was so enthusiastic that he dashed off a note to Orville Wright, saying: "It is significant that the achievement is realized through the aid of only three instruments which are not already the standard equipment of an airplane."[1] Those experiments, developing the techniques and equipment for flying by instrument, were considered by Jimmy Doolittle, in an incredibly distinguished career, as his "most significant contribution to aviation."[5]

In the same period, Harry Guggenheim had turned to the problem of making the general public more aware of the potential of aviation. He conceived of national "Air Tours" featuring famous fliers who would demonstrate the latest aircraft all around the country. The Fund sponsored these tours, which included such well-known people as Admiral Byrd and Floyd Bennett. They had just flown over the North Pole, and a natural successor was Charles Lindbergh.

The Guggenheim Fund's press release on June 28, 1927 read as follows: "Charles Lindbergh's airplane tour will be undertaken for the primary purpose of stimulating popular interest in the use of air transport. It will enable millions of people who have had an opportunity only to read and hear about the Colonel's remarkable achievement to see him and his plane in action."[2]

An unexpected outcome of Lindbergh's tour was his suggestion to make more roof-top markings of localities available for pilots on cross-country flights. "Guggenheim, promotion as well as air-minded, loved the idea and talked the Departments, of War, Navy, Commerce and the Post Office into cooperating. He also solicited support from major oil companies, railroads, businesses, civic groups and service clubs, plus the Ford Motor Company which enlisted the participation of its 7,600 dealers. When the campaign began, fewer than two thousand communities had air markings. When it finished, there were more than 8,000 carrying rooftop identification."[1]

Although the records available to this writer don't make it clear, it seems reasonable to think that the Fund's sixth university grant of $290,000 to the University of Washington had something to do with the close proximity of the Boeing Company there. That gift was, once again, to establish a School of Aeronautics, and was made on June 29, 1928.

As part of the efforts of Cal Tech's Dr. Millikan to convince Von Karman to accept the directorship of GALCIT, he told him of the Fund's interest in giving $250,000 in the summer of 1929 to create a research laboratory for airship studies in Akron, Ohio, as a branch of GALCIT. Von Karman would be in charge, operating from Pasadena, and someone of his choosing would be a resident director.

The Goodyear corporation had become involved with Zeppelin manufacture because of the US Navy's interest and because lighter-than-airships were regarded as promising for transoceanic travel. Harry Guggenheim saw that basic research was needed – a whole series of US-built rigid dirigibles had been destroyed by storms in midair – and the money he made available was given with the proviso that the city of Akron would match the grant with funds sufficient to run the center for five years. These arrangements did result in a large airship research institute being built at Akron Airport, which was attached to the University of Akron[4], and Dr. Theodore Troller, one of Von Karman's students in Aachen, became the Director of the Guggenheim Airship Institute for its entire existence,

In October, 1929, the Fund conducted a safe airplane competition that had been in planning for several years. A first prize of $100,000 was to be awarded to the designers of an airplane that would be impossible to "spin". Five additional prizes of $10,000 each would go to the first aircraft to meet "minimum flying requirements". The competition was held at Mitchel Field, Long Island, and a Curtis Tanager won first prize.[5] Twenty-seven aircraft had entered from three nations, and Daniel Guggenheim, in this seventies, went often to watch the test flying. "We must realize," he said in a published statement, "that the air age is already here. Once realized, our provincialism will fall away from us. Universal flying will make all of us neighbors, and as sure the steamboat and railroad are universal. The airplane will be."[1]

The last of the seven Daniel Guggenheim Schools of Aeronautics was established in the South Eastern United States on March 3, 1930 with the third largest grant of $300,000. It's site was the Georgia Institute of Technology.

But the Fund's leadership – the father and son – recognized that civil air transport needed their help. The Committee on Aeronautical Meteorology and the Full Flight Control Lab had made useful contributions, but airlines weren't making use of these advances, and that could only be done by advancing operations. The meteorology committee's chairman recommended to the Fund that it pay for a trial weather-reporting system to be used by the National Air Transportation Company on the New York-Chicago air route, which had both heavy traffic and bad weather. By the Fall of 1928, the Fund had made possible a radiotelegraph system operated twenty-four hours a day by the Department of Commerce, all along the route, providing point to point communication of weather information provided by the Weather Bureau.

Although the Guggenheims' old collaborator, Dr. Jerome Hunsaker, of AT&T's Bell Labs[*], was making good progress with instrument/equipment development and appeared to have an understanding with the pertinent government agencies, the New York-Chicago trial was in trouble with the Weather Bureau, because of rising costs. Sensing this, and becoming increasingly convinced as to the abilities of the Western Air Express Company, Harry Guggenheim used his support of the New York-Chicago trial to convince AT&T to also support a West Coast project of a similar nature between Los Angeles and San Francisco. The Fund proceeded to lend Western Air Express $180,000 to acquire three new Fokker F-10 trimotored airliners, capable of carrying twelve passengers. In addition, the Fund had notified the Commerce Department that it would grant Western Air Express $60,000 for a weather and communication system as a "model airline" project over the route, whose weather the meteorology committee chairman had pronounced variable enough to provide a fair test for the system.

But there was a small problem. "Harry Guggenheim wanted PT&T to make the service gratis for the year-long duration of the experiment, whereas the company was willing to donate only six weeks of service, after which it would charge commercial rates.[(3)]

To resolve the disagreement, Dr. Hunsaker "met with Guggenheim shortly after his arrival back in New York and asked him why the Fund would not pay for the communications service after the initial start-up period. Guggenheim answered that he thought that six weeks was not enough time to determine whether the system worked – that at least six months would be necessary. With a longer trial, more data could be obtained and a more persuasive case could be made to the airlines and the government that such a system could be extended to other airways. Finally, the Guggenheim Fund had been created to provide money for projects that neither the government nor private enterprise were in positions to support. This was obviously not the case, because AT&T had demonstrated that it had both the money and the facilities to back the experiment."[(3)]

"Hunsaker was convinced. He discussed the matter with one of AT&T's vice president's and reached an agreement that Pacific Telephone and Telegraph would not donate any services to the Guggenheim Fund but would instead function as an equal partner with the Guggenheim Fund and Western Air Express in an experiment for the government in the public interest. They further

---

[*] AT&T's interest in aviation had its genesis in Alexander Graham Bell's early collaborations with Samuel Langley and Glenn Curtis.

agreed that PT&T would provide its services gratis for at least a year in order to obtain a conclusive result of value to us and to the Air Transportation Companies."[3]

"With appropriate ceremonies and ample attention from the press, Western Air Express inaugurated flights over the route on the morning of May 26. Operations that day and over the next twelve months went smoothly, with the airline achieving a 99 percent on-time rate and in impeccable safety record. Passengers praised the airline for its performance and extolled the comfort and speed of the big Fokkers. The volume of traffic on the route, used not only by Western Air but also by Pacific Air Transport and Maddux Air Lines, climbed so rapidly that within a matter of months eighteen more reporting stations had to be added to the service, and forecasts were issued six times daily."[3]

Beyond the support given universities for education and research, the Guggenheim fund financed a number of more focused areas in aeronautics. Perhaps sensing that the conclusion of the Fund's efforts was growing nearer, in the midyear of 1930 the Fund made the following grants:

-$41,000 for various weather research projects, ranging from fog penetration to a study of North Atlantic weather

-$15,000 to the Harvard School of Business Administration for a study of commercial aviation's economic and industrial aspects. (The first Research Fellow appointed under this grant was Herbert Hoover, Jr., son of the president.)

-$60,000 to Syracuse University to establish a center for aerial photographic mapping and surveying

-$10,000 to the Northwestern University School of Law to help create an Air Law Institute

-$140,000 to the Library of Congress to endow a Chair of Aeronautics and stimulate the acquisition of historical aviation material.

Continuation by the Daniel and Florence Guggenheim Fund
But if the work of the Fund for the Promotion of Aeronautics was drawing to a close, a different challenge lay ahead. Charles Lindbergh had consulted with officials of the DuPont Company about the black powder used in solid propellant rockets – thinking of their use for flight. He had received a totally discouraging reply. In November of 1929, in Harry Guggenheim's living room at Falaise, they spoke of this depressing outcome. Harry's wife, Carol, having been silently reading a newspaper to that point, joined the conversation. According to Lindbergh, she read aloud to them an article about the rocket experiments of a Physics professor at Clark University, Dr. Robert H. Goddard. "When Carol finished, Harry was smiling. 'May be the answer to our problem,' he told Lindbergh. 'Why don't you check up on Goddard? Go have a talk with him, if you think it's worthwhile.'"[6] Lindbergh was interested enough to ask about Goddard in a visit to MIT, then to return with him to DuPont, and finally – empty handed – to the Carnegie Institution of Washington. The latter authorized a grant of $5,000 for rocket research, but a proposal for a considerably larger sum was unsuccessful.

During this time, i.e., in 1929, Harry Guggenheim was appointed United States Ambassador to Cuba. That undoubtedly had a role in the decision to liquidate the Daniel Guggenheim Fund for the Promotion of Aeronautics in the following year. When Lindbergh wrote to Harry asking if he would object to an approach to his father, Daniel Guggenheim, regarding the continuation of Goddard's work, Harry promptly told him to go ahead. It was, therefore, clear that when Lindberg visited the older man at Hempstead House, his request for financing was on a personal basis, that is, no foundation was involved. Lindbergh gave Daniel a summary of the status of Goddard's experiments. "The conversation was brief and to the best of Lindbergh's memory went about like this:

> Mr. Dan: 'Then you believe rockets have an important future?'

> Lindbergh: 'Probably. Of course one is never certain.'

> 'But you think so. And this professor, he looks like a pretty capable man?'

> 'As far as I can find out, Mr. Guggenheim, he knows more about rockets than anybody else in the country.'

> 'How much money does he need?'

> 'He'd like to have $25,000 a year for a four-year project.'

> 'Do you think he can accomplish enough to make it worth his time?'

> 'Well, it's taking a chance, but if we're ever going to get beyond the limits of airplanes and propellers, we'll probably have to go to rockets. It's a chance but, yes, I think it's worth taking.'

> 'All right, I'll give him the money.'"[6]

As the quotation shows – Daniel Guggenheim had let Robert Goddard be told that he would provide $25,000 a year for two years, with another two years to be allotted following a committee review of the work accomplished. But when Daniel died that year at age 74, delays in settling the estate kept the money from being available beyond the first two years.

Returning to Clark University from the test series in New Mexico where he had conducted over 100 firings and 31 successful flights, Dr. Goddard continued his rocket research on a smaller scale until 1934. At that time, having completed his tour as Ambassador to Cuba, (from 1929 to 1933) Harry Guggenheim resumed his presidency of the Daniel and Florence Guggenheim Fund, the oldest of the five foundations established by the Guggenheim family.

Following a meeting Harry Guggenheim had arranged for Robert Goddard, in May 1940, with representatives of the Army, the Army Air Corps and the Navy, these services turned down the opportunity to develop rockets as weapons. For the remainder of his life, Goddard worked on JATO, the jet assisted take off concept. After he died in August 1945, the Daniel and Florence Guggenheim Fund considered various ways to continue the momentum of Goddard's work. They ultimately commissioned G. Edward Pendray to survey government agencies and individuals distinguished for contributions to the field to determine the most pressing needs, the greatest opportunities and what a relatively small charitable foundation could do to help. The Pendray report on rockets and jet propulsion published in 1948 provided a useful picture of and welcome stimulus for the field.

Acting on the Pendray report's suggestion that schools should establish research centers where engineers could be trained for the space age which it had foreseen as being just ahead, the Foundation considered which campuses would offer appropriate sites. It decided to make grants of approximately $250,000 to both Cal Tech and Princeton in December 1948. The former was to be known as the Daniel and Florence Guggenheim Jet Propulsion Center (JPC) and the latter first the JPC at Princeton, later the Daniel and Florence Guggenheim Laboratories for the Aerospace Propulsion Sciences[6]. A major provision of the grants was to create a Robert H. Goddard Professorship at each institution to pursue rocket and jet propulsion technology. In 1955, at the end of what was considered a trial period, the Foundation made additional grants of $215,000 to the JPC at Cal Tech and of $205,000 to the center at Princeton. The scope of the Princeton Center was enlarged one more time, in July 1961, with still another grant from the Foundation of $225,000.

Harry Guggenheim's continuing interest in aviation was reflected in three more grants to universities. In 1950 Cornell University received $180,000 from the Foundation to establish the Daniel and Florence Guggenheim Aviation Safety Center, with headquarters in New York City. The famous names in aviation safety brought into the plans and operations of this Center included Cornell's T.P. Wright, the Flight Safety Foundation's Jerome Lederer, and the FAA's General E.R. Quesada. The Center's first report in 1951 indicated "gaps" in the 1500 aviation safety research projects active at that time. It was followed by a series of "Design Notes", which were widely disseminated. Another publication of the Center was the "Human Factors Bulletin". Foundation grants to the Center totaled $298,000 by 1962.

By the year 1953, Harry Guggenheim had been convinced by conferences with Columbia University's Dean of Engineering, J.R. Dunning and by the results of studies by Dr. Pendray that the aerospace field had unmet needs for advanced structural concepts appropriate for vehicles intended for supersonic flight. As a result, he established the Daniel and Florence Guggenheim Institute of Flight Structures at Columbia with a grant of $329,000 that year. He described its goals as follows: "To train exceptionally qualified students in the comparatively new field of air-flight structures, to conduct research in aircraft structure and design (especially for supersonic flight), to act as a national clearinghouse for technical information in the field, and to disseminate technical knowledge regarding flight structures."[6] An endowment was set up for fellowships, and the Institute conducted seminars and symposia in this field. To the original grant another $200,000 was added in 1960.

Finally, in this group of three, a Harvard-Guggenheim Center for Aviation Health and Safety was established during 1957 at Harvard University, as a unit of the Harvard School of Public Health, with a Foundation grant of $250,000. Both military and civilian, U.S. and foreign nationals were thereby able to partake of courses at the Guggenheim aviation health center leading to the certification which the American Medical Association requires of physicians who wish to practice flight surgery. In addition, research applicable to flight safety has been conducted there including, for example, aging effects on night vision, the influence of environmental extremes, pressure, temperature, etc. – on human performance, and other, similar, matters affecting flight safety, including astronaut selection.

The philanthropy of the Guggenheim family has also been a major factor in fields other than aerospace – notably, in music and the visual arts. But two of the five foundations bearing the family name, as has been noted here, in one case, the Daniel Guggenheim Foundation for the Promotion of Aeronautics was from its inception totally devoted to advancing aviation in the United States, and in the second case, the Daniel and Florence Guggenheim Foundation, continued the beneficial aerospace work of the first. Inspiration for the former and president of the latter, Harry Frank Guggenheim died on January 23, 1971, at the age of 81, after a lifetime of dedicated and effective service to his country.

Summary

The lives of Daniel and Harry Guggenheim, as public spirited men of vision and practicality, helped shape aviation and aerospace design, manufacturing and operations in the United States to a degree that is arguably unrivaled in their impact. They did this through generosity, dedication, good judgment and tough mindedness.

Although the totality of the sums of money they gave were not – could not be – huge, relative to the sums eventually spent in national aerospace developments, the late Prof. Martin Summerfield of Princeton University put it into perspective quite well when he said, "All the same, they were critical sums, in that they were made available at just the right time, under just the right circumstances. You might speak of them as the risk capital that has made possible everything that has happened since, the acorn from which the oak has grown."[6]

Partial Summary of Guggenheim Fund Activities

June 12, 1925	Grant of $500,000 given by Daniel Guggenheim to NYU to establish a Department of Aeronautics
January 16, 1926	Daniel Guggenheim Fund for the Promotion of Aeronautics formed
February 1926	Tour of European Aeronautical Facilities initiated by Harry Guggenheim and Admiral Hutchinson Cone; joined at the outset by Jerome C. Hunsaker (then of AT&T)
August 24, 1926	Grants of $305,000 and $195,000 given to California Institute of Technology and Stanford University, respectively, to establish Schools of Aeronautics
December 2, 1926	Grant of $78,000 given to the University of Michigan to establish a School of Aeronautics
January 16, 1927	Grant of $230,000 given to the Massachusetts Institute of Technology to establish a School of Aeronautics
June 1927	"Air Tours" featuring Admiral Byrd, Floyd Bennett and Charles Lindbergh are sponsored to bring aviation to public notice
August 1927	Daniel Guggenheim Committee on Aeronautical Meteorology formed
June 29, 1928	Grant of $290,000 given to the University of Washington to establish a School of Aeronautics
Fall 1928	Full Flight Laboratory, as a project of the Meteorology Committee, established at Mitchel Field, Long Island, NY
July 1929	Research lab for Airship Studies established in Akron, Ohio (as a branch of Cal Tech's Lab)
October 1929	Safe Airplane Competition organized with first prize of $100,000 for a spin-proof airplane and additional prizes of $10,000 for each of the first five designs to meet "minimum flying requirements"
March 3, 1930	Grant of 300,000 given to the Georgia Institute of Technology to establish a School of Aeronautics
Mid 1930	Focused grants totaling $266,000 given to Harvard, Syracuse, Northwestern Universities and the Library of Congress
Fall 1930	Daniel Guggenheim funds the rocket research of Dr. Robert H. Goddard
Fall 1930	The Daniel Guggenheim Fund is liquidated
Sept. 18, 1930	Daniel Guggenheim dies at age 74
1942	Widow of Daniel Guggenheim transfers title to Hempstead House to the Institute of the Aeronautical Sciences (now the AIAA).
1948	Daniel and Florence Guggenheim Jet Propulsion Center (JPC) formed at Cal Tech and the JPC at Princeton
1950	Daniel and Florence Guggenheim Foundation establishes Aviation Safety Center as part of Cornell University in New York City.
1953	Daniel and Florence Guggenheim Foundation establishes Institute of Flight Structures at Columbia University
1954	Daniel and Florence Guggenheim Foundation grants MIT $25,000 to help establish the Jerome Hunsaker Chair in Aeronautics
1957	Daniel and Florence Guggenheim Foundation established Center for Aviation Health and Safety at Harvard University
1960	Daniel and Florence Guggenheim Foundation provides the initial grant of $75,000 to establish the International Astronautical Academy
Jan. 23, 1971	Harry F. Guggenheim dies at age 81

References

1. "Daniel Guggenheim: The Man and the Medal", G. Edward Pendray, Editor, Robert J. Serling, Revision Editor, Guggenheim Medal Board of Award of the United Engineering Trustees, Inc. New York, 1981.

2. "Lindbergh", A. Scott Berg, G.P. Putnam's Sons, New York, 1998.

3. "Jerome C. Hunsaker and the Rise of American Aeronautics", William F. Trimble, Smithsonian Institution Press, Washington, 2002.

4. "The Wind and Beyond: Theodore von Karman, Pioneer in Aviation and Pathfinder in Space", Theodore Von Karman with Lee Edson, Little, Brown & Co., Boston, 1967.

5. "I Could Never Be So Lucky Again," General James H. Doolittle with Carroll V. Glines, Bantam Books, New York, 1991.

6. "Seed Money: The Guggenheim Story", Milton Lomask, Farrar, Straus & Company, New York, 1964

Part IV

Schools Accredited from the mid-1930s to the Present

Chapter 8
Department of Aerospace Engineering and
Engineering Mechanics
The University of Texas at Austin[*] -- A Brief History

Wallace T. Fowler, Distinguished Professor of Aerospace Engineering

THE EARLY PROGRAM

Aerospace Engineering at The University of Texas at Austin had its beginnings in the mid 1920s under the leadership of the late Professor Alexander Vallance. Courses in each of the areas of aerodynamics, structural analysis, mechanics of flight, and aircraft engines were made available as part of an optional program in mechanical engineering. Early support for the program by the Guggenheim Foundation was crucial to the development of these courses and the new degree program. The Department of Aeronautical Engineering awarded its first degree in 1927. Interest in aeronautics in general, which had been at a high peak in the late 1920s, following Lindbergh's crossing of the Atlantic, fell off rapidly with the advent of the depression of the early and middle 1930's. Consequently, the baccalaureate program in aeronautical engineering was discontinued in 1933-1934, although limited technical elective offerings in mechanical engineering continued to be given by Professors Vallance and Degler.

In 1937-1938, Professor Vallance left the university on a leave of absence. He never returned, resigning his position in 1938. As the late 1930's approached, the clouds of World War II, which were gathering in Europe, brought about a tremendous increase in the demand for military and transport aircraft. A new, expanded program of work in aeronautical engineering at the graduate level was initiated at The University of Texas in 1939-1940. This offering immediately received a favorable response from both engineering students and the aeronautical industry of the Southwest. In addition to the expansion of on-campus programs taking place early in 1941, there was a rapid growth throughout the state of Texas of both military and industrial aviation activities. This influx of industry brought still greater demands for trained engineers, and as a direct result, short training courses were established in industrial centers under the Engineering Defense Training Program.

WORLD WAR II

In 1940, as part of the changes preceding the entry of the US into World War II, a Civilian Pilot Training course was established under the auspices of the Civilian Aeronautical Administration.

Professor Venton L. Doughtie served as director of the program. Students who completed both ground and flight training in this program entered both the Army and Navy Air Corps. By May 15, 1942, 192 of the 225 graduates of this program had enlisted. This program produced over a thousand pilots and navigators.

Shortly after the entry of the United States into World War II in December 1941, the Engineering Defense Training (ESMWT) Program was expanded and renamed the Engineering, Science and Management War Training Program. Dean Willis R. Woolrich, acting as regional advisor, coordinated activities related to the national ESMWT program in the Southwest Region. Dr. Milton J. Thompson, from the University of Michigan, directed the work specifically relating to aeronautical engineering. Mr. Harry W. Brown, who had been brought to the campus early in 1941 as an instructor, assisted him. An outstanding contribution of the ESMWT program was the offering on the university campus of a full-time, one-semester course in the elements of aeronautical engineering.

With the rapidly growing interest in aeronautical engineering at The University of Texas, it soon became evident that further expansion of the program was indicated. It became obvious that the only effective way to develop a strong graduate program was to build upon the foundation of a strong undergraduate program. With this thought in mind, a more extensive course curriculum was developed and the formation of the new Department of Aeronautical Engineering was announced in 1942, with Dr. Milton Thompson as Chair, a position he held until 1966. Over 500 Bachelors degrees were awarded during this period. Also, during this time, The University of Texas was selected as one of only seven schools to conduct the Curtis-Wright Engineering Cadette Program. This was one of the first organized efforts to recruit women as engineers.

In 1945, Professor Thompson, Chair of the Aeronautical Engineering Department, along with Dr. C. Paul Bonner of Physics, organized the Defense Research Laboratory, a center for advanced research in military technology. Many engineering and science faculty members participated in this research, with the initial work being focused on missile guidance and control, radar, and underwater acoustics. Facilities were built for research into thermodynamics and flight mechanics. Faculty from Aeronautical Engineering worked with Electrical Engineering faculty on increasingly sophisticated guidance and radar systems.

After World War II, an off-campus research center was established eight miles north of campus on the site of a magnesium plant that had supported the nation's war effort. This center is the site for numerous research facilities, including the university's low speed wind tunnel.

Dr. Thompson served as Professor and as Associate Director of the Defense Research Laboratory until 1969. He guided the development of the Department of Aeronautical Engineering (which became Aerospace Engineering) for 24 years, from 1942 until 1966. Doctoral degrees in Aeronautical Engineering were offered for the first time in 1955 under his leadership. Dr. Thompson died while presenting a lecture to students in July 1971.

In the 1950's, the two departments, Aerospace Engineering and Engineering Mechanics, which make up today's Department of Aerospace Engineering and Engineering Mechanics, were separate. Faculty members from these two departments who continued into the 1960s were Drs. Harold Plass, Shao Wen Yuan, Milton J. Thompson, and Eugene A. Ripperger.

THE SPACE AGE

With the rapid expansion of the nation's space program, notably the race to the moon in the 1960's, a larger fraction of the curriculum was devoted to space-related subjects culminating today in the department offering space flight and atmospheric flight options to students. The driving force for the extension of the curriculum into space-related studies was Dr. Byron D. Tapley. Dr. Tapley received his Ph.D. in Engineering Mechanics from UT and joined the Department of Engineering Mechanics in 1960. He taught the first orbital mechanics courses and moved to the Aeronautical Engineering Department in 1964. He developed a research program focusing on low-thrust trajectory analysis and optimization. Under Dr. Tapley's leadership, a wide variety of space-related courses have been introduced over the past 40 years, resulting in an exceptionally strong program in space-related engineering, both at the undergraduate and graduate levels. During this period, the name of the department was changed to the Department of Aerospace Engineering. Dr Tapley became the Chair of the Aerospace Engineering Department in 1966, while he was still an Associate Professor.

Immense growth took place during the period from 1965 to 1969, primarily in Aerospace Engineering. This period also saw the merging of the two departments into a single unit, with many faculty members teaching courses in both disciplines. During this decade, thirty faculty members were added to the department and only two departed, for a net gain of twenty-eight faculty members. Faculty members added to the department(s) were Drs. Eric Becker, Anthony Bedford, John Bertin, Harry Calvit, Lyle Clark, Frank Collins, Roy Craig, John Dickerson, Richard Ensminger, Wallace Fowler, Charles Hickox, James Hill, David Hull, Glen Journay, Eugene Konecci, Lawrence Mack, Paul Nacozy, Henry Petroski, John Porter, Thomas Runge, Paul Russell, Bob Schutz, Ron Stearman, Morris Stern, Victor Szebehely, Byron Tapley, Enrico Volterra, John Westkaemper, Melvin Wilkov, and Ching Hsie Yew. Faculty members leaving the faculty during this decade were Harold Plass and Shao Wen Yuan.

In 1967, at the request of the Provost, the separate Departments of Engineering Mechanics and Aerospace Engineering were combined to become the Department of Aerospace Engineering and Engineering Mechanics. This combined department structure has survived until the present. Those coming from Engineering Mechanics were Drs. Clark, Ripperger, Volterra, Mack, Stern, Wilkov, Yew, Calvit, and Dickerson.

The period from 1970 to 1979 was a decade of slow attrition. During this ten-year span, ten faculty members were added and fourteen departed. Faculty members added during this decade were Drs. Dale Bettis, Roger Broucke, Graham Carey, Raynor Duncombe, Linda Hayes, Michael Macha, Tinsley Oden, Pol Spanos, Jason Speyer, and J.M. Summa. Faculty members departing were Drs. Calvit, Collins, Dickerson, Ensminger, Hickox, Hill, Journay, Konecci, Petroski, Runge, Russell, Summa, Thompson, and Volterra.

The period from 1980 to 1989 was a decade of slow growth. During this decade, thirteen new faculty members were added while ten departed. Those added to the faculty were Drs. Jeffrey Bennighof, David Dolling, George Dulikravich, John Kallinderis, Stelios Kyriakides, Kenneth Liechti, Parker Lamb, Hans Mark, Mark Mear, Richard Miksad, Greg Rodin, Philip Varghese, and Bong Wie. Faculty members departing during this decade were Drs. Bertin, Bettis, Dulikravich, Lamb, Macha, Nacozy, Ripperger, Spanos, Wie, and Wilkov.

Of special note in this decade was the addition of Dr. Hans Mark to the faculty. Dr. Mark came to the department after a distinguished career that included Director of the NASA Ames Research

Center, Deputy Administrator of NASA, Secretary of the Air Force, Director of the National Reconnaissance Office, and Chancellor of The University of Texas System. Even after he joined the faculty, his public service continued. He served as Director, Defense Research and Engineering for the Department of Defense during the Clinton administration.

The period from 1990 to 1999 was a decade of steady faculty size. During this decade, thirteen new faculty members were hired and twelve faculty members departed. Those added to the faculty were Drs. Maruthi Akella, Ivo Babuska, Robert Bishop, Noel Clemens, Clint Dawson, Leszek Demkowicz, David Goldstein, Glenn Lightsey, Steve Nerem, Richard Schapery, Jeff Shamma, Eric Taleff, and Mary Wheeler. Those leaving the department were Drs. Broucke, Clark, Mack, Miksad, Nerem, Shamma, Speyer, Stern, Szebehely, Taleff, Westkaemper, and Yew.

The period since 2000 has seen five faculty members hired and four faculty members depart. Those joining the faculty are Drs. Cesar Ocampo, K. Ravi-Chandar, Rui Hwang, Laxminarayan Raja, and Thomas Hughes. Those leaving the faculty were Drs. Bedford, Craig, Porter, and Schapery.

STRENGTHS AND EXTERNAL RECOGNITION

Faculty External Recognition - The Department of Aerospace Engineering and Engineering Mechanics has many notable strengths. Five faculty members from the department have been elected to membership in the National Academy of Engineering. They are Dr. Hans Mark, Dr. Byron Tapley, Dr. Tinsley Oden, Dr. Mary Wheeler, and the late Dr. Victor Szebehely. Dr. David S. Dolling is a Fellow of the Royal Aeronautical Society. Faculty members who are Fellows of the American Institute of Aeronautics and Astronautics (AIAA) are Dr. Byron Tapley, Dr. Roy Craig, Dr. David Dolling, and Dr. Wallace Fowler. Dr. Hans Mark is an Honorary Fellow of the AIAA, a step above Fellow. Faculty members who are Fellows of the American Society of Mechanical Engineering (ASME) are Dr. Tinsley Oden, Dr. Stelios Kyrikides, Dr. Kenneth Liechti, and Dr. Krishnaswamy Ravi-Chandar. Faculty members who are Fellows of the American Geophysical Union are Dr. Byron Tapley and Dr. Bob Schutz. Dr. Schutz is also a Fellow of the International Association of Geodesy. Dr. Ivo Babuska is a Fellow of both the International Academy of Mechanics and the US Academy of Mechanics.

Astronauts - There have been several department graduates who became astronauts. Department graduate and astronaut Alan Bean was the fourth person to walk on the moon. Robert Crippen was the pilot of STS-1, the first flight of the space shuttle Columbia. Michael Baker was a shuttle commander and headed the NASA team in Russia for a time. Paul Lockhart flew twice as shuttle pilot in 2003, and Stephanie Wilson has a flight scheduled for late 2003 or sometime in 2004.

RESEARCH AND TEACHING

Center for Space Research - Space-related research in the department was pioneered in the early 1960s by Dr. Byron Tapley. The early work was focused through an institute called IASOM (the Institute for Advanced Studies in Orbital Mechanics). This research expanded through the 1970s and in 1981, The University of Texas Center for Space Research was established. This center, first housed on campus, moved to a site northeast of campus in 1985 and then moved to the UT research campus eight miles northwest of the main campus in 1996. Currently the center is responsible for about 10% of the research funding of the entire College of Engineering. Research focuses on satellite geodesy, remote sensing, earth sensing, planetary exploration, precision orbit determination, and high precision satellite data reduction.

Institute for Computational Engineering and Sciences (ICES) - The Institute for Computational Engineering and Sciences (ICES), formerly known as TICAM, is an organized research center created to function as an interdisciplinary research center for faculty and graduate students in computational sciences and engineering, mathematical modeling, applied mathematics, software engineering, and computational visualization. Organizationally, ICES reports to the Vice President for Research, and draws faculty from fifteen participating academic departments. The Institute currently supports five research centers and numerous research groups, but new research units in distributed and grid computing, computational biology, biomedical science and engineering, computational materials research, and many others are planned over the next 4 years. It also supports the CAM Program, a graduate degree program leading to the M.S. and Ph.D. degrees in Computational and Applied Mathematics

Center for Aeromechanics Research - The Center for Aeromechanics Research conducts computational, analytical and experimental research in supersonic and hypersonic aerodynamics, high temperature gas dynamics, turbulence, combustion, laser diagnostics, aeroelasticity and structural dynamics, control of flexible structures, and flight structures. Funding from these sources provides support for graduate students engaged in MS and PhD research programs, undergraduate students, and postdoctoral fellows. The Center's goal is to produce work of the highest quality, which advances the state of engineering, meets the sponsors' needs, and is publishable in the most reputable national and international journals. Current work in the Aeroelasticity Laboratory and the Structural Dynamics Laboratory involves structures for new aircraft and system identification and control techniques for large, flexible structures such as Space Station components. Computational work in fluid mechanics includes non-equilibrium flows downstream of shock waves in supersonic nozzles, adaptive algorithms for aircraft with turbulent flow, direct turbulence simulation and smart algorithms for parallel computations; and in structural dynamics includes development of adaptive and parallel algorithms, and modeling of flexible multibody systems. The supersonic wind tunnel facilities include a Mach 5 blowdown wind tunnel. Recent investigations involve control and reduction of fluctuating pressure loads, flow in forward facing and parallel cavities, and unsteady interactions of shock waves with turbulent boundary layers. In the Flowfield Imaging Lab, laser diagnostic techniques are applied to the study of turbulent combustion and compressible turbulence. Non-intrusive measurement techniques for gas temperature, velocity, and composition using absorption spectroscopy, Raman and Rayleigh scattering are being developed.

Center for Mechanics of Solids, Structures & Materials – This center promotes research addressing fundamental and applied, issues in the broad field of mechanics of solids, structures and materials. Mechanics is playing an increasingly important role in new areas such as materials science and engineering, manufacturing and the durable design of a variety of microelectronics devices. The backgrounds and expertise of the Center's faculty bring together the necessary focus on theoretical, experimental and numerical approaches necessary to solve many of today's new mechanical problems. The Center's two main objectives are, first, to foster interaction between its members and develop the infrastructure required by its researchers; and second, to disseminate the results of the research to the public, increase the visibility of the activity and help expand interactions with industry and other research groups. The Center maintains the following laboratories: Solids and Structures Lab, Acropolis Computing Center, Composites Laboratory and the Mechanics of Materials Lab. The laboratories house modern testing facilities for examining the mechanical behavior of solids and materials at the structural level and the material macro, meso and micro levels. The group has its own dedicated computational facilities consisting of several workstations, graphics capabilities, data acquisition systems and digital image analysis systems.

Flight Simulators - In December 1969, a special edition Link GAT-1 flight simulator was delivered. Dr. Wallace Fowler used this simulator as the basis for a flight mechanics laboratory course that was integrated into the curriculum in the early 1970s. The simulator was also used in a laboratory attached to a senior elective in flight test engineering. The simulator was used instead of an actual aircraft for cost and liability reasons. This flight simulator, now over 30 years old, is still in use in the flight dynamics laboratory. However, in 2000, Dr. Robert Bishop reworked the flight dynamics laboratory, bringing in three additional simulators.

Low Speed Wind Tunnel - During the 1970s, a low speed wind tunnel with a 5 ' x 7' test section was constructed by faculty and students at the Balcones Research Center. The construction of the tunnel was supervised by Dr. John Westkaemper. The tunnel has been used for numerous aeronautical engineering research studies, architectural engineering studies, and for the aerodynamic laboratories.

Small Satellite Laboratory – In 2001, Dr. Glenn Lightsey established an on-campus laboratory in which undergraduate and graduate student teams could fabricate small satellites (nanoSats) and payloads for balloons and sounding rockets. He obtained a surplus tracking station from the US Air Force, which has been installed on top of the aerospace engineering building on campus. Sounding rocket and balloon payloads have already been flown successfully and students are currently building a pair of satellites for the USAF NanoSat program.

Teaching Excellence - In addition to its strong history of research excellence, the department has a long history of excellent teaching. Since 1957, there have been 10 faculty members from the department who have received the College of Engineering / General Dynamics / Lockheed Award for Teaching Excellence. These faculty are Drs. James Turnbow, Roy Craig, Richard Ensminger, Wallace Fowler, Victor Szebehely, Anthony Bedford, Eric Becker, Robert Bishop, Mark Mear, and Philip Varghese. Three departmental faculty have been selected for membership in the university-wide Academy of Distinguished Teachers (Drs. Fowler, Bedford, and Bishop). Three departmental faculty have also been recognized by ASEE/AIAA with the Leland Atwood Award (Drs. Fowler, Craig, and Bishop).

HIGH PROFILE RESEARCH PROJECTS

GRACE - The director of the Center for Space Research, Dr. Byron Tapley, is the principal investigator on the GRACE twin satellite mission (Gravity Recovery And Climate Experiment). The primary objective of GRACE is to improve knowledge of the earth's gravity field by several orders of magnitude. The twin GRACE satellites were launched from Plesetsk, Russia, in March 2002, and are performing well. The expected mission length for GRACE is five years.

ICESat - Dr. Bob Schutz, a professor in the ASE-EM department, also affiliated with the Center for Space Research, is the principal investigator for the GLAS (Geoscience Laser Altimeter System) instrument on the ICESat satellite, launched from Vandenberg Air Force Base in early 2003. The laser altimeter on ICESat is designed to provide a five-year thickness profile of earth's polar ice caps. Both GRACE and ICESat are part of NASA's Mission to Planet Earth.

DEPARTMENT LEADERSHIP

From 1942 to 1945, Dr. Milton J. Thompson led the effort to re-create a department of aeronautical engineering. The department was re-established in 1945 and Dr. Thompson served as its chairman until 1966. In 1966, Dr. Byron Tapley assumed the chairmanship and led the department through a period of explosive growth and the combining of the departments of Aerospace Engineering and Engineering Mechanics. Dr. Tapley resigned the chairmanship to lead an effort to expand the department's space research program in 1977. Dr. Victor Szebehely was the department chair from 1977 through 1981. Dr. J. Parker Lamb held the post of chairman from 1981 through 1988. In 1988, Dr. Richard Miksad became chairman. Dr. Miksad resigned the chairmanship in 1994 to become Dean of Engineering at the University of Virginia. Dr. David Dolling assumed the chairmanship in 1994. His term as chairman ended in 2003, and Dr. Robert Bishop assumed the chairmanship in 2003.

STUDENTS AND DEGREES

Aerospace Engineering Undergraduate Program - The enrollment in the Aerospace Engineering Program has fluctuated over the years in response to the pressures of the job market and national priorities. Figure 1 shows the total enrollment in the Aerospace Engineering (ASE) undergraduate program each year since 1963. Figure 2 shows the number of BS degrees in ASE awarded over the same time interval. Figure 3 shows the total graduate enrollment (MS + PhD) in the ASE programs since 1963, and Figures 4 and 5 show the numbers of MS ASE and PhD ASE graduates for the same time scale. Since 1963, the department has awarded 1942 BS ASE degrees, 760 MS ASE degrees, and 278 PhDs in ASE.

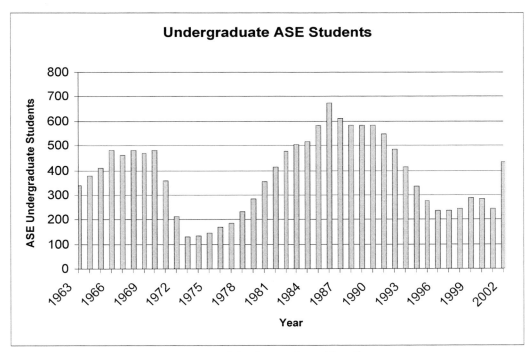

Figure 1 – ASE Undergraduate Enrollment

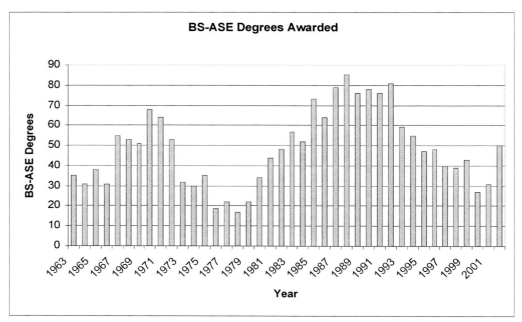

Figure 2 - BS Degrees in ASE Awarded

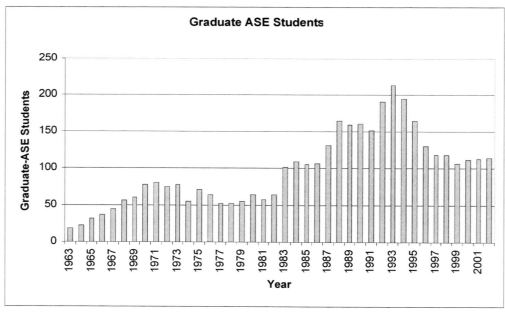

Figure 3 – Graduate Students in Aerospace Engineering

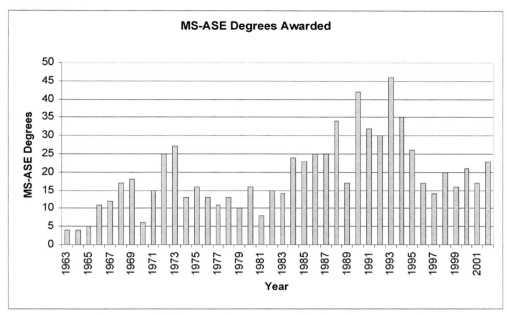

Figure 4 - MS Degrees in Aerospace Engineering Awarded

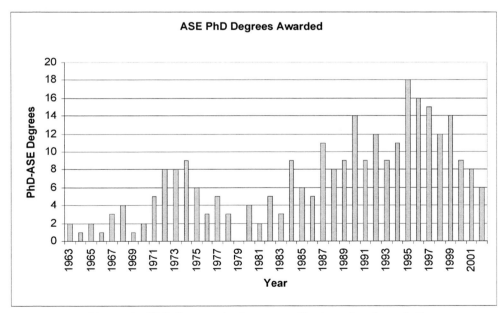

Figure 5 – PhD Degrees in Aerospace Engineering Awarded

Engineering Mechanics Graduate Program

The graduate program in Engineering Mechanics operates in parallel with the graduate program in Aerospace Engineering, with some faculty members producing students under both programs. Figure 6 shows the number of Engineering Mechanics graduate students enrolled each year since 1963.

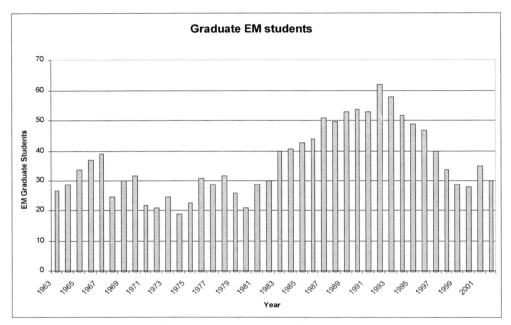

Figure 6 - EM Graduate Enrollment

Figures 7 and 8 show the number of MS and PhD degrees awarded in EM since 1963, respectively. There have been 194 MS degrees in Engineering Mechanics and 170 PhD degrees in Engineering Mechanics awarded by the department since 1963.

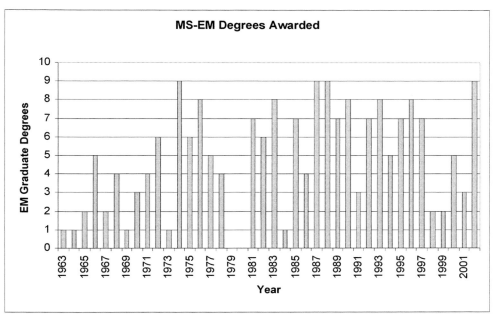

Figure 7 - Master of Science Degrees Awarded in Engineering Mechanics

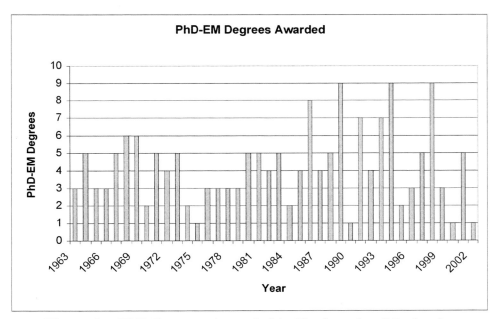

Figure 8 - PhD Degrees Awarded in Engineering Mechanics

CONCLUSION

The Department of Aerospace Engineering and Engineering Mechanics at UT Austin remains a strong program with major strengths in the areas of applied orbital mechanics, computational engineering sciences, and aeromechanics. It will continue to produce graduates who contribute strongly to the fields of aerospace engineering and computational engineering.

Chapter 9

The History of the Aerospace Engineering and Mechanics Department at the University of Minnesota

Amy Elizabeth Foster, MA
Graduate Student
Department of History
Auburn University
Auburn, AL

William L. Garrard, Ph.D.
Professor and Department Head
Department of Aerospace
Engineering and Mechanics
University of Minnesota
Minneapolis MN

1929 to 1958 - From Aeronautical Engineering to Mechanics

The University of Minnesota first offered courses in aeronautical engineering to undergraduates in mechanical engineering in 1926. This was 13 years after the first aeronautical engineering program in the U.S. was established at MIT. In early 1928, Ora M. Leland, Dean of the College of Engineering and Architecture, proposed to the Minnesota Board of Regents that an independent department of aeronautical engineering be established. He believed that "Minnesota is favorably located to become a center for this field of engineering for the Northwest." Leland recommended that the new curriculum continue much as it had from within the mechanical engineering department.

A special lectureship was given to John D. Akerman, who not only taught during the 1928-1929 school year, but also helped design the final form of the department. In the fall of 1929, the Department of Aeronautical Engineering at the University of Minnesota officially opened its doors to students. John Akerman, then an associate professor, served as its first department head, a position he would hold for nearly three decades.

Consistent with Akerman's background, the department's curriculum reflected the interests of industry. Born in Latvia, Akerman began his aeronautical studies at the Imperial Technical Institute in Moscow under the pioneer aerodynamicist Nickolai Joukowski. Akerman was also acquainted with Igor Sikorsky and maintained contact with Sikorsky after both immigrated to the USA. When World War I started, Akerman served as a pilot for the Russian Imperial Air Service. After the Bolshevik take over in 1917, he fled to France and served as a pilot in the French air force. He moved to the United States after the war in 1918.

Akerman's aeronautical interests led him to the University of Michigan, where he earned a bachelor's degree in aeronautical engineering in 1925. Akerman stayed at Michigan until 1927,

[1] Ora M. Leland to President L. D. Coffman, 4 February, 1928, College of Engineering and Architecture, Department of Aeronautics, 1929-1940 File, President's Office Papers, 1911-1945. University of Minnesota Archives.

doing coursework for a master's degree and working on a subsonic wind tunnel endowed by the new Guggenheim Fund for the Promotion of Aeronautics. He left Michigan for a position as chief design engineer at Hamilton Metal Plane Company in Milwaukee before finishing his master's degree. In 1928, Mohawk Aircraft Corporation, located in Minneapolis, hired Akerman as the chief engineer for their new low wing monoplane. It was this position that brought him to the Twin Cities and created the opportunity for Akerman to begin teaching at the university.[2]

The Aeronautical Engineering department at Minnesota offered eighteen courses its first year. These courses dealt primarily with hardware, pilot knowledge, structures, instruments, electrical systems, navigation, and communications. Charles Boehnlein of the Department of Mathematics was the professor for the more theoretical courses that dealt with aerodynamics. His three course series introduced the concepts of aerodynamic forces, stability, propeller theory, and laboratory practices. Professor Joseph Wise from the civil engineering department taught two classes on structures as applied to airframes and landing gear. Instructor Gustav Hoglund took responsibility for the laboratory courses, which covered airplane design, airplane parts and their construction, and airships.

Course offerings expanded during the 1930s with the addition of new faculty and new interests in industry. In the mid-1930s Akerman began studying the effects of high altitudes on pilots. He believed the next advancement in aircraft technology would be stratospheric flight "where high speeds are possible and bad weather in not encountered."[3] On October 23, 1934 Dr. Jean Piccard, a Swiss chemical engineer, and his balloon-piloting wife, Jeanette Piccard, ascended to 57,579 feet in a cloth balloon to record data on the stratosphere. The flight and the Piccards' possible contribution to his own project attracted Akerman's attention. Jean Piccard began experimenting with balloons in the early 1930s with his physicist brother, Auguste. With Dean Samuel Lind's approval, Akerman invited both Piccards to Minnesota, but only Jean became a faculty member—first as special lecturer, then as Professor in 1938. In addition to the stratospheric coursework, the faculty added courses on seaplanes in 1930 and dirigibles in 1931, both taught by Professor Wise.

Minnesota's Aeronautical Engineering department produced substantial numbers of talented engineers, thereby fulfilling industry's growing need for professional employees with formal aeronautical knowledge. In 1936, the program in Aeronautical Engineering became one of the first 10 programs accredited by the Engineering Council for Professional Development (the precursor of ABET). The Department has been in continuous accreditation since then. During the 1939-1940 school year, 3034 students were enrolled in aeronautical engineering programs across the United States and Canada, and 455 of those students studied at the University of Minnesota.

In 1926 Daniel Guggenheim organized the Daniel Guggenheim Fund for the Promotion of Aeronautics. Guggenheim intended the fund "to promote aeronautical education throughout the country, to assist in the extension of aeronautical science, and to further the development of commercial aircraft, particularly in its use as a regular means of transporting both goods and people."[4] The fund endowed seven of the ten schools that offered aeronautical engineering degrees at that time. Minnesota was not one of these. Since Minnesota received no Guggenheim funding, a major impetus for moving towards engineering science was missing. Instead, the department remained practice-oriented with little focus on the theoretical side of aeronautical engineering.

---

[3] Akerman, lecture notes, file J, Akerman Papers, University of Minnesota Archives.
[4] Harry F. Guggenheim, *Report of the Daniel Guggenheim Fund for the Promotion of Aeronautics, 1926 and 1927*, New York.

Those courses that were added in the 1930s mostly fell into the category of practice, not engineering science. In some cases, the faculty even dropped some science-based engineering courses.

In the late 1930s, the University administration began showing concern over the direction of the department. The department's and the school's competitiveness with other universities depended on a state-of-the-art, progressive curriculum. Consequently, Dean Samuel Lind of the new Institute of Technology (IT) felt obligated to evaluate his departments critically. He favored upgrading the curriculum, but the demands of WWII dictated that universities contribute to the war effort by training the largest number of scientists and engineers possible, as well as providing applied research. Lind thus postponed restructuring the curriculum until the war was over.

As one way of satisfying the country's demand for engineers during the war years, the University of Minnesota was one of seven universities participating in a unique educational experiment. One hundred and two young women from all parts of the country were selected to study aeronautical engineering at the University of Minnesota. These young women were employees of the Curtiss-Wright Corporation and were pledged to work in engineering departments of that company after graduation. They were called the Curtiss-Wright Cadettes. The Cadettes were intended to replenish Curtiss-Wright's dwindling supply of engineers. While in school, these women received room, board, tuition, and pocket money ($10/week). They were expected to work 40 hours/week with 30 hours of classroom instruction and 10 hours of supervised study for a period of 10 months.

The course of study for the Cadettes included drawing, structures, mechanics, aerodynamics, machine shop, materials testing, and aluminum fabrication. Those who doubted the engineering capabilities of women were soon proved wrong and before long the faculty found themselves teaching material far more advanced than originally envisioned. In the 10 months the students earned approximately two and one half years of college credit in engineering subjects.

Nearly 100 Cadettes completed the course and went on to Columbus, Ohio in December of 1943. The majority worked at Columbus until the end of the war. Some advanced to full engineering positions and some to supervisory jobs.

In 1946 the university began negotiations with the U. S. Government to acquire the idle Gopher Ordnance Works and its accompanying 8,000 acres of land south of the Twin Cities. The university finally purchased the installation in March 1948 for $1. This became the site of the Rosemount Aeronautical Laboratory (RAL), which would serve as the Aeronautical Engineering department's primary research facility for almost 15 years. Faculty members designed and installed a number of wind tunnels at Rosemount, including a hypersonic wind tunnel capable of producing speeds between Mach 7 and 11 and air temperatures of 3,000 degrees Fahrenheit. The RAL would be the site of significant research for both industry and the military. Graduates of the Department who worked at RAL developed a total temperature sensor for the Navy and then formed the Rosemount Engineering Company to produce it commercially. Rosemount Engineering became one of the world's largest suppliers of air data and other flow sensors. The company later divided into Rosemount Inc. and the Sensor Division of B.F. Goodrich Aerospace.

Dean Lind retired in July 1947. After a year-long search, the Board of Regents approved Athelstan Spilhaus, a professor of meteorology and the director of research at New York University, as the new Dean of the Institute of Technology. Spilhaus officially took office in January 1949 and brought with him a vision of developing the scientific foundation of Minnesota's engineering programs. He advocated a focus on fundamentals and aspired to leadership in theoretical research. He believed that a science-based curriculum and research were the principal components of a strong engineering program. In 1950, Spilhaus reported that within the Institute of Technology the "development of graduate instruction and research is emphasized."[5]

Akerman exerted little effort in support of Spilhaus's vision and eventually resisted it. The turbulent relationship between Akerman and Spilhaus hindered the development of engineering science in the Aeronautical Engineering department. In the spring of 1951, Dean Spilhaus's made his first major attempt to personally redesign the Aeronautical Engineering department. Enrollment had reached its lowest point at this time because of the mistaken perception of a surplus of engineers and the need for soldiers for the Korean War. The budget mirrored the drop in enrollment. Spilhaus pushed departments to reduce their costs as much as possible without risking their students' educations. As a means of eliminating the duplication of courses and thereby reducing costs, Spilhaus proposed making Aeronautical Engineering a division of Mechanical Engineering as it had been up until 1929. In this instance, Spilhaus was out of step with the rest of the country. Many aeronautical engineering programs started as options within mechanical engineering departments, as happened at Minnesota. However, by 1951, the trend was for independent departments of aeronautical engineering—Iowa State and Purdue being two such examples. Mechanical engineering coursework no longer met the technological and theoretical requirements of the aeronautical industry.

Despite decreases in enrollment and Spilhaus's perspective on the state of the department's curriculum and leadership, the Aeronautical Engineering department produced a number of notable graduates including Donald "Deke" Slayton, one of the original Mercury 7 Astronauts. However the curriculum was still extremely practice-oriented. Spilhaus refused to give up his quest to see Minnesota's engineering curriculum develop its strength in basic science, engineering science, and its "underlying principles," and an overhaul of the aeronautical course requirements was a priority for him. Spilhaus's second attempt to remedy the sluggishness he saw in the Aeronautical Engineering department came in May of 1957. He proposed a radical modification to the department—a merger with the Department of Mechanics and Materials and the subsequent removal of John Akerman as head of the combined department. Mechanics and Materials granted only graduate degrees, but taught undergraduate courses in mechanics. Aeronautical engineering was a professional department that granted mostly undergraduate degrees. Spilhaus intended the merger to combine the strengths of both departments and eliminate the weaknesses he saw in the aeronautical engineering program. Dr. Benjamin Lazan, Associate Dean of IT and head of Mechanics and Materials, took over as head of the combined departments in 1958.

In bringing John Akerman to the University in 1928, Dean Leland had hoped to create a program in aeronautical engineering with close ties to industry. The University of Minnesota Aeronautical Engineering department, however, was never able to establish relations with large aviation companies like those of the Guggenheim Schools. Except for some kit aircraft companies, all aircraft production companies in the state of Minnesota folded by 1931 because of financial failure or their inability to build working aircraft. Dean Leland's hope that Minnesota would become a major center for airplane manufacturing was not to become a reality.

---

[5] Athelstan Spilhaus, *The Biennial Report the President of the University of Minnesota approved and adopted by the Board of Regents 1948-1950*, 27 December 1950, Vol. LIII, no. 60,89.

1958 to 1992 - The Ascendancy of Mechanics

Dr. Lazan was a well-known researcher in the mechanics of materials whose orientation was toward engineering science. Lazan moved the department in the direction of engineering science by means of the faculty he hired. All faculty hires under Lazan were in the areas of solid or fluid mechanics with no faculty hired in aeronautical engineering. In the post-Sputnik era, research funds were re-directed to engineering science-oriented programs. During this time, the external funding base for the department changed from projects at RAL to projects on the main campus. The extensive facilities and personnel at RAL required large externally funded projects. Professor Rudolf Hermann was one of the world's leading experimental aerodynamicists and was instrumental in obtaining the large grants and contracts necessary for the existence of RAL. When Hermann left the University in 1962, there were few faculty who wished to continue the large aeronautical based projects at RAL. As a consequence of reduced funding, RAL closed its doors in the mid-1960s.

Dr. Lazan was forced to resign as Department Head in the early 1960s due illness; however, during his tenure as Department Head, he hired most of the faculty who shaped the Department in the 1960s and 1970s. Professor P. R. "Pat" Sethna replaced Lazan in 1966. Pat Sethna had received his Ph.D. from the University of Michigan in Engineering Mechanics and joined the University of Minnesota in 1956. His specialty was non-linear systems and he had a very strong applied mathematics orientation. Under his leadership the Department continued its orientation towards engineering science.

In 1972, the name of the Department was changed from Aeronautical Engineering to Aerospace Engineering and Mechanics. Up to that time courses had been designated Aero. or M&M (Mechanics and Materials) depending on whether they had been part of the old Aeronautical or Mechanics and Materials departments. The new course designator AEM (Aerospace Engineering and Mechanics) was adopted for all courses. Despite the name change, the department was still heavily oriented toward theoretical mechanics with an applied mathematics flavor.

In the early 1970s, the aerospace industry underwent a severe recession due to the end of the Vietnam War and the termination of the Apollo Program. Undergraduate enrollments plummeted. At the same time there was a substantial reallocation of funds from engineering, sciences, and liberal arts to the health sciences. Again there were proposals to merge the Aerospace Engineering department with Mechanical Engineering. This did not occur, but the department lost a number of faculty positions during this period and there were essentially no new faculty hires from the early 1970s to the early 1980s. During this period enrollments gradually increased but never reached the level of the late 1960s.

In the late 1970s expenditures in the aerospace field began to increase dramatically. This increase was driven primarily by increased defense spending. The demand for aerospace engineers rose sharply and enrollments in the Department grew. In 1989, enrollments peaked at about 600 undergraduates and the AEM Department had the largest undergraduate enrollment of any engineering department at the University of Minnesota. The number of faculty members also increased to a maximum of 21.

A number of new young faculty members were hired beginning in 1989, including the first two female faculty. These younger faculty members had, for the most part, a more applied orientation than did the more senior faculty members. This was to have important positive implications for the future. Nationwide, research funding priorities in engineering began to shift from theoretical to applied. This shift was due to the perception that theoretical engineering research performed at universities had produced relatively few practical benefits. Experimental and computational research began to replace "pencil and paper" studies. In the late 1980s and early 1990s, the number of research dollars per faculty member in the AEM Department was the lowest of all of the engineering departments at the University of Minnesota.

1992 to 2003 - A Balanced Department

In 1992, Pat Sethna retired and Professor William Garrard became Department Head. Garrard received his BS in Mechanical Engineering and Ph.D. in Engineering Mechanics at the University of Texas at Austin. He joined the faculty of the University of Minnesota in 1967. Despite having never taken a course in aerospace engineering, Garrard had a strong interest in the field and did his Ph.D. thesis on satellite attitude control. Garrard worked with the Honeywell Systems and Research Center in Minneapolis as a consultant and further developed his interests in the application of advanced control theory to control of aerospace vehicles.

During the 1990s, many of the faculty hired in the 1950s and early 1960s retired. Most of these had a theoretical mechanics orientation, and were replaced by faculty members with much more applied interests. Some of the remaining senior faculty members, realizing that national priorities for research favored more practical work, changed the focus of their research. The result was a large increase in funded research in the Department. In fact, the research dollar per faculty member in the AEM Department became one of the highest of all the engineering departments at the University of Minnesota. Research in computational fluid mechanics and hypersonics, experimental fluid mechanics, smart materials, and aerospace systems flourished.

With the end of the cold war and the dissolution of the Soviet Union in the early 1990s, the demand for aerospace engineers decreased and enrollment in the AEM Department declined. In addition Aerospace Engineering was no longer seen as a glamorous field and was eclipsed by computer science. During the 1990s, state support for higher education also decreased substantially as a percentage of the operating expenses of the University of Minnesota. This resulted in a series of retrenchments and reallocations at the University. The number of faculty members in AEM declined to 16, the same number as the lowest point in the early 1970s. As this is written in February of 2003, the state of Minnesota faces its largest deficit in history. The resolution of this deficit is not clear but its potential effect on the AEM Department and the University of Minnesota is likely to be severe. The survival and success of aerospace engineering at Minnesota, however, stand as evidence to the fact that the department has shown a vital sense of flexibility to change, a characteristic that should carry it well into the future of aerospace engineering.

Some Notable Early Faculty Members

H.W. Barlow. Professor H.W. Barlow came to the University as an instructor in September 1932. He was a native of Cleveland, Ohio, and obtained his BS degree from Purdue and a MS in Aeronautical Engineering at Minnesota. Professor Barlow worked with John Akerman on designing and building racing airplanes for Colonel Roscoe Turner, holder of several long-distance speed records from New York to Los Angeles and back and from England to Australia as well as a famous pylon racer. One of the pylon racers designed at the University of Minnesota is in the National Aerospace Museum. This aircraft was designed to be neutrally stable in order to increase performance and was a very early example of an aircraft purposely designed for reduced static stability, a feature prevalent in most current high performance aircraft. Another aircraft designed by Barlow and Akerman was a streamlined single-seat land monoplane that was expected to have a top speed of 400 miles per hour. Barlow later joined the faculty at Texas A&M.

Jean Piccard. Dr. Jean Piccard was already a world famous balloonist when he came to the University in 1936. He taught courses in stratospheric flight while doing research and conducting many pioneering balloon flights. Dr. Jean Piccard and his wife, Dr. Jeanette Piccard, made their first stratospheric flight in October 1934, in Detroit. Before coming to Minnesota, they were associated with the Bartol Research Foundation of the Franklin Institute. At that time, they were foremost among the five or six persons with scientific knowledge of the stratosphere. Jean's twin brother, Dr. Auguste Piccard, was also a stratospheric pioneer. Together the brothers designed the balloons and their gondolas for the first stratospheric flights attempted, and together they conducted those flights in Switzerland. Jeanette Piccard was the pilot of the famous Piccard balloon expeditions. She was the first licensed woman balloonist in the world and the first woman to ascend into the stratosphere.

One of Dr. Jean Piccard's first projects at the University was constructing an unmanned hydrogen-filled transparent cellophane balloon for ascents 10 to 14 miles into the stratosphere. The balloon was successfully flown on June 24, 1936. Three aeronautical engineering students—Harold Hatlestad of Minneapolis, who built the radio equipment for the flight, Robert Hatch of St. Paul, and Robert Silliman of Duluth—maintained radio contact with the balloon from the station on the roof of the University armory. Jean Barnhill was a graduate student who worked with Dr. Piccard and aeronautical engineering students Harold Larson and Lloyd Schumacher in cutting by hand the sixteen 33-foot long tapered gores that made up the 25-foot high balloon. The 'orange peel' gores were fastened together by a revolutionary product, inch-wide strips of cellophane covered with adhesive, called Scotch tape, developed by the Twin Cities' 3M Company. Jean Barnhill was the first woman to graduate from the University in Aeronautical Engineering, and was also a championship pilot in national air races. She married Robert Gilruth, another graduate of the Department. Robert Gilruth was later instrumental in development of swept back wind technology in the U.S.[6] and played a key role in the U. S. Space Program as Director of the NASA Johnson Manned Space Flight Center during the Apollo program.

Rudolf Hermann. The Rosemount Aeronautical Lab (RAL) provided an important research facility that helped the Aeronautical Engineering Department attract the kind of faculty members capable of advancing the department as a research entity. By 1960, RAL housed a continuous-flow transonic tunnel, continuous-flow and blow-down supersonic tunnels, and a high temperature hypersonic wind tunnel. The facilities at RAL attracted some top researchers to the Aeronautical Engineering Department at the University of Minnesota. One of these was Dr. Rudolf Hermann.

Rudolf Hermann earned his Ph.D. in physics from the University of Leipzig in 1929 and in 1935, he completed his Doktor habilitation (Dr. habil.), the second doctorate required of all professorial candidates in Germany. Hermann's first engineering position was in the Department of Applied Mechanics and Thermodynamics at the University of Leipzig from 1929 to 1933. In 1934 he took over as head of the supersonic wind tunnel division at Aachen, a position he kept until 1937.

In 1935 the Luftwaffe Technical Office introduced Wernher von Braun, the German rocket pioneer, to Rudolf Hermann who was working at Aachen as an assistant professor in addition to holding his position in the wind tunnel center. Von Braun's group had difficulty with the aerodynamic design of missile fins and turned to Hermann and his facilities at Aachen.

Because of the significant role supersonic aerodynamics played in rocket design and the distance of the Aachen lab from Peenemünde, von Braun felt that the rocket group needed its own supersonic wind tunnel and its own supersonic specialist. Hermann joined the Peenemünde group in April 1937 as Director of the Supersonic Wind Tunnel Laboratory of the Army Rocket Experimental Station. The construction of two supersonic tunnels was Hermann's priority. The

---

[6] John D. Anderson, *A History of Aerodynamics*, Cambridge University Press, Cambridge, 1998, page 427.

first tunnel was a 20-second, blow down tunnel with a 40-centimeter-wide test section and a maximum running speed of Mach 4.4. The second was an 18 x 18 centimeter continuous-flow tunnel with a maximum speed of Mach 3.1. The theoretical design of the De Laval nozzles used to accelerate the tunnel flows to supersonic velocities proved to be an extraordinarily complex task. Nevertheless, Hermann and his team perfected the designs for the testing facilities while providing novel methods for acquiring transonic and supersonic data, such as drop tests from an altitude of 7000 feet. Through these tests, Hermann and his staff gathered supersonic flight data on the aerodynamic design of the A-5, a redesigned A-3 rocket used to test guidance systems. The lessons learned from the study and testing of the A-5 were later incorporated into the design of the V-2 rocket. This experience gave Hermann the status of chief aerodynamicist for the V-2 rocket.

With the end of World War II, the Allied Powers sent representatives to occupied Germany to recruit the top scientists in a variety of fields for the benefit of science and weapons development at home. In the U.S., this operation was known as Project Paperclip. By the end of 1952, 544 German specialists were living and working in the United States because of Project Paperclip. As these scientists and engineers arrived in America, they were usually housed and put to work at military installations.

In 1945, Hermann was employed as a consultant with the Air Engineering Development Division at Wright Patterson Air Force Base in Dayton, Ohio. The American public was not told of the presence of German scientists and engineers working in the United States until early December of 1946. Newsweek magazine described the work of Hermann and his colleagues as follows: "As the war ended, [Dr. Rudolf Hermann] was building a 7,000-mile-an-hour wind tunnel in the Bavarian Alps. With six associates brought from Germany, Hermann is working on supersonic wind tunnels for the United States Army."

By 1948, some of the incoming Germans were being approved for work in American industry, and with that approval came essentially full freedom of choice. Scientists already in the United States were also being released for industry work. In fact, 516 of these German scientists and engineers and 1063 of their dependents obtained U. S. citizenship. Hermann was one of these.

In 1950, Hermann left Wright Air Force Base, and joined the faculty in the Department of Aeronautical Engineering at Minnesota. He brought knowledge and expertise in supersonic and hypersonic flight, subjects that were new to the curriculum. Hermann also taught mostly graduate level courses. The lack of graduate courses was a weak area in the Aeronautical Engineering department that was partially remedied by the addition of Hermann to the faculty.

Hermann served the University of Minnesota both as a teaching professor and researcher, much as he had in Germany. He and his family lived in one of the 25 staff houses on the grounds of RAL, where he was Technical Director of the Hypersonic Facilities. At RAL, Hermann conducted research on supersonic and hypersonic flow characteristics, rocket sleds, and ramjets, with much support and funding from the U. S. Air Force and Navy. Hermann was one of the top researchers in supersonic and hypersonic aerodynamics in the 1950s and 1960s.

In June 1962, Rudolf Hermann left the University of Minnesota to accept the position of Director of the newly founded aeronautical research laboratory at the University of Alabama in Huntsville, a neighboring facility to Marshall Spaceflight Center where Hermann's former collaborator from Peenemünde, Wernher von Braun, was in charge. During his time at Minnesota, he contributed to the Aeronautical Engineering program his knowledge and understanding of supersonic and hypersonic theory and an approach to engineering science at a time when the Institute of Technology was ready for change.

Helmut G. Heinrich. Dr. Helmut G. Heinrich was also one of the German scientists who came to America after the war as part of Project Paperclip. Educated at the Technical University of Stuttgart, "Doc" Heinrich, as his students and associates knew him, served as Chief of Aerodynamics at the Graf Zeppelin Institute in Germany during WW II. He was at Wright Airforce Base from 1946 until 1956 when he joined the faculty of the Aeronautical Engineering Department of the University of Minnesota. Professor Heinrich taught courses and conducted pioneering work on deployable aerodynamic deceleration devices, primarily parachutes. A number of undergraduate and graduate students worked on government contracts and grants under the direction of Dr. Heinrich. He invented the guide-surface parachute and several related devices that significantly improved parachute construction and performance. Heinrich developed supersonic parachutes that were considered for use in the Apollo program and his contributions to parachute systems were used for soft-landing scientific probes on Venus and Mars.

Dr. Heinrich died of a heart attack on March 7, 1979 in Houston where he had just received the first AIAA Aerodynamic Deceleration Systems Award. He was a fellow of the AIAA, a Fellow of the RAES, and a charter member of the AIAA Committee on Aerodynamic Deceleration Systems formed in 1965.

Note on Sources

The information in this paper was taken from the following four sources:

1. Foster, Amy Elizabeth, *Aeronautical Science 101, The Development of Engineering Science in Aeronautical Engineering Education at the University of Minnesota*, Masters Thesis, University of Minnesota, October, 2000.

2. Anonymous, "Fifty Years of Aeronautical Engineering, University of Minnesota 1929-1979", Department of Aerospace Engineering and Mechanics, University of Minnesota, 1979.

3. William L. Garrard, personal recollections.

4. *Aeronautical Research Facilities*, University of Minnesota, Rosemount Aeronautical Laboratories, Research Report 152, 1958

Fig. 1. First class of seniors taking flying lessons, 1933

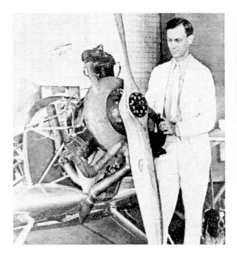

Fig 3. Mohawk first low-wing, twin-engine, cabin airplane that could fly on one motor

Fig. 2. John Akerman with his radical new tailless airplane.

Fig. 4. Mohawk Pinto on floats

Fig. 5. Engineers' Day display

Fig. 6. Cellophane strato-sphere balloon ascent in Memorial Stadium, 1930's.

Fig. 7a. Jean Piccard and John Akerman watching balloon ascend.

Figure 7b. Jean Piccard and John Akerman before launch

Fig. 8. Aerial view of the Rosemount Aeronautical Laboratories (RAL).

Fig. 9. Cadettes in the classroom, 1943.

Fig. 10. Aerial view of continuous flow facilities (Bldg 302) RAL.

Fig. 11. Demonstration by Drs. Hermann and Heinrich at RAL, late 1950's

Fig. 12. Supersonic test set up at RAL, 1950's

Fig. 13. Open throat test section (subsonic tunnel) with guide surface parachute, late 1950's.

Chapter 10

History of Department of Aerospace Engineering at Wichita State University

Dr. Melvin Snyder
Professor Emeritus, Department of Aerospace Engineering

The Department of Aerospace Engineering at Wichita State University offers the Bachelor of Science, Master of Science, and Doctor of Philosophy degrees. The faculty offers a complete program of study and is supported by state-of-the-art laboratories and research facilities. The department is supported by the National Institute for Aviation Research which is located on the same campus and provides support for the faculty and graduate student researchers.

HISTORY

The Aerospace Engineering Department is rather unique in that it was the first engineering course of study (originally Aeronautical Engineering) at the university rather than a later addition to the College of Engineering as in most universities.

In 1926, the Congregational Church turned over the facilities of Fairmount College to the city of Wichita, after citizens voted to support it as a municipal university. The task of converting the liberal arts college into a university involved setting up a public governing board – The Board of Trustees – and establishing four colleges within the University – the College of Liberal Arts and Sciences, the College of Education, the College of Fine Arts, and the College of Business Administration and Industry.

The words "and Industry" were included in the name of the Business Administration College so that the college could serve as home for freshmen and sophomore engineering courses. The basic courses in engineering drawing and statics could be taken along with chemistry, physics, mathematics, speech, English, and non-technical elective courses by students wishing to major in Civil, Mechanical, Electrical, and Chemical engineering. These students could transfer their credits to the University of Kansas or to Kansas State A. & M. College for the completion of their degree requirements.

From the time of World War I through the twenties, there was wild erratic growth of aviation throughout the nation, nowhere more than in Wichita, where the large number of aircraft companies changed the "Peerless Princess of the Plain" to the "Air Capital of the World." This burgeoning industry had to progress from barnstormers and pasture mechanics to professional pilots, trained engineers, and industry managers.

Responding to this community need, the Wichita University Board of Trustees approved, in early 1928, a four-year course in Aeronautics. A Bachelor of Science degree in Aeronautical

Engineering was to be awarded to the graduates of the four-year course. On October 4, 1928, Alexander A. Petroff, a graduate of the University of Michigan, was named the director of the School of Aeronautics of the university and Professor of Aeronautics.

Dr. E. J. Rodgers, later chairman of the Aeronautical Engineering department at Wichita State, described the life story of Mr. Petroff as follows:

"Alexander Petroff certainly can be ranked as one of the pioneers of the aeronautical engineering educational field. Alexander Petroff was born in Russia claiming Perm as his home town. He graduated from the Perm Gymnasium and entered the Military Academy at Moscow. On graduation he was made a lieutenant in the Russian Army during the first World War and fought near the Austrian lines. Later he attended the University of Perm, and, on graduation, became a member of the Russian Nationalist Army under General Kolchak. He also worked while in Russia as an engineer on the China Eastern Railroad in 1920 and assisted in the installation of an automatic telephone system in Siberia.

"He came to the United States in 1922 as a Russian refugee and attended the University of California for a short time holding various engineering positions. He migrated across the United States and ended up at the University of Michigan in Ann Arbor where he was employed in connection with the completion of a new aerodynamic laboratory and wind tunnel. He also pursued and was awarded the degree of Bachelor of Science in Aeronautical Engineering and later a Master's Degree from Michigan in the same area. From there he was appointed to the position of the Director of the Aeronautical Engineering program at Wichita University. He was one of the few holders of a Master's degree in aeronautical engineering at that time. After Mr. Petroff left Wichita University in 1933, he worked for the Curtis Wright Aircraft Company in New York, and the Hughes Aircraft Company in California."

In the fifties, Alex Petroff was persuaded by former students Dwane Wallace, Tom Salter, and Jerry Gerteis to return to Wichita to head up the research and development section of Cessna Aircraft, where he worked until his retirement. He served as the President of the Wichita Section of the American Rocket Society. In 1960 he worked with the U.S. State Department in Indonesia and Brazil establishing courses in aeronautical engineering. He died at the age of 74 on September 3, 1971, at his home in La Jolla, California.

One of Mr. Petroff's first major tasks at Wichita University was the construction of a wind tunnel. On December 6, 1928, the Board of Trustees appropriated $1,000 for the construction of the wind tunnel under the direction of Professor Petroff. He designed an open-return tunnel that was constructed of wood and lined with aircraft fabric. The tunnel had an open jet four feet in diameter. This open jet test section was enclosed in a small room also occupied by the student operators. The wind tunnel was constructed in the unfinished fourth floor of the Science Hall (now McKinley Hall).

The wind tunnel was powered by a 60-horsepower electric motor driving a wooden 4-bladed fixed-pitch propeller. The motor was started and controlled using a streetcar type controller. Motor speed could be varied from 400 to 860 rpm by use of the controller. Further adjustment of test section wind velocity was made by using variable vents downstream of the test section. By opening or closing these vents, more or less air bypassed the test section varying test section wind velocity. The balance-system was a three component balance.

113

The students of this time participated in a glider club, and there were aircraft and an engine test cell on campus. The Engineers Club was established and later supplanted by Tau Omega, the Aeronautical Honorary Fraternity, which was founded in 1927 at the University of Oklahoma. The second chapter was installed at Wichita University on May 14, 1932. Professor Petroff and students are listed as charter members of the Beta chapter of the first Aeronautical Engineering Honor Society.

At the Fourth Annual Commencement of the University of Wichita on Friday, May 30, 1930, five students received the first degrees of Bachelor of Science in Aeronautical Engineering: Virgil H. Adamson, Emerson Hayden Brooks, Jerry Howard Gerteis, Ernest Shih, and Byron Campbell Thayer. All of these men made their marks in the aircraft industry in Wichita and in California.

As the Great Depression deepened and aircraft companies closed, enrollment in the aeronautical engineering program decreased. Finally, in the mid-thirties, the program was closed. The aircraft industry almost disappeared as well. Clyde Cessna helped his nephew, Dwane Wallace (a Wichita University aeronautical engineering graduate), gain control of Cessna Aircraft Co., but Dwane had to enter air races to meet the company payroll. He later employed Wichita University classmates Tom Salter as chief engineer and Jerry Gerteis as assistant chief engineer. Walter Beech's Travelair Co. was bought (and later closed) by Curtis-Wright, and Walter started a new company, Beech Aircraft. The Stearman airplane company became a division of Boeing.

The engineering education effort remained in a state of limbo until the rapid expansion of the local aircraft industry during World War II. Wichita companies requested engineering courses for their workers. In response, Wichita University employed C. Kenneth Razak in June 1943. Razak had attended Kansas University and during summers had worked at the Stearman and Beech aircraft companies. He received the B.S. degree in 1939 from the University of Kansas and stayed as an Instructor. He earned his M.S. in 1941, and was subsequently promoted to Assistant Professor. He came to Wichita University as an Associate Professor in 1943.

He taught night classes for the engineers from the local plants, and during the day he and a student refurbished Professor Petroff's four-foot wind tunnel. This work included removing the thousands of tacks holding the fabric liner to the tunnel so a new plywood liner could be installed. The open-throat test section was changed to a closed-throat, and the Fairbanks-Morse balances were replaced with an automatic balance system of his own design and driven by sewing machine motors. Eventually (after the end of World War II) the fixed-pitch propeller was replaced by a P-38 electric variable-pitch propeller.

The refurbished wind tunnel made it possible for companies to test models of proposed aircraft design for the post-war period. These included numerous Cessna aircraft; an experimental constant angle-of-attitude plane for Boeing and its successors, the XL-15 and YL-15; and an attempt to test a model of the Buckminster Fuller Dymaxion house for Beech.

Ken Razak came to Wichita to teach aeronautical engineering courses because of the requests of Beech and Cessna aircraft companies. During his time at Wichita University he promoted the close relationship with the local aircraft industry that became a hallmark of engineering education at Wichita University and continues today at Wichita State University.

Academic Program

Through the thirties and into the post-war period, most engineering schools taught the art of engineering, and gradually evolved to the teaching of the science of engineering. The undergraduate program at Wichita University followed this pattern. In addition to satisfying the requirements of the Engineers' Council for Professional Development (ECPD) —now known as the Accreditation Board for Engineering and Technology (ABET) —it was necessary to satisfy the core requirements of the university. The requirements for the Bachelor of Science in Aeronautical Engineering degree in 1947 was 144 hours and included:

6 hours freshman English
2 hours speech
 12 hours humanities and social studies including the first course in political science
mathematics through differential equations
6 hours chemistry
6 hours physics
3 hours shop – welding, machine shop, foundry and pattern making
103 hours engineering courses, all required except 9 to 12 hours elective
4 hours physical education or ROTC

One of the first changes in the curriculum was the combining of the 3 hours of thermodynamics, 3 hours of fluid mechanics, and 3 hours of aerodynamics into two consecutive five hour courses of Aerothermodynamics.

One of the first steps in the transition from teaching the art of engineering to teaching the science of engineering was the employment, in 1954, of our first Ph.D. engineering professor, John Ruptash. (Photo 1) John was an energetic Canadian. He received his doctorate from the Institute of Aero-Physics at the University of Toronto. His first assignment was to take charge of the graduate program in Aerospace Engineering.

The graduate program required 30 hours, including thesis and a six to nine hour minor, usually in mathematics or physics. John Ruptash developed three courses that became the basic part of the graduate program: Aerodynamics of Incompressible Fluids, Aerodynamics of Compressible Fluids, and Aerodynamics of Viscous Fluids. The courses were taught at night and populated by engineers employed in the Wichita aircraft plants. These courses, or modifications of them, continue to be part of the graduate program. Many upper division and graduate engineering courses are still offered in the late afternoon and evening, and, in some cases, courses are offered in the early morning at aircraft plants.

A major factor that affected the department of Aeronautical Engineering was the addition of other engineering departments. In 1948, a four-year Industrial Engineering program was inaugurated. In 1952 the Departments of Civil Engineering and Mechanical Engineering were added and the School of Engineering was separated from the College of Business Administration and became the College of Engineering with Kenneth Razak as Dean. In 1957 the Department of Civil Engineering was closed because of decreasing enrollment, and, at about the same time, the Department of Electrical Engineering was established in response to demands from students.

In 1954 a committee on the basic curriculum proposed a new university core program, which was adopted by the faculty. Since that time there have been additional self studies of the general education program every four to eight years.

In 1957 a new university core program was adopted that required 45 hours, compared to the previous 29 to 36 hours, and distribution requirements spanned five general areas—math and general sciences, communications, humanities, social sciences, and physical activities. Thus, in 1958, the College of Engineering faced a new challenge as a result of the adoption by the University faculty of this university core curriculum that increased the number of non-technical elective courses from about 12 hours to a minimum of 24. Since the necessity to include this additional number of hours in the engineering program required that the engineering curriculum be revised, the faculty of the College of Engineering decided that this was an opportunity to examine the entire philosophy of engineering education at the University of Wichita.

The engineering faculty debated the problems and philosophy of engineering education and established the basis of a new program in engineering education. "Goals of Engineering Education" were debated and formulated, and then courses were designed from scratch to attain the goals. It was decided to require that all students should become engineers first and then specialize in their particular fields.

This new program consisted of 46 hours of University Core courses; including 24 hours of humanities and social sciences, 76 hours of Engineering Core courses, and 22 hours of department courses. The Aeronautical Engineering program was accredited by ECPD in 1952 and has been re-accredited at each evaluation interval to the present time.

The next structural change of the College resulted from the shift from municipal support to state support of the university. In spring, 1963, the bill making Wichita University a state university finally was approved by the State Legislature. Again, as in 1926, the citizens of Wichita voted. They overwhelmingly approved the incorporation of the University of Wichita into the state education system by a 96.7 percent majority in May 1963. On July 1, 1964, Wichita State University entered the state system. One immediate effect of this change was that the student enrollment increased by 38.6 percent.

When Wichita University became a part of the State system, the legislation provided that Wichita State University should be an Associate University of the University of Kansas, that further doctoral programs at Wichita should be developed cooperatively with University of Kansas, and that the degrees should be University of Kansas degrees.

In June 1966, the University of Kansas chancellor and the Wichita State president decided that cooperative doctoral programs at Wichita should be developed in selected areas. The WSU Graduate Council designated the Department of Aeronautical Engineering as the department most ready to begin planning doctoral programs.

The plan for the program was stated in a proposal for "new program of studies leading to the degree of Doctor of Philosophy to be offered jointly by the Department of Aerospace Engineering, University of Kansas, and the Department of Aeronautical Engineering, Wichita State University," dated August 1968, and revised January 1969.

The program was finally approved by the Kansas Board of Regents, and WSU embarked on a new venture. During the first year, there was an exchange of faculty. The committees for the students were composed of members from both institutions; they sometimes met midway on the Emporia State campus. While WSU was placing students into this program, the KU part of the program had practically no students. The KU Aerospace Department had started a Doctor of Engineering program into which most of their students were funneled. Consequently, after a few years of this unilateral usage of the program, the plan was altered so that each school ran its own program and granted its own degrees.

At the time of the launch of Sputnik by the USSR and the beginning of the U.S. space program, many of the aeronautical engineering departments throughout the country laid claim to this new area of engineering by changing their names to aerospace engineering. Partly because the Wichita area industry was involved with production of airplanes, i.e., winged flight vehicles; and partly because faculty members were aeronautical engineers, at WSU there was resistance to a name change, and the name remained the Department of Aeronautical Engineering—until 1992, that is. Bert Smith, the department chairman, was bothered by the fact that most other departments were titled Aerospace Engineering. Following the sage admonition to "...not be the first nor the last to adopt a new fad," Bert convinced a majority of the department faculty to vote to change the name.

The present undergraduate curriculum of the Department of Aerospace Engineering is:

9 hrs. Communication Classes
18 hrs. General Education Classes
34 hrs. Math and Natural Science Classes
13 hrs. Engineering Core
52 hrs. Engineering Courses required within the Department of Aerospace Engineering
9 hrs. Technical Electives

Physical Plant

At the end of World War II, when the four-year program was resumed, the facilities of the Aeronautical Engineering department consisted of the Aviation Building containing offices, a machine shop, two classrooms, and a simple materials testing lab; two engineering drawing labs; and the Petroff four-foot wind tunnel on the fourth floor of the Science Building. Fortunately, a large vacant field (roughly 20 acres) was available at the east side of the campus.

The first building erected on the East Campus was a triple Quonset hut that housed the shops. (Photo 2) When the machine and welding shops were moved to the East Campus Quonset hut in the summer of 1947, an aluminum foundry and a pattern making shop were added. Pattern making became the core of the wind tunnel model shop.

The Beech and Cessna companies decided that they needed a quality wind tunnel, and offered to finance one for the University. When Ken Razak recalls the day President Jardine asked him how much money he'd need for the construction. "I picked the largest number I could think of (and expect to get) and said $100,000." It turned out that wasn't nearly enough, but he did it anyway.

The new wind tunnel was built east of the Quonset hut, and a former PX building from Camp Crowder was built north of it. The latter one story frame building was used to house laboratories for electricity, internal-combustion engines, structures, and heat-power, as well as a classroom and a group of offices.

In designing the new wind tunnel, it was decided to have a 7 by 10 foot test section because of the extensive NACA experience with tunnels of that size. Because of the limited financing available, the tunnel was constructed of steel reinforced concrete. The lower part (roughly one-third of the tube) and the foundation were poured, and the upper part of the tube was formed by the gunite process. The designer put a rolled-channel ring every four feet along the tube to maintain the circular shape. A year later, there was a crack every four feet which was then waterproofed, so that many expansion joints were created rather than only one.

The balance system is a six-component pyramidal balance built in the university shop. Individual balances for each component made use of surplus electronic gear. The out-of-balance sensors and cursor drive motors were from B-17 autopilots. Originally, the balances were stopped at each test point and readings were recorded by hand. There have since been three generations of data acquisition hardware installed, and today, high speed data gathering, processing, and tabular and graphic presentation of results are all on-line.

The initial powerplants for the tunnel were Allison V-1710 (12 cylinder, liquid cooled) engines; the first used were removed from P-63 fighters. It required an engine operator in the powerplant room connected to the tunnel operator by intercom. Later, when sufficient transmission line capacity reached the East Campus, the engines were replaced by an electric motor.

Also located on the East Campus was a jet engine test cell with a J-33 turbojet mounted on a thrust stand and an aircraft test cell with a Jacobs radial engine mounted on a torque stand.

After the School of Engineering separated from the Business Administration College in 1952, Kenneth Razak was named Dean of the College of Engineering. A building for the College of Engineering was erected on the East Campus. Completed in 1953, the building provided office space for the faculty, three classrooms, three engineering drawing rooms, and two design rooms.

In 1964, when the university became state supported, the new dean Charles V. Jakowatz, promoted the effort to build new laboratories for the College of Engineering. (Photo 5)

The planning, financing, and construction of Wallace Hall was a great accomplishment. The new building was named to honor Dwane L. and Velma Lunt Wallace. (Photo 6) Dwane Wallace was a 1933 graduate of the Wichita University Aeronautical Engineering program. He became General Manager of the Cessna plant in January 1934, and from 1936 to 1975 he served as Chief Executive Officer of Cessna Aircraft Company.

The support of Dwane Wallace enabled the University to obtain sufficient state money to build this much-needed laboratory building. Also, in May 1976, the Dwane L. and Velma Lunt Wallace Fund was established at WSU to support scholarships and graduate fellowships for engineering students, and to purchase equipment needed by the College of Engineering.

Wallace Hall, dedicated April 22, 1977, added 78,204 square feet of space for laboratories, offices, and classrooms (including a lecture hall). (Photo 7) The Dean's offices and the Aeronautical and Electrical Engineering offices were moved into Wallace Hall. Most of the space was for thirty laboratories for the four departments. Also, a replacement for the old four-foot tunnel was created as part of the building. The building was wrapped around the Beech Wind Tunnel, and additional workrooms were provided for the tunnel personnel and customers. A super sonic laboratory was included with a vacuum-drive Mach 2 tunnel equipped with a 20 square inch test section and a research quality blowdown tunnel donated by the Boeing Company. Mach numbers of 1.2, 2.0, 3.0, and 4.0 could be obtained. A strength of materials lab and structures test floor was also included as well as a new engineering shop and wind tunnel model design and construction facility.

A major addition to the physical facilities occurred with the founding of the National Institute for Aviation Research. (Photo 8) Planning was done in 1985-1986. WSU President Warren Armstrong and his government relations director, Fred Sudermann, convinced the federal government that Wichita, Kansas, the "Air Capital of the World," was the place for such a research institute, resulting in financing in Spring 1986, with $5 million coming from the Army and $3 million coming from the Federal Aviation Administration. The plan was to make laboratories and equipment available to the WSU faculty and graduate students to work on problems in the aircraft industry. The faculty would be primarily engineers, but it was hoped that participants would come from physics, chemistry, business, and computer science as well.

The laboratory was built adjacent to the Beech 7x10 Wind Tunnel. The $7 million, 75,000 square foot building houses offices, conference rooms, seventeen seminar rooms, and laboratories, including:

Aerodynamic laboratories
- 2 ft by 3 ft Flow Visualization Water Tunnel
- Subsonic Wind Tunnels
  Walter Beech 7 ft by 10 ft Wind Tunnel
  3 ft by 4 ft Wind Tunnel
- Supersonic Wind Tunnels
  9 in by 9 in Pressure Blowdown Tunnel
  20 square inch Vacuum Drive Mach 2 Tunnel

Aging Aircraft Laboratory
  3,200 square foot high bay laboratory

Aircraft Icing Laboratory
  680 square foot lab

CAD/CAM Laboratory

Composites & Advanced Materials Laboratory
  7,420 square foot laboratory
    3 foot by 5 foot 1000°F, 400 psi. Autoclav oven, Enviromental Chamber, Hydraulic presses, Clean Room, Filament Winding Machines, etc.

Crash Dynamics Laboratory
  4500 square foot by 60 foot high bay lab
  VIA system floor-mounted impact sled
  Landing gear Drop Tower 20 ft high, 400-10,000 lb test load

Human Factors Laboratory

Structures Laboratory
  Eleven servo-hydraulic load units ranging from 5.5 kip to 220 kip capacity

Virtual Reality Center

Model Building Shop
  3-D Catia Cad/CAM controlled milling machines

Basic and Applied Research

Even before the end of World War II and the resumption of the four-year Aeronautical Engineering program, the local aircraft companies were using the four-foot wind tunnel to test models of aircraft being designed. Two airplanes which were produced, the designs of which were assisted by model testing in this period, were the Cessna Bobcat and the Model 190/195. From that time to the present, testing and research have been performed for aircraft companies from across the U.S. and some foreign countries; automobile and truck manufacturers; government agencies such as Office of Naval Research, NASA, FAA, U.S. Army; individual entrepreneurs; and, most importantly, for the education of students.

The first tests in the Beech 7 by 10 wind tunnel were drag tests of full-sized automobiles sponsored by the Nash Motor Company. (Photo 9) All body styles of the 1949 American car manufacturers were tested. The important result of these tests was that the wind tunnel became operational with an experienced crew of engineers and student technicians and with many of the "bugs" worked out of the system.

At this time, a working arrangement was developed with the Office of Naval Research (ONR) for the study of the use of energy to enable airfoils to develop higher lift coefficients. The result was more than a decade of sponsorship by the ONR boundary layer control and circulation control studies. Mississippi State University worked with area suction, and Wichita University investigated suction on the airfoils and blowing over the flap surfaces through finite-width slots. Much of the basic and applied research on boundary layer control was done at Wichita. Also, in June 1955, a short course was taught including lectures by Phillipe Poisson-Quinton of the French ONERA, and Professor Heinrich Helmbold and Frederick Wagner, two German scientists, who were members of the WU Engineering Research department and who had been supplied by the ONR through "Operation Paperclip." Another effort was the cooperative development, with Cessna, of a circulation-control flight demonstration airplane.

This research program was very important to the School of Engineering, not only for the research experience and training of the faculty and graduate students, but also for the reputation of the University in the academic community. During the fifties, a Department of Engineering Research was inaugurated. This department was composed of engineers and student technicians and was financed by "soft money," i.e., support from industrial and government funds.

In the mid-seventies there was considerable interest in the problems of energy sources and concern about the control of crude oil production by OPEC. Invited by the Kansas governor's office, a research proposal was submitted, which was included in the governor's budget and was funded by the Legislature.

A Wind Energy Center was set up utilizing both faculty and students from the Aeronautical and Electrical Engineering departments. The center worked with the Kansas Energy Office. Projects of the Wind Energy Center included:

- Survey of areas in Kansas suitable for "Wind Farms"
- Design and construction of a two-bladed wind turbine and tower that were tested on the campus
- Study of Himmelscamp flow on turbine blades
- Design and construction of a wind turbine grain bin dryer and installation of it at the Kansas State University grain lab

New support for wind energy came from NASA. WSU had done some excellent research in developing ailerons, flaps, and spoilers for modern airfoils. At first, it was developing spoiler controls for the Advanced Technology Light Twin (ATLIT) airplane and then testing the flaps for the new laminar flow airfoils that were being developed by NASA in the wind tunnel. This work was funded by the NASA Langley Research Center.

The U.S. government's development of aerodynamic wind turbines was being done for the Department of Energy by the Wind Energy Office of NASA at the Lewis Laboratory in Cleveland. The people there came to realize that one of the main problems was that of control of the output and speed of the turbines. It was suggested that perhaps spoilers or flaps could be used to control the turbines. They contacted the aerodynamics group at Langley and were referred to Wichita State University.

Thus began a long series of research contracts with NASA Wind Energy Office. WSU developed aileron controls for a 100-ft. wind turbine that was tested at the NASA test facility at Plum Brook, Ohio; converted the NASA PROP computer program to user-friendly WIND code (WIND-2 through WIND-15); and wind tunnel-tested tower components. A WSU staff member spent two summers at the Lewis Lab, and faculty members made numerous presentations of the research results, including a one-week course at von Karman Institute in Belgium and a three-week short course in New Zealand.

NASA is no longer in the wind energy business; the work is being done by the National Renewable Energy laboratory in Colorado. WSU aerospace engineering faculty and students continue to do testing for them and some of the commercial companies on next generation wind turbines. This work is being done by the National Institute for Aviation Research (NIAR). The former Wind Energy Laboratory was renamed the Center for Energy Studies in the mid-eighties; most of their studies are in the electrical engineering field. There are now wind farms in Kansas with plans to build more.

The Walter Beech 7' by 10' Wind Tunnel has been the center of the aerodynamic research and development work, particularly for local companies (e.g., the Learjet model, with modifications, has been in the tunnel for almost 30 test programs). Since the dedication of the tunnel in 1951, there have been many modifications to the data acquisition systems and improvements in the quality of flow. In 1991, a matching fund grant was secured from the National Science Foundation to refurbish and modernize the aerodynamic laboratory. A roof was built over the tunnel to protect it from the ravages of weather. The drive motor was removed – for the first time in about thirty years – and was completely overhauled and reinstalled. The liquid starting rheostat was replaced by a solid state system. Most important, a new state-of-the-art high-speed data acquisition and reduction computer system was installed.

Currently another modernization of the tunnel is underway. The balance system and the data acquisition computers will be replaced. The drive fan will be moved from the first tube downstream of the test section and will be installed in the back tube. New screens and an air exchanger will be installed. The cost of these modifications is between $5 and $6 million.

The addition of the National Institute for Aviation Research (NIAR) to the resources of the WSU Aeronautical Department was described earlier in this document. The following is a brief overview of the NIAR.

NIAR integrates university, government, and business entities in cooperative efforts to advance technologies for aviation and other industries. It is a high-tech research and development, testing, certification and learning center.

NIAR has been awarded the "center of excellence" distinction by prestigious state and national entities for technology advancement, including the Federal Aviation Administration (FAA) and the Kansas Technology Enterprise Corporation (KTEC). Operational funds come from the state through KTEC, WSU, and revenues generated by NIAR from federal and industry contracts, grants, and services.

As an FAA designated research center of excellence in both airworthiness assurance and general aviation, and as a KTEC Center of Excellence, NIAR conducts research in a number of specialties including aerodynamics, crash dynamics, composite materials, advanced materials, aircraft icing, structural components, propulsion, and human factors.

Built in 1990, there are currently 17 laboratories operating in NIAR's 74,000 square foot facility. At work are approximately 40 full time employees, 100 student assistants and graduate research assistants, as well as a large group of WSU faculty researchers, and other university associates.

One of the newest laboratories is the virtual reality center, the only one of its kind in Kansas. The 3-D virtual simulations projected on a 7' x 15' screen are used to enhance the efficiency and productivity of work and assembly lines in manufacturing plants, including maintenance tasks, related human factors, and ergonomics. NIAR's CAD CAM lab, one of the largest CATIA-based programs offered by any university in the country, provides the building blocks for simulations in the virtual reality center using CATIA and Maya 3-D computerized animation programming.

Congressional appropriations and funding from KTEC, WSU, and industry partners have paved the way for NIAR's newest addition, the Aging Aircraft Research Laboratory. This $1.1 million program is one of the few in the country that researches the effects of aging on general aviation aircraft. Congressional funding will also provide substantial upgrades of several key laboratories, including the Beech Wind Tunnel. Originally built in 1948, approximately $5 to $6 million in upgrades from federal and state funding will make the facility one of the premier low-speed wind tunnels in the nation.

Students and Alumni

The first five students to receive the bachelor of science degree in Aeronautical Engineering graduated in 1930. Six more BSAE degrees were granted in 1931, 1932, and 1933. The first post-war class of engineers graduated in Spring, 1948, and consisted of seven BS degrees in Aeronautical Engineering and one Master of Science. As of December 2002, the following number of degrees has been granted by the Department of Aeronautical Engineering and the Department of Aerospace Engineering:

1,016 Bachelor of Science
402 Master of Science
68 Doctor of Philosophy

The aerospace engineering students at WSU, unlike those at many other universities, have the advantage of being familiar with aircraft plants and, in many cases, working in those plants as junior engineers, technicians, assembly workers, etc. A number of students have worked as ferry pilots delivering new planes. Both undergraduate and graduate students are also employed by NIAR. In addition, many students are in the university Co-operative program and work summers and two or more semesters for NASA (particularly, NASA Houston), FAA, etc. Many students have Wallace and other university scholarships. In Fall 2002 there were 205 undergraduate and 84 graduate students enrolled in Aerospace Engineering.

Student organizations include Sigma Gamma Tau, aerospace engineering honorary; Tau Beta Pi, engineering honorary; Engineering Council, sponsor of the Engineering Open House and the annual Engineering Banquet; and student chapters of the professional societies AIAA, SAE, and SWE.

One of Dean William Wilhelm's (1979-2000) (Photo 10) first activities was to establish the Sam Bloomfield Distinguished Engineer Lecture Series. Each year, a distinguished practicing engineer spends a day on campus, meeting with students and faculty, and in the evening delivers a lecture in his area of expertise.

Dean Wilhelm also initiated the Dwane and Velma Lunt Wallace Outstanding Educator Awards. At the annual Engineering Banquet held in conjunction with the Engineering Open House, awards are given for excellence in teaching, research, continuing education, and for outstanding graduate teaching assistants. The Wallace Scholarship Program was also expanded under Dean Wilhelm's tenure.

The problems of rapid obsolescence of laboratories and high prices of replacements are difficult and continuing ones. Dean Wilhelm developed two programs to address these challenges. First, in 1989, he founded the Dean's Circle, a group of alumni pledged to raise an endowment fund to supply laboratory equipment. Second, he joined the KU and K-State Engineering Deans in obtaining, from the Board of Regents, the right to apply a surcharge to the tuition for engineering courses. This extra money goes into an engineering equipment fund.

An additional activity of the Dean's Circle is to designate one or two Distinguished Engineer Service Awards each year. Plaques commemorating these awards are placed in a position of honor on the wall in Wallace Hall.

The Dean's Circle also sponsors the Kansas BEST (Boosting Engineering and Science Technology) robotic competition for high schools. The winners compete in the Texas BEST national finals. The program has been quite successful as a student recruiting activity

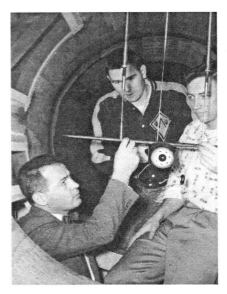

Photo 1 – John Rutpash and
Graduate Students in 4X4 Wind
Tunnel

Photo 2 – Buildings on East Campus

Photo 3– Dean Kenneth Razak (Term 1953-1965) and
faculty at new building site

Photos provided by Wichita State University Library Special Collections

124

Photo 4 – "New" Engineering Building

**Photo 5
Dean Charles Jakowatz
(Term - 1965-1979)**

**Photo 6
Dwane and Velma Lunt
Wallace, 1977 Dedication of
Wallace Hall**

Photo 7 - Wallace Hall – Completed 1977

Photos provided by Wichita State University Library Special Collections

Photo 8 - National Institute for Aviation Research Completed 1990

Photo 9 – 1949 Nash Wind Tunnel test

Photo 10 Dean William Wilhelm (Term 1979 – 2000)

Photos provided by Wichita State University Library Special Collections

CHAPTER 11

AERONAUTICAL AND AEROSPACE ENGINEERING AT N. C. STATE: A HISTORY

Prepared by

Frederick O. Smetana, Professor Emeritus
Mechanical and Aerospace Engineering
With contributions from Aerospace Faculty

World War I demonstrated that the 'airplane' would play a significant role in future military conflicts in the world. About 1920 a National Laboratory, established by the United States Congress for the theoretical and experimental study of the problems associated with advancing heavier-than-air manned flight, was located in nearby tidewater Virginia and provided properly prepared young people in North Carolina with a significant incentive to participate in this exciting adventure. This in turn led to vocal demands for an educational program at the North Carolina State University to so prepare its young people. The State's Land Grant institution responded by establishing a degree program in Aeronautical Engineering. Un-fortunately, existing records are quite scarce regarding the details but it appears that the program was authorized about 1923 and produced its first graduates in 1929. A small, hanger-like brick building was constructed to house faculty offices, laboratory space, and a lofting/drafting area. This building was known to be in use during the later 1930s but the date at which it was placed into service was not discovered.

Anecdotal evidence indicates that there were three faculty members in the Aeronautical Engineering program during the 1930s and around 10 graduates per year. A student chapter of the Institute of Aeronautical Sciences was chartered in 1937, according to the recently discovered charter. A small subsonic wind tunnel with a test section about 12" x 12" was located in the basement of the adjacent Mechanical Engineering building, Page Hall. The Engineering Mechanics program, responsible for the Engineering School service course in Fluid Mechanics, also used this facility.

WORLD WAR II

The onset of World War II changed everything. Faculty found new opportunities in the aviation industry and students enlisted or were drafted into the military services. The Army sent large numbers of men to the College for specialized training (ASTP, Army Specialized Training Program) in aeronautical-related topics. One of the instructors in this program was Robert W. Truitt, who had graduated from Elon College with a degree in Physics in 1941. Truitt turned the notes he had assembled for the program into a widely used introductory hardback textbook. In

1944 Truitt entered the U. S. Navy, leaving the N.C. State Aero program without permanent faculty, the ASTP program having by that time been discontinued.

When Truitt returned from Navy service in 1946 he was the only faculty member. Later that year, Professor L. L. Vaughn, Head of Mechanical Engineering, entered the small Aeronautical Engineering building and spotting Truitt working alone in his office said to him, "Boy, why don't you come on over with us?" And so, Aeronautical Engineering at N.C. State became the Aeronautical Option of Mechanical Engineering.

The first woman to graduate from an AE program in the USA, Kathrine Stinson, did so from NC State in 1941 with all the difficulties one would expect at that time. She was encouraged to "prove herself elsewhere" before entering the program at State. She did so, graduated, and went on to found the FAA and serve as its first Director. She knew several US presidents personally, was friends with Amelia Earhart, and was instrumental in the development of the aircraft transponder. In 1977 she received the Presidential Medal of Honor for her honorable career in US aviation. The street outside the Mechanical & Aerospace Building, Broughton Hall, is named Stinson Drive in her honor.

THE LATE 1940s AND EARLY 1950s

Beginning in the Fall of 1946 thousands of military veterans enrolled in N. C. State College under the G. I. bill. Four students were packed into dormitory rooms designed for two. Attrition was fierce. Two thirds of those who entered either flunked out or became sufficiently discouraged to drop out. At that time laboratory work was a much larger part of the curriculum than it is today. Aero students, for example, took courses in Pattern Making, Foundry, Welding, Steam Engines, and Instrumentation, in addition to junior and senior level wind tunnel labs. Laboratory space was so tight that only the top 100 junior students were permitted to continue in Mechanical Engineering. The remaining students were shunted into a laboratory-less program called Engineering General.

The open courtyard beside Page Hall and in front of the 1911 Building is today the "Court of the Carolinas". Prior to World War II this area was a drill field but after WW II twelve Quonset Huts were installed there, one of which was dedicated to the Aerospace Program.

The Aero Option required the equivalent of 160 semester hours in four years (actually 240 quarter hours). Classes were held until 10 PM and until noon on Saturday. Of the 160 hours, 16 were in ROTC (compulsory for non-veterans); 8 hours in Physical Education; 9 hours in drafting, kinematics, and descriptive geometry; 10 hours in laboratory; 23 hours in Mathematics; 10 hours in Chemistry and 10 in Physics; 6 hours in Thermodynamics; 9 in engineering science service courses; 15 in English and Economics; 9 in electives; and 35 hours in aeronautical-specific courses. The Aircraft Design class in the senior year met for seven hours each week (only 3 hours credit, however) in a Quonset Hut where each student could sit at a large drafting table. T-squares and triangles (not drafting machines) were used to prepare three-views of overall configurations and detail drawings of individual components. A mechanical desktop calculator supplemented slide rules for loads and stress calculations. This building was available to students and was much utilized after hours. The object of the program was to prepare students to take entry-level positions in industry without requiring any additional training.

Until 1949 Truitt taught all of the Aero-specific courses. In 1949 the department hired Philip L. Michel, who had been an aerodynamicist at Republic Aviation, to teach the design course and to offer some electives. While teaching at N. C. State, Truitt commuted to Duke University, some 30 miles away, to work on a Master's degree in Physics. Later he began to go to VPI during the summers to work on a Ph.D. in Engineering Mechanics. This was a time when few faculty members had a Ph.D. The emphasis was on faculty with professional experience, those who had actually done that which they were teaching.

In 1951 the Mechanical Engineering program moved from Page Hall to the newly completed Broughton Hall. As part of the financing for the new building, a subsonic wind tunnel was purchased. The tunnel has been painted but still uses the World War II Curtis Electric propeller with the same variable pitch mechanism to adjust speed. The facility is used extensively for teaching and research.

In the summer of 1951 while completing his doctoral requirements at VPI, Truitt was offered the headship of the Aeronautical Engineering department there and a full professorship. (He was still an Assistant Professor at N. C. State.) That left Michel as the only faculty member and in the summer of 1952 he too left to take the position of Chief of Aerodynamics at Sikorsky Helicopter. During the 1952-1953 academic year Dr. Fred Smetana, then a graduate student in the department, and Mr. Jack McCracken, another grad student, divided the teaching load for all the aero-specific undergraduate courses. In early 1953 the department brought in Robert M. Pinkerton from the Langley Research Center of NACA to supervise the two graduate theses and to take over the entire program during the fall of 1953.

**Figure 1 The NCSU Subsonic wind tunnel ,
used for teaching and research**

EXPANSION AFTER OCTOBER 1957

During 1949 and the first half of 1950 the dismal job market for Aeronautical Engineering graduates was limited almost exclusively to NACA. The outbreak of the Korean War in June of 1950 opened the market to additional jobs as the industry expanded to counter the Soviet MIGs. The cold war put pressure on the aeronautical industry, especially during the period 1953 to 1957 to develop techniques to analyze and cope with hypersonic flow. Budgetary pressures from the Eisenhower administration kept the increases in university research and enrollments to relatively modest levels. Professor Pinkerton was able to fill two additional faculty slots to assist in program instruction but these hires turned out to be transitory and by 1961 Professor Pinkerton was again by himself teaching the few Aero courses that remained.

The launching of Sputnik October 4, 1957 shocked the nation; as a consequence, the political leadership decided to fund the engineering education establishment to expand the number of engineers versed in the sciences needed to design vehicles to operate successfully in the cold voids of space, to travel to distant celestial bodies, and to return safely to earth. Because the space environment was so different from what the great majority of engineers in the industry knew, the clamor became deafening for scientists rather than engineers. The National Science Foundation funded many fellowships and Universities, including N. C. State. This started a revision of engineering programs to require doctorates of the faculty. Course offerings became science-based rather than expositions of current technologies, some laboratories were abandoned, and research became analytical or numerical rather than purely experimental.

At North Carolina State the Dean of Engineering was given the mandate and the resources to begin doctoral programs in all departments of the School. In 1961, the headship of the Mechanical Engineering department became vacant. Dean J. Harold Lampe persuaded Robert Truitt to return to Raleigh as Head of Mechanical Engineering with the mandate to build the Aerospace program to doctoral-granting capability. Truitt contacted former students and colleagues from N. C. State and VPI in an effort to locate qualified faculty candidates.

In the Fall of 1962 he was able to hire two of his former students who had recently completed their doctorates at the University of Southern California, Dr. F. O. Smetana and Dr. J. C. Williams III, and a former colleague from VPI, Dr. H. H. Hassan. In the next three years he was able to add Drs. J. N. Perkins and F. J. Hale with primary interests in aeronautical and space science and design.
The Aerospace program received accreditation in 1964 and the name of the Department was changed from Mechanical Engineering to Mechanical and Aerospace Engineering in 1966.

During the 60's NASA was given a mandate to put a man on the Moon and return safely to Earth. The Aerospace Engineering Program worked closely with NASA Langley Research Center in support of the space program. Dr. Truitt had succeeded in significantly enhancing experimental facilities at VPI during the 10 years he spent there. He was interested in doing the same thing at N. C. State. He was able to secure about $500,000 over a period of about six years toward that goal and asked Dr. Smetana to design and construct facilities that could be used in contemporary research and to instruct students in contemporary experimental research techniques. Among the facilities constructed or acquired were a supersonic blow-down wind tunnel shown in Figure 2, a continuous flow hypersonic wind tunnel, a transonic nozzle and arc jet facilities.

When the supersonic blow-down tunnel was first operated it had no muffler. The exhaust discharged normal to the street behind the building and was objectionably loud. The first effort to build a suitable muffler in-house resulted in an unsuitable two-story high cloud of fiberglass insulation that settled over a two-block area.

When the transonic nozzle was first run (using a roots blower train) there resulted a 120dB noise at 7 Hz. that caused coffee cups two floors above the lab to walk off desks. Another hasty muffler design activity was required.

During this period Drs. Williams and Perkins emphasized research on Fluid Dynamics and Aerodynamics, and Dr. Hassan established a research program in advanced propulsion systems with emphasis on Electrical Propulsion. This program was conducted in collaboration with NASA Langley Research Center. Dr. F. R. DeJarnette, who was specialized in Aerodynamic Heating and space vehicle design, joined the faculty in 1969.

In late 1971 Truitt discovered that he had cancer and he died in April 1972 at the age of 51. His death deprived the department of energetic leadership and significant clout in university. administration. Room 1402 of Broughton Hall, Truitt Auditorium, is named after him

NEW FOCUS

Following the first moon landing in 1969 federal funding declined for University research, particularly of the experimental variety, and aerospace engineering enrollments began to decline as well. Whereas the

Figure 2: The Supersonic Blow Down Wind Tunnel, still a valuable teaching and research tool

number of AE graduates had been as high as 45 per year in the mid 1960s they declined to a low of 9 or 10 by the mid 1970s before rebounding in the late 1970s. Federal agencies began to severely limit the size of grants and contracts and State funds for supplies and technicians to keep facilities operating also dried up. As a result, by the early 1980s all the facilities constructed 15 years earlier, except for the subsonic wind tunnel and the supersonic blow-down wind tunnel, were scrapped.

Faced with the above negatives, the situation demanded major actions on the part of the AE Faculty. Three major actions were taken. It was obvious that resources were not available to support and enhance the experimental facilities and it was decided to rely on facilities at NASA Langley and other Government installations.

The second step was to start a Flight Program. Dr. John Perkins initiated the program by modifying the senior design course so as to undertake the detailed design, construction, testing and finally flight testing a remotely piloted vehicle, RPV, during the senior year. Dr. Perkins was able to secure industrial support for the activity as well as permission from the University to build a paved runway on one of the University farms near Butner, NC. The runway was named later as Perkins Field. The design projects are reviewed each year by a group of professionals and the RPV's are flown at a picnic sponsored by the AIAA student branch and Sigma Gamma Tau. This activity is an important recruiting element for the Aerospace Engineering Program.

The third action was to start a Computational Fluid Dynamics (CFD) program. To implement this, three distinguished researchers were invited to teach a graduate course: Dr. John Rakich from NASA Ames Research Center, Dr. Fred Blottner from Sandia National Labs. and Dr. Pat Roach from Albuquerque, NM. Later, Drs. DeJarnette and Hassan taught a graduate course in CFD and negotiated a cooperative agreement with NASA Langley that lasted over 20 years. The CFD program at NCSU became a milestone defining the character of the Aerospace program at NC State.

During the mid-70's Dr. Todd Pierce joined the Faculty and taught courses in combustion. He left NCSU after four years and Dr. Robert Nagel was hired in 1980 to replace him. At the same time Dr. Wayland Griffith became the director of the College of Engineering Design Center and later joined the Aero Faculty. Dr. Williams then departed NC State having accepted a position as Department Head of Aerospace Engineering at Auburn University. During this same period Dr. Smetana led an effort sponsored by NASA involving a review of both NACA and NASA published research relevant to light aircraft. The research resulted in the publication of five NASA CR's and the development of computer codes to automate the computation of light aircraft stability derivatives and the resulting flight motions.

In the early 80's Dr. D. McRae joined the AE Faculty and took over teaching the CFD courses. In the basement of Broughton Hall, where the transonic facilities had been removed, a subsonic anechoic wind tunnel was built that was used with an existing anechoic chamber to support a noise program funded by NASA Glenn (Lewis). As the AE enrollment was rebounding, it became necessary to further increase the number of faculty. Dr. L. Silverberg with expertise in controls and E. Klang, with expertise in the areas of structures and composites, joined the Faculty during this period. In 1983 a graduate program in Aerospace Engineering was finally established. This made it possible to earn an M.S. and a Ph.D. in Aerospace Engineering at North Carolina State University.

The approval of the Ph.D. program signaled the start of the golden years for the AE Program. Shortly after approval of the Ph.D. Program, a solicitation from NASA was announced in the area of Hypersonic Aerodynamics. NCSU was one of six universities that received six year funding for the program with Dr. DeJarnette leading the effort. Resources from the program were used to hire Dr. Graham Candler, a young expert on non-equilibrium flows, and later Dr. Nd. Chokani, an expert on hypersonic transition. The Hypersonic program started as a joint NASA-Air Force-Navy program and one of the requirements of the Grant was that faculty and students sponsored by the Grant spend time at various Government labs. Dr. Griffith led the effort at NSWC at White Oaks where experiments were conducted at the Navy Hypersonic wind tunnel; Dr. McRae led the effort at AFRL in Dayton, while Drs. DeJarnette and Hassan led the effort at NASA Langley.

Another solicitation from NASA to establish University Space Centers was announced in 1988. Some of the AE Faculty teamed up with the NC State College of Textiles and North Carolina A&T University and submitted a proposal for one of the Centers. It was successful and the Center was named the Mars Mission Research Center because the thrust of the proposal was devoted to exploring technologies needed for the exploration of Mars. Examples of this were aerobreaking technology and surface rover research. The Center was led by Dr. Fred DeJarnette. Part of the funding from the Mars Mission Research Center was used to hire Drs. G. Walberg, who led the mission analysis activity, Dr. Gordon Lee who led the Spacecraft Controls work and Dr. F. G. Yuan who led the modeling effort in composites.

In 1991 an AE group, lead by Dr. Perkins, teamed with A&T State University and was awarded a NASA Space Grant. Shortly there after Dr. Hale retired and Dr. C. Hall was hired to replace him. Dr. Lee left the Mars Center to become an Assistant Dean of Research in the College of Engineering. In 1994 Dr. F. O. Smetana retired after a relation with the AE program that included not only his tenure as a faculty member, but also time in the 1940's when he was an undergraduate student in the Aero Option. Later, Dr. DeJarnette vacated the Directorship of the Mars Center to become Head of Mechanical and Aerospace Engineering and also Director of the NC Space Grant Consortium. Dr. Walberg succeeded him at Mars Mission. In 1993 Dr. G. Candler accepted a position at the University of Minnesota and in 1994 Dr. J. R. Edwards was hired to replace him. Also in 1994, Dr. W. Roberts joined the Faculty to work in the area of combustion and propulsion. In 1999 Drs. Perkins and Walberg retired and Dr. DeJarnette stepped aside as Head of Mechanical and Aerospace Engineering. Dr A. Gopalarathnam, who provided experience in aerodynamics and wing theory, was recruited to replace Dr. Perkins.

In 1999, reduced NASA funding led to the closing of the Mars Center, a blow to the faculty of the AE Program. The Faculty, nevertheless, adjusted to the new environment by diversifying their funding sources. Dr. Kara Peters, the first female tenure-track AE Faculty member joined the Aerospace program in 2000 and Dr. Nagel was appointed Director of the AE program in 2001.

In 2002, in an effort lead by Dr. Nagel, NC State joined five other universities to form a consortium that won the competition for the National Institute for Aerospace (NIA) from NASA Langley Research Center. This program promises to be a multi-million dollar per year activity with many new and innovative aspects regarding research, education, technology incubation, and outreach programs. That same year a Distance Learning program was initiated for the MS degree in Aerospace Engineering. The increased need for work force development coupled with enhanced research opportunities promises a great future for the Aerospace Program at the North Carolina State University. At present, the AE Program enjoys strong research funding and respectable graduate and undergraduate enrollment. It has an excellent nucleus of outstanding and productive young Faculty who are committed to excellence in both research and teaching.

Figure 3: Aerospace Senior Design, Build, & Fly

Chapter 12

HISTORY AND DEVELOPMENT OF AEROSPACE ENGINEERING AT THE UNIVERSITY OF OKLAHOMA

Tom J. Love[*] and Charles W. Bert[†]
School of Aerospace and Mechanical Engineering
The University of Oklahoma, Norman, Oklahoma 73019-1052

Abstract

The early history as well as the more recent development of the Aerospace Engineering Program at The University of Oklahoma is described. Topics included are the very successful participation in the AIAA, NASA, and industry student design competitions, descriptions of the outstanding careers of some of our graduates, enrollment trends, comparison of early and recent curricula, research highlights, and faculty activities.

Early History

The Aerospace Engineering program at The University of Oklahoma actually had its start in Mechanical Engineering when Assistant Professor Lester Clyde Lichty (later of Yale University) returned from his tour as an engineering officer with an army aviation squadron during World War I. Upon arriving back from service, he arranged for acquisition of a surplus Liberty aircraft engine. This was set up in the Mechanical Engineering laboratory and utilized for instruction for a number of years. In the early 1920's interest ran high in aviation and several elective courses were added. Mechanical Engineering offered the following courses: (1) Elements of Aeronautical Engineering, (2) Aeronautical Engines, (3) Theory of Propellers. Applied Mechanics offered: (1) Aerodynamics I, (2) Aerodynamics II, (3) Airplane Design.

Student interest continued to grow, and on February 16, 1928, the Aeronautical Engineering Honor Society, Tau Omega, was founded at The University of Oklahoma. Soon after, chapters were established at a number of midwestern schools, including Illinois Institute of Technology, The University of Minnesota, and Wichita State University in Kansas. Later Tau Omega merged with Alpha Gamma Rho, an honor society established in primarily the eastern schools in the United States. The new society was named Sigma Gamma Tau, and continues to be the primary Honor Society for Aerospace Engineering.

In 1929, an Aeronautical Engineering Option in Mechanical Engineering was established. By February 1930, there were fifteen students enrolled in this option. In March of that year, Tau Omega, under the supervision of Professor C. D. Case, constructed a glider with a 34 ft. wingspan. Prof. Case, who was a navy reserve pilot, piloted the glider during seven successful flights. In

[*] George L. Cross Research Professor, Emeritus. Fellow AIAA.
[†] Benjamin H. Perkinson Chair and George L. Cross Research Professor. Fellow AIAA.

1930, Carl Ritter, the first student to graduate with the Aeronautical Engineering Option, was working as an engineer with the NACA in Langley Field, Virginia.

In 1936, a wind tunnel was constructed as a WPA project. The tunnel had a test section of 4 ft x 6 ft, with a design air speed of 250 mph (see Fig. 1). It was a recirculating design with water-cooled straightening vanes and powered initially with a natural gas two-cylinder oil field engine. After WW II, this was replaced with a surplus Allison V-1710 engine from a P-38. In the 1960's an electric motor and a variable pitch propeller replaced the internal combustion engine. The original WPA project provided a brick building constructed as part of the tunnel. This included a room for the power source, a room for the test section and instrumentation, and a second story classroom used for lectures and data reduction. The initial design of the tunnel was done as a master's thesis by P. O. Tauson under the direction of Professor L. A. Comp. Professor Comp also later designed the balance system, providing measurements of lift, drag, and moments of pitch, roll, and yaw. In addition to its primary role as an instructional tool, this wind tunnel has been used in support of several industry contracts. All of the original aerodynamic studies for the Aero Commander series of twin-engine propeller-driven executive aircraft were conducted in this tunnel. Later it was used for low-speed aerodynamic studies of lifting body vehicles for the Martin Company. The tunnel, with an updated digital data acquisition system, was named the L. A. Comp Wind Tunnel in 1978 and is still in use.

In 1947, a Department of Aeronautical Engineering was established within the School of Mechanical Engineering for the purpose of administration of the Aeronautical option. A large hangar, constructed by the Navy during WW II and located at the University airport (Max Westheimer Field), was established as a laboratory. This included a small engine test stand and a small wind tunnel. On April 29, 1949, a tornado struck the airport and demolished the hangar. In February 1953, a brick building was completed on the location of the demolished hanger. It included two large rooms, providing for a machine shop, a laboratory/classroom area, and faculty offices. Six concrete engine test cells were constructed adjacent to the building.

In 1954, the School of Aeronautical Engineering was established. From 1951 through 1960, there were 184 BSAE, and 25 MSAE degrees granted.

As a reflection of the developing national interest in space exploration, the name of the school was changed to the School of Aeronautical and Space Engineering in 1959. A reorganization of the College of Engineering in 1963 resulted in the merger of the Aero Engineering program with Mechanical Engineering. At approximately the same time, the Aerospace Department Chairs Association published the result of a survey indicating a preference for the name Aerospace Engineering for the degree programs in Aero Engineering. The new department at the University of Oklahoma was identified as The School of Aerospace and Mechanical Engineering.

At the time of the merger, neither program offered the PhD degree. However, with an influx of qualified faculty, the School was approved to offer studies leading to the PhD in both Aerospace Engineering and Mechanical Engineering in 1964. An interdisciplinary program leading to the PhD degree had been established in the College of Engineering in 1957. Several candidates in that program, who had a sufficient Aerospace emphasis in their degree program in Engineering, were allowed to receive their doctorate in Aerospace Engineering. During the period from 1964 through 1972, 23 PhD degrees were awarded in Aerospace Engineering. During that same time period, the enrollments in the programs averaged 233 students in the undergraduate program, 20 in the masters, and 11 in the doctoral degree program.

The AE program and its predecessors are believed to have been accredited by the Engineers Council for Professional Development (ECPD) since its inception in the 1930's. ECPD was the predecessor of the present-day ABET (Accreditation Board for Engineering and Technology).

Both the Sigma Gamma Tau AE honorary society chapter and the AIAA Student Branch have been continually active. The AIAA student members have received numerous certificates of merit for their participation in the Minta Martin Student Paper Competitions at the Southwest Regional Student Conferences. Also they have received numerous national awards for their design projects (see section on Student Design Projects). The Student Branch was designated as the Most Outstanding Student Branch in the Southwest Region for Academic Year 1989.

Recent Enrollment Data

A listing of enrollments in the three degree level programs for the past twenty-five years is given in Table 1. The cyclic nature of the enrollments is obvious, but the increases in the undergraduate enrollments appears to be on an upswing in the past two years.

Table 1 Aerospace engineering headcount by graduate/undergraduate counts (Fall 1978 through Fall 2002)

Year	Bachelors	Masters	PhD	Totals
1978	111	7	4	122
1979	148	6	3	157
1980	163	3	2	168
1981	164	5	2	171
1982	188	7	3	198
1983	170	6	4	180
1984	165	5	2	172
1985	185	1	3	189
1986	209	7	3	219
1987	249	10	4	263
1988	243	10	3	256
1989	234	7	2	243
1990	249	16	2	267
1991	211	14	5	230
1992	157	13	4	174
1993	110	7	3	120
1994	91	9	3	103
1995	74	7	3	84
1996	73	9	5	87
1997	67	13	6	86
1998	99	5	7	111
1999	96	6	5	107
2000	78	5	6	89
2001	113	2	4	119
2002	138	2	5	145

Student Design Projects

For over a quarter of a century, OU Aerospace Engineering students have been entering various national design competitions and winning a significant number of awards for them. Prior to 1967, Professor L. A. Comp's aero design students were designing various flight vehicles, but they were not entered in national competitions. A complete chronological listing of the award year, name of the competition, name of the design project, award placing, and name of faculty adviser follows.

1974, AIAA Bendix Design Competition, The Sooner Shuttle, A Single-Stage-to-Orbit Vehicle, 3rd place, Professor Martin Jischke (now President of Purdue University)

1978-79 Bendix Design Competition, Super Sooner – Advanced Technology Agricultural Aircraft, 1st place, Professor Karl Bergey (now Professor Emeritus)

1979-80 AIAA Bendix Design Competition, Bandit General Aviation Airplane, 1st place, Professor Charles Bert (now Perkinson Chair Professor)

1983-84 AIAA Bendix Design Competition, Pegasus – Intermediate/Advanced Navy Jet Trainer Aircraft, 1st place, Professor Karl Bergey

1985-86 AIAA General Dynamics Corp. Team Aircraft Design Competition, Supersonic Cruise Executive Jet, Honorable Mention, Professor Karl Bergey

1986-87 Allied Design Competition, Space-Based Rescue Capsule for the Space Station, 3rd place, Professor Karl Bergey

1988 Allied Design Competition, An Antenna Array Satellite Using Rigid Inflatable Structure Technology, 2nd place, Professor Karl Bergey

1991-92 AIAA Air Breathing Propulsion Team Engine Design Competition, 3rd place, Professor Karl Bergey

1993-94 AIAA McDonnell Douglas Corp. Grad. Team Aircraft Design Competition, Navy Patrol Aircraft, 2nd place, Professor Dudley Smith (now at Bell Helicopter Textron, Ft. Worth)

1994-95 AIAA United Technology Corp./Pratt & Whitney Undergraduate Individual Aircraft Competition, 3rd place, Professor Dudley Smith

1998-99 NASA/FAA National General Aviation Design Competition, Best Retrofit Design (sponsored by the Aircraft Owners and Pilots Association (AOPA) Air Safety Foundation), Energy Absorbing Seat Design: an Innovative Approach, 1st place, Professor Karl Bergey

1999-2000 NASA/FAA National General Aviation Design Competition, Best Retrofit Design (sponsored by AOPA Air Safety Foundation), Innovative Aircraft Exhaust System, 1st place, Professor Karl Bergey

2000 AIAA Ground Test Engineering Competition, Impact Sled Tests of Energy-Absorbing Aircraft Seat Design, 2nd place, Professor Karl Bergey

2000-2001 AIAA Foundation Undergraduate Individual Aircraft Design Competition Award, Vortex Sport Plane, 2nd place, Professor Karl Bergey

2000-2001 AIAA Foundation Undergraduate Individual Aircraft Design Competition Award, SCX-1 Sport Plane, 3rd place, Professor Karl Bergey

2001 Experimental Aircraft Association National General Aviation Design Competition, Design It, Build It, Fly It Team Award, Design and Testing of a COUGAR Technology Demonstrator Aircraft, Professor Karl Bergey

2001-2002 AIAA Foundation Undergraduate Team Engine Design Competition, Airframe and Propulsion System Design for a High Altitude Long Endurance Unmanned Aerial Vehicle, 1st place, Professor Karl Bergey

Brief Career Descriptions of Some of Our Outstanding Graduates

Graduates of the University of Oklahoma's Aerospace Engineering program and its predecessors have gone into a wide variety of positions in industry, government (including both civilian and military), and academia. Many hold or have held positions of responsibility at the senior engineer, project engineer, group leader, department head, research associate, and junior faculty levels. Some of these have gone even further. In the interest of brevity, the following chronological listing of selected graduates was chosen on the basis of two criteria: (1) they must have been graduated with at least one degree in aerospace engineering from the University of Oklahoma (since the program was an option in mechanical engineering until about 1949, degrees earlier than that date were designated as mechanical engineering), (2) they achieved a position at the level of president, vice president, or technical manager in industry; general officer or colonel in the military; or senior faculty member in academia; or received a significant national award. Note: OU/CoE denotes The University of Oklahoma College of Engineering.

Jack Ridley, BSME, 1939
Colonel, USAF; test pilot/project engineer, X-1 rocket research aircraft (in which the sonic barrier was broken by Col. Chuck Yeager). Inducted posthumously into the Okla. Aviation Hall of Fame in 1961. See expanded biography below.

James Close, BSME, 1943
Engineer at Beech Aircraft; Founder & President, Close Bend, Inc., Tulsa, OK; donor of the L. A. Comp Chair, OU School of Aerospace & Mech. Engineering; OU/CoE Distinguished Graduates Society (1994)

Donald Malvern, BSME, 1946
President, McDonnell Aircraft, 1982-86; Vice President, McDonnell Douglas Corp., 1986-88; Fellow AIAA; AIAA Technical Management Award (1968); Oklahoma Aviation Hall of Fame (1987); OU/CoE Distinguished Graduates Soc. (1999); former member, OU/CoE Board of Visitors. See expanded biography below.

Oran Nicks, BSME, 1948
Team Leader, Scout Launch Vehicle, Vought Aircraft; Director, Lunar & Planetary Programs, NASA Headquarters, 1981-87; Deputy Assoc. Admin./Space Science & Applications and Assoc. Admin./Advanced Research and Technology, 1967-70; Deputy Director, NASA Langley Research

Center, 1970-80; organized the Space Research Center at Texas A&M Univ.; founding chair, Board of Governors, Texas Space Grant Consortium. See expanded biography below.

Jack Whitfield, BSAE, 1954
Engineer through President., Sverdrup Technology, Tullahoma, TN; Fellow AIAA; AIAA Simulation & Ground Testing Award; OU/CoE Distinguished Graduates Society (1993)

Jim Maupin, BSAE, 1956
He flew 100 combat missions in a F-80 Shooting Star in the Korean War before returning to OU for his BSAE and MSAE (1962) degrees. He was an engineer for the Air Force at Tinker AFB and later for the FAA Academy. For seven years he was Assistant Dean of the National Aircraft Accident Investigation School. He retired in 1986 and passed away in Feb. 2003.

Fred Haise, BSAE, 1959
NASA astronaut, Lunar Module Pilot for Apollo 13[1,2]; Vice President, Space Programs (1979-83), President, Space Station Support Division (1987-91), President, Grumman Technical Services (1992) until retirement, all at Grumman Corp.; Recipient of Presidential Medal of Freedom, AIAA Haley Astronautics award, NASA Distinguished Service Medal. Assoc. Fellow AIAA; Fellow of the Society of Experimental Test Pilots and American Astronautical Society. Oklahoma Aviation Hall of Fame (2000)

Robert Nerem, BSAE, 1959
Professor, Aeronautical & Astronautical Engineering and Associate Dean, College of Engineering, The Ohio State University; Professor & Chair, Department of Mechanical Engineering, University of Houston; Distinguished Professor of Bioengineering, Georgia Tech. Internationally known authority in biomechanical fluid mechanics and tissue engineering. One of the very few members of *both* the National Academy of Engineering and the Institute of Medicine. OU/CoE Distinguished Graduates Society (1994).

Robert Culp, BSAE, 1960
Faculty member and chair, Department of Aerospace Engineering Sciences, University of Colorado, Boulder; Fellow AIAA and American Astronautical Society (AAS); AAS Dirk Brouwer Award (1993); AIAA Mechanics & Control of Flight Award (1995); ASEE/AIAA John Leland Atwood Award (1995) for outstanding achievements in aerospace engineering education and administration.

James Abrahamson, MSAE, 1961
Director, Strategic Defense Initiative ("Star Wars"), Department of Defense, 1984-89. Retired as Lt. General, USAF in 1989. Executive Vice President, Corporate Development and President, Transportation Sector, Hughes Aircraft Co. (1989-92); Chairman, Oracle Corp. (1989-92); currently Chairman & CEO, Strat Com International LLC and Air Safety Consultants. Numerous military awards including the Distinguished Service Medal; NASA Distinguished Service Medal; three honorary doctorates; General Bernard Schriever Award (1984), Robert H. Goddard trophy for leadership and excellence in advancing space flight programs contributing to U.S. leadership in astronautics. OU/CoE Distinguished Graduates Society (1990), former member, OU/CoE Board of Visitors

Jamal Azar, BSAE, 1961
After receiving his Ph.D. from OU in 1965, he joined the faculty of the University of Tulsa in that same year. In 1974 he transferred to the Dept. of Petroleum Engineering where he was promoted to full professor and Director of the University of Tulsa Drilling Research Projects (a cooperative

research program supported by all major oil/gas companies and some service companies). His expertise in drilling engineering was recognized by the Society of Petroleum Engineers by awarding him the 1997 Distinguished Achievement Award and the 1998 Drilling Engineering Award. He retired in May 2002.

William Surbey, BSAE, 1963
Dynamics group leader, Aero Commander; Program manager, Citation executive aircraft, Cessna; President, Lear Fan Ltd.; Vice President, Northrop Corp., now retired

James ("Red") Hawkes, BSAE, 1964
Naval officer, Navy submarine service; senior engineer, United Nuclear Corporation, New Haven, CT; security analyst, portfolio manager, director of investment research, chief investment officer, and presently Chief Executive Officer, Eaton Vance, Boston. He has been with this major investment management firm for over 32 years.

Jerry Holmes, MSAE 1964
Maj. General, USAF (ret.); fighter pilot (135 combat missions in Vietnam); commander, AWACS, Tinker AFB; Deputy Chief of Staff - Logistics, Tactical Air Command; commander, NATO Airborne Early Warning Force, 1987-88; Senior Vice President & General Manager, Frontier Engrg., Inc., Stillwater, OK, 1992-96; currently Adjunct Professor, College of Engrg., The Univ. of Oklahoma. Elected to OU/CoE Distinguished Grad. Soc. (1997) & Okla. Aviation Hall of Fame (2001).

Donna Shirley, BSAE, 1965
Engineer through Manager, Jet Propulsion Lab. for 32 years. Manager, Mars Sojourner (Mars Pathfinder Project) and the Mars Exploration Program. Author of "Managing Martians" & "Managing Creativity". Holds three honorary doctorates; NASA Outstanding Leadership Medal; ASME Holley Medal; National Space Society Wernher von Braun Award; Okla. Aviation Hall of Fame (1998); OU/CoE Distinguished Graduates Society (1999); former member, OU/CoE Board of Visitors; Assoc. Fellow AIAA. Elected to the Oklahoma Women's Hall of Fame, 2003.

Carl Baerst, BSAE, 1967
Vice President, Allied Signal Engines and Systems. Member, OU/CoE Board of Visitors

Louis Duncan, BSAE, 1969
Pioneer in use of wind tunnels in automotive design. Now at Automotive Aerodynamics Inc., in North Carolina.

Robert Behler, BSAE, 1970 (MSAE, 1972)
Major General, USAF; SR-71 pilot; various command & staff positions in command, control, communications, & computer systems; Deputy Commander, Joint Command North, NATO, Stavanger, Norway; presently Commander, AF Command & Control, & Intelligence, Surveillance & Recon. Center, Langley AFB, VA.

John Baker, BSAE 1972
Lt. General, USAF; formerly Asst. Deputy Chief of Staff/Air & Space Operations, Headquarters, USAF; currently Vice Commander, Air Mobility Command, Scott AFB, IL.

Steven Fisher, BSAE, 1972
President, Space Data Division, Orbital Sciences Corp., retired; senior active member, OU/CoE Board of Visitors

Brian Argrow, BSAE, 1983 (Ph.D., 1989)
Asst. Professor, AME, Univ. of Oklahoma, 1989-92; Asst. through Assoc. Professor, Aerospace Engrg. Sciences, Univ. of Colorado, Boulder, 1992-present. Black Engineer of the Year (1988); W. M. Keck Foundation Engrg. Teaching Excellence Award (1995), President's Teaching Scholar (2000).

Three of the above have been selected for more extensive coverage:
Jack Ridley, representing military aerospace research
Donald Malvern, representing corporate management
Oran Nicks, representing both NASA and later university research

Jack Ridley was born in 1915 in Garvin, Oklahoma.[3-6] In his early schooling, Jack had a natural liking for mathematics and a definite aptitude for studying how machines worked. After high school, he entered the University of Oklahoma and its Army ROTC program. He was graduated in 1939 and received a commission in the Army in 1941. As a lieutenant he received his pilot's wings at Kelly Army Air Base, San Antonio, Texas in May 1942. Rather than being sent to a combat operational unit, Jack was first ordered to conduct acceptance tests of B-24 *Liberator* bombers at the Convair plant in Fort Worth, Texas. Later he was assigned to the B-32 and B-36 bomber projects.

With the war nearing its end, Jack was sent to the forerunner of the Air Force Institute of Technology at Wright Field, Dayton, Ohio and then to California Institute of Technology, Pasadena, California, where he received a master of science degree in Aeronautics in July 1945. He next assignment was to go through the Air Materiel Command's Flight Performance School, where he was graduated in 1946. Meanwhile the Army Air Force was assuming control of the revolutionary X-1 rocket-powered manned aircraft. Of the 125 AAF test pilots, Col. Albert Boyd, chief of the Flight Test Division, selected three for the project team to attempt the world's first supersonic flight: Capt. Charles ("Chuck") Yeager, 1st Lt. "Bob" Hoover, and Ridley. Yeager and Hoover were the pilots and Ridley was project engineer. See Fig. 2. Ridley was very knowledgeable about aerodynamics and was a pilot as well. He studied the phenomena encountered in passing through the transonic region and explained it to Yeager in pilot's terminology. As Yeager later said, "I trusted Jack with my life." In the vicinity of Mach 1, the aircraft would not respond to elevator control, due to a shock wave forming at the elevator hinge line. To solve the problem, Ridley dispensed with the trim tab and conceived the "flying tail" – now used on all supersonic aircraft. He also saved the day when the sonic barrier was broken (14 Oct. 1947) and Yeager had a broken arm, by providing a broom handle to provide sufficient leverage. He was awarded the Commendation Ribbon for his contributions.[6]

Later Jack worked with the operational development of the B-47 *Stratojet* bomber, the XF-92A research vehicle, the B-52 *Stratofortress*, and all of the first five X-series research aircraft. He was chief of the Test Engineering Branch and later Chief, Flight Test Engineering Lab. at the Air Force Flight Test Center (AFFTC). He is credited with creating AFFTC's basic philosophy still in use today and transplanting it to NATO as well. He was promoted to full colonel in 1956. Unfortunately, he died as a passenger on a C-47 which crashed into a mountain in Japan in 1957. In 1980, the Ridley Mission Control Center at Edwards AFB was dedicated to his memory.[6]

Donald Malvern was born in western Oklahoma and while he was in college, war broke out and he joined the U.S. Army Air Corps, serving as a pilot, flight engineer, and maintenance officer from 1943 – 1946. Upon graduation he went to work for McDonnell Aircraft Company starting a 42-year career in industry. He served in aerodynamics, as chief flight test engineer, and then as project manager for the F-4 *Phantom* program, for which he received the AIAA Technical Management Award. Later he was appointed vice president of the FX project which evolved into the well-known F-15 *Eagle*, for which he was awarded the Sylvanus Reed Aeronautics Award. Finally he moved into top management when he was appointed Executive Vice President and later President of McDonnell Aircraft. He retired as corporate Vice President of McDonnell Douglas Corporation in 1988. After retirement his services as a consultant were in high demand and he remained on the board of visitors of various aerospace organizations.

Oran Nicks spent two years in training as an aviation cadet and attended the Spartan College of Aeronautics in Tulsa, Oklahoma and received his BS degree from the University of Oklahoma in 1948. He received further training as an aeronautical engineer from the Universities of Texas and Southern California.

He was an engineer in industry for 12 years at North American Aviation in California and Chance Vought in Dallas, Texas. His duties included work in structures as well as aero-propulsion. He advanced to supervisor and project team leader at Vought for the *Scout* launch vehicle, which performed over 100 space missions for NASA and DOD.

In 1960, Nicks went to NASA Headquarters as Director of Lunar and Planetary Programs. In 1961-67 he had responsibility for thirty missions of *Ranger*, *Surveyor*, *Lunar Orbiter*, and *Mariner* to the Moon, Mars, and Venus. Later he served in two top-level management positions at the Deputy Associate and Associate Administrator level. As Deputy Director of NASA Langley Research Center, Hampton, VA, 1970-80, he was responsible for expanding aeronautics and space research in laminar flow, composite materials, reentry materials, and other high technologies.

In 1980 he retired from NASA and moved to Texas A&M University, where he conducted research and directed a wind tunnel facility and space research activities. He organized as well as directed the University's Space Research Center and led interdisciplinary research teams on regenerative life support systems, space power, and space transportation systems. He was instrumental in the formation of the Center for the Commercial Development of Space Power and chaired its management advisory board in 1987-92.

Nicks was very active at the state level in Texas. He led the organization of the Texas Space Grant Consortium consisting of 21 universities, 18 industry partners, and two state agencies. This was the first such consortium in the nation. The Governor of Texas appointed him to the Texas Space Commission in 1987, and he chaired the commission in 1991-92.

One of Oran's lifetime hobbies has been private and soaring flying. He owned a Cessna 182 and a Standard Class sailplane. He has flown in numerous regional and national competitions. He was active in the Soaring Society of America (SSA), serving on its board of directors for many years and as chair of its Technical Board. He invented a total energy sensor used on sailplanes throughout the world.

He was one of the leading proponents of the World Class Glider and was very proud of having spent three years in handcrafting such a beautiful glider from scratch for himself. Unfortunately, in 1998 he was killed in an accident in this very glider. It is ironic yet very fitting that he was posthumously awarded the Lilienthal Medal of the Federation Aeronautique International, nominated by the SSA and recommended by the International Gliding Commission. This medal is regarded as the highest soaring award in the world, and is named after Otto Lilienthal, who also was killed in an accident while conducting research on glider flight.[7]

Curricula: Thirty Years Ago vs. Today

The 1972 curriculum in AE is listed below. Note: AME denotes Aerospace & Mechanical Engineering.

1st Semester	Hr.	2nd Semester	Hr.
First Year			
English: Composition I	3	English: Composition II	3
Math.: Analyt. Geom.	2	Math.: Calculus II	3
Math.: Calculus I	3	Physics: General I	4
Chemistry: General	5	Engr.: Engr. Communications	2
Engr.: Introduction	2	Engr.: Engr. Analysis	2
	15		14
Second Year			
Math.: Calculus III	4	Math.: Engr. Math.	4
Physics: General II	4	History: U.S.	3
Pol. Sci.: U.S. Gov't.	3	Engr.: Rigid Body Mech.	3
Engr.: Thermodynamics	3	Engr.: Heat Transfer	
Engr.: Struct. & Props.		& Fluid Mechanics	3
of Materials	3	Engr.: Elec. Science	3
	17		16

Third Year

Engr.: Strength/Matls.	3	AME: Vibrating Systems	3
AME: Analysis	3	AME: Aero Structures	3
AME: Dynamic Analysis	3	AME: Gas Dynamics	3
AME: Measurements	3	Engr.: Numerical Methods	3
Elec. Engr. Elective	3	Humanistic Social Studies	3
Humanistic Social Studies	3		15
	18		

Fourth Year

AME: Gas Power Systems	3	Physics: Modern for Engrs.	3
AME: Aerodynamics	3	AME: Heat Transfer	3
AME: Stability & Control	3	AME: Vehicle Trajectory	
Technical Elective	3	Analysis	3
Humanistic Social Studies	3	Design Elective	3
	15	Experimental Elective	2
		Humanistic Social Studies	3
			17

In contrast, the recently approved 2003 curriculum in AE is listed below. An asterisk denotes a new course in this curriculum.

1st Semester	Hr.	2nd Semester	Hr.
Freshman Year			
English: Composition I	3	English: Composition II	3
Math.: Calc. & Analytic		Math.: Calc. & Analytic	
Geometry I	3	Geometry II	3
Chemistry: General	5	Physics: General I	4
Engr.: Introduction	2	*Computer Sci.: Programming	3
History: U.S.	3	for Non-Majors (AE Section)	
	16	Pol. Sci.: Amer. Gov't.	3
			16

Sophomore Year

Math.: Calculus & Analytic		Math.: Calc. & Analytic	
Geometry III	3	Geometry IV	3
Physics: General II	4	Math.: Ordinary DEs	3
AME: Intro. to AE	3	*Math.: Laboratory	1
Engr.: Rigid Body Mech.	3	*AME: Circuits & Sensors	3
Engr.: Thermodynamics	3	AME: Dynamics	3
	16	Engr.: Struct. & Props.	
		of Materials	3
			16

Junior Year

AME: Solid Mechanics	3	AME: Flight Mechanics	3
AME: Solid Mech. Lab.	2	AME: Aerosp. Structural	
AME: Aerodynamics	3	Analysis	3
AME: Wind Tunnel Lab.	2	*AME: Embedded Real Time	
AME: Control Systems	3	Systems	3
Elective: Social Science	3	AME: Interactive Engr.	
	16	Design Graphics	3
		AME: Experimental Elective[†]	2
		Elective: Artistic Forms	3
			17

[†] AME Robotics Lab. or AME Dynamics & Control Lab. recommended.

Senior Year

AME: Aerospace Systems		AME: Aerospace Systems	
Design I	3	Design II	3
*AME: Space Sci./Astrodyn.	3	*AME: Space Systems &	
AME: Aerosp. Propulsion		Mission Design	3
Systems	3	Technical Elective	3
AME: Flight Controls	3	Elective: Western Civ./Culture	3
Technical Elective††	3	Elective: Non-Western Civ./	
	15	Culture	3
			15

†† AME Systems Engineering recommended.

In comparing the 1972 and 2003 curricula, it is noted that the total number of hours (127) remains unchanged, as does the number of hours of Chemistry, Math., English, History, Political Science, and Humanistic/Social Studies electives. As for Physics, Modern Physics has been replaced by Space Science/Astrodynamics. Also, there are 3 additional hours of Technical Elective and 3 additional hours of Design in 2003. There are a total of 16 hours of new courses first introduced in this curriculum.

Research Highlights

Research topics have included low-speed drag reduction by compliant coatings, high-speed gas dynamics, composite materials and structures (especially sandwich structures), hypersonic flow (especially the waverider concept), radiative transfer, low-gravity combustion research, development of innovative concepts for aerospace plane applications and for general aviation aircraft, robotics for space exploration, structural optimization, and multidisciplinary optimization.

Research sponsors have included, in alphabetical order:
Aerospace Research Laboratory, Wright-Patterson AFB
Air Force Logistics Command, Tinker AFB
Air Force Office of Scientific Research
Army Aviation Materiel Labs., Ft. Eustis
Army Research Office
Federal Aviation Administration, Civil Aeromedical Institute
NASA: Headquarters and Dryden, Glenn, and Langley Research Centers
Office of Naval Research
Oklahoma Center for the Advancement of Science and Technology
Oklahoma Space Grant Consortium
Oklahoma Space Industry Development Authority
Piper Aircraft
Rockwell International

Faculty Activities

One measure of faculty quality is the extent of their activities on technical society committees and editorships. Two of our faculty members have served three-year terms as Associate Editor for the *AIAA Journal*. Four of our faculty members have served on the AIAA Thermophysics Committee (one of them as Chair). One faculty member has served on the AIAA Aeroacoustics, Applied Aerodynamics, and Atmospheric Flight Mechanics Technical Committees. One faculty member has served on both the AIAA Aircraft Design and General Aviation Technical Committees. One faculty member has served on both the AIAA Structural Dynamics and the Multidisciplinary Design Optimization Technical Committees. One faculty member has served on the AIAA Structures Technical Committee and has chaired the ASME Applied Mechanics Division's Composite Materials Committee. One faculty member has served on the AIAA Plasmadynamics and Lasers Technical Committee. One faculty member is currently President of the American Academy of Mechanics. One faculty member is currently on the editorial boards of *Composite Structures, International Journal of Structural Stability and Dynamics, Journal of Sandwich Structures and Materials,* and *Mechanics of Advanced Materials and Structures.*

Another measure of faculty quality is the recognitions and awards received. These include:
AIAA William T. Piper General Aviation Award – one emeritus
AIAA Fellow – one active and one emeritus
ASME Fellow – two active and one emeritus
AIAA Associate Fellow – three active & two emeriti

Fellow of American Academy of Mechanics, AAAS, Society of Engineering Science, and Society for Experimental Mechanics – one active

Concluding Remarks

As we look back over the past experiences of our AE program, including its faculty, facilities, and graduates, we believe significant contributions have been made to the greater good of our profession as well as our society. As we face the future, we look forward to a profession that will become even more systems and space oriented. We are changing our program to meet the challenge.

Acknowledgments

The authors acknowledge the continual encouragement of Professor S. R. Gollahalli, Director of the School of Aerospace and Mechanical Engineering. Information provided by Professors Karl Bergey, Martin Jischke, and Dudley Smith and Ms. Laura Kurz are acknowledged. Special credit is due to Ms. Lawana Cavins, Technical Communications Specialist, for major support on all aspects of this paper.

References

[1] *Apollo 13*, film, 1995.

[2] "Where Are They Now? The Homecoming", *People* magazine, July 24, 1995, pp. 160-162.

[3] *The Right Stuff*, the RKO Pictures film, 1983.

[4] Yeager, C. and Janos, L., *Yeager: An Autobiography*, Bantam Books, New York, 1985, various pages (see Index, Ridley mentioned on over two dozen pages).

[5] Hallion, R. P., *Test Pilots, The Frontiersmen of Flight*, revised ed., Smithsonian Institution Press, Washington, DC, 1988, pp. 198-201.

[6] Edwards Air Force Base Home Page, Nov. 1997 Cover story, http://www.edwards.af.mil/articles98/docs_html/splash/nov97/cover/ridley.html.

[7] *Soaring* (Magazine), Vol. 64, No. 11, Nov. 2000, p. 31.

**Fig. 1 The wind tunnel with a test in progress.
The late Professor L. A. Comp is on the far left**

**Fig. 2 The X-1
team in 1947.
Chuck Yeager
and Jack Ridley
are in the front
row and Bob
Hoover is behind
them.**

Chapter 13

A History of the University of Washington Department of Aeronautics and Astronautics 1917-2003

J. Lee,[*] D.S. Eberhardt,[†] R.E. Breidenthal,[‡] and A.P. Bruckner[§]
Department of Aeronautics & Astronautics
University of Washington, Box 352400
Seattle, WA 98195-2400

The Department

The University of Washington's Department of Aeronautics and Astronautics was one of the first aeronautical engineering departments in the nation, and one of the seven originally established with the help of the Guggenheim Fund for the Advancement of Aeronautics. It offers the only aerospace degree program in the Pacific Northwest, a region whose aerospace industry has been a major contributor to the technological development, economic vitality and the security of the United States. Educators and researchers in the Department over the years have made numerous contributions in all major areas of aerospace engineering. Graduates at all degree levels, have been successful and valued in industry at the local, national, and international levels, as well as in government organizations and institutions of higher learning.

Bill Boeing and the Early Years

In 1903, the year of the Wright Brothers' first powered flight, a man interested in establishing a timber business on the West Coast moved to Seattle after leaving Yale. Little did he know it at the time, but he was destined to change the face of aviation and the Pacific Northwest forever. His name was William E. Boeing. It is with this man that the story of aeronautics at the University of Washington begins.

The first airplane flight in Seattle took place March 11, 1910, when Charles K. Hamilton flew a Curtiss Reims Racer before a large crowd of eager onlookers at what was then called

Fig. 1   **March 11, 1910, the first airplane flight in Seattle, at The Meadows. Aircraft is a Curtiss Reims Racer. Pilot is Charles K. Hamilton.**

[*] Graduate Student
[†] Associate Professor
[‡] Professor
[§] Professor, Department Chair

The Meadows, a low-lying strip of land by the Duwamish River, south of downtown.[1] This location is now occupied by Boeing Field, otherwise known as King County Airport. It is not known whether Bill Boeing was present at this event, but what is certain is that he witnessed flying demonstrations in Los Angeles that same year, and was fascinated by what he saw.[2] For the next few years he tried to hitch a ride in an airplane, finally getting his wish on July 4, 1915, in a two-seater float plane on Lake Washington. Boeing caught the flying bug, and soon decided to start producing his own airplanes. Together with Navy Lieutenant Conrad Westervelt and Herb Munter, Boeing designed and built his first airplane, a float plane named the B&W. Shortly thereafter, on July 15, 1916, Boeing incorporated his aircraft manufacturing business as Pacific Aero Products Company, a name he changed to Boeing Airplane Company the following year. In early 1917, Boeing hired two students, Clairmont L. Egtvedt and Philip G. Johnson, from the University of Washington, to be his engineering staff. Though Egtvedt and Johnson were trained in mechanical engineering, and eventually became two of the most influential men in aviation history, their lack of formal background in aeronautics started Boeing thinking.[3]

To build a successful airplane company, Boeing realized that he needed trained aeronautical engineers and a facility to test new airplanes. Boeing devised a way to kill two birds with one stone. He decided to donate a wind tunnel to the University of Washington on the condition that the University develop an aeronautics curriculum.[4] Design and construction of the new wind tunnel started in 1917, supervised by Assistant Professor John W. Miller, then of the Civil Engineering department. Miller's interest in flight dated back to before the Wright brothers' famous

Fig. 2   The Boeing Wind Tunnel at the University of Washington (c.1918). Clairmont Egtvedt is third from left. This facility is still in use but with a modern 3'x3' wind tunnel inside.

flight. He had started experimenting with gliders in 1895 and developed his first powered airplane in 1909.[5] Miller later became the first person to take an aerial photograph of the University of Washington campus.[6]

As a first step toward fulfilling the University's end of the bargain, the Civil Engineering department offered an airplane structures class in the spring of 1917. Taught by Miller, this was the first aviation-related course offered at the University.[4] This class, however, was not destined to go any further. As a result of his work with Bill Boeing on the wind tunnel project, Miller resigned from the University in the summer of 1917 to become Chief Engineer at the newly renamed Boeing Airplane Company.[6]

Concurrent with this turn of events, the Mechanical Engineering department began a search for a new faculty member to implement and instruct a complete aeronautics curriculum. This search led to the hiring of Frank McKone for the 1917-1918 academic year. Per Boeing's specifications, McKone organized classes in basic aviation, aircraft design, aerial propulsion, and wind tunnel testing.[7] He taught these classes for just one year before leaving the University.[6]

The gap created by McKone's departure lasted for three months and was filled by McKone's predecessor, John W. Miller. The University had been desperate for an aeronautics professor, and with the end of World War I, Miller seems to have lost interest in work at Boeing. To sweeten the deal, Henry Suzzallo, president of the University, apparently promised Miller that, although he would initially be an Assistant Professor, he would be almost immediately promoted to the rank of Associate Professor. For some reason, this promise was not fulfilled, and Miller was not promoted. As a result, Miller again resigned from the University after teaching for only two academic quarters.[4, 6]

With Miller's departure, the University was again left with no one to teach aeronautics. As a result no classes on this topic were taught for the next two years. Little did anybody know at the time, but the man to fill this void was already right under their noses, teaching electrical engineering. As it turned out, this individual was destined to not only fill the aeronautics teaching void, but also to play a key role in the development of aeronautics at the University of Washington for the next 30 years; a man whose legacy continues to be felt even today. His name was Frederick Kurt Kirsten.

Enter Fred Kirsten

Born in Germany, Kirsten came to America in 1902 aboard an old sailing schooner.[8] At the encouragement of a friend, he began studying electrical engineering at the University of Washington, graduating *magna cum laude* in 1909.[4] In 1915, he joined the faculty as an Assistant Professor in Electrical Engineering. By 1923 Kirsten had been promoted to full Professor.

By nature Kirsten was an inventor. His most famous aeronautics-related invention was that of the cycloidal propeller, which Kirsten spent over 20 years trying to perfect for use on airplanes. At one point, he teamed up with Bill Boeing to further develop cycloidal propellers for both aeronautical and marine applications. Boeing put up $175,000 of his own money to start a company with Kirsten. However, the concept eventually proved to be impractical in aeronautical applications, and the Kirsten-Boeing company failed. It should, however, be noted that cycloidal propellers were viable as a marine propulsion system.[3] Even today, some tugboats are equipped with them.

Outside of his contributions to aeronautics, Kirsten invented everything from lights for airports to World War II air-raid sirens. Although he took out more than 100 patents, many of his non-aeronautical inventions[9] are now forgotten. There is, however, one exception, the "Kirsten Pipe." Kirsten had been a cigarette smoker and, while visiting his doctor about a persistent cough, he was told to quit smoking cigarettes. Kirsten went home and decided to design a pipe. The heat-absorbing aluminum stem of this pipe delivered a "delightfully cool, clean smoke." Kirsten demonstrated the pipe to his doctor who is reported to have told Kirsten that he could smoke it on one condition: that he made one for him. Kirsten started a company to manufacture his "perfect pipe,"[10] and over the years made a substantial amount of money selling them. They are still manufactured today, in Seattle, and the company is still in the family.[11]

Kirsten's personality was one of extreme confidence. According to one of his former students, "Kirsten was a great man, you could just ask him." An article in *The Daily*, the University's newspaper, stated that Kirsten was "about as meek as a General Sherman tank."[12]

Kirsten was extremely proud of his work and very dedicated to it. He could be the best of friends to those who took interest in his work, and a powerful adversary to those who criticized it. It was Kirsten's personality that gave him the driving force that would enable him to accomplish so much.

Founding a Department

Kirsten began teaching aeronautics courses at the beginning of the 1921-1922 academic year. He undoubtedly gave the courses a face-lift, but they basically remained the same as when Miller had taught them. Over the next four years very few changes would be made.

In early 1926, the University opened a dialog with Harry Guggenheim and the Guggenheim Fund for the Advancement of Aeronautics, in an attempt to procure funds to establish a school of aeronautics. At that time the Guggenheim Fund, founded by Harry's father Daniel, had already provided grants for similar schools at New York University, Caltech, MIT, Stanford, and the University of Michigan.[13] In its communications with Guggenheim, the University stressed its strong ties with Boeing and Naval Aviation, the Boeing Wind Tunnel on campus, and the promising development of Kirsten's cycloidal propeller.[41]

In 1927, a proposal for a $450,000 grant was submitted to the Guggenheims. They balked at the large amount of the request, but kept the matter under consideration. The University again approached the Guggenheims in 1928. This time, the University's approach was highly organized. After the first request, the Guggenheims had been given time to investigate the University's background and existing facilities. Bill Boeing wrote to the Guggenheims on behalf of the University. Most importantly, in the 1928 proposal, the University was asking only

Fig. 3   The Department's Founding Fathers. Left to Right: William E. Boeing, John W. Miller, Frederick K. Kirsten, Everett O. Eastwood

$290,000, just enough to construct a building to house the new department. The Guggenheims were sold. On June 15, 1928, the trustees of the Guggenheim Fund approved a grant of $290,000 for the construction of an aeronautics building on the University of Washington campus. This grant was contingent on receiving funding from the State of Washington to properly equip the new building. The state legislature approved this funding in early 1929.[13] Construction began soon thereafter.[14] The 1927-1928 academic year saw the first evidence of the preparations for the new department. Design work had begun on the new aeronautics building. John W. Miller, who had been serving as the secretary of the Guggenheim Fund board of trustees, had again returned to the University, and for the first time the aeronautics faculty consisted of more than one man. With two professors, the course offerings were expanded from five to eight. Support from the Boeing Airplane Company was evident, as some of these courses featured supplemental lectures from some of Boeing's best engineers, such as Claire Egtvedt and C.N. Monteith.[6] In July 1929, Professor Everett O. Eastwood was named the head of the department, thus marking its official beginning.[15] The building was completed in the spring of the following year.[16]

Eastwood was the very model of a "modern" mechanical engineer. He had a hand in almost everything. Educated at MIT, he joined the University of Washington in 1905 as the head of the Mechanical Engineering Department. He had developed the first master plan for the University of Washington campus, and served as the University's engineering consultant. Eastwood's appointment as head of the department seems to have originally been one of expedience. He was a competent administrator, and a good organizer; just the kind of person to get the new department up and running, but really nothing more. Eastwood never showed any particular interest in flight, either professionally or personally.[13] In fact, Eastwood never actually taught any aeronautics classes at all. This fact would later become a matter of some controversy. While these concerns led to two attempts to replace him as head of Aeronautics, he remained head of both departments until 1946, a year before his retirement from the University

Fig. 4   **Guggenheim Hall in 1931.**
(CF Todd Coll., PNW Coll. UW #14680)

On October 5, 1929, the Department of Aeronautical Engineering officially opened its doors.[17] There were four original faculty members: Miller, Kirsten, Eastwood, and Fred Eastman, who was hired earlier that year as an instructor, but would soon be given a professorship.[*] Twelve different courses were offered in the 1929-1930 academic year, leading to a degree in Aeronautical Engineering.

Fig. 5 Fred Eastman

The Daniel J. Guggenheim Aeronautics Hall was dedicated on April 11, 1930. In the words of the University of Washington's newspaper, *The Daily*, the Tudor-Gothic building was "architecturally perfect."[18] The building included room for six small instructional wind tunnels in the basement, only one of which was ever built. It is known as the Venturi Tunnel and is still in operation. The building housed a large laboratory with two full-sized airplanes and a number of aero-engines so that students could have hands-on experience with "the real thing." The building also sported a 355-seat auditorium which was designed not only to hold classes, but also to enable the Department to host large public lectures on aeronautics.

---

[*] Fred Eastman served the department until his retirement in 1970. He was Chair from 1946 to 1952. On February 10, 2003, he celebrated his 99th birthday.

Everybody wanted a piece of the new building. In addition to being the home of Aeronautical Engineering, the Civil and Electrical Engineering departments, the College of Engineering Administration, and the Engineering Library were also housed in Guggenheim Hall.[19]

The early 1930's were a time of evolution for the new department. Changes to the curriculum were constantly being made. In 1930, the Department awarded its first five Bachelor of Science in Aeronautical Engineering degrees; in 1931 there were 11 graduates and by 1933 this number had jumped to 28. However, the Depression took its toll, and from 1933 to 1939 an average of about 16 students graduated per year. Unlike other Guggenheim schools, such as Caltech and MIT, most of the Department's emphasis at this time was on teaching rather than research. The research that did go on was mostly applied research, such as Eastman's work on wind tunnel balances and Kirsten's cycloidal propellers.[13]

The Wind Tunnel Years

The years from the mid-1930's until 1960 can best be characterized as the Kirsten Wind Tunnel years. Although the tunnel still is used regularly today, it was this period that established it as a world-class facility. Throughout this era, faculty and students played a key role in operating the wind tunnel. Faculty had played a major part in its construction, and students were hired to operate the tunnel. In some cases, students who worked in the tunnel continued as technical staff and later were hired as faculty.

During the early 1930's, Fred Kirsten was eager to test his "Cycloplane," his cycloidal propeller aircraft concept. In 1934 he approached the Graduate Aeronautical Laboratories at the California Institute of Technology (GALCIT) to use their wind tunnel for this purpose. He was quoted a price of $200/day, which was much more than he could afford. At the time, GALCIT had the only wind tunnel of any practical size on the west coast. So, in 1935, Kirsten proposed a new wind tunnel for the University of Washington.[20]

Fig. 6 Kirsten with his "Cycloplane" in the wind tunnel that now bears his name.

The proposal, "An Aerodynamic Laboratory (Wind Tunnel) on the Campus of the University of Washington, Seattle, Washington," was for a wind tunnel with an 8x12-foot test section and a maximum speed of 250 mph. When completed, the tunnel would be one of only two capable of such speed. The proposal was for $120,000, of which $54,000 was requested of the federal Public Works Administration (PWA). The remaining $66,000 would come from other sources: $40,000 from the Washington State Budget Relief Administration, and $26,000 from Boeing, given as a loan against future rentals, at $15/hour.[20]

It is interesting to review some of the salient points made in the proposal. One was a barb aimed at GALCIT's high cost to other universities and its essentially holding a monopoly for testing on the west coast. Another was the training of students and staff for research. The wind tunnel would "allow them to contribute in considerable measure to the advancement of a new engineering field." Also it was pointed out that the department could not support graduate students because they had to go elsewhere to find facilities to do their research. Finally, it was added that Boeing was sending its work to the east coast and GALCIT at considerable cost. It is interesting to note that the tunnel was envisioned to operate with a student crew, as it still does today.

Construction began in January 1936 and, due to mild weather, progressed rapidly. The tunnel was completed in early autumn of that year. However, it needed much work before it would be ready for serious testing. In order to reduce cost, a decision was made to build a dual return tunnel that used two 500-hp motors rather than a single return with a 1000-hp motor. Much of the design, supervision and construction was done in house; an ingenious electromagnetic balance was designed by Fred Eastman and the 14 wooden fan blades were carved in Guggenheim Hall's machine shop. Kirsten's dedication was so great, he even had his son sanding fan blades.[10]

Fig. 7 The Kirsten Wind Tunnel under construction, March 1936.

No formal records appear to exist regarding the work done between the fall of 1936 and early 1939 but a "diary" indicates Boeing tested their model 307 Stratoliner extensively.[21] It appears that a shakedown period was going on concurrently with the Boeing tests. Notes indicate a few startup glitches, such as the loss of one set of seven blades due to a loose spinner. There are some interesting articles that were published in *The Daily*, the University's newspaper: one article, dated January 28, 1938 had the headline "Wind Tunnel Air Goes Wrong Way," and alluded to flow problems due to the merging of the two return streams. Formal test records did not start until March, 1939, when a test on the North American AT-6 "Texan" became the first entry in the official run log.[21]

The University of Washington Aeronautical Laboratory (UWAL), as it has come to be known, began testing furiously once it was open for commercial use. In 1939, a total of 21 tests programs were performed, which included Boeing, Lockheed, Davis, and Consolidated. Two vehicles that occupied the facility for much of the year were the Lockheed Constellation and the Boeing 307 Stratoliner. An interesting historical test from an aerodynamics standpoint was the testing of the Davis wing, used and tested on the Consolidated B-24 that year. The Davis wing was a poorly understood laminar flow wing, which performed exceptionally well in the wind tunnel but not in flight. It was not until years later that it was understood that the peculiar behavior in the operational environment was due to the flow becoming turbulent due to surface irregularities.

In 1940 the tunnel had 39 tests, with Boeing, Lockheed, North American, Consolidated, and Grumman all paying visits. The bulk of Boeing's testing was on the B-29, with some testing of upgrades to the B-17.[21] Lockheed tested the XP-49, which was an upgraded, pressurized version of their P-38. North American brought a model of their P-51 wing in secrecy from southern

Fig. 8   Famous early UWAL Tests. Clockwise from top left: Boeing Model 307 Stratoliner,

California. Throughout WWII, the tunnel saw constant action. The only breaks from military testing were in 1941 to perform post-collapse analysis of the Tacoma Narrows Bridge and then in 1942 to finally let Kirsten test his Cycloplane. It is somewhat of an irony that it took until 1942 for Kirsten to get testing time after starting with GALCIT in 1934. A big moment for UWAL occurred in July 1941, when, the construction loan from Boeing was paid off.[22] It took only two years of operation to fulfill the commitment of the $26,000 of testing to Boeing.

Military testing dominated the run logs of UWAL throughout the late 1940's. Some notable tests include the B-47 and the P-85 "Goblin". Boeing and McDonnell show up extensively in the tunnel logs. In 1948, the wind tunnel was officially named the Kirsten Wind Tunnel, after the man who had worked so hard to get it built.

As the 1950's progressed, Boeing started to become the exclusive customer in the tunnel. Aircraft such as the B-47, B-52, KC-135 and 707 were tested. By the end of the decade it is rare to find an entry in the run logs that is not a Boeing test. A strong relationship between Boeing and the Aeronautics Department blossomed during this period. For the next 30 years, UWAL would host most of Boeing's low-speed wind tunnel testing. nterestingly, Fred Kirsten, who had been so instrumental in procuring the wind tunnel for the department, was never its Director. Edmund L. Ryder, an instructor, was appointed as the first director, but held the post only briefly. He left the

University in 1937 to help start the Boeing Aeronautical School in Oakland, California. Ryder was replaced by Fred Eastman, who had designed the tunnel's balance. Sometime during the mid- to late 1940's the directorship of the tunnel transferred to James Dwinnell, a 1939 graduate of the department who had joined the faculty in 1941.

Up and Running

Although the Kirsten Wind Tunnel dominates the period from the mid 1930's to about 1960, there were other changes in the Department. In the late '30s the department began to offer flying courses under the sponsorship of the Civil Aeronautics Administration. In 1939 there were 350 applicants for 30 flight training spots.[23] After the U.S. entered WWII, the Navy sponsored the flight training. Throughout the war, the department trained 1200 people for the war effort, including 25 women, with E.O. Eastwood in charge.

Courses toward a Master of Aeronautical Engineering (MAE) degree were first offered during the 1946-47 academic year as a "fifth-year" program. However, the first graduates with MAE degrees did not appear until 1948. The Master's program grew until it represented approximately one third of the student body by the end of the 1950's. The Ph.D. degree was offered for the first time during the 1959-1960 academic year. Eight students entered this new program that year.

The size of the Department during this time can be best characterized by "slow growth". The Department had a regular faculty of five throughout the period 1935-1945. Eastwood stepped down as department head in 1946 and retired in 1947, and Kirsten retired in 1951 (he died shortly thereafter, in 1952, at the age of 67). A faculty position was added during the late 1940's reflecting the addition of the MAE program in 1946, and then two more during the next 10 years, probably due to the growth of the graduate program. Course offerings for undergraduates increased slightly during the period, while the graduate program increased course offerings dramatically during the late 1940's.[24]

Following the departure of James Dwinnell for Boeing in 1950, the management of UWAL was taken over by Robert G. Joppa, a 1945 graduate of the department. Joppa had stayed on after graduation to work at the wind tunnel, and in 1949 was hired as a part-time instructor and part-time research associate at UWAL. He went on to teach many courses, including flight testing and airplane design, until his retirement in 1988. In 1967 the helm of UWAL was handed to William H. Rae, Jr., ('53, M.S. '59) who had started as a research instructor in 1956. He continued as head of UWAL until his untimely death in 1993. Rae was co-author with Alan Pope of the 2nd edition of the well-known book *Low Speed Wind Tunnel Testing*, and was a founding member of the Subsonic Aerodynamic Testing Association (SATA).[25]

Another faculty member of note during this time was Victor Ganzer, a 1941 graduate of the department, who first worked at NACA on the P-38 and later at Boeing, where he contributed to the aerodynamic design of the B-47. In 1947 he accepted a faculty position in the department and taught here until his retirement in 1977. It was he who initiated the department's course in engineering flight testing. In 1953 Ganzer acquired a surplus NACA single-engine Fairchild 24W for the department. This airplane was later replaced by a Beech D18, which the department continued to own and operate for the flight test course until the late 1980's. Today, the course still exists, but an airplane is leased. From 1953 to 1957 Ganzer served as department head.

Finite Elements

During the 1950's new airplane configurations, using swept and low aspect ratio wings, strained the capabilities of classical structural analytical methods. At the same time, computing power was first becoming available to engineers. In the mid-1950's two senior structural engineers at Boeing, M.J. Turner and L.J. Topp; a visiting professor from Berkeley, R.W. Clough; and Harold C. Martin, from the UW Aeronautical Department, began a collaborative effort to make use of the computer in structural analysis. Their concept was to divide a complex wing into many simple triangular or rectangular pieces and construct a global solution for deformation and loads on the wing. Their pioneering research paper "Stiffness and Deflection Analysis of Complex Structures," was published in 1956, in the *Journal of the Aeronautical Sciences.*[26]

Martin continued to work on the method, expanding it to a wide variety of problems. Students of Martin also continued to expand the concepts, and courses were offered on the subject. In the abstract to the 1956 paper, Martin and his coauthors state: "Considerable extension of the material presented in the paper is possible." That prediction has come true beyond their wildest dreams. What they developed became known as the finite element method (FEM), and is the basis for a majority of the commercial structural analysis tools available today. FEM is also found in commercial tools for heat transfer and fluid dynamics.

Fig. 11 Harold C. Martin

Explosive Growth

In late 1960, with eight faculty members, the department initiated an external search for a new head. John H. Bollard, a member of the aeronautics faculty at Purdue, known for his work on aircraft and spacecraft structures, came to the forefront. Bollard, a native of New Zealand, impressed everyone with his expertise,

Fig. 9 Robert A. Joppa   Fig. 10 Victor Ganzer

enthusiasm, and people-skills. He was offered the position and accepted, taking up his new post in August 1961. He was only 33 years old.

Bollard immediately set out to build and expand the department. Times were good: the state's economy was healthy, support for the University in the Legislature was strong, and NASA was burgeoning as a result of President Kennedy's May 25, 1961 speech before Congress, committing the nation to land a man

Fig. 12 John H. Bollard

on the Moon before the end of the decade. They were heady days, indeed, and Bollard took full advantage of the situation. During the next nine years, with Bollard at the helm, the department doubled the size of its faculty, greatly increased the annual research budget, secured new equipment for the instructional labs, revamped the curriculum to include space-oriented courses, added the word "Astronautics" to the department's name, and initiated the planning for the Aerospace Research Laboratory (ARL) to be housed in a new building bearing that name. This laboratory was to be devoted to advanced, multidisciplinary aerospace engineering research. Bollard and Ganzer approached NASA for the funds to build the new facility and, in early 1966, began the search for its director. NASA granted the University $1.5 Million for the new building.

The search for the director of ARL netted the Department's most notable hire during this period, Abraham Hertzberg. Already well-known internationally for his work in high-energy gasdynamics and re-entry physics, and his development of shock tubes and shock tunnels, Hertzberg was head of the Aerodynamics Research Department at the Cornell Aeronautical Laboratory in Buffalo, NY. His name was first put forward by Arnold Goldburg, of the AVCO Everett Research Laboratory (Goldburg soon thereafter moved to Boeing to head the Flight Sciences laboratory at the Boeing Scientific Research Laboratory). Hertzberg was unhappy with his situation at Cornell, and was looking for a new position. The chemistry between him, the department's faculty, and Dean Charles H. Norris worked, and Hertzberg started his new post in the summer of 1966.

Fig. 13 Abraham Hertzberg

Initially, ARL, as a program, was located in Guggenheim Hall, as was Hertzberg; he had a staff of one, himself, but that didn't last long. Not only did he dive into the new building project but he also brought on board several new faculty, starting with David Russell and Walter Christiansen from JPL, both of whom would later go on to serve as department chairs. Construction began in 1967 and was completed in 1969. The new building was located adjacent to Guggenheim Hall. All the laboratory spaces in ARL were windowless because so much of the research that was to occupy it for the next two decades was oriented around gas-dynamic and chemical lasers, and other optical research. Three of the four floors of the building, plus the basement were occupied by research directed by Hertzberg and his colleagues. The third floor was assigned to the University's budding Bioengineering Program.

A spacious basement laboratory was devoted to a large shock tube installation. Other lab spaces housed Ludwieg tubes, chemical lasers, gasdynamic lasers, shear layer test facilities, and a picosecond laser facility that one of the new faculty used for bioengineering-related research. ARL became a truly interdisciplinary facility, with participating faculty drawn from Aeronautics and Astronautics, Nuclear, Mechanical, and Electrical Engineering, and Physics. The research income generated by Hertzberg and his colleagues soon overtook that generated by aero/astro faculty not associated with ARL.

One of Hertzberg's interests had been the application of lasers to controlled thermonuclear fusion, i.e., the laser heating of plasmas, and he initiated a small program in this area at ARL soon after the building was completed. He and collaborators John M. Dawson of Princeton University and Ray E. Kidder of Lawrence Livermore, and their colleagues, presented a paper titled "Controlled Fusion Using Long-Wavelength Laser Heating with Magnetic Confinement," at the Esfahan Symposium on Fundamental and Applied Laser Physics in Esfahan, Iran, in late summer 1971.[27] This seminal paper established the firm foundations of the fusion program at ARL, which continues to this day.

Bust and Recovery

As the 1960's came to a close, trouble was brewing in the aerospace industry. The first lunar landings had taken place but the Nixon Administration canceled the last four Apollo missions. Military spending declined with the winding down of the Vietnam War, inflation reared its head, and the economy began to weaken. Boeing's workforce started to decline from its high of 100,800 in 1967. In 1971 the Federal Government withdrew its funding of the Boeing Supersonic Transport program, and Boeing shed many more workers, ultimately reducing its payroll by more than 60,000. Because Boeing was the Seattle area's largest employer, the impact of these layoffs on the local economy was severe. It was in April 1971 that two realtors erected a billboard near SeaTac Airport, with the words "Will the last person leaving Seattle turn out the lights." The downturn in the aerospace industry, coupled with the faltering economy, was also felt state-wide, resulting in significant cuts to the University's state funding base.

Not surprisingly, enrollment in the department plummeted. In 1972, the graduating class had only 12 members, a number not seen since the very earliest days of the department more than 40 years previously. Gradually, however, enrollment began to rise, and by 1976 the department had 30 graduates. Despite the rocky start to the '70s, the number of faculty in the department, including non-tenure-track faculty, remained stable and even grew to 19 in 1975.

John Bollard stepped down as department Chair in 1976, and was replaced by Ellis Dill, who nine months later left to accept the position of Dean of Engineering at Rutgers University, a post he held for 25 years. David Russell assumed the helm of the department in 1977 and served in that role until 1992, when he was succeeded by Walter Christiansen. In 1975, the name of ARL was changed to Aerospace and Energetics Research Program (AERP), to better reflect the nature of the research that was being conducted there, and the name of the building was changed to Aerospace and Engineering Research Building AERB). Abe Hertzberg continued as Director until his retirement in 1993. He remained active in research, despite gradually failing health, until he passed away in March 2003.

Fig. 14 The Aerospace Research Laboratory (ARL), later renamed to Aerospace and Engineering Research Building (AERB).

In 1979, Bollard, together with James Mueller of the Department of Material Science, was appointed by

NASA to an independent, 12-member advisory committee formed to investigate problems with the thermal protection tiles on the Space Shuttle.[28] On the flight of the Shuttle Columbia from California to Cape Kennedy, atop its 747 carrier aircraft, more than 5000 of the tiles had fallen off. During two years of intensive research, Mueller, Bollard, and their colleagues discovered the fundamental initial cause of the tile attachment failure and the subsequent mechanics of detachment, and developed engineering solutions that were adopted by NASA. Their success resulted in special commendations from NASA, the State Legislature, the Governor, and the University.

Holding the Course

During the decades of the '80's and '90's, the Department was of relatively constant faculty size with student enrollment influenced by the roller coaster of the aerospace industry. This stable faculty population was fortunate, at least for those students who happened to enter during a lean period. By the time of their graduation, the job market was frequently strong with few applicants competing for many openings.

At the start of 1980, the Department had 20 faculty (including two supported on research) and 200 students. The aerospace industry reached a local minimum in the early 1980's, mimicking the previous decade, and then expanded rapidly. Classrooms were bursting by 1985, with strong demand for the Department's graduates. During this time undergraduate admission became highly competitive. In 1982, despite another budget crisis, the University created a new department, Applied Mathematics, which occupied the fourth floor of Guggenheim Hall. The faculty for this new entity were drawn from Arts and Sciences and Engineering. Three faculty from Aeronautics and Astronautics, Carl Pearson, Jirair Kevorkian, and Juris Vagners, were among them. About a decade later, Vagners rejoined Aeronautics and Astronautics.

Boeing's continued use of the Kirsten wind tunnel resulted in a decision to help the University modernize it. Gearing up for the 757 and 767 programs, Boeing donated $2,000,000 to upgrade the computer systems. A computer/operations room was added to the roof, using UWAL reserve funds, and two PDP 11/70 computers were purchased to process data.

The department gained a world-class authority in computational fluid dynamics when Robert McCormick from NASA Ames joined the faculty in 1981. He remained on the faculty for three years before returning to the Bay Area to accept a position at Stanford. In 1983 Hertzberg, together with colleagues Adam Bruckner (who would become department chair in 1998) and David Bogdanoff, created and developed the concept of the ram accelerator hypervelocity launcher. The initial successes of this work led to the establishment of similar laboratories throughout the world. Other notable research accomplishments came from the department's fluid dynamics and structures groups, in the areas of turbulence and vortex dynamics (Robert Breidenthal, Mitsuru Kurosaka, et al.), in fracture mechanics and composite materials (Kuen Lin et al.), and in multidisciplinary design optimization (Eli Livne).

A major event in the early 1990's was the voluntary fission of the Nuclear Engineering Department, due to a lack of students. The Dean assigned the newly homeless faculty to other departments within the College of Engineering. Thomas Jarboe, joined the Aeronautics and Astronautics Department, and was followed later by several other plasma researchers. Their work has emphasized the development of magnetic confinement fusion with an ultimate goal of commercial electricity production. Because of the large scale of the experiments, much of their research is conducted at an industrial site off campus, in Redmond, WA, across Lake Washington. This facility, known as the Redmond Plasma Physics Laboratory (RPPL), has been headed by

Alan Hoffman since its inception in 1992. Plasma research oriented toward space propulsion grew out of these efforts later in the decade.

Also in the early 1990's, UWAL went through a financial crisis, when Boeing decided to move its low-speed wind tunnel testing elsewhere in order to achieve higher Reynolds numbers. This left a large income void at a time when the staff was geared towards multiple shift support. The entire staff was transferred to other departments or chose to retire, as the tunnel's future was in doubt. During 1993-1994 there were only a handful of tests run by students and a temporary, part-time director. A "Last Wind" party was held and it was announced that the tunnel would close its doors forever. However, the Kirsten wind tunnel gained a second wind, and things began to pick up. The doors were kept open and tunnel operations were restructured to be more automated, with a leaner staff. A decade later, UWAL is operating at over 80% customer utilization with a wide variety of customers, including Boeing.

During the second half of the decade several faculty in the department (Juris Vagners, Scott Eberhardt, Eli Livne, and Uy-Loi Ly) began a research initiative in the area of unmanned autonomous vehicles (UAV), in collaboration with Insitu, a small company located in Bingen, WA, and with ARA, an Australian firm. Funding was provided by the Office of Naval Research. To demonstrate the capabilities of small UAV's, the group attempted the first crossing of the North Atlantic Ocean by a UAV in the summer of 1998. Four of the small aircraft were launched from Newfoundland in August. One of them, "Laima," named after a mythological goddess of good fortune, landed safely in Benbecula, a Scottish Island, 26 hours and 45 minutes after it had taken off. This remarkable feat marked the first transatlantic crossing by a UAV of any size, and did much to expand the department's activities and visibility in the UAV area. Laima was put on permanent display at Seattle's Museum of Flight the following year.

On the education front, as part of an expanded outreach effort, the department in 1997 began to offer a new course, AA101, Introduction to Air and Space Vehicles, aimed at non-engineering freshmen. Using balsa wood gliders, computer flight simulators, and water-bottle rockets, instead of mathematical equations, it introduced students to the rudiments of flight and rocketry, and their history. The class became very popular across the University's entire undergraduate population.

In autumn 1998, for the first time in more than 20 years, the department established an external Visiting Committee and formulated a strategic plan. Part of this plan was to restructure the department along lines that emphasize the systems aspects of aerospace engineering, namely aeronautical systems, space systems, and energy systems. The traditional areas of aerodynamics/fluids, structures, controls, and propulsion, plus plasma science, fell under these "umbrellas." The result was a greater degree of collaboration among the faculty, both within the department and with colleagues throughout the College of Engineering and the University. In addition, the undergraduate curriculum was restructured to require more prerequisites, provide more hands-on laboratory and design experiences, encourage collaborative learning, and expose students to systems concepts.

The New Millenium

At the beginning of the 2000-2001 academic year, there were 19 faculty (including three supported on research), 92 undergraduates, and 82 graduate students. True to form, at the beginning of the decade, the aerospace industry roller coaster began heading downward again, mirroring the downturn in the economy and the aftermath of September 11, 2001. Nonetheless, student demand increased, perhaps because other engineering fields were also relatively soft.

By 1999, faculty hired in the boom years of the 1960's were rapidly beginning to retire. Over a three year period, the department hired seven replacements, including the first woman, Kristi Morgansen, who was brought in with a Clare Boothe Luce endowed professorship. On average, the department faculty has not been this young since the go-go years of the space program in the 1960's.

Over the decades, the fraction of women students has slowly increased. The first one was Rose Lunn, who graduated in 1937 at the top of her class and later went on to an illustrious career at North American Aviation.[29] It is typical of the difficulties experienced by many professional women in those days that when Lunn was first hired, she was assigned secretarial duties. It took quite some effort on her part to convince her supervisors that she was capable of a lot more than typing! Even with strong recruitment efforts, the fraction of undergraduates in the department who are women is still only about 15%, roughly the same as in other aerospace programs nationwide.

The Department Makes Good

Early on, Boeing made an investment in aeronautics at the University. Over time, that investment began to pay off. In 1926, all but one member of Boeing's engineering department were University of Washington graduates. Even into the 1940's, the majority of Boeing's engineers came from the University. It is no exaggeration to say that the Boeing Company was built by University of Washington graduates. Almost every Boeing airplane project has had a University of Washington Aeronautics alum at the helm. Any history of the Boeing Company includes names of alumni such as Maynard Pennell ('31), George Martin ('31), George Snyder ('31), Jack Steiner ('40), William Hamilton ('41, M.S. '48), Joseph Sutter ('43), and Lynn Olason ('43).[29]

Boeing is not the only place where Aeronautics alumni made an impact. Scott Crossfield ('49, M.S. '50) was the first man to successfully break Mach 2, and was heavily involved in the design and testing of the X-15. George Jeffs ('45, M.S. '48) headed Rockwell's Apollo and Shuttle programs, and George Solomon ('49) became Executive Vice President of TRW. Robert Hage ('39) is best known as the co-author of the classic text, *Airplane Performance, Stability, and Control*. Dale Myers ('43) served as NASA Associate Administrator with responsibility for the Apollo, Skylab and Shuttle programs. Moustafa Chahine ('56, M.S. '57) was Chief Scientist at Caltech's Jet Propulsion Laboratory for many years. On the less technical side, Gregory "Pappy" Boyington ('34), leader of the famed Black Sheep Squadron during WW II, also earned an engineering degree from the Department.

Since 1978, three Aeronautics alumni, Jack Steiner, George Jeffs, and Joe Sutter, have been selected by the University of Washington as *Alumnus Summa Laude Dignatus*, the University of Washington's highest honor bestowed on its alumni.[30] Very few departments at the University can boast of this many alumni of such distinction.

**Fig. 15   Some distinguished University of Washington Aeronautics alumni.
Left to right: Jack Steiner, George Jeffs, Joe Sutter, Scott Crossfield.**

Acknowledgements

The authors are deeply indebted to their colleague, David A. Russell, for his very helpful suggestions on the manuscript.

References

[1] *The Seattle Times*, March 11-13, 1910. See also
http://www.historylink.org/output.CFM?file_ID=366

[2] Serling, R.J., Legend & Legacy, St. Martin's Press, New York, 1992, Ch.1. See also
http://www.historylink.org/_output.CFM?file_ID=2042

[3] Redding, R., and Yenne, B., *Boeing: Planemaker to the World*, Thunder Bay Press, reprinted ed., 1997.

[4] *A Century of Educating Engineers*, University of Washington College of Engineering, 1998.

[5] *The Daily*, University of Washington newspaper, February 12, 1936.

[6] University of Washington Archives, W.U. President, Accession No. 71-31, Box118, Folder 2.

[7] University of Washington Course Catalogs, 1917-1921, 1927.

[8] Levinson, M., "F.K. Kirsten, Illegal Immigrant Extraordinary: The Aeronautical Years, 1920-1938," *Journal of the West*, Vol. XXX, No 1, January 1991, pp. 18-29.

[9] *The Daily*, University of Washington, ca. February 1942.

[10] Richard Gahan (former son-in-law to Gene Kirsten), private communication, 2003.

[11] Kirsten Pipe Company, Seattle, WA, http://www.kirstenpipe.com/company.htm

[12] *The Daily*, University of Washington, February 25, 1943.

[13] Hallion, R.P., *Legacy of Flight: The Guggenheim Contribution to American Aviation*, University of Washington Press, Seattle, 1977.

[14] University of Washington Facilities Records,
http://www.washington.edu/admin/facserv/records/services.html

[15] *The Daily*, University of Washington, ca. July 1929.

[16] *The Seattle Times*, April 20, 1930

[17] *The Daily*, University of Washington, October 7, 1929.

[18] *The Daily,* University of Washington, ca. February 1930.

[19] Kramer, A., Sylvester, R., Colcor, J., and Seabloom, R., *Civil Engineering: 1898-1998,* University of Washington, 1998, p. 37.

[20] "Application of the University of Washington, for a grant to provide for the construction and equipping of an Aero Dynamic Laboratory (Wind Tunnel)," proposal submitted to the Federal Emergency Administration of Public Works, August 28, 1935.

[21] University of Washington Aeronautical Laboratory run logs, 1938-39.

[22] *The Daily,* University of Washington, ca. July 1941.

[23] *The Daily,* University of Washington, ca. 1940.

[24] University of Washington Course Catalog, 1935, 1940, 1945, 1950, 1955, 1960.

[25] Subsonic Aerodynamic Testing Association, http://www.sata.aero/

[26] Turner, M.J., Clough, R.W., Martin, H.C., and Topp, L.J., "Stiffness and Deflection Analysis of Complex Structures, " *Journal of the Aeronautical Sciences,* Vol. 23, No. 9, 1956, pp. 805-854.

[27] Dawson, J.M., Kruer, W.L., Hertzberg, A., Vlases, G.C., Ahlstrom, H.G., Steinhauer, L.C., Kidder, R.E., "Controlled Fusion Using Long-Wavelength Laser Heating with Magnetic Confinement," in *Fundamental and Applied Laser Physics,* Feld, M.S., Javan, A., and Kurnit, N.A., eds., Proceedings of the Esfahan Symposium, Esfahan, Iran, Aug. 29 – Sept. 5, 1971, Wiley, New York, 1973, pp.119-140.

[28] Illman, D.L., *Pathbreakers: A Century of Excellence in Science & Technology at the University of Washington,* Office of Research, University of Washington, 1996, pp. 197-200.

[29] Distinguished Alumni, Department of Aeronautics and Astronautics, University of Washington, http://www.aa.washington.edu/people/alumni/award/index.shtml.

[30] UW Alumni Association, Alumni Awards, http://www.cac.washington.edu/alumni/awards/asld/.

Chapter 14

A Historical Review of the Aerospace Engineering Curriculum at Oklahoma State University

E. A. Falk[1-] and A. S. Arena[2†]

Abstract

Oklahoma State University has enjoyed a long and storied affiliation with the aeronautical and space related engineering professions for almost 75 years. Historically, this affiliation has been bolstered by a strong industrial base in the State of Oklahoma related to aviation and the aerospace sciences, as well as through the activities of many nationally recognized native Oklahomans in aviation and space exploration. Even so, the development and success of the Aerospace Engineering program at the Oklahoma State cannot be strictly attributed to locale, but rather must be credited to the highest quality instruction, research, and extension. This paper focuses specifically on the history and development of the Aerospace Engineering curriculum at Oklahoma State. Topics discussed herein describe the birth, growth, and maturity of the Aerospace Engineering program, including its unique history, its close association with Mechanical Engineering, and developing national prominence.

Emergence of Engineering

The Agricultural and Mechanical College of the Territory of Oklahoma was established as a land-grant educational institution by the first territorial legislature of Oklahoma on December 24, 1890; merely eight months after the May 2 inauguration of the Territory that same year. Payne County was stipulated as the location for the new college, with Stillwater, the county seat, elected as the final construction site in the summer of 1891. Classes at the college commenced on December 14, 1891, with forty-five students (22 male, 23 female) and five faculty members meeting at various churches around Stillwater. At that time, each student enrolled in an identical curriculum, leading to a four-year Bachelor of Science (BS) degree. Required courses consisted of algebra, geometry, trigonometry, chemistry, physics, drawing, surveying, civil engineering, and mechanics; forming the early bases of a burgeoning engineering curriculum.

The Board of Regents made the first clear distinction between agricultural and engineering curricula at Oklahoma A & M College (as it was referred to then) when it ordered a mechanic arts department to be initiated during the 1898-1899 academic year. Richard E. Chandler was appointed the first full-time faculty member in the engineering division, and for one year he single-handedly taught all courses related to engineering and physics. Engineering courses under his direction included Blacksmith Shop, Machine Shop, Pipe Fitting, Mechanical Drawing,

[1-] Assistant Professor; School of Mechanical and Aerospace Engineering; 218 Engineering North

[2†] Maciula Professor in Engineering, School of Mechanical and Aerospace Engineering, 218 Engineering North

Statics, Hydrostatics, Moments of Inertia, Hydraulics, Kinematics, Theory of Structures, Steam Engines and Boilers, Thermodynamics, Machine Design, and Electrical Engineering. The title of Professor of Mechanical and Mining Engineering and Physics was formally given to Mr. Chandler in 1902. By this time, however, as many as three other engineers were assisting him in his teaching efforts; each engineer having a specific area of expertise, such as electrical engineering or civil engineering. Professor Chandler became Dean of the Engineering Division in 1910. He also founded the Oklahoma Society of Engineers, a precursor to the present day Oklahoma Society of Professional Engineers.

After 1898, engineering courses at the college were conducted in a separate "mechanical hall", apart from the agricultural and science departments, housing a 100 horsepower boiler with an attached 80 ft smokestack (thought to be the tallest structure in the Oklahoma Territory at the time). Engineering faculty and students played an important role in the construction of the mechanical hall, as well as many other physical improvements to the campus, including boiler, sewer, water, electricity, and pavement installation. Student laborers gained more than experience for their efforts toward campus improvement; often using the meager wage they earned to offset the $8.00/month room and board. Room and board expenses were a common source of student attrition from the college during its early years.

Despite having courses separately listed for Mechanical Engineering related topics, all engineering instruction prior to 1906 was offered as a single course of study. In 1906, however, engineering instruction above the junior level was separated into independent curricula for Civil Engineering and Mechanical and Electrical Engineering. Mechanical Engineering and Electrical Engineering were further separated into individual departments in 1909, with Electrical Engineering tasked to continue all physics-related instruction. A Master of Science degree was first offered for engineering in 1911. By 1914, all engineering disciplines had adopted the semester system and shared a common two-year general engineering program. After completing their first two years of study at the college, engineering students branched into a second two-year course plan, or Professional School (as it is now termed), in one of the stipulated engineering disciplines. The semester and Professional School concepts remain integral to today's engineering instruction at the University (despite a brief reversion to the quarter system after World War I). An engineering student in 1914 was required to complete 144 semester credit hours, not including military science or physical training, in order to graduate. Despite these rigorous requirements, nearly one-half of all graduates of the college studied engineering over the time period spanning from 1908-1911, with 211 BS degrees granted. By 1923, however, engineering enrollment had dropped relative to other programs in the college, accounting for only 15% of the total degrees granted over a two-year span.

Despite the early success of the engineering division at Oklahoma A & M College, few engineering graduates remained to work in Oklahoma. In fact, of the 29 men who graduated with engineering degrees prior to 1902, not one was employed in Oklahoma. The 1908 engineering class was unusual in that six of the nine graduates remained to work in the Territory. Of those graduates that remained in Oklahoma, most entered the public education system, or later on, worked in the expanding petroleum industry.

In the fall of 1919 engineering students from Oklahoma A & M College met with students from several other western academic institutions to form a new national engineering organization known as the "Guard of St. Patrick". Although never explicitly stated, the affiliation between the engineering profession and St. Patrick was initially attributed to the celebration and exhibition of engineering activities during the week of March 17. However, a second explanation suggests that the activities of Patrick, the patron saint of Ireland, who reportedly drove all the snakes out of Ireland, formed the first known "worm drive" and thereby established a connection with engineering. Regardless of the affiliation, after 1919 the college newspaper at Oklahoma A & M frequently referred to engineering students as "The Irish", a moniker that remained for many decades. In fact, to this day the top five graduating seniors in engineering at the University are selected to receive the "St. Patrick's Award" from the College of Engineering, Architecture, and Technology.

Mechanical Engineering Expands

While elements of mechanical engineering were firmly ensconced at Oklahoma A & M College as early at 1898, it was not until the appointment of Robert Lee Rhoades as the first official Head of the Department Mechanical Engineering in 1923 that the discipline began to fully settle in at the college. Ellis C. Baker replaced Rhoades in 1928, and subsequently built a solid group of Mechanical Engineering faculty. In particular, V. W. Young from Purdue University, R. E. Venn from Kansas State College, and V. L. Maleev from the Imperial Technical College were each hired by Baker prior to 1930. Baker also oversaw the establishment of several degree options associated with the Mechanical Engineering program. For example, along with several other engineering disciplines that began to offer options of specialization in their respective fields in 1929, Mechanical Engineering formally offered degree options in Aeronautical Engineering, Petroleum Engineering, and Refrigeration Engineering. The Aeronautical Engineering and Petroleum Engineering degree options were both two-year programs, beginning in the junior year. The first four-year curriculum for Mechanical Engineering with an Aeronautical Engineering degree option, as listed in the Course Bulletin of 1928-1929, is provided in Table 1.

The Aeronautical Engineering program grew rapidly during its early years. In fact, the program gained wide popularity due in part to a growing national and regional interests in aviation, spurred on by the celebrity of aviators such a Charles Lindberg and Oklahoman Wiley Post. Lindberg even visited Oklahoma in 1928, just one year after his historic transatlantic flight. He personally selected Waynoka as the site for the first transcontinental airport in Oklahoma. Similarly, Post's famous solo flight around the world occurred in 1933, just four years after the Aeronautical Engineering degree option became available at Oklahoma A & M. Other aviation developments in Oklahoma, such as the growth of the Spartan School of Aeronautics in Tulsa after 1928, and the selection of Oklahoma City in 1941 as a major aircraft and supply depot for the U. S. military (known today as Tinker AFB), continued to spur student interest in aeronautics and aviation through the 1940's.

The early Aeronautical Engineering degree option in Mechanical Engineering was nurtured through the efforts of Oklahoma A & M faculty members such as Professor V. W. Young. Professor Young initiated a local student chapter of the Institute of Aeronautical Sciences in 1943. He also greatly assisted in the development of a flight school at the college, beginning a civilian flight-training program under the guise of the Civil Aeronautics Authority in the fall of 1939. Many graduates from this flight program eventually became military pilots during World War II. The Civil Aeronautics Authority reported in 1942 that more pilots from Oklahoma A & M College had entered into military service than from any other U. S. school, and that not a single one of these pilots had "washed out" of the Army Air Corps pilot training program prior to that time. After World War II, a degree curriculum followed naturally from the Civil Aeronautics Administration flight-training program. The School of Flight was official formed in 1946, headed by Mechanical Engineering faculty member G. L. Rucker, but offering no separate degree from the Aeronautical Engineering degree option for Mechanical Engineering.

The advent of World War II significantly impacted both the college as a whole and the Department of Mechanical Engineering. For example, during World War II the Department of Mechanical Engineering provided a compressed degree program through which a four-year engineering education could be earned in six 12-week terms. The U. S. entrance into the war also significantly impacted the makeup of the Mechanical Engineering faculty. Prior to 1940, the faculty had consisted of a number of seasoned professors, as indicated previously; however, due to vacancies created by conscription at colleges throughout the States, twelve of the existing Mechanical Engineering faculty members departed from 1940-1945. Replacements were hired both during and after the war, including W. H. Easton, J. H. Boggs, R. E. Chapel, L. J. Fila, E. C. Fitch, M. A. Nobles, J. D. Parker, and G. H. Fila. Professor G. H. Fila was the first female Mechanical Engineering faculty member, hired in 1940; her tenure at the college lasted only one year. The position of Head of the Mechanical Engineering department also changed shortly after the war, as R. E. Venn succeeded E. C. Baker after his unexpected death in 1949.

Not surprisingly, many of the new faculty members hired during World War II were specifically engaged in the aeronautical engineering field, such as Professors L. J. Fila and R. E. Chapel, both of who were quite successful in soliciting research grants related to the aircraft industry. Professor Fitch also earned the first industrial research contract related to fluid power engineering in 1956. The subsequent Fluid Power Laboratory (FPL) would become enormously successful, productive, and nationally renowned. In fact, research productivity at the laboratory would eventually require a separate building and administrative staff apart from the Department of Mechanical Engineering. From the FPL, Prof. Fitch formally organized the Fluid Power Research Center (FPRC) in 1970, becoming the oldest research center in the college. The FPRC is responsible for establishing over one hundred industrial standards, patenting mechanical devices, and attracting external funding from over seventy-five industrial organizations in three continents. Professor C. M. Leonard, who was elected national President of Pi Tau Sigma, the Mechanical Engineering honor society, in 1956, also typified the success of the department during these years. Through his association, the student chapter of Pi Tau Sigma at the college hosted the National Convention of the society in Stillwater in 1957.

By the late 1950's, the exploration of space was becoming a reality as evidenced by successful orbital rocket programs, and the establishment of National Aeronautics and Space Administration (NASA). Oklahomans played a significant role in the growth of space exploration and development of space technology. Col. L Gordon Cooper, a native of Shawnee, OK, was one of the first men selected for the U.S. astronaut program, orbiting the Earth in both Mercury 9 and Gemini V flights. Similarly, Lt Gen. Thomas P. Strafford, a native of Weatherford, OK, became the pilot of Gemini VI, the commander of Gemini IX, and the commander of Apollo 10. In total, Oklahoma claims eight astronauts, including several participants in the Space Shuttle program. Two of Oklahoma's astronautics, Col. William R. Pogue and Col. Stuart A. Roosa, attended OSU, although neither graduated with degrees in Mechanical Engineering.

The Aeronautical Engineering degree option was fully accredited as a standalone curriculum, separate from the Mechanical Engineering degree program, in 1960. To reflect the broadening of the aeronautics field and the increasing importance of space-related engineering in Oklahoma, the name of the Aeronautical Engineering degree option at the college was officially changed to Aerospace Engineering in 1965. A sample of the curriculum for the Aerospace Engineering degree option in Mechanical Engineering at the time of the name change is included in Table 2.

The School of Mechanical Engineering was officially renamed the School of Mechanical and Aerospace Engineering (MAE) in 1968, in order to provide greater visibility to the Aerospace Engineering degree option. The nomenclature change followed suit with college, which was officially titled Oklahoma State University of Agriculture and Applied Science on May 15, 1957; this name was subsequently shortened to Oklahoma State University (OSU) in the years immediately following. Along with the name change, the position of MAE department Head was given to J. H. Boggs in 1958. Boggs served as Head until 1965, when E. L. Harrisberger replaced him.

During the 1960's and 1970's fields of specialization among the faculty of the MAE department successfully expanded into several new areas, particularly due to the appointment of several new faculty members such as G. W. Zumwalt, R. L. Lowery, D. K. McLaughlin, and P. M. Moretti. These energetic faculty members joined R. E. Chapel and L. J. Fila in augmenting departmental expertise in the areas of sonics, gas dynamics, vibrations, and instrumentation. Bolstered by the influx of faculty members, the School of Mechanical Engineering experienced a tremendous growth of sponsored research by 1961, such that nearly one-half of the research grants for the entire College of Engineering stemmed from the Mechanical Engineering department at that time.

During the 1960's a gradual shift occurred throughout the university in faculty expertise related to the area of fluid mechanics, moving from its traditional location in the School of Civil Engineering to the School of Mechanical and Aerospace Engineering. This shift was in large part due to the success of Professor Fitch's work in fluid power, but also bolstered by the hiring of J. E. Bose, K. N. Reid, W. G. Tiedermann, W. B. Brooks, and D. E. Boyd, each of who added unique expertise in various elements of fluid mechanics to the MAE faculty. Professor Zumwalt particularly championed teaching and research effort related to the Aerospace Engineering degree option, having designed and installed new wind tunnels in the mechanical engineering laboratory prior to his departure in 1968. A photo of the installed Zumwalt tunnel is shown in Fig. 1. He was also the first American Institute of Aeronautics and Astronautics (AIAA) faculty advisor for the OSU student chapter of the society, after a merger of the two antecedent societies, ARS and IAS. Professor Zumwalt was designated by the AIAA the as one of the nation's most outstanding advisors of a student chapter in 1965. Under his direction, the OSU student chapter of the AIAA received the Bendix Award, the highest honor that may be bestowed upon a student chapter.

Karl N. Reid succeeded E. Lee Harrisberger as the MAE department Head in 1972, and continued in that capacity until he was appointed to his present position of Dean of the College of Engineering, Architecture, and Technology (CEAT) in 1986. Lawrence L. Hoberock replaced Dean Reid as MAE department Head in 1986, and continues in that capacity to the present day. Prof. R. L. Swaim also joined the faculty in a full-time capacity in 1987, having previously served as Associate Dean for CEAT since 1978. Prof. Swaim played integral part in the development of several programs in both CEAT and MAE during the 1980s and early 1990s. For example, the cooperative education program in CEAT was nearly non-existent until Prof. Swaim's arrival from Purdue University in 1978. Prof. Swaim pushed for and successfully revived the cooperative education program, initiating what grew into a very successful program that until the mid 1980s boasted over 60 student participants per year from engineering and engineering technology. On December 12, 1989 Prof. Swaim also successfully initiated the Oklahoma State chapter of the National Aerospace Engineering Honor Society, Sigma Gamma Tau. The chapter inducted its first nine members on April 10, 1990. Since that time three people from the chapter have been nominated for the society's Honor Undergraduate Student Award, one of who won the award for the Southwest Region and competed nationally for the Ammon S. Andes National Award for Aerospace Engineering. The chapter had been recently inactive; it was reactivated as of 2001. Since 1990 the chapter has inducted well over 70 new members. During the 80's and 90's, MAE experienced a growth in faculty. A list of MAE faculty as well as department heads from the 1890's to today may be seen in Table 4.

Aerospace Engineering in the New Century

As described above, Oklahoma State University has had a thriving program in aeronautical and aerospace engineering for almost 75 years. Although accredited separately, the "option" terminology was used to reflect the commonality, and close working relationship, with the Mechanical Engineering curriculum. On June 30, 2000 the Oklahoma State University Board of Regents approved a degree designation from Bachelor of Science in Mechanical Engineering with an Aerospace Engineering option, to Bachelor of Science in Aerospace Engineering (BSAE). A degree designation change was necessary in order to eliminate confusion since the designation of "option" had become almost universally recognized as a non-accredited emphasis area within a discipline of engineering. At the time of the change, OSU was the only remaining university in the United States that used the title "Aerospace Engineering Option" instead of "Aerospace Engineering" for a fully accredited aerospace engineering program. The curriculum itself did not change with the new designation. The first BSAE degree was awarded to Aaron McClung in August of 2000. The current BSAE degree plan is provided in Table 3.

As illustrated in Table 3, the BSAE curriculum exposes the student to a broad range of fundamental science and engineering topics including: physics, chemistry, statics and dynamics, thermodynamics, fluid mechanics, heat transfer, materials, computational methods, and electrical and electronic circuits. In the junior and senior years, students are introduced to more specific topics including aerodynamics, propulsion systems, aircraft and spacecraft stability and control, aerospace structures, and aerospace vehicle design. Electives allow students to broaden, or specialize as desired.

The Aerospace Engineering curriculum has always demonstrated a commitment to hands-on and practical engineering and design experience. Early examples include multiple shop and design courses, as well as experimental labs. An example from the 1928-1929 curriculum, given in Table 1, is the Aero 423 course that provided training in design, construction, and testing of airplane engines from a theoretical and practical standpoint. Also, Aero 433 was a course in commercial application of aircraft, including the study of passengers and freight lines, local operating companies, and special uses of flying equipment with consideration of unit costs, and revenue, equipment regulation, etc. In recent history, a flight-test component was added to the required Aerospace Engineering curriculum. Twelve students enrolled in the first offering of the course in 1980, with Prof. F. Eckhart serving as both pilot and instructor. Typical flight test experiments during Prof. Eckhart's tenure were conducted in aircraft such as a Piper Twin Cherokee, the Cessna 172, and the Beechcraft Dutchess. A photograph of Prof. Eckhart, and the original 12 students in that course is shown in Fig. 2. Prof. Eckhart retired in 1999 at which time Prof. P. H. O'Donnell was hired as pilot and instructor for the course. Students now fly a series of 3-4 test flights making measurements in university aircraft. Figure 3 is a photograph of Prof. O'Donnell with his class of 2003 in front of one of the flight-test aircraft. Since 1997, all senior BSAE students have been required to participate in the AIAA Design-Build-Fly competition annually sponsored by the AIAA, Cessna, and the Office of Naval Research. The seniors form two teams that each design and build aircraft for the international contest. Oklahoma State students have enjoyed a marked level of recent success in the competition, including a 2nd place finish in 1999, 2nd and 5th place finishes in 2000, and 1st and 3rd place finishes in 2001. Figure 4 is a photo of both OSU teams placing in the 2001 competition, with aircraft and trophies.

It is also important to note that, due to its historically close association with the Mechanical Engineering program, OSU Aerospace Engineering students following the suggested curriculum need only complete two additional MAE courses to fulfill both BSAE and BSME degrees requirements for graduation. Some students also take advantage of the Department of Education Aviation and Space program and receive minors in Professional Pilot or Aviation Management.

Historical enrollment and graduation data for the OSU MAE department is illustrated in Figs. 5 and 6. Figure 5 shows the total undergraduate enrollment for Mechanical Engineering (BSME), Mechanical Engineering with an Aerospace Engineering Option (BSMEAERS), and Aerospace Engineering (BSAE) from 1985-2002. Despite a drop in enrollment in the mid 1990s, enrollment for the BSAE degree (previously BSMEAERS) at OSU has remained nearly constant at 180 students for almost two decades. Figure 6 shows the undergraduate graduation numbers for Mechanical Engineering (BSME), Mechanical Engineering with an Aerospace Engineering Option (BSMEAERS), and Aerospace Engineering (BSAE) from 1976-2002. BSAE graduates have numbered below 20 per year during the reported time period; however, 2002 provided a record number of BSAE student graduates. Combined, the enrollment and graduation data indicate the recent growth, increased retention rate, and popularity of the Aerospace Engineering program.

The popularity of the BSAE degree at OSU may be attributed to several factors, one of which is the significance of the aerospace industry in the Oklahoma economy.

Graduates of the BSAE program at OSU are primarily employed in Oklahoma and its neighboring states. Figure 7 illustrates the approximate geographic distribution of reporting MAE students graduating with Bachelor of Science degrees over the period of 1999-2002. About 70% of MAE graduates are, at least initially, employed in either Oklahoma or Texas; however, graduates have reported employment across 17 U. S. states, as well as several international countries. For those MAE students employed in Oklahoma and Texas, a large percentage are employed in the energy industry, including companies such as Conoco, Phillips Petroleum, Halliburton Energy Systems, and Exxon-Mobile. Aerospace Engineering graduates at OSU are typically employed by local Oklahoma and Texas companies such as The Nordam Group, Lockheed-Martin, NASA Johnson Space Flight Center, American Airlines, Raytheon, Aeromet, and the U. S. Air Force, located at Tinker AFB. Several Kansas-based companies also employ a large number of OSU BSAE graduates, including Cessna, Raytheon Aircraft, and The Boeing Company. The mean starting salary for MAE graduates reporting their employment plans upon graduation over the period of 1999-2002 was approximately $46,800, with a mean sign-on bonus of $2,800 for those receiving bonuses. Geographic distribution, mean salary, and sign-on bonus data specifically related to Aerospace Engineering graduates were not collected from 1999-2002; however, trends can be assumed to be equivalent to that of the nominal MAE graduate.

Aerospace Engineering Related Facilities at OSU

The MAE department at OSU has an extensive number of laboratories to support the experimental and hands-on design emphases of the undergraduate curriculum. A segment of the laboratories used by MAE undergraduates in Aerospace Engineering are detailed below.

1. Subsonic/sonic Wind Tunnel Laboratory

   Low turbulence, open return wind tunnel has a 15:1 contraction followed by an interchangeable test section with a 3 ft x 3 ft cross section. A 125 hp motor with a variable speed drive powers the centrifugal blower that drives the flow. The tunnel has an operating velocity range of approximately 0 – 185 ft/sec.

2. Zumwalt Wind Tunnel

   Open return wind tunnel has a 5.7:1 contraction followed by a test section with a 0.41 m x 0.61 m cross section. A 29.8 kW motor powers the centrifugal blower that drives the flow. The tunnel has an operating velocity range extending from approximately 11 – 49 m/s.

3. Aerospace Composites Laboratory

   Complete composites lay-up and curing equipment and facilities including clean room, and 8 ft x 4 ft x 4 ft, and an 8 ft x 8 ft x 6 ft curing ovens.

4. Aerospace Assembly Laboratory

   Large design and fabrication space to facilitate hands-on projects. Facilities and tools for construction of remotely piloted vehicles.

5.   <u>L. Andrew Maciula Aerospace Design Studios</u>
     Design studios to facilitate group communication and cooperation in the design of aircraft. Studios include computers with design and analysis software, white boards, and reconfigurable furniture.

6.   <u>Aerospace Engineering Laboratory</u>
     Used to conduct a wide range of aerospace related experiments including wind tunnel studies, rocket and jet propulsion. Approximately 40% of experiments are performed in the air in university aircraft

7.   <u>Aerospace Propulsion Laboratory</u>
     Axial-flow low-speed compression stage with 1 m diameter. Compressor consists of 1.5 stages (inlet-guide vane, rotor, and stator stages) driven by variable speed 50-hp electric motor with a maximum speed of 1800 RPM. A slip-ring assembly allows for rotating-frame measurements forward, between, and aft of the rotor blades. An auxiliary fan and terminating throttle provide both flow augmentation and restriction to operating at a variety of stage flow coefficients, including stall.

8.   <u>Aeroservoelasticity Laboratory</u>
     High-end computer workstations and software for computational studies of aerodynamics and vehicle flow/structure interactions.

9.   <u>Materials Science Laboratory</u>
     Wide variety of equipment for studying properties of materials, including computer-based metallograph, grinding, polishing, tensile testing, heat-treating ovens, microscopes.

References

10. Parcher, J. V., 1988, *A History of the Oklahoma State University College of Engineering, Architecture, and Technology*, Centennial Histories Series, Oklahoma State University, Stillwater, OK.
11. Oklahoma Aeronautics and Space Commission
12. Oklahoma Department of Commerce
13. Governor's office of Oklahoma
14. State of Oklahoma
15. Aerospace Department Chair Association
16. Aerospace Industries Association of America.
17. OSU Special Collections and Archives

Table 1. First Aeronautical Option in Mechanical Engineering curriculum, 1928-1929

Freshman Year		
Fall Semester	**Credits**	**Spring Semester**
Algebra	4.0	Engineering Drawing
English Composition	3.0	Analytical Geometry
Trigonometry	2.0	English
General Chem.	4.0	General Chem.
Descriptive Geometry	4.0	Surveying
Military Science	1.0	Military Science
Eng. Orientation	0.0	Eng. Orientation
Total	**18**	**Total**
Sophomore Year		
Fall Semester	**Credits**	**Spring Semester**
Calculus	5	Calculus
Physics	5	Physics
Pattern Making (Shop)	1	Statics
Foundry (Shop)	1	Machine Shop (Shop)
Kinematics	5	Machine Design
Military Science	1	Chemistry
		Military Science
Total	**18**	**Total**

Junior Year		
Fall Semester	**Credits**	**Spring Semester**
Power Plants	5	Power Plants
Machine Design	3	Machine Design
Strength of Materials	5	Kinetics
Forge (Shop)	1	Hydraulics
Adv. Eng. Shop (Shop)	1	Adv. Eng. Shop (Shop)
Elective	3	Manufacturing Processes
		Elective
Total	**18**	**Total**
Senior Year		
Fall Semester	**Credits**	**Spring Semester**
Electrical Engineering	5	Electrical Engineering
Thermodynamics	2	Thermodynamics
Economics	3	Engineering Economics
Speech	2	Spec. & Contr.
Plant Structures	3	Tech. Elective
Tech. Elective	3	Elective
		Inspection Trip
Total	**18**	**Total**

Additional Aeronautical Option Course Requirements:
Math 343-Theoretical Aeronautics: Aero 413-Aeronautics – General theory of the design of airplanesl: Aero 423
-Airplane Engine Design:Lecture/Lab-Design and construction of modern aeronautical engines: Aero 433
-Commercial Aeronautics- Commercial application of aircraft: Aero 473-Airplanes- Elementary course in air action
as applied to airplanes: Aero 483-Airplanes-An elementary course in airplane types, structures and motors.
Total units: 152

Table 2. Aerospace Option in Mechanical Engineering curriculum, 1965-1966

Freshman Year		
Course	**Credits**	**Course**
Analytics (Math)	3	Military or Air Science
Engineering Drawing	2	Beginning Analysis I (Math)
General Chemistry	4	Design Layout and Graphics
Orientation	1	General Physics
English Composition	5	Challenges in American Dem. Life
		Total Freshman Year
sophomore Year		
Course	**Credits**	**Course**
Beginning Analysis II (Math)	5	Differential Equations
General Physics	4	Fundamentals of Elec. Engineering I
Mechanics-Statics and Strength	5	Mechanics-Dynamics
Industrial Processes	2	Intro to Engineering Design
Military or Air Science	4	Challenges in American Dem. Life
		Total Sophomore Year
Junior Year		
Course	**Credits**	**Course**
Thermodynamics I	3	Thermodynamics II
Fundamentals of Elec. Engineering II	3	Heat Transfer and Fluid Mechanics

Course	Credits	Course
Dynamic Analysis	3	Design Stress Analysis
Introduction to Metal Behavior	3	Measurements and Instrumentation
Mechanics-Fluids	3	Seminar
Introductory Modern Physics	3	Aerospace Structures I
		Total Junior Year
Senior Year		
Course	**Credits**	**Course**
Humanities in Western Culture	4	Experimental Aerodynamics
Mechanical Engineering Lab	1	Aerodynamics II
Mechanical Metallurgy	3	Power Systems I
Aerospace Structures II	3	Tech Elective
Aerodynamics I	3	Tech Elective
Humanities in Western Culture	4	Eng. Economy, or Report Writing
		Total

Total Units 142

Table 3. Aerospace Engineering Curriculum, 2002-2003

Freshman Year		
Fall Semester	**Credits**	**Spring Semester**
Introduction to Engineering	3.0	Fortran
General Chemistry	3.0	Engineering Design
American History	3.0	General Physics
Calculus I	3.0	Calculus II
Freshman Composition I	3.0	Social Science Elective
		American Government
Total	**15.0**	**Total**
sophomore Year		
Fall Semester	**Credits**	**Spring Semester**
Statics	3.0	Material Sciences
Thermodynamics I	3.0	Dynamics
General Physics	3.0	Strength of Materials
Calculus II	3.0	Fluid Mechanics
Differential Equations	3.0	Electrical Science
		Science Elective
Total	**15.0**	**Total**
Junior Year		
Fall Semester	**Credits**	**Spring Semester**
Dynamics Systems	3.0	Mechanical Design I
Engineering Economic Analysis	3.0	Measurements and Instrumentation
Comp. Methods in Analysis & Design	3.0	Engineering Design

	Credits	
Compressible Fluid Flow	3.0	Mech./Aero. Engineering Elective
Engineering Statistics	3.0	Applied Aero. and Performance
Total	**15.0**	**Total**
Senior Year		
Fall Semester	**Credits**	**Spring Semester**
Social Science Elective	3.0	Social Science Elective
Aerospace Structures I	3.0	Mech./Aero. Engineering Elective
Gas Power Systems	3.0	Aerospace Vehicle Design
Aerospace Vehicle Stability and Control	3.0	Aerospace Engineering Laboratory
Advanced Social Science Elective	3.0	Technical Writing
Total	**15.0**	**Total**

Total units: 127

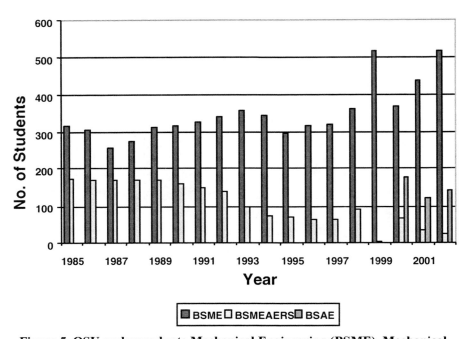

Figure 5. OSU undergraduate Mechanical Engineering (BSME), Mechanical Engineering with Aerospace Engineering Option (BSMEAERS) and Aerospace Engineering (BSAE) enrollment, 1985 – 2002.

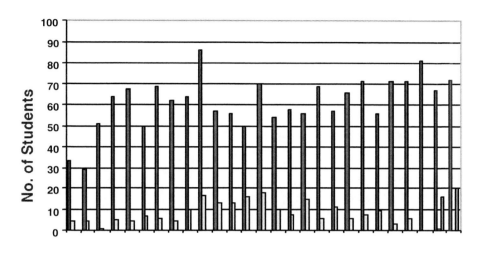

Figure 6. OSU undergraduate Mechanical Engineering (BSME), Mechanical Engineering with Aerospace Engineering Option (BSMEAERS) and Aerospace Engineering (BSAE) graduates, 1976 – 2002.

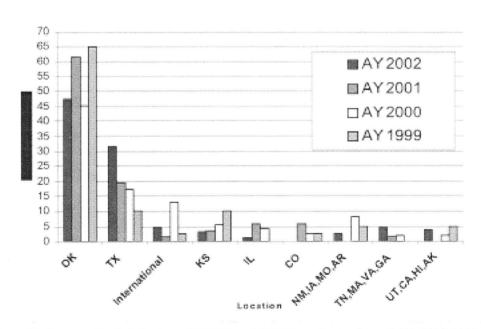

Figure 7. Geographic distribution of BSME and BSAE graduates from OSU, 1999-2002.

Appendix

Table 4. Heads and faculty members of the School of Mechanical and Aerospace
Engineering, 1898 – present

HEADS		
Name	**Appointed**	**Resigned**
Robert L. Rhoades	1923	1928
Ellis C. Baker	1928	1949
Rollo E. Venn	1949	1957
James H. Boggs	1958	1965
FACULTY		
Name	**Appointed**	**Resigned**
Henry J. Absher	1957	1958
Benjamin M. Aldrich	1941	1956
James J. Allen	1981	1984
Andrew S. Arena, Jr.	1993	Present
Glen L. Aupperle	1948	1951
Ellis C. Baker	1921	1949
Joseph H. Bell	1947	1948
Carl A. Bessey	1899	1900
Bruce A. Blackman	1947	1948
Thomas E. Blejwas	1978	1980
James H. Boggs	1943	1965
Samuel J. Boller	1923	1928
James E. Bose	1962	1974
Robert E. Bose	1965	1966
Donald E. Boyd	1969	1978
Frank R. Bradley	1910	1916
Merle C. Brady	1937	1940
Stanley O. Brauser	1948	1967

Martin Bretz	1941	1943
Edgar E. Brewer	1910	1916
William B. Brooks	1967	1973
Robert Brun	1967	1970
Wayne Buerer	1936	1949
Olden L. Burchett	1957	1960
ONeill J. Burchett	1959	1967
Brigitte Busch	1995	1996
Frank W. Chambers	1989	Present
Richard E. Chandler	1898	1914
Young Bae Chang	1991	2000
Raymond E. Chapel	1947	1983
Wilson R. Cherry	1941	1944
John C. Cluff	1930	1933
Alva G. Comer	1954	1971
Howard E. Conlon	1982	1998
Nicolae Constantin	1996	1997
Thomas A. Cook	1983	1988
Everett L. Cook	1967	1968
Neal A. Cook	1946	1950
David Cornell	1959	1961
Glen L. Corrigan	1946	1948
Jimmy E. Cox	1960	1962
Bob C. Crittendon	1950	1952
Bert S. Davenport	1948	1961
Ron D. Delahoussaye	1986	Present
Dean DeMoss	1965	1966
Ralph R. Denham	1943	1944
Ronald L. Dougherty	1985	1999
Earl L. Dowty	1967	1969
Peter Dransfield	1965	1967

Delcie R. Durham	1985	1988
William H. Easton	1942	1969
Lynn R. Ebbeson	1974	1982
Richard R. Ellis	1947	1951
Paul H. Evans	1943	1944
John P. Everett	1949	1951
Eric A. Falk	2001	Present
Bruce A. Feiertag	1984	Present
James H. Felgar	1905	1906
Gary B. Ferrell	1980	1984
Gertrude (Hill) Fila	1953	1954
Ladislaus J. Fila	1947	1978
Daniel E. Fisher	1999	Present
Ernest C. Fitch, Jr.	1953	1988
George E. French	1942	1943
Craig R. Friedrich	1982	1987
Charles R. Gerlach	1965	1966
Afshin J. Ghajar	1981	Present
Richard M. Gilmore	1953	1955
James K. Good	1980	Present
Larry D. Goss	1966	1971
Robert R. Graham	1937	1938
M. Clovis Green	1941	1941
Afif S. Halal	1989	1991
Gerald A. Hale	1929	1935
Syed Hamid	1976	1978
Vincent S. Haneman, Jr.	1966	1973
George J. Hanggi	1947	1950
Steven M. Harris	1985	1988
E. Lee Harrisberger	1963	1971
Howard M. Hawks	1948	1949

Donald R. Hawworth	1959	1966
John A. Herrington	1925	1926
Lawrence L. Hoberock	1987	Present
I. T. Hong	1988	Present
Robert R. Irwin	1940	1967
Everett C. Isbell, Jr.	1947	1950
Charles Jablow	1912	1920
J. L. Jones	1910	1912
Byron W. Jones	1975	1976
Milan K. Jovanovic	1958	1964
Bilgin Kaftanoglu	1981	1985
Ranga Komanduri	1989	Present
Harry L. Kent	1931	1939
Walter B. King	1941	1953
Sandford P. Kroeker	1935	1936
Edward J. Kunze	1914	1917

Name
E. Lee Harrisberger
Karl N. Reid, Jr.
Lawrence L. Hoberock

Name
Carroll M. Leonard
John M. Levosky
Steven Y. Liang
David G. Lilley
Edwin C. Lindly
Robert E. Little

Name		
Richard L. Lowery		
Hongbing Lu		
Don A. Lucca		
Loddie A. Maciula		
Vladimir L. Maleev		
Michael M. Mamoun		
Mack Martin		
Charles G. Martinson		
Clarence Murray, Jr.		
Amer Nasserharand		
William R. Mathews		
Georg F. Mauer		
Dennis K. McLaughlin		
Faye C. McQuiston		
Merlin L. Millett		
Timothy M. Minahen	1992	1993
Eduardo A. Misawa	1990	Present
Peter M. Moretti	1970	Present
B. N. Murali	1974	1975
J. Murali	1980	1981
Melvin A. Nobles	1954	1957
Brian ODell	1996	1996
Ronald R. Osborn	1965	1967
Prabhakar R. Pagilla	1996	Present
Ronald I. Panton	1966	1971
Jerald D. Parker	1955	1988
James R. Partin	1966	1972
Edward C. Pohlmann	1961	1963
C. Eric Price	1966	Present
William R. Qualls	1991	1999
Makaram Raghunandan	1995	1995

Karl W. Reber	1947	1948
Robert E. Reed	1965	1967
Troy D. Reed	1978	1982
Karl N. Reid, Jr.	1964	Present
Mary E. Reynolds	1965	1968
Robert L. Rhoades	1923	1928
Robert E. Roach	1948	1951
Allen M. Rowe, Jr.	1962	1976
Samit Roy	2001	Present
Glenn L. Rucker	1941	1946
Herbert A. Rundell	1953	1956
Richard P. Sauerhering	1906	1908
Virgil Scarth	1938	1943
Murel E. Schlapback	1959	1960
Alfred E. Schlemmer	1947	1950
Charles J. Schoene	1908	1910
Roger J. Schoeppel	1966	1977
Henry R. Sebesta	1966	1976
Adolph E. Shane	1900	1902
Yahya I. Sharaf-Eldeen	1982	1987
John J. Shelton	1984	Present
Yuh-Cheng Shiau	1989	1993
Stephen R. Sias	1981	1981
James D. Simpson	1965	1968
Charles W. Skinner	1910	1915
Frank H. Smith	1929	1931
Atmaram H. Soni	1966	1987
B. F. Spieth	1919	1920
Jeffrey D. Spitler	1990	Present
Arlo L. Steele	1937	1945
William O. Stephens	1940	1941

189

Robert L. Swaim	1978	1992
James H. Taylor	1978	1981
Robert Taylor	1993	1999
Bradley E. Thayer	1948	1950
George H. Thomas	1926	1929
Flint O. Thomas	1983	1988
Melvern F. Thomas	1902	1905
William G. Tiedermann	1968	1978
Warren C. Trent	1946	1947
Roger R. Tucker	1965	1966
James W. Turnbow	1940	1942
Lynn D. Tyler	1963	1964
C. Julian Vahlberg	1970	1972
Rollo E. Venn	1929	1968
Curtis Vickery	1989	1993
Robert K. Wattson	1946	1948
Newton P. Whaley, Jr.	1947	1949
John A. Wiebelt	1958	1985
Leroy A. Wilson	1921	1923
George S. Winchell	1947	1941
Anson T. Woods	1951	1957
Gary E. Young	1982	Present
Cline T. Young	1978	1983
Ronald K. Young	1947	1950
Vincent W. Young	1929	1951
Larry D. Zirkle	1970	1978
Glen W. Zumwalt	1959	1968
Y. H. Zurigat	1984	1989

Chapter 15

Aerospace Engineering at the University of Cincinnati: Co-operative Education and Research

A. Morrison, A. Hamed, W. Tabakof and G. Slater
Department of Aerospace Engineering & Engineering Mechanics
B. Dansberry, Division of Professional Practice
University of Cincinnati, Cincinnati, Ohio

Abstract

This paper gives a brief description of the first co-operative engineering college established in 1906 by Herman Schneider at the University of Cincinnati. It goes on to describe the establishment of the second aeronautical engineering department, and the first co-operative aerospace program in the United States, in 1929, by Bradley Jones. The paper also describes the establishment of a cutting-edge research program in the department in 1958 and the subsequent initiation of the graduate program.

Fig. 1 Herman Schneider

The First Co-operative Engineering College

At the beginning of the twentieth century, Cincinnati was over a century old and already had a well-established industrial base. It was also home to the eighty-year-old University of Cincinnati, an institution that was grounded in the classics and liberal arts but was moving towards the rapidly developing field of engineering. The first University catalog with a separate College of Engineering was published in 1900; Herman Schneider [Fig. 1] was named dean in 1906. The concept of Cooperative Education was conceived and implemented by Herman Schneider as a format of higher education in which the student alternates periods of academic study with periods of employment for the purposes of integrating classroom theory with work experience [1]. This combination of academic and practical education is designed to expand, enhance, and enrich the participating students' skills, abilities, and understanding of his/her chosen field at graduation. His resume was brief: he had graduated from Lehigh University as a civil engineer, practiced architecture, built bridges for the Oregon Short Line Railroad, and returned to Lehigh as an engineering professor.

Once there, Herman Schneider grew increasing concerned over what he perceived to be a gap between theory and practice in the subjects he taught. In 1901, he conceived the idea of incorporating the experience many of his engineering students received in the local mills immediately upon graduation and often over summer vacation into the curriculum of the college [2]. He had even found sponsors for his system among industries in Pittsburgh. Then the Carnegie Institute of Technology opened its doors and the co-operative education plan was no longer needed there. Cincinnati, however, was an ideal place to begin such an experiment in education: among its industries were a variety of machine tool companies which needed skilled labor to grow and prosper. At the time of his appointment in 1903, over thirty machine-tool factories were in operation and the city was in the process of constructing a new filtration plant and waterworks system which was being followed closely by engineers and municipal authorities across the nation. There were, as well, industrial leaders who were willing to lend their support, including Fred A. Geier, of Cincinnati Milling Machine, who was already a member of the university's Board of Directors.

In his work, Schneider had observed that students who had work experience were able to learn engineering theory better and more easily. He envisioned a military kind of school, with free tuition, in which students would spend mornings in the classroom and afternoons working in the factory. The program would emphasize developing their "business and social parts." Herman Schneider wanted to produce not only good engineers but well-rounded citizens. He therefore wasted no time in making his case for a new model of Engineering Education. In 1904, he was invited to write an article outlining his plan for Cooperative Education in the University News. The new president of the university, Dr. Charles Dabney, was favorably impressed with the concept. In October of 1905, Schneider presented his concept to the Cincinnati Society of Mechanical Engineers [3]. The presentation was well received and the media coverage caught the attention of John Manley, secretary of the Cincinnati Metal Trades Association [4]. Over the course of the next year, Manley and Schneider made the rounds of local factories, garnering support for their "cooperative plan." Prominent among the industrialists who gave support to Schneider's proposal were Frederick A. Geier, president of Cincinnati Milling Machine Company, and Ernest Du Brul, president of the Miller, Du Brul, Peters Company. Both were members of the University of Cincinnati's Board of Directors, with Du Brul also serving as chairman of the Board's Committee on the Engineering College.

By 1906, Schneider's efforts had resulted in a draft proposal implementing a six-year degree program employing his ideas. The proposal was passed by the university's Board of Directors on an experimental basis by a vote of five to four. However, co-operative education was nothing short of a revolutionary concept to which the Board gave its grudging support. They issued a statement: "We hereby grant the right to Dean Schneider to try, for one year, this cooperative idea of education at the University of Cincinnati, for the failure of which we will not assume responsibility." And so was born the College of Engineering's trademark, co-operative education.

The first incarnation of cooperative education manifested itself on the University of Cincinnati's campus in September 1906 with the arrival of twenty-seven "co-ops." This group of students was made up of twelve mechanical, twelve electrical, and three chemical engineering freshmen, who almost immediately dispelled the fears of some manufacturers that they would be "rah-rah" boys, and of some faculty that they would be a "gang of boilermakers" invading the campus [5]. For administrative purposes, the co-ops were divided into two sections. Specifically, there were six pairs of mechanical engineering students, six pairs of electrical engineering students, and one pair of chemical engineering students, with the remaining chemical engineering student having no "alternate". These students were scheduled to attend college every other week while their "shop partners" were at work at one of twelve participating companies. In each workplace there was a structured sequence of training, a gradation from simple to complex tasks, with consistently increasing responsibility. There was also rotation of duties to ensure a variety of experience. Continuity and smooth alternation of shop-work was made possible by personal supervision by Schneider and by conferences on Saturday morning between paired students.

The initial reviews of this experiment, labeled "the Cincinnati Plan [6]" in numerous articles, were so favorable that the University Board of Directors immediately set out to make the program a permanent addition to the University. Seventy students, out of an applicant pool of over 400, were selected for the second year of the program. By 1912, the University's co-op program consisted of 300 students with a total of fifty-five companies participating in the program. By the time the United States entered the First World War the number of students enrolled was up to 500, with eighty-five companies participating. In 1919 enrollment reached almost 800 students and 135 companies. By contrast, the number of traditional engineering students had dropped from 107 in 1906 to forty in 1912, and then twenty-five in 1917. The traditional program was scrapped entirely in 1920 [7]. By 1929, the year the Aeronautical Engineering Department was formed, the co-op program had evolved substantially from its first incarnations. Approximately 1800 undergraduates and over 300 companies participated in the program. Students alternated work and classes on a two-week cycle throughout the five years necessary to achieve a bachelors' degree.

Fig. 2 Bradley O. Jones, ca. 1918.

Bradley Jones and the Department of Aeronautical Engineering

The College of Engineering had been considering courses relating to aeronautics for some years, and with the interest in cross-country and transatlantic flight and the mounting realization that aeronautics was an important technology of the future, it was decided to offer a dedicated program of instruction. In 1927, using funds provided by an arrangement with the Jacob G. Schmidlapp Educational Fund, a Chair of Aeronautical Engineering was established and a search for a suitable candidate was underway.

One of the chief candidates under consideration from the beginning of the year long search was Major Bradley O. Jones, then employed at McCook Field in Dayton as a navigator and inventor [Fig. 2]. Bradley Jones had made national headlines, and certainly local ones, with the record nonstop flight in February 1926, from McCook Field in Dayton, Ohio to East Boston Airport in Massachusetts. This flight was the first nonstop flight to navigate solely by instrumentation: instrumentation that was developed by Bradley Jones.

Born in Boston in 1889, Bradley Jones entered the Massachusetts Institute of Technology (MIT) at the age of sixteen and graduated with a B.S. in physics in 1910. He embarked as first mate and navigator on a three-year cruise on the "Carnegie," a sailing ship funded by the Carnegie Institute and built of wood and brass, with no iron or steel that might affect a compass. The maps Bradley Jones charted of true vs. magnetic north variations were published and made available to mariners and aviators worldwide following the end of the voyage in 1913. He then took up teaching, moving from the University of Pittsburgh to Lehigh University to Norwich University. In the course of these activities he found time to finish his M.S. in Physics in 1914 at Norwich

University. He was drafted in 1917, but military records are sparse and there is no record of his activities. Judging from later statements he made to students, friends, and colleagues, it is likely that he instructed new pilots in the fundamentals of navigation and may have participated in balloon experiments. All that is known for sure is that when Bradley Jones entered the Reserves following the Armistice, it was with the rank of major.

Following the war, Bradley Jones taught Merchant Marine officer candidates and spent his summers barnstorming across America. According to him, in 1922 his commanding officer during the war, General Billy Mitchell, tracked him down and hired him as a civilian technical expert at McCook Field in Dayton, Ohio, where he was to lead navigation and instrumentation development. Mitchell was especially interested in developing instruments that would allow pilots to fly at night or in heavy cloud cover: At this time, all flights had to take place at low altitudes so that pilots could visually identify landmarks. They had to use sextants in order to provide a position fix, but the maritime sextant was quickly outdistanced by the speed at which aeroplanes could fly. Bradley Jones developed an aviation sextant that used a two-minute fix instead of a twenty to thirty minute one. It became the official Air Service navigation technique and was still in use by the Army Air Corps in World War II [8].

However, Bradley Jones was not finished. He spent the next year developing what the press called an "Earth Inductor Compass" as well as a flight indicator. Both of these were used in a 1923 flight to Boston. They flew at an altitude of 10,000 feet and were unable to see the ground during the flight. Despite this and strong winds, they were only five miles off course when they were able to verify their position visually [9]. It is worth noting that Charles Lindbergh insisted on having an earth inductance compass installed for his record-breaking flight from New York to Paris. He flew from San Diego to New York by following railroad tracks; when he reached New York, Pioneer Instruments installed the compass. His biography attributes his reaching the Irish coast just three miles off his planned route to the compass developed by Bradley Jones [10]. Clearly Jones was a far-seeing visionary with an extremely practical side: in his own words, "The engineer must be a far greater romanticist than the poet; for while the poet has no bounds for his whimsies, the scientist must always tie together his visions with clear reasons [11]."

This, then, was the candidate chosen by the University Board of Directors to head the first co-operative education program in Aeronautical Engineering in the country (and the second Aeronautical Engineering department). Bradley Jones was enthusiastic about this opportunity and threw himself, with characteristic energy, into developing a comprehensive program. He contacted Orville Wright: in a meeting, they developed the curriculum together, and were of the opinion that it would be much better than any taught at other colleges because of their combined practical experience with the mechanics and realities of flight navigation and aeroplane design [12]. Bradley Jones went on to write and publish two books following his appointment: *Avigation*, in 1931, and a textbook, *Elements of Practical Aerodynamics*, in 1936, both published by John Wiley & Sons.

The Early Years

In 1929, the first curriculum for the new Aeronautical Engineering Department [Table 1] was created. In keeping with the UC tradition, it combined a rigorous academic program interspersed with cooperative work experience into a five-year program of study. The new curriculum, as visualized by Bradley Jones in consultation with Orville Wright, emphasized hands-on experience but included key technical developments in modern aeronautics. All was not technical, however. Bradley Jones knew that in spite of the fact that aviation started just fifty miles up the road in Dayton, Ohio, much of the current technical achievement had moved to Europe, Germany in particular. Consequently a rigorous program in the German language was required of all the aeronautical students so that in the words of Jones, "students may avail themselves of books and reports in that language [13]."

The first two years of the five-year co-operative aeronautical curriculum were identical to the mechanical engineering program. Starting with the third year, however, courses in aerodynamics, aircraft structures, and, of course, German, replaced steam engineering and other more mundane mechanical courses of the time. By the end of the fifth year, students had covered most of the topics needed for a comprehensive aeronautical program, including propeller theory, materials, meteorology, instruments and equipment, air transportation (performance), and engine and aircraft design. Because students also participated in the co-op program (alternating eight weeks of work and seven weeks of school), they were employed in aeronautical concerns in the Cincinnati-Dayton area as well as in growing aeronautical industries around the country. The first Aeronautical Engineering class to be graduated, in 1932, consisted of five newly-minted engineers: Grant Clarence Adams, Edwin Hale, Daniel Loomis Pellet, Owen Cowley Stevens, and Philip Harold Stevenson.

Freshman	Sophomore	Pre-junior	Junior	Senior
Mathematics, Mechanics, Descriptive Geometry, Vector Algebra & Geometry, Engineering Drawing	Mathematics, Mechanics, Engineering Drawing, General Physics, Physics Lab	Mechanics, Differential Equations	Accounting, Economics	Management, Cultural Electives (2)
General Inorganic Chemistry, Chemistry Lab, Applied Chemistry	Engineering Materials, Metallurgy Lab	Heat Treatment Lab, Electrical Lab	Elementary Surveying, Meteorology	Indeterminate Structures
	Mechanism, Mechanism Drawing, Mechanical Lab	Mechanical Lab, Graphics, Automotive Engineering	Valve Gears, Experimental Engineering, Thermodynamics	Experimental Engineering
		Aerodynamics, Aeronautic Inspection Trips	Aeroplane Structures, Aeroplane Engines, Aeronautical Materials, Instruments & Equipment, Airships	Advanced Aeroplane Design, Aeroplane Engine Design, Propeller Design, Air Transportation, Seminar, Thesis
Coordination	Coordination	Coordination	Coordination	
English	English	English, German	English, German	English
Military Science	Military Science	Military Science*	Military Science*	Military Science*

Table 1. Aeronautical Engineering Curriculum, 1929

While records are scarce, companies where early aeronautical students spent their work assignments include the Glenn L. Martin Company and the Watson Airport Company as well as the Aeronautical Corporation of America and the All-Metal Aircraft Corporation, both aircraft manufacturers, and the Crosley Aircraft and LeBlond Aircraft Engine Companies, makers of aircraft engines [14]. Commercial aircraft companies operating out of Lunken Airport in Cincinnati were also good locations for co-op placement. These companies included Embry-Riddle, Universal Aviation, and Main Airlines [15]. In his 1929 article on the new Aeronautical Engineering program, Bradley Jo
nes himself commented that two fourth-year students were already working in the aeronautical field and that placement of future students was unlikely to be difficult [16]. Through the depression years of the 1930's and the build-up to World War II, the co-op program continued to evolve. Co-op employers prior to World War II included Alvey-Ferguson, Curtis-Wright, Goodrich Aviation, Goodyear Aircraft, Kellet Autogyro Corporation, U.S. Army Air Corps at

Wright Field, U.S. Bureau of Standards, and Wright Aeronautical in Lockland, Ohio. Parts manufacturers and tool-and-die companies such as National Automatic Tool, Cincinnati Shaper Company, and Republic Steel also trained students.

World War II

The start of the second World War followed hard on the heels of Dean Schneider's unexpected demise. The college, having successfully weathered the depression, was now faced with a mobilizing war effort and the effects of the draft - and all this under the guidance of a new and untried dean, Robert Gowdy. Dean Gowdy acted quickly to dispel anxiety over Dean Schneider's death: he said he would be "carrying on," in the tradition of Dean

Fig. 4 "Goodyear Girls" ca. 1943: two UC co-ops at the Goodyear Aircraft Factory

Schneider, rather than "taking over [17]." By 1940, although the United States had not officially entered the war, national service in defense of the country had already begun among the civilian faculty. Bradley Jones served on the Civil Aeronautics Bureau with other experts, preparing an expanded civilian pilot training program. He helped write several pilots' manuals for use in military training programs; one, Aerodynamics for Pilots, was published under his name [18].

Fig. 3 Ray Bisplinghoff (left), Class of '40 and a faculty member during World War II; later Dean of the School of Engineering at MIT.

The department's commitment to quality education was apparent even as early as 1940, with such students and faculty as Ray Bisplinghoff [Fig. 3] and Hans Liepmann (later the Theodore von Kármán Professor of Aeronautics at Caltech) on the rolls. Ray Bisplinghoff, for instance, was graduated in the Class of '40. While pursuing his Ph.D. in physics at UC (at the time, the department had no graduate studies), he joined the department faculty and taught for three years while on a graduate research fellowship. He was drafted in 1943 and served at the Bureau of Aeronautics in Washington, D.C. until 1946. After the war he was appointed at MIT as an assistant professor in aeronautical engineering. He spent sixteen years there teaching, conducting research, and co-authoring key textbooks, including the first textbooks to be published on aeroelasticity19. He also found the time to finish his Ph.D. at ETHZ, the Swiss Federal Institute of Technology, Zurich. As the space program gained momentum, Bisplinghoff took a leave of absence to serve as an assistant administrator at NASA, where he led the Advanced Research and Technology program. He returned to MIT to head the Department of Aeronautics and Astronautics, during which time MIT worked closely with NASA on the Apollo program; he was personally involved in the planning of Apollo missions 8, 9, 10, 11, and 12. He rounded out his tenure at MIT by serving as the Dean of the School of Engineering for two years. He died in 1985, having been recognized by his colleagues with honors ranging from Exceptional Civilian Service Medals and Distinguished Service Awards from federal agencies such as the U.S. Air Force, the Federal Aviation Administration, the National Science Foundation, and NASA, to invited lectures from the von Karman and the Wright Brothers Institutes, to election to the National Academy of Sciences, to honorary fellowship in the Royal Aeronautical Society.

The College of Engineering began to incorporate military training along with the rest of the university. Enrollment dropped even as the college took steps to accelerate its programs by eliminating vacations and co-op terms; by the time the United States entered the war in 1941, virtually all able-bodied men had been or were in the process of being drafted. One such undergraduate, Vincent Burinskas, had notations on each co-op section for a year that he was waiting to be called up. He was finally drafted by the Army Air Corps in 1943; amazingly, he returned in September 1945 and spent two more years finishing the program [20]. The department was also not immune to the draft. Although there were five faculty members at the time, some of whom were working towards graduate degrees, for several years two to three of them were working for the War Department: Bisplinghoff, Liepmann, and George Sibert. The catalogues of the war years show Lou Doty (later a professor in math and physics at UC) and Bradley Jones doing most of the teaching [21].

Another part of the war effort was civilian-military education. The department began a twelve-week training period for women undergraduates to be "Goodyear Girls" - airplane factory supervisors [Fig. 4]. As so often happened, the war opened up new opportunities for women: for the first time, they were granted access to all programs university-wide [22]. The department also continued research and development - war-related, along with rest of the university - such as testing aircraft propeller blades in the old wind tunnel in the UC hangar. The co-op experience reflected changes in society as well: Sydney Reiter, Class of '47, who worked at Wright Aeronautical (purchased by General Electric), relates the story of working on a segregated assembly line during his co-op assignment there. As the war effort gained momentum, the assembly lines showed the effect of the draft's attrition rate: due to the growing manpower shortage, the company announced the assembly lines would have to be combined in order to continue to operate at least one full-time. Many white employees walked off the job, although Sydney Reiter did not. The next day, all employees were told that if they didn't work, they would be fired. The assembly lines became, and remained, integrated [23].

The co-op program continued to evolve as a result of the changing conditions of World War II and the scientific advancements and increasing enrollment of the 1950's: following the war, enrollment climbed steadily as a result of the GI Bill and the release of young men from the armed forces. Temporary barracks-style housing was built to accommodate the influx of students in every available space on campus [Fig. 5]. As early as 1931 students had begun to call for a freshman year devoted solely to coursework prior to commencing co-op rotations. As more universities began to develop cooperative education programs, the time of each co-op rotation lengthened. By 1956, the fiftieth anniversary of co-op, students at the University of Cincinnati could expect to attend school exclusively through April of their freshman year. After that they rotated between work and school in eight-week intervals until returning to school full-time from January through June of their fifth year. Upon graduation, these students had compiled ninety-six weeks of work experience.

Dr. James Wade, a 1958 alumnus of the program and currently a faculty member in the department, recalls the curriculum and the faculty ca. 1956: "We diligently worked ourselves through the first two years without ever taking an aeronautical engineering course or really having any interaction with the faculty in the aeronautical engineering department. I do remember occasionally encountering a rather unusual looking avuncular old gentleman, who was always smoking and had a sequestered office on the seventh floor of Baldwin Hall. He eventually turned out to be Professor Bradley Jones [Fig. 6].

Fig. 6. Bradley Jones, ca. 1950

"The first two years of the aeronautical engineering curriculum consisted of preparatory courses in the basic sciences. However, the third year was the year of definition for most of the engineering majors because we had nothing but engineering courses. One in particular, 'Tech Mech,' a rigorous course in rigid body dynamics and the precursor to the present-day Mechanics III course, was known to cause many students to morph into six-year men. We also had our first course in our major: Basic Aerodynamics, taught by Professor Jones, from his text book *Elements of Practical Aerodynamics* [24]. His classes were always very interesting, especially when he told of some of his experiences as a navigator.

"On many occasions, he would bring students back to reality. For example, on one particular test, a student had calculated the landing speed of an aircraft to be 5000 mph; Professor Jones asked pointedly if the landing speed didn't seem to be totally outrageous, since there was not an airplane in the sky that fast in straight and level flight, and a landing speed of that magnitude should have made the student take notice. I myself was personally addressed in the uncomplimentary terms of being one of the 'biggest formula hounds' that Professor Jones had ever encountered. Being enshrined with such a moniker by Professor Jones was a descriptive remark that was consistent with the mindset of complete aquiescence towards the faculty by the student body. Professor Jones also taught a course in propeller design to fourth-year students. This was taught to my class in the '56 -'57 school year. The entire nature of aerospace was changing and I hazard to conjecture that our class of 1958 was probably the last class to take this course [25]."

As the nation began to cope with the growing pains of post-industrial status, job prospects were shaky, but UC aeronautical engineering graduates continued to find work, due in large

Fig. 5. Barracks on the quadrangle in front of Baldwin Hall, home of the College of Engineering

199

part to their co-operative work experience. In addition to many of the companies involved in the co-op program prior to 1941, students worked for such companies as Aeronco Aircraft, Bendix Products, Cessna Aircraft, Chance-Vaught, Fairchild Aircraft, General Electric, Goodyear Aircraft, McDonnell Aircraft, NACA Lewis (as well as other centers), North American Aviation, and Taylorcraft. Warner L. Stewart began a long tradition of working closely with the National Advisory Council on Aeronautics (NACA) as the first co-op at NACA Lewis Flight Propulsion Laboratory (known as NASA Glenn today) in 1948; he went on to become NASA Lewis' Director of Aeronautics, Director of Technical Services, and, finally, Director of the Lewis Research Academy [26]. Jerry Eastep, the first co-op to work at the newly reorganized Aeronautical Research Laboratory (ARL) at Wright-Patterson Air Force Base, arrived in 1957. He was followed by many others over the next forty-five years: the second co-op to arrive, James Snyder, found to his surprise that he was working next door to Hans von Ohein, inventor of the turbojet engine [27].

Fig. 7 Dr. Kroll with his Beechcraft Sundowner.

Other students of the time, such as Lawrence Lantzer and James Thomas (both Class of '61) went on to not only distinguished but unusual careers as well. For instance, Jim Thomas' cooperative assignment was at Chance Vought Aircraft in Grand Prairie, Texas; upon graduation, he joined the Air Force, where he became a test pilot. During his twenty-six year career, he flew over forty different kinds of planes, including the F-15, the U2 (test program), and the Lockheed F-117A (Stealth Fighter). He was one of the last pilots to fly the Douglas A1-E "Skyraider," a propeller airplane particularly suited to close-air-support missions [28], during the Vietnam War. He retired as a colonel in 1987 and moved onto Lockheed as a test pilot, where he flew until 1993. Larry Lantzer, too, became a test pilot, but for the Navy, where he graduated first in his class at the Navy Test Pilot School in Pax River; he went on to become the Air Department Head on the U.S.S. Constellation, and then the Squadron Commanding Officer for the Composite Squadron (jets, helicopters, and admirals' planes) at Pearl Harbor. No other class at UC had two test pilots; this is all the more remarkable considering that the Class of '61 had only seventeen members [29].

Aeronautical Engineering Gives Way to "Aerospace"
With the launch of Sputnik in 1957, surprising many in the Western world, the space race was on in earnest. Unfortunately for the department, Bradley Jones died unexpectedly that same year. This did not, however, deter the push to expand the newly-renamed aerospace program to encompass advanced degrees as well as expanding co-operative opportunities such as those begun at the Aeronautical Research Laboratories at Wright Field in Dayton, Ohio. It is interesting to note that although his death preceded the space program, a number of lectures given by Bradley Jones, and reported in the local press, focused on artificial satellites and space travel . In fact, in one article he predicted an unmanned flight via rocket to the moon within ten years: Bradley Jones was clearly prescient in terms of the future of aerospace [30].

In 1957 the Aeronautical Engineering Department at the University of Cincinnati had only three faculty to teach undergraduate courses, and offered a single graduate course. Ray H. Murray, the department head, specialized in gas dynamics; Robert Kroll's area was structures [Fig. 7]; and Odin Elnan concentrated on aircraft performance and dynamics. None had obtained a doctorate: at the time, it was very rare to find engineers holding advanced degrees. The department was fortunate, however, to retain their services for many years, including, in the case of Dr. Kroll, three turns as interim department head in as many decades [Appendix A].

In that same year the U.S. Army invited Dr. Widen Tabakoff to join Werner von Braun's research group in Huntsville, Alabama. Dr. Tabakoff, who was born in Bulgaria, went to study mechanical and aeronautical engineering in Prague in 1937 [Fig. 8]. When the German occupation began, he was forced to change from Karls University to the German University in Prague, where he obtained his Dipl. Ing. He was recommended to the University of Berlin to pursue his postgraduate education in aeronautical engineering; during the war, he worked in Berlin with von Braun's group in rocketry and Hans von Ohein's group in turbomachinery. He obtained his Dr. Ing. degree from the University of Berlin in 1945, continued to work in research and as an instructor there until 1947, and then went to Argentina to work with German scientists on supersonic aircraft. In 1955 he returned to Germany to work for Hoechst AG before joining von Braun's group in Alabama. As part of his duties, he traveled to Cincinnati to advise the U.S. Engineering Division Laboratory on rocket launch pad flame deflectors for the Saturn V complex. As a result of this assignment, he ended up spending time in Cincinnati. He was interested in aerospace education, so he contacted the department and became acquainted with the faculty. This resulted in the offer of a position from Professor Murray, then acting head, which Dr. Tabakoff accepted.

With the advent of the Space Age, a demand for a new type of advanced professional training was growing. Dr. Tabakoff and Professor Murray wanted to establish a doctoral program in the department but with only one Ph.D.-holding faculty, they were unable to obtain approval. They therefore developed a plan whereby faculty from related disciplines such as astronomy, physics, and mathematics would participate in the teaching and evaluation of the nascent graduate program in Aerospace Engineering. Drs. Paul Herget, Eugene Rabe, and Peter Musen from the Department of Astronomy, Dr. Russell Dunholter from Mathematics, and Dr. Tabakoff and Professor Murray from Aeronautical Engineering formed the "Institute of Space Sciences" and established M.S. and Ph.D. programs for this institute in 1959. The Aerospace Engineering Department established an M.S. program at this time; the first M.S., Leonard Beitch, was graduated in 1959.

Fig. 8 W. Tabakoff as a student in Prague, 1937

The objectives of the Institute were education and research directed towards the broad problems presented by the ever-increasing activities in the field of missiles and satellites. The educational program of the Institute was conducted entirely within the framework of the Graduate School of the University. Complete curricula were established in Dynamical Astronomy and in Aerospace EngineEngineering for the 1959-60 academic year [Table 2].

Dynamical Astronomy			Aerospace Engineering		
Practical and Spherical Astronomy	3 cr	Dr. Herget	Boundary Layer Theory & Heat Transfer	4 cr	Dr. Tabakoff
Orbit Calculation	6 cr	Dr. Herget	Theoretical Aerodynamics I	3 gr.	Mr. Murray
Numerical Analysis	6 cr	Dr. Herget	Rocket Flight Dynamics	3 gr.	Mr. Murray
Dynamical Astronomy	6 cr	Dr. Rabe	Advanced Propulsion	6 gr.	Dr. Tabakoff
Dynamical Astronomy II	8 cr	Dr. Musen			
Lunar Theory	8 cr	Dr. Musen			
Planetary Theory	8 cr	Dr. Rabe			

Table 2. Institute of Space Sciences Curricula, 1959

Innovations in Graduate Education

New courses in advanced propulsion, orbit mechanics and aerodynamic heating were added in 1958 as the curriculum followed the national trend towards an "aerospace" program. In addition to the new emphasis on space, it followed another national trend towards an "engineering science"-oriented program, moving away from design and hands-on experience. This move was necessitated by the new technical requirements of the aerospace industry and the increased computational power provided by the computer. During the next decade the department created a major computational program in fluid dynamics and solid mechanics and structures as part of the computational revolution in engineering.

As with the academic curriculum, the cooperative element of the newly named Aerospace Engineering Department expanded to include the growing list of space related businesses. Where co-ops once worked at NACA, now students held assignments at many of the National Aeronautics and Space Administration (NASA) centers. Those centers with the strongest contingent of UC co-op students included Goddard, Marshall, and the Manned Spaceflight Center

(Johnson). Companies like Chrysler Space Division, GE Space Power & Propulsion and Missiles Divisions, Lockheed Missiles and Space, Martin-Marietta, and Rockwell also participated.

Concurrent with this, many engineers who worked at General Electric Aircraft Engines (GEAE) in Evendale were finding their work demanded advanced research in propulsion and combustion as well as in structures and controls. This led to increased pressure to get approval for the doctoral portion of the graduate program within the Aeronautical Engineering Department. The University, however, turned down the proposal again due to the lack of advanced degrees in the department. Dr. Tabakoff spearheaded the recruitment and hiring of a fifth faculty member and new department head, Dr. Paul Harrington [Fig. 9], who came to UC from the Rensselaer Polytechnic Institute and had served as the first director of the Von Karman Institute in Brussels before joining Rensslaer.

Fig. 9 Paul Harrington

With the addition of Dr. Harrington in 1959 and the commitment by Professors Kroll and Elnan to obtain their doctoral degrees, approval was finally granted for the establishment of the Ph.D. program. Dr. Tabakoff was made the Graduate Director and was responsible for teaching many of the advanced graduate courses with an emphasis on propulsion. GEAE responded with significant interest, which resulted in very large course enrollments numbering over fifty, both in the propulsion and in the structures areas. Rodney Boudreaux, the department's second M.S. graduate (1960) and later a vice-president at RocketDyne, went on to become the department's first Ph.D. graduate five years later.

In 1965 Mr. Wysong, a graduate student from General Electric Aircraft Engines (GEAE), suggested that the department establish an Advanced Course in Engineering in cooperation with General Electric (GE), similar in nature to one at the Rensselaer Polytechnic Institute in Troy, New York. As Graduate Director, Dr. Tabakoff was asked to investigate the possibility, promptly traveled to New York, met with Professor J.V. Foa, and came back to the department with a complete program outline that established what was essentially a co-operative program at the graduate level: while attending classes at UC and the GE site, the student would cover his regular GE engineering assignment and thesis research would be developed in the area of his regular work. In addition to regular classroom study at UC, certain portions of the coursework were taught at GE by GE specialists and monitored by UC professors, some of whom occasionally taught sessions there. Sponsors of the program felt such advanced study in engineering had become a necessity with the rapid advances in space-age technology [31].

By 1966 the program was in place [Fig. 10] and continues to this day. Over 650 GEAE engineers have obtained their graduate degrees at the University of Cincinnati through the Advanced Engineering Program [32] and some have gone on to make significant contributions to aerospace research under GE's auspices. One such person is James Younghans (M.S. '78), Program Manager of GEAE's Preliminary Design, Performance and Operability Engineering; another is Thomas Wakeman (M.S. '71, Ph.D. '82), holder of thirty-nine patents, who led the effort to design UC's hot erosion tunnel. He was instrumental in the development of the revolutionary unducted fan engine (produced by GE, now found in the Smithsonian Institute), led the design team for the F110 nozzle for the F-16 fighter aircraft, and was the lead design engineer for the groundbreaking U.S. Navy ADEN Exhaust Nozzle, the first two-dimensional nozzle, still used as a reference today. His achievements spoke so well of his abilities that he was asked to lead the corporate-wide implementation of six-sigma, GE's patented method for designing in reliability [33].

Fig. 10 First GE-AE/UC graduates, 1969
L to R: David Amos; George Grant; Stav Perdumo, AC Supervisor; Bud Thomas; Bob Kraft; John Blanton, GE AE Program Manager; unknown.

Establishment of Research and Laboratory Infrastructure

Upon joining the department, Dr. Tabakoff started a vigorous externally-funded research program that continues to the present day; in addition, he built the propulsion and gas dynamics laboratories and infrastructure, and has supported a large number of graduate students and research staff who constituted the backbone of the department's graduate program for over fifteen years. He initially received U.S. Air Force funding to study the aerodynamics of munitions, and then moved to ramjet propulsion systems. In 1961 he was awarded a five-year Air Force contract entitled *Aerothermodynamic Investigations in High Speed Flow*, monitored by Colonel Andrew Boreske, Jr. A parallel theoretical research effort was initiated after calibrating the twenty-inch Mach 14 hypersonic blowdown tunnel at the Aerospace Research Laboratory (ARL) in Dayton, and the research results, encompassing topics such as shock interactions, hypersonic flow over cones and delta wings at high angles of attack, binary boundary layers and hypersonic wakes, were published in ARL reports and journal articles.

Fig. 11 Dr. Tabakoff with graduate student
Don Freund at the jet engine test facility

This project was important in establishing the department's reputation, in large part because it put together a highly competent research team of faculty and graduate research assistants and attracted many graduate students who went on to have distinguished careers. The faculty consisted of Drs. Ting Y. Li, Kinzo Hida, A. C. Jain, and Harou Oguru; the research associates were Robert Evans, Joseph Loch, Arnold Polak, and Maido Saarlas; and the research assistants were David Brown, Edson Goodrich, Barry W. Hannah, Werner Hoelmer, Phillip Kirk, Stewart Larson, James McDonel, Kenneth Token, David Will, and Frank Tepe, who acted as UC's staff supervisor at ARL during the last two years of the project. This group continued to contribute to aerospace research and education after graduation in various capacities. Mario Saarlas became the chair of the Aeronautical Engineering Department at the U.S. Naval Academy; Barry Hannah, the Navy's preeminent aerospace engineering leader for strategic nuclear re-entry systems, is currently head of the Re-entry Systems Branch, Strategic Systems Programs for the U.S. Navy; David Brown became director of UC's well-known Structure Dynamics Laboratory; Frank Tepe became Vice-President and Dean of Research at UC; Arnold Polak joined the department faculty as a professor; and Kenneth Token went on to a distinguished career at McDonnell Douglas, first as Head of Propulsion Thermodynamics, then as Deputy Program Manager of the F-15 project, and finally becoming the Director of Engineering at McDonnell Douglas' "Phantom Works."

At the same time, construction of the Jet Propulsion Laboratory was begun in 1958 in a large hangar on the northwest side of the main campus. An old wind tunnel used to study aircraft propellers during World War II was removed and combustion and heat transfer facilities were added, as well as a hot air mixing facility, cascade tunnels, and turbomachinery design facilities. The U.S. Army contributed seven 200 psig tanks of 3500 cubic feet; the laboratory was, at that time, fully supported with industrial and governmental research funds, the beginning of the concentration on cutting-edge research in the department. A large cascade water table was donated by GE, and a used Mach 3.5 supersonic wind tunnel in good operational condition was obtained from the U.S. Navy. Most of the work done in this facility was for advanced and/or graduate studies. Sadly, the hangar was torn down in 1983 to make room for the university bookstore.

The Propulsion Laboratory subsequently expanded into the newly completed Rhodes Hall in 1970 [Fig. 11], a 10,000 square foot space with a twenty-foot ceiling located on the ground floor and equipped with two extensive pipe delivery systems. A compressor plant was then purchased in order to supply the four 2,200 cubic foot 2,000 psig high-pressure tanks transferred to Dr. Tabakoff by the U.S. Air Force for the cost of moving them from Dayton to Cincinnati. This greatly enhanced the propulsion research capacity when added to the Army's seven tanks. He later transferred the titles of both the 2,000 psi and 200 psi air supply storage systems to the university in the nineties. The air supply system was and still is essential to operate all the facilities in the propulsion lab, including the aeroacoustic research facility, a 6" x 6" supersonic tunnel, compressor and turbine test stands, two subsonic tunnels and two erosion tunnels. In addition to the Rhodes Hall propulsion laboratory, the department maintains a facility at Center Hill that includes a bench scale TAPS combustor rig with laser diagnostics, a hot erosion wind tunnel. A number of laboratories are maintained in the Engineering Research Center, including computational fluid dynamics, computational modeling, non-destructive evaluation, a SQUID, and gas turbine diagnostics.

The Center of Excellence in Propulsion was established in 1968, when Dr. Tabakoff became director of Project Themis. UC's proposal to integrate the concepts of rocket propulsion, combustion, aerodynamics, and structural analysis into a light-weight high-output engine was one of two selected in the propulsion area from over 1300. Eleven faculty members from Aerospace Engineering, Material Science, and Chemical Engineering participated in this project on a variety of propulsion research topics including cascade aerodynamics, jet mixing, particulate flow in turbomachinery, gas turbine heat transfer, and high temperature propulsion system materials. Project Themis' contractual support continued for a decade and exceeded $6,000,000.

Among the large number of graduate students supported by Project Themis was Awatef Hamed, who joined the department faculty after completing her doctoral research in 1972. She was the first woman in the college of engineering faculty [Fig. 12] and later became head of the department in 2001. She continued research collaboration with Dr. Tabakoff in the related areas of turbomachinery particulate flows, erosion, and performance deterioration and retention. In addition to Project Themis, NASA's uninterrupted funding supported the research of Drs. Tabakoff and Hamed on radial turbine aerodynamics and heat transfer and turbomachinery erosion for more than two and a half decades. This research was widely disseminated through publications as well as professional engineering courses at the International Gas Turbine Institute of the ASME conferences in Houston, Montreal, and London. In addition Drs. Tabakoff, Hamed and Wenglarz (Allison Gas Turbine) offered a von Karman Institute lecture series in 1988, *Particulate Flows and Blade Erosion*, for which they received a NATO award [34].

Fig. 12 Dr. Hamed displays a new LDV system

The research results obtained under Project Themis had a direct and specific impact on the design of the T-800 engine which was developed to power the Army's next generation of rotorcafts. In addition, the funds provided under this project and the acquisition of government surplus equipment were instrumental in establishing the Propulsion Research Laboratory's experimental capabilities. The laboratory's well-known erosion facilities were an extension of the effort to study the effect of particulate flow on engine performance and, to this day, represent the leading edge in particulate flow theory and experiments. Dr. Tabakoff was later acknowledged for his work on this project under the United States Army Research Office's Selected Scientific Achievements.

In 1971, Dr. R.T. Davis, a highly respected fluid mechanics researcher at Virginia Polytechnic Institute, was appointed as department head following Dr. Harrington's death. His primary mandate was developing emerging areas of study and integrating them into the department curriculum. The major curriculum areas (Dynamics & Control, Fluid Dynamics, Propulsion, and Solid Mechanics) were enhanced by the addition of five new faculty members, including Distinguished University Professor Neil A. Armstrong [Fig. 13], who taught in the department from 1971 to 1979. The new faculty were primarily responsible for expanding the graduate-level courses and research productivity, as well as focusing on new research elements in Computational Fluid Dynamics.

The graduates of this era reflect both the depth of the research experiences available to students as well as the breadth of cooperative assignments [Appendix B]. Clark Beck (M.S. '69) began his career with his co-op employer, Wright-Patterson Air Force Base, where he became the Air Force expert in the area of extreme elevated temperature testing. Upon his retirement from the Air Force, he moved into higher education, where he served as the assistant dean of the College of Engineering at Wright State University. John Morrison, Class of '66, retired as a vice-president after three decades with Gulfstream. Edward Kraft, Class of '68, has had over thirty years of experience as a researcher, manager, and educator in the aerospace field. He is most known, however, for his integrated test and evaluation (IT&E) methodology, based on a systems approach to understanding the role of ground testing in the overall development of the flight system, to improve overall simulation and reduce both time and cost for developing flight systems. This approach underscores the complementary nature of computations, ground testing, and flight testing. He is currently the Executive Vice-President and COO of MicroCraft, Inc.

Fig. 13 Neil Armstrong at the annual student paper airplane contest.

Richard Johnson, Class of '73, started at Lockheed-Georgia as a co-op student in 1969. He returned to Lockheed following graduation as a structural engineer and participated in various engineering design and test programs for the C-130, C-141, and C-5 aircrafts. He moved to Gulfstream in 1981 as a structures technical specialist and has spent the past two decades in the design and certification of the Gulfstream III, IV, and V aircraft: he oversaw the latter, the first and most successful ultra-long-range executive jet aircraft, as Engineering Project Manager, then as Project Manager, and finally as Chief Engineer. The Gulfstream V received the Robert J. Collier Trophy in 1997 for its representation of outstanding achievement in U.S. aviation.

The tradition continues to this day. For instance, Dinesh Keskar received his M.S. and Ph.D. in Aerospace Engineering in 1977 and 1979, respectively. He worked as a research associate at NASA Langley in Flight Dynamics and Controls before joining Boeing in 1980. He developed techniques to conduct flight test data to obtain math models for the 737 - 767 flight simulators from 1980 to 1986. He is currently the president of Boeing India, a position he has held since 1995; during his tenure Boeing has maintained a one-hundred-percent market share in India. Yokichi Sugiyama, who received his Ph.D. in Aerospace Engineering in 1984, is currently the Director of the Third Research & Development Institute at the Japan Defense Agency, where he holds rank equivalent to that of a three-star general [35].

Winds of Change: Aerospace Education in the Post-Industrial Era

Throughout the 1980's, the Aerospace Engineering & Engineering Mechanics program was consistently ranked tenth in the nation by the Gourman Report. This brought new interest to the program in the late eighties when two major initiatives were undertaken: the NASA Health Monitoring Technology Center for Space Propulsion Systems, and the Ohio Aerospace Institute. Computational Fluid Dynamics activities were further strengthened when Dr. Stanley G. Rubin, then Associate Director of the Aerodynamics Laboratories at the Polytechnic Institute of Brooklyn, joined the faculty as department head in 1979, where he remained in the post until 1987. His influence, combined with existing research, led to the designation of the department by NASA as a "Center of Excellence in Computational Fluid Mechanics."

In 1980, following changes in the structure of the College of Engineering, the Engineering Science faculty were merged into the Aerospace Engineering Department. This significantly strengthened the faculty in the area of solid mechanics and dynamics, and the department name was changed to reflect the emerging Engineering Mechanics program. During approximately the same time period, there was a consolidation of research in biomechanics between the Colleges of Medicine and Engineering. The major component of biomechanics resided in the Aerospace Engineering & Engineering Mechanics Department from 1980 until 2000, when the biomechanical component became the Department of Biomedical Engineering. Thus, in two decades, the thrust of the department grew from well-established programs in fluid and solid mechanics, propulsion, and dynamics and control to encompass biomechanics and computational mechanics.

The department established a NASA Center in Computational Fluid Dynamics in 1980. This was one of only seven centers NASA funded nationwide; the others were Massachusetts Institute of Technology, Pennsylvania State University, Stanford University, Iowa State University, Michigan State University, and the University of North Carolina. The center received over $600,000 for five years and fifteen students were graduated under its auspices. During this time period, the Department of Energy oversaw a major research effort with Drs. Tabakoff and Hamed from 1980 to 1992. This multimillion dollar contract sought to quantify multiphase particulated flow with an emphasis on coal usage.

In 1987 Dr. Donald Stouffer, acting department head, spearheaded a winning proposal for a NASA University Space Engineering Research Center (USERC) training grant, which was awarded to the department in 1988, one of only nine to be funded by NASA under this initiative. The Health Monitoring Center for Space Propulsion Systems included eleven faculty from various departments such as Mechanical, Industrial, Materials Science, Electrical & Computer, and Aerospace Engineering to conduct research in micro-sensor arrays, fusion of distributed sensors, oxygen and hydrogen leak detection, turbo pump fault monitoring and simulation of fault effects, and digital imagine and diagnostics. Several private companies (RocketDyne, Aerojet, Pratt & Whitney and GEAE) were involved in the center's governance and research. The first director was the outgoing department head, Dr. Rubin. He held the post from 1988 to 1990. Dr. Bruce Walker was named Associate Director in 1989 and remained in that position until 1993. In 1990, Dr. Larry Cooper of NASA Lewis was named as director. Shortly thereafter, the Health Monitoring Center won an increase in NASA funding at its scheduled three-year review.

Members of the faculty were also involved in the founding and establishment of the Ohio Aerospace Institute (OAI). This collaborative research organization has grown to be an important part of research opportunity and funding in the state of Ohio, and has facilitated programs ranging from applications of aerospace technology in heart pump design and brain surgery to engineering photovoltaic equipment for the International Space Station. It counts among its corporate members General Electric, Goodrich, Boeing, Parker-Hannifin, Rolls-Royce Allison, Pratt & Whitney, and Honeywell, and is a rich source for cooperation in research in Ohio.

Dr. Stouffer also spearheaded an effort to write a winning proposal to support an Ohio Eminent Scholar in Experimental Fluid Mechanics. This effort culminated in 1994 when Dr. Miklos Sajben was appointed in the position, where he remained for five years. When he retired, Dr. Ephraim Gutmark was recruited for this position and is currently conducting groundbreaking research in flow control, jet noise mitigation, and pulse detonation engines.

In 1989 Dr. George Simitses was recruited from the Georgia Institute of Technology to become the new department head. During his tenure, he increased graduate enrollment, added faculty, and later served as the College's interim dean. Upon his departure in 1994, Dr. Gary Slater, an alumnus of the UC Aerospace Engineering program, stepped up to become head, where he remained until 2001. During his tenure, the department increased its research funding by over fifteen percent each year, revamped the curriculum in response to industrial and accreditation requests, and resisted several attempts by the College of Engineering to merge the Aerospace Engineering Department with other departments. The current head, Dr. Awatef Hamed, also an alumna of the Aerospace Engineering program, is a recognized expert for her research in aeropropulsion. Under her watch, the current research funding in the department has doubled to over $5,000,000 in FY2003, including support from such companies as General Electric, Ford Motor Co., Parker Hannifin, Halliburton, and Allison Rolls-Royce as well as government agencies such as NSF, NASA, the Department of Energy, the US Air Force, Army, and Navy, DAGSI, the Ohio Aerospace Institute, and the Ohio Department of Transportation.

By the early 1990's the winds of change were again blowing in the academic world. Many faculty, as well as industry employers, were beginning to recognize that the pendulum had swung too far towards science and not enough emphasis was being placed on real "engineering" and creative problem-solving skills. The aviation industry had changed from a number of small engineering firms to a few large corporate monoliths. Engineering in the modern, large corporate environment required the engineer to have more communication skills than in earlier days. In addition, the science of engineering was taught as a set of discrete topics with few relationships between courses. The nascent engineer had little training in cross-disciplinary problems which were prevalent in industry, and hence had difficulty adjusting to the unstructured nature of the problems to be faced.

With this backdrop, the department launched a major effort to evaluate its curriculum, and in particular to find out what people in industry and government felt were the critical skills needed but possibly lacking in current engineering graduates. With financial sponsorship from NASA and the Ohio Space Grant Consortium, the Department initiated a major study to update their program. The start was to survey a wide technical audience to see what aerospace leaders across the country viewed as critical needs for the modern engineer. To pick as broad an audience as possible, albeit one familiar with the aerospace industry, a survey was sent to all members of selected AIAA technical committees that dealt with topics typically contained in the aerospace program. Not surprisingly perhaps, the actual scientific know-how of the current graduates was generally not an issue. The survey confirmed what most faculty already knew: students, and hence young engineers, tended to treat courses individually, and generally saw little interconnection between the various components of the total curriculum and between disparate technical areas in aerospace.

Fig. 14 The Aerocats, UC's Aerospace team, with their student-designed and -built airplane at the SAE East Design Competition in April 2002, where they placed 8th in international competition after an on-the-runway emergency reconstruction.

To address these problems, the department faculty voted in 1989 to completely revamp the program. The key element was a new set of courses whose sole purpose was to focus on open-ended problem solving and the integration of elements from disciplinary courses. These courses started in the second (sophomore) year, and continued through the five-year program, culminating in the traditional "capstone design" project. Orthodox discipline courses were retained, but the new "integrated" courses required teams of students to examine trade-offs in a design of their creation; to write reports; and to give oral briefings on their projects. These courses covered topics such as application of probability and statistics to engineering with an introduction to six-sigma thinking and a comprehensive approach to aeronautical and space systems [36]. Recent experience has shown that, with this thread running through the entire curriculum, the current seniors are able to work together more effectively during the capstone design [Fig. 14]. However, curriculum issues are never static, and as with ABET criteria, continuous improvement and refinement of the program is sought. It goes without saying that in another twenty years the technical curriculum will evolve in new directions. However, problem-solving skills and creativity will remain key to the success of the graduating engineer.

The evolution of graduate student training has also not been static. In September 2000, UC was asked to join the GEAE University Strategic Alliance (USA) program as a partner in the Heat Transfer and Acoustics programs. Professors Gutmark and Orkwis took part initially, while Professors Tabakoff, Jeng and Nagy continued to work with GE on related programs. The USA program pools GEAE academic research resources and directs them to selected institutions that have an interest in and are dedicated to solving industrial research problems. Members of the coalition include such major research universities as Stanford University, Duke University, Ohio State University, Universität Aachen, and ETHZ (Switzerland). However, unlike most partner

Fig. 15 Class of 2006 tests student-designed wings in Nippert Stadium

institutions, UC graduate students have a unique advantage: many work directly with GE engineers at the Evendale site, thus resulting in a *de facto* graduate co-operative program. These students spend approximately half their day at UC and half at GEAE while classes are in session; during recess periods, students spend most of their time at GE. GE has also supported its investment in other ways, such as funding a new Beowulf cluster at UC as well as cascade and anechoic facilities.

Looking to the Future

In 2003, as we approach the 100th anniversary of Cooperative Education, the co-op program has changed only slightly since 1956. Aerospace students attend school full time through the first quarter of their sophomore year; the curriculum includes early design experience starting from the freshman year [Fig. 15]. At this point they are divided into two sections and begin alternating twelve-week terms of undergraduate work and co-operative assignments [Table 3]. All students serve their final work rotation the summer before their fifth year and spend the entire senior year at school. This represents a decrease from approximately two years to one and a half years of work experience as compared to a student in 1956, a necessary sacrifice to allow for the technical electives required to graduate qualified engineers considering today's rapid advances in technology.

Freshman	Sophomore	Pre-junior	Junior	Senior
Calculus I, II, III & Labs; Physics I, II & Labs; Chemistry I, II	Calculus IV & Lab; Physics III & Lab; Differential Eqns; Matrix Methods	Engineering Measurements Lab; Mechanics III; Basic Strength of Materials; Nature & Properties of Materials; Basic Thermodynamics	Basic Heat Transfer	Technical Options
				Technical Options
Computer Language; Graphic Fundamentals; Mechanics I	Mechanics II; Basic Fluid Mechanics		Integ. Spacecraft Engrg; Fund. Control Theory; Matrix Structural Analysis; Gas Dynamics; Flight Mechanics; Applied Aerodyn; Air-breathing Prop; Solids Lab	Aircraft Design I,II / Spacecraft Design I,II / Engine Design I,II; Rocket Propulsion; Fluids/Aerodyn Lab; Prop. Gas Dynamics Lab; Composites
Intro to Aerospace Engrg; Intro to Aircraft Engrg; Intro to Spacecraft Engrg	Probabilistic Engrg; Basic Integrated Engrg	Integ. Aircraft Engrg; Modeling & Sim. Phys. Systems; Aerospace Vehicle Performance; Fund. Aerodynamics; Vibrations; Num Meth Engrg Design		
English I, II, III	Foreign Language option		Foreign Language option	Foreign Language option
Humanities and Social Sciences	Humanities and Social Sciences		Humanities and Social Sciences	Humanities and Social Sciences
Professional Development	Co-op (24 wks)	Co-op (24 wks)	Co-op (24 wks)	Professional Development

Table 3. Aerospace Engineering curriculum, 2003

Fig. 16 UC Aerospace students conducting nano-satellite micro-gravity experiments on NASA's KC-135 zero-gravity aircraft Advisor: Dr. Trevor Williams

The aerospace co-op program at the University of Cincinnati currently has an active workforce of approximately eighty students each year out of a total enrollment of 240 undergraduates. Companies include the Aircraft Engines Division of General Electric, the Air Force Research Laboratories at Wright Patterson Air Force Base, Gulfstream Aerospace, Belcan Engineering, NASA, Lockheed, the Naval Surface Warfare Center, Naval Research Labs, and Pratt & Whitney. Over fifty other aerospace companies and government research laboratories have participated in recent years. As the department approaches its 75th anniversary in 2004, the faculty, students, and alumni can look back on a proud tradition of comprehensive education and advanced research that has resulted in over 2500 graduates. As one of very few aerospace programs with a mandatory co-operative element, current annual research funding over $5 million, world-class facilities in aeroacoustic and combustion diagnostics, non-destructive testing and evaluation, and aeropropulsion flow simulations, aerospace students from the University of Cincinnati [Fig. 16] will continue to be leaders in the field for the foreseeable future.

Acknowledgements

E. Sam Sovilla, Emeritus Professor of Professional Practice, University of Cincinnati, for unique background material on the co-operative program; Richard Bachman of Dayton, Ohio, for his tireless and indefatigable research on the subject of Bradley O. Jones; Anna Truman and Kevin Grace of the University of Cincinnati Archives, for their guidance in obtaining student and faculty data; Catherine Rafter and Thomas Curtis of the College of Engineering, University of Cincinnati, for providing the Herman Schneider Centennial program; and Bruce Walker, Associate Professor and Graduate Director and Prem Khosla, Professor, Aerospace Engineering & Engineering Mechanics, University of Cincinnati, for information on UC's NASA Health Monitoring Center for Space Propulsion System; and James E. Wade, Associate Professor of Aerospace Engineering & Engineering Mechanics, University of Cincinnati, for extensive and amusing personal reminiscences; Dorothy Byers, University of Cincinnati Engineering Library, for finding obscure and aged CoE periodicals; and Frank Tepe, for communication, networking, and proofreading.

Appendix A: Department Heads & Faculty

Year	1929-1957	1959-1970	1971-1978	1979-1987	1988-1994	1994-2001	2001-present
Head	Bradley Jones	Paul Harrington	R. Thomas Davis	Stanley Rubin	George Simitses	Gary Slater	Awatef Hamed
Faculty	Bisplinghoff Doty Dunhalter Elnan Liepmann Murray* Nolte Powell Sibert Wong *R. Murray Acting Head, 1957-1958 **Robert Kroll, Acting Head, 1958-1959; 1970-1971; 1978-1979 ***Donald Stouffer, Acting Head, 1987-	Adams Blitch Calico Clevenger Dale Elnan Ghia Hesse Jain Lee Li Loch Macke Mulholland Murovitch Murray Nagaraya Oguro Osborn Pasarello Peters Polak Saarlas Sheng Tabakoff Tepe Vitale Walchner Weidenhamer Wells	Adams Armstrong Benzakein Calico Clevenger Converse Da Silva Elnan Ghia Hamed Hope Kauffmann Kroll** Lee Nayfeh Pasarello Polak Savell Sheng Slater Sobel Tabakoff Wells Weidenhamer Werle Zipkin	Armstrong Becus Butler+ Da Silva Davis Disimile Elnan Gau Ghia Grood+ Hamed Khosla Kroll** Lin Nayfeh Osswald Polak Richardson Sheng Slater Stouffer*** Tabakoff Wade Young Zipkin +founded Biomedical Department	Abdallah Becus Butler+ Cooper Disimile Fan Ghia Grood+ Hamed Jeng Jinghong Khosla Kroll Lush Nayfeh Newberry Osswald Polak Richardson Rubin Slater Stouffer Tabakoff Wade Walker Williams Young	Abdallah Becus Butler Cooper Disimile Ghia Grood Hamed Jeng Khosla Kroll Nagy Nayfeh Newberry Orkwis Osswald Polak Richardson Rubin Sajben Simitses Stouffer Tabakoff Tabiei Wade Walker Williams Wysong Young	Abdallah Abot Becus Disimile Ghia Gutmark Jeng Khosla Nagy Nayfeh Orkwis Osswald Richardson Slater Tabakoff Tabiei Wade Walker Williams

Appendix B: Aerospace Engineering's Distinguished Alumni Award Recipients

1969
Raymond Bisplinghoff, ASE '40
L. Lee Burke, ASE '39
James L. Burridge, ASE '43
J.Russ Daniell, MAsE '36
Howard A. McGlasson, ASE '38
Thomas C. Raudebaugh, ASE '40
Robert R. Templeton, ASE '35

1970
Oscar W. Dillon, ASE '51

1971
Warner L. Stewart, ASE '49

1973
Walter Cornelius, AeroE '38

1976
Donald E. Hoak, AE '54

1977
George Fair, BSAE '57

1979
F. Robert Heyner, AE '53

1981
Clark E. Beck, ME '55, MSAsE '69

1983
Lawrence A. Lantzer, AE '61

1984
Linwood C. Wright, BSAE '44, MS '60

1985
Barry W. Hannah, BS '63, MS '65, PhD
'73

1986
Odin Elnan, BSAE '51, MSAE '56
George K. Grant, PhDAsE '73

1987
Leonard Beilch, MS '59, PhD '71

1988
Rodney A. Boudreaux, PhD '65

1989
Joseph A. Mancini, AE '35
Norman E. Nelson, AE '41

1990
Norman D. Malmuth, ASE '53

1991
Donald J. Dusa, MS '62

1992
Gregory M. Reck, ASE '65

1993
Kenneth H. Token, ASE '63

1994
Robert A. Calico, Jr., AE '66, MAS '68, Ph.D '71

1999
Edward M. Kraft, BS '68
Dinesh Keskar, MSASE '77, Ph.D. '79

2000
James Younghans, BSAE '63, MS '78
Richard Johnson, BS AE '73

2003
Thomas Wakeman, MAS '71, PhD '82

References

1   *Introduction to Professional Practice*, 3rd Edition. University of Cincinnati Div. of Professional Practice, Cincinnati, OH: McGraw-Hill, 2001

2   Park, Clyde W. *Ambassador to Industry: The Idea and Life of Herman Schneider*. Bobbs-Merrill, Indianapolis, IN: 1943

3   *American Machinist*. 19 Oct 1905

4   Burns, Paul. "Looking Back – 25 Years," *The Cooperative Engineer*, Oct 1930

5   Park, *Ambassador to Industry*, Bobbs-Merrill, 1943

6   Sovilla, E.Sam. "Co-op's 90 Year Odyssey," *75th CED Anniversary Magazine*, American Society for Engineering Education

7   Joerger, C. Albert. "College of Engineering," *UC Alumnus*, Spring 1956

8   Bachman, Richard. *Bradley Jones: Dayton's Forgotten Avation Expert*, Sycamore Creek Publications, Beavercreek, OH: unpublished proofs.

9   Boston Herald, 15 Oct 1923

10   Lindbergh, Charles A. *The Spirit of St. Louis*, New York: Charles Scribner's Sons, 1953, p.153

11   Bachman, *Bradley Jones:* unpublished proofs.

12   The Cincinnati Enquirer, 23 Jun 1929

13   Jones, Bradley. "The New Course in Aeronautic Engineering," *The Co-operative Engineer*, Oct 1929

14   Schuster, "Airplane Rib Structures," *The Co-operative Engineer*, Jun 1930

15   Schuster, "Development of Aviation in Cincinnati," *The Co-operative Engineer*, Jun 1931

16   Jones, Bradley. "The New Course in Aeronautic Engineering," *The Co-operative Engineer*, Oct 1929

17   Whissel, Pamela. *The First Cooperative College*, University of Cincinnati College of Engineering, Cincinnati, Ohio: 1993, p. 51

18   Jones, B. *Aerodynamics for Pilots*, U.S. Department of Commerce, 1940: Civil Aeronautics Bulletin #26

19   Bisplinghoff, Raymond S. *Aeroelasticity* (Reading, MA: Addison-Wesley, 1955); *Principles of Aeroelasticity* (New York: John Wiley & Sons, 1961); *Statics of Deformable Solids* (Reading, MA: Addison-Wesley, 1965)

20   College of Engineering enrollment records: handwritten index cards. University of Cincinnati, Cincinnati, OH: 1932-1969 inclusive

21   College Bulletins. Cincinnati, OH: University of Cincinnati, Vols. XXIII – XLI inclusive

22   Whissel, P. *The First Cooperative College*, Cincinnati, Ohio: 1993, p. 54

23   Reiter, Sydney. Personal reminiscences during a visit to the University of Cincinnati Department of Aerospace Engineering, 21 Jul 2003

24   Jones, B. *Elements of Practical Aerodynamics*, John Wiley & Sons: 1936.

25   Wade, James E. "Personal Reminiscences," manuscript, 14 March 2003

26   Stewart, Warner L. Email correspondence with A. Hamed, 28 Aug 2001

27   Snyder, James. Telephone correspondence with A. Morrison, Mar 21 2003

28   Gulledge, Frank. *THE PURPLE FOXES: Marine Medium Helicopter Squadron 364*, last updated 10 Jul 2003. 24 Jul 2003 <http://www.hmm-364.org/Ad.htm>

29   Tepe, Frank. "Some Additional Info for AsE History Article," email to A. Morrison and A. Hamed, 22 Jul 2003

30   Cincinnati Enquirer, Sunday, 9 Sep 1951

31   Cincinnati Post & Times-Star, 28 Mar 1968

32   Nelson, E. Telephone correspondence with A. Hamed, General Electric Aircraft Engines, 15 Mar 2003

33   Curtis, Thomas C. "Distinguished Alumni/ae," *University of Cincinnati College of Engineering website*, last updated 21 Apr 2003. 21 Jul 2003 http://www.eng.uc.edu/friendsalumni/alumniae/

34   Tabakoff, W., Hamed, A., and Wenglarz, S. "Particulate Flows and Blade Erosion," VKI LS 1987-88, Von Karman Institute of Fluid Mechanics, Brussels: May 24-27, 1988

35   Curtis,T. "Distinguished Alumni/ae," *University of Cincinnati College of Engineering website*

36   Walker, B.K., Jeng, S.M. , Orkwis, P.D., Slater, G.L., Khosla, P.K. and Simitses, G.J. "Redesigning an Aerospace Engineering Curriculum for the Twenty-First Century: Results of a Survey," *J. Engineering Education*, vol. 87, no. 4, pp. 481-487, October 1998.

Chapter 16

A HISTORY OF EDUCATION IN AEROSPACE ENGINEERING AT CORNELL UNIVERSITY

F.K. Moore, Profesor Emeritus, Sibley School of Mechanical and Aerospace Engineering

Introduction

At the turn of the last century, and until WWII, Aerospace Engineering meant the development of aircraft and airships for man to sail in the air as he had learned to do in the oceans. In general, the academic world at first paid little attention to this activity; The Wrights, Curtiss, Lanchester and others were certainly engineers, but not of the academic variety. However, readers of Horace Lamb's treasure chest of fluid mechanics theory, "Hydrodynamics" (first published in 1879), know that by the late 1800's academic research had already established a firm scientific basis for advances that would come decades later in the aerodynamics of flight. It would seem that Prandtl's school in Germany during the first few decades of the 20th century represented the real beginning of a powerful academic tradition in Aeronautical Engineering which was carried forward by Prandtl's students, notably Theodore von Karman who came to Caltech in 1928, and who indirectly but strongly influenced later events at Cornell.

During and soon after WWII, aeronautical interests broadened to include travel beyond the atmosphere, manned or unmanned, driven chiefly by military motives. Then, "Aerospace" Engineering perforce entered worlds of Physics and Chemistry far beyond that of the mechanics of an ideal fluid, and academic activities consequently dispersed and multiplied, so that it quickly became impossible for any academic department of Aerospace Engineering to embrace the totality of the subject of flight. At Cornell, and perhaps elsewhere, the response to this situation has been for Aerospace Engineering to form relationships with other departments for research and instructional purposes, to help form new academic entities, such as a Laboratory for Plasma Studies, and especially to rejoin Mechanical Engineering, with various Aerospace topics continuing to provide important but now highly diverse educational opportunities for faculty and students. This essay will describe how this historical process has worked out at Cornell; it was written with major contributions by Edwin L. Resler, Jr. and helpful suggestions by David Caughey, Frank Moon (these three are past Directors of the Schools to be described), by John Lumley, by Alice Anthony, and by Sidney Leibovich, who is the present Director of the Sibley School of Mechanical and Aerospace Engineering at Cornell.

Early History at Cornell, Before 1946

Ithaca was the scene of aircraft manufacturing during WWI; the Thomas-Morse company built 800 "Tommy Scout" trainers and many other more advanced aircraft there under government contract. Even so, there was only modest interest in education in aviation science at Cornell before 1946, despite urgings around 1900 by Robert Thurston, then Director of the School of Mechanical Engineering. The famous aviation pioneer Octave Chanute had given a lecture at Cornell in the 1890's, and a leading Cornell professor of Physics, Frederick Bedell, published a book "Airplane Characteristics" in 1918, based, he said, on certain courses he had given. One would have expected interest in aeronautics to have developed promptly in the School of Mechanical Engineering, but that happened rather belatedly. It is interesting that the subject of Marine Engineering was prominent in Mechanical Engineering, at least at Cornell, during those early years. In fact, a notable figure in aeronautics, W.F. Durand, whose later six-volume series was so important for aeronautical education, began his academic career in 1891 as professor of Marine Engineering in the Mechanical Engineering School at Cornell. He went on to Stanford University in 1905 to be head of Mechanical Engineering there, focusing on power-plant problems. But then he joined and chaired the newly-formed NACA in 1915, and thereafter became a national leader, teacher and author in the growing field of Aeronautics.

The first regular instruction in aeronautics at Cornell appeared in 1929, in the Sibley School of Mechanical Engineering. K.D. Wood, a young professor of Mechanics, began offering undergraduate elective courses in "Aerodynamics" and "Aircraft Design". In 1935, he published one of the first comprehensive books on aircraft aerodynamics, "Technical Aerodynamics". Also, in the years leading up to WWII, and during the war, Professor I. Katz in Mechanical Engineering taught aircraft engine design.

1946-1962

At the close of WWII, universities generally felt the need to acquire strength in Aeronautics, and each pursued that goal in a distinctive way. Some, like Caltech, by virtue of Karman's leadership, were already firmly established in that field. Cornell, however, began almost afresh. A new and unique "Graduate School of Aeronautical Engineering" (GSAE) was established at Cornell, to be directed by W.R. Sears, who had been a student of von Karman's, and who had been Chief aerodynamicist at Northrop Aviation during the war, designing the " Black Widow" fighter and the first "flying wing" (XB35).

At the same time, the Curtiss-Wright Corporation donated its laboratory in Buffalo, N.Y., including a major wind tunnel, to Cornell University. There was, for a time, a question how this facility should relate to Cornell's new venture in aeronautical education. It was soon decided that the "Cornell Aeronautical Laboratory" (CAL) would be an independent, not-for-profit corporation owned by Cornell University, but fully funded by contract research, with no programmatic or administrative ties to the new "Aero School" at Cornell. Relationships did nevertheless develop; certain professional exchanges occurred, and many Aero School graduates found employment in Buffalo. Notably, for many years the CAL provided an important stream of fellowship support for graduate students at Cornell, which did not have such support from the Guggenheim Foundation as did several other aeronautical engineering schools.

The GSAE at Cornell was uniquely independent; it was not part of Mechanical Engineering, for example. A graduate school only, its first degree followed a rigorous two-year Master's degree, requiring a research thesis. This degree was granted independently from the Graduate School, which is the academic entity normally managing graduate degree programs at Cornell. The result of this unusual and rather extreme independence was not academic isolation, as one might first suspect, but rather the reverse; students and faculty were free to develop a wide range of relationships around the Campus. As will be described subsequently, this happened to a spectacular degree, owing not only to institutional freedom, but to the talents and inclinations of

W.R. Sears and the fine faculty he assembled at first and in the years to follow. In the period 1946-1962, 100 students received the professional (2-year) Masters degree from the Aero School, and many others received the PhD with majors in Aeronautical Engineering.

In its first year, professorial faculty of the GSAE comprised, in addition to Bill Sears, Arthur Kantrowitz from NACA (Langley Field and the world of Eastman Jacobs), Y.H. Kuo from Caltech, and John Wild from Northrop Aviation. These were the aerodynamicists; and Fred Ocvirk and later Carlo Riparbelli covered structures. Al Flax was a Lecturer, visiting from CAL, covering the then new field of helicopters. Alice Anthony was secretary and administrator. A bit later, in 1951, the aerodynamicist Nicholas Rott came to GSAE from the ETH Zurich, Switzerland. Fig. 1 is a photo taken in 1946 or 7.

Fig. 1. The early faculty of the Graduate school of Aeronautical Engineering; Director William R. Sears is in the center, flanked by (from left to right) by David Sears, his son, and Professors Y.H. Kuo, John Wild, Arthur Kantrowitz, and Carlo Riparbelli.

Being a graduate school the GSAE offered no undergraduate program. However, Aero faculty taught courses in Thermodynamics and Statistical Mechanics for the undergraduates in Applied and Engineering Physics, which was also a new post-war department in the Engineering College with a very demanding and ambitious undergraduate program. Professor Sears himself taught a course in Continuum Mechanics to students in that department. Over the years, the Engineering Physics program came to provide many fine graduate students for the GSAE. This was an example of the productive, long-lasting, collegial relationships which were fostered by Sears and his faculty with other science departments at Cornell and which became a tradition of the School.

Most of the Aero students came from outside Cornell, except for graduates of the Engineering Physics program just mentioned. It should also be noted that the School of Mechanical Engineering at Cornell acquired distinction in the propulsion side of Aeronautics when D.G. Shepherd joined that School in 1951; he had been a key member of the British team that developed the "Whittle engine" during WWII. Many of his students became leaders in the Aircraft engine industry.

In these early years, research by GSAE faculty and students included aerodynamic problems of airplanes such as engine-wing interference, advances in boundary-layer theory, slender-body theory, supersonic wing theory, and unsteady aerodynamics of turbomachinery. Bill Sears was an enthusiastic pilot, flying his own aircraft--first a Mooney Mite and later a Twin Comanche. This enthusiasm influenced GSAE students, most of whom learned to fly in Ithaca, learning first hand about adverse yawing moment! Under the leadership of Kantrowitz, a vigorous and influential program of research in physical and chemical gasdynamics developed; wave processes were used to explore nonequilibrium kinetics of gases, laying the groundwork for the reentry studies which became of such absorbing national interest in the late '50's. Research support for the GSAE came from the Office of Naval Research, Air Force Office of Scientific Research, and NACA (NASA). The U.S. Army and the Cornell Aeronautical Laboratory also provided support in the forms of graduate fellowships, equipment and facilities.

Director Sears attracted a steady stream of distinguished visiting scientists from around the World to the GSAE, greatly adding to the vitality of the School's program of research and education. Some, like S.I. Pai and F.N. Frenkiel, stayed for a term or longer; extended visits were also made by von Karman, and by a succession of Japanese aeronautical scientists, Professors Imai, Kondo, Fukuda, and especially Professor Itiro Tani, who visited many times and became a great friend of the GSAE over the years. Among other notable visiting scholars were Professors Marble, Lighthill, Batchelor, Hayes, Liepmann, Goldstein, Spence, Burgers, G.I. Taylor, Alfven, Ferri , Laufer, Oxburgh, R.T. Jones, Bogdonoff, Atassi, and Lykoudis. Student experiences were of course greatly enriched by contacts with these world-famous scholars.

Research in the GSAE was much more communal than is usual in academia. Each student found help and encouragement from all the faculty and visiting scholars; there were no "turf" issues. A weekly "research conference" was a compulsory event for all; the shock-tube experimenter had to listen to the troubles of the boundary-layer theorist and vice versa, and all were encouraged to question and comment. The result of this style imposed by Sears was, as one could expect, an atmosphere of vitality, optimism (so important in research), and esprit de corps.

The late '50's were years of change. "Aeronautical" became "Aerospace", NACA had become NASA, and the GSAE, which had been instrumental in many of these changes, changed too. John Wild left to be Chief Engineer of the new Arnold Engineering Development Center of the Air Force in Tullahoma, Tennessee. The program of flow physics and chemistry led by Professor Kantrowitz led him to a deep involvement in the effort to develop heat protection for reentry vehicles, so vital for the ICBM and later for the Apollo vehicles of NASA. In 1956, he left Cornell to found and direct the AVCO Everett Research Laboratories, where those goals were successfully pursued with the help of the many Aero School graduates he employed. In the same year, his student, E.L. Resler Jr, returned to Cornell from the University of Maryland and assumed leadership of the flow physics program of the GSAE. In keeping with the outreach tradition of the School, Professor Resler helped teach the Graduate Physics Laboratory course in the Physics Department, Kinetic theory of Gases for the Electrical Engineering Department, and Thermodynamics for the Materials Science Department. Y.H. Kuo went home to China, where he had a distinguished subsequent career, recognized by a statue in his honor in Beijing. Professor Ocvirk returned to Mechanical Engineering at Cornell, where he became recognized for his work on the theory of short bearings. Al Flax went on to become Technical Director of CAL, Chief Scientist of the Air Force and head of the Institute for Defense Analysis in Washington, D.C.

N. Rott left GSAE in 1960 for a professorship at UCLA. S.F. Shen and P.C.T. deBoer arrived from U. Md. New junior members of the GSAE faculty were D.L. Turcotte, A.R. George, and A.R. Seebass. The Aero School had greatly extended its influence into the outside world of Aerospace, but, through new faculty additions, had nevertheless maintained and extended the School's great strength in experimental and theoretical flow physics, and continued to graduate students finely equipped for leadership and research in aerospace engineering and in other fields too, as we shall see.

1962-1972

Change continued apace during this time. In 1963, the Center for Applied Mathematics was founded at Cornell, with W.R. Sears as Director. The formation of this new center recognized a tradition at Cornell of strength in applied mathematics. This tradition had influenced and enriched the GSAE even as the GSAE enriched the programs of other departments. The late Professor G.S.S. Ludford in Theoretical and Applied Mechanics (T&AM) exemplified this collegial culture; he performed mathematical treatments of combustion and flows with electromagnetic fields, and taught and advised many students in the GSAE.

Thus, after 17 years, Bill Sears had relinquished his directorship of the GSAE, and was replaced by E.L. Resler, Jr. Professor Sears nevertheless continued his membership in the Aero School and his researches in fluid mechanics. Faculty and students studied the fluid mechanics of mantle convection and continental drift, helicopter noise, and the sonic boom thought to limit the feasibility of commercial flight at supersonic speeds. Research on unsteady boundary layers continued, and studies began on Hydrogen as an alternate fuel, and the mechanics of ferro-magnetic fluids.

This was a period of great interest at Cornell in magnetohydrodynamics (MHD), especially plasma wave processes in space, partly driven by the development of the radio telescope at Arecibo, P.R., and having also in mind promising possibilities for terrestrial power generation by MHD means. Aero School involvement in theoretical and experimental MHD research was instrumental in the founding in 1967 of the Laboratory of Plasma Studies (LPS) at Cornell, as part of a consortium including the University of Maryland and the Naval Research Laboratory. P.L. Auer came from ARPA in Washington to be a professor in GSAE and the first Director of the new LPS. Early successes of the LPS were in developing new plasma microwave devices, advances in experimental studies of plasma turbulence, and in flow laser research. Faculty and students of GSAE joined with those of Electrical Engineering, Chemistry, Engineering Physics, and Mechanical Engineering to achieve these results. As it evolved, the program of LPS turned toward problems of plasma containment for fusion. At Cornell, this effort was led by Professor R. Sudan of Electrical Engineering. Elsewhere, major efforts in this direction at Livermore, Sandia, and the University of Rochester came to be directed by veterans of the plasma studies consortium.

Meanwhile, high-temperature chemical kinetics and chemical lasers were being studied by faculty and students of GSAE in collaboration with Professor S. Bauer and associates in the Chemistry Department at Cornell. Shock-tube techniques were basic to this effort, and participating Chemistry students took the relevant Gasdynamics course in GSAE. The quantitative successes of this program have been widely influential.

Despite these important and successful initiatives, this was a period of contraction in the aerospace industry, and GSAE student enrollment was of course adversely affected. It did not help that Cornell University, in 1967, terminated its ownership of the Cornell Aeronautical Laboratory, ending the fellowship support that CAL had provided GSAE over the years. However, it was fortunate that in that same period, the traditional interests of Mechanical Engineering in fluid mechanics and thermal and combustion processes were reasserted at Cornell by the faculty additions of F.K. Moore from CAL, and a number of new younger faculty in the Thermal Engineering Department of that School, namely Professors Cool, Leibovich, Gouldin, Torrance and McLean. Various graduate research topics of aerospace interest were undertaken, largely under NASA sponsorship. Examples included studies of various vortex and wave processes , especially those underlying the phenomenon of "vortex breakdown". There were advances in combustion modeling, and studies of environmental fluid mechanics, as well as the flow-laser work already mentioned in connection with the LPS. The Aerospace-related content of the undergraduate curriculum of the Mechanical Engineering School was enhanced as well.

1972-Present

In 1972, the Mechanical Engineering and Aerospace Engineering Schools were combined, with E.L. Resler, Jr. as Director, to form the Sibley School of Mechanical and Aerospace Engineering (M&AE), with the traditional format of undergraduate program and graduate fields. At present, all faculty are members of both Mechanical and Aerospace graduate fields, as are certain members of Theoretical and Applied Mathematics. There is no undergraduate degree in Aerospace Engineering, although there is a popular upperclass concentration available, involving courses in aeronautics, spacecraft engineering, propulsion systems, flight vehicle dynamics and advanced fluid mechanics. Also, undergraduates may engage in Aerospace research projects in those same areas, for credit.

In 1974, W.R. Sears left for the University of Arizona, ending his long tenure at Cornell. However, in 1985, the Sibley School of M&AE established the "William R. Sears Lectureship" in Bill Sears' honor, and each year since then a distinguished figure in aerospace research has come to Cornell for a few days' visit, giving a formal lecture and consulting with faculty and students. Until a few years before his death in 2002, Professor Sears flew his airplane into Ithaca along with his wife Mabel to join in these festivities and to make the introduction of each "Sears Lecturer". Supported by a gift from Mabel Sears, the annual Sears Lectures will continue in the future as an important element of Cornell's program of aerospace education.

Soon after the merger forming the M&AE school, many other faculty changes significant for aerospace education at Cornell occurred. In furtherance of his laser research, Professor Cool joined the Engineering Physics department, and Professor Turcotte's studies of mantle convection led him to join the Geological Sciences Department, of which he subsequently became Chair. In 1974, D.A. Caughey joined the faculty from McDonnell Douglas Research Labs, bringing expertise in computational fluid dynamics, especially as applied to transonic wing design. Then in 1978, J.L. Lumley and Z. Warhaft joined the faculty, giving Cornell a particular eminence in turbulence research, enhanced by the subsequent arrivals of S.B. Pope in 1982 and L. Collins in 2002.

Of course, changes continue; faculty arrive and faculty retire (five Directors have followed Resler; S. Leibovich is the present Director); now the active M&AE faculty numbers 29, of whom about half are involved in research and teaching with a specific aerospace emphasis. Undergraduate classes are about 100, about one-quarter of whom elect an aerospace concentration. About 9 PhD's and 50 Master of Science and Master of Engineering degrees are awarded annually; again, typically one-quarter of these have an aerospace major. The aerospace involvement is strong and student interest is high, but Aerospace topics are widely dispersed; no person, nor any school, can embrace the entire aerospace field to the degree possible in 1946. Thus, we see a mosaic of discrete aerospace activities, often not formally identified as such, that reflect highly individual backgrounds and interests. This evolution is a natural one, but difficult to predict and plan for. After seeking outside advice, and conducting faculty "retreats", the M&AE school determined to make faculty additions expressing a stronger emphasis on issues of space technology, while still maintaining its tradition of distinction in the aeronautical sciences. Cornell's aerospace program would seem to properly reflect present realities, and should assure Cornell's continued leadership in aerospace education.

Certain important achievements of M&AE in the wide field of aerospace during the past few decades are outlined in the next few paragraphs, identifying the faculty responsible, who were of course assisted by graduate students and visiting scholars. This outline also serves to describe the present aerospace program of the M&AE School at Cornell. Faculty now emeritus are indicated by an asterisk.

Studies of the chemical kinetics of NOX production in internal combustion engines and in engines for high-altitude supersonic transports led to practical improvements covered by Cornell patents. This work was led by E.L. Resler, Jr*, who also initiated a program at the Glenn Center of NASA looking toward the application of wave-rotor principles to jet-engine design. An experimental study of fuel droplet combustion in microgravity conducted by T. Avedisian in a drop tower at Cornell has yielded important results and continues to do so. Also, experimental combustion research and modeling are long-term programs of F. Gouldin and E. Fisher and their students. Studies of the physics of secondary propulsion devices for space vehicles were carried out by I. Boyd (who recently left Cornell for the University of Michigan). Particularly significant advances in understanding turbulent combustion were achieved by S. Pope and his students through introduction of a computational method involving a probability density function (PDF). The results of this influential work have been incorporated in commercial CFD packages.

Other turbulence studies by Lumley* and Warhaft along with S. Leibovich and P. Holmes (then of Theoretical and Applied Mechanics (T&AM)) advanced understanding of coherent turbulent structures and the statistics of scalar quantities in turbulence. This program benefited from the work of many students and visiting scholars, and through Professor Lumley's leadership, a cooperative arrangement for studies of fluid mechanics and turbulence was established by Cornell and L'Ecole Centrale de Lyon, France. These various activities continue at a high level as reflected, for example, in the publication of Professor Pope's book "Turbulent Flows", and in recent studies of multi-phase turbulence undertaken by L. Collins.

An energetic experimental program concerning the dynamics of wake vortices was established by C.Williamson, who arrived in 1990. This program provides new understandings of the genesis and interaction of vortices, through the work of many graduate students and the many undergraduates who are attracted by the opportunity Professor Williamson provides to participate in this research. S. Leibovich advanced the theory of nonlinear waves in vortices, stability of vortices, and vortex breakdown, while advances in the theory of boundary-layer separation were made by S.F. Shen* and his students. P.C.T. deBoer* studied the prospects for general use of Hydrogen as a fuel, and more recently he has investigated novel thermodynamic cycles for refrigeration.

D. Caughey made remarkable advances in the application of CFD methods to the design of airfoils and wings for transonic aircraft. As shown in Figs 2a and b, use of CFD has clarified the old issue of shock formation, showing that smooth transonic flow can be achieved through design, but that small design changes can lead to shocks. Professor Caughey (with J.A. Liggett* of Civil and Environmental Engineering at Cornell) has recently published the first interactive fluid dynamics text, "A Computer-Based Textbook for Introductory Fluid Mechanics" to help students master the techniques of CFD. His teaching interests also include flight dynamics and, reviving a course begun years ago by D.G. Shepherd, aerospace propulsion.

Fig. 2a. Contours of constant Mach number computed by D.A. Caughey for flow about a certain supercritical airfoil shape, at a free-stream Mach number of 0.75 and 0 incidence.
For these "design" conditions, a smooth, shock-free recompression from supersonic (maximum Mach number about 1.2) to subsonic flow conditions on the after part of the upper surface.

Fig. 2b. Mach number contours computed for the same airfoil at a slightly lower stream Mach number of 0.74. In this off-design case, the supersonic zone has broken up into two smaller zones, each terminated by a weak shock.

A. George continued his research on the aerodynamic noise of helicopters, and in recent years has applied aerodynamic principles to the improvement of automotive design. Under his leadership, Cornell undergraduate teams have had great success in winning annual intercollegiate race-car design competitions, year after year. F. Moore* developed a successful model of rotating stall of axial-flow compressors, and he led in the formulation of the more general (Moore-Greitzer) model which describes gas-turbine dynamics when both surge and rotating stall are of concern. This model is the basis for current industrial designs for active and passive control of such undesirable behavior in aircraft engines.

M. Louge and students, collaborating with a colleague in T&AM, is engaged in an experimental study of the collective behavior of particles flowing in a microgravity environment. They are planning a related experiment to fly on the space shuttle. Also, a number of interrelated problems of dynamics and controls in space are being studied, especially by several younger faculty members. Controls and information processing are subjects of ongoing teaching and research by M. Psiaki, with applications to flight-path optimization and control of spacecraft, including global-positioning satellite systems. He also supervises model aircraft design projects and competitions for undergraduates. Dynamic controls, including robotics, are also studied and developed by R. D'Andrea and H. Lipson. In the past few years, D'Andrea has supervised the winning Cornell student team in an international competition for design of autonomous robots (the "RoboCup"). This work, and the techniques involved, are obviously of interest for space application. M. Campbell, in a new program, addresses problems of the control of satellite clusters which must maintain position with great precision in order to form antenna arrays in space. F. Moon studies the dynamics and stability of structures designed for space, such as robotic arms and antennas; these have mechanical properties quite different from those intended for terrestrial application. He also is concerned with failure phenomena, obviously of interest for space. E. Garcia studies novel opportunities for "smart materials" and "smart structures" which can adapt to changing needs in both aeronautics and space, perhaps "morphing" during the course of a mission.

Many important research activities in M&AE may not immediately relate to space technology but should do so in the future. For example, in addition to his important experimental and computational work in heat transfer, K. Torrance has contributed greatly to successful computer-generated imagery research at Cornell, through his theoretical and experimental studies of the radiative properties and behavior of material surfaces. One may reasonably expect this work to find application in future planetary explorations. Also, one may cite M. van der Meulen's research on the physical and bio-mechanical properties of bone as being potentially important for astronaut health.

The foregoing paragraphs describe certain research and project activities. M&AE also offers many courses relating to aerospace. In addition to the usual fundamental courses and the many project opportunities, the following specialized aerospace elective courses have been developed for undergraduates:

Introduction to Aeronautics, Spacecraft Engineering, Global Position System Theory and Design, Finite Element Analysis for Mechanical and Aerospace Design, Modeling and Simulation of Mechanical and Aerospace Systems, Aerospace Propulsion Systems, Dynamics of Flight Vehicles.

For graduate students, these aerospace courses are offered:

Foundations of Fluid Mechanics and Aerodynamics, Fluid Dynamics at High Reynolds Numbers, Stability of Fluid Flow, Physics of Fluids, Elements of Computational Aerodynamics, Multiphase Turbulence, Turbulent Reactive Flow, Turbulence and Turbulent Flows.

Cornell Graduates in Aerospace

Finally, a few remarks may be offered about the experiences of Cornell Aerospace graduates. The GSAE always had a strong international influence, continued in the School of M&AE, and many faculty and students have achieved distinction abroad: A Director of a Mechanics institute in China, a President of the Royal Aeronautical Society (UK), a Director of Canada's National Aeronautical Establishment, and a Director of the ETH (Zurich) all came from Cornell. Apparently, Cornell education in aerospace has inculcated the confidence to try and succeed in various things: Graduates designed a heart-assist device and then developed the first successful artificial heart; one graduate became a leader in California real estate development and joined the Cornell University Council; one became Editor-at-large of AIAA's "Aerospace America" magazine; two were co-inventors of a high-power flow laser; one became a Vice President of Aerospace Corporation; and one became an astronaut. Over the years, many graduates, of GSAE especially, have entered the defense industry, owing to their training in the flow physics and chemistry so important in such fields as missile defense. However, of the many students who received their PhD's from GSAE and MA&E, about fifty have become professors at Cornell and other institutions around the world, and two became Deans of Engineering at major universities. All these far-flung graduates have kept in touch with Alice ("Toni") Anthony who, as secretary and administrator of the GSAE and MS&AE schools from 1946 to 1993, took such good care of them all when they were young. Recognition of Cornell leadership has come in many forms; one might mention that nine past and present GSAE/M&AE faculty and seven graduates have been elected to the National Academy of Engineering, and two faculty joined the National Academy of Sciences as well. Also, two faculty have been elected to the American Academy of Arts and Sciences.

Chapter 17

GALCIT CONTRIBUTIONS TO EDUCATION AND DEVELOPMENTS IN AERONAUTICS

By Anatol Roshko
Graduate Aeronautical Laboratories
California Institute of Technology
Pasadena, CA 91125, USA

The principal contribution of GALCIT has, of course, been its graduates. They have distinguished themselves in the aerospace industry, in academia, in government institutions and the military. Its faculty is known for excellence in research and for leadership roles in national and international enterprises. Its official date of founding as the Guggenheim Aeronautical Laboratory at California Institute of Technology is 1928, just eight years after Caltech's official birth date in 1920, but the ideas and circumstances that led to the creation and success of both had their origins a quarter of a century earlier, in the vision and perseverance of a remarkable American scientist.

In 1903 George Ellery Hale was a distinguished astrophysicist at the University of Chicago, where he had founded the Yerkes Observatory. That year he came to Pasadena to open a site on Mount Wilson for a solar and astronomical observatory, to be funded by the new Carnegie Institution, and to include a 60-inch refractive telescope. Hale and his family quickly became established in the young, flourishing city whose climate and opportunities attracted wealthy families, many of whom took a philanthropic interest in its development. One of them, Amos G. Throop, had established Throop University in 1891, the other date from which Caltech calculates its origins. A "university" in name only, it soon changed that to Throop Polytechnic Institute, mainly a manual arts institute with units that ranged from an elementary school, a high school, courses in business and teacher training and a small College of Science and Technology. In about 1907 this College became the foundation for which Hale began to advocate and promote, from his position of prominence in the community and indeed in the nation, a "high grade institute of technology", concentrating on "doing one thing extremely well", which would "educate men broadly and at the same time make them into good engineers".

His ideas and enthusiasm attracted local philanthropists to the board of the College; they worked on funding for the new institute while Hale, also a trustee, was commissioned to seek out eminent scientists and persuade them to come to Pasadena to build the new school. With the advantage of national and international prominence for the observatories he had created, for his work on solar physics and for scientific journals he had established, and with connections at leading universities and at the Academy of Sciences, he was well positioned for his mission. But it was not accomplished easily or quickly; at first his targets were lured with short appointments, during winter, to lecture and discuss plans for establishing a center of science, while funding for laboratories was being sought. From winter to winter the association would become closer until finally it became a firm commitment. The first was Arthur A. Noyes, a physical chemist at M.I.T., whose course Hale had attended while an undergraduate; his first visit was in 1914. The second was Robert A. Millikan, an experimental physicist at the University of Chicago, famous for the "oil-drop experiment" for which he would later receive the Nobel Prize; he first came in 1917.

In the following years he and Noyes became more and more involved in planning and facilitating with Hale the new direction for Throop College but it was not until 1921, with a new laboratory for chemistry completed and one for physics on the way, that both Noyes and Millikan had finally parted from their eastern institutions and assumed full-time positions at the newly named California Institute of Technology (1920). The "triumvirate" of Hale, Millikan and Noyes (a.k.a. "thinker, tinker and stinker") are honored as the founders of the modern Caltech (Figure 1). Hale never joined the faculty; Noyes was content to devote himself to chemistry; Millikan became Chairman of the Executive Council (the position of President was recreated only after his retirement in 1946). The character and direction of the Institute after 1920 was shaped largely by R. A. Millikan, hence the title, "Millikan's School", of Judy Goodstein's readable history of Caltech[1] (from which quotations by Hale, Noyes and Millikan are re-quoted here).

Noyes and Millikan were scientists, so it is interesting to speculate why it is the California Institute of Technology and not the California Institute of Science. Hale, himself a scientist, tended to emphasize "engineering" and "technology" in expounding his ideas for a center of excellence in the West: ... the "prime object should be to educate men capable of undertaking vast projects". On the other hand, Noyes felt that... "The main field for educational institutions is research in pure science itself--a study of fundamental principles and phenomena, without immediate reference to practical application... Scientific investigation is the spring that feeds the stream of technical progress, and if the spring dries up the stream is sure to disappear." Millikan, a pragmatist, steered the new Institute on a course that embraced the views of both his colleagues. In fact, he leaned naturally to the views of Hale on the importance of technology in stimulating scientific research and he did not take kindly to "ivory-tower" attitudes. These views were of course invaluable in his presentations to wealthy entrepreneurs whose fortunes were tied to the development of Southern California.

In this vein, R. A. Millikan was already working with Hale in 1917 to bring Aeronautics to the still evolving College of Technology. By then it was obvious that the success of Europeans in having surpassed the United States in the design of aircraft was in large part due to the development of aeronautical science with theoretical foundations and strong experimental support. Hale, from his vantage point at the Academy of Sciences, saw the need for a similar elevation of the role of science in American aeronautics and he perceived his budding institute of technology in Pasadena as a good place to start. His search for a theoretician led to Harry Bateman. In 1917 Bateman was a lecturer at Johns Hopkins University, having immigrated to the U.S.A in 1910. A graduate of Trinity College, Cambridge in 1905, he was already well known in his field, with for some sixty published papers. The field was mathematical physics, with emphasis on mathematics, largely inspired by equations in physics including hydrodynamics. Millikan persuaded him to leave Johns Hopkins and accept a position as Professor of Aeronautical Research and Mathematical Physics, which he assumed in 1918, introducing courses in theoretical hydrodynamics and elasticity.

At the same time that Bateman was being courted, Hale persuaded one of Pasadena's millionaires to provide funds for an aeronautical laboratory with a wind tunnel (four feet by four feet, up to forty miles an hour) as its centerpiece. It was already built when someone was finally found to operate and use it. This was Albert A. Merrill, an early aviation pioneer who was a founding member of the Boston Aeronautical Society in 1894, learned to fly in Dayton in 1910, had acquired experience in airplane design and construction as well as wind-tunnel testing, but had no formal education in science or engineering. His initial appointment was Research Assistant to Bateman and later Instructor in Experimental Aeronautics and Accounting, the latter designation reflecting his training and profession. Self taught in the tradition of the early American inventors, Merrill knew a lot about practical aeronautics of the time. While Bateman was giving a course on the Design of Aerofoils (sic) and Streamline Bodies, Merrill was teaching students how to measure forces and moments in the wind tunnel and helping them design an airplane and a glider. These footholds on both the theoretical and practical sides of aeronautics became important when Millikan approached the Guggenheim Foundation for funding for a more ambitious program.

But preceding that proposal and certainly contributing to it was the presence of two young physicists who were involved in the nascent aeronautics program. One was Arthur L. ("Maj") Klein, a post-doc in physics who had started his undergraduate work in 1916 in Mechanical Engineering, later switched to Physics and completed work for a Ph.D. in 1924, with R. A. Millikan as advisor. His natural, engineering talent was to make things work and, in the years that followed, he played an important part in the "practical" side of GALCIT's development. The other was Millikan's son Clark B. Millikan, who had returned to Pasadena in 1924 with a bachelor's degree in physics from Yale. Enthusiastic about airplanes and aeronautics he enrolled for graduate work and a theoretical thesis with Bateman, while also working, with Klein, on the design and construction of a small airplane designed by Merrill.

In 1925 the aeronautics program at Caltech was still struggling. No degrees had been granted. Nevertheless it provided the basis on which R. A. Millikan made his appeal for funds from the newly established Daniel Guggenheim for the Promotion of Aeronautics. He emphasized the need for research and teaching in fundamentals while underscoring the practical activities already in place and pointing out Caltech's proximity to the aeronautics industry, which was developing in Southern California. In late 1926 the Guggenheim Fund granted $300,000 for the construction of a laboratory and establishment of a graduate school of aeronautics. Candidates considered for leadership of the program included Ludwig Prandtl, G. I. Taylor and Theodore von Kármán. Head of the prestigious Aerodynamics Institute at the Technical University in Aachen, with an impressive record of theoretical contributions to problems in physics and applied mechanics and with engineering contributions to German aeronautical technology, Kármán was the preferred one. He was persuaded to visit Pasadena in 1926 under the auspices of the Guggenheim Fund, to give some lectures and to advise on the plans being made for the new laboratory and school. He quickly became involved, pointing out, amongst other things, the inefficiency of the design of the large wind tunnel which was being planned, advising that it be closed- instead of open-circuit. Following a now well tested tradition he visited again in following years as Research Associate, dividing his time between Pasadena and Aachen, gradually taking over direction until becoming, in 1930, Director of the Guggenheim Aeronautical Laboratory at the California of Technology. The Laboratory was dedicated in 1928 on completion of the 10-foot (closed-return) wind tunnel (Figure 2) and the building which was constructed around it (Figure 3). The history of GALCIT is counted from that date. The acronym now stands for "Graduate Aeronautical Laboratories...", which includes the original Guggenheim Lab, the Kármán Laboratory of Fluid Mechanics and Jet Propulsion, and the Firestone Flight Sciences Laboratory, which were added in 1961 and 1962, resp.

In 1930, the full-time faculty included Kármán, Bateman, Klein and Millikan (Figure 4). Merrill, his important role ended, had transferred to Pasadena Junior College. (Later he helped in the design of a new, 3-by-4-ft, 200-mph wind tunnel that was installed in GALCIT in 1950. Named the Merrill Wind Tunnel, it is still in use for instruction and research.) Another very important, part-time member of the faculty was Arthur E. Raymond, an engineer at the Douglas

Aircraft Company in Santa Monica, who was recruited by R. A. Millikan to teach a course on airplane design, given on Saturdays and attended by faculty as well as students. The astuteness of this appointment became apparent in the symbiotic relation that developed between GALCIT and Douglas Aircraft. Faculty were actively engaged in consultation on the aircraft designs being tested in the wind tunnel, research problems were formulated and student employees were earning their support as well as experience for future careers. Arthur Raymond became chief engineer and eventually a Vice President of Douglas, in recognition of his key role in the development of the DC- series of aircraft, beginning with the DC-1 in 1932 through the DC-8 in 1968. The most famous of them, the DC-3, reached a total production of nearly 11,000. All of them, in scale model form, had many occupancy hours in the 10-foot wind tunnel.

The character and direction of GALCIT were now well defined. With Kármán's guidance and leadership, the school envisioned by Hale, stressing fundamental study and research, "...educating men broadly" in an environment of technological development and "...making them into good engineers", was in place. The program of instruction and research announced in Caltech's 1930 catalog was described as follows:

1. A comprehensive series of theoretical courses in
    aerodynamics and elasticity with the underlying
    mathematics and mechanics.
2. A group of practical courses in airplane design.
3. Experimental and theoretical researches on
    (a) the basic problems of flow in real fluids with
    (b) regard to the scientific foundations of technical hydro- and aero-dynamics.
    (c) practical problems in aerodynamics and
    (d) structures, especially as applied to aeronautics.

This has been the model up to the present, with occasional modification and extension to suit the times and the interests of the faculty. Thus, in the 2001-2002 catalog the section on Aims and Scope of Graduate Study in Aeronautics states succinctly that "The programs are designed to provide intense education in the foundations of the aeronautical sciences, with emphasis on research and the experimental method". Required full-year courses for the Master's degree include Fluid Mechanics, Solid Mechanics and Mathematics while advanced versions of the same are also requirements in the   PhD program. The present major areas of study and research are described in more detail under the following headings: Physics of Fluids; Computational Fluid Dynamics; Technical Fluid Mechanics (a.k.a. "Dirty Fluid Mechanics"); Mechanics of Materials; Computational Solid Mechanics; Mechanics of Fracture; Aeronautical Engineering and Propulsion; Jet Propulsion and Space Applications.

The astuteness of Millikan's choice of Kármán for Director of GALCIT was quickly apparent. His brilliance in science was matched by an unerring instinct for its application in technological opportunities. It was accompanied by an easy way with people, from beginning students to C.E.O.s and Generals that made him the inspirational leader that he was and the legendary figure that he is. In little more than a decade GALCIT became a world-famous institution.

Delightful accounts of those years and the traditions which Kármán established may be read in a book by Bill Sears[2] and a paper by him and Mabel Sears[3]. Those traditions include the weekly, informal Research Conference and the Aeronautics Seminar. In the Research Conference progress and problems in ongoing research are discussed with the faculty and other students, while the Aeronautics Seminar features mainly outside speakers on technological developments in aerospace and related fields. In addition, a weekly Fluid Mechanics and a Solid Mechanics seminar continue to play a vital role in the graduate experience at GALCIT.

The seminars provide a two-way street between GALCIT and the outside world. At one of them, in 1935, the seeds of JPL were planted. Probably at Kármán's suggestion, one of his graduate students, William Bollay, reviewed a paper on some rocket experiments by Sänger. The seminar came to the attention of John Parsons and Edward Forman, employees at an explosives factory, who wanted to build a rocket engine. Seeking advice, they contacted another student, Frank J. Malina, who got Kármán's blessing to get involved. As plans for a design and a test progressed, H. S. Tsien, A.M.O. Smith, and W. Arnold joined the group. Their first rocket test, in 1936, using oxygen gas and alcohol, was run in the Arroyo Seco ("Dry Gulch") in the foothills of Pasadena (Figure 5). Following suggestions of Parsons, the self-taught chemist in the group, more exotic, liquid-propellant fuels were developed. Later he and Malina developed a solid (potassium perchlorate and asphalt!) rocket, which became the basis for a full-fledged GALCIT project on jet-assisted-take-off (JATO), funded by the Air Corps in 1940. This Army Air Corps Jet Propulsion Research Project (with Kármán as director and Malina as chief engineer) was located in the Arroyo near the site of the original group's experiments and now the location of JPL (the Jet Propulsion Laboratory). In a paper[4] on the history of the project (used in making these notes) Malina explains why it is not called the Rocket Lab: the Air Corps still felt that "rocket" was a science-fiction word! Similarly, the company that was spun off to pursue commercial development was named the Aerojet Engineering Corporation. Responding to the challenge of Sputniks I and II, JPL launched the first American orbiter in 1958; it is now NASA's principal center for planetary exploration.

The origins of JPL described above serve to illustrate the environment and attitudes that Kármán created. Capable, enterprising students were attracted by the GALCIT program of study and research. Not bound to a course of studies based on established engineering practice, concentrating on fundamental mechanics and physics, graduate students developed the tools and confidence to quickly adapt to new technologies. They were given a lot of independence in developing their own interests and opportunities. Much of the research developed from student projects rather than defined contracts and grants. Sometimes it was the other way around--ideas that originated in frugally funded in-house work defined new projects for outside agencies. Much of this environment of student independence has been nurtured by succeeding Directors of GALCIT.

During the decade after 1930 the operation of the 10-ft wind tunnel and interaction with the engineers using it was assumed more and more by C. B. Millikan, together with Maj. Klein and Ernest E. Sechler. The latter had done his PhD (1934) research with Kármán on buckling strength of thin metal panels and became the sixth member of the faculty. He directed GALCIT's program of research on structures and solid mechanics for many years, and he played a unique role in identifying promising graduate applicants, especially during the large enrollments after World War II. By the time the tunnel was closed and removed, in 1997, some 800 tests had been run on aircraft models and about 200 on a wide variety of other vehicles and structures. The number of aircraft types in this count was about 200, including the DC- series already mentioned and a large number of well known World War II types, including Boeing's B-17 Fortress and B-29 Superfortress, North American's B-25 Mitchell (Figure 6) and P-51 Mustang (Figure 7). Post-war models included the DC-8 and DC-10, the Spruce Goose, the D-558 I (Skystreak) and D-558 II (Skyrocket), the XB-43, XB-46 and XP-92.

From his involvement with the development of some of these aircraft, Clark Millikan saw the need for wind-tunnel speeds higher than 200 mph and, in 1938, he wrote a proposal for a 12-ft high-speed wind tunnel (up to 600 mph) which he hoped would be built with government funding but it was realized as a privately owned facility of the large corporate users of the 10-ft tunnel. Completed in 1945 (Figure 8) at a site off campus and designated the Southern California Co-operative Wind Tunnel (CWT), it could be operated at pressures from 1/4 to 4 atmospheres. In 1955 it was upgraded to transonic capability but was shut down in 1960, "no longer economically viable", ironically because of new, government sponsored competitors like the Unitary Wind Tunnels at NASA. Over its 15-year life span, administered by Caltech with CBM as Director, the CWT was productive like its older relative on campus. Over 800 tests were performed, on scale models of a large number of famous aircraft including the DC-8, the B-58 Hustler, B-70 Valkyrie, F-100, F-104, F-106, C-130, as well as missiles like Sidewinder and Falcon.

By 1935, GALCIT had helped the US make up much of the lead in aviation that the Europeans had gained after the Wright brothers' flight. But in the area of compressible flow theory and experiment the Europeans were still far ahead. Again, Kármán played a leading role in getting this gap closed. What appears to be the first American paper on supersonic flow was published by him with Moore[5] in 1932; at the famous Volta Congress in 1935 Kármán presented a review on aerodynamic resistance at high speed[6]. These were followed by a long series of contributions, both theoretical and experimental, to high-speed aerodynamics across the whole range of Mach number. In 1940 Kármán initiated with H.S. Tsien the design and construction of a 2 1/2 inch supersonic wind tunnel, the first in the US. It was completed in 1942 by Allen E. Puckett, who was then a new graduate student from Harvard. The US had just entered World War II and the small facility became the prototype for a 3x3 ft supersonic wind tunnel at Aberdeen Proving Ground, and later for a somewhat smaller one at JPL. During the War, experiences with buffeting on high-speed military aircraft drew attention to mysterious effects that were called "compressibility burble" and "sonic barrier". To study these effects Hans Liepmann built the 2x20 inch Transonic Wind Tunnel, in which he and his students studied shock-wave boundary-layer interaction and other transonic phenomena. Liepmann had come to GALCIT a few years earlier with a PhD in Physics from the University of Zurich. He and Puckett taught wartime courses on high-speed aerodynamics to Lockheed engineers; their lectures developed into the first US textbook[7] on compressible flow. By 1946 Tsien was writing on hypersonic flow and in 1948 a wind tunnel at M=6 was in operation, supplemented soon after by a second leg at M=11.

Of the wind tunnels mentioned in the preceding, only the Merrill Tunnel is still in operation. Some of them have been replaced by more modern facilities designed by Hans Hornung, the present Director. The T-5 Hypervelocity Free-Piston Shock-Tube Wind Tunnel provides flow in a test section of 8 inch diameter for investigating real-gas effects at the temperatures and densities of re-entry speeds. It was operational in 1988. The Ludwieg-Tube Supersonic Wind Tunnel, up to M = 2.3 in an 8x8 inch test section, was built in 2001. The John W Lucas Adaptive Wind Tunnel, with 5x6 foot test section and speed up to 175 mph will be operational this year. Its "smart walls" will help compensate for the reduction in size from the 10 Foot Wind Tunnel which it replaces.

The preceding presentation highlights the importance of wind tunnels in the history of GALCIT but, of course, the experimental research program includes a large variety of other facilities and techniques. Thus the Graduate Aeronautical Laboratories contain other "Labs" as well as many special purpose facilities. These include a high-speed water tunnel and a free-surface water tunnel in the Hydrodynamics Laboratory; smaller water tunnels and a tow tank; a Supersonic Shear Layer Facility for investigating turbulent mixing with chemically reacting components; a Materials Dynamics Lab for high-strain loading, with application to crack propagation and fracture mechanics; an Explosion Dynamics Lab for experiments on ignition and detonation; facilities for studying combustion instabilities; a Vortex Dynamics (Heart) Lab and other biomechanics facilities; a Quantitative Flow Visualization Lab. There is also a computer equipped Laboratory for Spacecraft and Mission Design and, in cooperation with JPL, facilities for advanced propulsion design and development. The experimental point of view extends to the programs in computational fluid and solid mechanics, in which computational methods are developed for and applied to research problems ranging from flow-induced vibration to fatigue-crack initiation and propagation.

From the founding of GALCIT in 1928 through 2001 the total number of degrees granted to its graduates was 2291, composed as follows: 1311 MS, 439 AeE, 441 PhD. The total number of graduates receiving one or more of those degrees is about 1700. Figure 9 shows how the distribution in the three categories has changed over that period; during the war years large classes of civilian and military students produced a large number of Master's and Engineer's degrees. The majority of the GALCIT graduates have pursued careers in industry (about 60%) and academia (about 30%). More than half the industrial component is in aerospace companies, the remainder in a broad range of aerospace-related and other technology enterprises. In academia, GALCIT graduates have populated a large number of University departments of aerospace and mechanical engineering and, in several instances, have created new departments. In addition they are well represented in government research centers and in the military. In all categories, GALCIT graduates have provided leadership and innovation. They have been in demand, in good times and in bad, in aerospace and in other leading edge fields.

In closing, it is interesting to reflect on what has sustained this successful enterprise for 75 years. Clearly, the rapid rise and impact of GALCIT in its early years is due to Karman. More effectively than anyone else could have, he combined his immense talent and background with the opportunities that he found at Caltech and others that developed during his tenure. The continued success of GALCIT since then is, in large part, due to traditions and characteristics which were in place when he left. An interesting one is the name "Aeronautics", which has survived debate as to whether it should be changed to "Astronautics" or "Aerospace" or something else, and which has come to simply represent "a wide discipline encompassing a broad spectrum of basic as well as applied problems in fluid dynamics and mechanics of materials". It is not difficult to see why "Aeronautics" was chosen for the Laboratory but it is said that Karman would have preferred the name "Applied Mechanics" for the School. Leaving the name unchanged has no doubt help sustain traditions and continuity but perhaps even more important is the institution of "Director", which was created for Karman. Technically the title applies to the complex of laboratories in GALCIT but in practice the Director sets the style and carries forward the tradition of both Laboratory and School. There have been only four Directors (Figure 10): Theodore von Karman, Clark B. Millikan, Hans W. Liepmann and Hans Hornung. While bringing to GALCIT his own style, each of the successors of Karman has reaffirmed principles which have guided it since its founding, i.e. "...a tradition of research and teaching which stresses an appreciation for real applications in a very broad and deep base of fundamentals".[8]

ACKNOWLEDGEMENTS

Thanks for help with historical material and dates to Don Coles, Fred Culick, Hans Hornung, Hans Liepmann, Joe Shepherd and, especially, Gerry Landry.

REFERENCES

1. Goodstein, Judith R., (1991), Millikan's School, W.W. Norton Company.

2. Sears, William Rees, (1993) Stories from an American Life, (self published).

3. Sears, William Rees, and Sears, Mabel R., (1979) The Kármán Years at Caltech, in Ann. Rev. Fluid Mech., Vol.11, pp 1-10.

4. Malina, Frank J., (1983), "The Beginnings of Rocketry and JPL", in GALCIT: The First Fifty Years, F.E.C.Culick, (ed.), San Francisco Press, Inc.

5. von Kármán, T. and Moore, N.B., (1932), "The Resistance of Slender Bodies Moving with Supersonic Velocities with Special Reference to Projectiles" Trans ASME, Vol. 54, PP. 303-310.

6. von Kármán, T., (1935), " The Problem of Resistance in Compressible Fluids", Proc.5th Volta Congress, Rome, pp. 255-264.

7. Liepmann, H.W. and Puckett, A. E., (1947), " Aerodynamics of a compressible Fluid", John Wiley & Sons, Inc.

8. Liepmann, H.W, (1975) "Engineering and Research in GALCIT, unpublished memo.

Figure 1.
Noyes, Millikan and Hale (1929)
Oil painting in Caltech's
Athenaeum

Figure 2.
Entrance to GALCIT

Figure 3
Arthur (Maj) Klein and Clark Millikan
in 10-foot wind tunnel

Figure 4.
GALCIT group in front of 10-ft Wind
Tunnel. (1930) L-R: Visitors W. Tollmien
and R. Seiferth; wind-tunnel
superintendent, W. Bowen; Faculty C. B.
Millikan, H. Bateman, Th. von Karman
and A. L. Klein; student research
assistants F. L. Wattendorf and E. E.
Sechler; and three unidentified persons.

Figure 5.
GALCIT-Project No. l Rocket test team
(1936)
L-R: unidentified, A.M.O. Smith, Frank
J. Malina, Edward S. Forman, John W.
Parsons. Photo was taken at the site of a
rocket test stand in Pasadena's Arroyo
Seco.

Figure 6.
North American B-25 Mitchell
(1941)

Figure 8.
Southern California Cooperative (1945)
Wind Tunnel

Figure 7.
North American P-51 Mustang (1942)

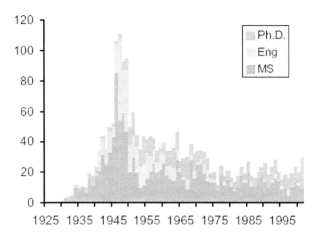

Figure 9
GALCIT Alumni
1928 to Present

Figure 10
GALCIT Directors
1930 - Present

Theodore von Kármán
1930 - 1949

Clark B. Millikan
1949 - 1966

Hans W. Liepmann
1972 - 1987

Hans G. Hornung
1987 - Present

Chapter 18
Aerospace Education and Research at Auburn University – From the Wright Brothers' Flight School to the Space Station

John E. Cochran, Jr. *
Auburn University, Auburn, AL 36849-5338

The history of aerospace education and research at Auburn University spans the period from the Wright Brothers' flight school in Montgomery, Alabama, to the present. This paper briefly recounts the major events of that period and provides information about some of the individuals who played significant roles.

The Early Years

The Wright Brother's achievement of manned, powered, heavier-than-air flight in 1903 and their later successes, described in 1908 issues of *Scientific American*,[1,2] provided the spark to ignite enthusiasm for aeronautics in faculty members and students of the Alabama Polytechnic Institute (API). Located in the small town of Auburn, Alabama, API would soon be intimately connected with flight and flying machines. Even before the Wright Brothers' great accomplish-ments, some independent researchers in Alabama were serious about the subject of aeronautics. In 1874, Lewis Archer Boswell, a physician from Eastaboga, near Talladega, Alabama, patented an "Improvement in the Aerial Propeller-Wheels."[3] Boswell continued to be actively interested in aeronautics until his death in 1909.

At least one *Scientific American* article about the Wright Brothers' historic flight was reprinted in the API student newspaper, the *Orange and Blue*, and a paper on aeronautics was presented at a meeting of the Engineering Society in 1908. In an interview some eighty years later, Robert Knapp, an API alumnus and later a U. S. Air Force General, (See Fig. 1.) claimed that the Wright Brothers had visited Auburn in 1907, stayed in his home (he was very young at the time), and met with engineering professors John J. Wilmore and M. Thomas Fullan. Knapp said the professors worked with the Wrights to redesign their aircraft so that it could be disassembled and transported in a wagon.[4]

Knapp's memories could not be verified. However, it is well documented that the Wright Brothers visited Alabama in 1910 in search of a site for a winter flying school. They found one near Montgomery, went back to get their airplane and returned to conduct instruction for three months later that year. In the fall of 1910, the *Opelika Daily News* (Opelika is a town adjoining Auburn) contained an announcement of instruction at API on "aeronautic construction and the principals [sic] of aviation," as part of a class in kinematics of machinery. Professor Fullan gave lectures on aviation around the state, and there were many "air minded" students and faculty on campus.[4]

*Professor and Head, Department of Aerospace Engineering; Associate Fellow, AIAA.

Just prior to the United States' involvement in World War I, a Reserve Officers Training Corps was formed at API. Service in World War I interrupted the studies of many API students and the lives of many alumni. According to Ref. 4, at least fifty students served in the army or navy air corps. During the war and for a while thereafter, a unit of the Student Army Training Corps provided training for around 1,700 students and funds for a shop and laboratory. After the war, the land used by the Wright Brothers for their flying school was the site for Maxwell Field. Robert Knapp, then a Lieutenant in the U. S. Army stationed at Maxwell Field in Montgomery, Alabama, helped encourage the interest of students in aeronautics. Knapp flew to and from Auburn to visit his brother and landed in a pasture owned by W. W. Webb, a veterinarian who had built two runways for his own airplane. Capt. Asa Duncan and Lt. Knapp flew sections of the first airmail route through Alabama in 1925. The student newspaper, the *Auburn Plainsman*, the alumni association's *Auburn Alumnews* and *The Auburn Engineer*, a School of Engineering publication, contain many articles on aeronautics during the period 1925-30. A visit by Charles Lindbergh to Birmingham in 1927 added to the considerable interest in aeronautics at API.[4]

First Aeronautical Program

In 1930, API President Bradford Knapp, who lead the reconstruction effort at the school beginning in 1928, and Dean of Engineering, John J. Wilmore, re-cruited and hired Volney C. Finch to establish an aeronautics program in the Department of Mechanical Engineering. Finch, a retired naval pilot had worked at the Naval Aircraft Factory and had done research in aircraft structures and engine cooling. A graduate of the Naval Academy, Finch had studied at MIT, had received a M.S. degree from the University of Washington, and had written four textbooks on aircraft design and engines.[4]

Alabama Polytechnic Institute became one of twelve universities or colleges in the United States offering a degree in Aeronautical Engineering.[4] The first aero-nautical courses taught at API were a junior course that covered aviation topics ranging from aerial navigation to meteorology and a senior course in aircraft design. Additional courses were added as an aeronautics option. As was to be the case for a number of years, the entire field of aeronautics was considered fair game. After all, the knowledge base was still relatively small. The December 1931 issue of *Southern Aviation*[5] contains an article describing a program, under Professor Volney C. Finch and Instructors Solon Dixon, (See Fig. 2.) and Victor W. Randecker, which included "airplane design, maintenance and operation of airplanes, and the business of commercial aviation." Since that time, although the titles have changed, there have been parallel programs in aeronautical engineering and aeronautical administration at Auburn.

The API catalogue for 1931-32[6] lists graduate courses in aeronautical engineering, including airplane structural design, aircraft instruments, advanced aerodynamics, aeronautical problems and theoretical aeronautics. The aeronautical labs were highly applied. Six types of airplane engines, including a 220 horsepower J-4 Wright Whirlwind, were used to study propulsion fundamentals. Airplane fuselage frames were used to test welded joints. A Vought U. O. two-place biplane with a radial air-cooled engine and a Boeing pursuit plane with a water-cooled engine were both used to demonstrate aeronautical principles and design techniques.[4,7]

The aeronautical engineering program produced thirteen graduates of the aeronautics option in 1932. A full four-year curriculum and a separate degree in aero-nautical engineering were in place by 1933. Marshall S. Cayley, a member of the first aeronautical engineering class of eleven students, later attended the Boeing School of Aeronautics, served in World War II and worked for

United Airlines for 32 years. Robert G. Pitts, another 1933 graduate, was to figure prominently in aerospace education at API.

The depression of the early 1930's caused great difficulties in all areas, but higher education in the southeast was particularly hard hit. Finch had plans for an engine test bed and wind tunnel in the mechanical and aeronautical building,[8] but, funds were not available. In fact, API ran out of money to pay employees and offered, as some relief, eggs, meat and produce, from the agricultural sector of the school, and script. After fighting for financial survival and winning many political battles, including keeping the aero-nautical program, Knapp returned to his home state as president of Texas Technological College.[4]

Rather than appoint a new president, the Board of Trustees established an administrative triumvirate of John J. Wilmore (Dean of Engineering), B. H. Crenshaw and L. N. Duncan. The three were actually able to function well enough together to bring some measure of financial stability. However, before that happened, another setback, this time to aeronautical education in particular, occurred in 1933. Finch, unable to support his family on what API could pay, accepted a position at Stanford University. Additionally, an aviation accident that year claimed the life of a student. In spite of these problems, aeronautics continued to be very popular and B. M. Cornell returned to API as a replacement for Finch. A true visionary, Cornell proposed, in 1934, a combination research center and recreational facility, including a golf course, on the site of the W. W. Webb airport, which was still privately owned. He wanted API to conduct interdisciplinary research in areas such as cotton fabrics for aircraft structures. Although Cornell's ideas were not included as part of the API planning, the aeronautics program, which included aeronautical business subjects, con-tined to be very popular and productive with 119 students in 1935. One of the 1935 graduates was Wilbur Funchess, who later became Col. Funchess during World War II. Funchess returned to API and served many years as the Director of Facilities.

After receiving a master's degree from the California Institute of Technology, Robert G. Pitts returned to Auburn as Cornell's assistant in 1935. With the help of students, he rebuilt wrecked aircraft to raise funds for the department. The 1937-38 senior design class also helped Pitts design and construct API's first wind tunnel. [4]

Cornell advocated the entrance of API into the Civilian Pilot Training Program (CPTP), but the privately owned Auburn-Opelika Airport (improved W. W. Web airport) was needed for the venture. Competition from Georgia Tech and the University of Alabama was a factor in convincing API President Duncan to take over ownership of the airport and engage API in the process of training civilian pilots. The payment to the former owners totaled about $375.

Cornell and Pitts began the ground school part of the CPTP in 1939 with twenty students. Alabama Air Service handled the flight instruction. Cornell and Pitts also helped instruct in the Tuskegee Institute ground school. Cornell convinced Duncan the API could do a better job of flight instruction and the Auburn School of Aviation was formed with Pitts in charge. During the next year, students from the advanced CPTP at Tuskegee Institute, some of the famous "Tuskegee Airmen," also received advanced flight instruction at the Auburn - Opelika Airport, because the field at Tuskegee had not been approved for aircraft heavier than Piper Cubs.

After a somewhat lengthy political process, federal funds for improving the airport were obtained in 1941. Over the course of the next several years, over 1,400 pilots trained at API.[4]

The Department of Aeronautical Engineering
1942-1960

Cornell retired in 1942 and Pitts became Head of the newly formed Department of Aeronautical Engineering, as well as Director of the Auburn School of Aviation and manager of the Auburn - Opelika Airport, positions he would hold for 35 years. A Cal Tech diploma and a picture of Theodore von Karman on Pitts' office wall in Wilmore Laboratory impressed visitors knowledgeable enough in aeronautics to recognize the world famous engineer. During the 1940s, Pitts contributed signifi-cantly to aviation in the southeast and especially to military aviation flight training as he helped design, build and develop numerous airfields and train pilots.

From its beginning until 1942, API was on the semester system.[8] "On June 8, 1942, the Alabama Polytechnic Institute began operation on a year-round Quarter System-four quarters of 12 weeks each-which permits students to graduate in three calendar years instead of four." Aeronautical engineering was a "critical occupation." The first quarter system AE curriculum consisted of 218 hours. Physical training was required, but students earned no credit for the course. The quarter system was retained until 2000. One of the initial reasons to retain the quarter system was that under it API could better handle the influx of students on the "GI Bill." The enrollment in the institute climbed to over 9,000 in the late 1940s. A co-op program was initiated for several purposes. First, the co-op program helped students earn money one quarter to pay their fees and expenses the next. Second, it provided practical experience. Third, it reduced the number of students on campus at any one time. In the opinion of many, the co-op program worked better on the quarter system.

Early engineering graduates of the Department of Aeronautical Engineering included Robert Hails, who rose to the rank of Lt. General in the U. S. Air Force. Hails, was instrumental in the development of the Head-Up Display (HUD) for military aircraft. He entered API in 1941 but left in 1943 to serve as a pilot in the U. S. Army Air Corps. Hails returned in 1946 and graduated in 1947. He recalls that a "water tunnel, or trough" was used to demonstrate the flow over airfoils. A contemporary of Hails, Robert H. Harris, rose through the ranks at General Electric to become a vice president. A 1943 graduate, Leroy Spearman, went to work for the NACA. He has continued to work at NASA's Langley Research Center in the areas of aerodynamics and foreign missile technology for well over fifty years. The AE curriculum in Hails' and Spearman's era included surveying, probably because of its importance in airport construction.

In 1945, Pitts developed and published a Master Plan for the Auburn-Opelika Airport.[8] He served on the State Aviation Commission, was active as a speaker to civic groups, and Junior Chamber of Commerce Man of the Year. In 1948, he had a hand in organizing the "Flying Farmers." The *Alumnews* for September 1951 contained an article entitled, "A Million Dollar Value - The Auburn Airport."[8] The airport was Class III, in the range of I to VI based on size. Pitts was involved in extension activities that included the Flying Farmers, an organization of farmers and others who advocated the use of aircraft in agriculture. The National Flying Farmers Meeting was held in Auburn in 1952.[9]

In 1955, Graham Newman, a sophomore in aeronautical engineering, wrote an article on the Department that appeared in *The Auburn Engineer*.[12] The engineering faculty members at that time were Pitts, R. B. Miller, R. R. Sanders, and W. G. Sherling, Jr. Aviation subjects were taught by Pitts, Sanders, and M. O. Williams. All the faculty members had degrees in engineering, but only two had masters degrees, Pitts (Cal Tech) and Sherling (Georgia Tech). A wind tunnel was in operation and there was an integrated relation-ship between aeronautical engineering, aviation administration, and flight training using the airport as a hands-on laboratory. Pratt & Whitney flew thirty-five students and faculty members to East Hartford, Connecticut for a plant trip. The enrollment was approximately 300 with around 50 in aeronautical administration.

Although a program that helped start many careers in aeronautics and astronautics, until 1959, the aero-nautical engineering program was not accredited by the Engineering Council on Professional Development (ECPD). The impetus for its accreditation was pro-vided by a serious academic problem. The electrical and mechanical engineering (EE and ME) programs lost accreditation in 1957, the same year the API football team won the national championship. At the request of President Ralph Draughon, the Alumni Association launched the Engineering Emergency Fund Drive with the goal of raising $250,000 to improve the School of Engineering. Dr. Roy B. Sewell '22 was chairman of the drive and trustee Dr. Frank Samford '14 was co-chairman. Both Sewell and Samford were great supporters of academics as well as athletics. Mr. Joe Sarver, executive director of the alumni association worked with Sewell, Samford, other alumni, friends, and industry to raise twice the goal.[12]

A board of consultants was hired to recommend actions that should be implemented to achieve re-accreditation and accreditation of additional engineering programs. They recommended five actions: (1) reorganize the curricula of the School of Engineering with more stringent requirements on credits in science, mathematics, and engineering sciences; (2) increase the amount of research done by engineering faculty members and students; (3) increase the teaching staff and decrease the teaching loads; (4) increase salaries and wages; and (5) provide additional space and equipment. Additionally, the ECPD required that the liberal arts component of the engineering curricula be substantially increased.

President Draughon hired one of the consultants, Dr. Fred H. Pumphrey, as the new dean of engineering. Pumphrey, a former API professor of electrical engineering, then at the University of Florida, lead the implementation of the recommendations. Under Pumphrey's leadership, the EE and ME programs were re-accredited and the AE program, which was ex-panding rapidly because of the increased emphasis on both aeronautics and astronautics, was accredited as an aerospace engineering program. The faculty of the School of Engineering increased by 40% and a new building, Dunstan Hall, primarily for EE, was constructed. Additional equipment was obtained for AE and ME operations in Wilmore Laboratories.[13]

Although the aeronautical engineering program was not accredited prior to 1959, it produced many graduates who contributed significantly to aeronautics and astronautics. Some are noted above. One of Auburn University's astronauts, Thomas K. Mattingly, received a degree in aeronautical engineering in 1958. According to Ms. Polly Martin,[7] her husband, Fred W. Martin, who joined the faculty in 1956, supervised Mattingly and other AE students in much of the construction of the low-speed wind tunnel. Mattingly flew missions in the Apollo program and was the "astronaut left behind" on the famous Apollo 13 mission and was instrumental in bringing the crew of the disabled craft back to earth safely. Later, Mattingly and Henry W. Hartsfield, a 1954 API graduate in physics flew together on STS-4.

In a 1981 *Auburn Plainsman* interview,[8] Mattingly recalled, "... In retrospect, to my experience, learning how to do with what you've got has been a far more potent lesson than all the far more theoretical esoteric subjects we could have covered."

Other examples of pre-accreditation successes are Ron Harris '58, who was a NASA engineer and administrator and later a Rocketdyne manager and Axel Roth '59, currently the Associate Director of George C. Marshall Space Flight Center, who has been a NASA engineer and administrator for over 30 years. Graduates of the aeronautical administration program during this period included John Stein, who became CEO of Golden Enterprises, makers of Golden Flake Potato Chips and other snack foods.

Although there was definitely something to be gained from practical engineering experiences like those of the late 1950s AE graduates, to achieve accreditation the aeronautical engineering program had to be improved. Part of this improvement appeared in the curriculum. Without question, much of the impetus for curriculum development was a very small object orbiting the Earth. In 1956, Martin had tried to get a course in ballistics and space flight included in the aeronautical engi-neering curriculum, but Pitts would not submit it to the curriculum committee because space flight was thought to be "...beyond the realm of achievement."[14] Shortly after the launch of Sputnik I, Pitts did submit Martin's proposal for the course, "Rocket Mechanics," and it was approved the next day.

In Fig. 3, Sherling and students inspect the "hot shot" tunnel. Pitts and Martin are shown in Figure 4 standing in a section of what was to become the pressure tank of the high-speed wind tunnel in 1959.

The Department of Aerospace Engineering
1960- 1969

With the support of alumni, industry, and the state, additional engineering faculty members were hired and research was emphasized along with instruction. In 1960, the Army Guided Missile Agency awarded a contract for special studies to the Aerospace Engineering Department. Under this contract, Branimir D. Djordjevic, a Yugoslavian engineer and military pilot who defected at the end of WW II, first developed instructional manuals for civilian Army engineers. During the next several years, Djordjevic wrote a seven-volume report on that covering most areas of aerodynamics and flight dynamics and a condensed handbook version of the report. Fred Martin and another new faculty member, James O. Nichols, as well as several graduate students, worked with Djordjevic. Djordjevic was of the European school and hence had amazing handwriting. He utilized colored chalk expertly and notes worthy of his lectures required a good set of colored pencils if the numerous axis systems of stability and control were to be appropriately identified.

By 1962, the aerospace engineering curriculum included more physics, mathematics, an introductory course in astronautics, boundary layer theory, gas dynamics and space propulsion systems.[18] Due to the demand for engineers with advanced degrees, especially at Redstone Arsenal, the Master of Science in Aerospace Engineering was approved in 1961.[20,21]

Funding for Auburn Research Foundation projects in the School of Engineering increased from about $35,000 in 1957 to more than $320,000 in 1961 and institutionally sponsored research was at the $100,000 level.[22] Sherling received a National Science Foundation grant in 1962 to fund the development of a hypersonic, "hot shot," tunnel. Martin was principal investigator on a $60,000 Air Force contract to investi-gate atmospheric re-entry trajectories.[23] Djordjevic was an associate investigator and Margaret Baskerville, a professor in the Department of Mathematics, was a consultant. After completing the project, Martin returned to VPI in 1965 to earn a Ph.D. in a related area. He returned in 1966.

The Master of Science program produced a number of graduates who later earned Ph.D.'s. and entered academia. Examples are Jewel B. Barlow, now a professor at the University of Maryland and John E. Burkhalter, W. A. Foster, Jr. and the author, now professors at Auburn.

During the early 1960's, astrodynamics was a new subject area. Although an undergraduate course in "astro" was taught in the AE department, graduate students in engineering, physics, and mathematics learned much of the classical methods of celestial mechanics, and the new area of satellites attitude dynamics, from a mathematics professor, Philip M. Fitzpatrick.[24] The author's love for the subject of astrodynamics is due largely to Fitzpatrick and his challenging sequence of courses that covered (very precisely!) the gamut from the dynamics of particles to the two-body problem to Lagrange's planetary equations to the application of Hamilton-Jacobi theory to satellite attitude dynamics problems.

Kenneth E. Harwell, a Cal Tech graduate (M.S. and Ph.D.) came to Auburn in 1963. One of his assign-ments was to lead the development of a doctoral program in aerospace engineering. A University of Alabama graduate (as was James O. Nichols), Harwell still worked hard at Auburn to develop not only the Ph.D. program, but also a world-class gas dynamics laboratory. Harwell's first Ph.D. student and the program's first graduate was Dwayne McCay, formerly a NASA engineer and manager and currently Vice-President for Research at the University of Tennessee at Knoxville. Harwell taught the author a lot more about gas dynamics and space propulsion than he can now recall.

With the expansion of the U. S. space program, Auburn's aerospace engineering program grew in terms of students and research. New faculty positions were funded and the subject of propulsion was one of the "hottest." Richard H. Sforzini was invited to come to Auburn as a Visiting Professor in 1966 after seven years with Thiokol Chemical Corporation. As Director of Engineering at Thiokol's Space Booster Division, Sforzini had led the development of the world's first three million pound thrust solid-propellant rocket motor ("solid rocket motor").[25] Prior to his work at Thiokol, Sforzini was a U. S. Army officer whose assignments included ordnance field maintenance and instruction at the U. S. Military Academy, and study that lead to the degree of Mechanical Engineer from MIT. Interestingly, the Army required him to specialize in automatic control.

Sforzini became a permanent faculty member after one year and remained on the faculty for an additional twenty years. He developed and taught propulsion courses and conducted research for the U.S. Army and NASA. The research that Sforzini and his graduate students did for NASA's Marshall Space Flight Center included the development of models and simulations of the internal ballistics of the solid rocket boosters for the Space Transportation System ("Space Shuttle"). A very important consideration in the successful use of the solid rocket boosters was the prediction of the degree of thrust difference in the two boosters, which, of course, would produce a yawing moment on the launch con-figuration. Sforzini was an excellent instructor who taught special extension courses for NASA and the U. S. Air Force and graduate courses in the Video-Based Instructional Engineering Outreach Program instituted at Auburn in 1984.

One of Sforzini's co-workers in solid rocket propulsion research was W. A. Foster, Jr., the second Ph.D. produced by the AE department. Foster, who is now a professor in the department, did his graduate work in structural dynamics under Malcolm A. Cutchins. However, in working with Sforzini he became well versed in solid rocket propulsion also. One of Sforzini's many students was Walt Woltosz, who later developed computer software for the handicapped including British astrophysicist Steven Hawking and pre-school children.[26]

Cutchins started his career at Auburn in 1967. Noted for his innovative teaching, he developed and taught many courses in structural dynamics and, in the latter years simulation. His students undoubtedly remember his multiple-choice exams, the correct answers of which would produce resultant answers like "War Eagle" (the Auburn battle cry).

During the early 1960's, the number of hours in the aerospace engineering curriculum was reduced from 240 to 228 by requiring calculus in high school and reducing the credit for some courses from five to four quarter hours. In 1969, the number of credit hours was reduced again to 208 by rewriting syllabi and changing the credit for many undergraduate courses from five or four hours to three hours. These changes were supposed to provide students with more time between classes to read and do other assignments. However, they also resulted in many students taking five or six major courses at the same time instead of three of four five-hour courses. Reducing the number of hours also involved dropping engineering drawing and labs such as machine tool and foundry. Thus, much of the tech-nology content was removed from the curriculum. It was assumed that students had taken engineering drawing in high school or could learn computer assisted design (CAD) software on their own.

 The enrollments in both aerospace engineering and aviation management grew during this time. The aerospace engineering enrollment was around 200, exclusive of freshmen. Students worked hard to stay in school, driven by both their career goals and the alternative of being drafted. Two years of ROTC were mandatory and the advanced program was an option many chose.

Djordjevic, died in 1967 due to complications following surgery. This was a great loss to the department and the author, who had been one of his graduate students.

The Lunar Landing and a Downturn in Enrollment

When the Eagle landed on the Moon, the world including Auburn University cheered! However, as all those in aerospace engineering know, that great achievement meant that the federal government could redirect funds from the space program to other areas, principally the United States' involvement in the Vietnam War and the Great Society. The downturn in spending on space flight lead to a decrease in aerospace employment and a corresponding decrease in enroll-ment in the aerospace engineering program at Auburn and across the nation. For the students who stayed in the program, this was actually beneficial due the smaller class sizes. One of these students was James Voss '72, an Army ROTC student and varsity wrestler. Voss' desire was to become an astronaut, a rare achievement for Army officers. His determination and perseverance, perhaps developed as a wrestler, helped him achieve his goal and flights as a mission specialist on the Space Shuttle and two Space Station Freedom missions.

The downturn in enrollment led, as is usually the case, to a decrease in the number of aerospace engineering faculty members. The author was one of the relatively new faculty members who stayed. Several faculty members who came to Auburn in 1968 had departed by 1973. Fortunately, when the NASA research declined, Department of Defense research was maintained at a high level and graduate school was a way to delay military service. In order to justify faculty positions, research was even more strongly encouraged. Collaboration of AE faculty on research projects, which had always been common (e.g., Sforzini and Foster) became more so. Cutchins and Burkhalter collaborated with Fred Martin on a major research project dealing with store separation from aircraft for the U. S. Air Force Research Lab at Elgin AFB. That project involved both experiments in the Auburn wind tunnels and theoretical research using finite aerodynamic elements (sources, sinks, vortices, etc.). Other research in the 1970's included, spacecraft attitude dynamics (Fitzpatrick and Cochran), missile launcher dynamics (Cochran), the development of an aerial seeding device for broadcasting pine seeds (Cutchins Burkhalter, and Foster), and spacecraft guidance and control (Art Bennett and Cochran). It is interesting to note that Bennet's contract was funded as a part of NASA's Comet and Asteroid Rendezvous and Docking (CARD) Mission, which was never flown. Thirty years later, NASA's NEAR Shoemaker spacecraft would accomplish things about which we only studied and dreamed.

A New Department Head and Star Wars

Robert G. Pitts retired in 1978 and his name was added to the Auburn-Opelika Airport. James C. Williams III, a VPI and UCLA graduate and North Carolina State faculty member, was named head of the department in 1980. Jim Nichols served as head during the interim. When Williams became head of the depart-ment, the organizational structure was changed. Gary Kiteley, Associate Professor of Aviation Management, who had been managing the airport, was made the Executive Director. Kiteley was also in charge of the Auburn School of Aviation that conducted flight training. Other aviation management faculty members were Bob Merritt, Leo Frandenburg, Bill Callan, Hal Decker, and Ollie Edwards. All these gentlemen were retired military officers with aviation and teaching experience.

In the early part of the 1980's, the author took a "sabbatical" as associate director of athletics. During that time, a Purdue graduate, Mario Innocenti, was the principal professor in the flight dynamics and control area. Enrollment was increasing. By 1984, the Strategic Defense Initiative had produced a boom in research. When the Challenger exploded in January of 1986, the winter quarter was just beginning and undergraduate enrollment (excluding freshmen) had increased to 230. By 1988, it was about 270. The peak in 1990 was 316.

Don Spring, a more than twenty-year civil service veteran of the U. S. Army Missile Command came to Auburn in 1983 to begin a second career as a professor. An expert in missiles and hypersonic aerodynamics, Spring taught courses in aerodynamics and was director of the wind tunnel facilities. Ron Jenkins, another Purdue graduate and excellent instructor in propulsion also joined the faculty in 1985.

A major research project during the middle and late 1980s was a "Stars Wars" guidance and control project for the U. S. Air Force. Guidance laws and simulations of exo-atmospheric interceptor concepts were developed. Other research of the period involved the characterization of wire rope vibration isolators and a study of the orbital lifetime of tethered satellites and free tethers. A visiting professor, Krishna Kumar from ITT Kampur, worked with Cutchins and Cochran on analytical models of wire rope. An officer in the U. S. Air Force, Ted Warnock did some seminal work on the tether lifetime problem utilizing neural networks to store data. There were no tethered satellites until several years later. Applied aerodynamics research and aerodynamic testing of FOG-M missile models were other projects that kept the aerodynamics faculty busy.

David A. Cicci was hired in 1988 to teach astrodynamics courses including orbital mechanics and orbit determination. Like Burkhalter and Cochran, he received his Ph.D. from the University of Texas at Austin. An excellent teacher, Cicci concentrated on improving the curriculum and engaged in some Star Wars research.

Robert S. "Steve" Gross also joined the AE faculty in 1988. Another excellent teacher, Steve has won all the teaching awards the department and college bestow at least once. His specialty is composite materials. On one research project, he applied his knowledge of com-posites to help College of Veterinarian Medicine professors develop artificial joints for animals.

The increased enrollment in aerospace engineering and the need for more space and better laboratories justified a new building and a contribution by Mr. John M. Harbert, III, a 1946 Auburn civil engineering graduate, made the construction possible. Two adjoining buildings, one providing aerospace engi-neering offices and laboratories, and the other classrooms for use by all engineering students, were built adjacent to the civil engineering building in the Harbert Engineering Center. Construction was started in 1989.

The fall of the Iron Curtain in 1989 began another downturn in the demand for aerospace engineers and another cycle of lower enrollment in aerospace engi-neering. The flow of research dollars through the Space Defense Initiative also dried up. However, an interesting applied research project with Hayes Targets (a part of Hayes International, later PEMCO) took up part of the slack. In that project, Burkhalter, Spring, Cochran, and Innocenti designed a maneuverable towed target, a control system for it and a simulation of the system of towing aircraft, towline, and target vehicle. Burkhalter created the aerodynamic design, which had "plus" canards and "X" wings. Spring conducted wind tunnel tests and Innocenti developed the control system. Tae Soo No, a graduate student (now a professor in South Korea), and Cochran developed the simulation using a lot of theory from previous tethered satellite research. This was apparently the first operational simulation for such a system. It was used to predict the behavior of the system prior to flight tests and to develop control logic for deployment, retrieval, and maneuvers. The guidance was a command type and a stability augmentation system (SAS) was included. The SAS channel for roll was especially important. Analytically predicted aerodynamic characteristics of the towline and the vehicle as well as data from wind tunnel tests of target models were used in the simulation.

Hayes Targets constructed the target and a limited full-scale wind tunnel test was conducted at Virginia Polytechnic Institute. A flight test was then conducted near Phoenix, Arizona without Auburn's assistance in Arizona. However, Hayes representatives were not confident enough to turn on the SAS during deploy-ment. Without it, the vehicle's motion was unstable, especially in roll and its motion was erratic. As a safety measure, the towline was cut and the vehicle crashed. Fortunately, the vehicle was not seriously damaged because it landed in a large cactus. Hayes included Auburn (Cochran) as a part of the team for the next test, possibly to have someone to blame if it failed too. The SAS was used for the second test and the vehicle flew very well. After deployment, it was maneuvered so that the towline and the vehicle formed a cone behind the towing aircraft as required. The retrieval was also successful. Apparently, this was the first time a remotely controlled towed vehicle was flown beside and above the towing aircraft and recovered successfully.[27] Flight test data was used to provide better estimates of aerodynamic coefficients that increased the fidelity of the simulation. This maneuverable towed system was a forerunner of decoys now flown off C-130s and other aircraft.

Post Cold War, the Space Station, and a Third Department Head

The beginning of the 1990's was marked by the flight of Col. Jim Voss on STS-44 as a Mission Specialist. Voss has since flown on several mission and helped assemble the initial version of the International Space Station.

A major event for the department was the move into the new Aerospace Engineering Building that had been begun five years earlier. The new building (See Fig. 5.) contained enough office space for all faculty members and many graduate students. It also provided what we thought at the time was enough laboratory space.

Another major event was the second change in department head in the history of the department. In 1992, after successfully guiding the department into the promised land of new facilities, Jim Williams returned to full time teaching. The author became department head with visions of the great opportunities that lay ahead. Enrollment in the engineering program was still decreasing, but demand for the aviation management program remained strong. Research funds were hard to get, but we had some plans. A flight simulator/air traffic control laboratory funded by a Federal Aviation Administration Airways Science grant, obtained by Williams and Kiteley, was added to the department assets.

Roy Hartfield (Ph.D., University of Virginia), with a specialty in non-obtrusive flow measurement joined the faculty in 1990. Hartfield is also a railroad enthusiast and in a former career was a contractor.

Jim Nichols retired in 1993, after 33 years of exemplary service as a faculty member, acting department head, program coordinator and principal design instructor. Ron Barrett, a enterprising graduate of the University of Kansas (twice) and the University of Maryland and a disciple of Jan Roskam was hired to teach airplane design, which he continues to do very well. At least equally as important, Barrett established an international reputation in the new field of adaptive aerospace structures and a laboratory in which un-manned aerial vehicles are currently being proto-typed. He received a Discover Magazine award in 1997 for his solid-state adaptive stabilator control for helicopter main rotors.[28]

When Innocenti returned to Italy (the University of Pisa) in 1993, his position was filled by John B. Lundberg, another University of Texas at Austin graduate. John added to the department's capability in astrodynamics and controls. An excellent teacher, he developed a GPS laboratory and began the involvement of Auburn AE students in the NASA Reduced Gravity Experiment program. In 1999, John took a position at the Naval Surface Warfare Center.

Jim Williams retired in 1997 to do more flying and travel with his wife. After retirement, he was still involved in engineering accreditation as a visitor for several years. His retirement resulted in the loss of much expertise in viscous aerodynamics.

Don Spring retired from his second career in 1998 and Wichita State graduate, Anwar Ahmed, filled his position. Ahmed's specialties are experimental aero-dynamics and fluid mechanics. He is especially interested in flow visualization and is very fond of water tunnels.

Academic Reallocation

The 1990s will be remembered as a truly difficult time for higher education in Alabama. There was what could be interpreted as a delayed reaction to the same difficulties (budget cuts, program closures, etc.) that higher education encountered in the late 1980s. Shortage of funds for public K-12 schools lead to a significant reduction in funding for colleges and universities. It is ironic that the governor who made the decision to cut funding to Auburn University was the Honorable Fob James, an Auburn alumnus in civil engineering and football star. James was of the opinion that higher education was inefficient.

In times of shortage of funds, extreme measures are often taken. In the opinion of some, the university had been overextended in terms of programs offered and services provided. An offer of early retirement incentives reduced the number of faculty members at the expense of losing veteran faculty who were still major contributors. The next step was to use the priority process, set up during the preceding four years before as part of a "continuous quality improvement" management emphasis, to cut programs and services and reallocate funds. As noted by numerous higher education experts, setting priorities within universities is always difficult.[29] It is doubly difficult for some programs that happen to have low enrollments caused by cyclic economic and political forces. Although by 1997 enrollment in aerospace engineering was increasing, that did not seem to matter much in the priority setting process.

In 1998, an election year, things reached a climax. When choices were forced, the aviation management program was deemed by some to be of lesser priority than other programs in the College of Engineering. Without the aviation management program, the aerospace programs were considered too small to justify a separate department. Thus, preliminary recommendations were made to phase out the aviation management program, which had 235 students, and merge the aerospace and mechanical departments. A significant effort, somewhat political and similar to those commonly used in response to proposals to close a military base, was mounted. Aerospace engineering and aviation management students, their parents, faculty, alumni, industry representatives and other constituents campaigned to retain the Department of Aerospace Engineering department and all its pro-grams. That effort was successful in most respects. The Department of Aerospace Engineering and all its programs were retained and future support of the department and all the affected programs was promised. However, in the end game, the aviation management program was considered more compatible with programs in the College of Business. It was moved there in 1999 as a part of the new Department of Aviation Management and Logistics.

Arguably, the major positive result of the reallocation exercise was the recognition by the Board of Trustees that tuition increases and major fund raising were required if Auburn University were to remain competitive and increase in prominence.

Today

As of October 2002, the aerospace engineering faculty consists of nine members. We have a staff of five. Bill Holbrook is a model maker extraordinary. Ginger Ware is our office administrator who keeps things running. Evia Vickerstaff keeps student records and assists Roy Hartfield and Steve Gross with student recruiting. Maxine Bryant keeps our books. Jim Lin is our Ph.D. electrical engineer.

Excluding freshmen, around 100 students are enrolled in the undergraduate program. The future looks bright since, this year about 160 aspiring freshmen have expressed the intent to enter the program. The relatively low student-to-faculty ratio in the undergraduate program will probably not remain low for long, but should continue to be low enough to allow for considerable student/faculty interaction. The graduate programs (Master of Aerospace Engineering, Master of Science, Doctor of Philosophy) have a combined enrollment of about 30.

Current research emphasizes are the application of genetic algorithms to design missiles, design and prototyping of micro aerial vehicles, experimental aerodynamics and fluid mechanics, applied structural dynamics, orbital mechanics, and modeling and simulation of flight vehicles and transportation systems. Even though much time is spent on research, the faculty members are all committed to providing excellence in instruction at all levels.

Plans are in effect to increase the undergraduate and, especially, graduate enrollments. Of course, this implies that research funding and faculty size must be increased also. The areas of modeling and simulation of transportation systems, dynamics and control, and aerodynamics will be emphasized in recruiting new faculty members.

Acknowledgements

Faculty members and alumni not mentioned explicitly herein, who have contributed to aerospace engineering education at Auburn are hereby acknow-ledged. The Master of Arts thesis by DiFante was an invaluable resource writing this paper. It provided a wealth of material and pointed the way to resources that provided information about the years following 1942.

References

1. *Scientific American*, August 29,1908, cover, pp.135-136.

2. *Scientific American*, September 26, 1908, pp. 208-209.

3. Boswell, Lewis Archer, "Improvement in Aerial Propeller-Wheels," U. S. Patent No. 155,218, September 22, 1874.

4. DiFante, Archangelo, Aviation at Auburn Uni-versity: An Introduction 1908-1941, Master of Arts Thesis, Auburn University, December 8, 1989.

5. "108 Students in Auburn Aeronautic School," *Southern Aviation*, December 1931, p. 29.

6. Catalogue of the Alabama Polytechnic Institute, 1931-1932, pp. 139-140.
7. Greer, J. A., and Turnipseed, C. L., "The Aeronautical Department," *The Auburn Engineer*, March 1931, pp. 139.

8. Finch, V. C., "Aeronautical Education at Auburn," *The Auburn Engineer*, February 1932, p. 116.

9. "Head of the Aeronautical School," *The Auburn Engineer*, February 1932, p. 119.

10. The Auburn Polytechnic Institute Bulletin, March 1942, Inside front cover.

11. "Robert G. Pitts File," Auburn University Library Special Collections.

12. "A Million Dollar Value - The Auburn Airport," *Alumnews*, September 1951, p. 1.

13. Newman, Graham, "Aeronautical Engineering at Auburn," *The Auburn Engineer*, December 1955, pp. 8-9 & 35.

14. Pitts, Robert G., "Aeronautical Engineering at Auburn," *The Auburn Engineer*, March 1959, p. 9.

15. "Aeronautical Department to Seek Accreditation," *Alumnews*, November 1959, p. 3.

16. "The Story Behind Re-Accreditation," *Alumnews*, October 1960, p. 5.

17. Martin, Polly, Private communication of auto-biographical material on Fred W. Martin, September 13, 2002.

18. Powers, Blake, "Auburn graduates backup crew for shuttle launch," *The Auburn Plainsman*, October 8, 1981, p. C-10.

19. Caption on photo of Branimir Djordjevic, *Montgomery Advertiser*, June 25, 1961.

20. Teer, Jerry, "Smarting In Success, Auburn Looks for Tougher Problems, *Sunday Ledger-Enquirer*, Columbus, GA, March 18, 1962, p. D-1.

21. Carson, E. Bruce, "The Undergraduate Program in Aerospace Engineering," *The Auburn Engineer*, December 1962, pp. 8 & 23.

22. "Two New Graduate Programs in Engineering – AU Offers ME Doctorate, Aerospace Master's," *Alumnews*, December 1961, p.1.

23. White, Bill, "The Master of Science Program in Aerospace Engineering," *The Auburn Engineer*, December 1962, p. 8 & 18.

24. Fitzpatrick, Philip M., Principles of Celestial Mechanics, Academic Press, New York, 1968.

25. Sforzini, Richard H., Private communication of material, July 2002.

26. Williams, John, "Tech Opens Stephen Hawking's Universe," *Business Week Online*, June 20, 2001.

27. "Hayes Officials Study New Uses for Software-Maneuverable Tow Targets," *Aviation Week and Space Technology*, June 24, 1991, p. 42.

28. "Whirly Like a Bird," *Discover*, July 1998, p. 57.

29. Benjamin, Roger, and Carroll, Steve, "The implications of the Changed Environment for Gover-nance in Higher Education," William G. Tierney, Ed., The Responsive University: Restructuring for High Performance, Johns Hopkins University Press, Baltimore, 1998, pp. 92-119.

Fig. 1 Brigadier General Knapp at age twenty-three (Logan, H. E. and Simms, J. D., **Auburn, a Pictorial History**, Donning, Norfolk, VA, 1981, p. 108)

Fig. 2 Solon Dixon, Aeronautical Engineering Instructor and student circa 1931 (Logan, H. E. and Simms, J. D., **Auburn, a Pictorial History**, Donning, Norfolk, VA, 1981, p. 135)

Fig. 3 W. G. "Bill" Sherling, with students, Bob Culberson and Wiley Robinson, inspecting the "Hot Shot" hypersonic tunnel (1962).

Fig. 4 Robert G. Pitts and Fred W. Martin in a section of the tank for the high-speed wind tunnel (1959).

Fig. 5 Auburn University Aerospace Engineering Building (2001).

CHAPTER 19
THREE QUARTERS OF A CENTURY OF AERO
AT IOWA STATE UNIVERSITY

Paul J. Hermann, Associate Professor Emeritus
Dept. of Aerospace Engineering, Iowa State University

Abstract

A review of the history of the development of an aerospace engineering curriculum and depart-ment at Iowa State University is presented. Significant events and people involved in that history are identified and several distinguished alumni are identified along with short surveys of their po-sitions and contributions to aerospace sciences and industry are indicated.

Early Years

The first courses in aeronautical engineering show up as elective courses in the Department of Mechanical Engineering at Iowa State College. Iowa State was, formally, Iowa State College of Agriculture and Mechanic Arts at the time. Iowa State is a state school established by the state in 1858 and it is the first school designated as a land grant institution under the provisions of the fed-eral Morrill Act of 1862. The college catalog for 1928-1929 showed four courses, one under-graduate course, "M.E. 482: Commercial Aviation" for 3 credits, and a sequence of three graduate courses, "M.E. 516a, 516b, and 516c: Aerodynamics and Airplanes" for 3 to 5 credits each. Pro-fessor Earl G. Smith was shown to be the professor in charge of the aeronautical engineering courses.

What the motivation was for having these courses has long since eased into the "great beyond" so it can only be speculation that those administratively responsible thought that aviation and aero-nautics were developing areas of technology and that Iowa State ought to make those areas to be areas of study available to engineering students who might wish those courses.   Dr. Raymond Hughes was president of the college, Professor Anson Marston was dean of the Division of Engi-neering, Professor Warren Meeker was head of the Department of Mechanical Engineering, and Professor Earl G. Smith was professor in charge. (It is of interest to this author that only President Hughes possessed a Ph.D.)

But, regardless of the motivation, Iowa State College introduced aeronautical engineering into its curricula. For the 1929-1930 catalog six additional undergraduate courses were introduced. The undergraduate aeronautical engineering courses listed were
    M.E. 433, Aeroplane Engine Construction
    M.E. 453, Aeroplane Construction
    M.E. 462, Aerodynamics
    M.E. 481, Aeroplane Designing
    M.E. 482, Commercial Aviation
    M.E. 489, Aeronautical Engineering
    M.E. 495, Aeroplane Engine Testing.

At the graduate level there were still M.E. 516a,b,c, Aerodynamics and Airplanes. Prof. Earl G. Smith was professor in charge.

(Here it is interesting to note that the word "aeroplane" is used for the undergraduate courses, and the word "airplane" is used for the graduate courses.)

A major change occurs for the college catalog of 1930-1931. There we see introduction of Professor William Alfred Bevan as professor in charge of the aeronautical engineering courses. Also, three additional courses at the graduate level are shown. They were

M.E. 519, Aircraft Instruments and Navigation
M.E. 529, Theory of Stability and Control
M.E. 549, Advanced Propeller Theory.
Professor Bevan is shown to be the professor in charge of the graduate courses.

It seems to be clear from this evidence that there was a vigorous effort being made in the Dept. of Mechanical Engr. to develop a set of aeronautical engineering elective courses for both undergraduate and graduate students that the students might select for their degrees in mechanical engineering. It is also quite clear that the people responsible for this effort were dedicated professional educators who were not developing these courses because of some idealistic whim to play around with some tom foolery toys of semi-crazy barnstormers and nutty military generals who liked to risk their lives in "stick and wire" Curtiss Jennys or bomb sitting duck battleships out of the water. These people had a clear objective to keep Iowa State College up there at the leading edge of a developing technology and they wanted to make that technology available to their engineering students.

Dr. Raymond M. Hughes was president of Iowa State College. He had received the A. B. degree from Miami University in 1893, the M. Sci. degree from Ohio State University in 1897, the LL. D. degree from Miami University in 1927, and was appointed president of Iowa State College on September 1, 1927.

Anson Marston was the first dean of engineering at Iowa State. He received the Civil Engineering degree from Cornell University in 1889, and had received honorary Doctor of Engineering degrees from University of Nebraska in 1925 and from Michigan State University in 1927. He had come to Iowa State in 1892 and was appointed the first dean of engineering in 1904. Dean Marston distinguished himself in the state of Iowa as a forward thinking engineer and was influential in the formation of the first Iowa Highway Commission. He was one of the first commissioners of the newly formed commission in about 1910. Dean Marston served as Dean of Engineering until 1932 and is one of the most revered members of the faculty of Iowa State College, now University.

Professor Warren H. Meeker received the Master of Engineering degree from Cornell University in 1891 and came to Iowa State in that year of 1891. He became professor and head of the Department of Mechanical Engineering in 1907 and served in that position until 1932.

Professor Earl G. Smith received the B. S. degree from the University of Missouri in 1903 and the Master of Engineering degree from Univ. of Missouri in 1905. He was appointed professor of mechanical engineering in 1925.

William Alfred Bevan received the Bachelor of Science degree in engineering from Iowa State in 1904, the Master of Science in engineering from Massachusetts Institute of Technology in 1921, and the Bachelor of Science in mechanical engineering from Purdue University in 1926. He had served previous stints as a member of the engineering faculty at Iowa State in 1904-1905, 1909-1912, and returned to Iowa State as professor of mechanical engineering in 1929. Professor Bevan had served in the United States armed forces during and after World War I and was involved in military aviation during that time. He had a reserve commission as Lieutenant Colonel in the U. S. Army when he returned to Iowa State in 1929.

Col. Bevan was active not only in teaching aeronautical engineering courses at Iowa State but also in aviation activities in the State of Iowa. Old, yellowing newspaper clippings attest to the fact that during the 1930s he made presentations to several aeronautical groups in the State of Iowa and that, in 1934, he was appointed by Governor Clyde Herring to an advisory position with the Iowa Aeronautics Commission.

In 1932 Dean Anson Marston stepped down as dean of engineering. His replacement was Professor Thomas R. Agg. Professor Agg had received the Bachelor of Science degree in electrical engineering from Iowa State College in 1905 and the Civil Engineering degree from Iowa State in 1914. He had been a member of the engineering faculty since 1913. Dean Agg demonstrated during his tenure as dean his commitment to the furthering and strengthening of the aeronautical engineering program in the Division of Engineering.

In 1936 Prof. Meeker stepped down as head of mechanical engineering. His replacement was Professor Mark P. Cleghorn. Prof. Cleghorn had received the B. S. in electrical engineering from Iowa State in 1902 and the M. E. from Iowa State in 1907. He had been a member of the mechanical engineering faculty at Iowa State since 1902.

Also in 1936 President Hughes stepped down as president of Iowa State College. His replacement was Dr. Charles E. Friley. Dr. Friley had received the B. S. from Texas A&M in 1919 and an A. M. from Columbia University in 1923. In 1929 he was awarded an honorary LL. D. degree by Hardin Simmons College. He had come to Iowa State as dean of industrial science in 1932. He served as president until 1953.

Thus, by 1937 the administrative leadership responsible for the aeronautical engineering program at Iowa State had completely changed from the original makeup in 1928. Friley was president, Agg was dean of engineering, Cleghorn was head of mechanical engineering and Bevan was professor in charge of aeronautical engineering, still in the Department of Mechanical Engineering. The indications are that this group of administrators was every bit as dedicated to developing an aeronautical engineering program at Iowa State as was the previous group.

As the decade of the 1930s wore on the indications are that the dedication was diminishing. One can speculate that the reason was Col. Bevan's health. He left the department and the college in 1940 and died in 1943.

But, the administration did go out to find a quality replacement for Col. Bevan. Wilbur C. Nelson came on board as assistant professor of aeronautical engineering in 1940. Prof. Nelson had received the B. S. from the University of Michigan in 1935 and the M. S., also from the University of Michigan, in 1937. He is

Wilbur C. Nelson, first department head.

shown to be professor in charge of the graduate courses in aeronautical engineering in the college catalog for 1941-1942.

The Curriculum and Department

The year 1941 turned out to be a very significant year for aeronautical engineering at Iowa State. On November 26, 1941 the State Board of Education, the predecessor of the present day Iowa State Board of Regents, the governing board for Iowa State, approved a new curriculum for the college, Aeronautical Engineering. The curriculum was to be administered in the Department of Mechanical Engineering and Assistant Professor Wilbur C. Nelson was appointed professor in charge. Eight days later, on Saturday, December 6, 1941, an article and photograph appeared on the front page of the college newspaper, The Iowa State Daily. Prof. Cleghorn, head of mechanical engineering, was quoted as saying that, "interest in these courses is now at a new high and we expect the demand to continue and to increase."

The next day, Sunday, December 7, 1941, the Japanese attacked Pearl Harbor, and on Monday, the United States was at war. Aeronautical engineering took on an entirely new and much more important position for the college, the state, and for the nation.

A year earlier, in 1940, Iowa State had taken on a mission for the Civil Aviation Authority (CAA). A program of pilot training for the Civilian Pilot Training program of the CAA was initiated with the Department of Mechanical Engineering being responsible for the program and with flight training taking place out at the Ames Municipal Airport. Now, after that historic Sunday, the trainees in the CPT program were looking forward to a much more serious future.

During war time many activities related to the war take place much more quickly than would otherwise be the case. That was true for aeronautical engineering at Iowa State. On June 5, 1942, the Iowa State faculty approved formation of a new department to conduct the curriculum and on June 30, the State Board of Education approved the new department. That is the true birth date of the Department of Aeronautical Engineering. Assistant professor Wilbur C. Nelson was promoted to professor and head of the Department of Aeronautical Engineering at that time. The effective beginning of the department and the curriculum was at the beginning of the school year of 1942-1943, in September of 1942.

The new department took quarters in a little brick building that had, for many years, served as the pattern shop for the mechanical engineering foundry. The building provided a departmental office, an office for the department head, an office for two, maybe three, faculty, one classroom, and a laboratory in the back that contained the new wind tunnel that Col. Bevan had designed and had constructed. The classroom had drafting tables rather than seats and students sat on stools.

The wind tunnel was an open throat closed circuit tunnel constructed of sheet metal. A wooden, four-bladed propeller was driven by an electric motor that was controlled from a control pedestal that looked as if it had been rescued from a Chicago street car. Test models were supported and suspended by wires from a platform above the test section upon which were five "meat scales" (as we called them) to measure left lift, right lift, left drag, right drag, and tail lift. The support wires extended into a pit underneath the test section and had plates that were immersed in oil in the pit, a system that provided for damping of the force measuring system. Two optical telescopes were mounted outside of the test section to provide for measuring of angle of attack of the wind tunnel airplane models. The entire wind tunnel system had been built for a cost of $5000 that had certainly been approved by the college administrative hierarchy.

When the department began its operation in the fall of 1942 the faculty consisted of Nelson, John M. Coan, assistant professor, and Lester G. Kelso, instructor. University records show that there were 97 undergraduate students, 48 sophomores, 23 juniors, and 26 seniors. Freshmen engineering students were not attributed to any department in those days. The enrollment history for the first five years of the department's operation is ugly: 97 undergraduates in 1942-1943, 40 undergraduates in 1943-1944, 13 undergraduates in 1944-1945, 76 undergraduates in 1945-1946, and 211 undergraduates in 1946-1947. The management problems for Nelson and Dean Agg had to be terrible!

John Michael Coan, Jr. was the structural analysis specialist. He had obtained the B. E. degree from Johns Hopkins University in 1935 and the M. E. from Johns Hopkins in 1938. He came to Iowa State in 1942.

Lester Grandville Kelso was more the aerodynamics specialist. He had his B. S. degree from the University of Washington in 1939 and he came to Iowa State in 1942.

For those first years of operation of the curriculum and department the curriculum was mostly a set of variations on the mechanical engineering degree with aeronautical engineering electives. The total number of quarter credits required to graduate was 209, 48 in the first year, 53 in the second year, 54 in the third year, and 54 in the fourth year. Those 209 quarter credits translate into 139 1/3 semester credits. Of those 209 credits 6 were for ROTC which was a requirement for all able bodied men enrolled in the college.

T. A. Wilson, of first graduating class, CEO of Boeing.

The first graduation from the new department was in March 1943. Our records show that there were thirteen graduates in that group. In that first group was probably our most famous alumnus, Thornton Arnold Wilson, Jr., T Wilson. T went on to work for Boeing and he eventually became the Chairman and Chief Executive Officer of the Boeing Company.

T was very generous to Iowa State during his lifetime. That was recognized by Iowa State University in 1993 when he was granted an honorary doctorate by the University, one of a very, very few that Iowa State has ever granted. He and his wife Grace have endowed a chair in engineering at Iowa State, a chair that is now being administered by the Department of Aerospace Engineering and Engineering Mechanics.

In July 1943 the Navy came to Iowa State College, and so did the Army. The war was on and the services needed officers for their expanding military units. The services wanted those officer candidates to have as much college education as they could get in the time available. The most important program for the Iowa State engineering division was the Navy V-12 program, a training program for general officers. A smaller program was the V-5 program for naval aviation candidates. The Army came to the campus with its ASTP program, Army Special Training Program. There was a sizeable ASTP presence on the campus but, as I recall, there wasn't the presence in engineering that was comparable with the Navy V-12 program.

The V-12 program brought in students from many places in the country, students who would have barely noticed Iowa State were it not that the Navy sent them to Ames. One of our most illustrious alumni came out of that group of V-12 students. John F. Yardley graduated in October 1944. John was from St. Louis and had been studying mechanical engineering at Washington University in St. Louis until he was drawn into military service and was sent to Iowa State. He went with McDonnell Aircraft Company in 1946 and, in the 1960s, he was deeply involved in the space program. He was, starting in 1958, project engineer at MAC for the design of the Mercury capsule. He went on to oversee the Gemini program and then became involved in Skylab. In 1974 John went "on loan so to speak" to NASA and became associate administrator for space transportation, i.e., the shuttle program. In 1981 he returned to McDonnell Douglas and became president of the McDonnell Douglas Astronautics Company. John retired in 1989 and he recently died, on June 26, 2001.

John Yardley, Navy V-12 grad., Pres., McDonnell Astronautics.

Arthur E. Bryson, Jr., V-12 grad. educator, prominent scientist.

Another of our most illustrious alumni who came out of the V-12 program was Arthur E. Bryson, Jr. Art graduated in December 1946. He went on to earn his doctorate in aeronautics at California Institute of Technology, and taught at Harvard and MIT where he made quite a name for himself in optimal control. Some might consider that he is the father of optimal control. Later he went to Stanford University where he was head of aerospace engineering. He is an honorary fellow in AIAA and has at least twice presented the Dryden Lecture at AIAA Aerospace Sciences Meetings.

Also during the war, probably in late 1942 or in 1943, Iowa State entered into a training program with the Curtiss-Wright Corporation to train young women to be engineering aides. These young women were taught basic skills of engineering calculations, use of the slide rule, elementary engineering drawing, and some basic principles of Mathematics, Statics, and Dynamics. The Department of Aeronautical Engineering was marginally involved with the teaching function and Mr. Gayle G. Carnes was brought onto the faculty to be one of the responsible faculty for the program.

In June 1945 the Board of Education approved a leave of absence for Wilbur Nelson to go to Johns Hopkins Applied Physics Laboratory "to do urgent war research in aerodynamics." When he returned is unclear, but it seems clear that Nelson was on campus again to start the 1945-1946 academic year. In May 1946 the Board then approved the resignation of Nelson who resigned to return to his native Michigan and to the University of Michigan to become director of aeronautical research. Eventually Nelson became head of the Department of Aerospace Engineering at the University of Michigan in Ann Arbor.

Carl N. Sanford,
dept. head 19467-1952.

Also in May 1946 the Board of Education approved the appointment of Assistant Professor Carl N. Sanford to be Professor and Head of the Department of Aeronautical Engineering at Iowa State. Sanford had obtained the B. S. in mechanical engineering from Oregon State College in 1928 and the M. S. from North Carolina State in 1940. He came to Iowa State as an assistant professor of mechanical engineering in 1945.

Carl ran into problems immediately. The war had ended and the GI Bill was coming into full force. The undergraduate student population in the department increased from 76 in 1945-1946 to 211 in 1946-1947. And, the faculty had gone. Nelson, Coan and Kelso were gone. Sanford was facing that crowd of students all alone!

What he did was cajole two instructors to come in and help. Ivan Jensen came from somewhere in South Dakota I believe and he came on board to handle the structures courses. The other new instructor was our illustrious alumnus, T Wilson. T had been at Boeing and came back to Iowa State to start, or continue, graduate work toward a master's degree. T once told me that he figured that Carl had given him, T, a raw deal. The three, Sanford, Jensen, and Wilson, didn't have any time to themselves or for personal advancement because of the crowd of GI bill students who had come back to finish their degree. Your present author was one of them.

Carl had to contend with an additional problem that was essentially unexpected. Dean Agg stepped down as dean at the end of June 1946. There was an interim dean until the next March and, on March 15, 1947, Dr. J. F. Downie-Smith took over as dean of engineering. Downie-Smith was a Scotsman. He had the B. Sc. From University of Glasgow in 1923, the M. Sc. from Georgia School of Technology in 1925, the M. E. from Virginia Polytechnique Institute in 1928, the S. M. from Harvard University in 1930, and the Sc. D. from Harvard in 1933. There are some of us who have the impression that Downie-Smith did not have the dedication to the aeronautical engineering curriculum that previous deans had displayed.

The curriculum Carl took over for the year 1946-1947 consisted of 210 quarter credits, 140 semester credits. Two years later, for the 1948-1949 catalog, it consisted of 216 quarter credits, 144 semester credits. There were still quite a few mechanical engineering courses included as requirements for the curriculum. The CPT flight training program had been terminated but flight training courses were still offered as optional additional courses in the department.

At some time just after the war the department obtained an airplane. Whether it was Nelson or Sanford who got it I don't know. It was an Ercoupe, an effort to "put an airplane in every garage." It had a "four banger" horizontally opposed engine and a two-bladed fixed-pitch propeller, tricycle landing gear, coupled aileron and rudder to eliminate rudder control for turns, and a "nonstall wing". That is an oxymoron, however, because any wing is going to stall. But the Ercoupe wing stalled from the inside out and the airplane would be losing lift and pitching forward before it was ever coaxed into a full stall. Possession of the airplane enabled Carl to create a flight test course and include that course in the graduation requirements.

For the next few years Carl seemed to have troubling times. The student population went down reaching a low of 46 for the year 1951-1952. He did get two or three faculty members from outside of Iowa State and he got two or three Iowa State graduates to take teaching positions. But, the faculty population seems to have been persistently turning over with people leaving and Carl having to struggle to find replacements. The post war job market was very poor so that the enthusiasm for the aeronautical engineering curriculum was poor.

Glenn Murphy,
dept. head, 1952-1955.

Then, as the academic year 1951-1952 came to an end, Carl Sanford became the victim of a political situation in the college. Dean Downie-Smith apparently did not find Sanford to be a suitable department head. In addition, a star member of the faculty of the Department of Theoretical and Applied Mechanics (T&AM), Dr. Glenn Murphy, was being proselytized by other schools to become a department head there. Murphy was, indeed, being looked upon as the heir-apparent for that position at Iowa State, but that position was still occupied by the original department head, Professor H. J. Gilkey, who was not scheduled to retire for another three or four years. The dean did not want to lose Murphy. So, he devised a solution: demote Sanford out of his position and appoint Murphy to be head of aeronautical engineering. How the dean could swing the deal I don't know, but he did.

Dr. Glenn Murphy became head of the Department of Aeronautical Engineering in 1952. Murphy had obtained the B. S. in civil engineering at the University of Colorado in 1929, and the M. S. in civil engineering from Colorado in 1930, the civil engineering (C. E.) degree from Colorado in 1937, and the M. S. in civil engineering from the University of Illinois in 1932. He obtained the Ph.D. from Iowa State in 1935. He had come to Iowa State in 1932 as a member of the faculty of T&AM and had advanced through the ranks to become professor of T&AM in 1941. He was the author of several technical books and was, clearly, a starring member of the faculty of the Division of Engineering at Iowa State. Later he became a prominent leader in the American Society for Engineering Education, ASEE.

During the early 1950s and through the tenure of Dr. Murphy as head two important names were added to the list of faculty for the Department of Aeronautical Engineering. Dr. Ernest W. Anderson had come to the department from the Mathematics Department. Ernie had obtained the B. S. in civil engineering from North Dakota State College in 1926. He came to Iowa State and the Department of Mathematics in 1928 and obtained the M. S. in mathematics in 1928 and the Ph. D. in mathematics in 1933. He was appointed professor of mathematics in 1947 and had a split appointment with the Engineering Experiment Station in the Division of Engineering. Ernie is shown to be a professor of aeronautical engineering in the college catalog for 1952-1953.

Ernest W. Anderson,
department head, 19855-1971.

Merlyn L. "Mike" Millett, Jr. had obtained the B. S. in aeronautical engineering from Iowa State in 1945. He stayed on to obtain the M. S. in aeronautical engineering in 1948 and then he went to California and Douglas Aircraft Company as an engineer. He returned to Iowa State in 1952 to be an assistant professor of aeronautical engineering and to take graduate work toward the Ph. D. He obtained the Ph. D. in engineering from Iowa State in 1957.

In 1955 Prof. H. J. Gilkey was required to step down as head of T&AM. Murphy was appointed to be his replacement. It is said that Murphy had wanted to merge the two departments so that he would be head of both T&AM and aeronautical engineering. But, Dean Downie-Smith and the university administration would not permit the merger. Years later, in 1990, that merger was edicted by the administration!

Ernie Anderson was appointed Head of the Department of Aeronautical Engineering in 1955 a position that he held until 1971. And, in 1955, the department traded the Ercoupe airplane for a four-place North American Navion. Mike Millett had been taking flight training and the bigger airplane provided for better demonstration flights, a bit more sophisticated flight testing, and for more convenience for any departmental business flying.

When Ernie took over he did so at the beginning of a long term period of enrollment expansion. A year earlier the college policy had been changed to assign freshmen who had identified a curriculum to the department conducting that curriculum. Thus, in 1955-1956 when Ernie took over the undergraduate enrollment was 199 72 of whom were freshmen—and there were only 7 seniors. The enrollment continued to increase each year to reach a peak of 509 for the year 1969-1970. Of those 227 were freshmen and 86 were seniors.

The reasons for the continuing interest in aeronautical/aerospace engineering were varied. There was increasing awareness of the military need for rocket powered missiles. There was increasing awareness of the big passenger transports that now regularly traversed the continent as well as traversing the Atlantic and Pacific oceans. There was increasing fascination with jet aircraft such as the B-47 and the "100-series" of fighter aircraft that came about because of the Korean war. And then there was space. The Russians beat us into space with their Sputnik in 1957 and space had, essentially over night, taken a major position of importance to the voters as well as the to the legislators and administrators of the government. President Kennedy's call for putting a man on the moon and returning him to earth before the end of the decade of the 1960s put a lot of focus on the problems of space. People got excited. And, the prime mover was that there were jobs! Aeronautical/aerospace engineering graduates were getting jobs.

Ernie faced much the same problem as had his predecessors: finding faculty to put in front of this expanding population of students. Mike Millett stayed on to be an enormous help to Ernie and for the 1956-1957 catalog only Anderson and Millett are listed as faculty. In 1956 Ernie got Fred Stuve to come down to Ames from North Dakota State in Fargo. And, in 1958 Dr. Cheng-Ting Hsu came to Ames from the University of Minnesota. Otherwise Ernie had to make do with graduate students. James D. Iversen and Lawrence H. Stein became stalwarts in the undergraduate classrooms along with John Havey. Dale Anderson came to Ames to take graduate work and to teach after getting his B. S. from Parks College in East St. Louis. These constitute just the beginning of a sizeable list of graduate students who manned the department's classrooms during the decade of the '60s. But, they did well. As the decade progressed it became clear that the aerospace companies liked the Iowa State graduates that they could hire.

In 1958 Iowa State College became Iowa State University and the Division of Engineering became the College of Engineering. It is probable that Ernie had little to do with causing those changes, however. But, he did have to do with changing the name of the department from Aeronautical Engineering to Aerospace Engineering. That occurred in November 1960 when the department became the Department of Aerospace Engineering.

In May 1961 the department granted a Bachelor of Science degree to its first female graduate, Beverly Birchmeier. Bev was the nemesis of the senior class that year because she was the one who always got the top grades. She received a Fulbright scholarship to go to England for a year, returned to get her M. S. in aerospace engineering in 1963, married Dr. Ned Audeh and they went to Alabama where she worked as an engineer and he got into engineering education. They have two children I believe.

A long term goal that Ernie had, other than to develop and maintain a strong undergraduate curriculum, was to develop a viable graduate program. The department had never had the authority to grant Ph. D. degrees in aeronautical engineering. The department could grant master of science and master of engineering degrees, but students could only join their aeronautics components with degrees granted by other departments such as Mathematics, T&AM, Electrical Engineering, etc.

In 1968 two aerospace engineering graduate students went out to NASA-Ames in Mountain View, CA to be the first in a kind of graduate student cooperative education program. Ernie and Jim Iversen had worked out the arrangement with NASA with the strong assistance of an alumnus, Dr. Vernon Rossow. The students were to go to NASA and work on a thesis project there. They then returned to Ames to stand their final examinations and be granted their Ph. D. degrees. The program was of enormous benefit to the aerospace engineering department and was responsible for a sizeable cadre of Iowa State graduates going onto the payroll of NASA-Ames in subsequent years. The "Iowa State Mafia" I called them.

The Iowa State graduate students tended strongly to get into the new and expanding area of computational fluid dynamics, CFD. Paul Kutler, Joe Steger, and Ron Bailey are the names of three recently prominent NASA scientists who went to NASA-Ames through the Iowa State program.

At Iowa State the leadership effort was with Dale Anderson, John Tannehill, and Dick Pletcher. Ernie, again with the assistance of Jim Iversen and others, had obtained a "training grant" from NASA that had the mission to develop Ph. D. graduates in the field of CFD. NASA considered the operation in the department to be a "center" but Iowa State administration did not recognize that designation for several years into the decade of the 1970s. An additional person added to the population of the center was Dr. Richard Hindman who came on board primarily to provide for the undergraduate teaching load that Anderson, Tannehill, and Pletcher would not be providing. Subsequently, over the years, other faculty have had part time appointments in the Iowa State CFD center.

In 1960 William D. "Bill" James, a 1958 B. S. graduate of the department, returned to Iowa State from Cornell University where he had obtained the M. S. in aerospace sciences. During the 1960s Bill had proposed and obtained funding for construction of a new wind tunnel for the department. It was an "open circuit" subsonic tunnel with provision for inserting smoke streams into the flow. The nominal test section is 3.00 ft x 2.50 ft and test section speeds up to 200 mph are possible. Bill's tunnel became available for serious research projects by the time that the department moved to its new quarters in Town Engineering Building in 1971. Bill himself used his tunnel to test drag characteristics of polygonal highway light poles for his Ph. D. dissertation.

Two projects that were conducted by Dr. Jim Iversen in Bill's tunnel concerned (1) wing vortex characteristics and modification for large aircraft, and (2) aeolian deposits downwind of craters on Mars. Jim also adapted the latter project for the Iowa Highway Commission to study snow deposits at interstate highway intersections.

Also during the 1960s Dr. Ernie, along with help from Dr. "Mike" Millett, worked out undergraduate cooperative education programs with McDonnell Aircraft Co. in St. Louis and with NASA-Edwards, now NASA-Dryden, in California. Later the program was extended to establish programs with several other companies. These co-op programs have been very valuable to our Iowa State University aerospace engineering students.

In 1964 Dr. Ernie, along with Jim Iversen, obtained an undergraduate research grant from the National Science Foundation. About ten undergraduates participated in the program with the result that several undergraduate papers were entered into the 1965 student papers competitions at St. Louis and at Dallas/Ft. Worth. Some AIAA sections throughout the country had established these student conferences and papers competitions to promote and reward student efforts in their respective schools. Iowa State was included in the invitations to the conferences conducted by the St. Louis Section and by the North Texas Section. Iowa State first participated in these conferences in 1965. John Tannehill won first place in the undergraduate division in St. Louis and Don Lewellyn won third place in the undergraduate division in Dallas, both with papers that had their origins in their NSF projects of the summer of 1964.

Iowa State then continued to participate in the student conferences during subsequent years and when the national AIAA took over the student conference programs throughout the nation in 1972 Iowa State has continued to be active in the Region V activities.

Vance D. Coffman, 1967 graduate, Chairman & CEO Lockheed-Martin.

In 1965 President James H. Hilton stepped down into retirement and Dr. W. R. "Bob" Parks took over as president. In 1967 another of our illustrious alums graduated, Vance D. Coffman. Vance is now chairman and chief executive officer of the Lockheed Martin Corporation.

In 1969, at the time of the first moon landing, it may be remembered that Armstrong and Schirra were having trouble with the computer overloading such that the landing on the moon was about to be aborted. An engineer at Johnson Spaceflight Center radioed up instructions about unloading the computer to permit the landing operations to continue. The landing was, as we know, successful. That engineer in Houston was Stephen G. Bales, Iowa State Aero. E., November 1965.

At some time in the mid-1960s the department again traded airplanes. The Navion was traded in for a Mooney. The Mooney had a low wing, a 200 hp engine, a constant speed propeller, and retractable gear.

In the year 1966-1967 the department conducted a series of courses at Collins Radio Co. in Cedar Rapids. The company was attempting to get into the business of manufacturing electronic equipment for aircraft flight control. The courses were a specially constructed set of courses that reviewed the entire gamut of aerodynamics, performance, stability, and control courses taught in the curriculum on campus. The sequence of courses was taught twice more, in 1973-1974 and in 1978-1979 after the company had been acquired by Rockwell International and had become the Collins Electronics Divisions of Rockwell International.

During the decade of the 1960s Dr. Ernie and Mike Millett had become more and more busy in planning a new engineering building that would house the Departments of Aerospace Engineering and of Civil Engineering. The building became a completed reality in the summer of 1971. But Ernie never operated as head of aerospace engineering out of the new building. He was required, by his age, to step down as administrative head at the end of the 1970-1971 academic year.

In the summer we got a new department head, Dr. Robert F. Brodsky, and a new home, which was subsequently named George R. Town Engineering Building for the dean of engineering who had retired a year earlier.

The dean who replaced Dean Town in 1969 was Dr. David R. Boylan. His first hire of a new department head was Dr. Bob Brodsky. Bob took over as head of aerospace engineering in July of 1971 and the department moved into its new quarters in the new engineering building in July and August. Brodsky had received the B. M. E. from Cornell University in 1947, the M. S. in Aero. Engr. from New York University in 1948 and the D. Sci. from New York University in 1950. He later received the M. S. from the University of New Mexico in 1957. He had been in engineering management at Aerojet Corporation before coming to Iowa State in 1971.

Robert F. Brodsky, department head, 1971-1980.

Bob took over at a time that might be called unpropitious. Enrollment in aerospace engineering plunged drastically. Undergraduate enrollment was at 509 for the year 1969-1970. It had dropped to 115 for the year 1974-1975. In the two year period from 1970 to 1972 total undergraduate enrollment dropped to be almost exactly the same in 1972 as the freshman enrollment in 1970.

The primary reason for the enrollment drop was, of course, lack of job opportunities. Beginning in 1969 job opportunities in the aerospace industry almost entirely evaporated. In addition, it was the time of great student unrest and sometimes violent opposition to the Viet Nam war, to the Selective Service, and to government and military institutions in general. Aerospace was associated with the military and with the war in the minds of students and they opted to avoid the aerospace engineering curriculum—in large numbers. Fortunately, Iowa State did not get shut down by the mob of student activists. There weren't enough of the militant activists to achieve a shut down, but they tried.

One thing that this drastic drop in enrollment did was to permit the new engineering management at Iowa State to reduce the number of "inbred" faculty in the department. The academic environment in colleges and universities had been slowly changing over the years and one principle that had come quite strongly forward was elimination of the "inbreeding" of faculty. Faculty were not to have their highest degree having been obtained from the institution at which they were employed. Brodsky felt the mandate that he should lay off those members of the departmental faculty who were untenured and had their degrees from Iowa State. I believe that there were seven of them who got their walking papers.

Another principle that was coming into force was that all faculty were to have an earned doctorate. This was certainly a restriction on hiring that had not been an institutional policy for the previous century. Bob then went out to hire new faculty who were not "inbred". Dr. Lennox Wilson and Dr. Thomas McDaniel were on board for the start of the 1973 academic year. Wilson had his Ph. D. from the University of Toronto in Canada and came to Iowa State from the University of Missouri where he had been professor of aerospace engineering. McDaniel had his Ph. D. from the University of Illinois and came to Iowa State from Dayton University in Ohio.

Then, on November 30, 1973, one of Ernie Anderson's goals was achieved. The department received approval for granting the Ph. D. in aerospace engineering. This meant that the department had fully matured and could grant the degrees B. S., M. S., M. E., and Ph. D.

One of Brodsky's pet projects was to get more intimate contact with aerospace institutions outside of Iowa State University. His idea was to "trade" personnel with commercial, government, and academic institutions for short periods of time. The first such "trade" took place with NASA-Ames Research Center. In November 1973 Dr. Gary Chapman of NASA-Ames came to Iowa State as a visiting professor of aerospace engineering and Dr. Jim Iversen went out to Mountain View, CA and NASA for a nine-month assignment.

For the 1975-1976 academic year Mr. David Hall, on leave from Battelle Memorial Institute in Columbus, Ohio, came to Iowa State as a visiting assistant professor of aerospace engineering.

In 1976 Mr. Victor Corsiglia, a research scientist from NASA-Ames came to Iowa State as a visiting professor of aerospace engineering for the academic year. Professor Bill James went out to Mountain View as the Iowa State part of the "trade" with NASA.

In December 1980 Dr. Pete Liepman was introduced as a visiting adjunct professor. Dr. Pete stayed for the winter and spring quarters of 1980-1981.

In February 1974 a distinguished alumnus was graduated. Kevin Peterson had been a co-operative education student at NASA-Edwards and he stayed with NASA after he graduated from Iowa State. Kevin is now director of NASA-Dryden Flight Research Center.

During 1975 members of the department faculty founded the I.O.W.A. Section of AIAA. The initial meeting of the new section was held in January 1976. The section serves as the institutional support for the AIAA Student Branch at Iowa State.

In May 1975 the first of what has become an annual event for the department occurred, the Iowa State Aerospace Engineering Spring Symposium and Student Conference. The concept and idea came from the fertile mind of Assistant Professor Bill

Kevin L. Petersen, 1974 grad., Director of NASA-Dryden.

265

James. Bill had observed the great benefits to the students who had participated in the NSF summer undergraduate research project in 1964. He proposed, and the departmental faculty accepted, the concept of "Senior Projects". Each senior was required to take a 1-credit course in each semester of his/her senior year for which he/she was to conduct a personal research project under the supervision and direction of a member of the faculty. The product of the student's work was to be a technical paper on the project. Then, in 1972, Bill proposed that the department conduct its own version of an AIAA student papers competition. Presenting a paper was optional—at first—and, surprisingly, four students opted to present their paper. There was a year end awards banquet to finish off the event and the speaker of the evening was Mrs. Ann Pelegreno who had made, in 1967, a circumnavigation of the globe following somewhat the path that Amelia Earhart was to have followed had she, Earhart, completed her flight. Ann and her husband, Dominick Pelegreno, a professor in the College of Education at Iowa State, were enthusiastic and active members of the Central Iowa Section of the Experimental Aircraft Association. Ann has written two volumes of a history of aviation in the State of Iowa and is still working on a third and last volume.

By 1975 Bill next proposed that there be a two-day event with the first day being a technical symposium and that the second day would be the student papers presentations. Engineers from industry and government would be invited to come to talk to the students about the "state of the art" of some aspect of aerospace engineering. Then these visitors would be oral judges at the student presentations on the second day.

The first of these extended events was held May 9-10, 1975. Bill was remarkably successful in getting industry and government cooperation. One of the six featured speakers at this first event was NASA's Richard Whitcomb of airfoil design fame. And, the banquet speaker was Alan Pope, practically the father of wind tunnel design.

After the first day's symposium presentations Bill organized a picnic out at Dr. Tom McDaniel's barn. Hamburgers, hot dogs, potato salad, beans, potato chips, beer (18-year olds could legally drink by then in Iowa), and soda pop were served, with very informal protocols. Then everyone gathered in the hay mow, sat around on bales of hay and Bill started a round of questions and answers between the students and visiting speakers. Alan Pope got so taken up with the process that he practically stole the master of ceremonies assignment right out from under Bill James.

The event has been a fixture for the department at the end of the academic year from 1975 on to the present. And, over the years students doing job interviews with companies found that one of the most interesting features of students' backgrounds for the interviewers has been their senior projects.

At the end of the 1979-1980 academic year Bob Brodsky resigned to go to TRW Systems in California. On July 1, 1980 Dr. Lennox N. Wilson took over as interim chair for a year and then as chair. Brodsky had headed the department through the drastic reduction in enrollment, then we began to see a recovery of the enrollment picture. Len Wilson inherited that enrollment expansion.

Len also inherited the problems of conversion of Iowa State University from the quarter system to the semester system. The conversion took place on July 1, 1981. Probably the big-

Lennox N. Wilson,
dept. chair 1980-1987.

gest problem that Len had was to accommodate the rapidly expanding enrollment. The undergraduate enrollment when Len took over in 1980 was 360. By 1984 the enrollment had peaked at 653. There were 130 seniors and 232 freshmen that year! The number of faculty was practically the same as it had been in 1974 when the enrollment was 115.

But aerospace engineering was not alone in being inundated by student enrollment. The entire College of Engineering was being inundated. Early in 1981 the faculty of the College of Engineering approved recommendations to the college and university administrations for management of the enrollment problems in the college. The college did receive approval for a management plan that put student advancement into the professional program of a curriculum to be dependent upon the student's achievement in a specified Basic Program. The plan did not limit admissions to the university but it did limit advancement into the sophomore, junior, and senior years. As a result of the grade point being used as the measure for admission into the professional program the quality of the students in the curriculum increased, attrition out of the curriculum decreased and the curriculum became somewhat an "elitist" curriculum. The plan was continued in operation until enrollment problems evaporated in the mid-1990s.

At some time in the 1980s Professor Bion Pierson became the founding editor of a technical publication, "Optimal Control Methods and Applications", a publication by John Wiley Company of England.

Throughout the long period of time from the accession of Dr. Ernie Anderson to be head in 1955 through the decade of the 1980s the department was characterized by a strong dedication of both administration and faculty to making the aerospace engineering curriculum a viable, up-to-date, high technology curriculum. The curriculum was tough and required a high number of credits for graduation. There was a lot of advanced theory in the courses and there was a strong emphasis on the use of theory in engineering design, high technology design.

Through the latter part of the 1950s compressible flow theory had filtered down from the graduate level into the junior level courses, sometimes with viscous flow theory also incorporated. Early in the 1960s space flight mechanics were introduced at the undergraduate level. Orbit mechanics and boost and reentry mechanics were emphasized, first in elective courses and then with some in curricular requirements. In 1972 a spacecraft systems design course became an available course and in 1973 a remote sensing course was added, both at the undergraduate level. And, automatic control systems analysis and conceptual design became features of the senior year, partly as required courses and partly as optional elective courses.

At the graduate level compressible flow theory continued to be emphasized, with increased sophistication. There was some attention given to hypersonic flow theory and to magnetohydrodynamics and Dr. Ernie introduced optimization and the theory of the calculus of variations into the graduate space flight mechanics courses. And then computational fluid dynamics began to enter into the picture. CFD soon took over the major attention of the department at the graduate level.

When modern, high speed digital computers began to enter into the technological picture in the country Iowa State aerospace engineering was right there to latch onto their use. Computational work began to enter into just about every graduate thesis, and, at the undergraduate level FORTRAN programming migrated down into the sophomore year. Our students showed varying proficiencies in programming when they graduated, but most of them had a pretty good grasp of the technology.

On July 1, 1986 Dr. Gordon Eaton took over as president of Iowa State University. He replaced Dr. "Bob" Parks. Eaton came to Iowa State from Texas A&M University and it would appear that when he came on board he did so with a mandate to change the university. Iowa State came under criticism for not going out to get its "fair share" of financial resources from the federal government. The school was criticized for not doing enough in the way of sponsored research. It appears that Eaton had the mandate to change those aspects of the school's previous administrations. In doing so Eaton ruffled a lot of feathers.

Eaton's policies and practices came to the Department of Aerospace Engineering at a stressful time. It was a time when the high undergraduate enrollment had been maintained at 600 or more for several years and the faculty were very busy just trying to keep up with the undergraduate teaching load. Something was going to have to give if the faculty were to spend more time soliciting and executing sponsored research. What gave was the time and attention given to undergraduate teaching. But, doing so was in contradiction with the departmental policies that had been in place in the department and college "since time immemorial". There was stress; there were many disagreements among the faculty. It was not a fun time!

In June 1987 the stress was accentuated and Wilson resigned his position. Dr. James D. Iversen was appointed to be the chair, first as interim chair and then as permanent chair. On June 30, 1988, Dean David R. Boylan resigned and Dr. David Kao was appointed dean of engineering. Dean Kao came to Iowa State from the University of Kentucky.

The Merged Department

During the first years of President Eaton's tenure the State of Iowa was experiencing serious economic problems. They prompted the Board of Regents to look for ways to reduce the costs of the three universities under the direction of the Board. The Board had the opinion that there was to too much duplication of effort among the universities and that there should be a reduction of those duplications. The president of the Board selected a typical political approach to get his proposals accepted. He had the Board hire a consulting firm to study the universities and make proposals for the more efficient operations of these institutions. If one is cynical it might be said that the consulting firm got rather firm suggestions from the Board as to what conclusions they should reach.

President Eaton then appointed a Strategic Planning Committee of the university faculty—typical: if we have a problem appoint a committee! Then, the new dean, Dean Kao, decided that he, too, should do something to anticipate what the Strategic Planning Committee might propose and what the consultants might propose. Economy of operation was one of the key factors in all of these considerations so Dean Kao came up with a plan that would involve several departmental mergers and permit a reduction of administrative people. When this plan was placed before the faculty it "flew like a lead balloon". Only one of the proposed mergers actually took place, that merging the Department of Aerospace Engineering with the Department of Engineering Science and Mechanics to form the Department of Aerospace Engineering and Engineering Mechanics (AEEM). That merger was formalized by the Board of Regents in their mid-July meeting in 1990.

At that meeting the Board also approved appointment of Dr. David K. Holger to be chairman of the merged department. In addition, at the same meeting, Holger was appointed chair of the presidential search committee. In April President Eaton had announced his resignation as president to be effective November 1, 1990. The search went rather quickly and in February it was announced that Dr. Martin Jischke had been appointed the new president to take office on July 1, 1991. He also became a tenured professor in the merged department. Jischke came to Iowa State from his position as chancellor of the University of Missouri at Rolla.

The merger of the two departments also became a management problem for Holger. He had to manage the department in the face of drastically reducing enrollments in the aerospace engineering curriculum and in the face of retiring faculty, particularly of teaching faculty of the aerospace engineering curriculum. The undergraduate enrollment in aerospace engineering was at 526 in 1990 and dropped to a low of 168 for the year 1996-1997.

In January 1994 Dean David Kao resigned as dean, effective June 30, 1994. After a year under an interim dean Dr. James Melsa took over as dean on July 1, 1995. Melsa immediately appointed Dr. Holger to be associate dean for academics and budget. On July 1, 1996 Dr. Thomas Rudolphi took over as chair of AEEM.

Iowa State University is one of the founding member institutions of the Iowa Space Grant Consortium and the current director, Dr. William Byrd, is a member of the AEEM department.

In 1994 Holger initiated an honorary lecture series for the department, a series named in honor of our distinguished alumnus and beneficiary, T. A. "T" Wilson. Wilson had, in 1993, been honored with an honorary doctorate by Iowa State. T himself presented the first Wilson Lecture in April of 1994. T and his wife, Grace, subsequently endowed a chair in the College of Engineering at Iowa State. That chair is currently administered in the Department of Aerospace Engineering and Engineering Mechanics and is occupied by Dr. Partha Sarkar.

Subsequent Wilson lecturers have been Vance D. Coffman, Lockheed-Martin Co., in April 1996; Eugene T. Covert, Emeritus T. Wilson Professor of Aeronautics at MIT in September 1997; Dr. Arthur Bryson, Paul Pigott Professor of Engineering Emeritus at Stanford University, in April 1999; and Norman Augustine, retired CEO and chairman of Lockheed Martin Co. in April 2000.

In October 1999 the department went through another move, from the George R. Town Engineering Building for the aerospace engineering faculty and from the Henry Black Engineering Building for the engineering mechanics faculty to the brand new Stanley and Helen Howe Hall.

Concluding Remarks

Now, in the academic year 2001-2002 the undergraduate enrollment is again increasing and replacement of retiring faculty is being extremely limited by state-imposed budget constraints. The department faculty can look to difficult times again as it tries to maintain a high quality degree program in aerospace engineering. The small degree program in engineering science has been dropped and will no longer be a part of the mission of the department. The name of the department has also been reverted back to Aerospace Engineering.

Bibliography

[1]   Hermann, Paul J., "Aeronautical/Aerospace Engineering at Iowa State, One Person's Story," Iowa State University, College of Engineering, Department of Aerospace Engineering and Engineering Mechanics, Ames, Iowa, 2001.
[2]   Pellegreno, Ann Holtgren, "IOWA Takes to the Air, Volume One, 1845-1918," Aerodrome Press, Story City, Iowa, 1980.

[3]   Pellegreno, Ann Holtgren, "IOWA Takes to the Air, Volume Two, 1919-1941," Aerodrome Press, Story City, Iowa, 1986.

* Associate Professor Emeritus, Dept. of Aerospace Engineering, Associate Fellow AIAA

Chapter 20

85 YEARS OF AERONAUTICAL/AEROSPACE ENGINEERING AT GEORGIA TECH: 1917-2002

R.G. Loewy, William R. T. Oakes Professor and Chair,
School of Aerospace Engineering
Georgia Institute of Technology, Atlanta, Georgia 30332-0150

ABSTRACT

Selected highlights of the first 85 years of Georgia Tech's programs, in aeronautics, at first, then in aerospace engineering education are reviewed here. This includes those events resulting in a free-standing department, or "School", as it is known on the Georgia Institute of Technology campus. Evidence is presented that research was a part of this School's program from its beginnings. Growth in all the important activities of such an academic enterprise is traced, accounting for some of its major sources of support and mentioning the names of some of the key individuals involved in bringing that growth about. Finally, brief indications as to the School's current strengths are alluded to.

INTRODUCTION

Formal instruction in aeronautical engineering in the U.S. may have had its start when, on May 2, 1890, Octave Chanute "delivered a lecture to the students of Sibley College" at Cornell University titled "Aerial Navigation."[1] The corresponding first university research in aeronautics may, similarly, have been documented with the thesis of Mr. A. J. Wells, titled, "An Investigation of Wind Pressures on Surfaces", in 1896, performed in a 30 in.^{2} wind tunnel with maximum speed of about 20 ft/sec[1] at MIT. Thesis supervisors were Mr. A.A. Merrell and Prof. Gaetano Lanza. Although informal courses and lectures in aviation were given earlier and elsewhere, it seems the first credit-bearing course in the US in aeronautical engineering was approved at the University of Michigan early in 1914[2] and taught as "Theory of Aviation" for two credit hours by Prof. Felix Pawlowski. These three pivotal events in aerospace engineering education are summarized here to provide a context for the counterpart formal events at the Georgia Institute of Technology.

HISTORY OF THE BEGINNINGS

On July 5, 1917, the U.S. War Department (predecessor of Department of the Army, for the younger readers) established one of eight U.S. Army Schools of Military Aeronautics at Georgia Tech – then known as "The Georgia School of Technology". This "ground school" for aviators continued for six months, during which time training was given to more than 1200 servicemen who were reported to have almost all been subsequently commissioned. The Army then established a special supply officers' school for aviation at Georgia Tech, which trained hundreds of men until the summer of 1918.[3] Presumably because of these earlier successes, the War Department selected Georgia Tech in January, 1921 as the site for establishing an Air Service Unit of the Reserve Officers Training Corps. Operations of this unit continued until 1928, when such training was discontinued in all college units. However, in September 1929, when Georgia Tech was the only institution of higher education in the Southeastern U.S. with a unit of the Naval Reserve Officers Training Corps, the Navy Department added a two-year ground school course to the NROTC unit for upperclassmen who expected to become naval aviators.

In all of this, the requisite instruction was provided by service officers with the appropriate technical expertise and Georgia Tech faculty members. Further, despite the fact that no separate department of aeronautics yet existed, courses in aeronautics and airplane design were offered at Georgia Tech as electives to upper-classmen.[3]

In keeping with both Georgia Tech's increasing interests in aviation and the nation's adulation of the man who had only recently flown – for the first time in history – non-stop and solo from New York to Paris, Charles A. Lindbergh was welcomed on Oct. 11, 1927 at Georgia Tech's Grant Field by a reception committee of notables and approximately 100,000 admirers. "We feel that we are honoring Tech in offering the field to so noted a man and to the city of Atlanta," President of Georgia Tech Brittain said. "Tech will be proud to assist in the welcome of Col Lindbergh, and we feel honored that we can be of service to him and to the general cause of aviation."[4]

Two years later, in October of 1929, correspondence from Captain Emory S. (Jerry) Land ,Construction Corps. US Navy - on leave with the Daniel Guggenheim Fund for the Promotion of Aeronautics - to Commander John H. Towers, Assistant Chief of the Navy's Bureau of Aeronautics, informed the latter of the Fund's interest in establishing an aeronautical engineering center in the South. It had already done so in the Northwest (U. of Washington), the West (CalTech and Stanford), the mid-West (U. of Michigan) and the Northeast (MIT and NYU). Commander Towers was, in fact, a Georgia Tech alumnus, and he wrote to President M.L. Brittain, suggesting that – should he desire to - he should "take up the matter directly with Captain Land."[5]

It may come as a surprise to those who consider "networking" a recent concept, that President Brittain not only immediately wired Captain Land, but also contacted Justice W. H. Black of New York (a former Georgian) and New York's Governor Franklin D. Roosevelt (who attended Georgia Tech football as his guest) asking them to write on behalf of his university. Both men did. Subsequently, a letter dated March 3, 1930 was received at Georgia Tech from Captain E.S. Land, as Vice President of the Guggenheim Fund, saying "the committee appointed by the President of the Fund has determined that the grant for the establishment of a southern aeronautical research center shall go to the Georgia School of Technology, Atlanta, Georgia." The letter enclosed a check for $300,000. This was the last of such grants, numbering seven, by the Guggenheim Fund for the support of aeronautical engineering in the U.S., and it was the second largest. (Fig. 1) A response, dated March 5, 1930 (without the likes of Federal Express, mind you) was sent to Captain Land reporting the action of Georgia Tech's Board of Trustees* whereby "the Daniel Guggenheim School of Aeronautics is hereby established as a distinct department of the Georgia School of Technology...."[6]

On the same day, President Brittain also wrote Captain Land of his intentions to inspect the buildings and equipment of the Guggenheim Schools already established at MIT and NYU,

* Dissolved later with the formation of the Board of Regents of the University System of Georgia

presumably to learn of their experiences in setting up a Guggenheim School of Aeronautics. Georgia Tech's planned apportionment of the grant was from ninety to one hundred thousand for a building, ten thousand additional for a wind tunnel, forty thousand for equipment, and one hundred fifty thousand for a permanent endowment, the interest from which would contribute to support for "the aeronautic faculty and expenses." Actually, $91,000 was spent on the building, $8,800 for the wind tunnel (and the construction company reportedly made a 10% profit) and $41,000 was spent on equipment. It is a commentary on the impact of inflation and the cost of computers and audio-visual aids, that the ninety-one thousand dollar Guggenheim building built on the Georgia Tech campus in 1930, was renovated in 1996 for more than two and one half million dollars.

In his letter of gratitude to the fund for its grant, President Brittain wrote "more than I can tell you will we appreciate your help in securing at the very outset the best man possible to head this new Engineering Department..." The man selected was MIT-educated NACA Langley researcher, Montgomery Knight, who led the School from 1930 until his untimely death in 1943. The Guggenheim building at Georgia Tech was dedicated on June 8, 1931 and first classes held as a separate department in September of that year with eighteen students, two faculty members and a budget of $10,000. In 1932 the School graduated its first aeronautics students; 13 in all.

Among Montgomery Knight's first actions was to design a low speed wind tunnel, which he patterned after the 10-foot GALCIT tunnel at Cal Tech. "The basic tunnel supporting structure consisted of steel frames tied together with longitudinal steel rods. Appropriate mesh screening was applied to this and the whole structure was sprayed with gunnite. Using long wooden forms which rotated about the tunnel center line (called screeds), the interior walls were made smooth. This basic tunnel shell was constructed by Rust Engineering of Birmingham, Alabama."[7] With a 200 HP drive and a two-bladed fan, top speed in the 9-foot test section was about 120 mph. Operations began in 1932. Improvements over time included changing to a variable pitch, three-bladed fan in the early forties and a four-bladed fan in the fifties. With the financial help of Lockheed Aircraft (now Lockheed-Martin) the test section was converted to a 7X9 foot rectangular section, the drive system up-graded with a 600 HP induction motor providing stepless speed control to 160 mph, and modern balance systems and mounting provisions added. (Fig. 2. The Georgia Tech 7x9 ft. tunnel drive fan as it is today.)

In addition to wind tunnels, Montgomery Knight had an active interest in rotorcraft. With Prof. Donnel W. Dutton, who would become the second director of the School, he designed, built and whirl-tested a one-bladed 10-foot radius, tip-jet driven rotor whose wood-bonded construction is quite similar to that of modern advanced filamentary composite rotor blades. Fig. 3 is a photograph of the full scale, counterbalanced, jet driven rotor blade, mounted for testing. These experiments began in 1936 and continued till 1944. Fig. 4 shows the jet driven rotor as part of a helicopter model, mounted in the Georgia Tech Wind Tunnel. In keeping with these research interests, a Pitcairn PCA-2 three-place, 300 HP autogyro was donated to the department for flight research (Fig. 5). It came to an ignominious end in 1935, when one T. Edward Moodie, inventor of the "Roadplane", readied the autogiro for a flight from a Candler Field runway to an intended landing at Georgia Tech's football field. He started the engine and, letting it "warm-up", he went into the hanger for coffee. When he returned to adjust the throttle setting, the brakes suddenly let go and the PCA-2 slowly taxied down the runway dragging Moodie as he tried to stop it. Although no one was hurt, the aircraft went into a ditch, a total loss.[8]

Among Georgia Tech's contributions to better understanding of rotary wing aerodynamics were Prof. Walter Castles Jr's theories in the early 50's, which provided one of the first representations of the rotor wake's effect on the non-uniform induced flow experienced by lifting rotors in forward flight, and Regents' Professor (Emeritus). Robin Gray's studies of the hovering rotor some years later, which established a model of the complex wake that is still used today.

ABBREVIATED RECORD OF GROWTH

Georgia Tech's growing program in aeronautical engineering broadened as it grew. The wind tunnel was used to conduct tests on speedboats, automobiles (Fig. 6) and skyscrapers, not just airplanes. Tech ran studies for the so-called Tower Place office building in Buckhead (a suburb of Atlanta) to see if wind conditions would cause glass windows to pop-out or create hazardous conditions. (They didn't.)[8]

The intent to transform Georgia Tech from an institution primarily devoted to teaching to a research university is well documented in Ref. 9, dated September 1954. It is stated (a little wistfully) there, "In a university which is committed both to teaching and creative scholarship, a faculty member is normally given a teaching load of 6 to 9 hours so that creative work can be carried on. Indeed in such institutions creative scholarship is expected of a large fraction of the faculty." This makes the research activities of Montgomery Knight, Donnel Dutton, and Walter Castles all the more remarkable, certainly by today's standards.

Donnell Dutton's stewardship of Georgia Tech's School of Aerospace Engineering from 1943 to 1963 was marked by supplementing what had become a solid, more or less classical undergraduate program with about 25 course offerings, to one including an early and substantial design component. With his guidance, both programs of instruction and research grew at a pace which can be inferred by comparing the statistics for the years 1950 and 1963 given in Figure 7. Under Prof. Dutton's direction, an MS program was given a solid foundation and the School's national reputation established.

In recognition of the School's expanding interests beyond the field of aeronautics, the School's name was changed from Aeronautical Engineering to Aerospace Engineering, effective July 1, 1962, much as was being done at many School's throughout the U.S. in this era. These years saw an expansion of enabling disciplines to a more widely based group driven by wider faculty interests, and the School's influence was, accordingly felt elsewhere on the campus, as will be mentioned in the following paragraphs.

Arnold Ducoffe became Director in 1963 and led the School until 1986. In 1967, the School was one of three on the campus to have its Institute-provided budget increased by $300,000 a year, as a result of competing successfully with a proposal to the university administration. This allowed new faculty members to be added, implementing Ducoffe's emphasis on research, building a new Combustion Lab and establishing a PhD program   .

In the early years under Dr. Arnold Ducoffe, much theoretical and experimental work was done on steady flow transition to turbulence in tubing systems and on unsteady flow in tubing systems with application to missile pressure lag problems. Application to the emerging field of biomedical engineering seemed a natural consequence. With this background, the School's Dr. Don P. Giddens developed, with his colleagues at Henrietta Egleston Hospital for Children and Emory University, means to quickly diagnose arterial disease in premature infants using a non-invasive technique. Giddens also worked with medical researchers at the University of Chicago doing basic research that identified where fatty deposits are likely to accumulate in blood vessels. "All this research grows out of basic research in unsteady three-dimensional fluid mechanics," Giddens said, harking back to earlier rotorcraft interests. "People in helicopter research are interested in the cyclic variations as a helicopter rotor turns, and I'm interested in cyclic variations as the heart beats."[8] This program eventually became a part of the nucleus for the interdisciplinary Bioengineering Center begun by Dr. Giddens (now GT's Dean of Engineering), which, in the years since has become a separate School of Biomedical Engineering and part of a unique joint program between Georgia Tech and Emory University Medical School.

In a similar sort of transition, analytical studies of rarefied gas phenomena, which added to the national reputation of the School in the early 1960's, played a role in having an Environmental Sciences Group join the School in 1964. Their research contributed significantly to knowledge of turbulence and wind shears at the 60-100 kilometer level of the atmosphere. This group eventually left the School to form the nucleus of a new School of Earth and Atmospheric Sciences at Georgia Tech, which continues to gain increasing nationwide recognition.

Following Sputnik and the transformation of the NACA into NASA, the School of Aerospace Engineering and Georgia Tech became the beneficiary of construction of four new buildings co-financed by NASA and the State's Higher Education Facilities Act. The first two, known at that time as "The NASA building" or Buildings #1 and #3, were joined at their juxtaposed corners, had been built at a cost of $1,023,700 and contained almost 49,000 square feet of laboratories and classrooms. They were dedicated on Dec 1, 1967 in a ceremony at which the principal speaker was James Webb, NASA Administrator. Building #2 was joined to the Guggenheim Building at a $90°$ angle at their respective ends, and contained among its laboratories an open-return, low turbulence wind tunnel with 2_ foot square test section, complete with 6-component balance system and used extensively for instruction including flow visualization work (Fig. 8). Erected at a cost of $1,716,000, with over 55,600 squ. ft. of floor area, Building #2 was dedicated in 1968 as the Montgomery Knight Building, as a memorial to the School's first director. Building's #1 and #3 are now known as the Space Science and Technology Building and Weber Building, respectively.

Building #4 in this new construction was to be an advanced combustion laboratory for instruction and research, with blow-out ceilings and walls for explosion protection, thick reinforced concrete walls for combustion cells, hoods to remove noxious effluents, and a large seismic-mass floor section for mounting vibration sensitive equipment. During the design stage, it was decided that it was inappropriate to site this building adjacent to Buildings #1, 2 and 3, as originally planned. It was, in fact, constructed at what was then the edge of the campus and became known as Building #102, the Combustion Lab.

THE PRESENT ERA

In 1982, the School won its first major national research competition, conducted by the U.S. Army Research Office (ARO) to establish a center of excellence in rotorcraft technology. The proposal effort was lead by Director Ducoffe and Regents' Professor (now Emeritus) Robin Gray, and the Georgia Tech center was known by the acronym CERWAT (for Center of Excellence in Rotary Wing Aircraft Technology). The first of three such national centers, the program continued under the Directorship of Prof. Daniel Schrage and the shortened name of CERT (for Center of Excellence in Rotorcraft Technology). In response to ARO initiatives, our School's proposals recompeted successfully and were renewed in 1987, 1992, and 1997; the last time under the broadened sponsorship of the NRTC (National Rotorcraft Technology Center; an agency funded by ARO, NASA and the three major U.S. rotorcraft manufacturers).

During the mid 80's, a major effort was made to establish a vigorous instructional and research program in the area of flight mechanics and controls. At the initiation of this effort, only one such course was offered. The faculty added Anthony Calise, JVR Prasad and later, Wassim Haddad and Panagiotis Tsiotras, who have been successful in growing an influential research program and increasing the number of flight mechanics and controls courses in the '02 catalog to 4 at the undergraduate level and 10 such graduate courses. During this same period, a strong program in the area of composite structures was initiated from within the structures area, and today these efforts involve interdisciplinary interactions with faculty in Mechanical Engineering and in Material Science and Engineering.

Following the untimely death of Arnold Ducoffe in 1986, and a two year period of transitional leadership provided by Robin Gray, Prof. Don Giddens was appointed School Director. From 1988 until 1992 the progress shown in Figure 7 continued, until Prof. Giddens was convinced to become Dean of Engineering at Johns Hopkins University. At that point, Prof. Alvin Pierce (now Emeritus) assumed the School's interim directorship. In 1993 the author joined the School as professor and the School's fifth director. In accordance with Institute faculty wishes, the position was renamed from "Director" to "Chair" in 1996.

In that year, the School won its second major research center contract, in the form of a MURI (Multidisciplinary University Research Initiative). Under the leadership of David S. Lewis Chair and Regents' Professor Ben Zinn, this 5- year program investigated the feasibility of "intelligent gas turbine engines", combining unsteady combustion and feedback control theory in experimental demonstrations requiring both real time signal processing and actuator developments never accomplished before that time. This project was a trail-blazer for the School in the area of turbo-machinery.

On April 18, 1996, a three year effort to renovate the original Guggenheim building culminated in a rededication ceremony. A gift to Georgia Tech by the AE class of 1939 had been the impetus for the project. The renovation increased the gross square footage of the building by 15%, mainly for expanding a design lab. Profs Suresh Menon, Jim Craig, and Stan Bailey (now Emeritus) had, roughly in parallel with renovation planning, induced the National Science Foundation (NSF) to provide a sophisticated computer network in the renovated building which would allow a number of diverse, individual computers to function as a large parallel processor. All told, the 1930's-building had been brought, by these means, into the start of the 21st century. Among the speakers at the rededication were Georgia Tech president Wayne Clough; former AE Director Donnell Dutton; Chairman of the Class of '39 Committee David S. Lewis; and the great grandson of the man for whom the building was named, whose own name is Daniel M. Guggenheim.

Beginning with the early 90's, an activity in the School lead by Prof. Daniel Schrage involved aerial robotics in various forms. Having student teams enter national competitions in which the challenge was to use flying models in increasingly autonomous operations was one. By 1998, the initial phase of a multi-phase program in Software Enabled Control (SEC) for Uninhabited Aerial Vehicles (UAV's) was initiated, under DARPA sponsorship, based largely on expertise developed in the aerial robotics activities. This research has continued to this writing, with increasing cooperation with industry partners.

The move to bring Design into the School's curriculum, begun by Donnel Dutton, was brought to fruition in 1997, when Prof. Schrage successfully proposed to its faculty that Design, as a disciplinary area, should be included in the PhD qualifying examinations alongside the more classical sub-disciplines, for graduate students specializing in Design. This practice continues to this writing with the addition of Boeing Associate Professor Dimitri Mavris, Director of the Aerospace System Design Lab (ASDL) and Associate Professor John Olds as Director of the Space Systems Design Lab (SSDL) to the School's faculty. These areas are among the most popular with aerospace engineering graduate students at Georgia Tech. They enjoy substantial research support from both government agencies and industry. ASDL, in particular, established very active and productive ties with several components of the turbomachinery community.

In 1999, to make room for Georgia Tech's burgeoning buildings and facilities in the life and environmental sciences, the School's Combustion Lab, mentioned as Building #102, earlier, was selected to be torn down. In its place as an aerospace program facility, the Institute erected a new combustion/propulsion laboratory in a new location, known as the North Avenue Research Campus. Demolition of the old building was delayed until the new one was ready to accept transfer of ongoing experiments, including, of course, their essential equipment. This new laboratory has four isolation rooms to exclude or contain light, dust, etc. (one contains an anachoic chamber) and four high pressure labs (one with high bay) the latter with positive pressure forced exhaust, 720 psi compressed air (room temperature and 995- F) and 1000 psi natural gas - making it eminently capable of supporting a wide range of instructional and research experiments in aerospace propulsion and power applications. The progression of buildings constructed at Georgia Tech to support its aerospace programs is indicated in Figure 9. The areas shown in SST/Weber includes about 1/3 occupied by other than the School of AE, at this writing, and represents possible expansion space.

Summary

The School of AE at Georgia Tech built rather steadily over 85 years on a solid foundation, making substantial contributions in its chosen fields even as these fields themselves expanded and grew. The five-year averages of degrees granted in the period 97-98 through 01-02 are approximately 40 BS, 44 MS and 16 PhD per year. The number of papers published by its faculty in refereed journals, in the same period, average 2.61 publications per faculty member per year. As one evidence of the quality of the School's programs and its student body, a few of its distinguished alumni are listed in Fig. 10. The current faculty of the School are listed in Fig. 11 according to the disciplinary committees of their choice. These committees are responsible for course content, curricular requirements, faculty recruiting, etc. in those areas. Names underlined show primary affiliation, those not underlined indicate secondary affiliation.

The faculty and staff of the School look forward to a future in which the - by now - well established traditions of teaching, research and service are enthusiastically continued.

REFERENCES

1. Ober, Shatswell, "The Story of Aeronautics at MIT 1895 to 1960", MIT Department of Aeronautics and
Astronautics, 1965.

2. Weeks, Robert P. "The First Fifty Years" the University of Michigan, Dept. of Aeronautical and Astronautical Engineering, 1964.

3. Brittain, M.L., "Proposal to the Daniel Guggenheim Fund for the Promotion of Aeronautics", Nov 20, 1929.

4. Georgia Tech Alumni Magazine, Spring 1998, 75th Anniversary Edition, page 24.

5. Correspondence dated October 24, 1929 Commander J.H. Towers, Bureau of Aeronautics, The Navy Department to M.L. Brittain; President, Georgia School of Technology

6. Excerpts from Minutes of the Board of Trustees, the Georgia School of Technology, March 5, 1930.

7. Harper, J.J., "History of The Georgia Institute of Technology Nine Foot Wind Tunnel", July 1988

8. Griessman, B.E., "The Big Step Nobody Noticed" Georgia Tech Alumni Magazine, Fall 1985.

9. "The Aims and Objectives of the Georgia Institute of Technology", A Report of The Committee on Educational Objectives and Methods, September 1954.

**Fig. 1.   Universities with
Guggenheim Schools of Aeronautics**

New York University		California Institute of Technology	
June 12, 1925	$500,000	August 24, 1926	$305,000
Leland Stanford Jr. University		University of Michigan	
August 24, 1926	$195,000	December 2, 1926	$78,000
Massachusetts Institute of Technology		University of Washington	
January 16, 1927	$230,000	June 29, 1928	$290,000
Georgia School of Technology			
March 3, 1930	$300,000		

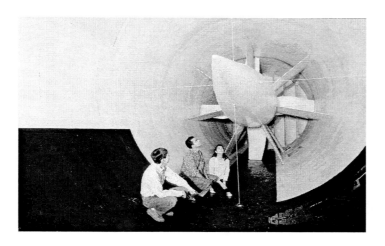

**Fig. 2
Low Speed
Wind
tunnel**

**Fig. 3. The Georgia
Tech single-bladed Drive
System in 2002 jet
driven rotor, mounted
for testing (Circa 1937)**

**Fig. 4. Model of the single-bladed,
jet drive helicopter model mounted in
the Georgia Tech Tunnel (Circa 1938)**

Fig. 5.
Georgia
Tech'sPitcairnPCA
-2
Autogyro at
Candler Field
(Circa 1938

Fig. 6. Georgia
Tech's "more
efficient, better
handling auto"
(Circa 1939)

Figure 7.
Growth Record School of Aerospace Engineering Georgia Institute of Technology

Year	Number of Faculty	Research Budget *	Total Budget
1931	2	$2,000	$10,000
1950	8	6,000	49,000
1963	11	20,000	150,000
1967	25	450,000	1,030,000
1988	25	3,500,000	5,500,000
1994	28	3,651,000	6,790,000
2002	30	10,000,000	15,000,000

* "then-year dollars"

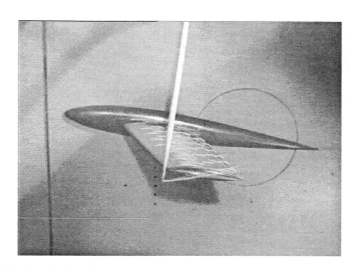

Fig. 8. Tufted semi-span wing in low turbulence tunnel

Figure 9. Record of Permanent Buildings Supporting the School of AE

Building	Year Open	Gross Area (ft^2)	Cost (then-year $)
Guggenheim	'31	12,900	91,100
Combustion Lab (Wing)[*]	'57	3,200	40,000
SST/Weber	'67	48,900	1,023,700
Montgomery Knight	'68	55,640	1,716,000
(Old) Combustion Lab	'68	10,800	25,300
Guggenheim Renovation	'96	2,000**	2,560,000
(New) Combustion Lab	'00	18,000	7,867,000

* Incorporated into M. Knight Bldg.
** area added to original bldg.

Fig. 10.
A Few Famous Alumni of Georgia Tech's
School of Aerospace Engineering

David S. Lewis '39
President, McDonald-Douglas Co.
 CEO, General Dynamics Corp.

Alan Pope '39
Director of Aerospace
Sandia National Laboratories

Robert Ormsby '45
President, Lockheed Aircraft Corp.

Capt. John Young, USN '52
Two Gemini Flights
One of 12 who walked on moon
Two Appollo Flights
Commander First Shuttle Flight
Total of Two Shuttle Flights
Assoc. Director, Johnson Space
            Center*

James Thompson '58
Deputy Administrator, NASA
President & COO, Orbital Sciences*

Admiral Richard H. Truly '59
Two Shuttle Flights,
Pilot, Stability Flight Tests, Shuttle
747 Combo
First Astronaut to be NASA
Administrator
Director, Renewable Energy Center,
DOE*

Hollis Harris '61
President Delta Airlines
President, Chairman and CEO,
Continental Airlines

President, Chairman and CEO,
Air Canada
President World Airways*

**Edward C. "Pete" Aldridge, Jr.
'62**
Secretary, US Air Force
President & CEO, The Aerospace
Corporation
Undersecretary, DoD*

Maj. Gen. Carl McNair, Jr. '63
Chief, U.S. Army Aviation Center
Special Asst. CEO, Government
Relations, DynCorp*

Lewis Jordan '67
Founder, CEO ValueJet Airlines
(now
Air Tran)

Roger Krone '78
Vice President & Treasurer,
McDonnell-Douglas Aerospace
Manager, US Army Programs
Boeing Company*

Larry Knauer '79
President, Space Propulsion &
Russian
Operations, Pratt & Whitney

**Fig. 11. SCHOOL OF AEROSPACE ENGINEERING
DISCIPLINE COMMITTEES (2002-2003)**

AERODYNAMICS & FLUID MECHANICS
Chair: P. K. Yeung

K. K. Ahuja*	S. Ruffin
J. Jagoda	L. Sankar
N. Komerath	J. Seitzman
S. Menon	M. Smith

PROPULSION &COMBUSTION
Chair: J. Seitzman

K.K. Ahuja*	S. Menon
J. Jagoda	J. Olds
T. Lieuwen	B. Zinn

FLIGHT MECHANICS & CONTROLS STRUCTURAL
Chair: J.V.R. Prasad

A. Calise	A. Pritchett±
J. Craig	D. Schrage
W. Haddad	P. Tsiotras
E. Johnson	Brian Stevens*

AEROELASTICITY & DYNAMICS
Chair: O. Bauchau

J. Craig	J.V.R. Prasad
S. Dancila	M. Ruzzene
S. Hanagud	D. Schrage
D. Hodges	M. Smith

STRUCTURAL MECHANICS & MATERIALS BEHAVIOR
Chair: E. Armanios

O. Bauchau	J. Holmes
J. Craig	M. Kamat
S. Dancila	G. Kardomateas
S. Hanagud	M. Ruzzene
D. Hodges	

SYSTEM DESIGN & OPTIMIZATION
Chair: D. Schrage

A. Calise	J. Olds
J. Craig	A. Pritchett±
M. Jenkins	J.V.R. Prasad
D. Mavris	

* Joint with GTRI   ± Joint with Industrial & Systems Engineering
NB Underlined denotes primary affiliation

Chapter 21
A Brief History of Aerospace Engineering
at the University of Virginia

Jim McDaniel, George Matthews, Jim May,
Earl Thornton, Bob Ribando and Chris Goyne

The Early Years of Engineering at UVa

The university opened officially in 1819, and the first engineering courses were offered in 1927. By 1836-1837 there were 17 students in a two-year curriculum in the "School of Civil Engineering." But the enrollment dwindled, and engineering disappeared from the University catalog by 1850. After the Civil War, under Professor Charles Scott Venable, engineering was revived, emphasizing Civil Engineering and Mechanics, Applied Mathematics, and Applied Chemistry. Major improvements and expansion of engineering took place under William Mynn Thornton who was a student from 1869 to 1873, a faculty member from 1873 to 1925, and the first Dean of Engineering from 1906 to 1925. By 1890 there was an average of 48 students a year. The 1908-09 catalog showed the thesis requirement for the first time. In 1924 the programs expanded from three to four years to allow broader studies in humanities, arts and social studies as was happening throughout the U.S. In 1935 the School of Engineering and Applied Science (SEAS) moved to the Thornton Hall complex then designed for a student enrollment of 300. Following WW II, SEAS was reorganized with extensive revision of the curricula, increasing numbers of students, and increasing research activity. In 1955 the first doctoral programs, Engineering Physics and Chemical Engineering, were established.[1]

Origins of Aeronautical Engineering

Mechanical engineering was recognized as a degree program in the 1890-91 catalog. It was a three year program and included courses in mechanics of machinery, boiler construction, steam engine design, hydraulic motors and, and mechanical drawing, with substantial laboratory work. The size of the ME faculty varied considerably, decreasing to one in 1933. Frederick T. Morse was appointed in 1933, a University Airport was acquired and opened in 1936, and reasonably steady growth ensued, except during WW II.

Aeronautical engineering courses were introduced in the early 1930s, and an aero option was added to ME in 1931. ME instruction featured five areas of concentration: heat, power, aircraft propulsion, refrigeration, and metallurgy. Aero option courses included aerodynamics, stability and control, structures and a wind tunnel laboratory. The aeronautical engineering program evolved from this option and was established as an independent department in 1956; its first bachelor's degrees, Bachelor of Science in Aeronautical Engineering, were awarded in 1957. New aero courses were added in jet propulsion, supersonic aerodynamics and aircraft design. Electives, including the military reserve courses, were allowed to broaden the program. A basic course in flight testing was also offered at the UVa airport in Milton, VA. With the help of engineers from NASA Langley, a supersonic blow-down tunnel was constructed. One hundred and eight undergraduate students were enrolled in the BAAE program in 1956-57.

In 1958 a Master of Science Degree in Aeronautical Engineering was initiated with the addition of a number of graduate courses. A large low-speed closed-return wind tunnel was constructed for laboratory instruction and graduate research. The aero faculty was greatly strengthened by the transfer of the Ordnance Research Laboratory (gas centrifuge project) faculty to SEAS from Physics in 1959. Five senior scientists, all holding Ph.D. degrees, joined the aero faculty. The Doctor of Science in Aeronautical Sciences was introduced in 1961. Opportunities for graduate research in advanced low-density gas dynamics were abundant under the new departmental faculty in the gas centrifuge project. Employees at NASA Langley enrolled in the Master's degree program spent summers enrolled in courses offered by UVa and guest faculty members.

Piper J-3 Cubs at Milton Airfield

The Changing Administrative Departments

Over the years the department offering the aero program changed several times due to administrative changes. As noted, in 1957 the program was offered in a stand-alone department devoted exclusively to the aero program, but over the years the program has been offered under departments that administered other programs as well. Throughout the period beginning in 1955 the number of aero faculty members increased, and the program generally prospered. Many of the faculty members hired during this early growth period have remained members of the faculty until recent times. The Aeronautical Engineering Department became Aerospace Engineering in 1963,

Aerospace Engineering and Applied Physics in 1966, and Engineering Science and Systems in 1974. The latter name reflected SEAS increased emphasis on the systems approach. During the 1968-69 academic year, the undergraduate enrollment peaked at 150 students in the 2nd, 3rd and 4th years. Aerospace Engineering joined the ME department to form the Mechanical and Aerospace Engineering Department in 1978. The department continued to offer separate undergraduate degrees, but combined the graduate degree programs.

The Aerospace Research Laboratory

By 1986 the vast majority of aerospace research within the department was associated with the gas centrifuge project. Thus the termination of the project by the Department of Energy was a major blow to the aerospace program. At about the same time, the renovation of the SEAS physical plant required the relocation of the recently developed supersonic combustion laboratory located in the basement of the Aerospace Mathematics Building (now Olsson Hall). Concurrent with these UVA changes was the development of the national interest in the hypersonic aerospace plane project. As a result, the gas centrifuge laboratory was converted into the Aerospace Research Laboratory. By the late 1980's significant investment of departmental, school, CIT, and NASA funds had led to the development of a unique supersonic combustion research facility. With this facility and continued support from the National Aerospace Research Plane project, supersonic combustion research flourished through the early 1990's. This research in hypersonic propulsion continues to flourish today with funding from the HYPER-X Program Office at NASA Langley. In the early 1990's, new laboratory facilities in laser technology and thermal structures were also developed in the ARL building.

The UVa AE program currently conducts leading-edge research in a number of areas which have given it international recognition. Research programs are well established in supersonic combustion for hypersonic propulsion systems, hypersonic aerodynamics of spacecraft entry,

UVa Supersonic Blow-Down Tunnel – mid 1950's

unsteady aerodynamics of flapping wing flight, low speed combustion and fire retardation, microgravity combustion, optical detectors, nanoscale mechanics and MEMS. Both graduate and undergraduate students participate in these projects. Undergraduate students are given the opportunity to join research groups over the summer as Research Assistants and are able to continue such research during their fourth year for their senior thesis topic. The senior thesis is required for all undergraduate engineering degree programs and has been found to give UVa graduates written and oral communication skills which serve them well throughout their careers.

Recent Aerospace Engineering Program Facts

Since 1990 AE graduates have accounted for 34% of the Department of Mechanical and Aerospace Engineering undergraduates and, during that period of time, 36% of the department's AE students have gone to graduate school upon graduation. UVa AE graduates have attended some of the best programs in the country. In 2001 one student received the prestigious Guggenheim fellowship at Princeton. In 2002 another AE undergraduate received full financial support at Stanford and this past year, two students began their graduate studies at Cal Tech with full financial support. One of these students also received a National Science Foundation fellowship for the duration of her studies for the doctoral degree.

The quality of the aerospace students at UVA is also exhibited in local awards that they receive. This past year, aerospace students won 3 out of 8 Harrison awards for the best undergraduate research accomplished during their senior theses. Four of this class also received Virginia Space Grant Consortium scholarships in recognition of their academic excellence. The capstone design courses in AE have been extremely successful in national design competitions. UVA has won 10 awards in the past 9 years in national airplane and spacecraft design competitions. In just the past three years, the UVA aircraft design has won two first place awards in NASA's General Aviation Design Competition. The design of Alaris, the 2001 aircraft design entry shown below, was featured in *Popular Science*. Each year, a number of the students in the aircraft design courses travel to the Experimental Aircraft Association's airshow in Oshkosh, Wisconsin, to receive their awards from NASA, FAA and Air Force officials. The aircraft design experiences give the students the opportunity to apply all their engineering knowledge to the design of a vehicle and continues to be the highlight for the graduating AE students.

The AIAA student chapter at UVA has always been very active. During their meetings, many outside aerospace experts have presented seminars about their experiences. Each year about 10 students attend the AIAA Mid-Atlantic Student Conference. Of the papers submitted for competition at this conference over the past five years, two students won first place in the Undergraduate Division and went on to present their papers at the national competition at the AIAA Aerospace Sciences Meeting in Reno, Nevada.

UVA's Alaris concept aircraft flies over Virginia.

     The AIAA student chapter has also been very active in extra-curricular activities. In April of 2001, 30 students and two faculty members traveled to Cape Canaveral to observe a space shuttle launch and to tour the Kennedy Space Center. In April of 2002, 15 students and their faculty advisor visited the Air Force Museum in Dayton, Ohio. In December of 2002, 7 students and their faculty advisor were invited to visit the USS John F. Kennedy. The group flew with a Navy rear admiral to the carrier located 150 miles off the coast of Cherry Point, NC. The group stayed on board for an entire day, touring the ship and observing operations on the flight deck.

Some Distinguished Alumni of AE

Year/Degree	Name	Position
1997 / B.S.	Eric Anderson	President and CEO, Space Ventures
1979 / B.S.	Michael Garrett	Vice President, Commercial Air Marketing, Boeing
1976 / M.S.	Robert Lindberg	V.P., Orbital Sciences; V.P., National Institute of Aerospace
1970 / Ph.D. 1966/B.S.	Alton Keel	Deputy Dir., Off. of Mgmt. & Budget, Ambassador to NATO
1964 / B.S.	Wesley Harris	Assoc. Admin. For Aeronautics, NASA; V.P. U.T.S.; Professor of Aeronautics and Astronautics, MIT
1961 / B.S.	A. Thomas Young	CEO, Lockheed-Martin Corp.

Chapter 22

Aeronautics and Aerospace Engineering
at The University of Alabama

Colgan H. Bryan, Professor Emeritus
Charles L. Karr, Department Head
Aerospace Engineering and Mechanics Department. The University of Alabama

Abstract

The Aeronautics and Aerospace Engineering program at The University of Alabama has a long and storied history. From its inception in the 1930's, this program has played an integral role in educating aeronautical and aerospace engineering students, in training pilots, in improving the economic development of the State of Alabama, and in conducting scientific research. The popularity of this program, like most aerospace engineering programs, has ridden the wave of defense spending and the economic times of the country. However, throughout its 70-year history, the aeronautics and aerospace engineering program at The University of Alabama has continued to be an important training ground for aerospace engineers from across the State, the region, and the nation. This chapter provides a glimpse into the history and the current status of this program.

Introduction

There has been an Aeronautical and Aerospace Engineering program at The University of Alabama (UA) since the early 1930's. Over its 70-year history, this program has graduated numerous students who have achieved successful careers in industry, education, the military, and government. These students have been able to utilize the technical education they received at UA in careers in a wide range of fields ranging from aerospace to ship building, from computers to business, and even in the medical field.

Aside from the job of educating future aerospace engineers, the aeronautical and aerospace engineering (AE) program at The University of Alabama has been involved with many, many research projects that have advanced the field of aerospace engineering, advanced the industrial development of the State of Alabama, and generally served the aerospace community. These research projects have covered the spectrum of the aerospace engineering field including aircraft design, propulsion systems, computer simulation, aircraft performance, risk analysis, structural design and testing, federal regulation, education and others. There is virtually no aspect of aerospace engineering that has not been investigated at The University of Alabama.

Over the years, the AE program has prospered and maintained accreditation under twelve University of Alabama presidents[1]. The AE program has enjoyed the support of these twelve distinguished gentlemen, and is currently anticipating the opportunity to advance under the new president of The University of Alabama, Dr. Robert Witt.

The State of Alabama is a key player in the aerospace industry, and has been for many years. Fort Rucker, located in the Southeast corner of the state, is the "Home of Army Aviation" [1]. In addition, NASA's Marshal Space Flight Center is located in the northern part of the state in Huntsville [2], as is Redstone Arsenal which is the home of the U.S. Army Missile Command. In 1958, the old National Advisory Committee for Aeronautics (NACA) became the National Aeronautics and Space Administration (NASA), and Huntsville became a focal point of space propulsion research. In fact, the demand for off-campus teaching in Huntsville increased to the point that the AE program began teaching and maintained off-campus classes until a school, The University of Alabama at Huntsville (UAH) was formed in 1969. Today, this university is a part of the University of Alabama system (UA, UAH, and the University of Alabama in Birmingham) and has taken on the primary role of continued education for engineers in the community. Originally, faculty from The University of Alabama (in Tuscaloosa) were utilized to handle teaching loads and faculty from the AE program were deeply involved with this endeavor into "distance education."

Naturally, the AE faculty at UA has consisted of dedicated educators and researchers. However, they have also been heavily involved in service to the aerospace engineering community. For instance, the Southeastern Section of AIAA (now the Ala-Miss Section) originated when its first meeting was held in 1949 at UA, largely through the efforts of John Hoover who later went to the University of Florida and worked until he retired some years ago.

The list of the alumni of the AE program is long and contains many prestigious individuals. Recently the College of Engineering created a list of Distinguished Engineering Fellows, and 22 graduates of the AE program were included in the list of 150.

Although it is impossible to detail the entire 70-year history of this program in such a concise form, this chapter represents an attempt to capture the flavor of this history, and to provide insight into the current program.

The Early Years

The 1930's was a decade of prosperity for the College of Engineering at UA. Lindbergh had crossed the Atlantic in 1927, and he returned to New York where he was greeted by the largest crowd to ever assemble there. Worldwide interest in aviation was increasing in spite of the Great Depression, and the College of Engineering responded to this growing interest in aviation to officially begin a new program of study in 1932. The program was called Aeronautical Engineering, and due to the support of UA and to the program's success, it has evolved into the current AE program.

---

[1] These include: Dr. Richard Foster (1941), Dr. George Denny (1942), Dr. Raymond Paty (1942-1947), Dr. John Gallalee (1947-1953), Dr. Oliver Carmichael (1953-1959), Dr. Frank Rose (1959-1969), Dr. David Mathews (1969-1980), Dr. Richard Thigpen (1980), Dr. Joab Thomas (1980-1989), Dr. Roger Sayers (1989-1996), Dr. Andrew Sorenson (1996-2002), and Dr. Barry Mason (2002).

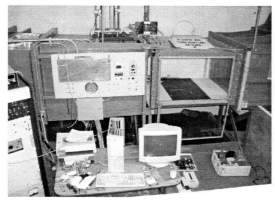

Figure 1: The low-speed wind tunnel was constructed in 1938, and is still in use today.

Figure 2: Students observing an experiment in the AE program's first wind tunnel, c. 1940

Dr. Frederick R. Maxwell, an experienced navy pilot and Electrical Engineering professor had been involved with teaching a class in navigation since 1929, and two years later (1931) fifty-nine students signed up for the as of yet unofficial aeronautical degree. By 1933, there were 165 students in the new program, including the first five recipients of the AE degree.

Prof. Leslie A. Walker, a 1915 graduate of UA and a navy pilot, handled the teaching duties during the early years. Courses were largely limited to airplane maintenance, aerodynamics, and meteorology. A single airplane was available for study. However, a new building, Hardaway Hall, named for Prof. Robert A. Hardaway, was completed in 1936. Professor Hardaway was the first full-time College of Engineering professor. Hardaway Hall later provided a home for the AE program, and laboratory space for a wind tunnel and other equipment.

In 1936 Professor Otto H. Lunde, a college trained aeronautical engineer and pilot was added to the faculty as Head of AE. Professor Lunde helped expand the curriculum to include more advanced courses in aerodynamics, aircraft engines, and aircraft design.

In 1938, a senior student in aeronautical engineering developed a design for a wind tunnel, and it was built of wood. This low-speed wind tunnel continues to be used to this day. In fact, it serves as a focal point when alumni visit the campus because virtually every graduate of the program has had the opportunity to run tests in this facility. The wind tunnel designer, Edward J. Finnell, later became the student recruiter for the College of Engineering, and served UA in this capacity until he retired and passed away several years ago.

One major mistake in the development of the AE program occurred in these early years. The Guggenheim Fund for the Promotion of Aeronautics offered to fund the construction of a wind tunnel. In addition, the proposed gift included a budget for the upkeep of the facility. Unfortunately, because money was so scarce as the Great Depression deepened, President Denny refused to erect a building to house the facility. Interestingly enough, Georgia Tech accepted the award in 1938, the facility was constructed, and Georgia Tech continues to reap the benefits of this contribution even today.

The War Years

During the depth of the Great Depression the United States of course began preparations to enter World War II. As luck would have it, UA and the City of Tuscaloosa had cooperated in building an airport with a paved runway in the early 1930's. As war preparations broke the depression, the airport enabled the AE Department to start one of the first collegiate civil pilot training programs in the country. The success of this program created a model for a national plan for pilot training. Ultimately, more than 200 pilots were trained in the civilian program. One of the key people involved with the program was Professor Walker from UA's AE program. In time, Professor Walker was borrowed by the Civil Aeronautics Authority (CAA) to initiate similar programs at other universities. Walker wrote a series of textbooks used in the new and expanded training programs.

The loss of Professor Walker was a mixed blessing in that it left UA short-handed, but allowed for the hiring of Professor Colgan Bryan. Professor Bryan had met Walker in Washington where he learned about the flight training at UA. Later Professor Bryan worked in a similar program at Roanoke, VA, sponsored by Roanoke College in Salem, Va. He was certified to teach ground subjects such as navigation, aerodynamics, aircraft engines, etc.; subjects vital to flight training. As preparation for the war increased many students were either drafted into or volunteered for military service. Due to the need to build and maintain air power, AE students were allowed to remain in school and graduate. Also, AE instructors and those involved in flight training were in critical demand.

Professor Bryan reported for duty in the fall of 1942. The war caused a major change in the distribution of students among the engineering departments. As the war approached AE enrollment increased from 148 students in 1937 to 268 in 1940, to 383 in 1943. Naturally, the Dean of the College of Engineering, Dean Davis, was worried about the ability of the faculty to handle the increased volume. When the United States entered the war in 1941, fully 40% of the engineering enrollment was in the AE program.

The AE program was faced with a daunting task to satisfy the need for pilot training and AE education for several reasons. These reasons included Professor Walker's death, Professor Maxwell's return to military duty, Professor Lundes' departure, and a greatly increased demand for an effective pilot training program. Professor Bryan met the challenge with the help and cooperation of the entire university. Professors were borrowed from other departments and temporary instructors were hired to satisfy the demand of the aviation community. In those days, Bryan and others worked long hours, seven days per week. Thousands of English and British soldiers completed pilot training at UA with only one accidental death: a student pilot crashed his airplane while showing his skill near the home of a girlfriend. Interestingly enough, Professor Bryan is still involved with educating young people at UA where he continues to teach the AE program's senior seminar.

As mentioned above, pilot training was a high priority for the program during these early years. The one paved runway at the local airport plus five practice fields served as adequate practice areas for pilot training. Thousands of acres of land owned by the Mental Health Department were made available for flying activities, and new airfields were made from cotton patches and cow pastures in the area by cutting the grass and removing the debris.

As the war activities decreased the AE program shifted its focus from pilot training to the scientific aspects of aircraft design. In 1948 Franz R. Steinbacher was appointed to be the head of the AE program. Graduate courses in aerodynamics, supersonic flow, and structures (plates and shells) were planned. Research was emphasized and a research contract from NACA for $5,000 was received. Professor Steinbacher left UA after a short stay but his plans for increasing research were continued by other faculty. The need for graduate education was stressed, and it remains an important part of the AE curriculum today.

1950 to 2003

Dr. Frank A. Rose became president of UA in 1958 and Dr. Vernher von Braun's group of rocket scientists arrived in Huntsville, AL, in 1950. The AE staff immediately became involved in space activities, and Dr. Rose fully supported their efforts. The low salaries of the day were increased and a very positive attitude prevailed. During a six year period from 1958-1964, the College of Engineering's budget increased 170%, and the number of faculty increased from 53 in 1957-1958 to 63 in 1958-1959 [3]. An 80% increase in faculty size between 1957 and 1963 lifted faculty morale, and the AE program enjoyed increased productivity. Research activities, graduate work, and laboratory facilities benefited immensely. The average salary of a full professor increased from about $5,000 in 1953 to $8,500 in 1958 and to $10,200 in 1963.

Professor Bryan became Chairman of the AE Department in 1952 after Professor Steinbacher resigned. One of the key contributions Professor Bryan made during this time was to recruit, hire, and maintain a very loyal and talented faculty. Many of these professors ended up spending their entire careers at UA and only recently retired. Professors Ray Hollub, Earl Bailey, and George Weeks have retired during the last five years, and continue to reside in the Tuscaloosa community. Professor Rey was eligible for retirement but passed away before he retired. These loyal and talented gentlemen educated many, many engineering students, most of whom continue to work in the industry today.

Faculty members in the AE program continue to receive many honors, engage in research, are active in professional groups, and support a well established department. Professor Bryan, at age of 93, continues to work as a part-time member of the faculty where he teaches the senior seminar in which students continue to benefit from his many years of experience.

During the last half-centry, the fate of the AE program was intricately woven with the Mechanical Engineering and Engineering Science and Mechanics programs at UA. At one time or another, these three programs have existed as three separate departments, as a single combined department, and as two departments of virtually every possible combination. Most recently, the Engineering Science and Mechanics Department was merged with the Aerospace Engineering Department, thereby forming the Aerospace Engineering and Mechanics (AEM) Department that thrives today.

The Current Program

The AEM Department at UA provides both undergraduate and graduate degrees. Specifically, the department offers the following degrees: (1) BS in aerospace engineering, (2) MS in aerospace engineering, (3) MS in engineering science and mechanics, and (4) PhD in engineering science and mechanics. In addition, the AEM Department offers an MSAE degree via distance education (the QUEST program [4]). Students receive video tapes of class lectures then interact with faculty and other students via the internet, faxes, and electronic mail. Aside from these degree programs, the department is responsible for teaching a variety of service courses for the college including the following undergraduate courses: (1) statics, (2) dynamics, (3) mechanics of materials, and (4) fluid mechanics, and the following graduate courses: (1) matrix and vector analysis and (2) partial differential equations.

The current faculty consists of thirteen full-time faculty and two temporary instructors; each of the fifteen have terminal degrees [5]. The qualifications of the faculty have changed dramatically over the years. In 1942 there was one faculty member with an earned PhD in the entire College of Engineering at UA, Dr. Taylor, who taught drawing. In 1957 only 14% of the faculty in the college held PhD's. By 1965 48% of the faculty in the college held terminal degrees. Like with most engineering programs, the educational attainment of the AE faculty has changed with time, reflecting more diversity to strengthen the AE curriculum and to allow for full coverage of the discipline for the students. The AEM Department also has two named professorships: (1) the AEM Cudworth Professor, Dr. Stan Jones and (2) the Jordan Chair (currently unfilled) named after William D. Jordan who was the head of the Engineering Mechanics Department at UA for over 20 years.

Of course, the collective qualifications of the faculty are not the only thing that has changed over the years; the curriculum has changed markedly. From the early years when pilot training was the focus, to the post-war years when aircraft design was placed at a premium, to the modern curriculum which includes classes on spacecraft [6], the curriculum in AE at UA has been constantly monitored and tweaked in an effort to provide students with the tools and knowledge to succeed in the current marketplace. Aside from adding and replacing individual courses, the AE program at UA has also acknowledged the interests of engineering technical societies (specifically the AIAA) in improving the image of the engineering profession and the desire for more and more engineers to become licensed.

Currently, every student completing a degree in AE is required to take the Fundamentals of Engineering Exam. In addition, UA has recognized the desire of employers to hire students with strong communication skills, the ability to work in teams, and a detailed knowledge of problem-solving and the design process. To this end, these aspects of a student's education have been addressed both in individual courses and across the curriculum. The process of monitoring and adjusting the curriculum has been an ongoing process, and the curriculum in the AEM Department is up-to-date and strong. Graduates of the program are well-prepared both to take jobs in industry and to attend graduate school; our graduates have gone on to succeed at prestigious graduate schools across the country.

Students in the undergraduate AE program are eligible to participate in at least two honors programs. First, they can participate in the University Honors Program. This program offers qualified undergraduate students a special academic challenge. Honors classes have limited enrollment, placing emphasis on interaction between students and faculty. Second, they can participate in the Computer-Based Honors Program [8]. This nationally acclaimed program focuses on getting students involved with faculty across the campus to work on various research projects. Many AE students have participated in both programs over the years. In addition to these two programs, the AEM Department has recently developed a program under the University Scholars Program by which students can complete both an undergraduate degree in aerospace engineering and a masters degree in either aerospace engineering or engineering science and mechanics via the use of dual course credit. This program is in its initial phases but should prove inviting to extremely bright entering freshmen.

Like with most things, the make-up of the student body in AE is also very different today than in the past. Currently, there are 91 students enrolled in the AE program. Of these, 20 (22%) are female. Of the six students who hold office in the student chapter of the AIAA, three are females (50%). This represents a marked improvement in increased diversity since the first female, Ms. Rose Rabinoyitz, graduated from the program before World War II.

Recruiting quality high school students is a continued emphasis of the AEM Department. This effort is based largely on providing prospective students with the kind of personal attention they can expect throughout their tenure at The University of Alabama. Of course, scholarships are an important part of any recruiting effort. The AEM Department regularly makes a number of scholarship offers including the Colgan Bryan scholarship.

The current AEM Department has made a concerted effort to increase the involvement of industrial partners in the process of educating students, impacting local industry, and making advances in scientific research. For instance, the department has a very active Industrial Advisory Board composed of thirty-eight members from industry and government [7]. This group meets twice per year and assists the AEM Department in:

- maintaining ABET accreditation;
- establishing an Endowed Scholarship Fund;
- provide laboratory equipment through donations, purchase or obtain surplus equipment as necessary to support the departmental programs;
- student recruitment;
- other tasks as deemed necessary to support the department.

Current Facilities

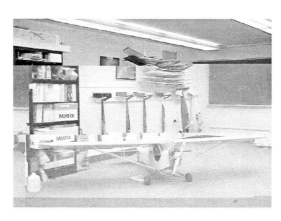

Figure 3: The Intelligent Control Lab is utilized in the development of small aircraft

Like at all research institutions, the AEM Department at UA continues to grow its research programs, and endeavors to tie these efforts to both undergraduate and graduate education. The faculty are involved with numerous research projects including: (1) a project with the FAA to develop a highly-accurate, very low-weight gyroscope for the FAA and (2) project NOVA which is a NASA-sponsored project to assist in the development of K-12 science and math teachers. These two projects provide an indication of the range of projects ongoing within the department.

Of course to perform quality research, universities must develop and maintain modern facilities. The AEM Department has a variety of modern laboratories used both for research and teaching. These facilities include:

- Jet Propulsion Lab;
- Aerospace Structures Lab;
- Computational Fluid Dynamics Lab;
- Experimental Aerodynamics Lab;
- Flight Dynamics Lab;
- Mechanics of Materials Lab;
- Structural Design and Testing Lab;
- Intelligent Control Lab;
- Experimental Stress Analysis Lab;
- Computational Mechanics Lab;
- Genetic Algorithms Lab.

Figure 4: The Experimental Aerodynamics Lab includes wind tunnels capable of achieving speeds of up to Mach 5

From its beginnings in which a single aircraft and one low-speed wind tunnel made up its core research and teaching facilities, the AEM Department has built facilities and infrastructure that allow it to conduct research and teach students to the highest standards. One interesting project of note involved Dr. Earl Bailey of AE and Dr. James Dudgeon of the Electrical Engineering Department. These two were involved with the development of an early flight simulator for the U.S. Army. Some of the same technologies used in the development of the simulator were later applied to the study of wind shear problems in aircraft. Their efforts applied the latest computer technology to the teaching of aircraft guidance and control. In addition, the test apparatus that was developed used color graphics and displays, a molded plastic cockpit and computer-generated navigational aids. Although this project may seem mundane by today's standards, the simulator they designed, built, and tested was at the leading edge of simulation research at the time, and their simulator remains a part of the current Flight Simulation Lab today.

There are, of course, numerous other projects in which the faculty in AEM are involved. For instance, the NASA Faculty Fellowship program for Marshall Space Flight Center is co-managed by UA and UAH. Also, the department is heavily involved with the management of the Alabama Space Consortium program.

The quest for better and better facilities is ongoing. The most recent construction project in the College of Engineering is a students' project building. This building will be used exclusively for students from across the college to work on projects for classes, research projects, and student organization-sponsored competitions. Students from the aerospace engineering program are looking forward to having a better place to work on the aircraft they produce as part of their senior projects.

Accreditation

Both the College of Engineering and the AE program in particular have an impressive record in the area of accreditation. In 1937 the Engineers' Council for Professional Development (ECPD) accredited six of the eight programs in the college; one of the six was the AE program at UA. This honor placed the College of Engineering among the top 12% of the schools in the United States. Of the eighteen southern schools examined by ECPD, UA had the highest number of accredited programs originally approved, providing national respectability and prestige. Perhaps more impressive is the fact that the AE program at UA has been continually accredited since 1937 (66 years), a source of strength, pride, and national prestige.

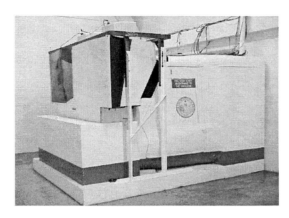

Figure 5: The Flight Dynamics Lab includes both helicopter and fixed-wing simulators.

The College of Engineering was most recently evaluated by the Accreditation Board for Engineering and Technology (ABET) in the Fall of 2001. At this time the AE program was again fully accredited.

The current AEM faculty is well aware of this history of maintaining its accreditation. Naturally, this is a great source of pride for everyone involved in the program including faculty, administrators, and students. This record is indicative of the emphasis placed on providing an excellent undergraduate education in AEM at UA. In addition, many of the faculty in the AEM, and in the College of Engineering, serve as ABET reviewers, thereby visiting and evaluating programs across the country.

Alumni

The list of alumni from the program is long and includes numerous prestigious individuals. To provide a flavor of the quality of the alumni, five are mentioned here. First, Mickey Blackwell was a 1962 graduate of UA. He went on to a distinguished career with Lockheed Martin Corporation where he retired as Executive Vice President.

Second, Lieutenant Colonel James Kelly was the first graduate of the distance education program in the department, and also the first alumnus to go into space. Kelly earned a master's degree in aerospace engineering in 1996 and piloted the space shuttle Discovery for the first time in March 2001. He was a U.S. Air Force Academy graduate in 1986 and an F-15 pilot until he was selected for astronaut training ten years later. Also, he spent 307 hours in space while serving as the Discovery pilot in 2001. Interestingly enough, Lt. Col. Kelly was scheduled to pilot the space shuttle a second time on March 1, 2003 – a mission that is subject to change based on the tragic loss of the space shuttle Columbia on Feb 1, 2003.

Third, Jack Lee (BSAE 1958) was appointed Deputy Director, NASA – George C. Marshall Space Flight Center in Huntsville in 1980 after serving with the Spacelab Program. He completed the Advance Management Program at the Harvard School of Business in 1985. He continues his involvement with AEM today as a member of the Industrial Advisory Board.

Fourth, Carl W. Albright, Jr. (BSAE 1967, J.D. 1970) is one of a number of AE graduates who have used their education to benefit the Tuscaloosa community. Mr. Albright served as Municipal Court Judge from 1975-1980, and served as President of Amsouth Bank Corporation until his recent death. His many respected professional activities and community activities are indicative of the strength of character of many alumni from the AE program.

Fifth, William Lawler is a prominent member of the UA class of 1962. He began his career with Boeing Aerospace in Seattle. While with Boeing, Mr. Lawler worked as the lead structural dynamics engineer for the Saturn V/Apollo launch vehicle; as the senior dynamicist for the Space Shuttle; and as the senior specialist engineer with the Assault Breaker missile program. After a brief stint with Northrop Grumman where he was deeply involved in the development of the B-2 "Stealth" Bomber, he returned to The Boeing Company as the vice president and general manager of Business Development and Strategy for the Military Aircraft and Missile Systems Group. He assumed his current post as the vice president and general manager of strategic operations and planning for the group in 2000.

It is tempting and somewhat ego-boosting to enumerate the most prestigious members of a program's alumni. However, it is also important to note that many of the alumni of the AE program have gone on to be successful (if not prestigious) members of the aerospace engineering field. This program has graduated many, many engineers who have gone on to be successful in industries across the State of Alabama, and across the United States. It is these engineers upon which the reputation of the program has been built.

Summary

Aerospace engineering has a rich and prominent history in the College of Engineering at UA. Over its seventy-plus-year existence, the program has educated numerous engineers, been a leader in scientific research, and contributed to the economic development of the State of Alabama and the Nation. Alumni of the program continue to be leaders in the fields of aerospace engineering, computer programming, business, medicine, and others.

Naturally, the program has changed over the years. This paper outlines some of the history of the program, and provides a flavor of the rich history that is AE at UA.

References

[1] Fort Rucker web page, (2003), http://www.armyflightschool.org/rucker.htm.

[2] NASA, Marshall Space Flight Center web page, (2003), http://www1.msfc.nasa.gov/NEWMSFC.

[3] Norrell, R. J., (1990). A Promising Field. Tuscaloosa, AL: The University of Alabama Press.

[4] Quality University Extended Site Telecourses at The University of Alabama web page, (2003), http://bama.disted.ua.edu/quest/default.htm.

[5] AEM at The University of Alabama web page, (2003), http://aem.eng.ua.edu/people/faculty_and_staff.html.

[6] AE Undergraduate Curriculum at The UA web page, (2003), http://aem.eng.ua.edu/undergraduate_programs/flowchart/_AEM/index.html.

[7] AEM Industrial Advisory Board web page, (2003), http://aem.eng.ua.edu/iab/.

[8] The University of Alabama Honors Program web page, (2003), http://bama.ua.edu/~uhp/

[9] College-Based Honors Program at UA web page, (2003), http://cbhp.ua.edu/

Chapter 23

A History of Aeronautics Education
at
Case Western Reserve University
(Case School of Applied Science, 1934 - 1947)
(Case Institute of Technology, 1947 - 1967)

The early years: 1934 – 1946, Case School of Applied Science

The Aeronautics program at the Case School of Applied Science was initiated by Professor Paul Hemke in 1934's as an option in Mechanical Engineering. Hemke hired John R. Weske who had been a student at Hanover in Germany and received his doctorate from Harvard, and turned over the program to him when Hemke left for R.P.I. in 1935. Weske inherited from Hemke along with some semblance of course structure, a rudimentary aeronautical laboratory, a good sized towing tank and water channel, and a very substantial return flow wind tunnel with an open working section of 3 feet by 4 feet cross section and a maximum velocity of about 150 feet/second. By 1936 the program consisted of several one-term courses, all taught by Weske: Aerodynamics I, Aerodynamics II, Aeronautics Laboratory, Airplane Design, and Aircraft Structures. There were graduate courses that were offered sporadically in the evenings: Advanced Fluid Mechanics and, toward the latter part of 1940-41, a course in Fluid Mechanics of Turbomachinery. As of the fall of 1937 Weske had two graduate student assistants who helped with the laboratory course and the general undergraduate fluid mechanics that was also under Weske. The laboratory assistants were given half-time tuition and a stipend of $60.00/ month, supplied by Case. These were awarded to the faculty, by the administration, on the perceived need or merit. There was at that time no contractual support for Aeronautics.

During this period at Case John Weske conceived and developed the constant temperature hot wire anemometer to eliminate the lag due to the time required to heat and cool the wire in response to turbulent fluctuations. The idea has been the basis of hot wire anemometers for more than fifty years. Once the difficulties of developing a suitable amplifier were solved, he had a reasonably stable system which he and undergraduate Frank Marble took to Washington to try in Schubauer's very low turbulence tunnel at the Bureau of Standards. The trial was a success, Schubauer was impressed and Hugh Dryden, Chief of Mechanics and Sound, paid a short visit on a Saturday afternoon.

The development of hot wire capability in the Case aero lab during the 1939 period increased the general interest in turbulence among students and faculty and Weske developed some ingenious student experiments for measuring turbulence levels and decay rates in wakes and jets. Weske put this technique to good use to investigate boundary layer control. To this end Frank Marble and Weske built a one foot square "low turbulence" tunnel in early 1939 with the capability of sucking off the existing boundary layer through a transverse slot and allowing the establishment of a fresh boundary layer. A larger 3 foot square tunnel followed and the new larger fan was designed and built by Marble summer '39. The tunnel and a first-class honey-combed inlet were finished by the beginning of the fall term, and were used in an investigation of the effects of free stream turbulence on boundary layer transition which constituted Marble's bachelors thesis.

In the spring of 1939 the U. S. government instituted the Civilian Pilot Training Program, a program designed to offer basic flight training to college students of the age to see military service in the event of our entry into the war. The program provided ground school and flight training sufficient to obtain a private pilots license. One such program was offered to Case and John Weske was asked to direct it. It was a significant effort to organize particularly because Cleveland was not cooperative in providing flying weather the winter and spring of 1939 - 40. The program did succeed, however, in providing about two dozen with pilot's certificates.

Marble decided to remain at Case for his Master's degree and began working on a new fan for the low- turbulence tunnel. Weske had sketched out a new fan design with sophisticated blading which would require new construction techniques. In addition to accurately contoured blades with good surface finish, Weske wanted to be able to measure the pressure distribution on the blade at several different radii. This necessitated making one blade with surface pressure taps and the "invention" of a slip seal arrangement to get the pressures out of the rotating fan. Working out the seal mechanism and construction of the final fan took well into 1941.

In the meantime industry and the military, aware of the high probability of entry into the war, became interested in encouraging academic institutions into areas where they saw needs. Early in 1941 a representative of Boeing met with Weske and Marble one evening and described their problems with pressure losses in elliptic ducts. There was essentially no reliable data of flow separation losses in bends and other contours of elliptic ducts. He suggested a contact with the Army Air Corps, Opie Chenoweth, who could supply funds in support of an appropriate program. This was a novel event for aeronautics at Case. The sum that Weske received was, $2,500, sufficient to support three students and to purchase the wood and plaster out of which the models were made by graduate student John Jacklitch. The effort went on for some 18 months, additional sums of money followed and constituted the first support for aeronautics at Case.

Aeronautics at Case did have a substantial association with the Guggenheim Airship Institute in Akron based primarily on Weske's friendship with Theodore Troller, its director. Troller had been a student of von Karman in Aachen and Karman had persuaded him to take the job. Arnold Kuethe spent several years at the Airship Institute after completing his work at Caltech, a fact which supports the high calibre of work done at that laboratory.

The United States entered WW II and this fact began to take its toll on both faculty and students. Almost immediately Professor Willey left Case for Lubrizol and Professor Burington, a first rate mathematician who was seriously involved in classical hydrodynamics, left for a position with the Naval Bureau of Ships.

Beginning in the fall of 1943 Frank Marble who had joined NACA Engine Research Laboratory in Cleveland after completing his master's thesis in 1942, taught aeronautics evening courses at Case in fluid mechanics. By 1944 - 45 the aeronautics program at Case had lost momentum. In 1945 - 46 Case arranged for Marble to give a course in Compressors and Gas Turbines at NACA where it would be convenient to people at NACA who were now developing an interest in gas turbines.

From 1934 through mid 40's, Case provided an excellent undergraduate education in Aeronautics. Although Case was not at that time a Research Institution, there was a substantial and growing body of research in fluid mechanics and aerodynamics to stimulate the interested student and to provide challenging problems for their Bachelors Theses which was a requirement at that time. Gradually Case's research program in aeronautics attracted notice both in the United States and in Europe. Its students were recognized by industry and fairly admitted to graduate schools. One student, John van Hamersveld, was the envy of many when he landed a job at Boeing! He became a mainstay on the design of the 707 and its predecessors. Case aeronautics did not have a high-level theoretician. Although several mathematicians were well versed in classical hydrodynamics, it was not a main interest of any. Due in large part to the generous and warm nature of John Weske, Aeronautics was probably the most student-oriented option within Engineering.

The middle years: 1947 – 1967, Case Institute of Technology

Further development of aeronautical activity at Case was fostered by Ray Bolz, who came to Case Institute of Technology in 1950, and became Chairman of the Mechanical Engineering Department in the 1952. Henry Burlage joined the Department in 1952, and was responsible for developing the Aircraft Propulsion and Aerodynamics Laboratory which housed a subsonic wind tunnel which evolved into the department's current wind tunnel, a variable Mach number supersonic tunnel (Mach numbers approximately 1.5 to 3.0, 5 inch square test section) whose test section is now part of a facility at NASA Glenn, a gas turbine test room with a free (not attached) roof that would be lifted by possible explosions, and a 9 square foot subsonic tunnel. Burlage left Case in 1960. The laboratory building was demolished in 1966 to make room for the Glennan Building, the present home of the department. Harold Elrod was a member of the Department during the 1950's after receiving his PhD with Howard Emmons at Harvard. After Ray Bolz became chairman he brought in Gustav Kuerti, Milt Swanson, and Isaac Greber (1959) to bolster the developing aeronautics activities. Kuerti had done significant work in theoretical mechanics and was the managing editor of the series of volumes, Advances in Mechanics. Swanson did theoretical work in fluid mechanics, and Greber had done important experiments in shock wave-boundary layer interactions. Greber's measurements became the standard test cases for validation of numerical computations of these interactions for many years.

A major connection with the aeronautics activity was the appointment of T. Keith Glennan as the first Administrator of NASA after NACA became NASA in 1958. Glennan was president of Case since 1947, and had been on leave to serve on the Atomic Energy Commission from 1950 to 1952 He went on leave again from Case as the NASA Admistrator until 1961 when he returned to Case, and left the presidency of Case in 1966. The building named in his honor was dedicated in 1969 and has been the home of aerospace education at Case to the present day.

Simon Ostrach left NASA Lewis Research Center in 1960 and he attracted Eli Reshotko from NASA Lewis Research Center 1964. Ostrach had done important work in boundary layers and heat transfer, and had been chairman of the Heat Transfer Committee of ASME. Reshotko had done important theoretical work in boundary layers and boundary layer stability. Because of the interest in magnetohydronamics at that time, Reshotko switched most of his efforts to plasma physics, and became Branch Chief of plasma physics at NASA Lewis. At Case he returned to his work in boundary layer stability and transition, and has continued to make major contributions to the understanding of those processes. Because of his leading role in that field he has been the chairman of the AIAA national committee on boundary layer transition since its organization about 30 years ago. In 1960 the engineering organization of Case was changed, dropping the departmental organization and melding the program into a Division of Engineering, with Ray Bolz as head of the Division. Within this framework degree programs were maintained in Chemical, Civil, Electrical and Mechanical Engineering. Groups were formed around research activities. At first these groups were informal. Later they became formal, with group heads, and by the end of the 1960's the new groups became Departments. The fluid mechanics and aeronautics faculty formerly of the Mechanical Engineering Department became the group and later the Department of Fluid, Thermal and Aerospace Sciences (FTAS), with Si Ostrach as the Head. This Department was originally responsible for a graduate degree program in Fluid and Thermal Sciences, but sufficient student demand caused it to develop an undergraduate degree program as well. This Department also shared responsibility for the undergraduate degree program in Mechanical Engineering with the Department of Structures, Solid Mechanics and Mechanical Design. The latter department was also the seat of important aeronautical research and instructional activity. Its faculty included Alex Mendelson and Stan Manson, who had come to Case from NASA Lewis, and Robert Scanlon who did important theoretical work in aeroelasticity. Manson's work was on the development of engineering models for predicting failure of metal structures, and Mendelson developed mathematical methods in elasticity.

In 1963 a Summer Faculty Fellowship program was organized between NASA Lewis, Case, and ASEE, with Walter Olson of NASA Lewis and Isaac Greber of Case as co-directors. By the time of the first summer session the program had grown to include two additional NASA center-university pairs. The program continues to the present day, and now involves 9 NASA center-university pairs. After Greber the Case directors have been Fred Lyman, who was another faculty member who had come from NASA Lewis, and then Joseph M. Prahl, the director from 1969 to the present time.

The 1960's saw additional faculty members brought into the department who had major research activities in aeronautics and were deeply involved with the development of the aeronautics educational program. These included Ted Morgan (1960-1970), who developed a shock tube facility for measuring non-equilibrium gas behavior behind shock waves, Lowell Domholdt (1960-1967) who developed a low turbulence water tunnel used by him and later by Prahl and Reshotko to investigate boundary layer stability, Ken Wiskind (1966-1970) who worked on turbulent flows, Fred Lyman (1967-1970)who came from NASA and worked on heat transfer and hest pipes, and Wilbert Lick (1967-1979) who worked on aerothermodynamics as well as wind driven circulation in the great lakes and modeling of sediment transport.

A very innovative research program, and method of funding research, was developed between NASA Lewis and FTAS. This stemmed from the efforts of John Evvard of NASA Lewis and Si Ostrach. (It began as a topic of a conversation at the reception at the first NASA summer faculty fellowship program). An amount of money, (about $100k) was awarded by NASA to FTAS for research broadly classified as fluid physics. Within this amount individual faculty members could submit very short (one or two pages) research proposals, which were given rapid reviews. This blanket grant procedure produced a large amount of significant research; it is a wonderful model that has not been copied, here or elsewhere.

The recent years: 1967 – present, Case Western Reserve University

In 1967 Case Institute of Technology and Western Reserve University federated, to form Case Western Reserve University, under the presidency of Robert W. Morse who followed Glennan as Case Institute of Technology's 5th president in 1966. The Case School of Engineering was established by the Board of Trustees of the new University, with Ray Bolz as Dean of Engineering, essentially continuing the position that he had held within Case Institute of Technology. Bolz continued as Dean until 1973, when he left CWRU. Under Ostrach's leadership FTAS grew to 10 faculty by 1970, with the addition of assistant professors, A. Dybbs, G. Janowitz, Y. Kamotani, J. Prahl, and J. T'ien. Si Ostrach resigned as Chairman of the Department in 1970, succeeded by Eli Reshotko. The research activities of the Department continued to be centered around its relationship with NASA Lewis. These included work by Ostrach and his students on the behavior of heat shields for spacecraft, theoretical and experimental work Reshotko on the effects of heat transfer and disturbances on boundary layer stability and transition, noise of jets by Greber, and other efforts.

In 1975 the Department of Fluid, Thermal and Aerospace Sciences and the members of the Department of Structures, Solid Mechanics and Mechanical Design who had been involved with the Mechanical Engineering program merged to form the Department of Mechanical and Aerospace Engineering. Eli Reshotko became chairman of the new department. In addition to the undergraduate degrees in Mechanical Engineering and in Fluid and Thermal Sciences, a degree in Aerospace Engineering was established in 1990 in response to demand by the undergraduate students. These degree programs, all accredited, have been offered to the present day.

In 1979 Isaac Greber succeeded Eli Reshotko as chairman of the department. T'ien continues to do significant work in combustion, and has been the principal investigator of combustion experiments on the space shuttle. After major earlier work on jets in cross flows with Greber, Kamotani and his students have done important experiments and numerical analyses of flows in microgravity, including experiments on the space shuttle. The 1980's saw the addition of Roger Quinn, who did pioneering work on dynamically stiffened aerospace structures and now works primarily in biologically inspired robotics. The robot control strategies are guided by neural control systems of insects, an application of cockroaches to aeronautics.

In 1984 Isaac Greber resigned from the chairmanship of the department, succeeded by Thomas P. Kicher. The research and educational activities in aeronautics continued unabated, including the close association with NASA Lewis. In 1998 Iwan Alexander joined the department, continuing his already well known work on the physics of flows in microgravity. In 1992 Joseph M. Prahl succeeded Tom Kicher as chairman of the department, a position that he holds to the present day.

Recognizing the growing importance of computation in aeronautical analysis and design and the absence of a local center of activities in this area Case joined with NASA Lewis in establishing the Institute for Computational Mechanics in Combustion (ICOMP) at NASA Lewis. ICOMP has brought in researchers in numerical methods from many institutions, to spend periods of time in residence. The visiting researchers interact with scientists at NASA and with faculty at Case. ICOMP was administered by a committee of NASA scientists and Case faculty, with Eli Reshotko as Chairman and Isaac Greber as a member of the committee and chairman of its computer sub-committee. ICOMP continues to function, but is now located at the offices of the Ohio Aerospace Institute (OAI).

Research on problems in microgravity had been going in the department for some time, primarily by Si Ostrach, Yasuhiro Kamotani and James T'ien and their students. The experiments in this field took advantage of the drop tower at NASA Glenn. In 1997 a center for this activity, the National Center for Microgravity Research in Fluid Mechanics and Combustion (NCMR) was established, with headquarters at NASA Glenn and at CWRU. The center is a joint effort of NASA Glenn and CWRU, with Si Ostrach as director. Iwan Alexander is chief scientist for fluids and James T'ien is chief scientist for combustion. The research staff of NCMR includes a group of scientists stationed at NASA Glenn, some of whom hold research professorships in the department. NCMR has interacted strongly with the educational as well as the research activities of the department. Students have worked with NCMR through research grants, summer internships, and projects in their laboratory courses and senior projects.

There has been a strong association of faculty and students with the space program, both in basic research and direct involvement with space flight and experiments. Joseph M. Prahl as been associated with the flight program, and was a backup mission specialist for STS-50/United States Microgravity Laboratory Mission 1 aboard the shuttle Columbia. Prahl, Ostrach, and several of the students have made training flights on the KC135 airplane (the "vomit comet"). Ostrach is second only to former astronaut and senator, John H. Glenn in holding the distinction of the oldest person to fly on the KC135.

The undergraduate students have had the advantage of association with research scientists at NASA Glenn in their courses and projects. This has included basing their course projects on research being carried out at NASA, working with NASA scientists on these projects, observing experiments on these projects at the NASA Glenn and at Plum Brook facilities, and making presentations of their work at NASA. The students have also had special flight test experiences, conducted specifically for them, at NASA and at Ohio State University.

The most recent additions to the faculty in the aeronautics are Chih-Jen Sung (PhD Princeton) who joined the department in 1999 and works in the combustion diagnostics, Edward White (PhD Arizona State) in 2000 who works in stability and bypass transition of boundary layers, and Alexis Abramson (PhD Berkeley), who joined the department in 2003 and works in the rapidly developing field of nanotechnology. White has made a major redevelopment of the department's subsonic wind tunnel for his boundary layer research program. The renovated tunnel now merits its low turbulence designation, and is now a modern facility producing important research results in by pass tr5ansition and receptivity.

At the present the department of Mechanical and Aerospace Engineering has grown to a faculty of 38, 17 full-time tenure track, 350 students, 250 undergraduates and 100 graduate students, and graduates 10 students a year with a BS degree in Aerospace Engineering, and 20 graduate students a year in mechanical and aerospace degree programs. The department's research support is approximately $8 million per year, of which $6 million/year is aeronautics and space research related.

CHAPTER 24

AERONAUTICS TO AEROSPACE
AT THE UNIVERSITY OF NOTRE DAME

Thomas J. Mueller[*] and Robert C. Nelson[†]
University of Notre Dame, Notre Dame, Indiana 46556

Abstract

The history of aeronautics at the University of Notre Dame, from Albert Zahm in the latter part of the nineteenth century, to the twenty-first century, is presented. Zahm experimented with gliders and wrote papers on stability and control before the Wright Brothers' successful flight in 1903. Because of his extensive education and early wind tunnel research, he has been regarded as America's first well-trained aeronautical engineer. He was also one of the driving forces behind the establishment of a national aeronautical research laboratory in 1915. The formal department of aeronautical engineering was established in 1935 and the first head was Professor F.N.M. Brown. Professor Brown and his colleagues pioneered the development of subsonic and supersonic smoke visualization in wind tunnels. In the 1950's these developments led to detailed smoke photographs of boundary layer transition and many other basic flow phenomena. From the 1960's to the present, the program has continued to evolve its undergraduate and graduate programs and add new faculty who greatly expanded the research topics studied. In 1991, the new Hessert Center for Aerospace Research building was dedicated and new experimental facilities added. The "Center for Flow Physics and Control" was officially made a center in the College of Engineering in 2001.

Before the Wright Brothers

Man's interest in flying began many centuries before the Wright Brothers' successful flight on December 17, 1903. From the beginning, there were emperors, kings, novelists, poets and all kinds of inventors interested in flying.[†] The experimentation in this area accelerated after the Montgolfier brothers made their first balloon voyage in 1783. Sir George Cayley produced the first design for the modern airplane configuration in 1799 and flew his full-sized gliders in 1809. In 1849, Cayley tested a boy-carrying glider and in 1853 made an uncontrolled flight of a man-carrying glider. Because of his experiments and publications he became known as the "Father of Aeronautics."[1]

It was during this period of aeronautical experimentation that the University of Notre Dame was founded by Reverend Edward Sorin in 1842. Notre Dame was the first Catholic university in America to establish a formal course of study in engineering when it created a program in Civil Engineering in 1873. The specialty of Mechanical Engineering was an outgrowth of this program in 1886.

Albert F. Zahm was an important early figure in aeronautics. Born in 1862, the eighth of fourteen children[2], Zahm entered the University of Notre Dame in 1879, enrolling in the Classical Course of studies. His experiments in aerodynamics began in 1880.[3] He built model airplanes and full size aircraft. In 1882, he tethered one of his aircraft to the ceiling of the old Science Hall with a 50-foot rope.[3] This aircraft was powered by a bicycle pedal mechanism that rotated a tractor propeller. The purpose of these experiments was to test the efficiency of the propulsion system, the lift from the wings and the action of the control system. Zahm also experimented with gliders. One of Zahm's glider experiments is depicted in a mural, Figure 1, which is in the LaFortune Student Center (also known as the old Science Hall).

Figure 1. Mural depicting one of Zahm's glider experiments at Notre Dame.

Zahm was an extremely bright student at the University and in 1883, while he was still an undergraduate, built a small hand-powered wind tunnel in an attempt to study the lift of various-shaped wings. This tunnel was made by removing the vibrating screens from a farmer's winnowing blower. It could have been the first wind tunnel built for research purposes in the United States. It was not the first wind tunnel ever built, however, since the wind tunnel was invented in England by Sir Frances H. Wenham in 1871. As a result of his experiments, he was among the first to favor slender concave surfaces (i.e. bird-shaped) for aircraft wings and propellers.[3]

A member of the class of 1883, Zahm remained at Notre Dame to complete an M.A. degree and to teach mathematics and mechanics. When the Mechanical Engineering program was established in 1886, Zahm was named professor in this program. A few years later, Zahm left Notre Dame to attend graduate school at Cornell University, where he received an M.S. degree in 1892. [3,4] It was there that he met Octave Chanute, who gave a seminar at Cornell on "Aerial Navigation."[4,5]

Correspondence between Zahm and Chanute relating to the physical and theoretical analysis of flight began in the spring of 1891.[4] In May of 1892, while he was teaching at Notre Dame, Zahm suggested that a meeting be called to provide serious discussions between those scientists and engineers interested in flight. He convinced Chanute that such a meeting could be held at the Chicago World's Fair to be held in 1893. Interest in this idea grew and during a meeting with Zahm on October 18, 1892, Chanute agreed to be the Chairman of the conference and oversee the general arrangements and Zahm agreed to be the General Secretary. Thus America's first International Conference on Aerial Navigation was held during the World's Fair in Chicago (often referred to as the Columbian Exposition of 1893).

Zahm delivered two papers of considerable significance at the Conference: "Atmospheric Gusts and their Relation to Flight" and "Stability of Aeroplanes and Flying Machines."[3,4,5] Dean Courtland D. Perkins of Princeton University, in his 1970 Wright Brothers lecture, credits Zahm for his contributions to flight stability.[6] Zahm, in the above cited paper, clearly defines the requirements for longitudinal static stability. In 1893, Zahm went to Johns Hopkins University to study physics, receiving his Ph. D. in 1898. He joined the faculty at Catholic University in 1895, and in 1901 built and tested what was then the largest "wind tunnel" in the United States.[7] This wind tunnel preceded the one built by the Wright brothers. He also developed instruments for the measurement of velocity and pressure in his wind tunnel. Zahm was interested in scientific research related to flight, while the Wright brothers were concerned with developmental problems.[2] Between December 1885 and December 1903, Zahm published fifteen papers on a variety of subjects in aeronautics.[5] When Catholic University's financial position became precarious in 1908, Zahm began lecturing at the Bureau of Standards. According to T.D. Crouch,[4] "In view of his superior methodology and the extent to which his studies, particularly in the area of skin friction, proved useful, Zahm deserved to be regarded as America's first well-trained aeronautical engineer."

Zahm had been aware of the fact that European countries, such as France, Italy and Germany, were more interested in flight than the United States.[2] Nine major wind tunnels had been built in these countries between 1903 and 1912. In January 1912, Zahm wrote an article "On the Need for an Aeronautical Laboratory in America."[5] He wrote fifteen other papers in 1912, including two that described European aeronautical laboratories.[5] He was the leading author of scientific and popular articles and leading spokesman for American aviation.

Zahm served as secretary to President Taft's committee to investigate the need for a national aeronautical laboratory. In 1913, he was named Secretary for the Langley Aeronautical Laboratory at the Smithsonian. The Advisory Committee of this laboratory commissioned Albert H. Zahm and Jerome C. Hunsaker to visit England, France and Germany to document the latest scientific developments in aeronautics. Hunsaker was a young naval officer who later was asked to design a course in aeronautical engineering at the Massachusetts Institute of Technology. The Zahm-Hunsaker final report presented to the Smithsonian on July 27, 1914, together with the start of World War I in August of 1914, convinced Congress to act. Congress increased appropriations for military aviation and in 1915 created the National Advisory Committee for Aeronautics (NACA).

Figure 2. Albert F. Zahm (right) with Glenn Curtiss (left) and Charles Manly (center). The flying machine is Langley's Aerodrome in an early stage of modification by Curtiss in 1914. [8, 9]

In 1914, Zahm was named chief research engineer for the Curtis Airplane Company, and in 1916 became the Director of the Aerodynamics Laboratory of the U.S. Navy. Figure 2 shows Zahm sitting on Langley's Aerodrome with Glenn Curtiss and Charles Manly in 1914. In 1929 he retired from this position and in 1930 the Guggenheim Chair of Aeronautics was created for him at the Library of Congress. He held the Guggenheim Chair until 1946 when he retired. His work was acknowledged by a commemorative envelope issued by the U.S. Post Office in 1938 for his glider experiments 50 years earlier in 1888.

The football program at Notre Dame also played a role in the development of commercial aviation. Notre Dame's legendary coach, Knute Rockne, died in a commercial airplane crash in Kansas on March 31, 1931. The accident investigation revealed that the wooden wing spars of the Fokker Trimotor F.10 failed due to dry rot. Figure 4 is a picture of a Fokker Trimotor F.10.

Figure 4. A Fokker Trimotor F.10 airplane.

Rockne's death caused a public outcry over the quality of passenger air service. TWA (then called Transcontinental & Western Air) responded by contracting with the Douglas company to build an all-metal transport that could take off fully loaded on one of its two engines and beat a Ford Trimotor (a clone of the Fokker Trimotor) from Santa Monica to Albuquerque. Douglas produced the experimental DC-1 in 1933, shown in Figure 5, and then went into production with the 14-passenger DC-2 version which started service with TWA in 1934. The 21 passenger DC-3 followed in 1936.[+]

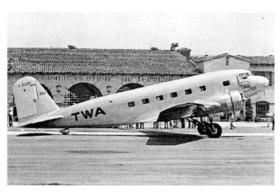

Figure 5. An all-metal DC-1 airplane produced by Douglas.

The New Program (1935-1944)

In the spring of 1935, President of the University Father O'Hara organized a conference on aeronautics to which he invited many well-known pilots and aviation enthusiasts from around the country. The purpose of the meeting was to discuss the beginning of an aviation program at the University of Notre Dame.

Captain Edward V. Rickenbacker, World War I ace, vice-president and later the president of Eastern Airlines, along with Major Alford G. Williams, U.S.M.C., manager of the Aviation Sales Department of the Gulf Refining Company of Pittsburgh, Merrill C. Meigs, publisher of the Chicago Evening American, Ben Smith of New York, and Lloyd Maxwell, a Chicago advertising man, were present as shown in Figure 6.[10] When it was decided that Notre Dame would have an aeronautical engineering program, Father O'Hara went to one of the four best universities in this field, the University of Michigan, to look for someone to head and develop this program. There he found Frank Newton Mithery Brown.

Figure 6. Aeronautics Conference at Notre Dame in 1935 (left to right) – Major Alford G. Williams, U.S.M.C. and manager of the aviation sales department of the Gulf Refining company of Pittsburgh; Capt. Edward V. Rickenbacker and Ben Smith, of New York; Rev. Father John O'Hara (standing); Merrill C Meigs, of Chicago.

F.N.M. Brown began his studies at the University of Michigan in aeronautical engineering in 1920 and studied there for two years. Then, in order to make enough money to continue his education, he went to work for the Ford Motor Company for two years where he moved up from laborer to assistant foreman. In 1924 he went back to the University for another year followed by a year at the Ford Motor Company as an engineer. He returned to the University of Michigan and completed the Bachelor's Degree in Aeronautical Engineering in 1927. He then mixed the next few years between working and completing his Master's Degree in Aeronautical Engineering at Michigan. He worked as a consultant designing aircraft; he worked as a flight test and research engineer at the U.S. Naval Air station at Mountain View, California, which is now the location of NASA Ames Research center; and he taught at Michigan.

In September 1935, Brown was named by Father O'Hara as Professor and Head of the Department of Aeronautical Engineering (see Figure 7). At that time he was the youngest person to be named a professor at the University. With the coming of Frank Brown to organize the aeronautical engineering department, the story from aeronautics to aerospace at Notre Dame begins. What follows is an abbreviated description of the evolution of the undergraduate curriculum, the research activities in the department, as well as other historically relevant facts, such as the development of research facilities, and a vision for the future.

Figure 7. Professor F.N.M. Brown, about 1936.

Shortly after the program began, J.A. McClean was added to the faculty. The first laboratory equipment installed was the wind tunnel, built in the summer of 1936. Since there was no internal combustion engine laboratory in the college, the department set up such a laboratory in the winter of 1936. In 1938, another University of Michigan graduate with his Master's Degree in Aeronautical engineering, Robert S. Eikenberry, joined the staff. Eikenberry, shown in Figure 8, was a skilled designer and engineer and supervised the design and building of much of the equipment in the original aeronautical engineering laboratory.

Figure 8. Professor Robert S. Eikenberry.

The curriculum in 1935-36 included a total of 159 credit hours, as shown in Figure 9. The senior year curriculum indicates a very strong orientation toward the practical engineering topics of the day. Airplane design, internal combustion engines, airport design, meteorology, and of course something that has always been important, English for Engineers. A student branch of the Institute of Aeronautical Sciences was established in 1937.

DEPARTMENT OF AERONAUTICAL ENGINEERING
1935-1936

SENIOR YEAR CURRICULUM

Philosophy 2c, 2d	The Philosophy of the Mind and Ethics	6
Civil Engr 4-4b	Plane Surveying	2
Metallurgy 108	Non-ferrous Metallography	3
English 21-22	English for Engineers	4
Elec. Engr. 17-18	Engineering Economics	4
Aero Engr. 7-8	Internal Combustion Engines	8
Aero Engr. 9-10	Airplane Design	10
Aero Engr. 11	Meteorology	2
Aero Engr. 13	Avigation	2
Aero Engr. 15	Airport Design	2
	Credit Hours	41

Total hours required for graduation - 159

Figure 9. Senior year curriculum, 1935-36.

This period was the beginning of six and a half decades of continuous research in aeronautics at Notre Dame. In the late 1930s, Frank Brown began his research on the development of low turbulence smoke visualization wind tunnels. He had experimented with different inlet contraction ratios, number of screens, and various other methods to reduce the turbulence intensity in the test section by using small tabletop wind tunnels. A two-dimensional smoke tunnel was operational in 1937, and a three-dimensional one was operational in 1940. By the end of the 1930s he had developed the ideas that would later come together to form the state-of-the-art low-speed flow visualization wind tunnels. However, although the design for a large three-dimensional smoke tunnel was complete in 1940, the actual construction had to wait because World War II began.

Figure 10. Glider built by Professor Eikenberry and students.[8]

Frank Brown invented the space time recorder, a navigational aid which was used by pilots during the war. Eikenberry was very interested in sailplane design and aircraft structures. He designed, and, with the help of his students, built a glider (shown in Figure 10). This glider was made with wood ribs covered with fabric and had a single seat open cockpit. The glider club support vehicle is shown in Figure 11.

Figure 11. Glider Club vehicle.

The first flight of the glider took place at the South Bend, Indiana airport on July 2, 1941.[11] Eikenberry piloted the glider at the National Soaring Meet in Elmira, New York, in the summer of 1941. It stayed aloft for six hours and took third place for endurance.

During World War II, the German Army attacked Fort Eber Emael in Belgium using troop gliders. The U.S. Army was surprised by this, since they didn't have any gliders. In fact, the Army had an eight-year prohibition on flying or even being a passenger in a glider for Army personnel. In order to provide trainers in sufficient quantity, the Army required civilian gliders.[12] One of these gliders was donated to the Army Air Corps by Notre Dame in July 1942.[13] It was designated "TG-21" (Training Glider #21) and was used to train pilots in California. This glider is listed in the Army Air Corps records as built by the students at Notre Dame.

Frank Brown spent much of World War II away from the University, working for the State Department, traveling to China and consulting on dirigible problems. While he was gone, the Navy basically took over the University for its V-12 program. There are many that have said that if the Navy had not used Notre Dame for the V-12 program, the University would have been in financial trouble. From the beginning of the program to 1941, there were an average of 7 students per year in the aeronautical engineering program, while during World War II the average number of students increased to seventeen.

After the War (1945-1963)

Eikenberry was in charge of the program during World War II, and when Brown returned after the war, the two began to build the large three-dimensional smoke tunnel which they designed before World War II. The curriculum changed to reflect the changes in the aviation industry. The number of credit hours decreased to 153 and the number of undergraduates increased to 31 per year from 1946 to 1949. Frank Brown and others in the department developed an internal combustion laboratory which they eventually turned over to the mechanical engineering program. During this period internal combustion engines were as important to the aeronautical engineering program as they were to mechanical engineering. The first design for an aeronautical engineering laboratory was made in March 1944. This design, Figure 12, shows the football stadium in relation to the proposed aeronautical engineering building and two runways. The thinking at the time was that if you were going to build an aeronautical engineering department, you needed to have an airport. This plan was not approved by the university.

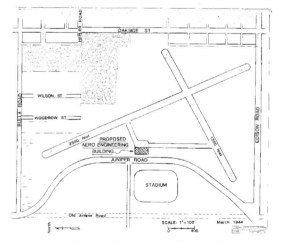

Figure 12. Proposed Aeronautical Engineering building and airport, 1944.[8]

Many of the concerns we have today with regard to research and teaching were also concerns of Brown and his contemporaries. Those faculty who are concerned about summer support will appreciate the following quote from Frank Brown: "I would like to know if you think there is any possibility of obtaining three months active duty pay in your department this summer?" (Taken from a letter to the Naval Bureau of Aeronautics, January 15, 1936). Another of Brown's quotes comes from a letter dated April 30, 1947 to University President John Cavanaugh. "It is recognized that graduate work in the Department of Aeronautical Engineering is desirable. It is further recognized that there must be no attempt to build the graduate division at the expense of the undergraduate division." Not many of the problems seen today are really new. During this period, the number of faculty ranged from three to five. The first graduate degree, an M.S., was granted in 1949.

The Department moved from the Engineering Building (Cushing Hall) to a separate building in the summer of 1947. The Aeronautical Building was a World War II surplus building of frame construction. It had a usable floor area of 10,000 sq. ft. It had several offices, two classrooms, a library and graduate student room, an instrument room, a darkroom, a senior drafting room, a junior drafting room, and a laboratory and shop. The shop included Clarence Smith ("Smitty"), a machinist and long-standing fixture who helped shape the character of the department.

In the 1950's, the department continued to evolve. In 1949, Vince Goddard, who had received his degree at Holy Cross College in physics and a degree in aeronautical engineering at MIT, came to Notre Dame and taught from 1949 to 1951 (see Figure 13). Goddard designed an indraft supersonic wind tunnel.[14] He left to go to the Douglas Aircraft Company, and he returned to Notre Dame about 1957. He had a very important part to play in the supersonic aerodynamics program at Notre Dame.

The curriculum in the 1956-57 version still had 153 hours, with all of the basic elements: airplane design, structures, laboratory, internal combustion engines and some options. Fundamentals of economics was an important course for engineers, as well as ethics. The undergraduate enrollment averaged about 20 students per year in the 50's.[15]

Figure 13. Professor Vincent P. Goddard.

Research in the 50's was dominated by the use of the flow visualization wind tunnels, both the subsonic and supersonic. Brown was very much interested in the magnus force on spinning cylinders and spheres, and flow around inclined bodies. He wrote research reports for the U.S. Navy and various companies such as Boeing and Bendix. He was doing sponsored research long before most people at Notre Dame. One of the major developments in the 50's was the building of the supersonic wind tunnel. Brown was quite an old horse trader from North Dakota and he built an entire laboratory largely on war-surplus equipment at minimal cost to the University. In fact, the supersonic wind tunnels are still driven by three sets of vacuum pumps and 125 horsepower motors that came off a World War II naval vessel. He originally obtained six of these units for $1500 and sold three to Bendix Corp. for $6600. With the money, he built the supersonic wind tunnel using the remaining three units.[15]

These were the early days of supersonic aerodynamics. It was a small wind tunnel, one that was later expanded by Goddard and his students to include a large inlet contraction to produce the first, and the only, supersonic smoke tunnel in the world. For many years, people didn't believe it actually worked. Figure 14 shows a simultaneous Smokeline-Schlieren photograph of a wedge at a Mach number of 1.38.[16]

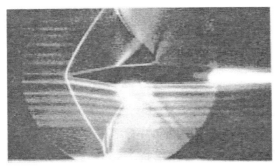

Figure 14. Simultaneous Smokeline-Schlieren photograph of the flow over a wedge at a Mach number of 1.38.

Brown started experimenting in 1937 with inlets and screens to reduce turbulence in the test section, which was an absolute necessity for flow visualization. He also experimented with methods of delivering smoke into the wind tunnel at the first screen. Most people in Europe and England put the smoke rake inside the tunnel and the disturbance produced by the rake in the tunnel increased the turbulence level and the system never worked. Frank Brown's ideas not only in the injection of smoke but also in the generation of smoke, were unique for his day. Figure 15 shows a drawing of the three-dimensional smoke tunnel in about 1959.[17] In the 1930's, smoke was generated by coking wheat straw, and then the smoke would go up through a pipe and into the smoke rake, producing a very dense smoke. Brown also used this method of generating smoke. The trouble was that it also produced a lot of tar, which clogged up the system. The tars deposited on the wind tunnel models and made quite a mess. In the 1960s, Brown developed a smoke generator which vaporized kerosene that eliminated the problems of the coked straw method.

They did a lot of work using an external, platform lift and drag balance which was designed and constructed by Eikenberry to measure lift and drag forces and also magnus forces on spinning bodies. Their entire laboratory- the wind tunnels, the smoke generator and rake, force balance, models, etc. – was designed and built at Notre Dame.

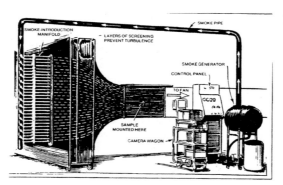

Figure 15. Drawing of the three-dimensional subsonic smoke tunnel in the 1950s.

Some of the most significant early research was visualization of the boundary layer. Figure 16 is of particular historical importance.[18] This is an axisymmetric body with which Brown worked. While he was doing this work, one of the smoke tubes impinged on the nose of the model and smoke was enveloped in the boundary layer. When he took this picture using a strobe light, he saw a wave pattern that eventually broke down. He was most likely the first person to ever visualize Tollmien-Schlichting waves in a wind tunnel. There were other people who were working on the same problem around that time, but it appears that he was the first one to photograph this phenomenon, around 1957.

There is a file of letters from almost every person of that day who was well known in fluid mechanics. Brown received a letter from Theodore von Karman from Paris. He had sent him some smoke photographs of a leading edge separation bubble with transition on an airfoil. Those who had seen his visual results showing the details of vortex shedding behind airfoils and sharp flat plates, and later boundary transition of axisymmetric bodies, were absolutely astounded, as was von Karman. At that time, most people didn't realize the value of flow visualization, but Brown was a very stubborn person and continued to improve his smoke tunnel system; that's what pioneers are all about! Schubauer and Klebanoff of the National Bureau of Standards were making point by point measurements with hot-wire anemometry in this region where transition was taking place on a flat plate. It sounds from the tone of their letter in response to Brown's sending them examples of his results that they were still a little skeptical of his visualization of transition.

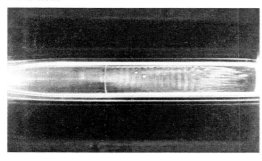

Figure 16. Smoke visualization of the boundary layer transition process on an ogive-nosed cylinder.

The trio of Brown, Eikenberry, and Goddard worked on several problems in the 1950's, and one of them was an early study of propellers. The smoke photograph shown in Figure 17 has been used in many textbooks.[19] Sometimes they gave credit to Brown, sometimes they didn't. The vortex off the tip of the blades is very clear. The renewed interest in propeller design for efficient subsonic aircraft increases the current interest in Brown's original work.

The magnus force produced by a spinning cylinder was also of great interest to Brown. He really was studying very basic phenomena in fluid mechanics, trying to understand them using the techniques he had available at that time. For the spinning sphere problem he found the best way to gain attention was by using a baseball.

Figure 17. Smoke visualization of the wake of a rotating propeller.[19]

His work on the spinning baseball was reported in articles in various magazines, for example, a smoke photograph of a spinning baseball appeared on the cover of the Saturday Review on April 5, 1958 (see Figure 18) and in a 1959 Popular Mechanics article in which he explained why the curve ball curves.[17] The dedication to this research by Frank Brown was noteworthy because most people, until the 50's, told him he was wasting his time trying to visualize complex flow phenomena. By the 1970's with people writing books and articles on flow visualization, everyone in the community realized how valuable this work was, how much it could help us understand basic phenomena, and how it could help formulate mathematical models and discover new phenomena. There have been numerous articles written on flow visualization and in these Frank Brown is given credit for most of the early contributions to smoke visualization in wind tunnels. Over 100 labs have used his tunnel smoke generator. [20, 21, 22]

Figure 18. April 5, 1958 cover of *Saturday Review* showing Frank Brown's spinning baseball.

Aerospace Engineering (1964-1969)

In 1964, the program name was changed to Aero-Space Engineering (not to be confused with Aerospace Sciences, which was the Air Force ROTC) by John D. Nicolaides, who joined the faculty as chairman in 1964. Thomas J. Mueller from United Aircraft Research Laboratories and Chuen-Yen Chow from the University of Michigan both joined the faculty in 1965. The 60's were a very exciting time in aerospace programs. The number of credit hours for the Bachelor's degree decreased to 144 and topics related to space were added (e.g., rocket propulsion and orbital mechanics). The undergraduate enrollment averaged about 30 students per year, with a peak of about 50 in 1969. There were about six Master's students per year and a Ph.D. program was established in 1965. The first Ph.D. was granted to Patrick J. Roache in 1967.

The department continued some of the research that Brown had originated in boundary layer transition, vortex shedding at the trailing edge of airfoils, and the effects of sound on vortex shedding from flat plates and airfoils. Brown and Goddard worked together on the trailing edge vortices and the control of the frequency of shedding by acoustical means. Unfortunately, these studies were not widely published.

New areas of research were introduced by the new faculty, such as the parafoil, computational fluid dynamics, altitude compensating rocket nozzle research, flight dynamics of missiles and aircraft, magneto-hydrodynamics, and separated flow research. In 1964, John Nicolaides was approached by an inventor of the parafoil from Florida. It so captured his imagination that from that time on he spent much of his time with the parafoil shown in Figure 19.

Figure 19. A parafoil plane.

In 1967, the aerospace building was enlarged because the enrollments had grown considerably. A metal addition was added to the north end of the Aero Building, at a cost of $40,000. In the late 60's the faculty had reached a peak of seven. The graduate program grew to 28 students.

The first digital computer on campus, the IBM 1620, arrived in 1959. With this Professor Eikenberry began his lifelong affair with computers. He was responsible for many innovations in programming this machine and those that followed. He was the author of the NDFOR compiler, which was a FORTRAN compiler well suited for educational settings that was eventually used in many schools. He also made major contributions to the area of missile flight dynamics in support of the research of Nicolaides.

Professor Albin A. Szewczyk joined the Mechanical Engineering Department in 1962 and began research in computational fluid dynamics in 1963, when the Computer Center obtained a

UNIVAC 1107. This research, which was supported by the Bendix Corporation, was published in 1966 as a Department Report and the Dissertation of David C. Thoman, and produced some of the earliest solutions for the incompressible flow around a cylinder.[23] These results were in agreement with available experimental data at the time. The only "computer" in the Aerospace Building was a Merchant Calculator which could add, subtract, multiply, divide, and take the square root.

When Professor Mueller joined the Aerospace Engineering Department in 1965, he brought research in compressible flow problems including basic separated flows and the complex flows in altitude compensating rocket nozzles, i.e. axisymmetric truncated plug nozzles and the planar aerospike type nozzle shown in Figure 20. This nozzle was being considered for the Space Shuttle main rocket engine but was not far enough along in development when the decision had to be made. His approach to the problems was both experimental and computational. His first Ph.D. student, Patrick Roache, produced one of the first numerical solutions to the compressible separated problems produced by a backstep and a rectangular cavity. After his graduation in 1967, Dr. Roache continued to refine and extend his dissertation until in 1972, he published the first book in Computational Fluid Dynamics,[24] which has become a classic.

Figure 20. Schlieren photograph of a planar truncated plug rocket nozzle flow field.

Aerospace and Mechanical Engineering, 1970-1989

The Department of Aero-Space Engineering merged with the Mechanical Engineering Department in 1969. Professor K.T. Yang became the chairman of the combined department. The two undergraduate programs, aerospace and mechanical engineering, maintained their separate curriculum, and that has continued to the present. After the merger and as a result of the significant decline in engineering enrollment over the entire country, a core curriculum was established in the College and Department that reduced the total credit hours for the B.S. degree to 128. The nationwide enrollment declined as a result of a downturn in the business cycle and the anti-technology and anti-military movements that were fueled by the Vietnam War. The successful Apollo mission landing men on the moon was unable to alter the mood in the country. Notre Dame had 49 senior aerospace engineering students in 1970 and only 9 by 1978. Research funding from NASA and the DOD all but disappeared during the early 1970s.

Notre Dame joined the Supersonic Tunnel Association (STA) in 1970. The first STA meeting at Notre Dame was held in September 1979, the second in April 1992, and a third in April 2002. In 1979, Notre Dame also joined the Subsonic Aerodynamic Testing Association (SATA) and hosted its annual meeting in 1991.

The AE and ME graduate programs have developed along discipline lines because of the common interest in aerodynamics, fluid mechanics, structures and design. Professor Hafiz Atassi, who joined the combined department in 1969, and Professor Szewczyk have been closely associated with the aerospace programs because of their teaching and research interests in fluid mechanics/aerodynamics.

Because it was clear in 1969 that it would be very difficult to obtain research funds for rocket propulsion related projects, Professor Mueller started to study separated flows with biomedical applications (e.g., arterial bifurcations). During the early 1970s, Professor Mueller and Professor

Jack Lloyd developed a biomedical research program to study the hemolytic and thrombogenic potential of artificial heart valves. This research was supported by NSF and the Indiana Heart Association. Both numerical and experimental studies were made. Figure 21 shows Professor Mueller and the tri-leaflet heart valve developed at Notre Dame in a mock circulatory system used for endurance testing.

Figure 21. Dr. Mueller observes a mock circulatory system experiment with the tri-leaflet artificial heart valve developed at Notre Dame.

During this period, Professor Goddard studied the drag reduction of deflectors on semi-truck cabs. He also completed his Doctoral Dissertation on the numerical solutions of the flow around an oscillating cylinder under the direction of Professor Szewczyk in 1971.

Figure 22. The atmospheric wind tunnel inlet with one of the four Turbulence Generating Boxes shown.

Professor Szewczyk studied the wind forces on and flow patterns around buildings, and was largely responsible for building the department's atmospheric wind tunnel shown in Figure 22. Arrays of jets perpendicular to the main flow are utilized as active devices on four sides of the turbulent generating box. Atmospheric wind tunnel studies were promoted by architects who were encountering problems with strong downdrafts from tall buildings that caused structural damage and affected pedestrian movement. [25]

In 1973, Professor Hafiz Atassi began his research in unsteady aerodynamics and Aeroacoustics with application to aircraft engine components and fan noise. His analytical and numerical work on the effects of nonuniform flows has continued to the present.[26,27] Professor Atassi, the Viola D. Hank Professor, is a recognized authority in this field and his research has been supported by NASA, U.S. Air Force Office of Scientific Research, Office of Naval Research, NSF, and Pratt & Whitney. Professor Atassi is an AIAA Fellow and has received the AIAA Aeroacoustic Award in 2000. He was also named Chevalier des Palmes Academiques by the French government.

Professor Robert C. Nelson came from Wright-Patterson AFB and joined the faculty in 1975. His research interests included airplane wake turbulence, aerodynamics, and flight mechanics. Dr.

Nelson and his students have examined aerodynamic phenomena associated with high angle of attack flight of slender wings. Their work includes fundamental studies of the flow structure of leading edge vortices before and after vortex breakdown.[28,29] Figure 23 is a flow visualization photograph of the leading edge vortices above a slender delta wing. A laser light sheet was used to illuminate the cross flow at various locations along the wing. The vortex core region and the secondary structures in the shear layer are clearly shown in this photograph. The core region is noticeable by the void in the smoke pattern.

Figure 23. Smoke visualization of the vortical structure over an 85° Delta Wing by Frank Payne and Robert Nelson.

The study of vortical flows above stationary wings led to an interest in how the structure and particularly vortex breakdown was affected by model motion. Large amplitude pitching motion studies revealed significant aerodynamic hysteresis in the measured aerodynamic loads. Visualization measurements showed that the hysteresis in the aerodynamic loads was related to the delay in vortex breakdown on the upstroke portion of the motion and in the re-establishment of the leading edge vortex after complete flow separation on the downstroke. From unsteady aerodynamics research Dr. Nelson became interested in the flight dynamics phenomenon called wing rock. Wing rock is an unwanted limit cycle rolling motion that affects many highly swept wing aircraft at high angles of attack. Nelson and Dr. Andrew Arena conducted a combined experimental and analytical investigation of the wing rock motion of slender flat plate wings. The data from these experiments coupled with the computational study provided an improved model of the flow mechanisms causing wing rock on slender delta wings.

In 1976, Professor Mueller began an interest in low Reynolds number airfoil aerodynamics that lasted for 25 years. His first sponsor was NASA Langley in 1977, and in 1980 the Naval Research Laboratory and Office of Naval Research helped him expand these studies. NASA supported this program for 12 years and the Navy supported it for 12 years. This project produced 14 M.S. students and two Ph.D.'s from 1977-1989. Over sixty journal and conference papers have been published from this research.[30] It also led to a series of conferences at Notre Dame on this subject. These conferences organized by Professor Mueller were held in 1985, 1989 and 2000.[31] In 2002, Mueller was elected to the grade of Fellow in the Royal Aeronautical Society of London.

Professor Szewczyk became the Department chair in 1978. Professor Stephen M. Batill joined the faculty that year. He came from the faculty of the U.S. Air Force Academy and his research interests included aerodynamics, aeroelasticity and aerospace structures.[32] His early research programs involved the development of design criteria for high-speed flow visualization in wind tunnels. This resulted in the development of a transonic test section for the Notre Dame facilities and eventually influenced the design of a major wind tunnel facility at Wright-Patterson AFB, the SARL facility. His research interests broadened to include dynamic system modeling and parameter identification, and this multidisciplinary perspective eventually resulted in a number of research efforts in systems design and automation.[33] Dr. Batill was also instrumental in the development of the aerospace engineering design education program at Notre Dame, and introduced the concepts of team-based, design-build-fly projects for all students. Every Notre Dame aerospace engineering student since spring 1986 has designed, built and tested a flight vehicle ranging from dirigibles and helicopters to fully autonomous UAVs. In 1988, Dr. Batill successfully led a team of students in the "Race for Space Software Chase" sponsored by Apple Computer, Inc. The team developed airplane design software called MacAirplane, which has been on exhibit in the National Air and Space Museum since 1988.

The ASEE/AIAA John Leland Atwood Award recognizes the accomplishments of a superior aerospace engineering educator and his/her contributions to teaching, research, and the profession. This award was presented to Professor Mueller in 1981, Professor Nelson in 1991 and Professor Batill in 2001.

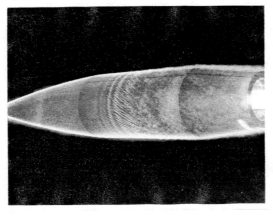

In 1979, Professors Mueller and Nelson continued the transition studies on spinning and non-spinning axisymmetric bodies started by Professor Brown in the 1950s and 1960s. On the spinning body at low Reynolds numbers both the viscosity-controlled Tollmien-Schlichting waves and the crossflow instability were observed to be present at the same time. This appears to be the first time these two phenomena had been visualized simultaneously, as shown in Figure 24.[34]

Figure 24. First visualization of simultaneous Tollmien-Schlichting waves and cross-flow vortices on a spinning axisymmetric body.

In 1985 Professor Patrick Dunn became a member of the faculty, coming from Argonne National Laboratory. He established the Particle Dynamics Laboratory, focusing on multi-phase flow research related to the dynamics of microparticles in various environments.[35,36] Initially he studied the electrostatic generation and dispersion of microdroplets and microparticles in a vacuum characteristic of low-Earth orbit (see Figure 25).

In 1989 he began to collaborate with Professor Raymond Brach in the Department on the experimental and theoretical aspects of microparticle deposition onto surfaces. This joint effort lead to seven years of funding from the Electric Power Research Institute and the Center for Indoor Air Research, resulting in over 40 journal publications and conference proceedings.

Figure 25. This unique microdroplet spray pattern was generated in the Particle Dynamics Laboratory using a tailored electric field.

A New Era, 1989-2000

Professor Flint Thomas came to Notre Dame from Oklahoma State University in 1987. Upon arriving at Notre Dame Professor Thomas was named an Office of Naval Research (ONR) Young Investigator and this supported his experimental investigations of shock wave / turbulent boundary layer interactions. For this work Professor Thomas developed new supersonic blow down wind tunnel facilities at Notre Dame. Professor Thomas has subsequently modified and utilized these facilities on a continuous basis for a wide range of projects in compressible transitional and turbulent flow.

These include studies of supersonic jet screech (supported by AFOSR & McDonald- Douglas) and more recently for studies focused on understanding noise source mechanisms in compressible subsonic axisymmetric jets (with co-investigator Prof. Thomas Corke) under support from the aeroacoustics branch at NASA Langley.

Prof. Thomas has also been involved in experiments involving free shear layer transition.[37,38] A unique feature of this experimental work has been the application of both novel polyspectral signal processing techniques and a nonlinear system model approach in order to provide quantitative measures of the complex sequence of nonlinear wave interactions that ultimately lead to fluid turbulence. As such, this work has sought to clarify the complex sequence of events that lie between the realms of linear stability theory and fully developed turbulence. In collaboration, Professors Thomas and R. C. Nelson have been involved in a range of research projects with each focused on the flow physics of high-lift systems used for commercial transport aircraft. This work has been supported by NASA Ames, NASA Langley and NASA Dryden Research Centers. The approach taken has been to "decompose" the high-lift system into a series of generic high-lift building block flows. These include wake development in pressure gradient, wake/boundary layer interactions, boundary layer relaminarization, and separated flow, among others. The work has focused upon understanding these building block flows in isolation, which is considered prerequisite to understanding their integrated behavior in actual high-lift systems. This work continues to be very well received by the aerodynamics community. In fact, one aspect of the high-lift work dealing with the identification of leading edge noise sources and their control recently received a Turning Goals Into Reality Award by NASA Langley Research Center.

Professor Mueller was appointed Department Chair in 1988 and the Roth-Gibson Professor in 1989. Professor Eric Jumper joined the faculty in 1989 after 21 years as a "Blue-Suit Scientist" with the United States Air Force. During his Air Force career, at various times he was on the faculty of both the Air Force Academy and the Air Force Institute of Technology. Just prior to joining the Notre Dame faculty Dr. Jumper was the Chief of the Laser Devices Division of the Air Force Weapons Laboratory; this Division contained all of the high power laser research and development in the Laboratory, and worked on problems related to laser beams interacting with turbulent airflows. It is not surprising that one of the first research areas that Dr. Jumper developed since coming to Notre Dame is in the area of laser interactions with the turbulent body of air surrounding an airborne laser platform; this field of study is known as Aero-Optics.[39]

The crux of the Aero-Optics problem is that when an otherwise collimated laser beam, whose wavefront is initially planar, encounters a turbulent, variable-index-of-refraction flow, it emerges with a distorted or aberrated wavefront. This wavefront aberration, which is rapidly time varying, interferes with the laser beam's ability to be focused. Dr. Jumper's Aero-Optics research group began their efforts with the development of the world's first, truly-high-speed wavefront sensor, capable of capturing time series of wavefronts at greater than 100 KHz. The sensor, known as the Small-Aperture Beam Technique or SABT sensor, is based on the discovery by M. Malley, G.W. Sutton, and Knicheloe, that aberrations due to a convecting turbulent flow are themselves convecting.[40] Up until that time, wavefront sensors were limited to bandwidths of about 1 KHz, which greatly limited the study of Aero-Optics, particularly for airborne-laser flow scenarios, which we have now learned require capturing wavefronts at 20 KHz. Notre Dame's continued Aero-Optics research has progressed to the point where it is now possible to contemplate applying Adaptive Optics techniques to the mitigation of the Aero-Optics problem.

Dr. Jumper's other area of research is in internal and external unsteady flows. Most recently, he and his students have been looking into the effect of upstream-propagating potential disturbances on turbine-engine components. This research contributes to a better understanding of

High Cycle Fatigue (HCF). HCF is responsible for a large percentage of unexpected turbojet and turbofan failures on commercial and military aircraft.[41]

Commander James D. Wetherbee (USN), Aero class of 1974, became Notre Dame's first astronaut (see Figure 26). On the first of several flights, he piloted the Space Shuttle Columbia STS-32, which was launched January 9, 1990. Two other Notre Dame aerospace engineering graduates have recently been named to the Astronaut Corps: Lt. Col. Kevin A Ford, class of 1982, and Major Michael T. Good, class of 1984, with a Master's degree in 1986. Both of these Astronauts in Training are in the U.S. Air Force. Although the undergraduate program is oriented more towards aeronautics than space, a large number of Notre Dame graduates have gone into space related activities either with NASA or NASA contractors.

Figure 26. Astronaut James D. Wetherbee.

The Hessert Center

By the late 70s it was quite apparent that the original Aerospace Laboratory building was nearing the end of its service life. A plan for a new building was developed as part of a proposal to the United States Army Research Office. At that time the Army was looking to fund several University Centers for Rotary Aircraft Technology. As part of the cost sharing included in the proposal the University agreed to renovate the Heat and Power Laboratory Building that prior to 1980 housed many of the laboratories for the Mechanical Engineering Program. However with the opening of the new Fitzpatrick Hall of Engineering building the Mechanical Engineering laboratories were relocated to Fitzpatrick Hall. Steve Batill developed the first plan on how the facilities for the Aerospace Laboratory could be included in the renovated Heat and Power building. Although our proposal was unsuccessful in the competition it did get us started in thinking about the requirements for a new laboratory.

Figure 27. Thomas and Marilyn Hessert.

Around 1987 the University Administration recognized the need to relocate the Aerospace Laboratory. Most Aerospace or Aeronautical Laboratories build in the 1940s were typically located some distance away from the main campus. This was true for the Notre Dame Aerospace Laboratory building as well. However with the large building program in the 60s and 70s the Aerospace Building was no longer in a remote location. The plans developed for the ARO proposal were revisited and requirements for a new building were developed that would allow Notre Dame to continue building its research and graduate program. It became apparent that the critical requirement for the new building was to recreate the excellent collaborative research environment of the old "Aero Shack." The excellent research environment that had existed in the old Aero Shack was due in large part to the layout of the laboratory facilities, graduate student and faculty offices. The old laboratory was an ideal integration of people and facilities. It was decided that the new building had to keep the faculty and graduate students offices together and in close proximity to the research facilities. Unfortunately the funding the University could provide at this time was only enough to renovate the Heat Power Building to accommodate the research facilities. The Aerospace faculty made the difficult decision to turn down the Administration's offer to relocate the Aerospace Laboratory.

About a year later, Mr. Thomas Hessert of our Engineering Advisor Council decided to see what he could do to move our building plans forward. A short time later Marilyn and Thomas Hessert (see figure 27) gave the University a significant gift to create in our opinion a world class University Aerospace Laboratory. The architects used our earlier plans and requirements to create the Hessert Center for Aerospace Research. The Hessert Center, shown in Figure 28, was dedicated in November 1991. The name was changed in 2003 to Hessert Laboratory for Aerospace Research. In addition to the two low turbulence subsonic wind tunnels, and the atmospheric tunnel, three new facilities were included in the new Hessert Center. An anechoic chamber was designed and built as part of the construction of the main lab, and the six sides of the chamber were treated by Eckel Industries, Inc., with fiberglass wedges that were paid for by the Office of Naval Research. Professor Mueller designed a low speed free-jet wind tunnel that was placed inside the anechoic chamber. Calibration of this facility indicated that the low frequency cutoff was about 100 Hz and the tunnel could produce speeds up to 100 ft/sec.[42]

Figure 28. Exterior view of the Hessert Center for Aerospace Research.

When this facility was finished, Professor Mueller began an aeroacoustics research program concerned with basic experiments to determine the sources of noise in propeller and airfoil flowfields.[43] This research, sponsored by ONR, continues to the present.

The Hessert Center Main Lab was configured to include a new blow-down supersonic facility. This facility has a 550 psi two stage compressor, twin tower dryers, a 2000 gallon trace heated storage tank outside of the building, and a flow control system designed by Professor Thomas.

A free-surface water tunnel purchased from Eidetics Corporation was assembled in the main lab. This facility has a speed range from 0 to about 15 inches/second, and has been very useful for flow visualization with colored dye and for very low Reynolds number airfoil aerodynamics.

In 1996, Professor Nelson was appointed Department Chair. His efforts to find the resources to continue the development of the Hessert facility and to recruit additional faculty have continued the tradition of excellence.

Professor Thomas C. Corke came to Notre Dame from the Illinois Institute of Technology in Chicago in 1999, and was named the Clark Professor. His initial research interests stemmed from his Ph.D. research, which was on turbulent boundary layers. That work dealt with the effects that suppressing the large-scale outer structure had on turbulence production near the wall (the so-called "burst" cycle), and on the viscous drag. This work became NASA/FED restricted from publication and presentation for three years. This led him to other research directions, in particular to the study of fluid instabilities and transition to turbulence. A particular approach that he has used in studying instabilities is to introduce controlled disturbances.[44, 45]Professor Corke and his students were the first to introduce controlled 3-D disturbances to document subharmonic resonance in boundary layer transition. This has been applied in many other flows including 3-D and supersonic boundary layers, jets and wakes. The techniques that were developed to control these instabilities resulted in a natural progression to general "flow control".

His present research interests involve various aspects of flow control as applied to a broad range of technologically relevant problems. These include control of transition to turbulence in boundary layers on hypersonic lifting bodies; control of transonic jets to reduce noise; separation control related to retreating blade stall on helicopter rotors, and on blades in the low-pressure turbine (LPT) stage of turbo-jet engines; and control of cross-flow instability on highly swept wings. Central to these problems has been the development of flow actuators which are based on locally ionizing the air to create a plasma over arrays of electrodes. This approach has a number of advantages such as being fully electronic and having a high frequency bandwidth that is necessary for high-speed applications. Another significant advantage of these actuators is that they can easily be incorporated into numerical simulations and therefore optimized and "reverse-engineered" for specific applications.

The 21st Century

Notre Dame is proud of its rich tradition in aerospace research and education and is optimistic about its future, and has developed a new center of excellence. The "Center for Flow Physics and Control" was made an official center in the College of Engineering in the summer of 2001. The Center builds on existing funded research and historical precedent of the original Aerolab and Hessert Center (renamed Hessert Laboratory in 2002), while consolidating common core interests and abilities to address larger focused efforts. It involves faculty in 5 departments. The overarching theme is flow diagnostics, prediction and control, in a process loop that involves computation, experiment and modeling. There is a strong emphasis on industry partnerships to identify flow control applications that can have an immediate impact, be technologically viable, and of economic interest. Core areas consist of aeroacoustics, aerooptics, fluid-structure interactions, multi-phase flows, and general flow control.

An example of the type of research that the Center is addressing is a ``Flow Control Designed Engine" which is the theme of a recent NASA URETI proposal. The concept is to focus on identifiable ``flow modules" which are basic but important elements of the internal and external flow field of turbo-jet engines. The rationale of this approach is to break down a complex system into simpler elements that can be more completely understood, modeled, and controlled. Such an engine could be more efficient, smaller, lighter, with a reduced part count, be substantially quieter, have a larger operation envelope, and cost less than present engines. This represents our vision of the type of a large focused effort, with a significant technological impact, at which the Center is aimed.

Acknowledgements

The authors recognize that there were other faculty who were at Notre Dame for less than ten years. Although these faculty are not mentioned specifically, they also contributed to the development of the program. The 1,532 B.S., 200 M.S., and 65 Ph.D. graduates also made significant contributions to this program.

Special thanks go to Charles R. Lamb of "The University Archives" for helping with the Albert Zahm material, and Professor Stephen M. Batill for his suggestions during the writing of this paper. Professors H. Atassi, T. Corke, P. Dunn, E. Jumper, A. Szewczyk, and F. Thomas all contributed to this paper. We would also like to thank John D. Anderson, Jr. for identifying Glenn Curtiss and Charles Manly in Figure 2.

References:

1) Wragg, D.W., "Flight Before Flying", Frederick Fell Publishers, Inc., New York, 1974.

2) Dethloff, H. and Snaples, L., "Who Was Albert F. Zahm?", AIAA Paper No. 2000-1049, presented at the 38th Aerospace Sciences Meeting & Exhibit, 10-13 January 2000, Reno, Nevada.

3) Hope, A.J., C.S.C., "Notre Dame: One Hundred Years", Icarus Press, Inc., South Bend, Indiana, 1978, pp. 210-211.

4) Crouch, T.D., "A Dream of Wings: Americans and the Airplane 1875-1905", 1981,W.W. Norton and Company, New York, pp. 78-83.

5) "Aeronautical Papers of Albert F. Zahm, Ph.D., 1885-1945," Volume I and Volume II, published by the University of Notre Dame, Notre Dame, Indiana, 1950.

6) Perkins, C.D., "Development of Airplane Stability and Control Technology", AIAA Journal of Aircraft, Vol. 7, No. 4, 1970.

7) Anderson, Jr., John D., "A History of Aerodynamics," Cambridge University Press, Cambridge, UK, 1997.

8) University of Notre Dame Archives.

9) Anderson, John D., Jr., private communication

10) The South Bend, Indiana News Times May 17, 1935.

11) News release, University of Notre Dame, undated, but additional information points to 1941.

12) Blacksten, Raul, Private communication, 2001.

13) Newspaper clipping, Catholic Herald Citizen, July 18, 1942.

14) The South Bend Tribune, February 21, 1951.

15) Annual Report of F.N.M. Brown, 1955/1956.

16) Goddard, V.P., McLaughlin, J.A., and F.N.M. Brown, "Visual Supersonic Flow Patterns by Means of Smoke Lines," Journal of the Aero/Space Sciences, Vol. 26, No. 11, November 1959, pp. 761-762.

17) Hicks, C.B., "The Strange Forces of the Air," Popular Mechanics Magazine, June 1959, pp. 124-127.

18) Brown, F.N.M, "The Organized Boundary Layer", Proceedings of the Sixth Annual Conference on Fluid Mechanics, University of Texas, Austin, Texas, pp. 331-349, September 1959.

19) Brown, F.N.M., "The Visible Wind", Proceedings of the Second Midwestern Conference on Fluid Mechanics, The Ohio State University, Columbus, Ohio, pp. 119-130, 1952.

20) Mueller, T.J., "Smoke Visualization of Subsonic and Supersonic Flows (The Legacy of F.N.M. Brown)", University of Notre Dame, UNDAS TN-3412-1, a Final Report for Contract AFOSR-77-3412, June 1978.

21) Mueller, T.J., "On the Historical Development of Apparatus and Techniques for Smoke Visualization of Subsonic and Supersonic Flows", AIAA Paper No. 80-0420, presented at the AIAA 11th Aerodynamic Testing Conference, Colorado Springs, Colorado, March 18-20, 1980.

22) Mueller, T.J., "History of Smoke Visualization in Wind Tunnels", Astronautics and Aeronautics, Vol. 1, pp. 50-54, 1983.

23) Thomas, D.C. and Szewczyk, A. "Time-Dependent Viscous Flow over a Circular Cylinder," The Physics of Fluids Supplement II, II-76 – II-86, 1969.

24) Roache, P.J., "Computational Fluid Dynamics", Hermosa Publishers, Albuquerque, New Mexico, 1972.

25) Owen, J.C., Szewczyk, A. A. and Bearman, P. W. 2000 Suppression of Karman Vortex Shedding. Gallery of Fluid Motion, The Physics of Fluids, 12, 9, pp. 1-13.

26) Atassi, H.M., "Unsteady Aerodynamics of Vortical Flows: Early and Recent Development," Aerodynamics and Aeroacoustics, Ed. K.Y. Fung, Ch. IV, pp. 119-169, World Scientific, 1994.

27) Golubev, V. and Atassi, H.M., "Interaction of Unsteady Swirling Flows in Annular Cascades; Part I: Evolution of Vortical Disturbances, " AIAA Journal, 38 (7), 2000, pp. 1142-1149.

28) Arena, Jr., A. S., and Nelson, R. C., "Measurement of Unsteady Surface Pressure on a Slender Wing Undergoing a Self-Induced Oscillation," Experiments in Fluids, Vol. 16, No. 6, pp. 414-416 1994.

29) Nelson, R.C. and Pelletier, A, "Unsteady Aerodynamics of Slender Wings on Aircraft Undergoing Large Amplitude Maneuvers," under revision for Progress in Aerospace Sciences, Elsevier Science Limited.

30) Pelletier, A. and Mueller, T.J., "Low Reynolds Number Aerodynamics of Low-Aspect-Ratio, Thin/Flat/Cambered Plate Wings", AIAA Journal of Aircraft, Vol. 37, No. 5, pp. 825-832, October 2000.

31) Mueller, T.J. (ed), "Fixed and Flapping Wing Aerodynamics for Micro Air Vehicle Applications," Progress in Aeronautics and Astronautics, AIAA, Reston, Va, 2001.

32) Batill, S.M., and Nelson, R.C., "Low Speed, Indraft Wind Tunnels", in Frontiers in Experimental Fluid Mechanics: Lecture Notes in Engineering, No. 46, Springer-Verlag, Berlin, pp. 25-94, 1989.

33) Batill, S.M., Sellar, R.S. and Stelmack, M., "A Framework for Multidisciplinary Design Based upon Response-Surface Approximations," AIAA Journal of Aircraft, Vol. 36, No. 1,pp. 287-297, 1999.

34) Mueller, T.J., Nelson, R.C., Kegelman, J.T., and Morkovin, M.V., "Smoke Visualization of Boundary Layer Transition on a Spinning Axisymmetric Body," AIAA Journal, Vol. 19, No. 12, pp. 1607-1608, December 1981.

35) Grace, J.M. and Dunn, P.F., "Droplet Motion in an Electrohydrodynamic Fine Spray," Experiments in Fluids, Vol. 20, pp. 153-164 1996.

36) Brach, R.M., Dunn P.F. and Li, X., "Experiments and Engineering Models of Microparticle Impact and Deposition," Journal of Adhesion, Vol. 74, pp. 227-282, 2000.

37) Thomas, F.O., and Chu, H.C., "Nonlinear Wave Coupling and Subharmonic Resonance in Planar Jet Shear Layer Transition," *Physics of Fluids A*, Vol. 5, No. 3, pp. 630-646, 1992.

38) Gordeyev, S.V. and Thomas, F.O., "Coherent Structure in the Turbulent Planar Jet. Part 1. Extraction of Proper Orthogonal Decomposition Eigenmodes and Their Self-Similarity," Journal of Fluid Mechanics, 414, 2000, pp. 145-194.

39) Jumper, E.J. and Fitzgerald, E.J., "Recent Advances in Aero-Optics," Progress in Aerospace Sciences, 37 (3), 2001, pp. 299-339.

40) Malley M, Sutton GW, Kincheloe N. "Beam-jitter Measurements of Turbulent Aero-Optical Path Differences," Appl. Opt. 1992:31:4440-3.

41) Falk, E.A., Jumper, E.J., Fabian, M.K. and Stermer, J., "Upstream-Propagating Potential Disturbances in the F109 Turbofan Engine Inlet Flow," Journal of Propulsion and Power, 17(2), pp. 262-269, March-April 2001.

42) Mueller, T.J., et al. "The Design of a Subsonic Low-Noise, Low-Turbulence Wind Tunnel for Acoustic Measurements." AIAA Paper No. 92-3883, presented at the AIAA 17th Aerospace Ground Testing Conference, Nashville, Tennessee, July 6-8, 1992.

43) Mueller, T.J. and Lynch, D.A., III, "An Anechoic Facility for Basic Aeroacoustic Research, Chapter 5 in "Aeroacoustic. Measurements," Experimental Fluid Mechanics Series, Springer-Verlag, Heidelberg, Germany, 2002.

44) Erturk, E. and Corke, T.C., "Boundary layer receptivity to free-stream sound on elliptic leading edges of flat plates," Journal of Fluid Mechanics, 429, pp. 1-21, 2000.

45) Erturk, E. and Corke, T.C., "Boundary layer leading-edge receptivity to sound at incident angles," Journal of Fluid Mechanics, 444, pp. 383-407, 2001.

Chapter 25
Of Aeronautics, Aerophysics, and Aerospace: A History of Aerospace Engineering at Mississippi State University

David H. Bridges[*]
Associate Professor of Aerospace Engineering
Mississippi State University

To Aero or Not to Aero?

Aeronautical engineering at Mississippi State began in a somewhat indeterminate manner. In 1932, the board of trustees for the college, in response to the urgings of various groups, issued a directive that Mississippi State College (MSC) begin offering courses in aeronautics. However, the wording of the directive was vague enough that Hugh Critz, president of MSC, decided to appoint a committee to interpret the directive to determine if that was in fact what the board had intended. The conclusion of the committee was "yes," and so the first students were admitted into the new aeronautical engineering program in September of 1933. Unfortunately the board did not provide any funding for the new program, which began as something of a foster child in the Mechanical Engineering department, as a subdivision of ME (Haug, p. 62; see "Note on Sources" at end of article).

The beginnings of aeronautical engineering at MSC were not much less definitive than the beginnings of engineering in general at the college. In 1878, a new College of Agricultural and Mechanical Arts had been established in Starkville, Mississippi, the new Mississippi A&M. The first president of the new college was former Confederate lieutenant general Stephen Dill Lee. When the new A&M college was created, however, funding was only provided for traditional liberal arts and agricultural education, and none for the "mechanical arts" (Haug, p. 9). In fact, according to Haug (p. 10), Lee actually resisted the addition of what he called "the mechanical feature." In 1888, Buz Walker joined the faculty as a professor of mathematics. Walker began to add education in the "mechanical arts," in the areas of surveying, statics, dynamics, elementary mechanisms, and fluid mechanics, in addition to such civil engineering topics as trusses, mechanics of materials, and railroad roadbed construction (Haug, p. 12). By 1892, instruction in the "mechanical arts" had become a permanent feature of the new college. The formal establishment of a School of Engineering at Mississippi A&M followed in 1902 by action of the board of trustees of the college. Walker was named the director of the school and shortly afterwards became the first Dean of Engineering at Mississippi A&M (Haug, p. 24). Mississippi A&M became Mississippi State College in 1932.

The urgings which prompted the board of trustees to create the new program in aeronautics may have come from diverse sources, but one of the leading proponents of the new program was Mitchell Robinson, who in 1932 became the new College Secretary, a position equivalent to today's Vice-President for Business Affairs (Barnett, p. 19). The late 1920s and early 1930s was

---

[*] P. O. Drawer A, Mississippi State, MS 39762. Senior Member, AIAA.

one of the golden ages of aviation, and Robinson could see the potential for growth in this new field, in both technical and commercial applications. To him, aircraft manufacturers, airlines and airports represented a new source of economic growth for the state, and he was determined that MSC would play a role. In fact, the new program that emerged was "Aeronautical Engineering and Commercial Aviation," and was open to both engineers and business majors (McKee, p. 36).

At that time, all engineering students at MSC had common programs for the first two years of study. Courses included algebra, trigonometry and calculus, as well as mechanisms, surveying, forge shop, and foundry. For the first year of the new aeronautical engineering program, juniors took Practical Aviation, which according to the catalog covered the material necessary in a Transport Pilot's ground school. They also took Simplified Aerodynamics, which included performance, stability and control and which reflected a common tendency of the time to lump all of aeronautics under the heading "Aerodynamics." Separate disciplines of aerodynamics as purely a study of fluid flow, aircraft performance, and stability and control would emerge later. Since as stated in the 1934-1935 MSC catalog the purpose of the program was "to train engineers rather than aerodynamicists; engineers who can take part in the practical work of designing, building and developing aircraft and aircraft engines on a scientific basis," juniors also took Airplane Welding and a second semester of Practical Aviation, which this time concerned itself with "advanced work in doping, gluing, and rigging." The next year, the first seniors would take Airfoils and Airscrew Theory, Airplane Design, Airplane Engine Laboratory, and Airplane Structures, in addition to other engineering and mathematics courses. Over on the business side, the Commercial Aviation students were taking Practical Aviation, Airport Management, Commercial Aviation, Aviation Law and Commerce Rules, Transport Aviation, and Airports and Civil Airways.

All of these courses were taught by two individuals. The first aeronautical engineering faculty member was Kenneth Withington, who had received his B.S. degree from Alabama Polytechnic Institute (now Auburn University) in 1931 and who would later receive an A.E. degree from API in 1936 (Barnett, p. 23). The second was Mason Sumter Camp, who had graduated from the Robertson School of Aviation in St. Louis in 1926. Camp's graduation certificate and his license were both signed by his instructor, one Charles A. Lindbergh. Camp had returned to Starkville prior to 1933 with the primary goal of establishing a flying school and airport in the area. Camp was hired as a flying instructor, but his duties included classroom work as well. He had the responsibility of teaching the junior level courses and Withington taught the senior level courses (Barnett, p. 23). Camp's formal status as a faculty member in aeronautical engineering would remain somewhat nebulous, as later history will demonstrate.

After one year of being located in the ME department, Aeronautical Engineering (hereinafter "AE") became a separate department starting with the 1934-1935 school year. Withington, who had been an assistant professor of mechanical engineering, was named department head. The aeronautics courses which had been taught as ME courses now had their own prefix, "Av" (at the time, "AE" was already taken by the Agricultural Engineering department). The new department still had no direct funding for faculty, offices, laboratories, or equipment. On loan from the U.S. Army Aviation Corps was a Curtiss P-1F pursuit plane, a Douglas O-2H observation plane, and a Liberty V-1650 aircraft engine, none of which was flight-worthy (Haug, p. 63; McKee, p. 36). Since there was no building to store these in, they were placed in a Dairy Judging Pavilion where they were used for instructional purposes. However, the pavilion remained in use, so the planes were damaged by visitors to the pavilion and by the rats that made homes in the aircraft and chewed the fabric and wood. Indeed, the conditions were so bad in general that Haug summarizes the state of the department in 1938 as follows: "From the beginning, courses were taught in a dairy barn by inexperienced faculty members without advanced degrees using borrowed, obsolete equipment" (Haug, p. 66).

Peacetime and Wartime Training

Despite its austere beginnings, the new aeronautics program managed to keep itself aloft. The first class received B.S. degrees in May of 1935. In the second group of graduates, those who received their degrees in 1936, was one Franklin Sproles Edwards, a future head of the AE department at MSC. Even with the lack of funding, the program was successful in attracting students. Department head Withington attributed this to his success at placing his graduates in jobs, with AE having the highest placement rate in the School of Engineering (Haug, p. 63).

The Airplane Design course that Edwards took in his senior year in 1936 had a slightly different description than the previous year. According to the new description, each student "design[ed] an airplane," which perhaps underscores the simplicity of airplane design in 1936 relative to today's complex aircraft. Topics listed for the course were "a detailed study of plan form, flaps, etc.; parasite drag; stability and control; and performance calculations for each design." The second semester of design was conducted "with reference to the Bureau of Air Commerce design regulations" and "design loads caused by the various maneuvers [were] calculated for each part."

Word was received in 1935 of possible federal funding for aviation training, a goal of Franklin Roosevelt's since he had been elected President in 1932 (Barnett, p. 38). In order to improve their chances of obtaining funding under this effort, the AE department conducted an extensive study of the state of aeronautics in the country, including a survey of aircraft manufacturers that included Pratt and Whitney, the Wright Aeronautical Corporation, and the Glenn-Martin Company. The survey presented information on the current status of aeronautics instruction at MSC and requested feedback on areas of possible improvements. The results of the study were used to make revisions in the program curriculum, an early example of external assessment of the program (Barnett, pp. 38-39).

In 1937 the AE department had the distinction of enrolling the first female student in the School of Engineering at MSC. Ms. Cora D. McDonald graduated from the program in 1940, becoming the first woman to receive an engineering degree from MSC. Part of the AE curriculum at the time was flight training, and some time after her graduation in 1940, during WWII, McDonald served as a ferry pilot (McKee, p. 45).

In 1938 it was announced that certain locations in the U. S. would be selected as experimental federal aviation schools, and MSC set out to land one of these schools. Even the president of the University, George D. Humphrey, exerted considerable effort to try to obtain one of the schools for MSC, contacting officials of the Civil Aeronautics Authority and members of Mississippi's congressional delegation. In January of 1939, the list of selected locations was announced, and MSC was not on the list. Humphrey continued his efforts, and the dean of the School of Engineering, L. L. Patterson, traveled to the University of Alabama with AE department head Withington to determine what could be done to improve the curriculum. The result of that study was the hiring of an additional faculty member, Thomas H. Dalehite, who received his B. S. in AE from MSC in 1939 (Barnett, pp. 47-51).

The federal aviation training program was expanded in 1939, and MSC's efforts finally paid off. In September of 1939 Humphrey was informed that MSC had been selected to participate in the program, and that the school to be established in Starkville would train up to forty students, making it one of the largest such training facilities in the region. The instruction consisted of seventy-two hours of classroom study and thirty-five hours of flight training. Withington, Camp and Dalehite were all examined and licensed as ground school and flight instructors (Barnett, pp. 52-55).

In 1940, AE finally obtained its first laboratory facility, although not the one that had been on the university books for construction since 1935. The basement of the Student Activities building was cleared out and converted into an engine laboratory. The facility did at least have two new engines, which had been donated by the Army, as well as some other used equipment (Barnett, p. 57).

The next significant event for AE occurred in 1941, when Franklin S. Edwards returned to MSC to join the AE faculty. Edwards had been working in industry since his graduation in 1936. Edwards remained at MSC until 1944, when he took leave to attend the California Institute of Technology. He received his M.S. in Aeronautics from Caltech in 1945 and returned to the AE faculty at MSC.

The war took its toll on the AE faculty. In 1942 Dalehite left on a military leave of absence and did not return to the department. In 1942 Withington requested that Camp be hired to assist in AE instruction, but his request was denied. Then in mid-February of 1942, Withington himself was called up to active duty in the U. S. Army Corps of Engineers. Camp was named acting department head and Edwards was put in charge of the pilot training program. Four student assistants were hired as temporary instructors, their pay coming from the remains of Withington's salary (Barnett, p. 64). However, by 1945 Camp was so involved with pilot training for the Army that his services to the college essentially ended, leaving Edwards as the sole AE faculty member after his return from Caltech (Haug, p. 78). The classroom load would not have been heavy, however. By the 1944-45 school year, enrollment at MSC had dropped from a peak of 2,327 in 1939-40 to a low of 404. Only twenty-one students received degrees from the college, and only six in engineering (Haug, p. 78).

The department and the university faced another significant challenge during the war. In 1943, the governor of Mississippi, Paul B. Johnson, Sr., proposed a committee to investigate the creation of a School of Aviation at MSC. This move met with violent opposition from MSC president Humphrey, who felt that the creation of such a school would endanger not only the chances of accreditation of the School of Engineering but also the chances of the college as a whole. According to Barnett, Humphrey "risked his position as President, verbally attacking the Governor of Mississippi, in order to preserve the past and future reputations of both the Department of Aeronautical Engineering and Mississippi State College" (Barnett, pp. 68-70). The proposal was eventually withdrawn.

Even with the war over, funding for the department did not improve. In 1946, Withington wrote to Dean Patterson, listing immediate needs as an exclusive departmental office and sufficient storage space. According to Withington, the department "has always been seriously handicapped." Unfortunately, Patterson's response struck Withington as inadequate, and

Withington submitted his resignation, effective July 1, 1946 (Barnett, p. 75). Edwards was named the department head.

The Research Feature

For many years there had been little motivation for MSC or its School of Engineering to become involved in research. Instruction had been its primary task, and some leaders had actually viewed publication as attempts by faculty to generate publicity for themselves. However, in the 1940s the School of Engineering began to re-evaluate this position. In 1941 the board of trustees authorized the creation of the Engineering Experiment Station, whose purpose was to determine "new or more economical uses for the natural resources of the state as they apply to engineering and industry." The demands of the war prevented the college from beginning a new program at that time, however, and the Station lay dormant until 1944 (Haug, p. 75).

In 1944, Dewey McCain, a professor of civil engineering, became assistant director of the Engineering Experiment Station, based in large part because of the research contract he had received from the Tennessee Valley Authority to study "Wood Joints made with Nails in Drilled Holes." This was the first externally-funded research conducted in the School of Engineering. His experience on this project opened his eyes to the value of research to the School of Engineering, and so he began advocating a greater role for research (Haug, pp. 82-83).

The efforts by various people to expand the research function in the School of Engineering began to pay off in 1948, when Dean Patterson hired Harold von Neufville Flinsch as a professor of civil engineering and associate director of the newly-renamed Engineering and Industrial Research Station (EIRS), with the understanding that Flinsch would become dean and director of EIRS upon Patterson's retirement. Flinsch had received the first Ph. D. in engineering at the University of Minnesota in 1941 for his work on the energy contained in ocean waves. Patterson retired as planned in 1949, and Flinsch became Dean of Engineering (Haug, p. 91).

Flinsch was given two primary duties as dean: increase the number of faculty with terminal degrees and expand the research programs in the School of Engineering. Budgetary and other constraints prevented rapid progress on the first goal. However, Flinsch could do something almost immediately about the second. Flinsch was a pilot and a sailplane enthusiast. He was familiar with sailplane research in Germany prior to WWII that had led to advances in aircraft technology, and had made particular note of the fact that such research could be done relatively inexpensively. He decided that if certain universities in Germany with limited means could conduct such research, there was no reason why Mississippi State College could not have a similar program. The effort to begin such a program brought him into contact with Dr. August Raspet, and the course of the history of the School of Engineering was changed (Haug, p. 94).

Raspet had graduated from the Carnegie Institute of Technology and gone on to receive a Ph. D. in physics from the University of Maryland in 1942. Like Flinsch, Raspet was a sailplane enthusiast and had achieved some recognition in this area. He served as director of the research phase of the Thunderstorm Project in 1946. In this project, he and other pilots would fly sailplanes into thunderstorms and the motions of the sailplanes would be tracked on radar, yielding information about the wind currents within the storms. Upon completing this project, he moved to New York, where he became the director of the Aerophysics Institute and began work on a grant from the Office of Naval Research (ONR). The purpose of this study was to use sailplanes to study airflow patterns over mountain ridges (Haug, pp. 94-95).

Flinsch contacted Raspet about moving to Mississippi State and beginning his own independent sailplane research program, which he did in February of 1949. From the beginning, Raspet was a very productive research scientist, but since he was essentially the only one in the School of Engineering, Flinsch had to figure out what to do with him. Since his degrees were in physics, he couldn't teach engineering courses. Also, since a heavy teaching load for each faculty member was one of the main impediments to research, Flinsch didn't necessarily want to put him in a classroom. Therefore he created the Department of Aerophysics just for Raspet and his sailplane research (Haug, p. 95).

The first project undertaken by Raspet upon his arrival at MSC seems to suggest a desire by Raspet to assimilate himself to his new surroundings. Raspet chose to study the aeronautics of bird flight by using his sailplanes to trail the local buzzards as they soared over the outskirts of Starkville. Raspet believed that an understanding of low-speed flight as illustrated by these birds would lead to improvements in low-speed aircraft performance, particularly as applied to short take-off and landing (STOL) aircraft currently under development. His preliminary results were sufficient that he was awarded a grant of $16,000 by the ONR to study the performance of soaring birds, the first large federal research grant ever received by the School of Engineering (Haug, p. 96). The research effort in engineering had literally taken off.

A Stearman biplane was purchased as a towplane (note: this aircraft is still owned and operated by the department). Several sailplanes were either purchased or donated, leading Flinsch to state that the Aerophysics Department was now ready to begin meteorological and aerodynamic studies that could not be performed anywhere else (Haug, p. 96). Interestingly, in order to perform their new duties, Raspet and his assistant, Mel Swartzberg, still had to be licensed as pilots, despite their considerable experience in sailplanes. This brought them into contact with Sumter Camp, the flight instructor, who recognized their abilities as pilots and was able to convince the FAA to reduce their required number of training hours (Barnett, p. 96).

Raspet's national reputation, and his presence at MSC, attracted other sailplane enthusiasts to Starkville. One of these was Dick Johnson, a native of California. Johnson came to MSC in 1950 and constructed his own sailplane, the RJ-5, under the direction of sailplane designer Harlan Ross and with the assistance of Raspet. Johnson used this sailplane to first win the 1950 national soaring championship, and then to set a new soaring distance record of 545 miles from Odessa, Texas to Salina, Kansas in August of that year. Johnson would go on to set numerous other sailplane records (Barnett, p. 100, p. 107).

Meanwhile, Back in the Classroom...

The Department of Aeronautical Engineering continued its slow but steady development. In the 1947-48 school year, one of the undergraduate courses, Applied Aerodynamics, mentioned "compressibility effects," the first such mention of the phenomena of compressible flow in the curriculum. In that school year AE also offered its first two graduate courses related to engineering. One was in Aircraft Propulsion Systems, which studied "the theory and design of propellers and rotor blades" and included "a comparative study of propellers, rotor blades, and jet propulsion" (the first mention of jets in the curriculum). The other new graduate course was Advanced Aircraft Structures Design and covered the topics of "shear lag, thin plates and shells under buckling, plastic bending, shear web theory, and torsion in single and multi-cell structures and open sections."

In 1947 Edwards hired Charles B. Cliett, a native of West Point, Miss., as a member of the AE faculty. Cliett had enrolled at MSC in the early 1940s but then entered the Navy during the war and participated in the V-12 program. Cliett received his B.S. from the Georgia Institute of Technology in 1945 and then spent two years completing his obligation to the Navy (Barnett, p. 79). Shortly after his arrival Cliett was put in charge of constructing the department's first wind tunnel. This was an open-circuit tunnel built in one of several "temporary buildings" on campus erected to meet the needs caused by the influx of post-war students, many studying on the GI Bill. The first AE course in laboratory methods was offered in the 1947-1948 school year and concentrated on wind tunnel measurements.

The School of Engineering finally received its long-desired accreditation by the Engineers' Council for Professional Development (ECPD) in 1952. This was due in part to achievements brought about by Flinsch, to revisions in curricula suggested by prior ECPD visits, and to the construction of Patterson Engineering Laboratories, completed in 1950. This building provided 47,000 square feet of space for offices, shops, and laboratories (Haug, p. 91). The AE department received office and laboratory space in the building, and moved in as soon as it was dedicated. Cliett had gone on leave for the 1949-50 school year to obtain his M.S. degree from Georgia Tech, and upon his return in the summer of 1950 was immediately placed in charge of building the new AE wind tunnel in Patterson Lab. This wind tunnel was a closed-circuit wind tunnel with an octagonal test section 3 ft high, 4 ft wide, and 5 ft long. The top speed was 150 ft/s, and this facility is still in use today.

At about the same time that Cliett was busy building the department's subsonic wind tunnel, work was proceeding on the first supersonic wind tunnel facility in the department. William Huntington, one of the first graduate students in AE, began his M. S. work on a blow-down supersonic tunnel facility, citing in his thesis[4] the necessity of generating experimental results to "generalize" the many theoretical solutions to supersonic flow problems. According to his thesis[4], work had begun on this facility in 1949. The nozzle was designed using the methods outlined in Puckett[5]. The test section was 4 inches high, 1 inch wide, and 4 inches long, and the facility was driven by a compressed air tank with a capacity of 200 ft^3 and an operating pressure of 150 psi. The design Mach number of the facility was 2.21, but measurements based on schlieren photographs of the shock wave on a wedge indicated that the Mach number actually reached was only 1.62. The tunnel was completed by January of 1951.

The necessity of studying supersonic flows cited by Huntington also manifested itself in the AE curriculum. One of the new courses in the catalog for the 1951-52 school year was Aerodynamics of Compressible Fluids, to be taken in the senior year. This was the first separate course dealing with compressible flow in the curriculum and Edwards was listed as the instructor. Based on the course description, the textbook probably used was *Aerodynamics of Supersonic Flight*[6] by Cliett's instructor and colleague at Georgia Tech, Alan Pope. Some new graduate courses were also listed, including Aerodynamics of Supersonic Flow, taught by Edwards and most probably guided by Liepmann and Puckett's *Aerodynamics of a Compressible Fluid*[7]; Theoretical Aerodynamics – Perfect Fluids, taught by Edwards and Cliett; and Theoretical Aerodynamics – Real Fluids, taught by Edwards. The "perfect fluids" course dealt with "the basic ideas of wing theory as developed from classical treatises on hydrodynamics," whereas the "real fluids" course dealt with "the aerodynamics of viscous fluids; boundary layer, heat transfer, fundamentals of boundary layer stability, turbulence, the fundamentals of isotropic turbulence, and experimental methods." According to Barnett, the AE graduate program bore its first fruit in 1951, when three individuals received M.S. degrees in aeronautical engineering: Glenn D. Bryant,

Joseph J. Cornish III, and William Huntington. Both Bryant and Cornish had been attracted to MSC by the idea of working with Raspet, and both continued working in the Aerophysics Department upon completing their M.S. degrees (Barnett, pp. 102-103). Cornish received the first Ph. D. granted by the School of Engineering in 1960.

The AE department received a new infusion of talent for the 1954-55 school year with the hiring of Leslie R. Hester. Hester had received his B.S. degree in AE from MSC in 1952 and gone to work for the EIRS on Raspet's ONR project (Barnett, p. 120). Upon completing his M.S. in AE in 1953, he joined the AE faculty. A new course, Aerodynamic Laboratory II, was added under Hester's instruction. This course dealt with aircraft performance measurements and aircraft stability and control. Indeed, by the 1955-56 school year, the areas of instruction seemed to have fairly well delineated between Edwards, Cliett, and Hester, with Edwards primarily responsible for aerodynamics, Cliett for structures, and Hester for flight mechanics. Hester taught the Aircraft Power Plants I & II courses (I dealt with internal combustion engines, II with jet propulsion) and both laboratory courses. Hester also assisted Cliett with the teaching of a new course, Vibrations and Flutter, which marked the first appearance of the discipline of vibrations as a separate course in the curriculum. In the 1957-58 school year, the senior design course was a two-semester course. The first semester dealt with topics such as layout, weight estimation, arrangement, balance, and performance estimation, while the second semester continued these topics through a detailed structural analysis of the wing. The next year, Hester's duties were extended further into the aerodynamics area when the graduate level viscous flow course was extended to two semesters, with Hester listed as the teacher for both. The first semester dealt with laminar flow, while the new second semester dealt with "the origins of turbulence; fundamentals of turbulent flow; turbulent boundary layers; turbulent flow through pipes; free turbulent flows; and determination of profile drag."

Smooth Sailing – and Rough

Raspet's early work in bird flight and associated soaring phenomena had already brought some recognition to MSC. However, in 1950, Raspet's research focus began to shift from bird flight to an area that would put Aerophysics and MSC squarely on the research map – boundary layer control (BLC). Raspet's early efforts in this area were in the promotion of natural laminar flow and the reduction of various surface roughness elements, such as gaps at panel edges or protruding fixtures, to decrease drag (Haug, p. 113). Efforts to produce natural laminar flow included the use of distributed suction, and the degree of success achieved warranted inclusion in the volume by Thwaites on incompressible flow[8]. In 1951, Raspet discovered another use of distributed suction: preventing separation of the boundary layer in large adverse pressure gradients. It was this discovery that would in Haug's words "shape the research program of the Aerophysics Department and provide it with its first national recognition for engineering research" (p. 114). Other researchers had experimented with boundary layer suction, but in those cases it was usually applied at a single location, usually through a slot. Raspet hit upon the idea of putting thousands of small holes in a wing surface and then using a small suction motor to pump air out of the interior of the wing. With this method applied to various aircraft, extremely large lift coefficients resulting in extremely small stall velocities could be obtained. The most dramatic results came when distributed-suction BLC was applied to an L-19 "Bird Dog" reconnaissance aircraft. The airplane could take off in 50 ft, had a minimum flight airspeed of 30 kts, and stalled at just 24 kts. Its maximum climb angle was 45 degrees and its maximum lift coefficient was 5. The potential for the application of this technology to STOL aircraft was obvious, and both the Navy and the Army were interested in pursuing this avenue of research.

Unfortunately for Raspet and the Aerophysics Department, support from MSC was almost exactly in inverse relationship to the progress made in research. During the initial formation of the Aerophysics Department, it had received strong support from MSC President Fred Mitchell. Mitchell's health began to fail, however, and by 1952 he had to be hospitalized in New Orleans. Most of his administrative duties fell to Benjamin Hilbun, his administrative assistant since 1949. Hilbun became acting president in 1953 and then president in 1954. Hilbun did not hold an advanced degree nor had he spent any of his working life outside of Starkville. Hilbun was not a supporter of research or the Aerophysics Department, and the budget allocated to the department by the college was cut from $32,000 in 1951 to $2,300 in 1952. By the time Hilbun retired in 1960, college funding had only been increased back to $3,400, even though during that time the Aerophysics Department had generated over $2 million in outside funding. The situation was so bad that when in the mid-1950s Bill Lear, out of gratitude for Raspet's efforts to improve the aerodynamics of Lear's *Lodestar*, offered a generous scholarship fund for research in Aerophysics, Hilbun turned him down (Haug, pp. 110-112). According to Barnett (p. 119), Hilbun had just rejected $100,000 per year in support of Raspet's research. The one bright spot in this period, from an administrative point of view, was that Raspet was formally named Head of the Aerophysics Department in 1953 (Barnett, p. 120).

The problems with funding did not end on campus. In 1954, Flinsch asked Graduate School Dean Herbert Drennon not to publicize the amounts of the contracts that Aerophysics had received from ONR. The fear was that if members of the board of trustees and state legislators friendly to the University of Mississippi found out, they would try to reduce the state's appropriation to MSC by the amount received from the ONR (Haug, p. 112). Flinsch himself did not last much longer. Shortly after his arrival at MSC, he had insulted Hilbun, who at that time was the college registrar. In 1957, now-President Hilbun called Flinsch into his office one day and summarily fired him (McKee, p. 72).

By 1957, Raspet had developed a proposal for an Aerophysics Institute to be built at Mississippi State at a cost of $500,000 and began attempting to obtain support for this proposal from outside sources, since it was obvious none was forthcoming from the college. This would take a few years. Meanwhile, research work continued and by 1960, according to Haug, "thirteen graduate students studying under Raspet and other faculty members associated with him had written master's theses on aspects of viscous flow, and researchers associated with Raspet's group had published seventy-four technical papers on sailplanes, fluid flow, and control of the boundary layer" (pp. 114-115). Joseph Cornish, David Murphree, Donald Boatwright, Cliett, and Bryant – all of whom would contribute significantly to the development of AE at Mississippi State – published papers or notes on aerodynamics and/or boundary layer control while working with Raspet (Haug, p. 115).

As fate would have it, 1960 was a traumatic year for both the Aerophysics Department and the Department of Aeronautical Engineering. In April of that year Raspet was demonstrating a Piper Super Cub equipped with BLC for Lowell L. Meyer, a representative of the Chance-Vought Aircraft Corporation, when something went wrong, and the airplane crashed, killing both Raspet and Meyer. Harry Simrall, who had replaced Flinsch as Dean of Engineering, was in the state capitol lobbying for the Aerophysics Institute funding when he heard the news. The shocked Mississippi legislature voted soon afterward to approve the $500,000 in funding, and the Raspet Flight Research Laboratory opened its doors at the Starkville airport in the summer of 1964. Cornish was named director to succeed Raspet (Haug, p. 115, 117). Cornish had been working on his Ph. D. under Raspet and had completed writing his dissertation, which Raspet had read shortly before the crash. However, no one else was familiar with the results and so an outside review board was convened, eventually granting Cornish his degree, the first Ph. D. awarded by the School of Engineering (Barnett, pp. 158-159). Upon completion of the degree, Cornish was also named Head of the Aerophysics Department (Barnett, p. 163).

On October 18, 1960, AE department head Franklin "Tiny" Edwards suffered a massive stroke and died the next morning. Edwards had been one of the earliest students of AE at Mississippi State and had been on the faculty since 1941, except for the year he had taken off to get his M.S. at Caltech. According to numerous accounts, Edwards had helped Aerophysics through some of the difficult years in the 1950s. Charles Cliett, AE faculty member since 1947, was named the new department head.

A *MARVEL* to Behold

As discussed above, both the Navy and the Army were interested in STOL aircraft and in the application of BLC technology to such aircraft. Beginning in the mid-1950s Raspet and his group began an effort to develop a STOL aircraft which would combine several different technologies. The first step was the realization that the turbulent slipstream of a propeller on a standard tractor-type airplane contributed significantly to boundary layer separation on the wing and the conclusion that a pusher configuration would be more suitable. Next, since this was to be a STOL aircraft, a way had to be found to increase the static and low-speed thrust of the propeller. This called for the ducted fan approach. After considering the available aircraft, an Anderson Greenwood AG-14 pusher aircraft (serial number 4) was purchased in 1955. A ring duct was added around the propeller and flight tests were conducted to determine the changes in aircraft performance. It was found that the duct did in fact increase the static and low-speed propeller thrust, but it decreased thrust at higher speeds and added to the overall drag of the aircraft at higher speeds as well. It was decided to try to design the duct so that it could replace some other aircraft component by performing that component's function.

Next, the issue of BLC control was addressed. The idea was still to prevent separation of the boundary layer in large adverse pressure gradients. One way to avoid separation was to avoid disturbances to the flow, such as those created by flaps. Therefore, a variable camber wing was designed. A large crank inside the wing turned and distorted the wing surface, changing its camber and therefore producing a similar result to the deployment of a flap. By using a variable camber wing, a smooth wing surface could be maintained, without the joints, rivets, and other hardware that would be necessary for a wing and flap of standard construction. However, in order to have a flexible surface, the wing would have to be made of fiberglass instead of metal. Finally, distributed-suction BLC was also to be applied to the wing, to provide further protection against boundary layer separation.

The new airplane was designed under the Mississippi State *MARVEL* Program. The "*MARVEL*" acronym, which stood for "Mississippi Aerophysics Research Vehicle with Extended Latitude," was coined by Joseph Cornish. The Raspet group decided that in order to prove the concept and gain some experience, it would be wise to design and build a smaller, less powerful version of the proposed airplane. A second AG-14 (s/n 2) was acquired in January of 1957. The cockpit module, engine and landing gear were retained, but everything else was replaced. A new fuselage fairing was fitted around the cockpit. The fairing extended back to the tail, where the ducted fan was now located. The propeller was driven by a fiberglass extension of the engine driveshaft. The propeller shroud, a ring with an airfoil section, now served the function of the tail surfaces. Two articulated segments on the sides served as rudders, and four segments, two on the top and two on the bottom, served as elevators. The main wing had the variable camber mechanism and perforations for BLC. All of the fiberglass components were built by the Parsons Corporation of Traverse City, Michigan. This aircraft was dubbed the *Marvellette*, and in January of 1960 received the U. S. Army designation XAZ-1, having been built under a contract with the Army signed in 1959.

The main drawback of the *Marvellette* was that it was overweight and underpowered. Tests were conducted with this aircraft until March of 1964, during which landing speeds of 31 kts were demonstrated. During this period work began on the primary aircraft, the actual *MARVEL*. The power plant of the new airplane was a 250 HP Allison T-63-A-5 turboprop. In its construction fiberglass was used almost exclusively, with stainless steel only used in selected areas for reinforcement, making the *MARVEL* the world's first all-composite aircraft. It had a wing span of just over 26 ft, a length of 23.25 ft, a wing area of 106 ft^2, and a maximum take-off weight of 1,890 lb. The first flight of the *MARVEL*, U. S. Army designation XV-11A, took place on December 1, 1965. Unfortunately, the BLC system did not provide enough suction to achieve maximum performance from BLC, and the expected results were not quite obtained. Nonetheless, a minimum take-off distance of 125 ft was demonstrated, as was a maximum lift coefficient of approximately 3.5.

Unfortunately for the *MARVEL*, external developments would intrude and cut short its usefulness as a research vehicle. During the 1960s the Department of Defense came under the direction of Robert McNamara, who brought a different perspective to defense spending, particularly on research. According to Haug (p. 117), McNamara directed each branch of the service to define itself in terms of its mission. To obtain funding, research projects had to be geared toward helping a service achieve its mission. By the mid-1960s, the Army's mission no longer required research on fixed-wing aircraft, and neither the Air Force nor the Navy wanted slow, STOL-type aircraft. Therefore, defense spending on STOL aircraft dried up, and the *MARVEL* project's funding was terminated in 1968, although flight testing would continue until 1970. An entire life cycle for research funding into STOL technology had come and gone.

Growing Our Own

The AE program underwent some fairly significant modifications beginning in 1960. For the first time the discipline of aircraft stability and control was separated out of aerodynamics and taught as a separate course in the 1960-61 school year. Also in that year the two aircraft power plant courses were dropped, and a single Internal Aerodynamics course, dealing essentially with jet propulsion, was substituted. Interestingly, Internal Aerodynamics was one of three electives that could be chosen, along with Structural Dynamics and another structures course, meaning that no aircraft propulsion course was *required*, a situation which continued for several years. Graduate courses in Applied Elasticity, Advanced Strength of Materials, and Elastic Stability were added.

In 1962 the department's name was changed to the Department of Aerospace Engineering. All courses now carried an "AsE" prefix (now "ASE"). The name change reflected the addition of two new courses related to space: Space Mechanics, and Propulsion Systems, which dealt primarily with rockets. Both of these new courses were taught by Hester. In 1959, Associate Professor Walter Carnes joined the ASE faculty. Carnes, who had received his M.S. from Georgia Tech, left in 1961 for the University of Illinois, where he received his Ph. D. in Mechanics in

1963, and then returned to MSU. In the 1964-65 school year, the courses of statics, dynamics, mechanics of materials, fluid mechanics, and vibrations were listed for the first time as engineering mechanics (EM) courses, separate from any particular department. There was some thought at the time of establishing a separate EM department, but that did not take place, although the separate EM courses would remain. The 1964-65 school year was significant also in that the School of Engineering had become the College of Engineering of Mississippi State University. The ASE department began offering a two-semester senior design sequence, but that would be short-lived.

When Charlie Cliett became department head in 1960, he was soon confronted with the difficulty of recruiting faculty with advanced degrees to Mississippi State, which had become a university in 1957. Cliett realized that the department's best chance of obtaining talent was to identify promising graduates and graduate students, hire them as lecturers for a year or two, and then send them off to obtain their Ph. D. degrees, supplementing their stipends and promising them jobs when they finished. Cliett was able to recruit a few notable faculty members in this way. His summation of the situation was "we have to grow our own."

The 1960s saw the growth of the experimental facilities of the ASE department. In 1963, A. George Bennett, Jr. returned to MSU after completing his M.S. at the University of Illinois. Bennett had acquired his B.S. in ASE from MSU in 1959 and had gone to work for Douglas Aircraft in Long Beach, CA. Cliett recruited Bennett to get his M.S. and then come to MSU as an assistant professor. When Bennett arrived at MSU, his first assignment was to oversee the assembly of a large supersonic tunnel that had been donated to ASE. This was a blow-down tunnel with a 9 in by 18 in test section that had been used in a structures group at NASA Langley Research Center for flutter testing. The components included a settling chamber, nozzle blocks for Mach numbers between 1 and 5, and a vacuum sphere with a capacity of 5,500 ft^3. The only component missing was a high-pressure tank, and this was constructed by Babcock and Wilcox, the steam boiler manufacturers, who had a plant in West Point, Miss.. The new tunnel was located in Patterson Laboratories, with the tank standing just outside the back door.

The supersonic tunnel took its place alongside another new facility in Patterson Lab, a high-speed water tunnel facility designed and built by Aerophysics graduate student Graham Wells[9]. Wells toured 19 different water tunnel facilities in North America and Europe before building the MSU tunnel, constructing it from fiberglass. It had an 11-inch diameter test section, a 13.2:1 contraction ratio, and was designed for a top test section speed of 75 ft/s. Because of the power requirements for such a large speed, the first power plant for the water tunnel was a 250-HP Buick V-8 car engine. The tunnel may have been one of the few in the world with a gear shift, mounted on a tunnel support just downstream of the test section. Subsequent modifications included replacing the V-8 with an electric motor that reduced the top test section speed to 50 ft/s.

Another of Cliett's early recruits made a significant impact on MSU's aeronautical engineering program almost immediately on his return. David Murphree had worked as an undergraduate for the Raspet group until he completed his B.S. in AE in 1960. When he left to get his graduate degrees, Cliett promised him a job if he came back. Murphree completed his Ph. D. in physics at the University of Wisconsin in 1965 and did return to MSU. His presence was felt almost immediately, as new courses in Fundamentals of Plasma Dynamics, Hypersonic Aerodynamics, and Magnetoplasmadynamics sprang up almost overnight. These courses reflected Murphree's interest in the subject of re-entry flows, as did one of his early research projects. In 1966 Murphree received a National Science Foundation (NSF) Research Initiation Grant for the study of plasmas and aerodynamic processes occurring in an arc-driven, hypersonic plasma tunnel, along with the development of the necessary instrumentation[10]. According to the first annual report for this grant, the work was divided into six areas, with the first three being related to the assembly of a Mach 12 plasma tunnel system that had been donated to the department. This was a small hypersonic facility that had a 7-inch diameter, 10-inch long test section that was to be used in general to study problems of hypersonic aerodynamics, and in particular to study variable-area diffusers for use in Mach 12 flow[11]. This facility made use of the vacuum sphere that had come with the supersonic wind tunnel. The other three areas covered by the NSF grant were related to fundamental properties of plasma flows, including the determination of the velocities of such flows. Research in the hypersonic tunnel did not proceed very far, but the work done in preparing for its use, both in hardware and in theoretical development, would pay off significantly in a few short years.

Murphree was not the only new faculty member with new courses. In 1961 Joe F. Thompson received his B.S. in physics from MSU. In 1963 he completed an M.S. in ASE, working at RFRL under Cornish on the problem of turbulent vortex structure. After working on the Apollo program for NASA in Huntsville for just over a year, he was hired as an assistant professor in ASE in 1964, the third of Cliett's recruits. Thompson introduced two graduate courses for the 1966-67 school year: Compressible Viscous Flow, and Aerothermochemistry, which was essentially a course in combustion.

The 1960s saw a number of important additions to the ASE curriculum. A graduate course in Advanced Guidance and Control of Aerospace Vehicles had been added for the 1965-66 school year. In the 1968-69 school year, first-semester juniors were required to take IE 1111 Computer Programming – the first appearance of FORTRAN in the ASE curriculum. The three-hour laboratory course was taught for the year by the Industrial Engineering (IE) department. The next year it was moved to a new Department of Computer Science and was given the designation CS 1011, a number it carried for several years. It was also moved to the first semester of the sophomore year in ASE. The 1968-69 school year also saw the renaming of Space Mechanics to Astrodynamics I and the addition of a new graduate-level Astrodynamics II course. In the 1969-70 school year a course in Fundamentals of Rotary Wing Analysis was taught for the first time. The establishment of this course paralleled research on helicopter aerodynamics being conducted at RFRL, primarily under Donald Boatwright. Boatwright made helicopter propeller downwash measurements that are still used in standard helicopter texts today[12, 13].

Tailspin and a Calculated Recovery

As can be seen from the history of the departments to this point, Aeronautical/Aerospace Engineering and Aerophysics had separate beginnings. During the early years of Aerophysics, the relationship between the two entities was fairly informal. Aeronautical engineering would supply faculty to assist with Aerophysics research as needs arose, and graduate students in aeronautical engineering would do their research on Aerophysics projects. As noted above, Edwards as head of AE had helped out Aerophysics financially from time to time. In the late 1950s, steps were taken to make the relationship between the two units more formal. In 1959, Raspet, Cornish and Bryant were named research professors of aeronautical engineering (assistant research professors in the cases of Cornish and Bryant), and Edwards, Cliett and Hester were named associates of the Aerophysics staff (Barnett, p. 143). Upon Raspet's death in 1960, Cornish was named director of the flight laboratory and then head of Aerophysics upon the completion of his Ph. D.. However, Cornish left in 1964 to begin a long and distinguished career with Lockheed. The directorship of the RFRL fell to Sean Roberts, who was acting head of Aerophysics until 1967, when Aerophysics and ASE were merged into a single department, the Department of Aerophysics and Aerospace

Engineering. Cliett was head of the combined department, and Roberts remained as director of RFRL and Research Professor of Aerophysics and Aerospace Engineering.

The year 1969 was a high point in aerospace engineering around the country. Armstrong and Aldrin walked on the moon. Unfortunately, the decline in aerospace engineering after that peak, both nationally and at MSU, was extremely sharp. As noted earlier, the redirection of research funding by the military ended the *MARVEL* project in 1970. The end of the Apollo program in 1971 marked the beginning of a massive decline in the aerospace industry, a decline reflected in ASE enrollment at MSU and around the country. The year 1971 also saw the resignation of Sean Roberts as director of RFRL. Thus both ASE and RFRL found themselves at perhaps the lowest points in their histories. By this point the ASE faculty had grown rather large, too large to be supported by the student enrollment in ASE at the time. RFRL had little external support and certainly not enough to maintain its operation. Cliett convened the ASE faculty and presented them with two options: begin generating funded research or see the ASE department reduced greatly in size, if not eliminated altogether. The faculty eventually decided to concentrate on three areas: general aviation research to be carried on at RFRL; magnetohydrodynamics (MHD), a promising area being pursued by the group headed by David Murphree, and a relatively new area – computational fluid dynamics, or CFD, under the direction of Joe Thompson. (Haug, p. 122).

Thompson had returned to MSU after completing a Ph. D. in CFD at Georgia Tech in 1970. Thompson's efforts soon paid off. In 1974, Thompson and ASE graduate student Frank C. Thames, along with MSU mathematics professor C. Wayne Mastin, published the first work in the area that would put MSU on the CFD map – numerical grid generation[14]. Thompson and the others discovered a method for distributing the grid points in a boundary-conforming grid that made use of the solution of an elliptic partial differential equation. This was a major breakthrough in CFD technology and would lead to the publication of the first textbook in the area of numerical grid generation[15], co-authored by Thompson, Mastin, and Z. U. A. Warsi, a graduate of the University of Lucknow in India who joined the ASE faculty in 1971 after working for four years as a researcher at RFRL. Warsi would become one of the most prolific authors in the department and would eventually publish a graduate-level textbook on fluid mechanics[16]. The breakthrough in numerical grid generation would also lead to numerous short courses and other conferences that would draw attention to the CFD work being done at MSU and would attract other researchers in the area. The first course in Numerical Fluid Mechanics at MSU was offered by ASE in the 1973-1974 school year.

Prior to Thompson's return, another of Cliett's recruits, George Bennett, came back to MSU in the fall of 1969 after completing his Ph. D. at the University of Illinois. At Illinois Bennett had been a classmate of Bob Liebeck, who was doing his dissertation work on inverse-design techniques for high-lift airfoil sections. Shortly after his return to MSU, Bennett began a course on Applied Airfoil and Wing Theory, which made use of the methods developed by Liebeck and others for airfoil and wing design in incompressible, subsonic, and supersonic flows. Bennett also instituted a course in Advanced Performance which dealt with advanced aircraft configurations and advanced performance analysis techniques.

The last of Cliett's recruits returned in 1969. John C. McWhorter, III had received his B.S. in mechanical engineering from MSU in 1964. He had then obtained an M.S. in ASE in 1965, performing wind tunnel and flight research under Sean Roberts. After teaching engineering mechanics courses at MSU for a couple of years, McWhorter left to obtain his Ph. D. in Mechanics at the University of Illinois, in the same program where Carnes had obtained his Ph. D. a few years before. When McWhorter returned to MSU, he was listed in the catalog as a member of the ASE faculty, but his individual listing in the catalog referred to him as Assistant Professor of Engineering Mechanics. McWhorter taught mechanics courses and ASE structures courses for several years and in August of 1979 became coordinator of the engineering mechanics (EM) program at MSU. ASE and EM had always been connected, with the direction of the EM program being housed in ASE. In the 1970s, however, Cliett took steps to make the connection more formal, giving ASE primary responsibility for teaching the EM courses and thus providing ASE with a larger enrollment base. This was another example of Cliett's skill in protecting ASE during the dark days of the 1970s.

That engineering students shall take thermodynamics is a maxim of almost biblical proportions, and ASE students were no exception. For many years ASE students had taken the ME thermodynamics course. However, the long-recognized fact that most ASE graduates did not need a working knowledge of the steam tables, but many did need a deeper understanding of compressible flow, was acted upon in the mid-1970s. In the 1975-76 school year, a new course called Gas Dynamics was introduced, replacing the ME thermodynamics course. The content of Gas Dynamics was very similar to the thermodynamics course just removed, and was in fact about four-fifths classical thermodynamics and one-fifth compressible flow. The new topics were an introduction to isentropic compressible flow, flow through converging-diverging nozzles, and normal shocks, preparing the students for more advanced studies later in the curriculum when they took Compressible Aerodynamics. The Gas Dynamics course is significant in that with its adoption, the ASE curriculum took on the form that it would keep for close to twenty years, with very few changes. Students started with two years of calculus, two semesters of chemistry, two semesters of engineering graphics, and three semesters of physics. They took the engineering mechanics courses of Statics, Dynamics, Mechanics of Materials, Fluid Mechanics, and Vibrations, along with Gas Dynamics and the FORTRAN course discussed earlier. These were followed (or in some cases paralleled) by a three-course sequence in flight mechanics (Performance, Static Stability and Control, Dynamic Stability and Control), a three-course sequence in aerodynamics and propulsion (Incompressible Aerodynamics, Compressible Aerodynamics, and Propulsion), and a three-course sequence in structures. Two lab courses and a one-semester Systems Design course, along with three technical electives, an ME heat transfer course, a couple of math courses and a number of humanities courses, completed the ASE undergraduate degree requirements.

David Murphree's efforts in developing expertise in high-energy flows paid off in the mid-1970s. In 1976, Mississippi State University was selected by the federal Energy Research and Development Administration, a product of the Arab oil embargo of the early 1970s, as a research center for magnetohydrodynamic (MHD) technology, along with the University of Tennessee Space Institute, the Massachusetts Institute of Technology, and Stanford University. Early on the MSU group discovered its niche when Murphree and W. Steve Shepard, a 1967 addition to the ASE faculty, realized that the best way to determine the composition of the high-temperature, high-energy gases in the MHD facilities was to use laser spectroscopy as a non-intrusive diagnostic tool. The development of instrumentation for studying MHD flows became the focus of efforts at MSU. By 1980, the U. S. Department of Energy had provided over $4 million in funding for MHD research, the largest amount received to that time for a research project. In 1980, the MHD Energy Center was organized as a separate unit that reported directly to the director of EIRS. Murphree and Shepard became the director and associate director, respectively, effective January 1, 1980. All of the employees in the ASE department who were being funded by the DOE contract for MHD work were transferred to the MHD Energy Center. In 1985, Murphree left the university to found the Institute for Technology Development, a state-wide effort that was funded by a $20 million grant from the state of Mississippi. Shepard was named Director of the

MHD Center. Later the MHD Center at MSU would change its name to the Diagnostic Instrumentation and Analysis Laboratory, or DIAL (Haug, p. 123; McKee, p. 144).

In 1978, one other significant change in the ASE program took place. Since the merger of Aerophysics with Aerospace Engineering in 1967, the department had been known as the Department of Aerophysics and Aerospace Engineering. In 1978, department head Cliett wrote a letter to university president James McComas, petitioning that the name be changed to the Department of Aerospace Engineering. Cliett pointed out that RFRL had acquired a national reputation of its own, without a direct tie to "Aerophysics" as such. The one-page letter, which essentially served as the application and approval form, carried the signatures of Cliett, Willie L. McDaniel, director of EIRS; Harry Simrall, Dean of Engineering; J. C. McKee, Vice-President for Research and Graduate Studies; Robert E. Wolverton, Vice-President for Academic Affairs; and James D. McComas, President of the university. The change took effect July 1, 1978. While the RFRL would certainly perpetuate the memory of August Raspet in a manner befitting his efforts, the last vestige of the separate department that Dean Flinsch had created so that Raspet could conduct his research unimpeded had been removed.

The *MARVEL* Becomes a Phoenix, and the Computational Feature

In 1974, Ernest J. Cross, who had acquired his Ph. D. from the University of Texas in 1968 and worked at the U. S. Air Force laboratories at Wright-Patterson AFB, became director of RFRL upon his retirement from the Air Force. During his tenure as director, RFRL was involved in tip vortex research for NASA using a Cessna AgWagon and in cooling drag measurements on a Navy T-34B trainer, among other projects related to general aviation. After Cross left to become head of the Department of Aerospace Engineering at Texas A&M University in 1979, he was succeeded as director by George Bennett (McKee, p. 109). Bennett oversaw a new study of BLC performed for Lockheed. The effects of spanwise blowing on boundary layer separation were studied on a Caproni sailplane equipped with a small jet engine that extended its time aloft. The blowing increased the maximum lift coefficient at high angles of attack. The glider was also used for laminar flow research using a "glove" in the shape of a natural laminar flow airfoil attached to the wing of the glider (Haug, p. 125).

Following completion of its flight testing in 1970, the *MARVEL* had been placed in storage. In 1981, however, it was returned to service when RFRL received a contract from Brico Limited to perform research on a desert utility observation aircraft. The requirements were that the aircraft be able to take off from sand and possess low infra-red and radar signatures. The *MARVEL* seemed to be an ideal candidate, having been designed as an STOL aircraft. However, some significant modifications were required. Because of the BLC and variable camber technology applied to the wing, there was no room for fuel inside the wing, and so as originally constructed the airplane's fuel tank was small, thus limiting the range of the aircraft – and was located under the pilot's seat. It was also recognized that suction BLC would not work very well in a sandy climate. For these reasons a completely new wing was designed and built in a very short time. The spar was constructed from fiberglass in New Mexico, the wing skins were built of Kevlar in Florida, and the wing was assembled at RFRL. The airplane was tested in Saudi Arabia in 1982, where it was determined that the landing gear design was not appropriate for sand operations.

Even though the *MARVEL* test was not completely successful, it opened the door for new efforts at RFRL. From the beginning RFRL had been involved in constructing structural components from fiberglass. With each new project, expertise in this area grew. After Bennett became director of RFRL, efforts in the area of fiberglass construction and the more general area of composites were increased. In particular, the work concentrated on rapid prototyping, which consisted of constructing molds using computer-controlled milling machines and then using the molds to fabricate components from composite materials such as fiberglass. These efforts were rewarded in 1986 when the Honda Research and Development Corporation began a program at RFRL to develop a turbojet-powered aircraft constructed completely from composite materials. The project began with the replacement of the horizontal stabilizer and rudder on a Beechcraft A-36 with new components built from composite materials. Later the main wings were also replaced. Finally, a completely new aircraft with twin turbojet engines mounted on top of a high-wing configuration was completed at the end of 1992 and flight-tested in 1993. The airplane was constructed in a new building built by Honda at RFRL and called the Honda Annex. At the end of the testing program, the building and a Rockwell Turbo Commander twin turboprop, used as a chase plane for the Honda projects, were turned over to the university. By the end of their 10 years at MSU, Honda had been the source for over $16 million in research and development funds. During the years of the Honda project, work at RFRL on rapid prototyping continued and efforts at nondestructive testing and evaluation of composite structures using acoustic techniques were initiated. The composite materials fabrication capabilities were exercised in an unusual fashion in 1991, when the senior design class designed and built a one-third replica of the National Aerospace Plane, or NASP. The 30-foot long model was completed in one semester, on the day it was scheduled to be shipped to Oshkosh for its premier.

Joe Thompson's response to Cliett's challenge would also be rewarded. In January of 1981, David Whitfield joined the CFD efforts at MSU. Whitfield was a noted CFD researcher who came to MSU from the Arnold Engineering Development Corporation (AEDC) at Tullahoma, TN. While research in CFD was going on in ASE, work in designing integrated circuits for specific applications was being done in the Department of Electrical Engineering at MSU, under the direction of Donald Trotter. These two efforts were merged into one when in 1987 the Air Force Office of Scientific Research provided $5 million to establish the Research Center for Advanced Scientific Computing at MSU (Haug, p. 124). The purpose of RCASC was twofold: the CFD group would develop codes to solve the equations of fluid mechanics, and the digital circuits group would develop computer chips that were "hard-wired" to run the CFD codes. RCASC as a separate entity would be short-lived, however. In 1988 Thompson, Whitfield and Trotter submitted a proposal for a National Science Foundation Engineering Research Center. The decision to fund the center was announced in January of 1990, and the center became the NSF Engineering Research Center for Computational Field Simulation, one of only 19 similar institutions around the United States. The new center was dedicated on May 10, 1991, with Joe Thompson as the founding director (Haug, p. 125).

The standard life cycle of an NSF ERC was 11 years. After the end of that period, the Center was to become a stand-alone operation, no longer dependent on NSF funding. During its 11 years under the NSF umbrella, MSU's ERC saw an enormous increase in CFD work. Bharat Soni, a noted researcher in the area of unstructured grid solutions, joined the ASE faculty as an ERC researcher. Dave Whitfield established the Simulation Center, a group within the ERC that developed new flow solvers and applied them to complicated geometries. Much of the SimCenter's work was funded by the Office of Naval Research. Work on very large scale integrated circuits (VLSI circuits) also continued, although breakthroughs in parallel processing technology meant that chips hard-wired to solve fluid mechanics problems were no longer necessary. Other work, particularly research in scientific visualization and the graphic display of data, became important components of the ERC's efforts. Periodic reviews of the ERC by the NSF regularly commented on the quality of the work done at the ERC and how well the ERC was operated.

Changing of the Guard

In 1991, two veterans of the ASE department, Charles Cliett and Leslie Hester, retired. At the time of his retirement, Cliett had served the department for almost 43 years, 31 of them as department head. He had certainly left his mark on the department. Three of his "recruits" had become directors of research units: Murphree, who was the founding director of the MHD Center, later DIAL; Bennett, who became head of RFRL; and Thompson, founding director of the ERC. The fourth recruit, McWhorter, became interim department head upon Cliett's retirement and was later named department head. There is no doubt that Cliett had an eye for talent, and the dramatic increase in the research output of the department, reflected in the growth of work at RFRL and the establishment of the MHD Center / DIAL and the NSF ERC, demonstrate well his abilities of motivation. Hester's contributions had been numerous, and the continued operations of the various ASE facilities in Patterson Laboratories were due in very large part to his efforts.

After a successful ABET accreditation visit in the spring of 1993, McWhorter embarked on a major revision of the ASE curriculum. As discussed before, the curriculum had really not seen much modification since the mid-1970s. The review of the program was motivated in large part by the downturn in ASE enrollments that began in the early 1990s and rivaled the downturn of the 1970s in its severity. The magnitude of the decline can be understood by a look at the enrollments in Flight Mechanics I – Aircraft Performance. This course was the first ASE course ASE students took, in the second semester of the sophomore year. In 1992 enrollment in Flight I peaked at 50 students. The enrollments for the following years tell the story: 1993, 25 students; 1994, 12 students; 1995; 6 students. McWhorter and the faculty realized that dramatic steps needed to be taken to make the program attractive to students both in Mississippi and around the country, if it were to survive. It was recognized that the department possessed unique resources in RFRL and the ERC, and that it needed to figure out ways to incorporate these resources into the curriculum.

The faculty began looking closely at both individual courses and the curriculum as a whole. One deficiency that had been noted for a number of years was the lack of contact with students early in their programs. As just stated, the first ASE course was not taken until the second semester of the sophomore year. The necessity of establishing a relationship with students early on was recognized as a key to student retention. Therefore a three-semester Introduction to Aerospace Engineering sequence, beginning in the first semester of the freshman year, was established. This course had multiple purposes. In addition to introducing students to basic concepts of aerospace engineering, it was also designed to teach basic computer skills, such as word processing and spreadsheets, use of mathematical toolkits such as Mathcad, and proficiency in Unigraphics, a computer solid modeling and analysis program. This sequence thus replaced the traditional Engineering Graphics classes that had been in the curriculum since the beginning (the FORTRAN course had been dropped as a requirement a few years earlier). In order to make room for the third semester, the course in Performance was dropped, but the material was included in the freshman sequence. Other courses were revised to make use of the skills acquired by the students in the freshman sequence. Another course that was dropped was the ME heat transfer course that had also been part of the curriculum for a number of years. This course was removed to decrease the total number of hours required for graduation. The Gas Dynamics course was modified to include an introduction to heat transfer and was renamed Aerothermodynamics. The placement of other courses in the curriculum was changed, with both the structures sequence and the laboratory course sequence beginning a semester earlier. Finally, a new two-semester design course was put together and will be taught for the first time in the fall of 2003. In addition, new recruiting efforts were begun. The *MARVEL* was pressed into service once again, this time as a traveling static display and attention-getter that was carried all the way to the annual fly-in at Oshkosh. Along the way, ASE enrollments began to recover, with approximately 25 students currently in the junior level courses, the largest such enrollments since 1992.

The end of the 1990s and the beginning of the 21st century saw further turnovers in personnel. In 1997 Shepard retired as the director of DIAL, around the time of the completion of its new 55,000 ft^2 building, funded by $8 million in grants from the Department of Energy and the state of Mississippi. At the time of his retirement, MHD/DIAL had received over $70 million in research funding since its inception in 1976. RFRL celebrated its 50th anniversary in 1999. Bennett retired as director of RFRL and ASE faculty member in 2001 and was succeeded by David Lawrence, formerly CEO of Tracor Flight Systems and a former associate director of RFRL. In the summer of 2002 the department was racked by a number of departures. Bharat Soni left the ERC to pursue opportunities at the University of Alabama at Birmingham. Dave Whitfield, along with ME faculty member Roger Briley and a number of research personnel from the ERC, left to start a new simulation center at the University of Tennessee at Chattanooga. Four other ASE faculty members were offered positions at UTC, but chose to remain at MSU. A major impact of the departures of Whitfield and Soni from ASE was financial, since both had brought in significant research funding and had also provided significant released-salary and overhead funds. Then, also in the summer of 2002, MSU offered an early retirement option, and McWhorter and Warsi chose to exercise this option. Because of budget cutbacks at the university, these faculty members could not be replaced immediately. In very short order ASE had lost four faculty members and was in significant danger of losing the positions as well. Boyd Gatlin, who had joined the ASE faculty after completing his Ph. D. in CFD in 1987 under Dave Whitfield, became the interim department head. Since being named interim head, Gatlin has instituted a number of efforts aimed at preserving the ASE department. A search for a new permanent department head is currently underway.

Conclusions

The history of aerospace engineering at Mississippi State University has seemingly followed a cycle of crisis and exertions to survive the crisis that has appeared to repeat itself every ten years or so. There were the initial struggles to survive in the early 1930s, followed by the difficulties of the war years in the mid-1940s. The mid-1950s saw the problems in funding faced by Raspet with the university administration. The cycle then shifted slightly, with the next crisis coming in 1960 with the deaths of Raspet and Edwards. The early 1970s saw the near-demise of RFRL and the department because of declines in research funding and in enrollment, followed by a similar crisis in the early 1990s. Each time the department has responded with renewed effort, rising to the challenges and overcoming the obstacles. The first years of the 21st century have brought their own challenges, with significant faculty retirements and departures and the lack of the immediate ability to replace all who have left. It is to be hoped that those of us who currently comprise the ASE faculty at Mississippi State University will respond as our forerunners did, with equal amounts of imagination, inspiration, and determination.

Note on Sources

In 1992 C. James Haug wrote a history of the College of Engineering at MSU for its centennial celebration[1]. In 2001, Cory Barnett wrote a master's thesis in history on the first 27 years of the ASE department[2]. In 2002, Chester McKee, a former MSU electrical engineering faculty member and the first Vice-President for Research at MSU, published a second history of engineering at MSU that consisted primarily of his personal recollections[3]. These three works provided significant contributions to this article. In an attempt to give these authors reasonable credit for their works and to also keep the reference list manageable, the author has made use of the author/page citation method in the text for these three works. All of the rest of the references are cited in standard endnote format. The author would like to thank the current and former members of the ASE faculty for their assistance in providing information and in reviewing the text, in particular Charles Cliett, George Bennett, Joe Thompson, and John McWhorter. The author was privileged to study under these individuals in the early 1980s and then to serve alongside them beginning in 1993, and is grateful for their contributions to his professional development.

References

[1]Haug, C. J., *The Mechanical Feature: 100 Years of Engineering at Mississippi State University*, Jackson, MS: University Press of Mississippi, 1992.

[2]Barnett, J. C., "Building Rockets on the Moon: A History of the Department of Aerospace Engineering at Mississippi State University, 1933-1960," M. A. Thesis, Department of History, Mississippi State University, December 2001.

[3]McKee, J. C., *From Slide Rules to PC's: 10^2 Years of the College of Engineering at Mississippi State University*, published by the author, 2002.

[4]Huntington, W. S., "The Design and Construction of a Low-Cost Intermittent-Type Supersonic Wind Tunnel at Mississippi State College," M. S. Thesis, Dept. of Aeronautical Engineering, Mississippi State College, January 1951.

[5]Puckett, A., "Supersonic Nozzle Design," *J. Applied Mech.*, vol. 13, no. 4 (Dec. 1946), pp. A-265-A-270.

[6]Pope, A., *Aerodynamics of Supersonic Flight: An Introduction*, New York: Pitman Publishing Corporation, 1950.

[7]Liepmann, H. and Puckett, A., *Aerodynamics of a Compressible Fluid*, New York: John Wiley & Sons, 1947.

[8]Thwaites, B., ed., *Incompressible Aerodynamics*, New York: Oxford University Press, 1960, pp. 231-232.

[9]Wells, W. G., "Design Principles for a High Speed Fiberglass Water Tunnel," Mississippi State University Department of Aerophysics Research Report No. 51, April 1964.

[10]Murphree, D. L., "Research Initiation – Variable Area Diffusers for Mach 12 Flow: July 1, 1966 – June 30, 1967," National Science Foundation Research Initiation Grant GK-904, Annual Report No. 1, Department of Aerophysics and Aerospace Engineering, Mississippi State University.

[11]Murphree, D. L., and Williford, J. R., "Description and Predicted Performance of the Hypersonic Plasma Tunnel Facility at Mississippi State University," Department of Aerophysics and Aerospace Engineering Research Report No. 82, October 1967.

[12]Boatwright, D., "Measurements of Velocity Components in the Wake of a Full-Scale Helicopter Rotor in Hover," U. S. Army Air Mobility Research and Development Laboratory Technical Report No. 72-33, August 1972 (also Mississippi State University Aerophysics and Aerospace Engineering Report No. 72-59).

[13]Prouty, R. W., *Helicopter Performance, Stability, and Control*, Malabar, Florida: Robert E. Krieger Publishing Company, 1990.

[14]Thompson, J. F., Thames, F. C., and Mastin, C. W., "Automatic Numerical Generation of Body-Fitted Curvilinear Coordinate System for Field Containing Any Number of Arbitrary Two-Dimensional Bodies," *J. Computational Physics*, vol. 15, no. 3, July 1974, pp. 299-319.

[15]Thompson, J. F., Warsi, Z. U. A., and Mastin, C. W., *Numerical Grid Generation: Foundations and Applications*, North-Holland, 1985.

[16]Warsi, Z. U. A., *Fluid Dynamics: Theoretical and Computational Approaches*, CRC Press, 1993; 2nd ed., 1998.

FIGURES

Fig. 1 Kenneth Withington, first head of Aeronatical Engineering, 1933-1946

Fig. 2 August Raspet

Fig. 4 *Marvellette* STOL Research

Fig. 3 L-19 Boundary Layer Control Research Aircraft

Fig. 5 The *MARVEL*

Fig. 6 Raspet Flight Research

Fig. 7 Charles B. Cliett, Department Head. 1960-1991

Fig. 8 George Bennett (left) oversees assembly of supersonic wind tunnel, ca. 1964

Fig. 9 Students working in the subsonic wind tunnel in the 1970s

Fig. 10 The Honda Jet, with the Honda Annex in the background

Fig. 11 The Honda Jet in flight

Fig. 12 MSU Engineering Research Center

Fig. 13 Diagnostic Instrumentation and Analysis Laboratory (DIAL)

Chapter 26

Dreams of Flight Fueled Aeronautic Advances at Catholic University of America[1]

Catherine Lee, Senior Writer/Editor
Office of Public Affairs, The Catholic University of America

In 1901, two years before Wilbur and Orville Wright made history in North Carolina by flying their spindly aircraft a distance of 121 feet, an associate professor of mechanics in Washington, D.C., set out to make his own dream of manned flight come true.

With financial help from a businessman, Professor Albert Francis Zahm set up the first wind tunnel at a U.S. college or university. The wind tunnel — housed in a one-story wooden shed with tall windows and a metal roof — became the hub of an innovative aerodynamics research laboratory.

The wind-tunnel apparatus is long gone, but the building that housed it still stands at The Catholic University of America, where the professor taught from 1895 to 1908. The building — originally erected behind McMahon Hall and later moved to a spot behind the university's power plant on John McCormack Road — is a reminder of both Zahm's pioneering work in aeronautical engineering and CUA's role in an exhilarating era of U.S. aviation history.

As the United States commemorates the 100th anniversary of flight on Dec. 17, Catholic University also celebrates its significant contributions to the field of aeronautics. Zahm, who corresponded with the Wright brothers by letter around the time of their historic experiments, was the first in a succession of distinguished CUA professors and graduates whose work has enhanced the field of aviation.

Zahm's initially cordial relationship with the Wright brothers would turn acrimonious after the professor sided with aviation pioneer Glenn Curtiss in Curtiss' patent dispute with the Wrights. The dispute, which centered on the airplane control system developed by the Wrights, eventually overshadowed Zahm's achievements — an unfortunate development, given the professor's contributions to the field of aeronautics, says author and former U.S. Air Force historian Richard P. Hallion

"Albert Zahm was a significant pioneer in the field of aeronautics," says Hallion, author of the book *Taking Flight: Inventing the Aerial Age From Antiquity Through the First World War.* Hallion adds that "for a time, Catholic University was the gold standard in aeronautical research."

At left from top: The building that housed Zahm's wind tunnel was built with financial help from entrepreneur and aviation enthusiast Hugo Mattullath. The wooden wind tunnel was 40 feet long by 6 feet wide and tall. Hanging in Zahm's laboratory were devices of different shapes used to determine the best configuration for an aircraft's hull.

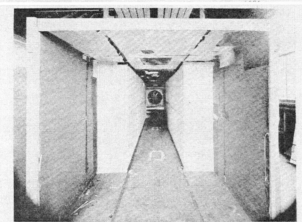

2. ZAHM'S TUNNEL WITH INTAKE CONE REMOVED

With his CUA wind-tunnel experiments noted by newspapers around the country, Zahm became a leading aviation expert in the United States. In 1912, he proposed that the government set up a national aeronautics laboratory. His report on European aeronautical research facilities influenced Charles D. Walcott, then secretary of the Smithsonian Institution, to throw his support behind formation of the National Advisory Committee for Aeronautics — the precursor of NASA. In 1917, NACA founded the Langley Memorial Aeronautical Laboratory in Hampton, Va. — still one of NASA's leading aerospace research centers.

In 1935, CUA established its Department of Aeronautical Engineering, which was part of the School of Engineering and Architecture. The department's first chair was Louis H. Crook, an affable inventor and a protégé of Zahm. When Crook died in 1952, German-born researcher Max M. Munk — internationally respected for his work in aerodynamics — took over the department.

BALANCE AND GROUP OF TEST MODELS USED IN ZAHM'S TUNNEL

At the department's peak in the year 1950, the university awarded bachelor's degrees in aeronautical engineering to 37 CUA graduates who left the university to take jobs with NASA, the military and the aviation industry. By 1960, however, CUA had merged the departments of mechanical and aeronautical engineering. At that time, most of the aeronautical engineering faculty members transferred to the physics and mechanical engineering departments.

But CUA's interest in flight didn't disappear. Responding to developments in the U.S. space program, the university created master's and doctoral programs in aerospace engineering in 1960. From then until the early 1970s, when CUA disbanded the program because of funding cuts, the program was popular with NASA employees who could take courses part time at the university.

In 1996, with ties to NASA well established, the CUA physics department set up the Institute of Astrophysics and Computational Science at the space agency's Goddard Space Flight Center in Greenbelt, Md. The institute develops research and educational programs and promotes closer cooperation between CUA and government agencies and industry. The relationship between NASA and the university has led to the hiring of a significant number of CUA graduates for space agency jobs. Several of those graduates have had distinguished careers at NASA.

Daniel R. Mulville, who earned his CUA doctorate in 1974, retired last February as NASA's second in command after 16 years with the space agency. CUA honored him in 2001 with an Alumni Achievement Award. Colleen Hartman, who earned her doctoral and master's degrees at CUA, was director of NASA's Solar System Exploration Division before leaving in October to become assistant administrator of the National Oceanic and Atmospheric Administration's National Environmental Satellite, Data and Information Service.

Gwendolyn Brown, who earned a CUA bachelor's degree in 1987, has been nominated by President George W. Bush to be chief financial officer for NASA, where she has served as deputy chief financial officer for financial management.

Portrait of the Scientist as a Young Man

Albert F. Zahm was born on Jan. 5, 1862, in New Lexington, Ohio. He came from a religious family; his older brother, John, was a priest who studied and taught at the University of Notre Dame. Following in John's footsteps, Albert enrolled at Notre Dame in 1879 as a classics major but quickly became absorbed in the pursuit of science. He earned his bachelor's and master's degrees at Notre Dame and taught there in the 1880s and early 1890s.

Anecdotes about Zahm indicate his interest in discerning the mechanics of flight. At Notre Dame, he once suspended a cable between two buildings and dangled a brave student volunteer wearing a crude set of wings from the line, according to a 1953 *New York Times* story. In the article, Zahm described a Greek class at Notre Dame that fueled his interest in flight: "... the professor told the story of Daedalus and Icarus, the two mythological characters who fashioned wings for themselves so they could fly. I decided then and there to find a method of flight."

In 1893 — the same year that Zahm started work on his doctorate in physics at Johns Hopkins University — he and Octave Chanute, noted engineer and aviation enthusiast, helped to organize the first International Conference on Aerial Navigation, held in Chicago the first four days of August. Historian Hallion says that the conference was critical because "it brought together leading people in aeronautics at a time when the momentum in aeronautics was shifting from Europe to America."

A diary kept by Zahm in the months prior to the conference indicates that he could be disdainful of other researchers who didn't meet his standard of commitment to the new field of aviation. He noted in the diary — now part of the National Air and Space Museum Archives collection — that a gentleman from Norfolk, Va., had requested a conference admission card but didn't actually plan to attend the sessions. "He is evidently an imbecile," Zahm wrote in his diary.

Zahm's letters — part of his collection at the University of Notre Dame Archives — include correspondence between the scientist and Chanute from 1891 to 1909. It was a time when scientists were trying to understand the most rudimentary principles of flight and atmospheric resistance by watching birds in motion and measuring the effect of drag on objects dropped from the Eiffel Tower in Paris.

Zahm had been conducting experiments using birds since at least 1891. In a handwritten letter dated April 8, 1904, Chanute told Zahm that he was "chuckling over your proposal of borrowing a tame buzzard and towing him by the beak."

Major Developments in the Wind Tunnel

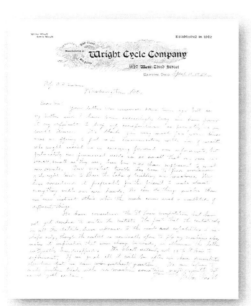

In handwritten letters such as this one, Wilbur Wright described the brothers' work to Zahm.

In 1895, when Zahm started working at Catholic University, "he represented a new kind of aeronautical researcher who emerged at the end of the 19th century," says Hallion.

"Zahm was a trained scientist who appreciated that research should be done in a systematic fashion," says the historian. "His wind tunnel was, in a sense, ground zero for American aeronautical research and development."

The wind tunnel and the building that housed it were constructed by the National Transit Co. in the winter of 1901 with funding from Hugo Mattullath, an inventor and businessman who had sought Zahm's services to conduct research related to the inventor's design of a large twin-hull seaplane.

Built to Zahm's specifications, the wooden wind tunnel was 40 feet long by 6 feet square with a 5-foot suction fan at the rear of the tunnel, according to a paper in the Notre Dame collection. The fan was driven by a 12-horsepower motor that drew air through it at any speed up to 25 miles an hour.

Zahm developed new instruments to measure air resistance in the wind tunnel. Taking exact measurements and carefully recording wind-tunnel data, Zahm reached pioneering conclusions about the role of air friction in flight and the optimum shape of an aircraft's hull.

In an Oct. 20, 1904, paper that is part of the Notre Dame collection, Chanute describes Zahm's work as "little short of revolutionary. Physicists will need to revise the formulae heretofore employed for computing the reactions of air in motion."

Zahm proved that air friction over a plane's surface — known as skin friction — was a major cause of drag. Prior to Zahm's experiments at CUA, scientists thought that skin friction did not contribute significantly to friction drag, the force that slows down a body in motion.

That work and Zahm's tests involving spindle and fish shapes led to the determination that the blunt torpedo shape for a plane's hull was the most efficient, an observation that was quickly adopted in the construction of new aircraft.

An Influential Figure

During his years in Washington, where he resided until a year before his death in 1954, Zahm lived at the Cosmos Club, Washington's social headquarters for the city's leading intelligentsia. By the early 1900s, he had established himself as a major figure at the club — a gathering place for leaders in academia, government and industry who were interested in aviation, according to historian Hallion.

With his CUA professorship and growing influence as an authority in aeronautics, Zahm was in a position to help inventors like the Wright brothers, then bicycle shop proprietors in Dayton, Ohio. In a letter to the professor dated April 10, 1904, Wilbur Wright alluded to Zahm's connections to influential people.

"We thank you very much for your kindness in offering to put us in communication with men of wealth who might assist us in carrying forward our experiments," Wright said in the note written on Wright Cycle Co. stationery, now in the Notre Dame collection. Wright politely declined Zahm's offer, saying that "our resources, small as they are, have been sufficient to meet our wants."

Written primarily by Wilber Wright, the letters to the professor were friendly and informative about work on the brother's "machines," as they referred to their planes. But in 1910, a letter from Wilbur hints at the root of the disagreement that eventually would sour the brothers' relationship with Zahm.

"Naturally we regret that you will be lined up against us even in a professional capacity as confidential advisor in the legal struggle," wrote Wilbur in another letter in the Notre Dame collection, dated Jan. 29, 1910. "But we do not think that such service carried out in a spirit of fairness need interrupt the friendship which has always existed between us."

Earlier Zahm reportedly had offered his services as a scientific adviser to the Wrights in their lawsuit against aviation pioneer and entrepreneur Curtiss, but the Wrights had declined his assistance. Zahm, in turn, offered his technical know-how to Curtiss.

Afterward, Zahm would be severely critical of the Wrights. But in 1951, nearing the end of his life and mindful of the historic significance of the Wrights' letters, Zahm sent them to a priest at Notre Dame for safekeeping, with a note penned in the shaky handwriting of an old man. A lifelong bachelor, Zahm died on July 23, 1954, at the age of 92.

A Colorful Successor

Professor Emeritus Gabriel Boehler, who taught at Catholic University from 1951 to 1998, and Thomas M. Clancy, who earned his CUA bachelor's and master's degrees in 1949 and 1950, are longtime friends. On a recent morning, they sit in the den of Boehler's Northwest Washington home reminiscing about CUA Professor Louis H. Crook, who continued Zahm's tradition of excellence in aeronautical engineering and dedicated his life to the university.

Left: Aeronautical engineering alumnus Thomas M. Clancy, left, and Professor Emeritus Gabriel Boehler peruse an old Cardinal yearbook. Right: Remembered as an inventor, Louis H. Crook was a protégé of Zahm's.

Boehler and Clancy recalled Crook as a tinkerer and an inventor. "He was always working on something," says Clancy, who lives in Bethesda, Md., and retired in 1994 as assistant director of system planning and analysis at the Office of Naval Technology in Arlington, Va.

Crook was born in Havre de Grace, Md., on June 16, 1887 — the same year Catholic University was founded. According to a 1949 *Washington Post* article, Crook spent much of his life at CUA where his father, Nicholas Crook, was hired as the university's general manager when the professor was a child.

After graduating from high school, he enrolled at CUA, where he earned his bachelor's degree in 1909. He started teaching at the university shortly after graduation.

By Crook's graduation day, Zahm had left the university to become chief engineer of the Curtiss Aircraft Co., but the two scientists continued their association. From 1917 to 1924, Crook worked part time as an assistant to Zahm, who by then was director of the Aerodynamical Laboratory for the U.S. Navy. (In 1924, Crook also managed to squeeze in an unbeaten season as coach of CUA's freshman football team.)

Together Crook and Zahm conducted research and wrote several critical publications, including *Airplane Stress Analysis*, the first complete textbook on the subject.

Appointed chair of the brand-new aeronautical engineering department in 1935, Crook had an office on the second floor of the two-story, red-brick aeronautical engineering building, which was torn down in the late 1950s to make way for the parking lot of Pangborn Hall. Alongside his office was a newer wind tunnel put together after Zahm's was dismantled, where students conducted experiments with model airplanes.

Crook also had put together a wind tunnel inside an old tobacco barn at his home in Bethesda, Md. Students and colleagues were welcome to stop by and observe or conduct experiments. Clancy, who now lives about a half mile from Crook's old house, remembers doing research for his master's thesis inside the barn.

At his home wind tunnel, Crook carried out research in the 1930s and 1940s, designing, building and flying a 1,700-pound prototype of an experimental two-engine warplane.

Crook's sudden death from a heart attack on Nov. 19, 1952, at the age of 65, stunned the CUA community; classes were suspended two days later so that students and fellow faculty members could attend his funeral at a nearby Catholic church.

Clancy, who took classes with Crook, remembers that the professor — thoroughly engrossed in his teaching — sometimes had trouble wrapping up a class even though the allotted time had elapsed and an afternoon baseball game was starting outside in the old CUA stadium.

"He'd say, 'Just one more graph, one more graph,' " says Clancy. "Meanwhile we could hear people cheering outside. We'd start sliding our feet on the floor, which was always covered with dirt that would blow in from outdoors. Finally he'd give up, and say, 'Alright, alright, go home.' "

Crook to Munk

German scientist Max Munk, right, chair of CUA's aeronautical engineering department in the 1950s, confers with Boehler about the design of an airplane wing.

In contrast to Crook — the hands-on inventor — his successor was an esoteric thinker, according to those who knew him. Max Munk was born on Oct. 22, 1890, in Hamburg, Germany. After graduating from high school, he apprenticed for a year at a machine shop. In Germany, he earned his bachelor's degree in mechanical engineering at the University of Hanover and his doctorate at the University of Göttingen.

At Göttingen, Munk studied with the legendary Bavarian engineer Ludwig Prandtl, who made major contributions to the understanding of turbulence and the function of a wing in producing lift.

In 1921, Munk immigrated to the United States — among the first of a succession of scientists who left Germany after World War I. Upon his arrival, he accepted a position as a researcher at the Langley Memorial Aeronautical Laboratory — the NACA facility envisioned by Zahm. There he distinguished himself by solving the problem of correlating results from wind-tunnel model experiments with measurements from actual planes.

Munk recommended compressing the air in a wind tunnel to 20 atmospheres (a standard unit for measuring atmospheric pressure) and using a 1/20-scale model airplane. Munk's argument that test results should then correspond with data from a full-scale plane at normal atmospheric pressure proved correct when the Langley wind tunnel became operational in 1923.

Munk left NACA in 1926 to work at Catholic University, but his genius is recalled in dozens of scientific papers he wrote while at the national aeronautics agency.

Aerospace entrepreneur Charles H. Kaman, who earned a CUA bachelor's degree in 1940, describes Munk as "a very knowledgeable scientist whose lectures were always a revelation."

The CUA alumnus, whose Kaman Aircraft Corp. developed the first gas turbine-powered helicopter and is now a $1 billion company, recalls that Munk used to supervise students conducting wind-tunnel experiments in the late afternoon or early evening when classes were over for the day.

CUA graduate and former lecturer Clancy also remembers consulting with Munk about the design of the model airplanes that he and other students used in their wind tunnel experiments.

A student would mount a model made of Philippine mahogany on a balance in the wind tunnel. The balance was an arrangement of rods similar to a trapeze that held the model in place as air blew over it. The balance also was a scientific instrument that measured the aerodynamic forces of lift — the air force acting on a wing that causes a plane to move upward — and drag — the force holding a plane back.

Clancy also recalls pleasant evenings at Munk's home in Brookland, where the professor would teach a graduate course to the sound of classical music playing on a stereo.

Munk was appointed chair of the aeronautical engineering department after Crook died in 1952 and served until 1958 when the department was merged with mechanical engineering. He retired in 1959 but continued teaching at the university as a part-time lecturer. Munk, like Zahm a lifelong bachelor, died on June 3, 1986, at the age of 95.

From Aeronautics to Aerospace

With the disbanding of the aeronautical engineering department and increasingly closer ties to NASA, research at Catholic University began to shift from aeronautics to aerospace in the early 1960s. Originally founded as a graduate institution, CUA offered a variety of programs in physics and engineering that appealed to NASA and Goddard Space Flight Center scientists who wanted to pursue advanced degrees.

Established in 1959 in nearby Greenbelt, Md., Goddard is a major NASA laboratory for developing and operating unmanned spacecraft.

"By the early 1960s, many of the senior-level engineers at Goddard were Catholic University graduate students," says Professor Emeritus Boehler.

One of those students was retired NASA aerospace engineer Steve Paddack, who earned his bachelor's, master's and doctoral degrees at CUA between 1959 and 1973. His doctoral dissertation focused on a space phenomenon now called the Paddack Effect in his honor. The phenomenon involves the way that solar radiation pressure induces small celestial bodies to spin and burst, eventually driving them out of the solar system. Paddack describes the effect as "a vacuum cleaner of space debris."

Left: Three-time alumnus Steve Paddack, who's been honored for his cutting-edge work at NASA, reflects the shift at CUA from aeronautics to aerospace engineering. Above: The COBE mission, which Paddack helped to manage, studied the conditions of the early universe.

Now an aerospace engineering consultant who lives in Northwest Washington, Paddack has been honored by NASA for his significant contributions to the aerospace field. He also received CUA's Distinguished Service Medal in 1990 and the university's Alumni Achievement Award in 1995. He is a member of the CUA Alumni Association's Board of Governors and of the board's executive committee.

The scientist says that by the time he earned his CUA bachelor's degree in 1959, he was straddling two fields — aeronautics and aerospace engineering. Among the courses that he took as an undergraduate were Aerodynamics with Munk and Aircraft Stability and Control with Boehler. Later Paddack found that what he was studying at CUA often matched what he was doing professionally at Goddard.

Paddack, who participated in a number of significant space flight missions while at Goddard, recalls that Munk had a profound effect on him: "He taught me the fundamentals of aeronautical engineering. But he also taught me a way of life."

The aerospace engineer says that Munk had high standards for his students, tempered by a wisdom about life that he revealed in conversational sayings. Paddack says that over the years he's incorporated a couple of Munk's truisms into his own conversation.

"Munk used to say, 'Never measure a person by his shortcomings but rather by his merits,'" says Paddack.

The three-time CUA alumnus recalls being upset with Munk when the professor gave him an "F" on a thermodynamics exam. Paddack had done the exam problems correctly but hadn't converted his calculations from the Fahrenheit scale to the absolute temperature scale. Paddack says he had to repeat the course but he never forgot the importance of checking his work.

After earning his bachelor's degree, Paddack moved to the West Coast and worked for the Boeing Co. in Seattle for two years. In 1961, he returned to Washington and started working at Goddard where he helped to manage several missions, including the Cosmic Background Explorer or COBE project, designed to study the physical conditions of the very early universe. As chief of Goddard's Advanced Missions Analysis Office from 1981 to 1995, he was a member of a NASA "think tank" that designed and started new projects.

Paddack, who is training to be a docent at the new National Air and Space Museum annex at Washington Dulles International Airport, says he's had "the most wonderful career imaginable. I couldn't believe that people actually paid me for what I did."

Paddack's words are similar in sentiment to those of Munk, who, in 1981, at the age of 90, said, "It makes me happy to remember that long ago, at the beginning of aviation, heaven granted me the opportunity to make significant contributions to aerodynamics."

Chapter 27

A Brief History of Aerospace Engineering at the Virginia Polytechnic Institute and State University

Robert W. Walters, Professor and Department Head
Jane Echols Johnston, Program Support Technician

The Early Years (circa 1913-1947)

In many ways, the roots of the current Aerospace and Ocean Engineering (AOE) department at Virginia Tech can be traced back to a time prior to 1913, when according to old timers, a flying field was set up at a farmer's pasture about a mile south of Blacksburg. During the 1920's, aviation grew in popularity and development such that the Civil Aeronautics Authority (CAA) of the federal government began a project to create a national airway system. Blacksburg was strategically located close to the CAA approved route between Washington, D.C. and Nashville, Tennessee and was considered to be a desired location for an emergency landing field in the national airway system.

On the afternoon of November 11, 1926, President Julian A. Burruss of V.P.I. appointed a special committee of faculty members to study the aeronautics situation at the college and to consider possible sites for a landing field. The committee was appointed with the definite understanding that the college did not intend to teach flyers, but would concentrate on aeronautical engineering. At least one farmer that owned level land within a mile of V.P.I. thought that defying the law of gravitation through aviation was immoral and refused to lease his property because he "did not want anything to do with such goings-on." Finally the problem was resolved and during the last week of July, 1929 engineers of the Virginia Highway Department began work on the land which today constitutes the Virginia Tech airport.

Early flying field prior to 1913

At that time the V.P.I. Airport was the only one in the United States to be owned or operated by a college or university. Most of the early airplanes at the airport were open cockpit biplanes. Enterprising pilots charged a cent per pound for passenger weight for a ten-minute flight or a flat rate of $3. If the passenger wanted a bit of stunt flying (loop-the-loop, barrel roll, spin, etc.) the charge was $5.

During the 1928-1929 academic year President Burruss was appointed to membership on the permanent committee on aeronautics of the Association of Land Grant

Colleges and Universities of America. At its 43rd annual convention, the V.P.I. Dean of Engineering, Earl B. Norris, was chosen secretary of the Aeronautical Section of the association. Fifty colleges and universities were represented at the conference. The nation's institutions of higher learning promoted Aeronautical education at the time and the academic world recognized V.P.I.'s role.

Faculty taught courses in Aeronautics in 1929 as an Aeronautical option under the Mechanical Engineering curriculum. The first Bachelor of Science degree in Mechanical Engineering with an Aeronautical Engineering option was granted in 1937. Interestingly, the 1938 Course Catalog under the Aeronautical Engineering option states, *"The technical studies in aeronautical engineering are exceedingly difficult and only a few of the fundamental subjects can be mastered by undergraduate students."* In addition to a typical mechanical engineering background in Statics, Dynamics, Materials, Fluid Mechanics, etc., fourth year students were required to take courses in Airplane Design, Aerodynamics, Principles of Aeronautics, Airplane Engines and an Aerodynamics Laboratory course. As a side note, the entire academic year fee in 1938 was $100 for tuition and $260 for board, room and laundry.

In 1941, the Aeronautical Engineering Department was formed as a separate engineering curriculum and granted its first degree in 1942 to Edwin Faunce Burner. In 1943, the number of students was large enough to warrant a separate discipline and the Department of Aeronautical Engineering was created with Professor Leon (Lee) Z. Seltzer taking the role as the first Department Head. Professor Seltzer served as department head until 1947 when he left V.P.I. to accept the position of Professor and Chairman of the Department of Aeronautical Engineering at West Virginia University.

Virginia Tech Airport in 1942.

During his tenure, the entire equipment inventory consisted of a three-foot diameter open throat, single return wind tunnel located in McBryde Hall. The tunnel was later renamed the "Green Monster."

In the 1943 time frame, V.P.I. had approximately 3300 total students. Approximately 150 were women, predominately studying the Home Economics curriculum. The first woman graduate of the Aeronautical Engineering department was Jane Hardcastle (B.S. 1944) and was noted by Professor Seltzer as a top student. Perhaps our most famous alumnus, Dr. Christopher Kraft (former head of the NASA Johnson Space Center) was also a member of that graduating class. The catalog from that era states, *"Only those students who demonstrate ability in the elementary aeronautical engineering courses and advanced mathematics are allowed to continue with the fourth year curriculum in aeronautical engineering."* Clearly, high standards of excellence were expected of the students back then just as they are today.

The Japanese attack on Pearl Harbor before the end of the first trimester of Professor Seltzer's tenure had a large impact on the department. The curriculum was accelerated and the army called up the ROTC students. Seniors reported first followed by the juniors. V.P.I. became a part of the Army Specialized Training Program and the biggest problem faced was the shortage of faculty for the enormous student class load. During the remainder of the war, Professor Seltzer taught 27-30 credit hours per quarter! By 1944, the curriculum had expanded to include courses in Airplane Propellers, Airplane Structures and Applied Air Loads.

During this same time period, the College had a flight-training program at the Blacksburg airport composed mainly of Fleet model biplanes. One interesting feature of these planes (or more accurately a lack of feature) was that they did not have brakes. Apparently, it was quite an experience to make a crosswind landing on the narrow airstrip without the aid of brakes. Moreover, Fleet biplanes would occasionally be flown to campus and landed on the Drill Field for full dress parades. The only things that fly around the Drill Field today are Frisbees and footballs. Back in that day, students were not even allowed to walk across the Drill Field, a tradition that has long since passed.

Post World War II Era (1947-1967)

In 1947, Professor Arthur E. Rowland became Department Head serving until 1951. Under his leadership, the curriculum expanded to include courses in Rotary Wing Aircraft and Aircraft Performance. Compressible flow was added to the Aerodynamics II course. In addition, students were required to take a 9 quarter credit (6 semester hours) summer term after the sophomore year that had once been part of the senior year. In 1951, a third course in Aerodynamics and a fourth course in Structures had been added to the senior year to bring the total hours required for graduation to 225 quarter credits (=150 semester credits), significantly more than the 136 credits required today.

Dr. Robert W. Truitt headed the department from 1951 to 1961. Under his dynamic leadership, new wind tunnels were added and the department moved into the newly built Randolph Hall. Courses in Vibration and Flutter, Boundary Layer Theory and Hypersonic Aerodynamics were added to the curriculum and students were (strongly) encouraged to write papers for IAS (Institute of Aerospace Sciences - now AIAA). V.P.I. students set records by taking awards in as many as three IAS meetings in one year. The department also began to offer advanced degrees including the Ph.D. degree in this period. Sadly, Dr. Truitt passed away in 1972 at the age of 51.

In 1958, Charlie L. Yates became the first black student to graduate from V.P.I. earning a degree in Mechanical Engineering. Dr. Yates received his Ph.D. from the John Hopkins University and later joined the faculty of the Aerospace and Ocean Engineering Department. Charlie retired in 2000 and is currently a Professor Emeritus of the department. In 2002, Peddrew-Yates Residence Hall was named in the honor of Dr. Yates and Irving L. Peddrew III (the first black student to enroll at V.P.I. and a man who spent his career in the aerospace industry).

The dawning of the space age brought about significant changes of emphasis. In 1961 the department changed its name to the Aerospace Engineering Department under the direction of its new head, Dr. James B. Eades, Jr. Dr. Eades had previously served on the faculty for 13 years. The facilities were significantly expanded to include the six-by-six foot stability wind tunnel graciously donated by NASA, two supersonic blow-down tunnels, and a "plasma-jet." An instrumentation lab was also added and an analog computer became available to the faculty and students. The laboratory work consisted primarily of aerodynamic investigations in the speed ranges from 150 miles per hour up to a Mach number of 4.75.

The 1962 Course Catalog shows that the curriculum expanded substantially to include 18 undergraduate course offerings and 17 graduate courses. A wide range of courses were offered with additions in Hypersonics, Aeroelasticity, Magnetoaerodynamics, Mechanics of Space Flight, Molecular Flow of Gases, and Energy Systems for Space Operations to name a few. This list is amazing considering that only five faculty members served in the Aerospace Engineering Department that year. A minimum of 223 quarter credits (149 semester credits) were required for graduation in 1962.

University records show that in 1963, there were 12 Bachelor of Science degrees, 10 Master of Science degrees and 1 Ph.D. awarded by the department. By contrast, in 1972, 34 B.S. degrees, 4 M.S. degrees and 8 Ph.D. degrees were awarded.

In looking back, we found a hand-written note that read *"Dr. J.B. Eades, Jr. called his secretary and told her that he would be in the Spherical Dynamics Lab (pool hall) for an hour. When Dean Whittemore called and asked for Dr. Eades, she told him, 'He is in the Spherical Dynamics Lab'"*. Apparently it satisfied the Dean and possibly a tactic that is worth remembering. Unfortunately, our current Dean will probably read this and I (RWW) will have to try something new.

A bulletin written by the senior class of 1962 also offers some insights into the times. In that document the following hints written for the benefit of entering students can be found:

1. You must learn that you will have to do things that you do not want to do.
2. You must learn to wait for rewards.
3. You must recognize that learning does not end in college.
4. You must be realistic about your relations with other people – be cooperative.
5. You must become more self-confident.
6. You must have clear in your mind what it is <u>you</u> want to do.

Finally, one other section in that document reads,

"It is through the medium of these men (the faculty) and machines that you will receive the training necessary for a career in the Aerospace field. The fact still remains, however, that the most important contribution to your education must come from <u>you</u> – the student."

The students of that time really had a great grip on the educational process and something that we need to continually reinforce today. As a side note, V.P.I. students were not allowed to own or ride in automobiles in the expansion Post-War era (1945-1955).

The Modern Era
(1967-Present)

Rollout of the 80' mock up of the National Aerospace Plane (NASP).

Although dating back 36 years, we refer to this as the modern era primarily because in 1968 women were given the right to wear pants on campus. Dr. Fred DeJarnette served as Interim Department Head of Aerospace Engineering at V.P.I. from 1967-1969 prior to joining NCSU. One personal comment (from RWW) about Dr. DeJarnette is that he is an excellent teacher and writes so fast on the chalkboard that he can almost disappear in a cloud of chalk dust - literally.

Dr. Joseph A. Schetz took over as Department Head in 1969 serving for an amazing 24 years until 1993. Dr. Schetz currently holds the Fred D. Durham Chair and remains as an extremely valuable faculty member to this day. Both the university and the department went through many changes during his tenure as Department Head. In 1970, V.P.I. officially changed its name to the Virginia Polytechnic Institute and State University and began its expansion from approximately 5,000 to over 25,000 students. In 1972 the department changed its name to the Aerospace and Ocean Engineering (AOE) Department drawing on the synergy that exists between the two disciplines. At first, a joint AOE degree was offered. Later, a separate accredited OE degree was offered in parallel to the AE degree. Under the direction of Dr. Schetz, the faculty tripled in size and undergraduate students were given significant roles in sponsored research programs. In addition, the department's record in winning AIAA student papers continued to grow and its exemplary record in winning student design competitions began.

After leading the department for almost a quarter of a century, Dr. Schetz decided to devote his efforts to research and teaching. Dr. Bernard Grossman began his nine year tenure as Department Head having served on the faculty for eleven years. One of Dr. Grossman's goals was to build on the design program and to emphasize the multi-disciplinary aspects of aerospace and ocean engineering in both the undergraduate and graduate programs. The Multidisciplinary Analysis and Design (MAD) Center for Advanced Vehicles began under his direction and resulted in numerous awards and graduate student fellowships. In 2002 Dr. Grossman stepped down from his position to assume the role of Vice President of Education at the newly formed National Institute of Aerospace under contract from the NASA Langley Research Center.

As a testament to the outstanding faculty members in the department (especially Dr. William Mason and Mr. Nathan Kirschbaum), Virginia Tech students placed first, second, third or received honorable mention in each AIAA undergraduate design competition over the 12 year period spanning 1989-2001 (with seven first place teams). Dr. James Marchman also played a key role in establishing international student design teams that contribute to the success of the department. Dr. Fred Lutze, the most senior faculty member to ever serve the department is retiring this year. His work with student organizations including AIAA and DBF (Design-Build-Fly) and his academic and graduate advising for over 30 years denote special historical recognition.

In 1999, the AOE department was named an exemplary department for its excellence in effectively linking research with teaching with particular emphasis on innovative undergraduate programs. In the past three years, numerous other awards have been given to the faculty and students of this department including our three-time World Champion Human-Powered Submarine Team.

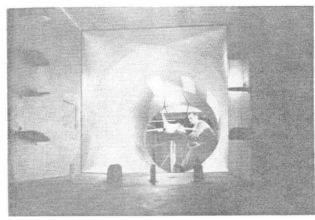

Model in test section of the Stability Wind Tunnel.

In August of 2002, Dr. Robert (Bob) W. Walters assumed the position of Professor and Department Head of Aerospace and Ocean Engineering. Long before ever having stepped foot on the Virginia Tech campus, Bob had been strongly influenced by the Aerospace program in Blacksburg. His Ph.D. advisor, Dr. Hassan A. Hassan had been a Professor here before joining NCSU. Moreover, Dr. Fred DeJarnette (Interim Department Head) was one of his major professors as was Dr. James Williams, a 1951 graduate of Aerospace Engineering at Virginia Tech who later when on to assume the position of Department Head of Aerospace Engineering at Auburn.

Today, our curriculum covers fundamental courses in structures, fluid mechanics, dynamics and control, ship design and optimization techniques as well as a wide range of advanced courses in these areas. We have a world class faculty with Ph.D. degrees from many fine institutions including MIT, Princeton, Stanford, Georgia Tech, Purdue and Michigan to name a few. The first female faculty member, Dr. Naira Hovakimyan (Ph.D. from the Russian Academy of Sciences and currently at Georgia Tech) will be joining the department in the fall of this year. We wonder if the faculty from the 40's and 50's could ever have imagined this possibility!

Our facilities are first-rate, including a flight simulation laboratory with a 2F122A Operational Flight Trainer donated by Naval Air Station Oceana valued at $13 million. We also operate one of the largest university owned Stability Wind tunnels in the world with a replacement value exceeding $15 million donated by NASA. In addition we have developed a Satellite Tracking laboratory that will be used to monitor the HokieSat, a student built satellite that will be launched into the ionosphere by NASA Goddard with the purpose of performing scintillation measurements. The AOE faculty have also developed unique diagnostic equipment including Laser Doppler Velocimeters (even credit card sized LDV's), a diode array velocimeter for which a patent has been obtained and many other one-of-a-kind devices. Moreover, we have a new Hypersonic Wind Tunnel and have been able to obtain Mach 10 flow in Randolph Hall this past year.

For many years, our enrollment data has basically tracked with ADCA data. This year, we have 95 sophomores, awarded 70 B.S. degrees in Aerospace and Ocean Engineering, 31 Masters degrees and 11 Ph.D. degrees. Enrollment is on the rise, which we expect to continue for some time to come.

Research in the department is also on the rise with a 16% increase over the previous year. Our faculty members not only perform research in the basic and applied aerospace sciences spanning an amazingly wide variety of topics but also have extended their efforts to include work in bioengineering, automotive engineering, nanotechnology, and information technology. We are also focusing a great deal of effort on Autonomous Vehicular Systems leveraging our expertise in vehicle design that has been developed over the years.

Notable Alumni

The success of our graduates has truly been amazing and it is not possible to list them all in this article. At the risk of excluding some remarkable people, we have decided to list a very few.

Dr. Christopher C. Kraft Jr. ('45) – Director of NASA's Lyndon B. Johnson Space Center;
Mr. Paul Holloway ('60) – Director of the NASA Langley Research Center;
Dr. William Grossmann, Jr. ('58, '62, '64) - Vice President of Technology and Chief Scientist, SAIC;
Dr. Fred DeJarnette ('65) – Director of the Mars Mission Research Center; VPI Department Head;
Dr. Douglas Dwoyer ('64, '68, '75) - Associate Director of the NASA Langley Research Center;
Mr. John B. McKay ('50) – NASA X-15 Test Pilot,
Mr. Jerry C. South ('59, '59) – NASA Langley Chief Scientist;
Dr. C. Howard Robins, Jr. ('58) – NASA Deputy Associate Administrator for Space Systems Development
Dr. James Williams ('51) – Aerospace Department Head at Auburn University;
Mr. Toby Bright ('77) – Vice President of Boeing Commercial Airplane Group;
Dr. Joseph W. Meredith, Jr. ('69) – President, Virginia Tech Corporate Research Center;
Mr. Norris E. Mitchell ('58) - President and Owner, Garden-Homes Realtors;
Mr. Nicholas J. Moga ('76) – President, PCSS;
Mr. Philip R. Compton ('47, '50) – Douglas Aircraft/NASA;
Mr. Robert J. Hanley ('79) – Deputy, U.S. Navy Airworthiness Office;
Mr. Marc W. Sheffler ('73) – Director of Apache Integrated Product Teams, Boeing;
Dr. Thomas F. Swean, Jr. ('72) – Program Manager, Ocean Engineering Office of Naval Research;
Mr. Kevin Crofton ('82) – Vice President for CMP, Lam Research Corporation;
Mr. Larry Marshall ('66) – Senior Research Fellow, E. I. DuPont;
Mr. Lester W. Roane ('58) – Chief Engineer, H.P. White Laboratory;
Dr. Robert H. Tolson ('58, '63) – NASA Langley Chief Scientist;
Dr. Thomas H. Thornton ('55, '58) – Division Director, JPL;
Mr. John W. Boyd, Jr. ('47) – NASA Associate Administrator for Management;
Mr. Robert Warrington ('68) – Dean of Engineering, Michigan Technological University.

Acknowledgements

The authors wish to acknowledge Lynn Nystrom in the College of Engineering for her helpful information and suggestions. We also want to thank each of the previous department heads whose notes and materials made writing this brief history possible. In many cases, we used text written by them verbatim or made very slight changes to their descriptions.

Chapter 28

The History of
Aerospace Engineering At West Virginia University

Dr. Donald W. Lyons and Dr. Ever Barbero
Department of Mechanical & Aerospace Engineering
West Virginia University

West Virginia University first added Aeronautical Engineering as an option within the Department of Mechanical Engineering in 1938, and established a separate Department of Aeronautical Engineering in 1944. The first Department Chair was Professor Harold Wickersham. Professor Henry W. Woolard took over as acting chair of the Aerospace Engineering Department in 1947. The first BSAE students, Joann Berry and Clyde Cokely graduated in 1948. The Aerospace Engineering Program received accreditation by the Engineering Accreditation Board in 1953. The first MSAE was awarded to Lu-Chung Chang in 1961. The first Ph.D. in AE was awarded to Jerry L. Gester in 1967. Professor Leon Seltzer was appointed as Chair in 1949 and served until 1963. Professor Ben Urich served as acting chair in the period 1963-1964 and Professor Jerome B. Fanacci was appointed as Chair in 1964 and served until 1980. In 1981, the Aerospace Engineering Department was merged with the Mechanical Engineering Department to form the Department of Mechanical and Aerospace Engineering (MAE). The Aerospace Engineering education and research programs were strengthened by the merger and have continued to grow and be enhanced to this time.

The first Chair of the combined Mechanical and Aerospace Engineering Department was Dr. Severino L. Koh. In 1983, Dr. Russell R. (Rex) Haynes became acting Chair followed by Acting Chair Dr. Richard E. (Dick) Walters in 1984-1985. Dr. Donald W. Lyons was appointed as Chair in 1985 and served until 2001. Dr. Gary Morris served as Interim Chair for part of the next year until Dr. Ever Barbero was selected to be the permanent Chair of the Department in 2002.

In 1961, all of the Departments of the College of Engineering moved into a new building, the Engineering Sciences Building, that still stands as the core of the current Engineering Building Complex. In addition to large mechanical shops and laboratories, the building has an attached building housing several wind tunnels and a separate aerospace propulsion laboratory. A civilian pilot training program was established in 1939. The program served 1,082 student pilots between 1942 and 1945. The flight training program was terminated in 1967 but the aviation ground school course continues till this day.

The WVU Aerospace Engineering Program has responsibilities in the three areas; teaching, research, and service. Over the years there has been major growth in the area of research. Faculty and students have conducted many major programs of research which have had and continue to have a major impact on the science and technology of aerospace engineering.

Figure 1. Dr. Jerry Fanucci and Dr. John Loth stand before the STOL aircraft designed , built , and tested by them and their students at WVU.

One of the first major research programs began in 1971, lead by Professors Jerome Fanucci and John Loth, sponsored by the Office of Naval Research. The program was to study very short takeoff and landing (VSTOL) aerodynamics. The program included the design and construction of a technology demonstration aircraft shown in Figure 1. The aircraft was physically built in the WVU hanger and flight tested under the supervision of WVU faculty and students. The Department still has the demonstration aircraft. Although it is no longer airworthy, it is used as a demonstration for current students and visitors.

A more recent major aircraft research program is directed by Professor Marcello Napolitano. For the past eight years, with major funding from the Department of Defense and NASA, he has been developing neural network based damage tolerant computer control systems for aircraft. To test the developments he has constructed a small fleet of remote control airplanes modeled after the common new high performance commercial and military aircraft. These remote controlled aircraft, typically with 8 to 10 feet wing spans, are being regularly flown as test platforms for new aircraft control technologies. WVU students and engineers are shown with one of these aircraft in Figure 2.

Figure 2. Dr. Marcello Napolitano (back right) with students and research scientists and one of their fleet of remote controlled aircraft for control system testing.

Another recent major research area is the development of alternative fuel technology for aircraft. The program which has been ongoing since 1990, has major funding from the US Department of Energy and other agencies. The program involves many faculty and students and is directed by Professors Donald Lyons and John Loth. One result of the program is the conversion of a Cessna

Figure 3. Dr. John Loth and Doctoral Student Rob Bond beside the alternative fueled aircraft they developed and flight tested.

150 aircraft to operate on alcohol fuel. The aircraft, shown in Figure 3, is currently operable on an experimental basis on either alcohol fuel or conventional aviation fuel, switchable in flight

Another recent project, directed by Dr. John Kuhlman, is the development of methods for magnetic control of fluid flow in microgravity environments. Students tested their experimental designs using the NASA Johnson Space Flight Center microgravity test bed aircraft shown in Figure 4.

Figure 4. Dr. John Kuhlman (back right) with students and their microgravity fluid flow experiment flown for testing on the NASA Weightless Wonder zero gravity test.

The Aerospace Program has hundreds of alumni and a long list of very distinguished alumni. In 1988 the Department established an Academy of Distinguished Alumni of Aerospace Engineering to recognize some of the most distinguished alumni and a few honorary alumni who were individuals with a long association to the program and a record of major contribution to the fields. A list of the members of this Academy is shown in Table 1.

Table 1. WVU Academy of Distinguished Alumni
Of Aerospace Engineering

Name	WVU Degrees	Example Position
Dr. David Anderson	BSAE 1954	Scientific & Technical Advisor, USAF
Mr. Darius N. Brant	MSAE 1961, MSAE 1963	Mgr. of Aerothermophysics Martin Marietta Corp
Dr. Subrto Chandra	MSAE 1973, Ph.D. 1975	Dir. of Research & Dev., Florida Solar Energy Ctr.
Mr. William S. Clapper	BSAE 1969	Mgr. Bus. Plan & Mkt. Dev., Gen. Elec. Aircraft Eng
Mr. Ralph C. Cokeley	BSAE 1948, BSME 1948	Test Pilot, Lockheed Corp.
Dr.Thomas A. Csencsitz	Ph.D. Eng 1973	Assoc. Dir. Med. Edu., Orlando Reg. Med. Ctr.
Mr. Larry W. Dooley	BSAE 1962, MSTAM 1963	Group Engineer, Bell Helicopter
Dr. Jerome B. Fanucci	Honorary Member	Chairman, Aerospace Eng. at WVU
Dr. Michael E. Fourney	BSAE 1962, MSTAM 1963	Chair Mech. Engr. & Aero. Eng., Univ. of Maryland
Mr. Daniel J. Holt	BSAE 1969, MSAE 1972	Editor in Chief, Aerospace Engineering (SAE)
Mr. C. Neil Jubeck	BSAE 1960, MSAE 1962	Tech. Dir. Rotary Wing, Directorate Naval Air Test Ctr.
Mr. Harry L. Lemasters	BSAE 1965	V.P. of Com. Bus., Hamilton Stand. Div. United Tech
Mr. Richard E. Longhouse	BSAE 1964	Supv. Of Advanced Sus. Sys.Delphi Chassis Div. GM
Mr. Jon A. McBride	Honorary Member	Pilot & Commander, Challenger Space Shuttle Miss.
Mr. Thomas V. Murphy	BSAE 1957, MSAE 1969	President, AAI Corporation
Mr. George K. Oss	BSAE 1958	President & CEO, Versar Laboratories
Mr. Leon Z. Seltzer	Honorary Member	Chairman, Aerospace Eng. at WVU
Mr. Robert E. Walter	BSAE 1959, MSAE 1961	Team Leader, Boeing Corp.
Dr. Richard E. Walters	BSAE 1956, Ph.D. 1967	Professor & Assoc. Chair, MAE Dept. WVU
Mr. Chester L. Whitehair	BSAE 1959	V.P. for Space Launch, Operations Aerospace Corp.
Dr. F. David Wilkin	BSAE 1967, MSAE 1969	President, Virginia Highlands Community College
Mr. Calvin F. Wilson	BSAE 1957	Dir. of Eng., Piper Aircraft
Dr. James D. Wilson	BSAE 1968, MSAE 1970	Science Consultant, U.S. House of Representatives

Table 2. Current Faculty
WVU MAE Department

Bajura, Dick	Professor & Director NRCCE	Morris, Gary	Professor
Banta, Larry	Assoc. Prof. & Assoc. Chair	Mucino, Victor	Professor
Barbero, Ever	Professor & Chair	Mukherjee, Nilay	Assistant Professor
Campa, Giampiero	Research Assistant Professor	Napolitano, Marcello	Professor
Celik, Ismail	Professor	Norman, Tim	Professor
Clark, Nigel	Professor	Palmer, Mike	Professor
Dean, Russel K.	Professor & Assoc. Provost	Perhinschi, Mario	Assistant Professor
Gautam, Mridul	Professor	Prucz, Jacky	Assoc. Professor & Assoc. Chair.
Huebsch, Wade	Assistant Professor	Seanor, Brad	Research Assist. Professor
Johnson, Eric	Professor	Shoukry, Samir	Professor
Kang, Bruce	Assoc. Professor	Sivaneri, Nithi	Professor
Kuhlman, John	Professor	Smirnov, Andrei	Assistant Professor
Lewellen, David	Research Assoc. Professor	Smith, James	Professor
Liu, Xingbo	Research Assistant Professor	Sneckenberger, John	Professor
Long, Thomas	Professor	Stanley, Chalres	Professor
Loth, John	Professor	Thompson, Gregory	Assistant Professor
Lyons, Donald	Professor and Director CAFEE	Wayne, Scott	Research Assistant Professor
Means, Ken	Professor	Yavuz, Ibrahim	Research Assistant Professor

The Mechanical and Aerospace Engineering Department currently has 26 full time tenure track faculty, 3 jointly appointed tenure track faculty, and 8 full time research faculty, and almost all provide support in some way for both the Aerospace Engineering program and the Mechanical Engineering Program. A list of the current faculty is presented in Table 2.

The Department offers BS, MS and Ph.D. degrees in both Aerospace Engineering and Mechanical Engineering and also a unique nine semester long bachelors program offering students an opportunity to simultaneously receive a bachelor's degree in both AE and ME at the same time. The Department currently has 400 undergraduate students (sophomore through senior year) and 165 graduate students. Approximately a quarter of these undergraduate students choose the dual degree program which allows them to obtain both a BSAE and BSME degree in approximately four and one half years. Approximately one third of the remaining students are majoring in Aerospace Engineering and two thirds in Mechanical Engineering.

Chapter 29

Aerospace Engineering at San Jose State University

Richard D. Desautel, Professor and Head, Aerospace Engineering

The Aeronautics Program became the first four-year engineering-type program at San Jose State College in 1938. The Department of Engineering was established in 1946 with a single faculty member! Chartered to provide 4-year undergraduate education, the campus (and entire Cal State University system) had to fight a political battle with the University of California system when the California Master plan was established in the 1950's in order to be allowed to offer graduate (MA and MS) degree programs. Mechanical Engineering was established in 1958 as a separate program. In the 1980's, the ME program hosted significant faculty research in aerospace applications in rotorcraft dynamics and control and in hypersonic flow measurements (shock tunnel). In 1985, Engineering Dean Jay Pinson asked Professor Dick Desautel of Mechanical Engineering to develop an AE program and Department from the 'ground up." Dr. Desautel proceeded to do so while remaining in ME and dividing duties between ME and AE through the middle of 1988.

The BSAE degree program "opened its doors" to students in the Spring 1987 semester with a fully defined BS curriculum and initiation of 10 laboratories. Several students immediately transferred into the major from both aeronautics (a technology program) and mechanical engineering, with the result that the first student graduated in Spring 1989! A full-time faculty team was quickly developed beginning with Dr. Nikos Mourtos (Fall 1987), then Jeanine Hunter (Lecturer, Fall 1989) and Dr Henry Pernicka (Fall 1990). In this team of four (including Dr. Desautel), two faculty had significant aerospace industry experience and two were recent PhDs. Also, two of the four were private pilots. The team had disciplinary representation in aerodynamics, propulsion, dynamics and control, and flight mechanics including rocket launch and reentry.

The distinctive characteristics of the Program at its inception were:
- Innovative, broad curriculum
    Breadth - inclusion of a full layer of engineering sciences between math/science/engineering basics and aerospace applications;
    Breadth - specialization to either aircraft or spacecraft only occurs in the yearlong senior project, i.e., innovatively packaged applications courses integrated both aeronautical and astronautical applications;
    Practice - industry-like processes and environment in the senior design course.
- Innovative, advanced laboratories
    Approach – goal was to implement as many theoretical concepts in physical laboratory experiences as possible;
    Character – modern, advanced, innovative experiments (rigid body dynamics, shock tunnel, satellite bus subsystems, electro-optical and laser sensors.

The distinctive philosophy was one of a broad integrated treatment of aerospace engineering education integrating analysis, computation, design, and experimentation supported by wide-ranging hands-on laboratory experience.

By 1991 (four years from first enrollments), the accomplishments of the program were an enrollment growth to nearly 400 majors, establishment of the ten laboratories, formation of the faculty team augmented by valuable adjunct faculty from industry, and full ABET accreditation of the BSAE program. There also occurred the design and development of the practice-oriented MSAE Program that was implemented in 1992. This program began as primarily a series of satellite subsystems courses that expanded to include aeronautical disciplines as well. By California's Master Plan, SJSU is not chartered to grant doctorate degrees. The Aerospace Program was housed in a new Department of Aerospace Engineering. The Department included an AIAA Student Chapter and a Sigma Gamma Tau Honor Society Chapter.

The tremendous enrollment growth in the program's first few years was attributed to concentration of 2nd and 3rd generation aerospace industry families in the bay area. Lockheed, Space Systems Loral (formerly Ford Aerospace), and a concentration of smaller firms and defense electronics industries had build a heritage of families involved in aerospace from the 1950s up to the redefinition of Silicon Valley as computer related in the 1980s and 1990s. When the SJSU program opened, it was the only undergraduate aerospace (aeronautics and astronautics) engineering program from Seattle to Boulder to Los Angeles. That remains the case up through 2003.

The rapid development and innovations of the BSAE program – curriculum and laboratories - were reported to enthusiastic response in a series of papers presented at ASEE and IEEE education sessions. Awarding of the 1993 Leland Atwood Award for Outstanding Aerospace Educator to Dr. Desautel recognized the results. Although an individual award, clearly the outstanding success of the new program was due to the faculty team that supported it.

When the cold war ended in 1991, the nation's undergraduate aerospace engineering programs suffered a 67% decrease in enrollments; SJSU's program followed this trend with about a year delay. As is typical of aerospace engineering programs, this cycle precipitated several mergers across the country between aerospace programs and allied programs, most commonly mechanical engineering. SJSU's Aerospace Engineering and Mechanical Engineering departments were merged in 1996 into the Mechanical and Aerospace Engineering Department. At its nadir, the SJSU AE program decreased to 90 students in the major. It rebounded by the end of the decade, and was 140 in AY 2002/2003.

Notable achievements of the Program include national award-winning student papers, good showings in the annual SAE aircraft fly off competition, and successful development of a microsatellite (Spartnik) by undergraduates. Most Program alumni have launched their careers in bay-area industry and established themselves in R&D and leadership positions. Many have gone on to earn their MS degrees at SJSU and other institutions, and a few have achieved PhD degrees at institutions such as UC Davis and Stanford. Program faculty have also provided important service to the College of Engineering and University in leadership in the SJSU Teaching & Learning Program, Dean Search Committees and Chair positions.

References
J. Hunter and D. Desautel, "Undergraduate Astronautics Initiatives at San Jose State

University," J. Engineering Education, Vol. 82, No.1, January 1993

D. Desautel, et al, "Development and Integration of Modern Laboratories in Aerospace
Education," AIAA Paper 92-4022, AIAA 17th Ground testing Conference, July 1992

D. Desautel, "An Example of New, Modern B.S. and M.S. Programs in Aerospace Engineering,"
American Society for Engineering Education Annual Conference &
Proceedings, June 1989

D. Desautel, "Development of a Combined Research / Instructional Gas Dynamics
Laboratory," American Society for Engineering Education Annual Conference &
Proceedings, June 1988

SJSU Aerospace Engineering Faculty
1985 – Present

Dick Desautel Professor	Ph.D. - Stanford	1985 – present	Gas dynamics; flight mechanics; sensors
Jeanine Hunter Lecturer	MS - Purdue	1989 – 1995	Dynamics & control; space systems design
Nikos Mourtos Professor	Ph.D. - Stanford	1987 – present	Fluid mechanics, Aerodynamics, Propulsion, Aircraft design
Periklis Papadopoulos Associate Professor	Ph.D. - Stanford	2002 – present	Space transportation vehicle design, Computational fluid dynamics & geometry, Multi-physics, multi-phase modeling.
Henry Pernicka	Ph.D. - Purdue	1990 – 2001	Dynamics & Controls; astrodynamics; space systems design

Dick Desautel   Nikos Mourtos   Henry Pernicka

SJSU Aerospace Engineering Instructional Laboratory Facilities
(As of June 2003)

Location / Name	Courses Served	Example Facilities & Activities	Area (sq ft)	Director
E 107 - Aerodynamics	AE 162, AE 170AB	1x1ft subsonic wind tunnel smoke tunnel aircraft model fabrication LDV & hot-wire stations	1357	Mourtos
E 131 – Space Mission Analysis & Design	AE 110	AIAA design competitions CFD and computational mechanics Space transportation systems design	400	Papadopoulos
E 164 - Gas Dynamics	AE 114, AE 164, AE 110, satellite test	6x6in supersonic blowdown tunnel hypersonic shock tunnel w/schlieren vibration and vacuum testing LDV & hot-wire instrumentation Gyroscope and EO experiments Heat Pipe station Photographic darkroom Satellite electrical power experiment	2622	Desautel
E 236 - Space Engineering	AE 170AB, satellite fabrication, storage & operations	4-bay Class 100 clean room electronics assembly and test bay software development bay Mission Operations stations	1318	Desautel
E 240 - Aircraft Design	AE 170AB	Computer stations with AAA, ProE MatLab, Sub/super 2D Specialized bibliographic resources	1342	Mourtos
E 272 - Spacecraft Design	AE 170AB	Computer stations with Sub/super 2D, STK, POHOP, SINDA, MatLab, ProE Specialized bibliographic resources	1975	Desautel

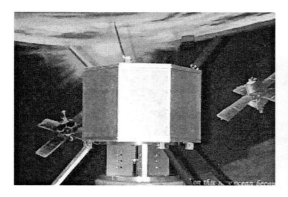

378

Chapter 30

HISTORY OF AEROSPACE EDUCATION AND RESEARCH
AT PURDUE UNIVERSITY: 1910 - 2002

A. F. Grandt, Jr., AIAA Fellow, Professor

W. A. Gustafson, Professor Emeritus

School of Aeronautics and Astronautics, Purdue University

Abstract

This paper summarizes how Purdue University has developed into one of the world's leading institutions of aerospace education and research. It is estimated that during the past fifty years, Purdue has awarded 6 % of all B.S. degrees and 7 % of all Ph.D.'s in aerospace engineering in the United States. These alumni have led significant advances in research and development, headed major aerospace corporations and government agencies, and have established an amazing record for space exploration. Over one third of all U.S. manned space flights have had at least one crew member who was a Purdue graduate (including the first and last men to step foot on the moon), and a fourth of all U.S. space missions have been flown by a graduate from the School of Aeronautics and Astronautics.

Early History

Although it may surprise some that a small community in northwest Indiana has played a leading role in developing air and space travel, the Lafayette area has a long tradition with exploring the frontiers of human progress. Just a few miles southwest from the current Purdue campus, Fort Ouiatenon was established as a French trading post with the Wea Indians in 1717, and became an important center for the Indian and white man to exchange respective cultures and technologies. The fact that these interchanges were not always peaceful is also evident in local history, as the Battle of Tippecanoe occurred a few miles northeast on November 7, 1811. This was the last large conflict between organized Indians and white men east of the Mississippi River, and was a milestone in bringing peaceful settlement to the Northwest Territory.

The Lafayette area also has a rich aerospace tradition. It was, for example, site of the first U.S. airmail delivery on August 17, 1859, accomplished by a hot air balloon piloted by John Wise of Lancaster, Pennsylvania, and directed by U.S. postmaster Thomas Wood of Lafayette [1]. This first airmail consisted of 123 letters and 23 circulars, and traveled approximately 25 miles before the balloon was forced to land from lack of buoyancy. Mr. Wise also conducted experiments during this short flight to measure the presence of ozone in the upper atmosphere. Thus, 10 years before Purdue University's founding in 1869, Lafayette already had a history of experimentation with air travel and with using that new technology for scientific exploration.

Community interest in aviation continued when Purdue was established across the Wabash River in what was to become West Lafayette. The Purdue Aero Club was organized in 1910 under the direction of Professor Cicero B. Veal of mechanical engineering, and the community's first aircraft demonstration was held on the campus June 13, 1911 (Figure 1). Sponsored by the Purdue Alumni Association and a local newspaper, this "Aviation Day" attracted 17,000 people to see two flimsy biplanes land on the Purdue athletic field [1,2]. Other flights to campus in the next few years continued to draw large crowds (Figure 2).

The first Purdue graduate to become an aviator was J. Clifford Turpin (class of 1908), who was taught to fly by Orville Wright. Turpin set an altitude record of 9,400 feet in 1911, establishing an alumni tradition that was continued 55 years later, when an X-2 aircraft flown by Captain Iven C. Kincheloe (BS '49) flew to 126,000 feet in 1956. (That record was subsequently surpassed by alumni Neil A. Armstrong (BS '55) and Eugene A. Cernan (BS '56) during their 1969 and 1972 flights to the moon.) Lieutenant George W. Haskins was the first alumnus to land an aircraft on campus in 1919, when he flew from Dayton, Ohio with a resolution from Dayton alumni proposing that Purdue establish a School of Aviation Engineering [1].

Purdue began limited education in aeronautical engineering during the 1921-22 academic year with four elective courses offered by the School of Mechanical Engineering. Professor Martin L. Thornburg, a 1915 ME graduate and veteran of the Air Service, was in charge of instruction [1]. An aerodynamics laboratory was soon established and equipped with a fully assembled airplane and operating engines (Figure 3). When Professor Thornburg left in 1924, responsibility for the new aeronautical courses was given to Professors Elbert F. Burton and Alan C. Staley. They were later followed by Major William A. Bevan of the Air Service in 1926, and by Captain George Haskins in 1929 (Figure 4). Although a formal four-year aeronautical curriculum was not available until the 1940's, many civil and mechanical engineering students took the aeronautical electives, commonly referred to as the senior aeronautical option, and entered the new aeronautical industry during the 1920's and 30's. Donovan R. Berlin (BS '21) is one early graduate who designed a number of important aircraft during the first half of the 20th century, including the P-36, P-40, and P-48 of WW II fame. He later worked on the Navy's FH-1 Phantom and the Army's CH-47 Chinook helicopter, and was awarded an honorary doctorate by Purdue University in 1953 for his significant contributions to the aircraft industry.

Purdue became the first U.S. university to offer college credit for flight training in 1930, and opened the nation's first college-owned airport in 1934 [1,2]. Although somewhat controversial among faculty, the concept of providing academic credit for flight training was actively promoted by Purdue President Edward C. Elliott. President Elliott was later responsible for bringing Amelia Earhart to Purdue as a "Counselor on Careers for Women," a staff position she held from 1935 until her disappearance in 1937 (Figure 5). Purdue was instrumental in funding Earhart's ill-fated "Flying Laboratory," the Lockheed Electra that she attempted to fly around the world in 1937. The University library houses an extensive Earhart collection, which continues to be studied by those seeking to solve the mystery of her final flight.

Active involvement in flight training continued during the 1930's, and Purdue was an important military flight-training center during World War II. Training of aviation technicians was started in 1954 and a two-year professional pilot program was created in 1956. A general aviation flight technology course was created in 1964, and a B.S. program in professional piloting approved in 1964. These pilot training and aircraft maintenance programs continue in the current Department of Aviation Technology. The focus of this paper, however, is on Purdue's aerospace engineering programs.

Professor Haskins returned to the Air Corps in 1937, and the aeronautical engineering programs were taken over and expanded by three key individuals: Professors Karl D. Wood (Figure 6) and Joseph Liston (Figure 7) who joined the faculty in 1937 and by Professor Elmer F. Bruhn (Figure 8) in 1941. Professor

Wood came to Purdue after many years at Cornell University, where he had published a comprehensive book on airplane design in 1934 [3]. Professor Wood was a strong advocate of education that balanced theory, technical analysis, testing, and design, and although he left Purdue in 1944 to head the Aeronautical Engineering Department at Colorado University, his seven-year tenure left a permanent imprint on Purdue. Professor Liston remained for 35 years until retiring in 1972. He was responsible for developing excellent propulsion courses and laboratories.

Professors Wood and Liston strengthened and expanded offerings in theoretical aerodynamics, airplane design, and aircraft engine design. A weakness in the structural area was ably addressed by recruitment of Professor Elmer F. Bruhn in January of 1941. Professor Bruhn taught 5 years at the Colorado School of Mines, followed by 12 years in industry with the North American Aviation and Vought-Sikorsky Aircraft Companies. One of Professor Bruhn's goals for returning to academia was to prepare an aircraft structural design text [4]. The first version of that comprehensive volume was completed in 1943, and with subsequent revisions in 1965 and 1973, remains in print, having sold over 100,000 copies, and continues to be a mainstay in industry.

WW II Developments

Since Purdue had taught aeronautical engineering for 20 years, and had greatly expanded coursework after the arrival of Professors Wood and Liston, Dean of Engineering A. A. Potter had decided by mid-1941 that aeronautical engineering should play a key role in Purdue's growing war training effort. Under his leadership, the School of Mechanical Engineering changed its name to the School of Mechanical and Aeronautical Engineering in 1942, and began a four-year B. S. curriculum in aeronautical engineering. This program required 160 2/3 credits, and students could also obtain a B.S. in mechanical engineering with an additional semester of work. An M.S. degree in aeronautical engineering was also offered at this time. The first students to pursue this "official" aeronautical engineering degree began study in 1940, and since the University went on an accelerated three-semester-per-year schedule during the war, received their diplomas in August of 1943.

Following the nation's formal entry in WW II after the Pearl Harbor attack of December 1941, Purdue pursued a number of aeronautical engineering efforts to support the war effort. The **Air Corps Cadet Aeronautical Engineering Program** was an extensive 12-week course given to groups of 50 students in January and in April of 1941. The rigorous curriculum was based on the fact that all students were engineering graduates with strong academic records. A large majority held mechanical engineering degrees, obtained from some 40-odd universities. These cadets were trained for Air Corps operations and aircraft maintenance. After completing the three-month Purdue program, they went to Chanute Field at Rantoul, Illinois, for several weeks of intensive practical training, and were then commissioned as Army Air Corps officers.

The **Curtiss-Wright Cadette Programs** resulted from a decision by the Curtiss-Wright Airplane Corporation to train young women for technical positions normally held by men at that time. Cornell, Iowa State, the University of Minnesota, the University of Texas, Rensselaer Polytechnic Institute, Pennsylvania State College, and Purdue participated in what became known as the Curtiss-Wright Cadette Training Program. The first 100 Cadettes who arrived at Purdue on February 12, 1943, were college graduates selected from throughout the country. Another group of 116 young women arrived on July 4, 1944. The first Cadette curriculum was common to all seven participating colleges, and consisted of two 22–week-long terms heavy in drafting, materials processing, and testing (Figure 9). The second program was shortened to six months with two twelve-week terms, and the age limit was lowered to 18 by the date of plant induction.

The U.S. Navy designed The Navy College Training program, usually referred to as the **Navy V-12 Program**, to solve its own critical staffing problems. The V-12 program began at Purdue on July 5, 1943, with 1263 men. Successive enrollments raised the total to 2730 at Purdue, with approximately 400 of these men receiving B.S. degrees. The V-12 program was conducted in 16-week terms, and by giving a semi-term in the fall of 1943, Purdue was able to provide a single calendar for both civilian and Navy students. Separate options were offered in the structures and the engines areas. The structures option required 12 credit hours of aerodynamics and 13 credits in aircraft structural theory, laboratory testing, materials and processes, structural design, and a course in vibrations and flutter. The engine option emphasized courses dealing with theory, testing, and design of aircraft power plants.

Post World War II Programs

At its spring 1945 meeting, the board of trustees approved an independent School of Aeronautics effective July 1, 1945, with Professor Elmer F. Bruhn as Acting Head. Professor Bruhn was widely known for his textbook on airplane structures. The broad title "School of Aeronautics" was selected because degrees were to be offered in both aeronautical engineering and air transportation.

Three new staff had joined the School of Mechanical and Aeronautical Engineering several months previously in anticipation of this official action. These new faculty included Mr. Grove Webster, who had been general manager of the Purdue Aeronautics Corporation, Dr. Paul Stanley, who had been director of the ground school for the Purdue Aeronautics Corporation, and Mr. Edward Cushman, who came from the Allison Company in Indianapolis. Mr. Webster joined the school with the rank of associate, and Dr. Stanley and Mr. Cushman as instructors.

The first tasks were to prepare detailed curricula for both the air transportation and aeronautical engineering programs, and to inform the public of Purdue's post-war aeronautical offerings. A pamphlet was published listing the air transportation and aeronautical engineering curricula. It described the relationship of the School of Aeronautics and the Purdue Aeronautics Corporation, as well as the various sections of the Aeronautical Engineering and Air Transportation departments. The proposed post-war curriculum in aeronautical engineering was primarily the thinking of Professors Bruhn and Liston, while the air transportation curriculum was due to Professors Bruhn, Webster and Stanley.

Air Transportation shared a common freshman year with other engineering curricula, followed by three options: airport management and operations, flight and flight operations, and traffic and administration. The airport management and operations option included aircraft maintenance and shop work to allow students to qualify for aircraft and engine mechanic's certificates. The flight and flight operations option provided all ground and flight courses needed for a professional pilot's rating.

The goal of the post-war Aeronautical Engineering program was to provide a well-grounded understanding of how a flight vehicle is designed to meet given performance and operating requirements. The common freshman-engineering curriculum was followed by a summer session devoted to shop work, descriptive geometry, and drafting. The next one and a half years consisted of courses in mathematics, physics, engineering mechanics, drafting, economics, and thermodynamics. The second half of the junior year required engineering mechanics, aerodynamics, thermodynamics, materials testing and mechanisms. The final year allowed the choice of one of four options: airplane design, aircraft power plants, airline engineering, or production management.

There were only two professors and two instructors available in 1945, so a hiring effort began immediately, and by 1949 there were ten professors and four instructors on the faculty. The post war surge of students produced rapidly increasing enrollment, and by Fall of 1947 there were about 700 students enrolled in the two degree programs. By that time a new building had been constructed at the university airport, located adjacent to campus, to provide additional space beyond the two original campus buildings.

Graduate education in the School of Aeronautics began with a Master's Degree program in Aeronautical Engineering in 1946. Ph. D study was approved for aerodynamics and propulsion in 1948, followed by the structures area in the early 1950's. The new school's first Ph.D. was awarded to R. L. Duncan in 1950 for his work with Professor M. T. Zucrow on the performance of gas turbines. A Master's Degree program in Air Transportation came into existence after 1950.

The 50's Decade

Professor Bruhn decided to return to full-time teaching in 1950, and Milton U. Clauser (Figure 10) was selected to be the next School Head. Dr. Clauser came from the Douglas Aircraft Corporation in El Segundo, California, where he had been head of mechanical design. During the early 1950's, the School enrollment began to drop as the WW II veterans graduated, alleviating some of the pressure felt during the immediate post war years.

The aircraft internal combustion engine laboratory was well established and being used by undergraduate and graduate students (Figure 7). This laboratory was developed during the 1940's by Professor Joseph Liston, and consisted of two test cells where students could run engineering tests of reciprocating-type aircraft engines. By 1950 one of the test cells had been modified to use a small Westinghouse axial flow turbojet engine. The jet propulsion area was developed under Maurice J. Zucrow's leadership, but most of that program moved to Mechanical Engineering in 1953, leaving only a small part in Aeronautical Engineering.

At this time the aerodynamics laboratory had five wind tunnels in operation. The latest of these was a large subsonic tunnel with a capability of 350-400 miles per hour (Figure 10) designed by Professor G. M. Palmer, and built mainly of plywood with student help. The test section was approximately 11.5 square feet, and it used a 400 H.P. electric motor. This tunnel was upgraded by means of a large grant from The Boeing Company in the early 1990's and is still in use. Other tunnels in use at that time included a Japanese variable-density wind tunnel that had been confiscated and brought to this country after the war. Its use was rather limited because of its low power (100 H.P.). It had a 15-inch throat and was capable of speeds up to 300 mph. Another 110-mph tunnel that was originally powered by a Dodge auto engine, was modified and lengthened by Professor Palmer and driven by a 50 H.P. electric motor. This tunnel was used primarily in the required wind tunnel courses for aero students. Two other small wind tunnels, including a smoke tunnel, were moved from campus to the airport facility in 1948. A water table was added in 1953 for demonstrating compressible and viscous flow. At about the same time a shock tube was added for study and research in high-speed flow.

The structures laboratory consisted primarily of load-applying equipment and instrumentation involved in measuring the effect of various loads on flight vehicles (Figure 11). The following equipment were used: a 60,000-lb. Tinius Olsen testing machine, a 60,000-lb. Tate Emery testing machine, a Rockwell hardness tester, several tension dynamometers, strain indicators, load cells, oscilloscopes, vibration meters, vibration analyzers and velocity pick-ups.

The Aeronautical Engineering program was modified in 1954 to include a theoretical option for students desiring a stronger background in mathematics and physics. Also, because of the loss of some of the propulsion capability, the previous airplane design and power plant options were combined to form the airplane and power plant option. About this time, Professor Clauser left the university, and Dr. Harold De Groff became the new Head.

By the mid 1950's, the Air Transportation enrollment was dwindling, so a decision was made to close this program effective with the 1958 graduating class. Some aspects of this field were transferred, however, to a new Transportation Studies program in the Department of Industrial Management. The School's name was then changed to the School of Aeronautical Engineering in July 1956.

With space flight coming closer to reality, the School began to initiate new courses that reflected that interest. In the fall of 1957, Angelo Miele offered a course entitled "A preliminary approach to the mechanics of terrestrial and extraterrestrial flight." By the following year, Hsu Lo offered a course on orbit mechanics. This was followed by a course on astrophysics given by Paul Lykoudis. Research in these areas included work on chemical and radiative aerodynamics, plasma jets and high temperature materials and structures.

A New Era

In 1960, the School of Aeronautical Engineering and the Division of Engineering Sciences were merged, and the School name changed to include both of these areas. The Engineering Sciences department had a small building on the main campus, so in 1960 most of the faculty from both programs were housed in that building, but the laboratory buildings at the airport were still retained. The Division of Engineering Sciences had been created in 1954 to offer advanced engineering mechanics with a strong basis in mathematics and physics. Since the Aeronautical program had been moving in that direction, it was felt that the two curricula could function well together. The expectation was that they would find common ground and merge into a single program. Although there was cooperation in developing basis courses at the sophomore level, the faculty could not agree to a single unifying curriculum that satisfied all aspects of the School's mission. Thus, two separate degrees were maintained until the Engineering Sciences program was terminated in 1973.

During the early years of this era, Professor De Groff became involved with a small company in the Research Park, and took a two-year leave of absence to devote more time to that project. Professor Paul Stanley became the acting Head during that time period. De Groff later came back to the Dean of Engineering Office, and the School had a series of acting heads until Hsu Lo was named the Head in November of 1967. John Bogdanoff was Head from 1971 to 1972.

A new type of graduate program was established in June 1963. Titled Master of Science in Astronautics, this new degree was developed in conjunction with the U.S. Air Force Academy. A select group of 12 to 15 cadets who had advanced credit beyond that necessary for graduation were brought to the Purdue campus immediately after graduation from the Academy. They took three courses during the summer session, followed by a special one-month course in August, and then 15 credit hours during the fall semester with graduation in January. The original program involved a thesis, but that was soon deleted in favor of additional courses due to time constraints. After a few years, the program was lengthened to include the entire academic year, and provided courses in hypersonics, electronics, propulsion, along with choices in a major and minor area. This program lasted about twelve years and was highly successful in that it enabled Air Force personnel to acquire a Master's degree relatively quickly before going on to flight training. Although not intended to be an astronaut-training program, seven graduates subsequently did become astronauts. In addition, Purdue students from a variety of other disciplines have also been selected for space flight over the years, and there are now twenty-two Purdue graduates who have been astronauts. The university was greatly saddened in January of 1967 to learn that an Apollo spacecraft fire on the ground had killed two Purdue graduates, Gus Grissom and Roger Chaffee. Grissom had been involved in other space flights, but this was to be Chaffee's first. Two campus buildings were named in their honor. Grissom Hall was remodeled in 1967, and became the primary home for Aeronautical Engineering and for Industrial Engineering.

The School also began to participate in the Engineering Co-op program in 1964. This course allowed five work sessions alternating with study semesters beginning after the freshman year. The program took five years to complete and included all summer sessions during that time. This co-op effort is still quite active, and provides an excellent method of coordinating industrial experience with academic training.

During this time period, several new laboratories were created to complement existing ones. An IBM 1620 digital computer was acquired, and all students had to learn programming. Other laboratories provided test facilities in experimental engineering sciences, magneto-fluid mechanics, material research, random environments and composite materials.

Transition and Growth

As the Engineering Sciences program was phasing out, the School was renamed the School of Aeronautics and Astronautics, the title retained to the present time. Professor Bruce A. Reese was appointed the new school Head in 1973, transferring from the School of Mechanical Engineering. This was the first step in bringing a stronger propulsion influence back to the aero curriculum.

The early 1970's were a time of reduced student enrollment due to a downturn in the aerospace industry. This came about because of a mid-east oil embargo, which caused increased oil prices, produced a reduction in air travel and a reduction in new commercial airplane orders. Military spending also decreased, and the end of the Apollo missions curtailed spending in the space field. Curriculum changes made programs more flexible to attract students, and to facilitate a continuation directly to a Master's degree. Basic courses were required in aerodynamics, structures, thermodynamics, airplane performance, and control systems. The student could then choose a major and a minor area and also have 12 credit hours for technical electives over a wide range of courses. The major and minor areas were: (1) aerodynamics, (2) flight mechanics, astronautics, guidance and control, (3) propulsion, and (4) structures and materials. There were also a few courses in air transportation that had been developed recently by a new faculty member. Graduate education had been growing continuously for several years, particularly in control systems and in space mechanics, and was accompanied by a steady rise in sponsored research from various government agencies and private industry. The 1970 to 1980 decade saw sponsored research grow from $350,000 to $500,000 per year. Faculty members diversified their research interests to include wind tunnel testing of buildings, automobiles, and ships. Research was also done in bio-engineering, aerodynamic noise, composite materials, and seismic resistance of fossil-fuel power plants.

After a low undergraduate enrollment of about 200 students (excluding freshmen) in 1972, the enrollment began to steadily increase throughout the decade, reaching 500 in 1980, and increasing to 570 students by 1989. Graduate enrollment followed a somewhat similar pattern, but the changes were not as drastic, since as industrial jobs became more difficult to find, more students did graduate work.

Professor Henry Yang became the new School Head in 1980. He had come to Purdue in 1968 and was very active in research in finite element analysis.

Under his leadership the School grew in student numbers and in research expenditures, so that by 1985, the sponsored research budget had reached $1,500,000 per year. The structure of the undergraduate curriculum remained the same, but a number of new courses were introduced which broadened the range of interests. Some of these new courses were: fatigue of structures and materials, flight dynamics laboratory, unsteady aerodynamics, digital flight control, transonic aerodynamics, nonlinear systems, advanced composite materials, optimal trajectories, low gravity fluid mechanics, and optimal systems design. During this time period, aerospace design received increased emphasis by means of a program supported by Lockheed Missiles and Space Company at the senior level, and another design course at the sophomore level. Design projects were developed with Lockheed engineers, who visited the campus twice a semester to evaluate the student's work. Those engineers assisted with the final design evaluation at the end of the semester, and awarded prize money to cover expenses of preparing reports. Airplane design was also helped by means of funding from the University Space Research Association (USRA) and later from the Thiokol Corporation.

In 1984, Professor Yang became the Dean of Engineering at Purdue, and held that position for ten years before becoming Chancellor at the University of the University of California at Santa Barbara. His successor as School Head was Alten "Skip" Grandt who joined the faculty in 1979 in the structures and materials area. By 1990, the undergraduate enrollment had reached a maximum and was decreasing rapidly. The graduate enrollment did not show such a significant change, and sponsored research spending continued to increase, reaching a yearly value of $2,300,000 in 1993. The undergraduate curriculum remained relatively unchanged in the late 1980's and 1990's, although a number of new courses were developed to represent the faculty interests. New laboratory developments during this period included a laser laboratory, a guidance and control laboratory, a composite materials laboratory, and rocket combustion and propulsion laboratory. In 1993, John P. Sullivan became the School Head, and Thomas N. Farris replaced him in 1998.

Current Status of the School

As of the Fall semester of 2002, the School of Aeronautics and Astronautics had 22 faculty, 407 undergraduate students (excluding freshmen), and 157 graduate students (90 M.S. and 67 Ph.D.). The separate Department of Aviation Technology numbered 33 faculty and 610 undergraduate students. For comparison, the Purdue Schools of Engineering enrollment is approximately 8000 students with 275 faculty, and the total Purdue West Lafayette campus enrollment is approximately 38,000.

Current Purdue President Martin C. Jischke is formally a member of the Aeronautics and Astronautics faculty. Other faculty play research leadership roles in many technical disciplines as evidenced by current external research expenditures of more than $4 million annually. Many faculty serve on prestigious editorial boards, 7 are fellows of at least one major society (AIAA, AAS, ASME, SEM), an additional 6 are associate fellows of the AIAA, and one emeritus professor is a member of the National Academy of Engineering.

The present Aeronautics and Astronautics undergraduate curriculum has two areas of concentration, aeronautics, and astronautics. These two curricula contain many of the same basis courses, but differ in terms of required courses in areas such as dynamics and control, and propulsion. The astronautics curriculum contains rocket propulsion, spacecraft attitude dynamics, and spacecraft design, while the aeronautics curriculum contains jet propulsion, flight dynamics and control, and aircraft design. Both curricula require 15 credit hours of major and minor area electives that may be chosen to emphasize a student's particular interest.

The School is involved in research in all disciplines. Aerodynamics research includes computational methods, separated flows on bodies at high angles of attack, aerodynamics of rotors and propellers, boundary layers, wakes, and jets in V/STOL applications and noise, experimental methods using laser systems, and boundary layer transition. Laboratory equipment includes the Boeing subsonic wind tunnel with a 4x6 ft. test-section, along with three smaller wind tunnels and a water table. The Boeing Compressible flow Laboratory includes a 2-inch blow-down wind tunnel, and a shock tube. Also in use are a 4 inch Mach 4 Ludweig tube, and a 9.5 inch Mach 6 Ludweig tube with a quiet flow test section.

The Control Systems Laboratory includes high-end work stations to develop methods and software for the analysis and design of complex dynamical systems. Undergraduate laboratories include a two-degree-of-freedom helicopter experiment, and a three-degree-of-freedom experiment to simulate the attitude dynamics of a flexible spacecraft. A remotely piloted vehicle is under development to perform a variety of dynamics and control experiments.

The propulsion area has an Advanced Propellants and Combustion Laboratory composed of two reinforced concrete test cells. One cell contains a rocket thrust stand handling thrusts to 1000 lbf. The other cell is for hybrid rocket combustion studies for a variety of nontoxic hypergolic propellants. The newly renovated High Pressure Laboratory in the Zucrow Laboratory is shared with Mechanical Engineering. This facility has two test cells for testing propulsion devices to 10,000 lbf thrust. One cell is devoted to air-breathing propulsion, and one to rocket propulsion. The Energy Conversion Laboratory located at the Aerospace Sciences Laboratory has four work areas: propellant area, electrochemistry area, physical energy conversion area, and the catalysis area.

Structures and Materials research includes work in composite materials, computational structural mechanics, damage tolerance analysis, experimental structural analysis, aeroelasticity, tribology, manufacturing, wave propagation, and optimal design methods. The McDonnell Douglas Composite Materials Laboratory contains equipment for fabrication and testing of composite laminates. The laboratory includes an autoclave, a hot press, an En Tec filament-winding machine, a water jet cutting machine, an x-ray and an ultrasonic C-scan system. The Fatigue and Fracture Laboratory is equipped for research directed at evaluating the damage tolerant properties of materials and components. Two computer controlled electro-hydraulic test machines and associated equipment are used to measure fracture loads, and to study fatigue crack formation and propagation. The Structural Dynamics Laboratory has the latest equipment for recording ultra-dynamic events, such as Norland and Nicolet recorders, a one-million-frame-per-second camera, an impact gun, and data acquisition equipment for the study of impact to structures and stress waves. The Tribology and Materials Processing Laboratory is maintained jointly with the Center for Materials Processing and Tribology. A test machine is available for the study of fretting fatigue at room and elevated temperatures, a frictional apparatus for high and low speed sliding indentation, lapping and polishing equipment, a vibration isolation table, and a variety of equipment for measurements. A piezo-electric based load frame is available to perform high frequency fretting fatigue experiments related to HCF of aircraft engines.

Alumni Accomplishments

An outstanding academic program needs quality students to succeed. As Indiana's only state supported engineering program, Purdue has an obligation to admit qualified state residents, but also attracts a number of out-of-state students, so that undergraduate enrollment is more than 50% out-of-state. Few other disciplines enjoy a student body with such an avid interest in their profession. Indeed, the aerospace mystique has a powerful motivational influence for learning, and plays a most effective role in outreach activities directed toward encouraging educationally disadvantaged students to higher levels of achievement. Thus, aerospace engineering students tend to be among the most capable and enthusiastic students on campus.

Alumni achievements are the standard by which any academic program is ultimately judged, and in this regard Purdue is widely recognized. The School inaugurated the "Outstanding Aerospace Engineer" designation in 1999 by recognizing its 58 alumni who had previously been awarded an Honorary Doctorate, a Distinguished Engineering Alumni Award, or who had served as a NASA astronaut. Twenty-one additional alumni have been selected for this honor as of December 2002. These alumni have demonstrated exemplary service to industry, government, academia, or other endeavors that reflect the value of an aerospace engineering degree.

Perhaps the most well known alumni are those who have become astronauts and have participated in space flights, which include Mercury, Gemini, Apollo, space shuttle, Mir space station, and the International Space Station. These individuals are listed below.

Purdue Astronauts (Aeronautics and Astronautics)

Neil A. Armstrong, John E. Blaha, Roy D. Bridges, Mark N. Brown, John H. Casper, Roger B. Chaffee, Richard O. Covey, Guy S. Gardner, Gregory J. Harbaugh, Gary E. Payton, Mark L. Polansky, Loren J. Shriver, Janice Voss, Charles D. Walker

Purdue Astronuats, (Other disciplines)

Eugene A. Cernan, Andrew J. Feustel, Virgil I. Grissom, Michael J. McCulley, Jerry L. Ross, Mary E. Weber, Donald E. Williams, David A. Wolf

Some specific spaceflight accomplishments from these Purdue alumni are summarized below.

- As of November 2002, 52 of 143 (36.4%) total US manned flights have had at least one Purdue alumnus as a crewmember (5 flights had crews with 2 Purdue alumni).

- Two of the seven longest US space flights (aboard Russian space station MIR) were by Purdue alumni (John Blaha and David Wolf)

- Purdue alumni were the first (Neil Armstrong) and last (Gene Cernan) humans to step on the moon.

- In April of 2002, Purdue alumnus Jerry Ross became the first human to be launched into space 7 times. Ross also has the most US space walks (9) and longest accumulated US space walk time (58 hours 18 minutes).

- By the end of 2002, Purdue alumni have flown 610 days in space, believed to be the longest total accumulated by alumni from any US university.

Summary

The goal of this paper is to summarize the history of aerospace activity at Purdue University. This is a many-faceted story (see Reference 5 for further details), as Purdue has grown to world prominence and leadership in aerospace education and research. Although the School of Aeronautics and Astronautics was not formally established as a separate academic unit until July 1, 1945, aeronautical education and research at Purdue began much earlier. The Purdue Aero Club was organized on campus in 1910, and was instrumental in bringing the first aircraft to the Lafayette community in 1911. Aeronautical engineering courses were first offered by the School of Mechanical Engineering in 1921, and extensive World War II programs led to the first aeronautical engineering degrees awarded by the School of Mechanical and Aeronautical Engineering in August of 1943. Through December of 2002, Purdue University has produced 6,011 B. S. degrees, 1,331 M. S. degrees, and 444 Ph.D.'s in aerospace engineering. These alumni have made many significant contributions to aerospace engineering and other technical and non-technical fields.

As the School marks the end of the first century of manned flight, it looks forward to strengthening its leadership in the aerospace arena. It has a dynamic faculty involved with a broad spectrum of basic research issues of national importance. This faculty is backed by an excellent support staff and well-equipped, modern laboratories and computational facilities. The School enjoys an enthusiastic and talented student body that is tremendously excited by the opportunity to further advance aerospace technology. Most of all, the School has a tradition of excellence and accomplishment earned by prior generations of faculty and alumni that encourages the current members to higher levels of achievement.

Purdue University relishes the continued challenge to provide students with the aerospace technology needed for the second century of manned flight. Indeed, it looks forward to helping government and industry shape the future aerospace opportunities for our nation and the world. Purdue faces the next hundred years of flight with confidence and optimism. It is proud of its accomplishments, delights in those of its alumni, and looks forward to continuing to play a leading role in providing the world with undreamed-of opportunities for air and space travel.

References

1. H. B. Knoll, *The Story of Purdue Engineering*, Purdue University Studies, 1963.

2. R. Kriebel, "Old Lafayette," newspaper columns of July 15 and October 7, 1990, *Journal and Courier*, Lafayette, Indiana.

3. K. D. Wood, *Airplane Design, A Textbook on Airplane Layout and Stress Analysis Calculations with Particular Emphasis on Economics of Design*, published by the author, 1st edition, 1934.

4. E. F. Bruhn, *Analysis and Design of Aircraft Structures*, 1st edition, 1941. Subsequent revisions titled *Analysis & Design of Flight Vehicle Structures*, Tri-State Offset Company, 1965, revised 1973.

5. A. F. Grandt, Jr., W. A. Gustafson, and L. T. Cargnino, *One Small Step: The History of Aerospace Engineering at Purdue University*, School of Aeronautics and Astronautics, Purdue University, 1995 (392 pages).

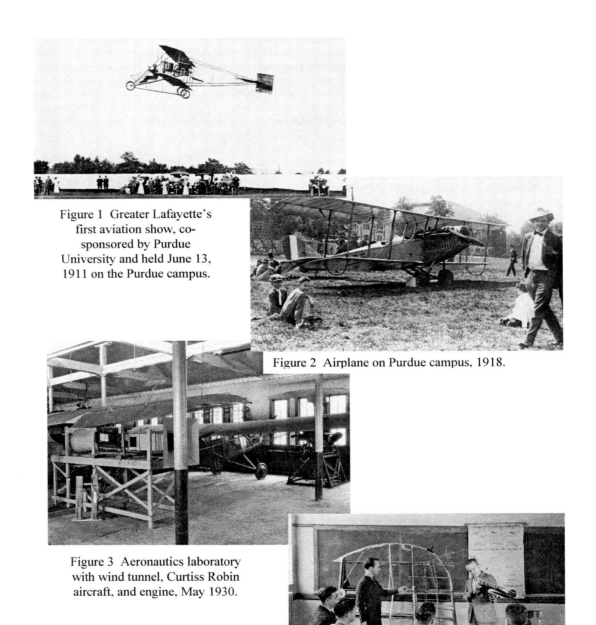

Figure 1 Greater Lafayette's first aviation show, co-sponsored by Purdue University and held June 13, 1911 on the Purdue campus.

Figure 2 Airplane on Purdue campus, 1918.

Figure 3 Aeronautics laboratory with wind tunnel, Curtiss Robin aircraft, and engine, May 1930.

Figure 4 Professor George Haskins (standing right) with aeronautical engineering class, September 1934.

Figure 5 Amelia Earhart, "Counselor on Careers for Women" at Purdue from 1935-37 with students, September 1936.

Figure 6 Professor K. D. Wood with student and wind tunnel, February 1943.

Figure 7 Professor J. Liston (second from left) and students examining aircraft engines, circa 1943.

Figure 8 Professor E. F. Bruhn and structural analysis class, 1942. Note slide rules and shear flow formula on black board.

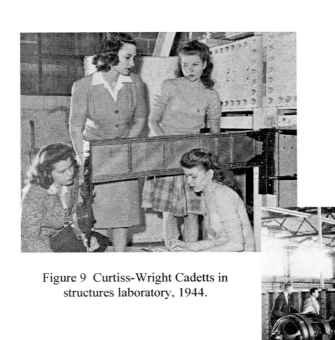

Figure 9 Curtiss-Wright Cadetts in structures laboratory, 1944.

Figure 10 Large subsonic wind tunnel constructed under direction of Professor Palmer (center), 1951. Tunnel was renovated and renamed the Boeing Wind Tunnel in 1992. Also shown are Professors DeGroff (left) and Clauser (right)

Figure 11 Structures laboratory, circa 1945.

Chapter 31

Texas A&M University
Department of Aerospace Engineering

History of the Department

Texas A&M University recognized the importance of the air age as early as 1928 when it introduced an aerodynamics course into the mechanical engineering curriculum. By the early 1930s, several faculty with backgrounds in aeronautical engineering were added, and the course offerings were expanded to include three additional courses in "Aeroplane Design." Interest continued to grow, and in 1938 the Board of Regents of the Agricultural and Mechanical College of Texas approved the creation of the Department of Aeronautical Engineering. However, it was not until the fall of 1940 that the department was established with Dr. Howard W. Barlow as the first head. Offices for the new department were located in an old laundry building, and the departmental objective was to graduate engineers with a firm foundation of theoretical and applied principles in aeronautical engineering.

The first class (1941) consisted of seventeen mechanical engineering seniors who took ten hours of elective work in the new Aero Department. The second class, the class of 1942, and the third class, the class of 1943, were approximately the same size (19) and consisted of ME students converted to aeronautical engineering by the reality of war. In 1944 the first full class of the department was graduated. During the 1942-1943 school year, the Aeronautical Engineering Department became the first in the Southwest (and fourteenth in the nation) to be accredited by the Engineering Council for Professional Development (ECPD), the forerunner of ABET.

In the post-World War II era, research became an important component of the department; and the first graduate degree was awarded by the department in 1949. Most of the early research in the department was in the field of structures and conducted by such well-known individuals as Ben Hammner and Norm Abramson (later head of Southwest Research). In the early 1950s, Fred Weick directed the department's Personal Aircraft Center (forerunner of the Flight Mechanics Laboratory) and conducted far reaching studies on stall-spin phenomena and the prevention of aircraft accidents. He also designed and supervised the construction of the AG-I, the first airplane specifically designed for agricultural spraying. Over six hundred pilots tested this plane, and with their suggestions two improved versions, the AG-II and AG-III were designed. One of the students who worked on the design of these planes was Leland Snow. After graduation he founded Snow Aircraft and later Air Tractor, which combined have produced over 2,000 agricultural airplanes. It is noted that Fred Weick in 1989 received the Daniel Gugenheim Medal for his "notable achievements in the advancement of aeronautics."

In 1954 the Aeronautical Engineering Department moved into the new "Engineering Building." While this building also housed the Dean's offices and several other engineering departments, it did provide space for new wind tunnel and structural mechanics laboratories. In 1956 Alfred E. Cronk, from the University of Minnesota, was appointed Head of the department, a post he held for the next twenty-two years. Under his leadership, many changes occurred. The 7x10 wind tunnel at the campus airport, which was originally constructed in 1943, was converted to a closed-circuit facility with a sensitive six-component balance system and full instrumentation. Today this facility, now named the Oran Nicks Low Speed Wind Tunnel, is one of the finest large low speed wind tunnels in the southwest. With the development of astronautical and space engineering as well as aeronautical, the department started to unofficially call itself aerospace engineering in 1959; and in 1963 it officially became the Aerospace Engineering Department. Subsequently, the curriculum was modernized and the faculty and facilities expanded. Notable among the additions was Dr. W. P. Jones, former director of Aerodynamics for AGARD-NATO and former head of the National Physical Laboratory in England, who joined the faculty in 1967 as one of the first individuals to hold the rank of Distinguished Professor at Texas A&M. Under his guidance, the aerodynamics research within the department grew rapidly. The graduate program was also expanded and the first Ph.D. degree from the department was awarded in 1969. During the three decades prior to 1975, the number of faculty was generally about six to ten. During the period from 1965 to 1975, the faculty included Profs. Cronk, Carlson, Chevalier, Jones, Haisler, Lowy, Norton, Parker, Rand, Rao, Rodenberger, Stricklin, Sweet, and Thomas.

In the immediate post-Apollo era, the enrollment in the department followed national trends and decreased significantly. However, recovery began about 1975 and the number of students increased until it peaked in 1987 at 1,054 undergraduate students (including freshman). In 1984, enrollment management policies were instituted in all engineering programs at Texas A&M University, and since the late 1980s upper-class enrollments have been relatively steady. In 2002, the undergraduate enrollment was approximately 500. During the late 1960's and early 1970's, graduate enrollment was a modest 20-25 students. In early 1980, the department made a decision to increase graduate enrollment with an increase from 20 to 40 graduate students in 1983, and to eighty graduate students by 1991. Since that time, the graduate enrollment has grown steadily until the current steady-state of about 90 graduate students. These enrollment trends are shown in Figures 1 and 2.

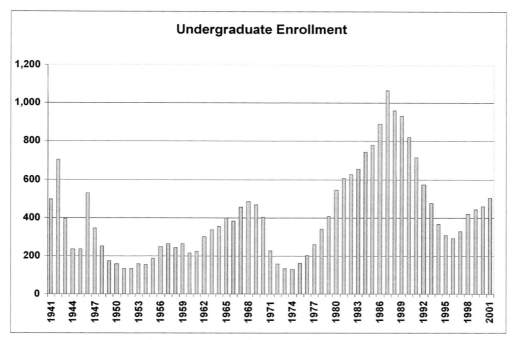

Figure 1. Undergraduate Enrollment History

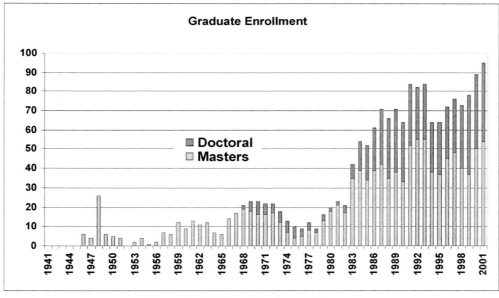

Figure 2. Graduate Enrollment History

By the mid to late 1970s, a number of faculty had left the department (Jones, Rand, Rao, Rodenberger, Stricklin, and Sweet), and additional faculty in the fluid and solid mechanics area joined the department during the 1975 to 1985 time period including Profs. Allen, Ahmed, Cross, Hall, Harris, Horn, Highsmith, Jenson, Kinra, Korkan, Macha, Miley, Ostowari, Pollock, Porteiro, and vonLavante. Many of the faculty that were added during the early 1980s were in the solid mechanics area and the department rapidly became known for its composite materials research. Also during this period, the department expanded its flight mechanics program with the addition of Profs. Chilton and Ward, and the expansion of the Flight Mechanics Laboratory (with 5 aircraft used for instruction and research). By 1995, additional faculty were also added to further strengthen the solid mechanics faculty (including Profs. Lagoudas, Strouboulis, and Whitcomb).

While the department had changed its name from Aeronautical Engineering to Aerospace Engineering in 1963, it was not until the mid-eighties that a substantial change in curriculum was made to include more aerospace engineering topics, and to add new faculty in the space dynamics and control area. Notably among these was chaired professor, Dr. John Junkins. During the late eighties and early nineties, the faculty in the dynamics, control and space-related area was expanded with the addition of Profs. Chilton, Vadali, Kurdila, Crassidis, Alfriend, and Valasek. Prof. Strganac was added in 1989 to provide expertise in the aeroelasticity area. After Profs. Chilton, Kurdila and Crassidis left the department; they were replaced in early 2000 by Profs. Hurtado and Mortari.

During the past five years, the fluid mechanics faculty has been significantly strengthened by the addition of Profs. Rediniotis, Cizmas, Girimaji and Bowersox. Additional faculty have also been added in the solid mechanics area (Boyd and Talreja). Overall, the number of faculty has grown appreciably during the past twenty years, from ten in the mid-eighties, to 17 in the mid-nineties, to a present faculty of 21.

The department has always had a strong reputation in the undergraduate design area. This began in the early days with Prof. Stan Lowy who produced many winning teams in the AIAA design competition. With the retirement of Prof. Lowy in the late 1970s, Prof. Chevalier undertook the teaching of the aircraft design course sequence. With Chevalier's retirement, Prof. McElmurry joined the faculty after retiring from NASA-JSC. He continued teaching the aircraft design courses until 1997 when Prof. Valasek joined the faculty. Prof. Valasek implemented a design-test-build-fly approach in which students designed an aircraft, built and tested a wind tunnel model in the 7' x 10' low speed wind tunnel, then built an RC model (typically 6 ft wingspan), and finally flew the RC model. Prof. Carlson took over the aircraft design course in 2001 with the assistance of Prof. Lund who joined the faculty from his career at Boeing. The design-test-build-fly concept has continued, although the hand-built wind tunnel models have been replaced by models created with a rapid prototyping machine. The design sequence was expanded in the late 1990's with a rocket design option headed by Prof. Pollock. Students design, build, test and fly rockets which typically have a mass of 15-20 kg.

As a result of the growth in the academic and research programs, the department in 1991 moved into the new H. R. Bright Building. This relocation, which was overseen by the department head, Dr. Walter Haisler, essentially quadrupled the classroom, laboratory, and office space available to the department and involved the construction of new undergraduate and graduate laboratories. Currently the department utilizes the basement, first, sixth and seventh floors of the H. R. Bright Building which provides approximately 25,000 square feet of laboratory space; plus departmental classrooms, and office space for all faculty, graduate students and staff. Instructional laboratories were brought to state-of-the-art with the acquisition of modern wind tunnels, water tunnels, a flight simulator, materials and structures testing equipment, an operational jet engine, an undergraduate and graduate computer laboratory, and other laboratory equipment. A more complete listing of current facilities is provided in a later section.

Figure 3: H.R. Bright Building

Since the mid-1960s, the department has experienced a steady growth and expansion in its research and graduate programs. Some of the significant contributions of these programs to the aerospace community are listed below. Starting with just a few research faculty in the mid-1960s and a research volume of approximately $500,000 annually, external research funding has now grown to more than $5M in 2003. While the early (1940s-50s) research programs focused on structural mechanics, the 1960's-70s saw a focus on fluid mechanics and aerodynamics, and a decreased emphasis on solid mechanics. The period from 1980 through 1990 saw an expansion again in the solid mechanics area, particularly in composite materials (from two to seven faculty). In the mid 1980's to 1990's, the department expanded its teaching and research capabilities into the flight mechanics and space dynamics and controls area. Commensurate with increased research activities beginning in the mid 1980's, Figure 2 shows a similar increase in graduate enrollment, from twenty in 1982 to eighty in 1992. In recent years, the ratio of doctoral to masters students has been increasing, and presently doctoral students make up almost 50% of the total graduate student body. Faculty are active in scholarly activities and publish on average three plus referred journal publications per year. Details of current and recent research programs are provided in a later section.

Significant Contributions of the Department to the Aerospace Community

The faculty and students of the Department of Aerospace Engineering have over the years made many significant contributions to aerospace technology and the aerospace community. In addition, the students have regularly performed very well in national and international competitions. Some of these achievements are listed in this section.

Aero/fluid Dynamics

- Space Shuttle Orbiter – Test of configurations from the straight wing orbiter, versions with jet engines, through final version. Building of the model used around the country.

Testing of Orbiter and 747 combination. Testing of Shuttle in launch tower, particularly after the Challenger accident. Testing of escape procedures from the Orbiter. Tile roughness effects on heating and roughness effects on landing and low speed performance. (Oran Nicks Low Speed Wind Tunnel)

- Computational prediction methods for the thermal behavior of high altitude balloon packages and the night-day trajectory performance of high altitude balloons (THERMTRAJ). Modified versions still in use today.
- Vortex box unsteady lifting surface method. Used for helicopter aerodynamics and vortex/hover analyses.
- Transonic airfoil and wing design methods (TRANDES, TRANSEP, ZEBRAINV)
- Chemical and thermal nonequilibrium radiating flowfield analysis methods
- Development of intelligent fluid diagnostic instrumentation and microprobes
- Studies of wind effects on high rise building and offshore structures
- Development of methods for controlling oil spills in the open ocean
- Methods for hydrodynamic propulsion and navigation
- Modeling and closed loop control of complex flows
- Fuel powered and thermoelectric SMA actuators
- Reconfigurable synthetic jet actuators for closed loop hingeless flow control and separation control
- Active skin for turbulent drag reduction
- Smart systems for meso and micro-air vehicles.
- Unsteady methods for predicting performance and clocking of large turbines
- Methods for the prediction of stall-flutter in turbo machinery
- Rotor/stator interaction in multi-stage compressors and turbines
- Reheat turbines and turbine-combustors
- Aeroacoustic prediction methods
- Turbulent mixing and combustion
- Analysis of the physics of turbulence in complex fluid flows

Solid Mechanics

- Pioneering efforts (1960s) in the development of finite element software tools for static and dynamic, nonlinear, inelastic response of shells of revolution (SNASOR, DYNAPLAS)
- Nonlinear FEM analysis of Apollo aft heat shield
- Practical finite element methods, numerical nonlinear structural analysis algorithms for metals and composites, and constitutive modeling of materials
- New algorithms for solving large deformation and large strain structural problems. Constitutive models for characterizing inelastic and creep behavior of materials at high temperature. Predictions methods for the dynamic response of large space structures.
- Studies of the material degradation of large space structures
- Composite material characterization
- Analytical models to characterize the response of composites
- Micromechanics of composite materials
- Constitutive modeling and applications of shape memory alloys
- Damage mechanics of composite materials
- The effect of ice and foam impact on shuttle tiles
- Fatigue and aging of aircraft structures.
- Methods for the structural analysis of high altitude balloons
- Constitutive modeling -- especially viscoelasticity, shape memory alloys, progressive damage, and damage mechanics for both metals and composite materials

- Smart materials, SMA technology and MEMS
- Textile composites
- Computational FEM, error estimation, grid optimization, and global/local stress analysis methods
- Nano mechanics

Dynamics and Control
- Fundamental stall spin studies
- Design and construction of the first airplane designed specifically for agricultural aviation
- X-38 studies of separation effects from B-52
- General aviation GPS system
- Intelligent air traffic management and control
- Formation flight control of UAVs
- Fault tolerant adaptive control
- Spacecraft navigation and control
- Actively involved in Star Wars research during 1987-1992. Our focus was on rapid maneuvering of space structures and vibration suppression. Some of the projects we worked on:
  - R2P2: Rapid Retargeting and Precision Pointing Experiment
  - ASTREX: Advanced Space Structures Technology Research
- Spacecraft dynamics and control and nonlinear optimal control/guidance
- Spacecraft attitude controls using Lyapunov theory for maneuvering spacecraft
- Numerical optimization schemes for determining optimal maneuver torques for spacecraft.
- Spacecraft formation flying
- Optimal guidance of interplanetary vehicles
- TechSat21 and Power Sail Projects
- New electro-optical sensing technologies
  - StarNav – a self-calibrating active pixel stellar camera for real-time star pattern recognition attitude estimation.
  - VisNav – an analog vision sensor and a unique structured light approach to proximity navigation.
  - SunNav – a sensor for autonomous orbit navigation
  This work has culminated in a flight experiment (StarNav I) for the Space Shuttle STS-107, the attitude sensor (StarNav II) and the NASA New Millenium spacecraft EO-3 GIFTS, and a relative navigation sensor (VisNav) for autonomous aerial refueling.
- Development of short courses in flight test engineering
- Designed, built, and tested an on-board experiment for NASA's Long Duration Exposure Facility
- Designed, built, tested a wind tunnel test apparatus that permits unique investigations of aircraft limit cycle oscillations and related nonlinear instabilities
- Development and validation of analysis tools for nonlinear aeroelastic responses (with and without control surfaces)

Awards in Student Competitions
- First Place, 1973 AIAA Bendix Design Competition
- First Place, 1974 AIAA Bendix Design Competition
- First Place, 1974 International Astronautical Federation Contest
- Honorable Mention, 1983 AIAA Bendix Design Competition

- Second Place, 1984 AIAA United Technologies Individual Student Aircraft Design Competition
- First Place, 1990 AIAA Allied Team Space Design Competition
- Second Place, 1998 AIAA/Cessna Aircraft/ONR Undergraduate Design/Build/Fly Competition
- Third Place, 2000 AIAA Foundation Undergraduate Team Airplane Design Competition
- First Place, 2001 AIAA Region IV Student Conference Undergraduate Team Design Competition
- First Place, 2003 AIAA Region IV Student Conference Undergraduate Team Design Competition
- Innumerable First Place winners during the past thirty years, AIAA Region IV and National Student Paper Competitions

The Department Today

Today, the Department of Aerospace Engineering is primarily located in the H. R. Bright Building, but it also has the Oran Nicks Wind Tunnel at Easterwood Airport, a flight dynamics laboratory at the Riverside Campus, and access to numerous facilities in the Space Technology Center. In the fall of 2002, the department had twenty-one tenure track faculty, 497 undergraduate students, and 98 graduate students. Approximately half of the graduate students are enrolled in the doctoral program. The department conducts over $5.5 million in research annually. Faculty are active in scholarly publication with an average of over three journal publications per faculty member. Many of the faculty have been recognized nationally for their educational and research contributions. Two of the faculty are members of the National Academy of Engineering (Junkins and Alfriend). Two of the faculty have been awarded the AIAA Mechanics and Flight Control Award, and the AAS Dirk Brouwer Award (Junkins and Alfriend). Junkins has also received the Frank J. Malina International Academy of Astronautics Medal, and the Pendray Aerospace Literature Award. Three of the faculty (Junkins, Carlson and Alfriend) are Fellows of AIAA and three (Haisler, Alfriend and Junkins) are also Fellows in the American Astronautical Society. Three faculty (Haisler, Junkins, and Carlson) have each won the AIAA/ASEE J. Leland Atwood Award for contributions to education. Lagoudas is a fellow of ASME and Haisler and Carlson are Fellows of ASEE. Nine additional faculty are Associate Fellows of AIAA.

Mission

The overall mission of the Texas A&M University Aerospace Engineering Department is:
- To provide a quality undergraduate and graduate aerospace engineering education.
- To advance the engineering and science knowledge base through research.
- To assist industry in technical applications and innovations.
- To serve the aerospace profession through leadership in these areas.

Faculty

The faculty (July, 2003) consists of eleven full Professors, seven Associate Professors, and three Assistant Professors; and these are listed in Table 1. The faculty are generally grouped in three discipline areas: Aerodynamics and Fluid Mechanics (5 faculty), Solid Mechanics and Materials (8 faculty), and Flight Mechanics, Aeroelasticity and Dynamics and Control (8 faculty). Professor Alfriend will move to a Research Professor status September, 2003. Prof. Amine Benzerga will join the faculty as an Assistant Professor January 2004 in the area of nano solid mechanics. The department is presently advertising to fill a faculty position in connection with the newly created Texas Institute for Intelligent Bio-Nano Materials and Structures for Aerospace

Vehicles (TiiMS) and the associated NASA URETI. In addition, the department, as part of TAMU's Strategic Reinvestment Plan has been designated as a Signature Program by the College of Engineering which provides a number of additional faculty positions. It is anticipated that the department will receive approximately ten new faculty positions over the next four years (four in 2003-04). The new faculty will strengthen all of our teaching and research areas; but will focus on new research initiatives in the area of space technology, advanced materials, modeling and control in aero/fluid dynamics, and design, safety/security and integration. In addition, future research will focus on nanotechnology (nanoscale modeling, simulation, testing, multiscale integration), air transportation systems (traffic management, aviation noise and emissions, avionics), next generation spacecraft (propulsion and power, energy sources, launch vehicles), and autonomous vehicles (UAVs and UCAVs).

Professors	Associate Professors	Assistant Professors
Terry Alfriend, Wisenbaker Chair and NAE member	Rodney Bowersox	Paul Cizmas
Leland Carlson, Dir. of Undergraduate Programs	James Boyd	Johnny Hurtado
Walter Haisler, Interim Head and Dir. of Graduate Programs	Sharath Girimaji	John Valasek
John Junkins, Regents Professor, George Eppright Chair, Dir. of Center for Mechanics & Control and member of NAE	Daniele Mortari	
Vikram Kinra, General Dynamics Professor	Thomas Pollock	
Dimitris Lagoudas, Ford Professor and Dir. of TiiMS	Othon Rediniotis	
John Slattery	Thomas Strganac	
Theofanis Strouboulis		
Ramesh Talreja, Tenneco Professor		
S. Rao Vadali, Stewart & Stevenson Professor		
John Whitcomb, Dir. of Center for Composites		

Table 1. Aerospace Engineering Faculty (July 2003)

Undergraduate Program

The primary educational objectives of the TAMU undergraduate Aerospace Engineering program are:

- Using a high quality faculty, provide a comprehensive aerospace engineering education that develops in students the fundamental skills necessary for the design, synthesis, analysis, and research development of aircraft, spacecraft and other high technology flight systems.
- Prepare students for the aerospace engineering profession and related fields by developing in them the attributes needed so that they can contribute successfully to society and to the engineering profession now and in the future.

The 134 semester credit hour undergraduate program offers a balanced and solid foundation in aerospace engineering. After a common freshman year, the sophomore year contains topics common to many fields of engineering as well as an introduction to aerospace engineering and modern numerical methods. The junior year has sequences in the fundamental areas of aerospace engineering – aerodynamics, structures and materials, dynamics and control, and propulsion. These along with technical electives in the senior year provide a strong fundamental basis for advanced study and specialization. Design, while contained throughout the curriculum, is particularly emphasized in the senior year through a required design elective in either wing design, jet engine design, or structural design and a two-semester design-build-test-build-fly (or simulate) capstone design sequence. During the first semester of the project, students working in small teams use integrated analysis and design tools to design an aerospace system such as an aircraft, rocket, or spacecraft subject to specific goals, objectives, and constraints. Then during the second semester, they plan and oversee the construction of accurate models, conduct wind tunnel and bench tests, analyze the results, and then for aircraft and rockets build models that actually fly. For spacecraft, special simulation techniques are used to test the flight design. The overall curriculum, which integrates theory, analysis, and design, will prepare a student for either graduate school or entry into industry. For details concerning the curriculum and courses, see the departmental website at http://aero.tamu.edu/.

Graduate Program

The Department of Aerospace Engineering offers four graduate degrees: Master of Engineering, Master of Science, Doctor of Engineering and Doctor of Philosophy. The Master of Science degree is suited for research or design-oriented students, and has two options: thesis and non-thesis. The thesis option requires a minimum of 32 credit hours; the non-thesis option requires 36. The Ph.D. program is intended to prepare well-qualified students for management, research and teaching careers in industry, government and academia. It requires a minimum of 96 credit hours beyond the bachelor's degree or 64 beyond the master's degree. A minimum of 48 credit hours of formal course work is required beyond the bachelor's degree. The rest involves intensive research. A dissertation is mandatory. Doctoral students are required to pass a qualifying examination within one year of beginning the Ph.D. degree. The Master of Engineering and Doctor of Engineering degrees are intended for students planning careers in industry or small business as engineering specialists or technical managers. Both degrees combine elements of engineering and business administration. The Master of Engineering degree requires a minimum of 30 credit hours. The Doctor of Engineering degree includes advanced courses in analysis and synthesis of engineering systems, business administration and humanities and social sciences. The DE program consists of 96 credit hours beyond the bachelor's degree, 16 of which may be earned through a one-year internship.

Average GRE scores for admitted students in 2002 was 1,310 (verbal + quantitative) and 1,430 (quantitative + analytical). The department offers graduate students competitive stipend rates starting at $1,500 per month, with full medical benefits, and tuition/fees are often provided by grants and contracts. Students may specialize in the broad areas of aerodynamics and fluid mechanics, structures and materials, flight mechanics, and spacecraft dynamics and control. In July 2003, a new graduate program in Materials Science and Engineering (MSE) was approved; and new students are expected in the program starting in Fall 2003.

Research Centers

Several centers provide a focus for interdisciplinary research in the department. These include:

- *Center for Mechanics of Composites.* CMC is an interdisciplinary research effort focusing on mechanics of composites and utilizing faculty from Aerospace, Mechanical and Civil Engineering.
- *Center for Mechanics and Control.* This center focuses on space technology, navigation and sensor systems, next generation spacecraft, and related topics.

- *Texas Institute for Intelligent Bio-Nano Materials and Structures for Aerospace Vehicles (TiiMS).* TiiMS was established in 2002 and utilizes expertise in biotechnology, nanotechnology, biomaterials and aerospace engineering to develop the next generation of bio-nano materials and structures for aerospace vehicles. TiiMS is a NASA University Research, Engineering and Technology Institute (URETI), one of only seven URETIs chosen throughout the United States.
- *Space Technology Center (STC).* STC is a NASA Research Partnership Center created under a cooperative agreement with NASA's Marshall Space Flight Center to provide low-cost access to internal and external platforms on the ISS and the Space Shuttle. STC supports industry's development of commercial payloads for those platforms, and also supports development of commercial payloads involving other space platforms.
- *Aerospace Vehicle Systems Institute (AVSI).* AVSI is a cooperative of aerospace companies and the Department of Defense working together both to improve and to reduce the costs of complex subsystems in aircraft. AVSI's mission is to lead and facilitate cooperation between industrial organizations, academic institutions, government agencies and to accelerate development of "faster, cheaper, better" vehicle systems, architectures, tools and processes.
- *Center for Aerospace Design, Safety/Security & Integration (ADSSI).* This Center forms an umbrella for coordinating interdisciplinary research activities across the campus in relevant areas associated with aerospace design, safety and security and integration. The new national Homeland Security initiative provides newly emerging opportunities that require multidisciplinary research, systems approach and transfer of innovation to practical products.

Facilities and Laboratories

The department has extensive laboratory and research facilities, and the majority of these are located in the H.R. Bright Building. Among these are the Aero and Fluid Dynamics Lab for undergraduate education and graduate research, the Damping Lab that examines the properties of metal matrix composites in the space environment, the Dynamics and Control Lab for verifying control laws for large flexible space structures, and the Flight Mechanics Lab which uses manned and uninhabited flight vehicles to study the dynamics, control, and flow about full-scale aircraft. The Space Technology Center's 15,000 SF of space hardware fabrication and test facilities includes a Class-10,000 clean room, thermal-vacuum chambers (24" bell jar and 7' x 20' chamber), 8K vibe table, small board-level shaker, and EMI and acoustic chambers.

The department also has a Flight Simulator Laboratory containing a full size two-seat flight simulator that is currently used for control law development associated with intelligent air traffic management, a large Water Tunnel Laboratory for detailed flow visualization, and a Materials and Testing Lab for the processing, study, and evaluation of high-temperature metal matrix composite materials. The Structural Dynamics Testing Lab that can perform dynamic testing of components and assemblies complements the latter, and the Wave Propagation Lab provides nondestructive evaluation of adhesive joints, composites, thin coatings, multi-layered media and granular media.

Instructional and graduate research laboratory equipment includes:
- A 7' x 10' closed circuit low speed wind tunnel (The Oran Nicks Low Speed Wind Tunnel facility)
- Closed circuit 3' x 4' and 2' x 3' subsonic wind tunnels with 200 feet per second capability
- A 2' x 3' x 6' long water tunnel, 4 fps
- Jet engine test set
- 3-D laser Doppler velocimetry

- 3 X 5 surface water table
- Mach 2+ supersonic wind tunnel
- Schlieren system
- Hele-shaw flow apparatus
- Flow visualization bench
- Vortex flow apparatus
- 2-D smoke tunnel
- Two wind tunnels, each 12" x 12", one open return and one closed return
- Three hydraulic MTS testing machines, with high temperature environmental control
- Instron testing machine
- Creep testing machine
- Hot isostatic press (HIP)
- Rockwell hardness tester and Charpy impact tester
- Shaker table
- FFT/frequency spectrum analyzer
- Engineering flight simulator
- Space structures ground test laboratory

Laboratories include:
- Flight Mechanics Laboratory
- Aero-Fluid Dynamics Laboratory
- Dynamics & Controls Laboratory
- Materials & Testing Laboratory
- Propulsion Laboratory
- Main and Satellite Shop
- Structural Dynamics Testing Laboratory
- Water Tunnel Laboratory
- Wave Propagation Laboratory
- Computer Laboratory

Current Research

In 2003, the department had research projects totaling approximately $5.5M, involving 21 faculty, 90 graduate students, 8 staff and 11 post-doctoral researchers in more than 100 active research projects. Current research topics include the following:

Flight Mechanics, Dynamics & Control
- Formation control of spacecraft, autonomous aircraft and autonomous underwater vehicles
- Novel sensor and navigation systems using the stars (StarNav), the sun (SunNav) and active structured lights (VisNav) to provide accurate solutions to a variety of navigation problems
- Air-traffic management and control systems
- Nonlinear analysis of aeroelastic vehicles
- Cooperative control of ground and space robotic systems
- Rocket design and spacecraft-component design
- Attitude estimation
- Orbit theory and modeling of orbital debris
- Fault-tolerant robust adaptive control of ascent and re-entry vehicles

Solid and Structural Mechanics
- Damage mechanics of polymer matrix composites
- Thermomechanical response of metals and composites

- Mechanics of nonlinear viscoelastic media such as solid rocket propellant and toughened composites
- Fundamentals of experimental mechanics, including Moir interferometry and ultrasound techniques
- Adaptive Finite Element research
- Active materials and smart structures
- Nano materials and structures

AeroFluids
- Non-equilibrium hypersonic flows around re-entry vehicles
- Multidisciplinary design methods for aircraft
- Two- and three-dimensional transonic wing design methods
- Active blade-vortex interaction noise suppression
- Massive separated flows on airfoils and wings at high angles of attack
- Reacting and detonating flows, tip vortex flows and viscous flows in nozzles and ducts
- Application of SMART materials to flow diagnostics, neural networks, fuzzy logic and multi-resolution analysis
- Convergence acceleration for two- and three-dimensional Euler and Navier-Stokes calculations
- Adaptive methods for unsteady flow simulations
- Scientific visualization for complex multidimensional volumetric data
- Analysis and modeling of turbulent mixing and combustion at all speeds
- Fundamental analysis of the physics of turbulence
- Mathematical models for CFD
- Methods for compressor/turbine flows

In Closure

As we celebrate the 100th anniversary of the Wright brothers' historic flight, it is appropriate to note that keeping alive the flames of imagination and innovation is the hallmark of aerospace engineering. The U.S. aerospace industry has shaped the past century for America as well as the world. Indeed, our national security, economic growth and quality of life depend on a variety of aerospace products and technologies. Global mobility, space-based communications, precision farming, advanced medical devices, and a myriad of other benefits we enjoy in modern society, are impacts of efforts in the aerospace fields. The minds that make these efforts are shaped in the aerospace and other engineering disciplines. The Department of Aerospace Engineering is proud to be a participant in educating and training these minds.

The challenge faced by the Department of Aerospace Engineering is to maintain a high level of its core disciplines as well as to keep itself at the frontlines of new developments in the aerospace field. Our history shows that since the establishment of the Department in 1940, we have succeeded in meeting this challenge. The aerospace field continues to lead technological developments, in spite of occasional setbacks, and new opportunities continue to emerge at the horizon.

Chapter 32
Aeronautical and Aerospace Engineering at the University of Florida

Richard L. Fearn[*] and Wei Shyy[†]
University of Florida

ABSTRACT

In one sense aeronautical/aerospace engineering has come full circle at the University of Florida. An aeronautical option in the mechanical engineering department started in 1941. Aeronautical engineering became a separate department in 1946 with a faculty of three, leaving mechanical engineering with a faculty of five. After fifty-six years of parallel developments as separate academic departments, aerospace engineering and mechanical engineering reunited in 2002 to form the Department of Mechanical and Aerospace Engineering. At merger, the two departments were approximately the same size, and the new department has more than 50 faculty members, 700 undergraduate students and 250 fulltime graduate students.

Introduction

The growth and development of the aeronautical/aerospace engineering program at the University of Florida took place within a wider context. Global and national events of the period since 1941 are common knowledge, but regional and local events may not be. Information from the following sources is used throughout this paper without additional citation. Histories of Florida and of the University of Florida are given by Gannon (1) and Proctor and Langley (2), respectively. Various publications by the college and departments over the years provided more local history. Interesting information about the aeronautical/aerospace engineering program was also obtained from the University of Florida Archives including old university catalogs, office files of university presidents and college deans, and files of old photographs. More recent photographs were obtained from college and departmental files. Some of the following material is anecdotal and depends on the memory of those telling the stories.

Describing the aeronautical/aerospace program at the University of Florida is complicated by the various mergers, separations and name changes of the department within which the aero program has resided. Table 1 provides dates associated with the beginnings of the program and subsequent departmental changes.

Perhaps it is useful to view a measure of the size of the aerospace engineering program at the University of Florida before describing its development. A history of the annual number of graduates, shown in Figure 1, provides some indication of the teaching effort but only a hint of the research effort within the program. The numbers of graduate degrees in aerospace engineering are estimated after 1972 when there were several degree programs within the department.

[*] Professor Emeritus, P. O. Box 116250, University of Florida, Gainesville, FL 32611
[†] Professor, P. O. Box 116250, University of Florida, Gainesville, FL 32611

The Early Years

Although the University of Florida is a land-grant university and traces its founding to 1853, it was 1909-1910 when the current name and location in Gainesville were established. The College of Engineering was one of the four colleges formed at that time, offering curricula in civil, mechanical and electrical engineering. There were twenty-seven faculty members at the university including five in engineering, and 181 students enrolled including forty-seven in engineering. Elsewhere in the state, Henry Flagler was busy spending his Standard Oil fortune extending a railroad from Miami to Key West. This engineering marvel was completed in 1913. The Naval Aeronautical Center at Pensacola was established in 1914. During World War I, the university campus was used for military training programs. The war also focused on the need for an efficient transportation system and led to federal subsidization of state road construction. Air passenger service between Key West and Havana, Cuba was initiated in 1918 by Aeromarine.

Access by rail and automobile contributed to the Florida land boom of the 1920s. During the early part of the decade, roads improved to a degree that 2.5 million tourists visited Florida in 1925, most of them arriving by automobile. The Florida land boom was showing signs of strain by 1926 when a September hurricane that caused 400 deaths and left 50,000 homeless in south Florida, effectively ending the boom. An even more devastating hurricane in 1928 caused an estimated 2,000 deaths in the communities near Lake Okeechobee. The stock market crash of 1929 and the national depression followed in short order.

John Tigert was inaugurated president of the University of Florida in 1928, the year that graduate programs leading to a master's degree in chemical, civil, electrical and mechanical engineering were started. Severe spending restrictions were imposed at the university including prohibition of all travel, but a new football stadium seating 22,000 was completed in 1930. During 1933, the university had not been allowed to hire or replace employees, purchase equipment or authorize travel. There were sixteen faculty and 336 students in engineering that year.

Joseph Weil became Dean of Engineering at the University of Florida in 1936 and guided the growth of the college for the next 28 years. At the time Weil became dean, the primary business of the college was teaching, and research played a relatively minor role. Weil emphasized industrial and federally supported research as a means to develop the resources of the College.

Air-conditioning played an important role in the economic development of Florida where summer heat and humidity can be extreme. In 1929 Willis Carrier introduced the modern prototype air-conditioner. Over the next thirty years air-conditioning made its way from public transportation and urban buildings to private homes and vehicles, making Florida summers tolerable for millions of new residents. It is an interesting aside to note the role of a Floridian in the development of air-conditioning. John Gorrie, a physician in Apalachicola, was awarded a patent in 1851 for a machine that produced ice to cool sick rooms and hospitals.

Hard economic times ended with the military buildup for World War II. During the war years, college resources were directed to supporting the war effort in training and research. Fewer than 1,000 students remained on campus, but that was offset by several armed services training programs including the Civilian Pilot Training Program offered by the Civil Aeronautics Authority and administered by the College of Engineering. Shifts in engineering curricula were made to stress material having war significance. This included the mechanical engineering curricula placing more emphasis on aeronautics. The enhanced resources for research in the college of engineering were devoted almost entirely to the war effort and contributed to the development of the mortar type radio proximity fuse, radar, and weather tracking by radio, a forerunner of radar weather tracking.

A phenomenal enrollment boom occurred after World War II as veterans took advantage of the GI Bill. Fall enrollments climbed from 1,500 in 1945 to 6,300 in 1946 to 8,800 in 1947 to 10,200 in 1948. Staff size in engineering grew at a comparable rate. Facilities that had been declared inadequate for a prewar enrollment of 3,500 students were serving almost three times that number. War surplus buildings became the quick fix for student housing as well as classrooms and laboratories, and some remained on campus for many years. In 1947, the year of official transition of the University of Florida to a coeducational university, there was no medical school, no community college system, and no public university in the state south of Gainesville.

The University of Florida had its share of social conflict during the 15 years after the war. Integration was not accomplished at the University of Florida until it received a specific court order about four years after the 1954 United States Supreme Court decision that segregated school systems were unconstitutional. Once forced to do so, however, the university and community were integrated without violence. In the late 1950s, after Joe McCarthy had been censured by the United States Senate, a special Florida Senate committee was formed with Charley Johns as chairman to root out communists, homosexuals and other un-American elements from state universities. More than a dozen faculty members at the university were investigated and forced to resign.

Aeronautics first appears in the course listing of the Mechanical Engineering department at the University of Florida in 1928, the year that Robert Thompson, founding head of the Aeronautical Engineering department, entered the university as a freshman. Thompson earned his bachelor's degree in Mechanical Engineering in 1932 and joined the mechanical engineering faculty as a part-time instructor and operator in charge of the mechanical engineering laboratory while he worked on his master's degree.

In 1939, the Head of Mechanical Engineering Department, N. C. Ebaugh, wrote a brief report recommending that an aeronautical engineering degree not be developed at that time at the University of Florida. He pointed out that of the twelve degree programs in aeronautical engineering offered in the United States, six were endowed by the Daniel Guggenheim Foundation. Low enrollment in the aeronautics option of mechanical engineering (about twelve students per year) and the expense of expanding engineering facilities were cited as reasons for not offering a special degree in aeronautical engineering. Thompson was by then a full-time member of the faculty with the rank of assistant professor. He was involved about half-time with initiating a flight training program for the Civil Aeronautics Authority including serving as ground school instructor. Additionally, he was given responsibility for building up the aviation work in the Department of Mechanical Engineering, and aeronautical engineering became an option in mechanical engineering in 1941. During the next few years, Thompson supervised the construction of a small (18-inch by 18-inch test section) low-speed wind tunnel, a six-component balance system (see Figure 2) and acquired a small smoke tunnel for flow visualization.

The Aeronautical Engineering department, with a separate degree Bachelor of Science in Aeronautical Engineering, was formed in 1946 with three faculty members including Robert Thompson as "Head Professor of Aeronautical Engineering". The other two founding members of the faculty were Ford Prescott and Sam Goethe, both graduates of the mechanical engineering program at the University of Florida and listed as research engineers. The aeronautical engineering program started as a home-grown product. Thompson remained head of the new department for ten years, but Prescott appears on the faculty list for only the first year, and Goethe, for two years. The Bachelor of Aeronautical Engineering degree was accredited by the Engineer's Council for Professional Development by 1949. A graduate program leading to a master's degree was offered by the new department, and the first master's degree in aeronautical engineering was awarded in 1950 to Charles Pearson who shows up on the faculty list during his graduate studies. In a list of thirty-two graduate theses for the College of Engineering in 1951, Thompson is named as the director for six. For the first five years of the department's existence, the faculty consisted of Thompson and one or two interim members, often his graduate students.

Although aeronautical engineering was probably housed originally in the old Benton Hall and Benton Annex with the other engineering departments, it moved in 1950 to an airplane hangar that had been used during the war for training activities (3). The "hangar building" was shared with chemical engineering and was located west of the football stadium on the current site of the O'Connell Center sports dome (see Figure 3).

John Hoover joined the department in 1951 as an associate professor. Hoover earned a bachelor's degree from Alabama Polytechnic Institute (Auburn University) and a master's degree from Georgia Institute of Technology, both in aeronautical engineering. He was director of academics for the Southern Aviation School during the war years, worked for a couple of years in the aircraft industry and was an associate professor of aerospace engineering at the University of Alabama before joining the University of Florida. Hoover taught and advised students in the department for more than thirty years (Figure 4). Many of our alumni have fond memories of him and credit his influence as contributing to their personal and professional maturity. Hoover was also active in engineering professional societies, including the American Society for Engineering Education where he chaired the Aeronautical Division, and the Florida Engineering Society.

Thompson and Hoover, and sometimes one other interim instructor, shared the teaching and research duties for undergraduate and graduate courses in aeronautical engineering until 1955 when William (Bill) Miller joined the faculty as an associate professor. Miller had earned an MSAE degree from Massachusetts Institute of Technology and was a retired naval officer. By the time Thompson resigned in 1957, the department had granted 107 bachelor's degrees and 9 master's degrees. This averages to about the same number of undergraduates as cited by Ebaugh in his report in 1939 recommending against starting a new aeronautical engineering department. During the two-year period ending in 1954, the six departments of the college of engineering awarded a total of 278 bachelor's degrees, 34 master's degrees and one PhD. Aeronautical engineering accounted for 13 bachelors degrees compared to 51 in mechanical engineering and 75 in electrical engineering.

In 1957, Raymond Doll was hired to replace Thompson as head of the department. Doll was an aeronautical engineer and retired naval officer. It seems reasonable to assume that he was responsible for the biennial report for the department submitted at the end of that academic year. It begins, "The Aeronautical Engineering Department is still inadequately housed in the Hanger Building and has only the barest minimum of laboratory facilities required for teaching future aeronautical engineers in this space age. The facilities on hand are not suitable for prosecuting research projects at any level". Doll appears to have a tendency toward plain speaking; he is on the faculty list for only one academic year. The fledgling department of Aerospace Engineering appeared not to be participating in the phenomenal post-war growth of the university and college.

The arrival of David Williams in 1957 marks the beginning of a transition from the early years to a viable research-oriented department. Williams received his PhD in physics at New York University in 1938, and his experience included working as a physicist for the National Advisory Committee for Aeronautics during the war years, a year as Associate Professor at the University of Michigan, followed by ten years as a research scientist at Battelle Memorial Institute in Columbus, Ohio where he contributed to the development of the xerography process. Williams' research interests ranged over many topics including hypersonic flow, orbital mechanics, crystal growth and composite materials, but his focus was on applied optics and spectrometry. He was also noted as an outstanding classroom teacher during his twenty-year career at the University of Florida (see Figure 5).

The department had hired several of its own students as interim instructors after they received their master's degree. Except for Thompson, they usually stayed only a year or two before moving on. In 1958, Allen Ross joined the faculty in such a position. Ross made his career at University of Florida, growing with the department over the next 30 years. He continued his graduate work part time, earned his PhD in 1971 and moved to the university's Graduate Engineering Center at Eglin Air Force Base in the panhandle of Florida. He quickly established a productive research program and continued to be active as a consultant after his retirement from the university in 1985.

Establishing a Viable Research Program

Although some of the major events of the 1960s and early 1970s, such as the civil rights movement, directly impacted the university, and some, such as the war in Vietnam, caused major disruptions of personal lives for some of our undergraduate students and recent alumni, only one branch of this turbulent river of history resonated with the aeronautical/aerospace engineering program at the University of Florida, the space program.

The launch of Sputnik I on October 14, 1957 focused the attention and competitive spirit of our country on space. The Air Force Missile Test Center at Cape Canaveral had been launching missiles since 1950. Echo, a satellite that one could see easily in orbit was launched in 1960. John Glenn's flight into orbit on February 20, 1962 raised the level of interest even more. Facilities of large private companies such as Martin (Orlando), Pratt & Whitney (West Palm Beach), Lockheed (Dunedin), Boeing, Northrop and Douglas (Cocoa) were moving into Florida. Many of the brighter young people who were entering the university to study engineering wanted to participate in the excitement of the space program. This phase of rapid growth and development of aerospace engineering at University of Florida almost coincides with the Apollo program from 1963 to 1972 to land humans on the moon and bring them back safely to earth.

Guidelines for restructuring engineering education in the United States were also in place. In 1955, the "Report of the Committee on Evaluation of Engineering Education" (known as the Grinter Report) was published by the American Society for Engineering Education. It stated that the technical objective of engineering education includes "mastery of the fundamental scientific principles associated with any branch of engineering" and recommended that basic science (including mathematics, physics and chemistry) constitute about one-fourth of the undergraduate program. At that time, L. E. Grinter, chairman of the committee, was Dean of the Graduate School at the University of Florida. This report had significant influence on engineering programs nationwide and guided the transformation that was about to take place in aeronautical engineering at the University of Florida.

By the early 1960s, the college of engineering at the University of Florida had grown and developed to offer the bachelor's and master's degrees in nine fields of engineering and a PhD in six. Undergraduate engineering enrollment was about 2,000 and graduate enrollment, about 300. It was recognized by the university that to successfully compete for competent faculty there must be adequate space and support for research activities as well as adequate salaries. Federal dollars were matched by state funds to construct seven new buildings to house engineering departments. The need was so great for aeronautical engineering that a special "surge" building was built to house the department temporarily from 1964 until the new aerospace engineering building was completed in 1967.

Mark Clarkson was hired as chairman of the aeronautical engineering department in 1961. He earned a bachelor's degree in aeronautical engineering from the University of Minnesota in 1939. After working in the aircraft industry from graduation through the war years, he became a research engineer and mathematician at the Defense Research Laboratory at University of Texas and earned his PhD in mathematics from the University of Texas in 1952. After a year at University of Manchester in England on a postdoctoral research fellowship, he returned to work at Chance Vought Aircraft Corporation in Dallas, Texas. He was supervisor of the Aerophysics Group at Chance Vought when he left to head the aeronautical engineering department at the University of Florida. Clarkson's research interests were initially plasma diagnostics, but evolved through the years to extracting aerodynamic coefficients from test data to flow visualization by the time of his retirement in 1986 (see Figure 6). Clarkson arrived at University of Florida to join Hoover, Williams, Miller and Ross as the faculty of the aeronautical engineering program. It was finally time for rapid growth and development of this program at University of Florida.

It was probably inevitable that one of the first acts of the transformation was to change the name of the department and curriculum from aeronautical to aerospace engineering. Approval from the state board of regents was received about the time of John Glenn's flight aboard Friendship 7. Substantial changes began during the next two years with the addition of two senior faculty members, Knox Millsaps and Bernard (Bernie) Leadon.

Clarkson liked to tell the story of how he had seen in a newspaper article that Millsaps was resigning as Chief Scientist and Executive Director of the Air Force Office of Scientific Research. Informal contacts indicated that Millsaps might be interested in joining the aerospace engineering faculty at the University of Florida. Before opening more formal negotiations, Clarkson called a contact in Washington to inquire if he should be aware of any circumstances about Millsaps' resignation. The reply was that although Millsaps had not been forced to resign these positions, the decision for him to leave had been mutual because, "Millsaps doesn't just call a spade a spade, he calls a spade a bloody shovel". Millsaps earned his PhD from the California Institute of Technology in 1943 in Applied Mathematics and Theoretical Physics. Three years earlier, he had received a BS degree from Auburn University in English literature; he reportedly chose Cal Tech for graduate school so that he could study drums under Sol Goodman. Before joining the University of Florida in 1963, Millsaps had also been on the faculty at Ohio State (aeronautical engineering), Auburn University (physics) and Massachusetts Institute of Technology (mechanical engineering). Millsaps primary love was classical problems in fluid mechanics and heat transfer, but his interests also included electric propulsion, mechanical vibrations and statistical theory of gases.

Leadon received his MS degree from the University of Minnesota in 1942 and worked during the war years at Cornell Aeronautical Laboratory. After the war, he returned to the university where he taught while working on his PhD, graduating in 1955. He was a senior staff scientist at

General Dynamics Corporation for several years before joining the University of Florida in 1964. This is also the year the department finally moved out of the old hanger building that Doll had referred to in his 1939 report. Leadon's research interests were primarily in experimental fluid dynamics including shock phenomena, turbulent mixing, sensors and wind engineering (see Figure 7).

A graduate degree program leading to a PhD was initiated in 1964. In keeping with local tradition, bright students were recruited from the undergraduate program for graduate studies, and three were hired as part-time faculty during the latter stages of their studies. The first PhD was earned by Roland C. Anderson in 1965, with the guidance of Millsaps. Anderson remained on the faculty of the University of Florida until his retirement in 1990, and he was active professionally until his death in 1997. Kynric (Ken) Pell and Dennis Keefer both studied under Clarkson's supervision and received their degrees in 1967. Pell went to the University of Wyoming after graduation where he has served as Head of Mechanical Engineering and Dean of the College of Engineering. Keefer remained on the University of Florida faculty for a few years before moving to University of Tennessee Space Institute where he is the B .H. Goethert Professor of Engineering Sciences. Another student from this era took a slightly different route to the aerospace engineering faculty. After receiving his BS degree in aeronautical engineering at the University of Florida in 1960, James E. (Ed) Milton changed colleges to begin graduate studies in physics where he earned a PhD in 1966, then returned to aerospace engineering as a member of the faculty. He moved to the University of Florida Graduate Engineering and Research Center at Eglin Air Force Base in 1972 where he served as its director until his retirement in 1998.

By 1965 the aerospace engineering faculty on the main campus had tripled in size to fifteen, but that included at least four graduate students who received their PhDs within the next year or two and accepted positions elsewhere. The department moved into its new building in 1967, and Millsaps left the University of Florida to accept a position as Head of Mechanical Engineering at Colorado State University in 1968. Faculty size was about ten as the boom times ended.

Job prospects for graduating aerospace engineers were grim in 1971 and undergraduate enrollment was plummeting. A report on engineering programs in Florida commissioned by the state board of regents precipitated a reorganization of the college in which the department of Aerospace Engineering was merged with the department of Engineering Science and Mechanics, with Lawrence (Larry) Malvern serving as interim chairman. Malvern earned his PhD from Brown University in 1949 and had been a Professor at Michigan State when he joined the Engineering Science and Mechanics department at the University of Florida in 1969. Malvern did pioneering work in plastic wave propagation analysis, and several generations of graduate students studied his text "Introduction to the Mechanics of a Continuous Medium". The merger increased the breadth and depth of the aerospace program, particularly in the area of structures, and provided increased potential for collaborative research among faculty members. The combined faculty size at merger was thirty one, and the engineering science and mechanics branch brought with it responsibility for teaching the service courses statics, dynamics and mechanics of materials. The name chosen for the new department was Engineering Science, Mechanics and Aerospace Engineering.

Disney World opened in Orlando in 1971, the HP 35 calculator was introduced in 1972 and by 1973 the slide rule had disappeared from campus.

Building a Foundation

Tourism became the foundation of Florida's economy, increasing from 5 million visitors in 1950 to 40 million in 1990. Retirees seeking milder winter climates sought affordable housing. By 1980, Florida had more mobile homes than any other state with over 12 percent of the population housed in 760,000 mobile homes and 25 percent of the population classified as elderly. More than 800,000 Cuban exiles have arrived in the United States since the fall of Bastista in 1959, most of them settling in Florida. The vast majority of this growth took place in areas of Florida south of Gainesville where the University of Florida is located.

In 1967, a Federal District Court ordered reapportionment in Florida, shifting political power in the state to the south. During the next decade, although per capita funding of higher education remained among the lowest in the nation, four new public universities were created south of the University of Florida in large population centers. Building a foundation of excellence at the University of Florida would require augmenting state funding by establishing a healthy endowment fund and successful competition on the national level for funded research. One wise investment the state legislature made in higher education was to provide matching funds for large private gifts to state universities.

During the next twenty-three years the department continued to build the foundation for a strong research and teaching program in aerospace engineering. Millsaps returned from Colorado State in 1973 to chair the newly formed department. A year later, an additional merger occurred with the Coastal and Oceanographic Engineering department. Extending the name list was beyond the tolerance of the faculty, and the name Engineering Sciences was chosen for the department. A couple of years later, Coastal and Oceanographic Engineering was restored as an independent department, but Engineering Sciences remained the name for the home department for the aerospace engineering program. Millsap's strategy to guide the department through its next phase of development was to leverage the winter weather and his personal contacts to bring well-known senior people who were nearing retirement age to the University of Florida and then to seek their help in recruiting competent young faculty. Millsaps' health began to fail in the early 1980s and he resigned as chairman in 1986, but continued to teach until his death in 1989. Martin (Marty) Eisenberg became chairman of the department in 1986. Eisenberg had been a member of the Engineering Science and Mechanics department before the merger in 1972. He earned his doctoral degree in mechanics of solids from Yale in 1967 after receiving a bachelor's degree in aeronautical engineering and a master's degree in aeronautics and astronautics from New York University. In 1987, the department's name was changed again, this time to Aerospace Engineering, Mechanics & Engineering Science.

Karl Pohlhausen had retired to Gainesville in 1967 to be near the Millsaps family. When he came to the United States in 1946 as a participant of Operation Paperclip, Pohlhausen was well known for his earlier work in fluid mechanics, particularly boundary-layer theory. He and Millsaps (see Figure 8) began their collaboration at Wright Field in 1948 and continued to work together at the University of Florida until Pohlhausen's death in 1980. Although their primary interest was forced and free heat convection in fluids, they made excursions into other fields such as the metabolism of glucose and the rail gun.

Raymond L. Bisplinghoff visited the University of Florida during winter semesters from 1978 until his death in 1985, teaching courses on aeroelasticity and presenting seminars on technical and national policy topics. Charles E. (Chuck) Taylor joined the faculty in 1981 to strengthen the area of optical stress analysis. He is currently Professor Emeritus and active in the Society of

Experimental Mechanics. Hans von Ohain spent winter semesters on campus during the period from 1982 to 1985, providing undergraduates a special treat for their propulsion course, and reminding them that "inventions are the result of a playful mind". One of those taking his course was Millsaps son, Knox Millsaps, Jr., who is currently a professor at the Naval Postgraduate School. Daniel (Dan) Drucker joined the faculty in 1984, was active in teaching continuum mechanics in our graduate program and in research until his retirement in 1994, and was consulted often on both technical and engineering education topics by the faculty throughout the college and university until his death in 2001. Harry N. (Norm) Cotter, retired vice president for engineering at Pratt & Whitney Aircraft, was an adjunct professor from 1984 to 1992 and taught the undergraduate propulsion course several times. Chia-Shun (Gus) Yih joined the faculty in 1987, teaching the joys of fluid mechanics in our graduate program and continuing his research until his retirement in 1990. Those of this group in residence and other colleagues had lunch together on Fridays and referred to themselves as the "lunch bunch" (see Figure 9).

Research projects underway in the department during the 1970s, and 1980s of direct interest to the aerospace engineering community included the following examples. Damage and wave propagation in composite materials by ballistic impact were investigated by Lawrence Malvern, C. T. Sun, Robert Sierakowski and Allen Ross. Research on compression failure in composite plates and fracture toughness in stitched graphite/epoxy composites was conducted by Bhavani Sankar (see Figure 10). Bernard Leadon worked on the effect of boundary-layer separation on vehicle stability and control, and shock tubes. Ulric Kurzweg studied flow stability and time dependent heat transfer and received patents on a process for thermal pumping. Additional research topics of this period that attracted significant student participation include high-angle-of-attack aerodynamics, fluctuating wind pressures on tall buildings, turbine-blade cooling, combustion, vortex breakdown, computational fluid mechanics, turbulent jets in a cross flow and their effects on vertical take-off and landing aircraft, optical information processing, projectile aerodynamics, structural optimization, control of large space structures and sensor development.

Pursuit of Excellence

Since the late 1980s, the department has shifted its strategy to hiring mostly junior and mid-career faculty. A number of them have become productive and professionally visible at the University of Florida or elsewhere. For example, six younger members of the departmental faculty have won National Science Foundation CAREER or Office of Naval Research Young Investigator awards.

Overall research funding for the department has experienced substantial growth, reaching a level of about $3.5 million dollars for a faculty of about 25 just before the recent merger with mechanical engineering. Faculty and students form working groups to pursue topics of both research and undergraduate teaching interests. Research groups in aerospace engineering include computational fluid mechanics, gas dynamics, experimental combustion and propulsion, pressure and temperature sensitive paints, composite materials, dynamics and control, engineering optimization, and micro-electro-mechanical systems (MEMS). Most faculty members participate in joint projects, and regularly produce publications collaboratively. Our faculty and students are now regularly winning research and teaching awards in professional societies.

During the last several years, the group approach has also contributed noticeably to the University of Florida's success in yielding nationally recognized activities. Brief descriptions of two research projects of the many underway are given as examples of current activity. In early

1996, shortly after Wei Shyy became the department chair, a few faculty members, including Rafi Haftka, David Jenkins, Wei Shyy, Norman Fitz-Coy and Bruce Carroll, wondered what "new" aerospace topics would make sense for the department to promote. The topic should be scientifically and technically interesting, and multidisciplinary. Haftka suggested micro air vehicles (MAV) which received an enthusiastic reaction from the group. A special-topics course was offered the same year by Jenkins, and collaborative activities began. Shyy suggested that the group focus on the flexible-wing concept, and with Jenkins and some students, began to pursue studies in membrane wing aerodynamics. These efforts were before the term MAV became trendy in the technical community. Shortly after the University of Florida started its activities, experts such as Peter Lissaman, Rick Foch, and program managers such as Jim McMichael visited the University of Florida either to offer advice or to learn about our activities. An annual international flight competition began in 1997 with much publicity. Later, Peter Ifju and his students joined this effort. Ifju's interest in windsurfing and his talent in fabrication technologies have significantly advanced vehicle development (see Figure 11), and the University of Florida group has succeeded in winning flight competitions since 1999. The group has also produced scholarly works, and won grants and recognition within the MAV community.

The University of Florida's aerospace engineering faculty, leading a consortium of several universities, won a major project in the Reusable Launching Vehicle (RLV) area under a University Research, Engineering and Technology Institute (URETI) competition. The project has the potential of ten years support from NASA with an initial phase of fifteen million dollars already committed. The goal is to develop scientific foundations for key aspects related to a future RLV, including hypersonics, combined air-breathing and rocket cycles, multidisciplinary analysis and optimization, computational tool development, materials and life expectancy, vehicle health management, and educational outreach. URETI was the first broad-ranged, nationwide competition organized by NASA. More than ten faculty members from the University of Florida are active participants in the effort. Different aspects of our URETI activities have been covered recently in news media such as the New York Times, St. Petersburg Times and Christian Science Monitor.

On the teaching side, broad revisions to the curriculum, especially the undergraduate program, were implemented in the 1990s. The State of Florida requested that its public universities standardize the number of credits required for all bachelor's degrees. In response, the aerospace engineering program not only trimmed the semester-hour credits to 128, but has also gone through substantial revisions of the curriculum. Table 2 summarizes the curriculum structure early in the program (1948), at the time of the first merger (1972) and currently. While there are clear shifts in topics of emphasis through the years, there is an interesting consistency in the overall spirit, namely, a combination of mathematics, basic and engineering science, and the aerospace engineering discipline. Additionally, there have been changes in textbooks and teaching methods within topics, usually an increase in mathematical sophistication and increased use of numerical methods. A modern aerospace engineer needs a strong foundation in mathematics and science, breadth in other engineering disciplines and depth in a chosen specialty for technical competence. Additional studies in areas of "people skills" are increasingly important for multidisciplinary approaches to problem solving. Graduate studies are an important path to a suitable level of competence in aerospace engineering.

A number of graduates from the University of Florida's Aerospace Engineering degree programs have become strong contributors to their chosen professions. We offer a few examples of those who have also continued to contribute to the aerospace program at the University of Florida by visiting periodically to offer advice and encouragement. John D. Anderson, Jr. (BSAE

1959) has established a distinguished career as an aerospace educator and researcher. He is now curator of the National Air & Space Museum. Anderson gave the Distinguished Alumni Lecture at the golden anniversary celebration of the Aerospace Engineering, Mechanics & Engineering Science department in 1996. Could his first technical paper have been "Rocket Fuels", published in the May 1959 issue of the student publication, Florida Engineer? William Powers (BSAE 1963) became a professor of aerospace engineering at the University of Michigan before joining Ford Motor Company where he held several positions including vice president for research before his retirement in 2001. He established an endowed professorship in aerospace engineering at the University of Florida, and earned distinction as a member of National Academy of Engineering, as well as a senior spokesman for the automotive and aerospace professions. Donald Daniel (BSAE 1964, MS 1965, PhD 1973) became Executive Director and Chief Scientist of the Air Force Research Lab, then Deputy Assistant Secretary for the Air Force, the highest ranked career civil servant in the US Air Force, before retiring in 2002. William Sellers, III (BSAE 1973, MS 1975) and Robert Garcia (BSAE, 1986) have become branch leaders at NASA Langley Research Center and Marshall Space Flight Center, respectively. Darryl Van Dorn (BSAE 1962) became the Director of NASA and Commercial Delta Programs for McDonnell Douglas, overseeing activities of Delta rocket launch services for a host of customers. Carl Meece (BSAE 1969) became Director of Pratt & Whitney's Turbine Component Center, responsible for all aspects of the air-breathing engine's turbine technology for the company.

Merger with Mechanical Engineering

With encouragement by the new dean of the college, Pramod Khargonekar, faculty members of the Aerospace Engineering, Mechanics and Engineering Science department, and of the Mechanical Engineering department agreed to merge, creating the department of Mechanical and Aerospace Engineering. The two departments were of comparable size, and Shyy became chairman of the new department when it was formed in 2002. This merger reunited the programs of aeronautical/aerospace engineering and mechanical engineering after a separation of 54 years. Currently the Mechanical and Aerospace Engineering department has more than 50 faculty members, 700 undergraduate students and 250 full-time graduate students.

The rationale for this action came from the following realization: Aerospace Engineering has focused on an interest in all aspects related to flight including fluids mechanics, structural mechanics, and dynamics and control, while Engineering Science/Mechanics has historically been a home for nontraditional and developing curricula in engineering, most recently, biomechanics. Mechanical Engineering is broad in its scope, encompassing all aspects related to vehicles, machinery, thermal and fluid engineering, and energy. Faculty and students from both departments regularly interact with each other, and often publish scholarly works in the same journals. Intellectually, rapid advances in science and technology have significantly broadened the scope of engineering in all disciplines while blurring the boundaries between them. Pragmatically, available financial resources limit the ability to expand and modernize the teaching and research missions.

Wide ranging actions began immediately after the merger, including building renovation, new faculty searches, unified graduate course requirements and Ph.D. qualifying exams. The goal, simply stated, is to maximize faculty and student interactions and to minimize administrative barriers.

References

1. Gannon, Michael, ed., "The New History of Florida", University Press of Florida, 1996.
2. Proctor, Samuel and Langley, Wright, "Gator History", South Star Publishing Co., 1986.
3. Block, Seymour S., Chemical Engineering at the University of Florida, Chemical Engineering Department, University of Florida, 1999.

Table 1. Departmental mergers and "firsts"

1928	First course in aeronautics offered in the mechanical engineering department.
1941	Option in aeronautical engineering offered in mechanical engineering.
1946	Aeronautical Engineering department founded.
1948	First graduating class in aeronautical engineering.
1950	First master's degree in aeronautical engineering.
1961	Department name changed to Aerospace Engineering.
1965	First PhD in aerospace engineering.
1972	Merged with Engineering Science and Mechanics department to form the Engineering Science, Mechanics and Aerospace Engineering department.
1974	Merged with Coastal and Oceanographic Engineering department to form the Engineering Sciences department.
1976	Coastal and Oceanographic Engineering restored as separate department.
1987	Changed name to Aerospace Engineering, Mechanics & Engineering Science department.
2002	Merged with Mechanical Engineering department to form the Mechanical and Aerospace Engineering department.

Table 2. Sample aeronautical/aerospace engineering curricula

Topic	1948	1973*	2003
English including composition	8	6	6
Humanities and social science	16	14	15
Math through differential & integral calculus	16	13_	12
Additional engineering math including num. meth.	0	8_	10
Chemistry with lab	8	5_	4
Physics with lab	8	8	8
Science elective	0	2_	3
Physical education or military science	4	2	0
Elective	6	6_	0
Computer programming	0	1_	2
Intro. to engineering	0	_	0
Engineering drawing _ computer-aided graphics	4	2	3
Elementary design	3	0	0
Mechanisms & kinematics	3	0	0
Statics, dynamics & mechanics of materials	9	6_	9
Thermodynamics	3	3_	3
Electrical engineering	6	3_	4
Materials behavior & selection	3	2	3
Engineering lab courses	9	2	5
Manufacturing operations	6	0	1
Specifications & industrial safety	3	0	0
Elements of aeronautics & astronautics	0	2	0
Fluid mechanics	0	2_	3
Aerodynamics	6	12	3
Astrodynamics	0	0	3
Structures	6	9_	3
Stability & control of aircraft	0	0	3
Propulsion	5	3_	3
Control systems	0	0	3
Electives restricted to approved technical or aerospace courses	5	12_	12
Professional development	2	_	1
airplane design _ aerospace design	6	4	6
Total semester hours	145	134_	128

*   Quarter system being used this year, credits converted to semester hours.

_   Denotes change in nomenclature between 1946 and 1973, but topics seem comparable.

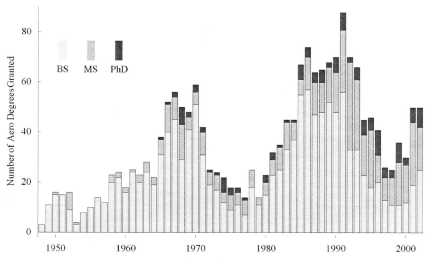

Figure 1 History of the number of graduates in Aeronautica l/ aerospace engineering from the University of Florida

Figure 2 Low Speed wind tunnel and balance system with R. A. on upper deck Thompson

Figure 3 Looking over the football stadium and press boxes toward temporary buildings, the old ROTC building and the hanger building

Figure 4 John Hoover with students
in design class

Figure 5   Dave Williams in the
early 1970's

Figure 6 Mark Clarkson with
undergraduate
Paul(Dean) Johnson in the lab in the mid
1980's

Figure 7 Bernie Leadon and student
constructing a wing tunnel for
studying wind loads on buildings

Figure 8. Karl Pohlhausen and
Knox Millsaps in early 1970's

Figure 9. Lunch bunch one day in 1982 – Knox Millsaps, Chuck Taylor, Joe Hammack, Stan Ulam (mathematics), Prabhat Hajela and Hans von Ohain: Knox Millsaps, Jr. in front.

Figure 10. Bhavani Sankar curing fiber composites in an autoclave.

Figure 11b. Close-up of MAV

Figure   11a. David Jenkins ready to take control of MAV after Peter Ifju launches

Chapter 33
History of The University of Kansas
Aerospace Engineering Department

Jan Roskam, Ackers Distinguished Professor of Aerospace Engineering

1. Academic History

The aeronautical engineering degree program at The University of Kansas (KU), was officially authorized in 1941. Before that time various aeronautical engineering topics were taught at one time or another since the year 1911. Airplane aerodynamics and airplane design were taught as part of the Mechanical Engineering curriculum starting in 1928.

In the 1931-1932 catalog four professional options are listed under Mechanical Engineering: aeronautics, design, petroleum and power. Each of these options consisted of six courses. The courses required for the aeronautical option are listed in Table 1.

Table 1 Courses Required for the Aeronautics Option in 1931-1932

Title	Semester Hrs	Instructor	Title	Semester Hrs	Instructor
Aeronautics	3	Hay	Airplane Design	3	Hay
Aerodynamics Laboratory	1.5	Hay, Baker	Aero. Constr. Laboratory	1.5	Hay, Baker
Aircraft Welding	1	Smith	Thesis	3	Hay & Staff

Note the course in welding. In those days the primary structure of many airplanes consisted of welded steel tubular trusses covered with some type of fabric.

In 1936 the KU School of Engineering applied to the Engineer's Council for Professional Development (ECPD*) for accreditation of various programs. The following year the Council made the following statement about the undergraduate program in Mechanical Engineering and its four options:

"The curriculum of mechanical engineering, with its present staff and equipment, should offer at most two options, one in industrial management and the other in power or heat engineering. Electives, to a limited extent, may be available in aeronautics and heating and ventilating. The petroleum option should be discontinued."

This could be interpreted as saying that the Aeronautical program was inadequate. There are no ECPD records which indicate that the aeronautical program at KU was accredited before 1949.

Between 1936 and 1949 the only change made in the aeronautical option was the addition of a course in Statically Indeterminate Structures taught by the Civil Engineering Department.

The first windtunnel at KU was built in the 1929-1931 period with a budget of $2,000 (sic!). This open-section windtunnel was located underneath the seats on the West side of the football stadium.

KU students were active building and flying their own airplanes already in 1911. The idea of design, test, build, fly has a long tradition at KU. Of particular interest is the joined wing airplane designed and built by four students and a professional welder. Figure 1 shows this airplane which was powered by a 90 hp air-cooled engine via a drive shaft to the pusher propeller. The airplane was tested in the new windtunnel and then built and flown by the students. One can imagine the legal problems with the latter events if this would be attempted in 2003!

The students learned a lot about drive shaft vibrations and aeroelastic deformation. The airplane was first flown by Louis Cogwell and next by Bill Wells who reported that when he tried to roll to the right the airplane actually rolled to the left. The reason was: wing torsional

Figure 1 First Design, Test, Build, Fly Project at KU

deformation as a result of the imposed aileron loads. By stiffening the wing this problem was solved. The shaft vibration problem was solved with the help of automobile brake drum components.

By 1940 rumors of war abounded. The KU administration became convinced that the university should participate in the many activities needed to help the aeronautical part of the war effort. A flight training program was started and by 1941 a degree program in aeronautical engineering was officially established with the blessing from the State of Kansas Board of Regents.

In 1942 a complete AE curriculum (see Table 2) had been established albeit with a freshman year common to all engineering departments. Over the ensuing years the AE curriculum was adjusted to reflect changes in technology and changes in industry expectations. The most recent AE curriculum for 2002 is shown in Table 3. The evolution of credit hour requirements from 1944 to 2002 is indicated in Table 4.

he first acting chairman of the department was the Dean of the School of Engineering, J.J. Jakosky. He hired Edward Brush as potential chairman. However, Brush resigned for unknown reasons in 1942. The next chairman was Harry Stillwell who in turn left in 1944 to head up the AE department at The University of Illinois. Table 5 shows the names of all AE department chairmen. Table 6 lists all AE faculty members over the years.

At first the AE department had one room in Marvin Hall. In 1943 the department moved into the so-called Quonset Hut which was officially known as the Engineering Experiment Station. The AE department stayed there until 1962 when it was moved to the second floor of the (then) new engineering building, Learned Hall. It has been there up to the present.

Between 1962 and 1967 the Engineering Mechanics part of Civil Engineering and the AE department were combined and operated as the Department of Mechanics and Aerospace Engineering with Kenneth Deemer as the chairman.

In 1964 David L. Kohlman joined the faculty. In 1967 it was decided to make the AE department a separate department again and David Kohlman became its chairman. Dr. Kohlman brought in Dr. Roskam and Dr. Smith from Boeing and Dr. Lan from NYU. Vincent Muirhead obtained an NSF grant with matching funds from KU to build a new windtunnel in Learned Hall. That windtunnel is still in operation in 2002.

Table 2 Aeronautical Engineering Curriculum in 1942

Common Freshman Year

First Semester	Hrs	Second Semester	Hrs
Math. 2a, College Algebra	3	Math. 4, Analytical Geometry	5
Math. 3, Plane Trigonometry	2	Engl. 2E, Rhetoric II	2
Engl. 1E, Rhetoric I	3	Chem. 3E, Inorg. Chem. and Qual. Anal.	4
Chem. 2E, Inorganic Chemistry	4	Engr. Dr. 2, Machine Drawing	2
Engr. Dr. 1, Lettering and F.H. Draw	2	Engr. Dr. 3, Descriptive Geometry	3
C.E. 5, Engineering Lectures	1	M.C. 8, Metal Working	1
Gym. or ROTC	--	Gym. or ROTC	--
Total	15	Total	17

Sophomore Year

First Semester	Hrs	Second Semester	Hrs
Math. 5E, Calculus I	4	Math. 7E, Calculus II	4
Phys. 7a, General Engrg. Physics	5	Phys. 7b, General Engrg. Physics	5
Econ. 1E, Introductory Economics	3	A.M. 1, Statics	2
M.C. 1, Foundry Practice	1	M.C. 2, 6, Pattern and Mach. Tool Work	2
A.E. 1, Aeronautics	2	M.E. 3, Mechanisms	3
A.E. 2, Navigation and Meteorology	3	M.E. 154, Heating and Air Conditioning	2
Total	18	Total	18

Junior Year

First Semester	Hrs	Second Semester	Hrs
M.E. 151, Thermodynamics	3	M.E. 150, Machine Design	5
A.M. 50, Dynamics	3	A.M. 55, Hydraulics	3
A.M. 51, Strength of Materials	4	M.E. 159, I.C. Engines	3
A.M. 52, Testing of Materials	1	Engl. 56, Technocal Report II	0.5
M.C. 50, Heat Treatment	1	A.E. 101, Aerodynamics II	3
Engl. 59, Advanced Composition	3	A.E. 102, Aerodynamics Laboratory I	2
Engl. 6, Technical Report I	0.5	A.E. 105, Aircraft Matl's and Proc.	2
A.E. 100, Aerodynamics I	3		
Total	18.5	Total	18.5

Senior Year

First Semester	Hrs	Second Semester	Hrs
E.E. 71, Direct Currents	3	A.E. 151, Airplane Design II	5
E.E. 91, Electrical Laboratory	1	M.E. 53, Seminar	0.5
A.E. 162, Aero Structures	3	A.E. 166, Aero Engine Laboratory	1.5
A.E. 163, Aero Structures Laboratory	2	C.E. 267, Statically Indeterminate Struct.	3
A.E. 150, Airplane Design I	3	E.E. 72, Alternating Currents	3
C.E. 56, Industrial Administration	3	Nontechnical option	2
Nontechnical option	3	Technical option	2
Total	18	Total	17

Grand Total   140 hours

Table 3 Aerospace Engineering Curriculum in 2002

Freshman Year

First Semester	Hrs	Second Semester	Hrs
Math. 121, Calculus I	5	Math 122, Calculus II	5
Engl. 101, Composition	3	Engl. 102, Composition and Literature	3
Chem. 184, Chemistry I	5	Phys. 211, Physics I	4
AE 245, Introd. To Aerospace Engrg.	3	HSS* Elective	3
AE 290, Aerospace Colloquiem	0.2	CPE 121, Fortran	3
		AE 291, Aerospace Colloquiem	0.3
Total 16.2		Total	18.3

Sophomore Year

First Semester	Hrs	Second Semester	Hrs
Math. 250, Math. Of Engrg. Systems	5	Math 124, Calculus III	3
CE 301, Statics and Dynamics	5	AE 445, Aerodynamics	3
Phys. 212, Physics II	4	Phys. 351, Physics III	3
AE 345, Fluid Mechanics	2	ME 312, Thermodynamics	3
AE 290, Aerospace Colloquiem	0.2	CE 310, Strength of Materials	4
		AE 291, Aerospace Colloquiem	0.3
Total 16.2		Total	16.3

Junior Year

First Semester	Hrs	Second Semester	Hrs
Ae 507, Aero Structures I	3	AE 508, Aero Structures II	3
AE 550, Dynamics of Flight I	3	AE 551, Dynamics of Flight II	4
AE 571, Reciprocating Engines	3	AE 572, Jet Propulsion	3
AE 545, Aerodynamics	5	AE 421, Computer Graphics	4
EECS 319 Circuits	4	AE 430, Aero Instrumentation	3
AE 290, Aerospace Colloquiem	0.2	AE 291, Aerospace Colloquiem	0.3
Total 18.2		Total	17.3

Senior Year

First Semester	Hrs	Second Semester	Hrs
AE 521, Aircraft Design I	4	AE 522, 523 or 524 Design II	4
AE 510, Materials and Manufacturing	4	TE/HSS* Electives	12
TE/HSS* Electives	9	AE 291, Aerospace Colloquiem	0.3
AE 590, Senior Seminar	1		
AE 290, Aerospace Colloquiem	0.2		
Total	18.2	Total	16.3

Grand Total 137 hours

* TE stands for Technical Electives, HSS stands for Humanity or Social Science Electives
    TE electives must total 10 hours, HSS electives must total 14 hours

Table 4 Evolution of AE Credit Hour Requirements

Year	1944	1961	1968	1988	2002
Credits taken in AE Dept.	31.5	35	34	53	54
Credits taken in Engrg. outside AE Dept.	39	27	20	17	19
Credits taken in other Engrg. Subjects	15	7	5	5	0
Credits taken outside Engrg.	54.5	71	75	62	64
Total credits required	140	140	134	137	137

Table 5 AE Department Chairmen since 1941

1941 - 1942	J.J. Jakosky (Acting)
1942 - 1944	Harry S. Stillwell
1944 - 1952	W.M. Simpson
1952 - 1961	Ammon S. Andes
1961 - 1967	Kenneth Deemer
1967 - 1973	David L. Kohlman
1973 - 1976	Jan Roskam
1976 - 1988	Vincent U. Muirhead
1988 - 2000	David R. Downing
1991 - 1992	Saeed Farokhi (Acting)
2000 - present	Mark S. Ewing

Table 6 AE Department Faculty Members since 1941

Edward E. Brush	1941	Chuan Tau Lan	1968
Harry S. Stillwell	1942	Howard W. Smith	1970
Robert W. McCloy	1942	William Schweikard	1979
William H. Simpson	1944	Paul E. Fortin	1981
Reid B. Lyford	1945	David R. Downing	1981
Ammon S. Andes	1946	John C. Ogg	1982
Harry W. Johnson	1950	Saeed Farokhi	1984
Edwin K. Parks	1952	James E. Locke	1988
James B. Tiedemann	1953	David R. Ellis	1989
Laddy G. Kimbrel	1959	Ray R. Taghavi	1991
Ferdinand Bates	1960	Mark S. Ewing	1992
Costas J. Choliasmenos	1960	Tae Lim	1992
Vincent U. Muirhead	1961	Richard D. Hale	1998
David L. Kohlman	1964	Trevor Sorenson	2000
Leroy Devan	1965		
Theodore Bratanow	1967		
Jan Roskam	1967		

Circa 1968, the department was granted the privilege to add a PhD degree program in addition to the already existing MSAE program. In 1970 the department added the Master of Engineering (MEAE) and Doctor of Engineering (DEAE) degrees. These differ from the classical MSAE/PhDAE programs by adding a significant design, management and industrial internship requirement. The research components associated with the MSAE/PhDAE programs are slightly reduced in the MSAE/DEAE programs. The rationale was to address the needs of industry.

The number of AE students has fluctuated considerably over calendar time. Figure 2 shows the BSAE degree history. Major dips can be correlated to general downturns in the aerospace industry.

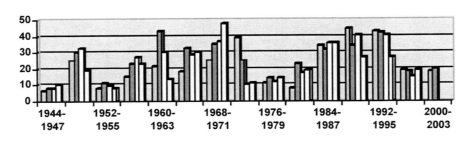

Figure 2 BSAE Degrees Granted During the 1944-2001 Period

Figure 3 indicates the number of MSAE/MEAE degrees granted.

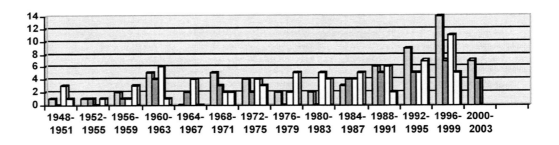

Figure 3 MSAE/MEAE Degrees Granted During the 1948-2001 Period

Figure 4 shows the number of PhDAE/DEAE degrees granted.

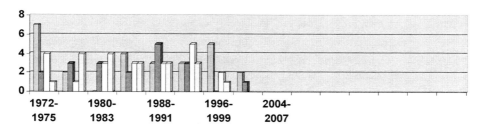

Figure 4 PhDAE/MEAE Degrees Granted During the 1948-2001 Period

2. History of Department Research

Before 1968 research projects were conducted by individual faculty members through the University Office of Research. Starting with 1968 all department research was funneled through the Flight Research Laboratory (FRL). In 2000 the department also formed a Flight Test Center at the Lawrence Municipal Airport.

2.1 Research before the Flight Research Laboratory

Fred Bates started not only research in meteorology but was instrumental in establishing the KU Department of Meteorology. He also initiated a research program in thunderstorms, established the first severe weather warning station in Kansas City. Gus Choliosmenos and Vincent Muirhead expanded this research in the area of tornados. Muirhead even built a tornado simulation facility which generated spectacular pictures of laboratory tornadoes and which helped explain how real tornadoes come about.

Muirhead also designed and supervised construction of the subsonic windtunnel which is still operational in Learned Hall.

The first airplanes were bought by Norman Hoecker. In 1952 Norman acquired a Cessna 120 which was damaged. He rebuilt it into a flight-worthy airplane. This airplane and an old unflyable Stinson were traded for a Cessna 172 which allowed the department to fly up to four people.
In 1955 Norman bought a USAF surplus Beech Model 18 which he and Beech Aircraft Corporation refurbished. This became the first airplane to be used by a KU Chancellor.
In 1973 Cessna donated a new Cessna 172 to the department, followed in 1990 by a Cessna TR182.

2.2 Flight Research Laboratory

The University of Kansas Flight Research Laboratory (FRL) was organized by Dr. Roskam and Dr. Kohlman in 1968. During the years 1968 through 1996 a large number of design and flight test research projects were carried to fruition with many faculty members and students (graduate and undergraduate) involved. These programs were mostly funded by NASA. Typical examples were:

1. The Redhawk* (see Figure 5) program aimed at developing advanced flap and roll control systems for light aircraft.
2. The ATLIT* (Advanced Technology Light Twin, see Figure 6) program aimed at developing advanced airfoil, high lift and ride control systems for light twins.
3. The SSSA* (Separate Surface Stability Augmentation, see Figure 7) program aimed at developing advanced stability augmentation systems without feedback to the pilot.
4. Under a USNavy underwater program an actuator test facility was developed and delivered to the USNavy.
5. Under a Teledyne Brown funded program three autopilots for supersonic target missiles were developed and delivered.
6. Under a NASA funded program a Beranek test facility was constructde and used to test the sound transmission characteristics of many aircraft materials. Much of that testing was carried out on behalf of the aircraft industry in Wichita.
7. Dr. Lan and Dr. Roskam collaborated on the development of a computer code for the determination of steady state aeroelastic effects. This resulted in a NASA certificate of Recognition.
8. Dr. Kohlman and Dr. Roskam collaborated on a program to develop a low cost Parameter Identification system for light aircraft. This program resulted in a spin-off company: Kohlman Systems Research of Lawrence, KS.
9. Dr. Roskam with initial grant support from General Dynamics, and with considerable help from several graduate students was instrumental in the development of the Advanced Aircraft Analysis (AAA) code. This code allows preliminary designers to size and analyze the performance, stability, control, weight, balance and cost of a new airplane on a PC

---

* Reference 2 provides details about these programs.

Figure 5 KU/NASA/Cessna Redhawk High Lift and Spoiler Research Airplane

Figure 6 KU/NASA/Piper/Robertson ATLIT High Lift and Ride Quality Research Airplane

Figure 7 KU/NASA/Beech/Boeing SSSA Research Airplane

Dr. Lan has been instrumental in developing a variety of CFD codes. His first major contribution was the development of a quasi-vortex lattice method to determine and explain the effect of upper-surface and over-the-wing-blowing. Very notable is his VORSTAB code which allows inclusion of nonlinear viscous effects and vortex bursting effects in the predicting of stability derivatives. Recently Dr. Lan has focused on dynamic aerodynamics and post-stall maneuvering. To predict nonlinear dynamic stability and control characteristics, experimental and flight data may now be analyzed through a fuzzy-logic algorithm to capture all the fine details, including the interdependence of longitudinal and lateral-directional dynamic aerodynamics. He has also used Navier-Stokes solvers to improve wind-tunnel interference corrections in both static and dynamic tunnel testing.

Dr. Taghavi has conducted   extensive parametric studies on the aeroacoustics and noise control of supersonic multiple-jet mixer ejectors in support of the NASA High Speed Civil Transport Program. He has conducted experimental studies into twin supersonic-jet coupling, effects of nozzle exit geometry, boundary layer swirl, screech tone and artificial excitation of supersonic jets. His collaboration with Dr. Farokhi in supersonic flow control studies has resulted in a patent for a "supersonic vortex generator" which has been tested on thrust vectoring nozzles at NASA Langley.

Dr. Taghavi is collaborating with Dr. Lan in the assessment and correction of windtunnel wall and support interference methodologies. He has also collaborated with NASA Glenn in the evaluation of JP-8 as a rocket propellant.

Dr. Farokhi has conducted extensive research in active flow control, turbulence modeling in curved ducts, hydrodynamic stability analysis of swirling jets, propulsion installation and flight test of pusher propellers, smart supersonic vortex generators, transition duct aerodynamics, pusher propeller noise and aerodynamics of wind turbines. Dr. Farokhi's research at KU has received external funding from NASA-Lewis , NASA-Langley, GE Aircraft Engines, Kansas Corporation Commission and DOD. A pusher propeller flight test program was conducted with Cessna Aircraft and involved a modified Citation Jet airplane. In 1999 Dr. Farokhi initiated a Joint MS-Diploma Program between KU-Aerospace Engineering and the Von Karman Institute for Fluid Dynamics in Belgium.

Dr. Downing has been involved in a variety of flight dynamics, navigation, guidance and control research projects funded by NASA Langley, NASA Dryden and NASA EPSCoR. Research topics have included takeoff monitoring systems, agility metrics, ride quality enhancement systems, nonlinear control design, optimal and hybrid control. He has been the Director of the Kansas NASA EPSCoR program since 1996. Dr. Downing is the driving force behind the new Kansas Flight Test Center described in Section 2.3.

Dr. Hale is active in research for composite materials and structures, ultrasonic nondestructive evaluation, manufacturing, and engineering education. He is currently funded to investigate nonlinear steered fiber composites producible with fiber placement, and three-dimensionally reinforced textile composites. Dr. Hale holds two U.S. patents for the Boeing composite design process, with further national and international patents in review.   Flight Research Laboratory improvements realized under Dr. Hale's guidance include equipment for composite manufacture, nondestructive evaluation, CNC machining and rapid prototyping. The latter includes a fused deposition modeling machine, or 3D printer, which enables solid geometry CAD files to be converted to solid ABS plastic components. Such capabilities enable students to literally print aircraft models to scale

Dr. Ewing is involved in a variety of research programs related to the reduction and control of structural vibrations to cut unwanted interior aircraft noise.

Before Dr. Sorensen joined the faculty there was very little "space" to the "aerospace" in the name of the department. Dr. Sorensen has not only given a significant boost to aerospace course offerings (including a course on spacecraft design) but he has also started a vigorous research program into various innovative satellite development programs.

Several research programs conducted through the Flight Research Laboratory have resulted in "spin-off" companies. Examples are:

438

1. Roskam Aviation and Engineering Corporation which went on to make major contributions to the design and development of several Learjet designs, in particular the Model 36, the SIAI-Marchetti S-211 jet trainer and to the Piaggio P-180 Avanti three-surface, twin turboprop pusher executive airplane.

2. Aerotech Engineering and Research which is a research and development company started by Dr. S. Farokhi. Aerotech became active in the Federal Government's Small Business and Innovative Research (SBIR) program.

3. Kohlman Systems Research (started by Dr. D.L. Kohlman) which became one of the foremost companies in the world for flight testing and simulator model development.

4. DARcorporation (started by Dr. J. Roskam) which went on to develop and market the Advanced Aircraft Analysis software and is engaged in design and development activities on a wide variety of aircraft programs

2.3 Kansas Flight Test Center

In 2000 the Kansas Flight Test Center (FTC) was formed under the leadership of Dr. D.R. Downing. The FTC will support the undergraduate, graduate and professional education missions of the Department of Aerospace Engineering. It will do so by offering flight test courses and by conducting and involving students in a broad range of avionics, flight testing and navigation-guidance and control research activities. The Center will also serve the university, state and professional communities by providing flight test planning and operational services. The Center was jump-started by a $300,000 gift from Cessna Aircraft Company. In 2002 the Center received a very significant boost through a gift of $2,000,000 from Walter and Jane Garrison. Part of these funds are designated to purchase and instrument an aircraft to be used as a flying laboratory for advanced flight systems in manned and unmanned aerospace vehicles. The remainder of the gift will be used to expand and equip the office and laboratory space of the Kansas Flight Test Center at the Lawrence Municipal Airport.

3. History of the Aerospace Engineering Short Course Program

In 1976 Dr. Roskam taught his first short course in Wichita, Kansas on the subject of General Aviation Feedback Control technology. This short course was sponsored by NASA and attracted 21 engineers from Cessna, Beech, Learjet and Boeing. As a result of the surrounding publicity KU received requests from various parts of the country to hold similar courses on a regular basis.

Consequently, in 1977, under the auspices of the newly established Aerospace Engineering Short Course Program the first public short course was offered. This short course program has been very successful and in 2002 it comprised more than 60 short courses per year, conducted by over 25 instructors. These courses annually attract well over 1600 engineers. Courses have been conducted in the USA, Canada, Australia, Singapore, Brazil, England, Germany, France, The Netherlands, Norway, Switzerland, Italy, Mexico and Sweden.

Most of the KU AE faculty are also short course instructors. Many short course instructors come from other universities or from industry. The program is part of the KU Continuing Education Division. Since 1981 the program has been under the leadership of Jan Thomas Barron the Senior Program Manager.

Recently, several short courses are offered in a truly long-distance fashion. Several are being taught through the medium of streaming video. A typical example is the series of courses being conducted on behalf of EMBRAER in San Jose dos Campos in Brazil.

The plan is to expand the program into space: Dr. Trevor Sorensen and Dr. Roy Myose (from WSU) will co-teach the first space short course in Fundamentals of Astrodynamics in 2003.

For information on current and future Aerospace Short Courses visit the website at: www.kuce.org/aero.

4. Notable KU AE Graduates

Following is a list of notable graduates of the AE program at The University of Kansas. They are listed alphabetically.

Jack Abercrombie, Retired Head, Aerodynamics, McDonnell-Douglas St. Louis
Ron Barrett, Professor of Aerospace Engineering, The University of Auburn
John Brizendine, Retired President, Lockheed Aeronautics Company
Sam Bruner, Director Advanced Concepts Engineering, Raytheon Aircraft
Marc Bouliane, Director Marketing, Bombardier Business Jets
Kees van Dam, Professor of Aerospace Engineering, The University of California, Davis
Linda Drake, General Manager, Launch Vehicle Division, Aerospace Corporation
General Joe Engle, Retired Astronaut
Steve Erickson, Manager UAV Programs, Lockheed Skunk Works
Dick Etherington, Retired Director of Engineering, Gates Learjet Corporation
Walt Garrison, Retired, President and CEO, CDI
Donna Gerren, Professor of Aerospace Engineering, The University of Colorado
Charles Guthrie, Director UAV Systems and Fuel Cell Systems, Boeing Phantom Works
Malcolm Harned (†), President, Cessna Aircraft Company
George Hill, Captain USN
Bruce Holmes, Manager, NASA Langley General Aviation programs
Harry Johnson, Retired Manager General Aviation Programs, NASA Headquarters
Vicki Johnson, Technical Director, Boeing 767 Tanker Conversions
David L. Kohlman, Chairman of the Board, Kohlman Systems Research
Sudhir Mehrotra, President, Vigyan Associates
Alan R. Mulally, President and CEO, Boeing Commercial Airplane Company
Ton Peschier, Vice-President Engineering, De Schelde Shipbuilding Co., The Netherlands
Ronald Renz, President, Alligator Inc.
Wendell Ridder, Captain USN
Douglas Shane, Director Flight Test Operations, Scaled Composites
Trevor Sorensen, Professor of Aerospace Engineering, The University of Kansas
Ray Taghavi, Professor of Aerospace Engineering, The University of Kansas
James Thiele, President and CEO, American Blimp Corporation
Robert Waner, Director of Engineering, Boeing Wichita
William Wentz, Retired Director NIAR
Kenneth Wernicke, Retired Director of Engineering, Bell Helicopter Co.

List of References

1. Downing, D.R.; Proceedings of the Alumni Reunion 1944 – 1994; University of Kansas, School of Engineering, 1994.
2. Roskam, Jan; Roskam's Airplane War Stories; DARcorporation, 120 East Ninth Street, Suite 2, Lawrence, KS, 66044.

Chapter 34

Aerospace Engineering Science at the University of Colorado at Boulder

Brian Argrow and Robert Culp
Department of Aerospace Engineering Sciences
University of Colorado

Introduction

The history of the Department of Aerospace Engineering Sciences (AES) at the University of Colorado at Boulder encompasses much of the 20th Century. For the following discussion, we divide this history into five eras: 1) Aeronautical Origins, 1930-1946; 2) Department Establishment, 1946-1963; 3) Aerospace Engineering Arrives, 1963-1984; 4) The Space Initiative, 1984-1997; 5) Aerospace Engineering Education Re-Engineered, 1997-present. The following discussion revisits the people, decisions, and outcomes that lead to AES as we know it today. The discussion in Aerospace Engineering Education Re-Engineered focuses on our current program to re-establish the hands-on and design focus of pre-Cold War engineering using a new state-of-the-art teaching laboratory and new teaching methods. We conclude with a summary of the current status of the Department.

Aeronautical Origins

On February 11, 1929 the University of Colorado Aero Club, with twenty members, held its first meeting. Its goal was to promote aviation, which included flying airplanes, and studying airplanes and their design. In that same year, in his biennial report, the Dean of Engineering noted there was a growing demand for a course in aeronautical engineering. This led to immediate action, as the 1930-31 College of Engineering catalog listed an aeronautical engineering option in Mechanical Engineering, supported by courses in aerodynamics, airplane stress and analysis, airplane engines and instruments, airplane transportation, and navigation.

The assistant professor in charge of the aeronautical option was Norman Parker. In 1939 he built the first aeronautical laboratory, featuring the first wind tunnel in the region. This same year the federal government established a Civilian Pilot Training program, with an annual enrollment of 40, in conjunction with the aeronautical option. Both the aeronautical laboratory and the aeronautical option curriculum expanded rapidly in the next two years. In addition, during the war the Navy, in what was named the V-12 program, sent annual classes of about 40 through this option, providing the Navy with both pilots and aeronautical engineers.

In 1943, because of the growing demand both by the Navy V-12 program and the civilian students, the University Regents approved the BS degree in aeronautical engineering with options in engines and structures. Norman Parker, as Head of the Department of Mechanical and Aeronautical Engineering, assured the continued growth of the aeronautical engineering degree.

In 1944, the University Regents appropriated $4,000 for remodeling and $32,000 for new equipment to upgrade the aeronautical laboratories, keeping the laboratories among the most modern in the nation. A key step in raising the program to national prominence was the appointment of the nationally renowned authority of aerodynamics and aircraft design, Karl Dawson Wood. He had been head of aeronautics at Purdue, and had wind tunnel aerodynamics experience at Cornell, Purdue, Cal Tech, and the Consolidated Aircraft Corporation.

The Department is Established

In 1946, the Regents established the Aeronautical Engineering Department as a separate department and appointed K. D. Wood as its first Head. Three outstanding faculty members were responsible for the Department's reputation at this time. K. D. Wood, the nation's leading authority on airplane design, had two of the field's recognized standard texts, *Airplane Design*, and *Technical Aerodynamics*. Harold W. "Hap" Sibert joined the faculty in 1946 from the University of Cincinnati. He was one of the most respected experts in aerodynamics and structures, and author of the text, *High-Speed Aerodynamics*. Franklin P. Durham came to the Department in 1947 from Pratt and Whitney Aircraft Company. He was a nationally known authority in aircraft power plants including jet engines, and perhaps the nation's top rocket scientist. He authored two texts, *Aircraft Jet Power Plants* and *Thermodynamics*.

Figure 1 From the left, K. D. Wood, the first Chair of the Department of Aeronautical Engineering, accompanied by Robert Culp, who was later the 13th Chair, and Culp's PhD advisor Adolf Busemann, circa 1970.

These three stars, and the influence of their texts, led the Massachusetts Institute of Technology in 1954 to list the University of Colorado in its "Big Ten" of aeronautical engineering programs. By 1954, the undergraduate enrollment was 198, and grew to 334 in 1957. Research expanded into several areas of rocket engine and nozzle development, aerodynamic heating at high speeds, and materials, supported by the U.S. Naval Ordinance Test Station, Marquardt Aircraft, the Martin Company, USAF Wright Field Research Center, and other government agencies.

Figure 2 A photograph of the Department of Aeronautical Engineering's 1949 senior class, a gift to Department Chair K. D. Wood.

Aerospace Engineering Arrives

In the last half-century, three strong presidents dramatically advanced the thrust of the University of Colorado. Quigg Newton in the late fifties put the University on the road to becoming a premier research university. Arnold Weber in the early eighties proclaimed his Space Initiative with the goal of making the University the leading space university in the nation. E. Gordon Gee in the late eighties renewed Weber's Space Initiative and put money and resources behind the drive to become the number one space university.

In 1962, Quigg Newton brought Max S. Peters from the University of Illinois to be the new Dean of Engineering. With the new dean the College won an award of $15,000,000 from the National Science Foundation to build a new Engineering Center.

For aeronautical engineering, a new Department Chair, Mahinder S. Uberoi, (PhD, Johns Hopkins, 1953) came from the University of Michigan in 1963, when the Department was renamed the Department of the Aerospace Engineering Sciences. Professor Uberoi had a reputation in magnetohydrodynamics, turbulent flow, and other basic sciences of fluid flow. The faculty members he added were, for the most part, strong in basic sciences and secondarily interested in applications to engineering.

In 1963, the Department added the greatest star in its history: Adolf Busemann (cf. Figure 1), the father of supersonic aerodynamics, the inventor of the swept wing, the creator of conical flow theory, and a genius with seminal research in nearly every field of fluid flow. After joining the faculty he continued his research in fluids, and ventured into new areas of space trajectories, propulsion, and structures. He was awarded the Prandtl Ring and gave the Prandtl Lecture in 1965. Professor Busemann remained active even after his retirement until his death in 1986.

As Chair, Professor Uberoi added faculty in basic research areas in the fields of fluids, controls, and even biology. His theory was to bring in researchers strong in the basic sciences, and have them apply their work to engineering. Hence, he hired three biologists to build a bioengineering program.

During the sixties and seventies this approach worked well. The Department grew to twenty-nine faculty and nearly twenty postdoctoral assistants. In 1975, Uberoi's reign as Department Chair ended, and the Department entered a ten-year period of acting and short-term chairs. The absence of leadership and other turmoil in the Department took its toll. By 1984 there were only nine regular faculty members.

The Space Initiative

In 1984, University President Arnold Weber launched his Space Initiative. He was determined to make the University of Colorado the outstanding space university of the nation. He brought Don Hearth, former Director of the NASA Langley Research Center, to Boulder as the University's space czar. Dean of Engineering A. Richard Seebass worked with Don Hearth to implement this strategy in engineering.

In 1985, Arnold Weber left to become president of Northwestern University. The next University President, E. Gordon Gee, adopted the Space Initiative as his own, and announced that it would feature sizeable resources poured into aerospace engineering. George Morgenthaler was brought from Martin Marietta to Chair the department, and George Born was hired from the University of Texas to establish the Colorado Center for Astrodynamics Research. Over the next ten years, 1985-89 with Morgenthaler as Chair, and 1989-1996 with Robert D. Culp as Chair, twenty new faculty members were added and three million dollars of new offices and laboratories were built. Gordon Gee proclaimed the Aerospace Engineering Sciences Department to be "the gem in the crown" of his space university.

The University of Colorado, Boulder has produced 15 astronauts who have completed at least one space flight. Of these, the Department's eight astronauts, including Scott Carpenter, one of the original "Mercury 7," are a testament to its space heritage. Figure 3 includes the photographs of the eight astronauts and their connection to the Department.

Research developed massively with the new faculty. From a low of a few hundred thousand dollars in 1984, annual research expenditures approached $10 million by 1990 and remains over $10 million today. These expenditures are generated primarily from the three research centers established in the eighties: the previously mentioned Colorado Center for Astrodynamics Research , the Center for Aerospace Structures, established in 1986, and BioServe Space Technologies, established in 1987. An additional large contributor to the research expenditures is the Department's atmospheric scientists who are affiliated with various research centers throughout the University. The Department led the University in research expenditures for several years in the nineties, and has remained the top engineering department and one of the top four departments in the University since 1987. AES is consistently ranked nationally in the top aerospace engineering departments in terms of research expenditures.

Figure 3 Astronauts from the Department of Aerospace Engineering Sciences. Top row from left: Vance Brand (BS 1960), M. Scott Carpenter (BS 1962), Kalpana Chawla (PhD 1988), Takao Doi (Postdoc 1987-88). Bottom row from left: Richard Hieb (BS 1979), Marsha Ivins (BS 1973), Ellison Onizuka (BS, MS 1969), James Voss (MS 1974).

Aerospace Engineering Education Re-Engineered

Around 1990, Dean of the College of Engineering and Applied Sciences, A. Richard Seebass, lead the Integrated Teaching and Learning (ITL) Initiative. This was the beginning of a College-wide effort to return to the hands-on, experiential roots of engineering education, while recognizing the necessity for teamwork, communications skills, and a multidisciplinary environment. After 13 years as Dean of the College of Engineering and Applied Sciences, Professor Seebass served one term as AES Chair, from 1996-1999. His great interest was in revamping the AES curriculum to take advantage of the new Integrated Teaching and Learning Laboratory, a unique, state-of-the-art product of the ITL Initiative, shown in **Error! Reference source not found.**, along with the "ITL Wheel" that illustrates the integrative and multidisciplinary program plan. This complete curriculum overhaul coincided with the establishment of the new Accreditation Board for Engineering and Technology (ABET) guidelines that are currently enforced. Professor Seebass had been largely responsible for the new ABET rules by helping to organize the research engineering colleges in a movement to bring greater freedom and recognition of the research colleges and their abilities. His efforts culminated in 1997 with the Department's Aerospace Engineering Curriculum 2000.[1]

Quoting Seely,[3] in his discussion of the history of education in American engineering colleges:

> *"Recent efforts to re-emphasize design in engineering schools and develop a better balance with engineering science fit into a history that extends further into the past than two decades ... the changes being proposed in the 1990s seek to undo an earlier "re-engineering" of engineering education in the United States, an effort that dominated the first half of this century. Those earlier changes culminated in a substantial reworking of engineering education in the period 1945-1965, and brought into place the style that current reformers wish to overturn, or at least modify. It was only after World War II that American engineering colleges completely embraced engineering science as the foundation of engineering education. That decision led to sharp reductions in the time and coursework devoted to practical skills such as drafting, surveying,*

and other traditional features of engineering curricula. Replacing them were courses in fundamental sciences, mathematics and engineering science."

The lesson here is:[3] "A good engineer ... must strike a balance between knowing and doing." The recognition of this balance was the impetus for the re-engineered curriculum that is the Curriculum 2000; a curriculum with renewed emphasis on design and hands-on learning to balance the theory of the engineering sciences. Horizontal integration of engineering science topics with hands-on and design experiences is a priority. This is within a learning environment where communications and teamwork development is ubiquitous. Specifically, we have:[2]

- Established a core curriculum
- Integrated the material in this core
- Made the curriculum relevant to applications
- Made it experiential, i.e., "hands-on"
- Integrated communication and teamwork skills into all courses
- Provided more curricular choice at the upper division
- Implemented continuous improvement procedures

Near the completion of the Curriculum 2000 planning, ABET announced new guidelines and criteria for program accreditation. The 1997 aerospace engineering program criteria proposed by the American Institute of Aeronautics and Astronautics (http://www.aiaa.org) helped to finalize the first iteration of the new curriculum. The outcomes-based assessment plan of the Curriculum 2000 was in the spirit of that to be used by ABET evaluators. In 1999, the Curriculum 2000 was successfully reviewed by ABET.

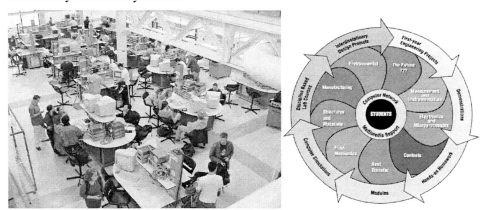

Figure 4 A laboratory plaza in the Integrated Teaching and Learning Laboratory and the ITLL wheel that captures the program plan and organization of the facility.

Summary: Aerospace Engineering Sciences Today

The following is a compilation of recent enrollment, faculty, and research expenditures data; most of it coincides with the start of the University's Space Initiative that was discussed earlier. From 1943 through 2002, aerospace engineering has produced 2226 BS degrees, 720 MS degrees, and 220 PhD degrees. The Regents authorized the PhD degree in 1948, however the Department granted its first PhD in 1965.

Undergraduate enrollments reached an all-time high of 660 in 1988 and the characteristic cyclic behavior of the aerospace industry, usually reflected in these enrollments, is evident in Figure 5. As expected, the number of degrees conferred over this same time period reflects the trends of the overall enrollment as shown in Figure 6. Graduate enrollments and degrees also follow these trends, with an expected 4-5 year phase shift. Reflecting the steady research expenditures over the past seven years, Figure 7 shows a relatively stable MS/ME and PhD production rate.

As shown in Figure 8, since the mid-nineties, the number of AES faculty has been constant at about 24. Combined with the research expenditures shown in Figure 7, the average research expenditures per faculty has ranged from $379,000 to $480,000 during this time period. The Department continues to thrive on its multidisciplinary diversity, with research expertise covering the traditional aeronautical, astronautical, and atmopsheric sciences. Our strategic plan builds on current strengths and will take us into relatively new areas such as bio-astronautics, GPS technologies, massive aerospace computations and simulations, and unmanned aerospace vehicles.

Starting with K. D. Wood, the first Chair of the Department of Aeronautical Engineering, Table 1 lists the 13 professors who have served a total of 15 terms as Chair of the Department. The faculty of the Department of Aerospace Engineering Sciences continue the legacy started early in the 20th Century. We look forward to the challenges of the 21st Century. Our makeup and interests continue to change as we collectively adapt to the technical and societal needs of our country and the world.

Table 1 AES Department Chairs

Chair	Term
Karl D. Wood	1947-1956
Franklin P. Durham	1965-1957
Karl D. Wood	1957-1962
Charles A. Hutchinson (Acting)	1962-1963
Mahinder S. Uberoi	1963-1975
Louis C. Garby	1975-1976
Franklin Essenberg	1976-1979
Klaus Timmerhaus (Acting)	1979-1980
George Inger	1980-1983
Marvin Luttges (Acting)	1983-1984
Klaus Timmerhaus (Acting)	1984-1985
George Morgenthaler	1985-1989
Robert D. Culp	1989-1996
A. Richard Seebass	1996-1999
Charbel Farhat	1999-present

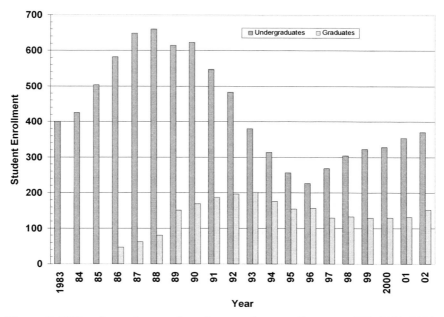

Figure 5 AES undergraduate and graduate student enrollment per FY, 1983-2002.

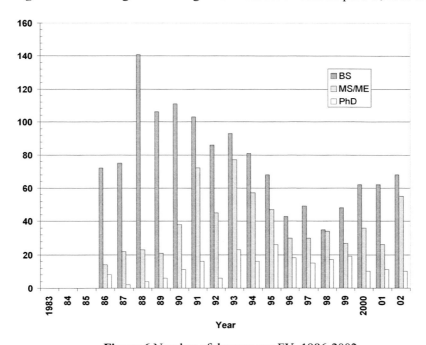

Figure 6 Number of degrees per FY, 1986-2002

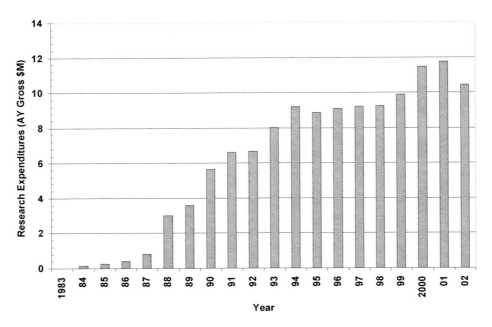

Figure 7 AES research expenditures per FY, 1984-2002.

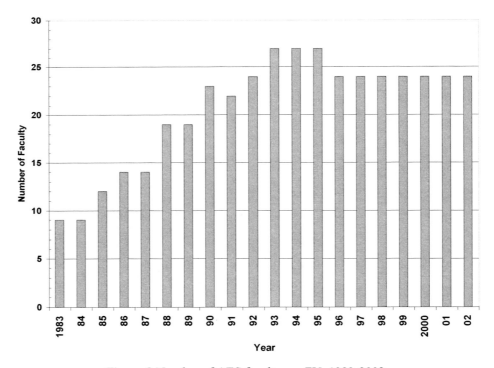

Figure 8 Number of AES faculty per FY, 1983-2002.

References

1. Argrow, B. M., "Proactive Teaching and Learning in the Aerospace Curriculum 2000," Proceedings of the 2002 ASEE Annual Conference & Exposition, Montreal, June 2002.

2. Seebass, A. R. and Peterson, L. D., "Aerospace Engineering 2000: An Integrated, Hands-On Curriculum," *Frontiers of Computational Fluid Dynamics* 1998, ed. Caughey, D. A. and Hafez, M. M., World Scientific, pp. 449-464, 1998.

3. Seely, B. E., "The Other Re-engineering of Engineering Education, 1900-1965," *Journal of Engineering Education*, **88** (3), Jul. 1999, pp. 285-294.

Chapter 35

The History of Aeronautical/Aerospace Engineering at Penn State

Barnes W. McCormick, Professor Emeritus and Dennis K. McLaughlin,
Professor and Head, Department of Aerospace Engineering, The
Pennsylvania State University

Introduction

Aerospace Engineering at the Pennsylvania State University began with one 3-credit course offered by Mechanical Engineering for the first time around 1928. The course description found in the 1928-1929 Catalog reads:

Figure 1
Aeronautical "Pioneers" Class of 1943
Front row: Frederick Young, Donald Steva, Albert Yackle, Samuel Schnure, Donald Wion; Back Row: John Scheppman, Boris Osojnak, Luther Boyer, Louis Borges

ME 56 Airplanes (3) A brief survey of aerodynamic principles underlying the design and performance of the modern airplane followed by a study of the materials commonly used, and the propulsive power plant. Elective Sr. me. Prerequisite: me 7 (elementary mechanics)

In the next year's catalog this course is renumbered as ME 456 and a preliminary course, Airplanes (2) is added to be taken in the fall semester of the senior year. These courses remained until around 1942 when an expanded option in Aeronautical Engineering was formally offered by ME.. This option, however, was never accredited. The first class pursuing this option graduated in 1943 and the names and hometowns of those pioneers are presented here.

First Graduates to Complete the Aeronautical Engineering Option in ME   October 21, 1943

Louis J. Borges, Jr., Royalton
Luther J. Boyer, Ringtown
Charles E. Duke, Hershey
Clinton H. Fitzgerald, Freeland
Joseph E. Greiner, Erie
Warren E. Herr, Hershey
Lloyd E. Hill, Apollo
Bernard A. Koval,.South Waverly
Howard E. Kugel, Zelienople
William W. McKenna, Norvelt
Herbert E. Means, New Woodstock, N. Y.
John F. Melzer, Erie
Boris M. Osojnsk, Smithton
Robert F. Painter, Erie
Edmund G. Pinger, Freeland
Thomas H. Randall, Jr., Arlington, Va.
William J. Scarborough, Scranton

John C. Scheppman, Brookline
Samuel Ausben Schnure, Milton
Harry K. Search, Jr., Wilkes-Barre
Earl D. Shank, Waynesboro
William R. Slivka, McKees Rocks
Earl N. Stauffer, East Petersburg
Donald G. Steva, Erie
Kendrick C.Taylor, Bellefonte
Glenn D. Walters, Altoona
Morris W. Hazelton
Donald A. Wion, Bellefonte
Albert R. Yackle, Willow Grove
Franklin D.Yeapie, Jr., Roselie Park, N. J
James H. Yeardley, Uniontow
Frederick A. Young, Oil City

A number of the first class are still quite active. They have collectively endowed a scholarship and return regularly for reunions. A group photo taken in 2002 is presented in **Figure 1** and the Department salutes their devotion to "Dear Old State".

Figure 2
Dean Harry P. Hammond

Founding and Growth during the 40's

During this time, Penn State was known as the Pennsylvania State College with a School of Engineering. Its president was Ralph Dorn Hetzel and the present student union building is named the HUB, or Hetzel Union Building. The Dean of Engineering in the early 40's was Harry P. Hammond for whom the large engineering building that houses the present Department of Aerospace Engineering is named. A photograph of Dean Hammond is shown in **Figure 2**. The courses and laboratory required by the ME Option were nearly identical to those adopted when a separate Department of Aeronautical Engineering was formed in 1945. The ME Department operated a 3' by 4' subsonic wind tunnel capable of approximately 135 mph. The degree required 150 semester hours including 17 credits in machine and airplane detailed design. Subject areas included aerodynamics, airplane engines and airplane structures. Nine credits of mathematics with calculus through differential equations was required. Also, shop courses in foundry, sheet metal working and use of machine tools were required. Two instructors, Associate Professor David J. Peery and Assistant Professor David. J. Gildea, taught most of the Aeronautical Engineering courses.

Figure 3
Dr. David J. Peery
The First Department Head

The Department of Aeronautical Engineering was formed in 1945 with Dr. Peery (**Figure 3**) as Head of the Department. It was housed in Engineering D built as a temporary building during the First World War and still standing today.

The Department's wind tunnel was on a second floor balcony of the ME Lab. The number of required hours was reduced to 149 semester credit hours. Several students came in 1945 from Bloomsburg State Teachers College, members of the US Navy's V-5 and V-12 programs. They joined students already at Penn State in the military programs. These members of the military were housed in fraternity houses that were taken over by the Navy and given Barracks numbers. For example, Barracks 22 had been the Kappa Sigma house. One of those students coming from Bloomsburg in 1945 was Barnes McCormick who was to become the Department's Head twenty-four years later. A photograph showing the complex of Engineering buildings is presented in **Figure 4**. Although taken around 1930, this is the way the area looked in 1945 when the Department was formed. The current Hammond Building now lies along this block of College Avenue between the street and the older Engineering buildings.

Figure 4 College Avenue in the 1930's.
Aeronautical Engineering was housed in Eng D in 1943

During the time when the Department was formed, World War II was still being fought. The cover of the commencement shown in **Figure 5** was typical of the wartime environment. Inside the cover the following instructions were to be found.

"The Civilian Defense Authorities request that in the event of an Air Raid Alarm, you are to remain seated and not attempt to leave the building unless otherwise instructed. Air Raid wardens, Auxiliary Police, and Auxiliary Firemen will report to their stations when notified. Members of these units on leaving the building are requested to wear their arm bancs or show their identification cards to the wardens on duty at the exit."

Samuel Hoffman who had been the Chief Engineer for Lycoming Motors joined David Peery and David Gildea during the period 1946-47. Sam left Lycoming Motors to join academia when his physician advised him to find an environment that was less stressful. In the academic year 1947-48, two instructors, Charles Duke and John Montgomery joined the trio of Professors. Dr. Perry and Mr. Hoffman were Professors while Mr. David Gildea was an Associate Professor and Charles Duke and John Montgomery were Instructors.

By 1953, the Institution was still named The Pennsylvania State College but the faculty continued to grow in number. The Professors were David Peery, Samuel Hoffman and a newcomer, Martin Lessen. Another new member was Henry Lew, an Associate Professor. Instructors were John Fox and Jerome Fanucci. Both of these Instructors

Figure 5
Commencement Program 1943

would become Department Heads many years later, John Fox at the University of Mississippi and Jerry Fanucci at the University of West Virginia. Barnes McCormick, a graduate student and Research Associate at the Ordnance Research Laboratory also taught occasionally.

In 1954 the institution was renamed The Pennsylvania State University and the School of Engineering became the College of Engineering. At that time, the faculty consisted of Dr. Peery and Dr. Lew as Professors, Dr. Charles Duke as an Associate Professor and John Fox, Jerry Fannucci and Richard Mathieu as Instructors. Dick Mathieu would later distinguish himself as the Senior Professor at the US Naval Academy.

Figure 6
Harold Hipsch,
Head 1955-1957

Struggles during the 50's

Dr. Harold Hipsch (**Figure 6**), a graduate of Cal Tech was named Head of the Department when Dr. Peery left the University in 1955 to join the faculty at the University of Michigan. Dr. Lessen also left for the University of Rochester, and Dr. Lew joined The General Electric Co. in Valley Forge, PA where he became Manager of Gas Dynamics. Dr. Peery later went to Convair in San Diego and then to Lockheed Space & Missiles in Sunnyvale. He died in 1979 and a Memorial Scholarship was established by his wife, Joanna, in memory of the first Department Head and founder of this Department. Dave Peery also wrote one of the outstanding texts that has endured the test of time,

"Aircraft Structures". During this time, Dr. George F. Wislicenus was named to a joint appointment as the Director of the Garfield Thomas Water Tunnel and a Professor of Aeronautical Engineering. Charlie Duke, Jerry Fanucci, John Fox, and Dick Mathieu remained in the Department but Barney McCormick left ORL to join the Piasecki Helicopter Company. In 1956 Charles Duke left to join HRB, and in 1957 Lloyd LeBlanc came as an instructor.

In 1957, the Department suffered the tragic loss of its Head, Dr. Harold Hipsch from cancer and a year later, Dr. Irving Michelson was named Head of the Department with the remaining faculty being Dr. Wislicenus and Messrs. Fox, Mathieu, and Leblanc. Otis E. Lancaster, the George Westinghouse Professor of Engineering Education, joined Drs. Michelson and Wislicenus in 1959. Assistant Professor Dr. Huon Li and Instructors Fox, Mathieu, LeBlanc, Marek Jakubowski., and Bernard Carson. Joint appointments with ORL included Assistant Professor of Engineering Research J. William Holl, Associate Professor Barnes McCormick (who returned from the University of Wichita), Assistant Professor John Lumley and Research Associate Joseph Eisenhuth..

A Few Interesting Trivia

Figure 7
Wind Tunnel Balance built by
Bill Brown in this Building

Early in its existence, the Department's wind tunnel, located on the balcony of the ME lab, was driven by an Oldsmobile 88 engine and it's "hydramatic" transmission salvaged from a car wreck in front of the Autoport Motel. As the tunnel was brought up to speed, it would literally shift gears. It was cooled by tap water, the flow of which was adjusted by students after feeling the temperature of the discharge. The balance was mechanical and built by one of the Department's machinists, Mr. Bill Brown. Bill made a significant contribution to the world of aviation by inventing and building the first model airplane gasoline engine. An article appeared in The Philadelphia Evening Bulletin, May 26, 1931, about Bill and his engine. Bill was just a senior at Frankford High School. Later in life, Bill also invented the miniature CO_2 engine which became popular for powering indoor models. The small building in which Bill built the balance is pictured in **Figure 7.** It was located just behind the ME Building. To the left of it is the all-wooden building that housed Engineering Extension and in the background to the right is the foundry. All of these buildings have been torn down but not before the small one housed the Penn State Amateur Radio Club for several years. A small park, called Foundry Park, now occupies most of this area and a small cupola monument marks the spot where a time-capsule was buried.

Figure 8
Dr. George F. Wislicenus
Department Head 1962-1969

The Rescue by George Wislicenus

In 1961 John Fox, now an Associate Professor, became acting Department Head after Dr. Michelson left for The University of Illinois-Chicago. In 1962, Dr. Wislicenus was named as Head of the Department while continuing to direct the Water Tunnel. Many Water Tunnel staff joined the Department including Budugur Lakshminarayana, William Holl, Barnes McCormick, Maurice Sevik, Ed Rodgers, and Roger Arndt. This move may have saved the Department as it had become void of faculty and students during Dr. Michelson's tenure and would probably have been dissolved or merged with another Department by Dr. Merritt Williamson who was the Dean at the time. Dr. Wislicenus (**Figure 8**) was a renowned Mechanical Engineer in turbomachinery and fluid mechanics and, as a result, the Department tended to develop in this direction during his tenure. He supported the acquisition of an airplane in 1962 that was used to teach a course in techniques of flight testing. This airplane, a Cherokee 160, was donated to the Department by the Piper Aircraft Corporation located in Lock Haven, PA, thanks to the help of the Vice-President of Engineering for Piper, Mr. Walter Jameneau who was at that time a member of the Department's Industrial and Professional Advisory Committee (IPAC). The readers may be interested to know that the "J" in the designation for the classic Piper Cub J3 stands for "Jameneau" who was one of Mr. Piper's early associates. Walter strongly supported the activity to introduce applied aeronautics into the curriculum and a few years later replaced the Cherokee 160 with a Cherokee 180 with IFR capability.

Figure 9
Dr. Barnes W. McCormick
Department Head
1969-1985

In July, 1969 Professor Maurice Sevik was named Assistant Director of the Ordnance Research Lab, and Director of the Garfield Thomas Water Tunnel, assuming the position once held by Dr. Wislicenus.

Faculty In the 70's and 80's

In 1969, Dr. Wislicenus retired and moved to Tucson, AZ. Before he retired however, several new Assistant Professors joined the Department including Dr. Hendrik Tennekes, Dr. Winfred Phillips, Dr. Thomas York, Dr. John Allen, and Dr. Marshall Kaplan. Following Dr. Wislicenus' retirement, Dean Nunzio J. (Joe) Palladino named **Dr. Barnes McCormick** as the new Department Head.

Dr. McCormick (Barney) (**Figure 9**) served as the Head for 16 years. At the time of his appointment to lead the department Dr. McCormick had a long history with Penn

State. Initially an undergraduate student, he was the second Ph.D. graduate of the Penn State Department of Aeronautical Engineering. He spent 1955 to mid 1958 as
an engineer and ultimately as chief of aerodynamics of the Vertol Company (later to become Boeing Helicopters), and as head of Aeronautical Engineering at the University of Wichita (for 1958 – 59). He joined the Department faculty in 1959 as a joint appointee with the Ordnance Research Lab (later to be named ARL).

Dr. McCormick was (and still is) an international expert in helicopter aerodynamics and wake vortices (those produced from underwater propellers as well as from the tips of aircraft wings.)

During McCormick's tenure as Head, the department strengthened the applied aerodynamics research, much of which he personally directed. He also brought considerable renown to the department with the authorship and publication of two outstanding textbooks. The first, concerned with V/STOL flight was published in 1967 and is still in print. The second, a text on aerodynamics was published in 1977. At present this text has had 13 printings and is used internationally as an undergraduate text in aeronautics.

Significant Research in the 50's through 70's

The faculty during the tenures of both Dr. Wislicenus and Dr. McCormick performed significant research. Dr. Holl's work on the fundamentals of cavitation is recognized internationally. Research on marine propellers by Drs. McCormick and Eisenhuth led to a doubling of the silent speed of torpedoes, a fact that was classified for many years. Dr. Phillips played a major part, together with two other faculty from ME and the Hershey Medical Center, in developing the University's artificial heart. Dr. McCormick guided some of the first in-flight research done on wake turbulence. His student team obtained the World's first detailed measurements of the wake geometry and decay from a full-scale airplane in flight, both near and far -field. The Army L-19 used in this research is shown in **Figure 10** with the truss attached to the right wing supporting a vortex meter that could be traversed in flight.

Figure 10
Army L-19 with Vortex Meter
and Porous Wing Tip

During this period from 1969 until 1985 there were many significant

Figure 11
The Department's "Fleet" N907PS and N911PS

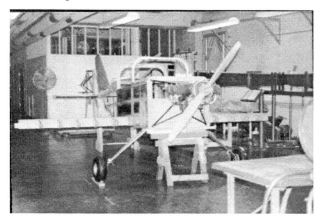

Figure 12 PSU-1

changes with an emphasis being placed on developing aeronautics and astronautics. Flight training using the Department's airplane was subsidized and three faculty members joined Drs. McCormick and Smith as pilots. Research developed that utilized flight-testing and, as a result, a second airplane was procured from Piper, a Cherokee Arrow. Permission was obtained to use the airplanes for professionally related travel at an audited rate in order to keep the five faculty pilots current. Still, the preponderance of the use was for academic instruction, research and University travels. A photograph of N907PS and N911PS are shown in formation in **Figure 11**. Many Alumni of the 70's will remember **PSU-1** pictured in **Figure 12**. This project was undertaken to introduce some realism into the airplane design course. The purpose was to design and build a light two-place airplane. The project was completely supported by industry with the aluminum, engine, wheels, brakes, and propeller all being donated. Each student was given a mini-course in sheet metal working, educated as to how the airplane got to its current state and then asked to design and build at least one component of the structure. The airplane was designed under Part 23 of the FAR's. After some evolutionary learning the quality of the work became excellent with each part being give a dye penetrate check and then zinc chromated. A standard design handbook was prepared and hand drawings were made, numbered and filed in accordance with industry practice. Admittedly, the project was time-consuming but valuable and stretched over about ten years. Unfortunately PSU-1 was never completed as the principal faculty directing the project were obligated to give their time to other projects.

During Dr. McCormick's tenure as Head, there were several personnel changes. In 1974 Dr. John Lumley was named Evan Pugh Professor of Aerospace Engineering. Win Phillips left to become Department Head of Mechanical Engineering at Purdue University. Marshall Kaplan resigned to do private consulting work and John Allen left to pursue research in oceanography. Dr. Lumley became the first person to resign from an Evan Pugh Chair to accept a similar distinguished appointment at Cornell University. Dr. Tennekes returned to his native Holland to become the Chief Scientist for their equivalent to our NOAA. Dr. Roger Arndt went to the Univ. of Minnesota in the late 1970's to become Director of the Saint Anthony Falls Laboratory. Dr. Maurice left PSU and in 1972 went to the David Taylor Model Basin in Carderock, MD, where he was Head of Acoustics. Dr. Blaine R. Parkin was hired from Convair (San Diego) as Professor of Aerospace Engineeirng and replaced Dr. Sevik as Director of the Garfield Thomas Water Tunnel.

In 1963 Professor Wislicenus hired a young PhD graduate from the University of Liverpool who had become an accomplished researcher in the area of turbomachinery (Dr. Wislicenus' area of expertise). Thus Dr. "Bud" Lakshminarayana began a faculty career that would eventually span 37 years. Bud was instrumental in developing the turbomachinery laboratory at PSU, which comprised numerous experimental research facilities. He taught numerous courses in the department, and authored a graduate level textbook on turbomachinery. In 1986 he was awarded the Evan Pugh Professorship, being only the second Aerospace faculty member to hold this honor. Bud, who died recently, is shown in **Figure 13**. He is sorely missed by those of us who were fortunate enough to have known him.

**Figure 13
Dr. Budugur
Lakshminarayana**

Enrollment Fluctuations

The undergraduate enrollment swings that occurred at Penn State, and almost every Aerospace Engineering program in the country, had a far-reaching impact on the development of the department. **Figure 14** shows a historical record of the number of BS graduates from 1968 to the present (2003). The first rapid decline began in 1970 and resulted in several graduating classes whose total numbers were in the teens. With such a situation, Dr. McCormick had limited opportunities to hire new faculty except to fill occasional positions made vacant by resignations of standing faculty. (Professors Skip Smith and Philip Morris were two such hires.) Then in the late 1970's, a stupendous growth began in the undergraduate population that soon left the department faced with graduating class sizes of around 150 students with only a dozen full time tenure track faculty. The College administration responded by providing several new faculty positions. During that period Professor McCormick hired Professors Michael Micci, Robert Melton and Mark Maughmer, all of whom have served the faculty for over 20 years and are still with the Department.

In addition to the new faculty positions, the College administration instituted enrollment controls during 1986-87, which limited the upper division class size to approximately 115. The smaller classes, together with the increased number and breadth of expertise of our faculty, improved the quality of student education considerably.

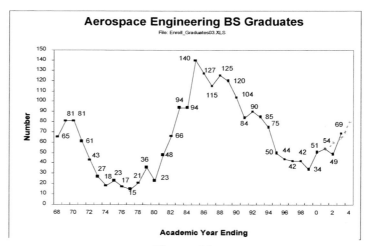

**Figure 14
Record of the Number of BS Graduates per Year
in our Department**

460

Boeing Professor

During the mid 1980's, Dr. McCormick encouraged Boeing Vertol (later to become Boeing Helicopters) to endow a distinguished professorship. In 1985, he stepped aside from department head responsibilities to become the first Boeing Professor of Aerospace Engineering. Professor Thomas York, an expert in plasma dynamics, served as Acting Head during the 1985-86 academic year. During this period he continued to take advantage of the faculty expansion demands that were driven by the large enrollments. He hired Professors David Jensen, George Dulikravich, and William Pritchard, the latter being a joint appointee with the Mathematics Department, and started the initial recruitment of Cengiz Camci who joined the department the following year.

New Department Head, 1986

In 1986, Dean Wilbur Meier appointed **Dennis McLaughlin** as Professor and the new Department Head of Aerospace Engineering **(Figure 15)**. Dr. McLaughlin, originally from Canada, held a BS degree from the University of Manitoba and graduate degrees including the Ph.D. from MIT. He had previously held a faculty position (Assistant to Full Professor) at Oklahoma State University and 5 years as a Group Manager at Dynamics Technology, Inc., in Torrance, CA, where he directed research and development in several aeronautical and oceanographic applications. At Oklahoma State Dr. McLaughlin worked with graduate students to develop a unique low Reynolds number jet noise facility. He continued to pursue research in experimental aeroacoustics upon joining Penn State.

Figure 15
Dr. Dennis McLaughlin
Department Head, 1986-Present

Big Changes in the 1980's and 1990's

With the substantial growth in the enrollments and the research activity in the late 1980's, two new administrative positions were created. Beginning in the fall 1987, Dr. J. William Holl was appointed to the new position, Director of Graduate Studies, and Dr. Hubert Smith was appointed Director of Undergraduate Studies. Following the retirements of Dr. Holl in 1991, and Dr. Smith in 1999, Dr. Anthony Amos became Director of Graduate Studies and Dr. Robert Melton became Director of Undergraduate Studies, a position he currently holds. Dr. George Lesieutre assumed the position of Director of Graduate Studies following Dr. Amos's retirement in 2000.

In 1992, Professor Philip J. Morris was named Boeing Professor of Aerospace Engineering replacing Barnes McCormick. In 1997, the Boeing Company of Seattle, WA, committed an additional $250,000 to augment the professorship with the new name, the Boeing/A.D. Welliver Professorship in Aerospace Engineering in memory of Albertus "Bert" Welliver, ME'56, one of Boeing's most distinguished leaders.

Faculty Additions

The late 1980's and the 1990's provided ample opportunity to add new tenure track faculty in the department. From the start of Dr. McLaughlin's term as department head, to the present, thirteen new faculty joined the department. These faculty are listed in Table 1 (b) which is a compilation of all faculty who have served following the headship of Dr. Wislicenus. Faculty prior to the Wislicenus era are tabulated in Table 1 (a).

Table 1 (a) Aerospace Engineering Faculty (Years of Service) Prior to Wislicenus Era

David J. Peery (1943-1955) DH	John Fox (1953 - ??)
David J. Gildea (1943 - ??)	Jerome Fanucci (1953 - ??)
Samuel Hoffman (1946 - ??)	Richard Mathieu (1954 – 1959 ?)
Charles Duke (1947 – 1956)	Harold Hipsch (1955 – 1957) DH 1955-57
John Montgomery (1947 - ??)	George F. Wislicenus ('55 - '69) DH 1961-69
Barnes W. McCormick(1951-92) DH 1969-85	Irving Michaelson (1958 – 1961) DH 1958-60
J. William Holl (1963 – 1991)	Huon Li (1958 - ??)
Martin Lessen (1953 – 1955)	John L. Lumley (1959 - 1977)
Henry Lew (1953 – 1955)	Otis E. Lancaster (1959 – 1975)

Table 1 (b) Aerospace Engineering Faculty (Years of Service) Following Wislicenus Era

Wislicenus Era (hires)	
Budugur Lakshminarayana (1963 - 2000)	York Era (hires)
Maurice Sevik (1965 – 1972)	David Jensen (1986 - 1995)
Herbert B. Kingsbury (1966? – 1968)	George Dulikravich (1986 – 1999)
Hendrik Tennekes (1966? – 1977)	William Pritchard (1986– 1989) joint w/ Math
Fred R. Payne (1966? – 1967)	
David P. Hoult (196? – 1967)	
Edward Rodgers (1966 - ?)	
S. Charles Wakstein (1966 – 1967)	McLaughlin Era (hires)
Roger E. Arndt (1967 - 1977)	Dennis K. McLaughlin (1986–present) DH
John S. Allen, Jr. (1968 – 1973)	Cengiz Camci (1986 – present)
Marshall H. Kaplan (1968 - 1981)	Wayne Pauley (1988 – 1995)
Winfred M. Phillips (1968 - 1979)	Roger Thompson (1988 – 1995)
Thomas M. York (1968 – 1987)	Anthony K. Amos (1989 – 2000)
McCormick Era (hires)	Lyle N. Long (1989 – present)
Hubert C. Smith (1968 – 1999)	George A. Lesieutre (1989 – present)
Joseph Eisenhuth (1969 – 1986) joint w/ARL	Michael M. Rieschman (1989 – 1997)
Zachary Sherman (1969 – 1972)	Edward C. Smith (1992 – present)
Blaine R. Parkin (1972 – 1989) joint w/ ORL	Farhan Gandhi (1995 – present)
Philip J. Morris (1977 – present)	David Spencer (1999 – present)
Gilbert H. Hoffman (1979 - 1981) joint w/ ARL	Kenneth S. Brentner (2000 – present)
Robert G. Melton (1981 – present)	Joseph Horn (2000 – present)
Michael M. Micci (1981 – present)	Deborah A. Levin (2000–present)
Mark D. Maughmer (1984 – present)	

Figure 16 --Faculty 1998
Front Row L-R: Dennis McLaughlin, George Dulikravich, Barnes McCormick, J. William Holl, Michael Micci, Robert Melton, Philip Morris, Farhan Gandhi
Back Row L-R: Edward Smith, Joseph Eisenhuth, Mark Maughmer, Lyle Long, Hubert Smith, Anthony Amos, George Lesieutre, Cengiz Camci

Growth in Graduate Program

In the 1960's and 70's the department had strong research activity in hydrodynamics, underwater propulsion, turbomachinery and applied aerodynamics. These had been the strengths of Drs. Wislicenus and McCormick. With such an influx of new faculty, the base of research in the department was significantly broadened. Professors Lesieutre, Ed Smith and Farhan Gandhi brought expertise in structural dynamics that had been absent since the time of David Perry. Aeroacoustics grew from a small, one faculty member program (Dr. Morris), with the additions of Drs. McLaughlin, Long and Brentner, to a major strength of the department. The largest research contract has produced the interdisciplinary Rotorcraft Center of Excellence with no fewer than eight faculty participating in the projects. Professor Edward Smith is the director and Dr. Lyle Long is administrative director. Finally, the astronautics area has been gradually solidified with Dr. Micci's novel space propulsion research, Drs. Melton and Spencer's astrodynamics activity, and Dr. Levin's hypersonics and space environment research.

Research Areas in the 21 st Century

The research expertise of the Penn State Aero Department in the 21st Century is summarized in the block diagram of **Figure 17**.

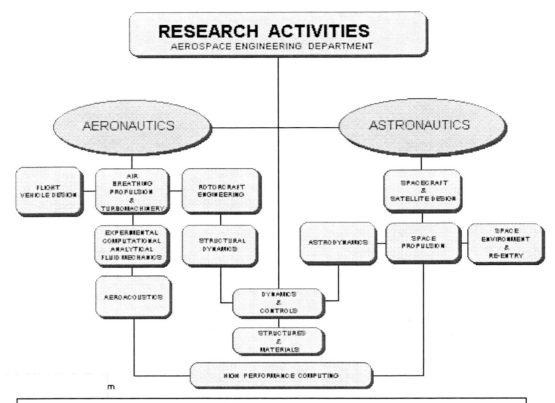

Figure 17   Research in the PSU Aerospace Engineering Department in the 21 st Century

Our faculty members pursue research that is more or less aligned with our teaching. In a broad sense, this research can be divided into aeronautics and astronautics activities. Research also proceeds along the disciplinary lines of aerodynamics, propulsion, structures, dynamics & control, and high-performance computing. Vehicle design research involves the appropriate integration of these technologies using optimization tools from systems engineering. Research on high-priority topics includes quieting airplanes, autonomous flight control, small surveillance vehicles enabled by MEMS sensors, and morphing aircraft structures.

Numbers of Graduate Students

With the influx of new faculty in the late 1980's, stronger emphasis on graduate student recruiting resulted in significant growth. **Figure 18** shows a graph of the MS and PhD enrollments for the past 20 years. Note also that following the strong growth to 1995, there has been a gradual decrease to more moderate levels associated with the

years of low number of BS degree students nationwide. The graduate student population is approximately equally divided between MS and PhD candidates. Additionally approximately one-half of the graduate students are US citizens and one-half are international students from about a dozen countries.

Penn State PhD Enrollment

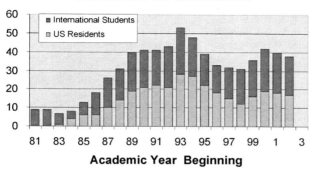

Penn State MS Enrollment

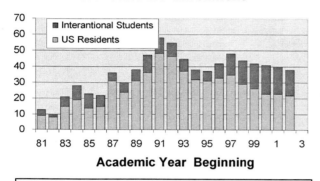

Figure 18   Record of Graduate Student Enrollments for Past 20 Years

Aero GSA (Graduate Student Association) was formed in 1994 to promote interaction among Aerospace Engineering graduate students and faculty and to address graduate student concerns. Activities include organization of departmental seminars, fall orientations, and participation in IPAC reviews of the department.

External Support for Research

Certainly a major factor in the strong increase in the graduate student population has been the steady increase in the research contract funding base. **Figure 19** shows a graph of the annual expenditures in research for the past 20 years. Grants provided for the sole purpose of acquiring new equipment are shown separately on this graph.

Equipment Grants

It is apparent from Figure 19 that there were several years in which substantial numbers of specialized equipment grants were awarded to our department through our faculty's proposal writing efforts. These grants were awarded by NSF and the US Department of Defense. Not shown on the graph are the substantial funds provided by State-matching programs that allowed significant expenditures to be made on equipment predominantly for instructional laboratories. At the same time strong efforts have been made to integrate the teaching and research so these facilities are also heavily used in research.

Facilities Development in the 80's and 90's

The decade beginning in 1986 was the most conducive one for facility development in the department. The combination of new space, adequate budget and

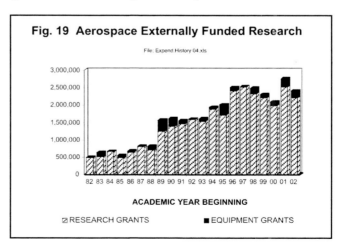

committed faculty was the right mix of ingredients to make tremendous progress in facility development. The first round of these efforts occurred in the late 1980's and early 1990's. The activity started with the dismantling of the department's large wind tunnel (test section size approximately 4 ft. x 5 ft.) for relocating to the Academic Projects Building on the east end of campus. Although somewhat inconvenient for students, the much larger room allowed us to rebuild the tunnel to produce a much higher quality facility. Upon taking apart the 30 year old tunnel we discovered that replacement of all but the fan-motor drive was appropriate. The fabrication took two years, almost all work done by an undergraduate student team under the direction of senior student Rick Auhl. Following this tunnel project Rick joined the department staff and eventually became the laboratory director, a position he currently holds. Eventually the wind tunnel served dual use with undergraduate student projects and Prof. Maughmer's low to moderate Reynolds number airfoil studies. The latter soon demonstrated the outstanding low turbulence levels and overall flow quality of the facility.

When the large wind tunnel vacated the Hammond Building the space allowed a large expansion of the turbomachinery lab with the eventual occupation of four large facilities: the existing single-stage compressor and the new turbine facility, the three-stage compressor and the automobile torque converter facility. All of these facilities were heavily used for a decade by students supervised by Profs. Lakshminarayana and Camci.

There was considerable facilities development underway in other areas of the department during the early 19990's. Professors' Jensen and Lesieutre added several composites fabrication and dynamics test specimen facilities working cooperatively with the Engineering Science and Mechanics Department. Prof. Pauley developed a low-turbulence water channel facility that has seen considerable use in research and laboratory instruction (in the Hammond Building). Prof. McLaughlin developed a new supersonic shear layer facility in conjunction with the low Reynold's number jet noise facility that came originally from Oklahoma State University. He also worked with Dr. Don Thompson of ARL to develop a centrifugal fan noise facility in the anechoic chamber of the Center of Acoustics and Vibration, CAV, (in Hammond). Finally, Prof. Micci built an impressive space propulsion lab housed in the Propulsion Engineering Research Center (also on the east end of the campus). Among several experiments, Micci's development of the microwave heated propulsion unit drew national attention for innovation.

Besides the laboratory (physical) facilities discussed above, the faculty made impressive advances in computer facilities for both research and instruction. The majority of these workstations and "clusters" of parallel computers were purchased with external grant support. Led by Professor Lyle Long, a 1993 winner of the IEEE Gordon Bell Prize for fast computer program development, the department has assembled a number of Beowulf clusters. There has been strong interaction in most of this computer development with the instructional program with the introduction of several new computational courses.

For the remainder of the 1990's very productive use was made of the aforementioned facilities. Research activity expanded partly because of this expanded capability. Perhaps this contributed to a citation from the Institute for Scientific Information website (2001) that ranked our department fifth in the country in terms of number of publications (from 1994-98), and second overall in publications per faculty member.

Aside from the computer facilities area, the department did not develop significant new facilities during the last half of the 1990's until the department was provided with a major space acquisition. This came to the department as a result of two departments obtaining new buildings

(Engineering Science and Industrial Engineering) and vacating Hammond. The new space allowed the department to significantly expand both the structures and the aerodynamics labs. In the latter case, we were able to install/fabricate a new wind tunnel (2 ft. x 3 ft. test section) that is dedicated primarily to instruction. We were able to acquire the majority of this tunnel from the University of Southern California (surplus) as they were upgrading to a tunnel donated by Nissan Company. During the last several years we have also been developing two additional major facilities: a hot jet facility, installed in the CAV anechoic chamber and a helicopter rotor test facility. The facilities were developed by modernizing a surplus jet noise plenum, obtained from Syracuse University and the rotor facility, obtained from Boeing Helicopters. During the same time period we sent the torque converter facility to the University of Toledo and the single-stage compressor to Oklahoma State University when our use of those capabilities diminished.

In summary, the department's laboratory and computer facilities are able to provide our students with an excellent capability to pursue state-of-the-art research on a wide variety of topics. At the same time there has been a gradual process of involving undergraduate students in projects that interrelate to the graduate students' research. The learning environment is something we are proud of.

Centers of Excellence

Rotorcraft Center of Excellence Established, 1995

The National Rotorcraft Technology Center, a collaborative program between NASA, the US Army, and industry, awarded to Penn State Aerospace Engineering Department a National Rotorcraft Center of Excellence (in 1995). The Center's research program performed by a multi-disciplinary technical team consisting of researchers from several departments and other PSU research centers focused on rotorcraft and vehicle dynamics, smart composite structures, aeroacoustics and aerodynamics, and drive train technology. Professor Edward Smith serves as center director and Lyle Long is administrative director.

Institute of High Performance Computing Applications (IHPCA)

Professor Lyle Long was appointed Director of The Institute for High Performance Computing Applications and Philip Morris appointed as one of five Associate Directors upon the center's inception (in 1999). The center developed (with an NSF grant) a graduate Minor in High Performance Computing that has served as a model for other schools. This Minor is very popular with students in disciplines such as aerospace, chemistry, physics, mathematics, petroleum engineering, mechanical engineering, nuclear engineering, chemical engineering, and acoustics. Since 1999 we have awarded 28 degrees, and there are 40 more students currently enrolled in the program. This HPC graduate minor program was started and is administered by the Aerospace Engineering department.

Center for Acoustics and Vibration (CAV)

Research in acoustic and vibration is one of Penn State's enduring strengths; our program is one of the largest and most respected of its kind at a major research university. Dr. Gary Koopmann (professor of Mechanical Engineering) is the Director of the CAV, and Dr. George Lesieutre (professor of Aerospace Engineering) is Associate Director. The center consists of faculty, graduate students and staff in nine laboratories throughout the College of Engineering. These labs perform both disciplinary and cross-disciplinary research in areas related to acoustics and vibration. The Center strengthens basic and applied research in related engineering areas, fosters graduate education in acoustics and vibration engineering as well as providing a base for technology transfer to industry.

UNDERGRADUATE STUDENT PROJECTS:

During the 1990's, the aerospace engineering faculty began to integrate more engineering project work into the undergraduate curriculum. The commonality of the activity described here is the aspect of design-build-test, providing students with more "real world" experiences to accompany their theoretically based coursework.

Sailplane course (Flight Vehicle Design and Fabrication)

The longest lasting design-build-test experience, the "sailplane course," has evolved into an official multi-year elective course, in which students participate in the design and construction of an advanced composite sailplane. The goal of this project course is to provide a "hands-on" environment in which students learn about aerospace engineering in a cooperative manner that provides opportunities for mentoring and leadership training. The program provides students with an integrated education in total vehicle design, construction, testing, and project management. By its very nature, this project includes extensive interdisciplinary involvement in aerodynamics, structures, materials, stability and control, fabrication and flight-testing. The coursework is vertically integrated to give freshmen and sophomores experience in various aerospace engineering principles by working with juniors and seniors on design projects and components of the full-scale sailplane. Students are allowed to explore implications that their designs will have in manufacturing the final product by building their designs in the dedicated laboratory. The course was originally started by Profs. Mark Maughmer and David Jensen. Dr. Maughmer kept it going (and healthy) following Jensen's departure for Brigham Young University.

Flying Boat Project

The flying boat project has evolved from research activity in the aerodynamics of large wing-in-ground-effect vehicles. An independent study course was created to allow a team of students to work on the project for an extended period of time. The team's current design is for a recreational vehicle that will be towed behind a speedboat. It is envisioned that the Penn State flying boat will fit a new recreational market competing with towed water sports, such as water skiing and parasailing, and self-propelled watercraft. The students are working toward a goal of developing a design to the point where it could be commercialized by licensing to a recreational watercraft manufacturer. In the student development work, composite fabrication and testing forms the basis of model fabrication at several model scales. The smaller scale models are wind tunnel tested within the activity of the aerodynamics laboratory course, and the large-scale radio-controlled models undergo field tests by being towed behind a boat at a local lake.

Rocket Project

A sounding rocket project for undergraduate students was initiated by the department in fall of 2001, and offers the students a challenging hands-on space-related learning experience. The students build rockets and incorporate commercially available flight computers as the payload.

The flight computer contains a microcontroller as well as various sensors to measure acceleration and pressure. The 2003 flight was a milestone, having incorporated our program into a new incentive at NASA's Wallops Island Center. The Penn State Phoenix rocket is a pioneer in an outreach program NASA Wallops is initiating.

Uninhabited Air Vehicles

Uninhabited air vehicles (UAV's) are becoming a very important part of aerospace engineering. During the 2002-2003 academic year the department offered a first-year seminar in radio-controlled aircraft. Students each built an electric-powered radio-controlled (r/c) aircraft, and learned to fly them. This activity will expand in the 2003-2004 academic year with students building simple electric r/c aircraft, but then using them to complete a reconnaissance mission. They will then be provided with a larger vehicle that can carry microprocessors, telemetry equipment, GPS systems, and sensors. The students will have to assemble the systems, use them to accomplish a more complex mission, and measure the performance of the vehicle. Profs. Long and Maughmer initiated this activity.

ALUMNI

Outstanding Engineering Alumni (OEA)

During the first half of its existence, the Department graduated many who later came to distinguish themselves in all facets of the profession of aeronautical and astronautical engineering. Space limitations allow a listing of only a few of these alumni who we are proud to call "Nittany Lions".

Louis Borges	1944 BS (ME) with Aero Option
Mary Ilgen Douglas	1946 BS, PSU Woman of the Year, Chief Commercial Flight Testing,
Karl Bergey	1948 BS, Chairman Bergey Windpower Co., Inc. and Prof. Emeritus of Aerospace and Mechanical Engr., Univ. of Oklahoma
Barnes McCormick	1948 BS, 1949 MS, 1954 PhD
Paul Weitz	1954 BS, 30 days in Skylab, Command Pilot for 1st flight of the Challenger, Chief, Astronaut's Office (**Figure 20**)
James Marley	1957 BS, President, AMP, Inc
Ernest (Jim) Cross	1959 BS, Head, Dept of Aero., Texas A&M, Dean of Engineering, Old Dominion Univ.
Jerome Schutzler	1961 BS, Co-Founder, PDA, Inc., Re-Entry Transparencies
Niles Kenyon	1964 BS, President, Conair, Inc.
Ardell Anderson	1964 BS, Boeing executive, then United Technologies Executive
Guion Bluford (**Figure 21**)	1964 BS, Shuttle Mission Specialist, Nation's first black Astronaut
John McKeown	1965 BS, Executive, U.S. Navy BuAir
Glenn Spacht	1968 BS, Grumman VP, Manager, X-29

Outstanding Engineering Alumni (OEA) Continued

Andrew Logan	1966 MS, VP, McDonnell-Douglas Helicopters, inventor of NOTAR
Paul Leamer	1967 BS, Director X-35 Liftfan Program, Lockheed Martin Aeronautics Company
James Hargrave	1969 BS, VP, Navy Helo Program, Lockheed Martin Federal Systems
Robert Cenker	1970 BS, 1973 MS, Shuttle Mission Specialist

Figure 20
Astronaut Paul Weitz

Figure 21
Astronaut Guion Bluford

Honorary Lecture Series

In November 2000, the **Barnes McCormick Honorary Alumni Lecture Series** was initiated. This lecture is the first of a new series, which will annually honor outstanding achievements of distinguished alumni of the Aerospace Engineering Department. Professor McCormick was an especially appropriate alumnus to be the first recipient of this award. **Samuel L. Venneri** (BS 1969) was selected as the second McCormick Honorary Alumni awardee in October 2001. Mr. Venneri received a MS in Engineering Science from George Washington University in 1975. Mr. Venneri was appointed NASA's Associate Administrator for Aerospace Technology in February 2000, while retaining his previous position as Chief Technologist. The 3rd recipient of the McCormick Honorary Alumni Lecture Series was **Daniel Mooney** (BS 1990, Civil). Mr. Mooney is Vice President of Product Development for Boeing Commercial Airplanes, where he leads the group responsible for developing new and derivative commercial airplane products and features.

Aerospace Scholarship History

Enhancement of our student (College and Departmental) scholarships and fellowships during the late 1990's. Departmental scholarships include:

David J. Peery Memorial Scholarship ('84)
Donald and Jayne Steva Scholarship (first awarded '97)
James Reynolds Norris Memorial Scholarship
  (first awarded '98)
Mary Ilgen Memorial Scholarship (first awarded '99)
John and Brenda Myers Scholarship (first awarded in 2000)
John Pierre Hemler Scholarship (first established in 2002)

Richard W. Leonard Scholarships
  (first awarded '97)
Lou Borges Scholarships (first
  awarded '98)
Carl A. Shollenberger Memorial
  Scholarship (first awarded '88)
Aero Pioneers Class of 1944
  Scholarships (first awarded '99)

Finally, in closing, Figure 22 presents a recent photo of the Department's present faculty

Figure 22
2003 faculty in the Department of Aerospace Engineering, The Pennsylvania State University
Front Row, L to R: Cemgiz Camci, Lyle Long, Edward Smith, Phillip Morris, Deborah Levin, George Lesieutre, David Spencer, Joseph Horn
Back Row, L ro R: Kenneth Brentner, Robert Melton, Mark Maughmer, Barnes McCormick, Dennis McLaughlin, Michael Micci, Farhan Ghandi, Goetz Bramesfeld, Richard Auhl, ,

Summary

The co-authors have served the Penn State Department of Aerospace Engineering in the department head position for a combined total of 33 years. And one of us (BWM) has over 40 years of total service. Throughout this time, it has been a distinct pleasure working with the outstanding faculty who have served this Department with such distinction. Also, behind the scenes, have been the excellent staff members who also contribute so much, but who are too numerous to mention. It is appropriate, however, to single out Ms. Sheila Corl and Ms. Tammy Besecker, who have both served our Department so superbly for over twenty years, much of the time as the lead staff administrator(s) in the Department. Ms. Besecker deserves a special thanks for assembling much of the material in this Department History.

Chapter 36

Department of Aeronautical and Astronautical Engineering
The Ohio State University

Prologue

When the Board of Trustees of The Ohio State University selected Garvin L. Von Eschen to organize and develop the newly authorized Department of Aeronautical Engineering, they could not have known what a serendipitous choice they had made. Over the next 33 years, Von Eschen led his department with a clear vision, strong will, and a graceful style as he laid down the foundation of the Aeronautical Engineering Program at OSU. He organized the department, hired the first faculty, developed the curricula, assembled hardware for laboratory facilities, taught, and mentored undergraduate and graduate students. He was a leader, both of his faculty and of his peer group of national Aerospace Engineering Department Chairs. Von Eschen continued his Herculean efforts for three decades, and when he retired in June of 1979, he left a legacy of a highly rated program, competent faculty, and quality laboratories. He had the respect, admiration and love of the many hundreds of students, staff, and colleagues that were influenced by his life's work as Chair of the Aeronautical and Astronautical Engineering Department.

This history begins, appropriately with a brief review of the Von Eschen Era, from 1946 to 1979. Professor Frank McLean Mallett, one of the first hires of Von Eschen, provided the first portion of this era, covering the early years up to 1969. Mallett provides detail and a unique perspective of those early years, written at Von Eschen's request as OSU was preparing for its 100th anniversary. The years following the Von Eschen Era are divided into two distinct periods as a result of a College of Engineering reorganization. Any edits resulting in inconsistencies or errors are the responsibility of this author. In 1994, the College reconfigured its departments, reducing them from 12 units of varying sizes, to just 8 departments and a School of Architecture. In that process the Aeronautical and Astronautical Engineering Department was combined with the Engineering Mechanics and the Aviation Department to form the Department of Aerospace Engineering, Applied Mechanics, and Aviation. The 1994 date, therefore, is the latest year when there was Department status for the Program. Thereafter, the Aeronautical and Astronautical Engineering Program existed as a component of a larger entity.

The history that follows touches lightly on the curriculum, faculty, students, and research over the many years of the Department. In some cases too briefly, but in all cases without regard to the budgeting matters that weigh into operations and administration of academic departments. These details must await another time.

Gerald M. Gregorek, Professor
Aerospace Engineering
The Ohio State University
September 2003

The Von Eschen Era   1946 to 1979

Introduction

The determination of the scope of a history of an academic department of a university is a problem with no easy solution. In this brief account, emphasis has been placed on the early years, since this period is apt to be of the greatest interest and at the same time the least known.

One of the difficult aspects of the determination of the scope is the question of the faculty. At one extreme, complete biographies could be included. In the interest of brevity, this account omits all biographies. Generally, those faculty members who served for more that a few years are the ones mentioned in this text, but a complete listing, with dates of service, is provided in Appendix 1.

A similar question arises in connection with students. It is a temptation to mention some of the most outstanding, but since the graduates of the department range through the whole spectrum from prominent to obscure there seems to be no satisfactory criterion for inclusion that will not risk omitting some who are perhaps as worthy of mention as some of those included. Reluctantly, it was decided not to mention any individuals except the earliest graduates.

Realizing how little material has been included in this history, even after many hours of search, the writer finds himself hoping that more information can be collected and saved for the use of some future writer who will address himself to the history of Aeronautical and Astronautical Engineering at Ohio State University.

The Beginning and Some Milestones

If the date of authorization may be taken as the beginning of a department, then 8 March 1943 is the birth date of the Department of Aeronautical and Astronautical Engineering at the The Ohio State University. It was on this date that the Board of Trustees approved the establishment in the College of Engineering of a new department, the Department of Aeronautical Engineering. Due to war-time complications, the actual organization of the department was delayed for three years.

Although wartime conditions complicated the beginnings of the department, it was World War II that caused the University to reverse its earlier policy of not entering the field of aeronautical engineering instruction. In 1942, a committee was appointed to study the whole question of aviation and aeronautical science. This "University Policy Committee" presented a report dated 21 December 1942 that included six recommendations. The first of five parts of the third of the recommendations was to develop an undergraduate curriculum in aeronautical engineering. Graduate work and research in this field were included in the fourth recommendation.

Some of those interested apparently anticipated the action of the committee as well as the approval by the Trustees, because the 1942-1943 Engineering College Bulletin listed a curriculum in Aeronautical Engineering.

It was not until 1946 that the department really came into existence. Garvin L. Von Eschen was brought from the University of Minnesota to become the chairman and to organize the department. Edward Miller was the first to join him, being hired as an Instructor working on a Master of Science degree.

Also during the first year, Knox Millsaps, at that time at Wright-Patterson Air Force Base, was recruited for the new staff. The department was given office space in the southeast corner on the first floor of Robinson Laboratory, also a small war-surplus hut located just east of Robinson where part of the Physics building now stands. This hut was used to store war-surplus equipment, which Professor Von Eschen was beginning to accumulate.

During the first year, the department had a few undergraduate students, completing work previously started, and two graduate students. As the post-war rush of students continued, the second year of the department saw a considerable increase in enrollment. A number of veterans, whose previous education has been interrupted at a various stages, enrolled in Aeronautical Engineering. A temporary building, standing on part of the same ground now occupied by the Civil-Aero Building and Hitchcock Hall, was assigned to the department. The building was of two-story wooden construction with three classrooms on the first floor and office on the second floor. It was called "Engineering Annex B" and was one of three war-surplus buildings shipped to the campus in pieces and reassembled on a site that had previously been open land along Neil Avenue between West Woodruff and Nineteenth Avenues. Two more instructors were added to the teaching staff, Frank Mallett and Lloyd Yates.

The department used Annex B until March 1957, when space was assigned in part of the second floor of Townshend Hall so that the temporary buildings could be removed to make way for the construction of the Civil-Aero Building. The new building was occupied on 19 December 1959. Many of the students and faculty members helped assemble tables and chairs for the classrooms in the last week of the year, and the rooms were ready for classes for the Winter Quarter, 1960.

In the meantime, consideration was being given to the name of the department. The successful launching of artificial satellites had brought attention to the astronautical part of the department's activities. Many departments of Aeronautical Engineering in other universities had changed their names to Aerospace Engineering or some other such name that reflected the dual nature of the field. In April of 1960, the department requested that its name be changed to the "Department of Aeronautical and Astronautical Engineering." With the approval of this request, the department gained its present designation and began to grant the degree of Bachelor of Aeronautical and Astronautical Engineering.

The Faculty

In the second year of the department's operation with its own staff, Professors Von Eschen and Millsaps with the three instructors, Mallett, Miller, and Yates, comprised the teaching staff, as mentioned herein. In addition, Carroll Pierce, the Director of the School of Aviation at Don Scott Field, was attached to the department and assisted with some of the instruction. For the academic year 1948 – 49, Mallett and Miller became Assistant Professors and Arthur Tifford joined the department and remained a member for ten years. The following year brought Bruno Boley to the staff for a three-year stay.

From the beginning, the faculty has represented a strong mixture of analytical and experimental interests, in many cases both being represented in the same individuals. Also established from the beginning was the department's interest in undergraduate students, who have not been neglected in favor of graduate work and research. A good balance between teaching and research has always been a characteristic of this department.

Until about 1957, the size of the faculty, including instructors, averaged about seven. This figure increased in the late fifties and during the sixties has averaged about eleven, with some of these only part-time, devoting the rest of their time to research projects.

Of the present staff, Professors Von Eschen and Mallett remain from the early years, as well as Professors Rudolph Edse and John Lee, who joined the department in 1951 and 1953, respectively. Professor Lee Petrie started as an instructor in 1959, as did Professor Gerald Gregorek in 1960. Also in 1960, Professor Buford Gatewood joined the department. Professor Robert Nerem started as an instructor in 1962. In 1964, Professor Odus Burggraf returned to the department in which he had once been an undergraduate student, and Professor Ting Li joined the department. The two most recent members of the faculty are Professors Cecil Bailey, 1967, and Professor Jeffrey Young, 1969. In the summers of 1967 and 1969, Keith Stewartson of the University of London was a visiting professor, and it is planned that he will continue to serve in future summers.

In Appendix One is a list of past and present members of the faculty, including instructors, with dates. The first year given is the date of joining the department, a second year shows the date of leaving.

The Course of Study

The Early Years
Between the authorization of the department and its actual organization under Professor Von Eschen a curriculum in Aeronautical Engineering was offered under the auspices of the Department of Mechanical Engineering. Examination of college bulletins for 1942-3, 1943-4, 1944-5, and 1945-6 shows a curriculum in Aeronautical Engineering for these years. The (nominal) department is also listed under "Departments of Instruction" with two courses offered; 601, Aerodynamics and 610, Aircraft Stress Analysis. The curriculum resembled that in Mechanical Engineering, but included the other relevant courses. In the bulletins for 1944-5 and 1945-6, the fourth year is marked "tentative" rather than "not offered".

The bulletins for the first year of the actual organization of the department, 1946-7, lists nine courses including 950, Research. In addition to the courses 601, 610, 701, 710 and 711 mentioned above, AE 713, Aeronautical Laboratory and two graduate level courses, 810 and 812, are shown.

Three-Hour Courses and Five Options
The Engineering College Bulletin for 1947-8 shows the first full-fledged curriculum for Aeronautical Engineering. Under "Course of Instruction" were listed 59 courses. Of the 33 in the 600-700 number range, twelve were marked as not being offered that year, but all were offered the following year. Listed separately was a fourth year program for the old four-year curriculum. The five-year curriculum was listed in its entirety. The first aeronautical course was not taken until the spring quarter of the third year. This first course, A.E. 640, introduced the student to the nature of aircraft and their components. The more analytical work commenced in the fourth year. In the summer between the third and fourth years the student was given a choice between flight training and "summer experience" on a job.

For the fourth and fifth years, the student chose from five options, Aerodynamics, Flutter and Vibration, Propulsion, Stability and Control, and Structures. Some courses were taken by all, the remaining course selections were governed by the choice of options. Most of the courses were for three hours of credit, and covered a wide range of subjects, as partly reflected by the options. All students took a laboratory course and an aircraft design course.

In the academic year, 1948-49 two courses were added that presaged the space age. Called Exterior Ballistics I and II, they went somewhat beyond the classical exterior ballistics in being concerned with orbits of hypothetical artificial satellites. These courses were joined by three more the following year, when Exterior Ballistics III and Missile Ballistics I and II were added. Courses devoted entirely to viscous aerodynamics and to superaerodynamics (very low density, or high altitude, aerodynamics) were also added. By the next year, the total number of courses listed had risen to 71. This listing remained for two more years.

Five-Hour Courses and Three Options
The first major change in the curriculum went into effect in the academic year 1953-54. A consequence of the three-hour course pattern was that the students were studying too many different subjects at once, often as many as six. For this reason, the department decided to make most of the undergraduate courses five hours each. This necessitated a considerable revision of the course structure, and reduced the total number of courses to 61. At the same time the five options were consolidated into three, Aerodynamics, Propulsion, and Structures. Flutter and Stability and Control were not, however, removed from the curriculum. The student's introduction to his major still came with A.E. 640 in the spring quarter of this third year. There was no change in the mathematics required after the Calculus, although the previous year there had been introduced A.E. 691, a course that was designed to show aeronautical applications of mathematics.

For the academic year 1956-57 a change was made in the introductory course. A new course, A.E. 642 was placed in the autumn quarter of the third year, with an aerodynamic course in the spring quarter. The following year 642 was moved to winter, but it still provided the student with an earlier start on his major than had prevailed earlier.

In 1957-58 and additional course in Mathematics was required. Previously, the three courses beyond the calculus had been, respectively, ordinary differential equations, partial differential equations, and a combination of vector analysis and complex variable. It was decided that one three-course was not enough for these last two subjects, and a separate course in each was required.

The Professional Division
The academic year 1959-60 was the first year for several changes. At the College level, the separation into the "pre-professional" and "professional" divisions was adopted. With the professional division including the last three years, a curriculum for the first two years became possible that was more nearly common for all departments in the college. Included in the common first two years was a change in the Mathematics requirements. Instead of starting with "College Algebra", the first course listed was the course that had been third, the analytical geometry and beginning calculus course. With two quarters thus gained, ordinary differential equations and statistics were then included in the winter and spring quarters of the sophomore year. The statistics course was dropped in 1967.

These changes, of course, were college-wide. For the department it meant that the four third year courses could be reduced to three, since the ordinary differential equations would already have been taken. The applications course 691, was also dropped at this time. Another change resulted in three Aero courses being included in the third year, the introductory course in the autumn, aerodynamics moved to winter and a beginning structures course moved forward to the spring.

More noticeable changes also took place at this same time. The options were dropped and more use made of technical electives. The basic course pattern was changed when almost all of the undergraduate courses were changed to four hours. The five-hour courses had proved to provide too little flexibility in curriculum arrange, as a given subject could be assigned only five hours or ten, or other multiple of five. In some cases five was more than needed, in other cases, five was not enough, but ten too many. Since the five-hour pattern seemed to be the opposite of the old three-hour pattern, the obvious compromise was tried and has been quite satisfactory. It provides for a basic student load of four courses for a total of sixteen hours.

PROFESSIONAL DIVISION CURRICULUM 1959 -60 (Pre-Engineering Division)

Year 1		Hours	Year 2		Hours
Autumn	Math 440	5	Autumn	Math 543	5
	Calculus			Calculus	
	Chemistry 404	4		Physics 531	5
	General			Mechanics	
	Eng. Dr. 440	3		Chem. 421	3/0
	Orthographic Projection			Quantitative Analysis	
	English. 416	3		Basic Education Requirement	3/6
	Survey of Eng. 401	1		Military or Air Science	2
	Military Or Air Sc	2			
	Phys. Ed. 401	1			
Winter	Math 541	5	Winter	Math 544	5
	Calculus			Differential Equations	
	Chem 405	4		Physics 532	5
	General			Heat, Light, Sound	
	Eng. Dr. 441	3		Chem. 422	3/0
	Principles			Quantitative Analysis	
	English. 417	3		Basic Education Requirement	3/6
	Military or Air Science	2		Military or Air Science	2
	Health Ed. 400	1			
	Phys. Ed. 402	1			
Spring	Math 542	5	Spring	Math 546	3
	Calculus			Statistics	
	Chem 406	4		Physics 533	5
	General and Qualitative Analysis			Electricity and Magnetism	
	Eng. Dr. 442	3		Eng. Mech. 521	5
	Working Drawings And Graphics			Statistics	
	English. 418	3		Basic Education Requirements	3
	Survey of Eng. 402	1		Military or Air Science	2
	Military or Air Science	2			
	Phys. Ed. 403	1			

PROFESSIONAL DIVISION CURRICULUM 1959-60

Year 3		Hours		Year 4		Hours
Autumn	Math 622	3		Autumn	Mech. E. 611	3
	Vector Analysis				Heat Transfer	
	Eng. Mech. 602	5			Aero. E. 706	4
	Strength of Materials				Ideal Aerodynamics	
	Aero. E. 642	4			Aero. E. 705	4
	Introductory Aeronautics				Aerothermochemistry	
Winter	Math 609	3			Aero. E. 710	4
	Fourier Series and Boundary Value Problems				Aircraft Structures	
	Physics 614	3		Winter	Elec. E. 642	4
	Introduction to Modern Physics				Electrical Engineering	
	Met. E. 611	4			Aero. E. 707	4
	Production and Properties of Structural Materials				Compressible Aerodynamics	
	Aero. E. 673	4			Aero. E. 716	4
	Applied Aerodynamics				Unsteady Aerodynamics	
Spring	Math 624	3		Spring	Elec. E. 644	4
	Complex Variables				Industrial Electronics and Control	
	Chem 689	4			Aero. E. 760	4
	Chemical Equilibrium				Propulsion I	
	Eng. Mech. 617	5			Aero. E. 775	4
	Dynamics				Aerodynamics of Viscous Fluids I	
	Aero. E. 610	4				
	Aircraft Stress Analysis					

Year 5

		Hours
Autumn	Aero. E. 713	4
	Aeronautical Laboratory	
	Aero E. 762	4
	Propulsion II	
	Aero. E. 724	4
	Aircraft Stability and Control	
	Aero. E. 776	4
	Aerodynamics of Viscous Fluids II	
	Aero. E. 790	1
	Seminar	
Winter	Aero. E. 740	4
	Preliminary Design of Aircraft	
	Aero. E. 754	4
	Aeroelasticity I	
	Aero. E. 791	1
	Seminar	
	Technical Elective	8
Spring	Aero. E. 731	4
	Aircraft Design Laboratory	
	Aero. E. 777	4
	Superaerodynamics	
	Aero. E. 792	1
	Seminar	
	Technical Elective	8

In addition, the Basic Education Requirements were to be met during the second, third, and fourth years to a total of forty hours, five in Natural Science, and at least fifteen each in the Social Sciences and Humanities for the remaining thirty-five.

The New Four-Year Program

In 1969, effective with the Summer Quarter, the College of Engineering left the five-year bachelor's curriculum and went back to a four-year program. The story of this change belongs in the College history, but its effect on the department can be briefly related. In general, the reduction was effected by "cutting off at the top", i.e. the more advanced courses were no longer required of the undergraduates. In addition, it was necessary to start the work of the department in the sophomore year, which was done by placing one course each in the winter and spring quarters. The mathematics requirements beyond the Calculus were reduced from fourteen to nine hours, the difference being made up by the addition of a new course in the department, AAE 480, Mathematical Methods in Aeronautical and Astronautical Engineering which has the advantage that various topics can be taken up as needed and related directly to their uses in aeronautical and astronautical courses. As part of the change, the degree awarded under the new program is Bachelor of Science in Aeronautical and Astronautical Engineering.

THE FOUR-YEAR CURRICULUM 1969 -70

Year 1		Hours	Year 2		
Autumn	Math 151 Calculus and Analytic Geometry	5	*Autumn*	Math 254 Calculus and Analytic Geometry	5
	Physics 131 Particles and Motion	5		Chem. 204 Principles of Chemistry	4
	Engl. 101 Composition and Reading	3		Eng. Mech. 210 Statistics	4
	University College 100 Freshman survey	1		Eng. Gr. 200 Computer Utilization	3
	Phys. Ed. 101	1		National Defense Option ROTC or Academic	
	Health Ed. 101 Hygiene National Defense Option ROTC or Academic	1			
Winter	Math 152 Calculus and Analytic Geometry	5	*Winter*	Math 415 Ordinary and Partial Differential Equations	5
	Physics 132 Waves and Quanta	5		Chem. 205 Principles of Chemistry	4
	English 102 Composition and Reading	3		Eng. Mech. 420 Strength of Materials	4
	Phys. Ed. 102	1		Aero-Astro. E. 400 Elements of Aeronautics and Astronautics	4
	National Defense Option ROTC or Academic			National Defense Option ROTC or Academic	
Spring	Math 153 Calculus and Analytic Geometry	5	*Spring*	Aero-Astro. E. 401 Elements of Aeronautics and Astronautics	4
	Physics 133 Particle Systems, Electrodynamics	5		Eng. Mech. 510 Dynamics	4
	Eng. Gr. 110 General Engineering Graphics	4		Met. E. 300 Materials Science	4
	Phys. Ed. 103	1		National Defense Option ROTC or Academic	
	National Defense Option ROTC or Academic				

Year 3			Year 4		
Autumn	Math 416	4	*Autumn*	Aero-Astro E. 520	4
	Vector Analysis and Complex Variables			Flight Vehicle Dynamics	
	Aero-Astro. E. 505	4		Aero-Astro. E. 570	4
	Thermodynamics			Viscous Flow and Heat Transfer	
	Aero-Astro. E. 660	4		Aero-Astro. E. 641	4
	Classical Aerodynamics			Structural Design of Flight Vehicle Components	
	Elec. E. 500	4		Aero-Astro. E. 695	1
	Electrical Engineering			Senior Seminar	
Winter	Aero-Astro. E. 603	4	*Winter*	Aero-Astro. E. 710	4
	One Dimensional Gasdynamics			Aeronautical Laboratory	
	Aero-Astro. E. 480	4		Electives	8
	Mathematical Methods in Aeronautical and Astronautical Engineering				
	Aero-Astro. E. 500	4			
	Flight Vehicle Structures I				
	Elec. E. 520	4			
	Electron Devices and Controls				
Spring	Aero-Astro. E. 750	4	*Spring*	Aero-Astro. E. 715	4
	Principles of Flight Vehicle Propulsion			Preliminary Design of Flight Vehicles	
	Aero-Astro. E. 661	4		Electives	4
	Compressible Dynamics				
	Aero-Astro. E. 640	4	*Electives*	*Basic Education Requirements are scheduled to provide 40 credit hours throughout the 4 year program*	
	Flight Vehicle Structures II				
	Engl. 305	3			
	Technical Writing				

The department has been accredited by the Engineers' Council for Professional Development, the official accrediting agency for engineering colleges, since 1949.

The Students

Earliest Graduates
The first graduates of the new department were Nancy Ann Ewing and William A. Gunn, who received the BAE degree on 6 June 1947. There were no additional bachelor's degrees awarded until the following June commencement when a class of thirteen received their degrees, with one more at the end of the 1948 summer quarter, and two more in December.

The first Master of Science degree from the department was awarded to Frank Mallett on 19 March 1948. Shao Yung Tung received this degree 11 June 1948 with three more, Edward Miller, Paul Rowe, and Karl Wein, on 3 September 1948. Mr. Wein was the first to receive his degree from this department through the extension center at Wright-Patterson Air Force Base. The department's first Doctor of Philosophy was Sheng To Chu, who received his degree on 10 June 1955.

Enrollment
The number of students enrolled in the curriculum before the actual organization of the department has been commented on above. Autumn quarter enrollments for the years 1946-1969 inclusive provide the following enrollment information: . The average for the four post-war years, 1946-

49, was about 196 in total enrollment, all classes. With the post-war rush ended, the average in the fifties was 180. This figure swelled to 311 for the sixties perhaps reflecting a post-Sputnik growth in interest. The enrollment figures for upperclassmen, omitting freshmen and sophomores, shows an average of 66 for the first four years, about 77 for the fifties, and about 110 for the sixties. It is notable that autumn of 1969 shows the largest enrollment of any year (469), and by a considerable margin.

Enrollment figures for graduate students indicate a similar upward trend, averaging 32 in the fifties and 82 in the sixties. In 1969 there were 103 graduate students enrolled in the AAE program. The total number of graduate degrees granted by the department, as of the end of the Summer Quarter 1969, included 168 students with the Master of Science degree and with 27 the Doctor of Philosophy.

Student Organizations

A student branch of the Institute of the Aeronautical Sciences was established in 1946, the first year of the department. The name has since been changed to the American Institute of Aeronautics and Astronautics. The purpose of the student branch is to provide the students with opportunities to get acquainted with each other, the faculty, and with the nature of the profession they plan to enter.

In addition to the student branch there is a chapter of the honorary society in Aeronautical and Astronautical field, Sigma Gamma Tau. This society was formed by the merger of two earlier organizations with essentially the same purpose, Tau Omega, founded in 1927, and Gamma Alpha Rho, founded in 1945. A chapter of Tau Omega had been installed at Ohio State on 9 December 1950. Sixteen students were initiated as charter members, along with two faculty members, Boley and Mallett, who had not previously been members of the society.

In February of 1953, the two societies met in joint convention to effect the merger. After considerable lack of agreement about which name or what name to use for the new joint society, one of the members of the Ohio State delegation proposed the name Sigma (for summation, or the joining) Gamma (for Gamma Alpha Rho) Tau (for Tau Omega). This suggestion met with immediate approval. Besides providing recognition for outstanding students, the chapter provides service for the department, especially in liaison between the department and the student.

The Department Laboratories

Since its authorization in 1943 and its initial staffing in 1946, the Department of Aeronautical and Astronautical Engineering has been developing its research and graduate training capabilities. Active construction of the Aerodynamic Laboratory was initiated in 1949 and the first wind tunnel tests were conducted in 1950. The Rocket Research Laboratory, originally developed near the end of World War II, was transferred from the Chemistry Department to the Department of Aeronautical Engineering in 1951. The two research laboratories are now combined and known as the Aeronautical and Astronautical Research Laboratory (AARL) located at the University Airport, Don Scott Field.

The creation of facilities for research from nothing forced the Aerodynamic Laboratory to facility-development types of contract support during the early years of operation. Although some facility-oriented research continues, increasing emphasis is continually being place on more basic studies utilizing the apparatus and intellectual talents accumulated and developed until most studies now being conducted can be classified as basic. The research program of the Rocket

Laboratory of the department was initiated on a more basic level from the beginning because of the Laboratory and some personnel from the Chemistry Department.

Much of the early Rocket Laboratory work under Director Rudolph Edse, dealt with premixed propellants and oxidizers, particularly hydrogen and oxygen. Research on detonations and related pressure oscillations, and the chemical kinetics of combustion with various gases have been performed for many years. The feasibility of supersonic combustion has also been explored in recent studies.

In the Aerodynamic Laboratory, originally located in a hangar at the University Airport, the 12-inch blow down supersonic wind tunnel was first operated in 1950. In 1952, the 12-inch transonic wind tunnel began operating as a model tunnel for the Air Force 16- foot tunnel at Tullahoma. In the ensuing years, various research studies were conducted and to facilitate data handling, an analog computer was installed in 1954; partly funded by North American Aviation, it was the first use of such a computer for the "on-line" reduction of wind tunnel data. Professor John Lee, the Director of the Aerodynamic Laboratory, started hypersonic flow studies in 1954. A 12-inch wind tunnel was developed as the first in the country capable of continuous operation with temperatures from 1000 degrees F to 2400 degrees F and Mach numbers up to 15. Since then research studies have been conducted on a variety of hypersonic flow problems. In 1962, development was begun on the use of electric arc discharges for driving a shock tube. Research has since been conducted on a variety of hypervelocity problems, including gas radiation, utilizing that high-energy shock tube. Over the same period, research studies were conducted with a continuous arc-heated (plasma) tunnel particularly in conjunction with electron beams and ion beams.

The experimental facilities of the Department are currently being used to conduct basic research and to train graduate and undergraduate students in experimental techniques. From the viewpoint of equipment, these facilities are among the most advanced to be found in any university.

Plate I – Engineering Annex B The home of the Aerodynamics Department for its first ten years, viewed from 19ᵗʰ and Neil Ave., 1956. Plate II below shows the same geographical location four years later.

Plate II – The Civil-Aeronautical Engineering Building, viewed from 19th and Neil Avenue, 1960. The name was changed to Bolz Hall in the 1990's.

Plate III – The Rocket Research Laboratory at Olentangy River Road and Lane Ave. The Lab shared space with the horse pastures, 1955.

Plate IV – The Astronautical and Astronautical Research Laboratory located on the OSU Airport Facility, 1966.

Professor Mallett ended his brief history in 1969. By that year, Von Eschen has chosen the faculty, refined the academic curriculum, and overseen the construction of a new building for the AAE Department on campus and the Aeronautical and Astronautical Research Laboratory at the OSU Airport. He presides over a vibrant department with increasing numbers of undergraduate and graduate students attracted to the AAE Program by its productive faculty, excellent research facilities, and the broad range of research efforts underway.

However, the times are changing, and the next decade, Von Eschen's last, will present challenges to both Von Eschen and the AAE faculty.

The Seventies

As the decade of the Seventies opened, the Apollo Moon Program was winding down, the jumbo jets had been designed and were entering commercial service, the military was completing the design and implementation of the missile deterrent, and the third generation of jet flights was already in flight test. This confluence of completed aerospace programs, when coupled to the debilitating effects of the waning Vietnam War and the associated campus unrest produced a downturn in the aerospace industry that reached into academia.

Cutbacks in the space, commercial, and military budgets followed with a direct impact on the research support given to the Universities. Companies and government agencies reduced job opportunities for both undergraduate and graduate students. The public was entertained by the billboard on the outskirts of Seattle admonishing the last one to leave the city to "turn out the lights." It was inevitable that student enrollment would follow this downward trend.

The AAE Program

Chairman Von Eschen responded to the times. To offset the lower funding from the reduced number of research projects, he tightened the budget. Travel and long-distance phone service was curtailed as well as other administrative reductions. Professor Foster joined the faculty in September 1970 to increase the faculty number to thirteen. The next faculty additions were not to be until 1981, with the hiring of .Professors Hayrani Oz and Mike Bragg.

The four-year program was in place and the last of the five-year students had graduated by 1973. Their graduations accelerated the enrollment decline. By 1976, the total student enrollment had declined to about 150, one fourth of the more than 600 students enrolled in the AAE Program in 1969. With the student enrollment decreasing, the teaching loads for the faculty were becoming dangerously low. Even at that time, the possibility of combining AAE with another department existed. To engage the AAE faculty, Von Eschen decided to "farm out" the faculty to overloaded departments in need of help in the College of Engineering. Both the Engineering Graphics and Mechanical Engineering Departments used AAE faculty. It was unorthodox, if not amusing, to have an AAE professor teaching Mechanical Engineering thermal dynamics and heat transfer courses to Electrical Engineering students.

Research Activity

The faculty research adjusted to the new era. At AARL, with the research into Hypersonic Aerodynamics almost eliminated, Professor John Lee moved his experimental work to the transonic regime and developed two new facilities – a high Reynolds number transonic wind tunnel, and a low turbulence transonic tunnel specifically designed for airfoil testing. Professor Lee Petrie maintained his diagnostic efforts with non-intrusive measurements in plasma flow and Professor Nerem continued his transition from high temperature plasmas to the study of blood plasma. Professor Gregorek developed a flight test capability for general aviation aircraft and used OSU aircraft for several NASA flight test programs. Professor Edse focused his research on detonations and supersonic combustion of Hydrogen.

On campus, Professor Burggraf continued his partnership with Visiting Professor Stewartson as they developed a new triple deck theory for boundary layer analysis. Professor Bailey, brought in to assist Professor Gatewood's work on thermally stressed structures, moved into more analytical work after a detailed examination of Hamilton's Law of Varying Action. Bailey developed a unique generalization of Hamilton's Law that bypassed the differential equations of motion completely. With his approach, he was able to demonstrate direct solutions to non-holonomic, non-conservative, non-stationary, linear and non-linear systems.

In 1976, a center that was to have a major impact on department research, the General Aviation Airfoil Design and Analysis Center (GAADAC) was established at AARL. This NASA supported center used computational methods developed at the NASA Langley Research Center to design advanced airfoils specifically for the manufacturers of general aviation aircraft. The novel concept allowed the NASA personnel to focus on the technology development of airfoil design methods. Aircraft manufacturers went to GAADAC for specific airfoil designs for advanced aircraft. Professor Gregorek was the Director of GAADAC, with Senior Researchers Kenneth Korkan, Ph.D. and Richard Freuler, Ph.D. assisting.

486

Faculty Changes

The number of faculty was reduced during this period, first with the job change of Professor Young in 1974, then with the sudden death of Professor Mallett and the retirement of Professor Li in 1975. Professor Gatewood, and Chair Von Eschen retired in 1978 and 1979, respectively. Professor Nerem moved up to Associate Dean for Research in the Graduate School also in 1979. These major faculty losses would become the next Chairman's challenge and opportunity to enhance the Program as the Department entered the next decade.

The End of an Era

Professor Von Eschen retired in June 1979. In a quiet dignified exit, he carried the thanks and best wishes of the many hundreds of students and colleagues that profited by his lifelong dedication to the Aeronautical and Astronautical Engineering Department of the Ohio State University. His legacy was a highly rated undergraduate and graduate program with research facilities and faculty that were among the best in the nation. He was, and would be, greatly missed.

II. *The Aeronautical and Astronautical Engineering Department 1979-1994*

Introduction

For the first 33 years of its existence, the Aeronautical Engineering Department- later to be renamed the Aeronautical and Astronautical Engineering Department- had the stability and vision of a single Chairman, Garvin Von Eschen. The next fifteen years brought three different Chairs to the leadership position. Stuart L. Petrie, an OSU graduate, replaced Von Eschen in 1978 and was faced with rapidly increased enrollments and only seven faculty to teach and conduct research. Petrie left in 1976 to take an executive position in industry. Professor Thomas M. York, appointed after a national search, came to OSU in 1987 from Pennsylvania State University with expertise in the space sciences. With enrollment near its peak, York was able to increase the faculty to thirteen, stressing and initiating space research programs. Professor Gerald M. Gregorek, who served one year as Acting Chair during the search for Petrie's successor, replaced Professor York in 1991 as York returned to teaching roles. Professor Gregorek was Chair during decreasing enrollments and major changes within the College of Engineering.

The AAE Program and Enrollment Management

The curriculum did not change significantly during this period, although both Chairs, Petrie and York, made efforts to increase the visibility of the Astronautic arm of the program. One of the major efforts was to provide a better separation of the two areas in the senior year. The design sequence was offered with aircraft or spacecraft design task, and elective courses oriented toward space flight were made available to the students.

In the early Eighties, Dean Glower of the College of Engineering initiated an enrollment management procedure to reduce the number of engineering students entering the Engineering College. College Enrollment was heading toward 7500 and as a Land Grant University, OSU was required to accept all students with valid diplomas from Ohio high schools. State funding for faculty limited the faculty so an Enrollment Management Plan was put in place to reduce the number of students in the upper years. The plan required a certain Grade Point Hour Ratio to be attained prior to being accepted into the upper classes of a particular discipline. Most of the departments in the College adopted a PHR of 2.3 for admittance to their discipline in the junior years. For the AAE Department, which had recovered from the 1976 low of 150 students, this meant about a 30% reduction in the students as they moved from the sophomore to the junior year. By 1982, there were 125 students taking the Sophomore AAE 201 course and these numbers were increasing. Sophomore enrollment peaked at 186 in 1987. From this high, however, the enrollment drifted downward again so that by the end of this period (1984) the entering sophomores were down to 24 and the BSAAE graduates numbered 35.

In 1982, Chairman Petrie convened the Industrial Advisory Board for the AAE Department. This board, made up of seven experienced executives of aircraft and engine companies and aerospace researchers, produced a document that led to a Ten-Year Plan presented to the Dean in 1983. They identified weaknesses, opportunities, and necessities to move the AAE Program to a higher level. Their comments included the need for modern structural methods such as finite element analysis, and to include composite structures, a course with increased emphasis on design and the system or disciplinary approach on design, and to encourage the use of personal computers. The Advisory Board agreed with the need to limit the class size, but felt that the number of faculty should increase. A comment worth noting even today was made by this panel: some university people had the view that the Aeronautical and Astronautical field was a mature field with a limited future. The committee disagreed and was unanimous that we were on the threshold of new and exciting developments.

Chairman Petrie used these comments in the Ten Year Plan, approved by the faculty in the fall of 1983 to work toward a faculty of thirteen. As a reference, the Dean had set a student to faculty ratio of 15 as an ideal target for the College. In 1983, the College average for other departments varied from 19 to 21. AAE average at that time was 31. Thirteen AAE faculty would move this ratio to the College average. The Dean approved this increase in the faculty in 1983.

The Faculty

When Professor Petrie stepped in as Chairman in the summer of 1979, Von Eschen's retirement had reduced the number of faculty to seven. Petrie immediately set about to increase the number with the addition of two new tenure track Assistant Professors. Appointed in the fall of 1981 were Professors Bragg and Oz. Bragg had just graduated from OSU with a strong research effort in aircraft icing. Oz arrived with specialties in space structures and dynamics. A few years later, in 1984, Nesrin Sarigul and Gary Yip were appointed as Assistant Professors in response to the Ten-Year Plan and the suggestions from the Advisory Board. Professor Sarigul was an expert in composite structures and Professor Yip specialized in high temperature gases. Professor Cecil Bailey retired in 1985, accepting Emeritus Professor status but remaining active. Professor Richard Bodonyi, who obtained his doctorate under Burggraf during the Seventies, joined the faculty in 1986 from a teaching position at the University of Indiana and Purdue in Indianapolis. He joined the faculty as an Associate Professor to work with Professor Burggraf on theoretical aerodynamics.

As Professor York assumed the Chairmanship, there were eleven faculty members, a number that increased immediately to twelve with the addition of Rama Yedavalli, coming to OSU as an Associate Professor with an excellent reputation in robust controls. Professor York searched for experienced faculty to assist in developing research areas, and in 1989, made appointments of Dr.

James N. Scott as an Associate Professor in Computational Fluid Mechanics and Dr. Peter Turchi as a Professor for Astronautics. Scott had worked many years at NASA and was teaching at the University of Dayton before he came to OSU. Turchi was a Director of a Research and Development Laboratory when he was attracted to OSU. Two young Assistant Professors were added in 1989, Hacene Bouadi and Mo-How Shen, to pick up the structures area left open by the early departure of Assistant Professor Sarigul. Faculty losses during this period included the departure of Yip in 1988 to return to California, Sarigul in 1989 to take a faculty slot at University of California, Professor Burggraf in 1989 to retirement as an active Professor Emeritus and Professor Michael Bragg in 1990 to accept another Associate Professor position at the University of Illinois.

Research Activities

As the Eighties began, the initiatives started during the lows of the past decade matured. The loss of the research efforts of Nerem and of Edse that were in the process of phasing out were replaced by new programs. On campus, Professor Burggraf was continuing his boundary layer work, with one of the more interesting applications of viscous flows to the study of severe storms, as both he and Professor Foster studied the fluid dynamics of tornadoes and hurricanes. Professor Bodonyi, joining the theoretical group in 1986, worked with high Reynolds number viscous/inviscid interactions, boundary layer stability, and separation. Professor Bailey, after his retirement in 1985, continued his work with Hamilton's Law, and Professor Oz worked with Bailey's concepts, as well as continuing his structural dynamics investigations.

At the Aeronautical and Astronautical Research Laboratory, the space research efforts begun by Professor York continued with instrumentation developed for studies of the ionosphere. By 1990, Professor Turchi had assembled a Gigawatt Pulse line consisting of banks of capacitors, and was initiating research projects with application to magneto-plasma-dynamic thrusters. The General Aviation Airfoil Design and Analysis Center was flourishing. Airfoils had been designed for many aircraft, propeller, and helicopter companies with projects that featured natural laminar flow designs, supercritical designs, and multi-element, high lift systems. These projects were ongoing and many involved wind tunnel tests in AARL facilities. As a direct result of these capabilities, considerable work was done on airfoils for wind turbine rotors – both for horizontal and vertical axis machines. The analytic work conducted in support of GAADAC led to computational studies of aircraft icing that expanded to wind tunnel and flight tests of aircraft in the icing environment. Similarly, the work with propeller airfoils naturally developed into the analytic prediction of the noise generated by propellers. In turn, this effort led to aero-acoustic propeller design and subsequent flight test of several AARL propeller designs.

Flight research peaked in the early eighties. On one particular day in September 1980, for example, there were three AARL teams of faculty, staff, and students conducting flight operations off-site: in Oklahoma performing near-field and far-field acoustic measurements with a Rockwell AeroCommander Twin Turboprop, in Canada evaluating icing on a US Army helicopter and in Virginia documenting the extent of laminar flow on the wing of a Bellanca Skyrocket. This flight activity was not sustained however, and ended by 1985.

New starts occurred as the decade progressed, with an original project that determined the performance of an engine test cell. Modeling an engine test cell enclosing a working jet engine began a long-term association with General Electric Aircraft Engines. Using simulators of GE jet engines designed by Professor Lee, scale models of test engine cells have been calibrated for noise and flow distortion since the early eighties. Dr. Richard Freuler conducted these tests and then monitored the full-scale validations of these model tests on site.

Another "new" start in the latter part of the Eighties was the selection of OSU as a NASA Hypersonic Research and Training Center with Professor Lee as Director. The Center re-vitalized the hypersonic facilities at AARL and led to work on Hypersonic Wave Riders and other experimental studies.

New Wind Tunnels

In 1991, with the support of the US Air Force and partial funding by the TrueSports Racing Team, AARL took over the operations of the 7'x10' Subsonic Wind Tunnel at Port Columbus Airport. This facility owned by the US Air Force, but leased to McDonnell Douglas as the tenant in the large Columbus facility fabricating parts for the McDonnell Douglas C-17, was not being utilized by the McDonnell team. The wind tunnel, with a top speed of 300 knots, was one of the last of the 7'x10' Subsonic Wind Tunnels built originally for North American Aviation in 1962. After several months of contractual effort, the facility was designated AARL-East and AARL researchers proceeded to perform aerodynamic tests on race cars, predominantly for the TrueSports Racing Team. AARL-East utilized a moving belt, provided by the racing team that correctly simulated the racecar environment. Other Air Force, NASA and Industry aerodynamic studies were conducted in this facility.

This 7'x10' closed return wind tunnel was the second of the low speed wind tunnels obtained during this period. Earlier, in 1983, a 3'x5' open-circuit, low turbulence level, subsonic wind tunnel capable of 150 ft/sec was donated by Battelle Memorial Institute. This wind tunnel has seen heavy use from its first operation and that use continues to this day.

Reorganization

In 1993, the President of The Ohio State University, Dr. E. Gordon Gee felt the administrations of the several Colleges of the University required streamlining. The Engineering College had 12 departments and a School of Architecture with 3 departments. The President, supported by the Provost, felt that this College of Engineering organization with many departments – some as small as eleven faculty, others with more than 50 faculty – would operate better if there were fewer, more balanced departments. With the threat of withholding funding for non-compliance, the faculty of the College of Engineering spent the academic year re-organizing. The College organization that emerged had eight departments and a School of Architecture with no departments. The AAE Department was combined with the Engineering Mechanics Department each with 12 faculty members and the Aviation Department to bring together a total of 27 faculty. Other groupings coupled the small Welding Engineering Department with the Industrial and Systems Department to produce the Industrial Systems and Welding Engineering Department while the Civil Engineering Department was grouped with the Engineering Graphics Department and Department of Geodesic Science (a transfer from the College of Arts and Sciences). The large departments: Computer and Information Science, Electrical Engineering, and Mechanical Engineering stood alone. The Material Science and Engineering Department had undergone an earlier combination with Ceramic Engineering joining with Metallurgical so it, and the Chemical Engineering Department, were left untouched.

On July 1, 1994, the Department of Aeronautical and Astronautical Engineering ceased to exist. From this date forward, the Aeronautical and Astronautical Engineering Program would reside in the Department of Aerospace Engineering, Applied Mechanics and Aviation (AAA). The College of Engineering as befits the problem solving character of its engineers, was the only college in the university to complete President Gee's reorganization desires.

III. 1994 to the Present

Introduction

In October 1994, the new organizational structure of the College of Engineering was in place. The University defines an administrative division of a Department as a Section. Therefore, the new AAA Department had three sections, each with a Head. A Chair of the Department was selected by the Dean after a suitable search and input from the entire faculty. The Chairman determined the head of each section based on faculty recommendations. Chairs serve 4 years, Section Heads serve 2 years. As the 1994 school year started, Professor Gregorek was appointed Chairman with Professor Hayrani Oz as AE Head, Professor June Lee as AM Head, and Professor Nawal Taneja as AV Head.

This organizational structure and basic interests did appear to have merit. Aerospace Engineering deals with fluids, but also with structures, so the structures efforts could always use more support and research activity from the Applied Mechanics Section. Also, AM had a large service load since the statics, dynamics and strength of materials were taught to most of the Engineering College by this section, and this service load could balance the decreasing teaching load of the AE Program. The Aviation Department, though with few faculty (3) but with many adjuncts, operated the OSU Airport, a major reliever airport, and taught flight training with ongoing research in human factors. Finally, the AE Section had both an undergraduate and graduate program, the AM Section only a graduate program and the AV Section only an undergraduate program. The possibility of developing a Master's Program for the AV section with the help of AE faculty existed, on the other hand, the AE undergraduates would have the opportunity for a Master's or Doctorate in Applied Mechanics.

While these potential synergetic activities were possible with this unique grouping, this synergism did not occur. The three sections were located in three different buildings. There was insufficient office space to consolidate the faculty under one roof. There was a possibility to bring the AE and AM Sections together, but was vehemently rejected. As a result, the hope for congenial integration of the faculty did not occur. The net result was that after three years, the AM Section voted to leave the AAA Department, and become part of the Mechanical Engineering Department. In October 1998, the entire faculty of the Applied Mechanics Section was absorbed into the ME Department. The ME Department took over the service course load of the AM Section and eliminated the Engineering Mechanics Graduate Program.

The Aerospace Engineering Section and the Aviation Section were reconstituted, at this time into the Department of Aerospace Engineering and Aviation (AEA).

Faculty and More Changes

With the retirement of Professor John D. Lee in 1993, the number of faculty in the AAE Program was now eleven. New appointments were soon to come, with Assistant Professor Reza Abhari in 1995 and Professor Michael G. Dunn in 1996. These two additions were the result of the acquisition by the AAA Department of the Gas Turbine Laboratory of the Calspan Corporation of Buffalo, New York. In 1994, this laboratory with excellent experimental facilities for gas turbine research and directed by Dr. Dunn, was about to be downsized, and Dr. Dunn was seeking a University as a home for his Laboratory.

Chairman Gregorek, assisted by OSU Vice President for Research Edward Hayes, gathered sufficient funding in 1995 to transfer much of the Calspan equipment and selected staff to OSU. The innovative funding process utilized two $750,000 five-year loans, one from The Ohio State University, the other from the State of Ohio's Department of Development to be re-paid from the indirect charges from the funded research projects conducted by the Gas Turbine Laboratory (GTL). The entire indirect charges of 46% were to be applied for loan reduction at the order from Vice President Hayes, an unusual concession necessary to enable the loans to be retired in a minimum time. As part of the acquisition process, Dr. Dunn was appointed Professor of Aerospace Engineering and Director of GTL. Reza Abhari was appointed Assistant Professor, jointly with the AE Program (75%), and the Mechanical Engineering Department (25%), he served as Associate Director of GTL.

In 1997, Professor Alfred B.O. Soboyejo received a one-third appointment to the AEA Department to assist in our structural courses; his other appointment was with the Agricultural Engineering Department. When Professor Abhari left for a Chair position in Switzerland, the faculty numbered 11.

Professor Gregorek stepped down as Chairman of the Aerospace Engineering and Aviation Department in 1999, to return to teaching and research. After an internal search for a Chair, Dean David Ashley selected Professor Nawal Taneja, a faculty member from the Aviation Section to head the AEA Department. Professor Taneja has an international reputation in Airline Operations, a Bachelor Degree and Doctorate from the University of London and a Master's from the Massachusetts Institute of Technology, all in the area of aeronautical engineering and economics. Professor Taneja appointed Professor Michael Foster as the AE Section Head.

When Dean Ashley left for the Chancellor position at a new California University, being formed at Merced, Professor James Williams, a faculty member from the Material Science and Engineering Department, replaced him in July 2001. In an effort to consolidate the aero propulsion activities of the College of Engineering, Williams moved three faculty of the AE section to the Mechanical Engineering Department. By October 2002, Professors Dunn and Shen, and a new AE appointment, Assistant Professor Igor Adamovich, (just hired to replace the vacancy left by Professor Turchi's recent return to industry) had moved to the ME Department, along with the Gas Turbine Laboratory facilities and staff.

The Present AAE Curriculum

Except for minor tinkering, the AAE curriculum had not been changed for many years. In 1995, the Provost charged the faculty with re-evaluating its curriculum. He had observed that Bachelor Degree Programs were requiring more than four years to complete. In the quarter system, with each quarter assuming 15 hours of coursework, 4 years at 3 quarters per year totals 180 credit hours. Many of our College of Engineering Programs required more than 200 credit hours for the Bachelors Degree. The Provost, feeling some pressure from parents as well as students, ordered revisions of all Bachelors Programs to reduce the required credit hours to make four-year graduations more probable.

The faculty revamped the AAE curriculum, and implemented it in the fall quarter of 1997. The reduction dropped an advanced aerodynamics course and reduced hours in several others.

Although the AAE Program still requires 189 total credits, this present organization of courses introduces several novel features. The three course sophomore introduction series has been reduced to two courses, freeing up spring for the Thermodynamic Course, and a new course, team taught by an AAE faculty member and a Mathematics professor to demonstrate the application of differential equations to aerospace engineering. The Junior Year has a new AAE course in linear systems and a new course series, one credit hour each quarter of the Third Year, to demonstrate the integration of the many elements of aerospace engineering. Student teams work on design tasks coordinated with their class work, with the teams presenting their efforts at the end of the Spring Quarter. The Senior Year has been completely altered. This last year has a two hour laboratory each quarter, taking up problems in aerodynamics and aero elasticity, propulsion and gas dynamics, and structures, dynamics and controls of flight vehicles – essentially aerodynamic, propulsion and structure, labs. A three-course sequence in design – conceptual, preliminary and structural design completes the required senior courses. In addition, each quarter of the senior year, the student can choose an elective in aerodynamics, propulsion, structures, and/or flight mechanics to obtain additional strength in areas of interest and to emphasize aeronautics or astronautics, as he or she desires.

The Present

The AAE faculty are entering a new era, and with only eight full-time members, the lowest number since 1976. In the near term, the demands of teaching undergraduate and graduate courses and of conducting research to support the graduate programs will be great. Experimental research activity has suffered from the dual loss of the Gas Turbine Laboratory and that of the large 7'x10' subsonic wind tunnel at AARL – East, demolished earlier. To replace these losses, the faculty are resolutely searching for research opportunities at nearby Air Force and NASA facilities, as well as with industry. Fortunately, the AARL transonic and subsonic facilities are engaged to assist their search.

On a more positive note, the student undergraduate enrollment has been increasing, whether due to the revised curriculum or the national scene. In 2002, the sophomore enrollment, the best indicator of the undergraduate program status, had risen to 82 for the Fall Quarter introductory AAE 200 course. In June of 2002, the Department awarded 23 BSAAE degrees. While these numbers do not compare with the peak enrollment year of 1987 with 186 entering sophomores and 82 BSAAE degrees, the statistics are very much improved over those of 1996. In that year, only 23 sophomores were enrolled in the AAE 200 class. Two years later, the number of graduates was 11, the lowest since the early fifties. If the 77 freshmen who have indicated Aerospace Engineering as a choice do, indeed, enroll into AAE as sophomores, the total AAE enrollment for the 2002-2003 year is 228. This surge in enrollment may even require a return to enrollment management, with a proposed limit of 55 students being contemplated as a limit for entering into upper divisions, a somewhat ironic consideration.

And so this chapter ends with the list of 28 faculty that have served in the AAE Program since 1970. They have worked through the decades that seem to alternate between up times and down times. Looking back, the even number decades – like the sixties and eighties – were periods of high enrollment and research activity. The opposite is true of the downturns of the seventies and nineties.

The tenor and history of this first decade of the new century will be written by future aeronautical and astronautical engineers.

Appendix One

Teaching Personnel
1946 - 1969

Cecil D. Bailey, 1967-
Willard P. Bergren, 1949-50
Bruno Boley, 1949-52
Loren E. Bollinger, 1958-66
Odus R. Burggraf, 1964-

Sheng To Chu, 1952-61
Ralph G. Dale, 1958-64
Rudolph Edse, 1951-
William E. Eiselstein, 1950-51
E. Stokes Fishburne, 1960-67

Buford E. Gatewood, 1960-
Alva R. Glaser, 1963-66
Gerald M. Gregorek, 1960-
William H. Lane, 1957-58
John D. Lee, 1953-

Ting Y. Li, 1964-
Frank McL. Mallett, 1947-
Edward L. Miller, 1946-50
Knox T. Millsaps, 1946-49
John A. Murphy, 1959-61

Theodore C. Nark, Jr., 1958-60
Robert Nerem, 1962-
Carroll J. Pierce, 1947-52
Stuart L. Petrie, 1959-
Keith Stewartson- summers, 1968, 1969

Francis Sturms, 1957-58
Matthew A. Sutton, 1952-58
Richard E. Thomas 1958 - 64
Robert N. Thurston 1949 - 50
Arthur N. Tifford 1948 - 58

Benjamin H. Ulrich, Jr. 1964 - 66
Garvin L. VonEschen, Chairman 1946 -
Lloyd D. Yates 1947 - 50
Jeffrey O. Young 1969 -
Vincent P. Zimnoch 1951 - 52

Appendix Two

Faculty
1970 - 2003

Reza Abhari 1995 - 98
Cecil D. Bailey 1967 - 85
Richard D. Bodonyi 1986 -
Hacene Bouadi 1989 - 94
Michael R. Bragg 1981 - 90

Odus R. Burggraf 1964 - 89
Michael G. Dunn 1996 - 2002
Rudolph Edse 1951 - 1996
Michael R. Foster 1970 -
Buford E. Gatewood 1960 - 78

Gerald M. Gregorek 1960 -
Joseph H. Haritonidis 1990 -
John D. Lee 1953-93
Ting Y. Li 1964 - 75
Frank McL. Mallett 1947-75

Robert Nerem 1962 - 79
Hayrani Oz 1981 -
Stuart L. Petrie 1959 - 87
Nesrin Sarigul 1984 - 89
James N. Scott 1989 -

Mo-How Shen 1984 - 2002
Alfred B. O. Soboyejo 1997 -
Peter Turchi 1989 - 2000
Garvin L. VonEschen 1946 - 1979
Rama K. Yedavalli 1987 -

Gary Yip 1984 - 1988
Thomas M. York 1987 -
Jeffrey O. Young 1969 - 74

CHAPTER 37
AERONAUTICAL AND ASTRONAUTICAL ENGINEERING EDUCATION AT THE UNIVERSITY OF ILLINOIS

Kenneth R. Sivier, AIAA Associate Fellow, Associate Professor Emeritus
Diane E. Jeffers, Coordinator of External Relations
Department of Aeronautical and Astronautical Engineering
University of Illinois, Urbana IL

Abstract

This paper traces the development of aeronautical and astronautical engineering education at the University of Illinois. It follows its development from World War I, to the modest beginning in 1944 as the Department of Aeronautical Engineering, through its times of growth in the 1960s and 1970s when it became the Department of Aeronautical and Astronautical Engineering, to its present status of vigor and world-class reputation. The paper describes the development of the laboratories, and the concurrent development of the research programs, from the construction of a low-speed wind tunnel to the present array of 18 laboratories for analytical, computational, and experimental instruction and research. Finally, stories from alumni and emeritus faculty, about the curriculum, faculty and students, provide a sense of the character of the department.

1. In the Beginning; Through 1949

The Department of Aeronautical Engineering (AE) was established during the 1944-45 academic year. But before 1944, there had been education in aeronautics at the University of Illinois. The following is taken from a history of the department written by Professor Robert W. McCloy for the department's 25th anniversary. "… aeronautical (education) activity started with the establishment of a School of Military Aeronautics during World War I. Instruction was in the Armory, Men's Old Gym Annex, and in the Aircraft Engine Building (now [1969] the annex to the Men's Old Gym Annex). In February 1920, Matthew R. Riddell was appointed Assistant Professor of Aeronautic Engineering in the Department of Mechanical Engineering. Courses were offered in aerodynamics and aircraft structures. In September 1942, an aeronautical option was established in civil engineering. … aircraft structural testing had been going on in Talbot Laboratory since 1940."

"In 1944, President Willard established a faculty committee … to study the relative merits of offering aeronautical engineering courses through optional programs in the various existing engineering departments as compared to a complete offering in a separate Department of Aeronautical Engineering. At the same time a seven-man Aviation Advisory Board was appointed. It was made up of outstanding men in the aircraft industry. They recommended the development of a broad program in aeronautical education and research. The recommendations of the (University's) Aviation Advisory Committee, to establish a separate Aeronautical Engineering Department, was the culmination of (early) aeronautical activity at the University of Illinois."

Professor Henry. S. 'Shel' Stillwell, of the University of Kansas, was a member of the Aviation Advisory Board. In the fall of 1944, Professor Stillwell (see Picture 1) was appointed

Associate Professor and Head of the Department Aeronautical Engineering. (See the note, at the end of the paper, about Professor Stillwell's early career.) He assumed his duties at the University on November 1, 1944. Classes were first offered in the winter 1944 semester. The enrollment, recorded on December 1, 1944, consisted of 24 students; 19 freshmen, one first-semester sophomore, three second-semester sophomores (one of those was Allen Ormsbee [B.S. '46] who, in 1951, became an Assistant Professor in the department), and one first-semester junior.

Ormsbee remembers that: "In the mid 40s, engineering programs at the University of Illinois were on a different schedule from that of the rest of the campus, operating three full semesters a year, starting in late fall (the beginning of the winter semester). This was to accommodate the needs of the Navy V-12 students who comprised about 90 percent of the engineering student body. The end of World War II was still a year away and the whole campus swarmed with students in uniform from all branches of the military service. Many were assigned to campus for short-term non-degree programs, but the V-12 students were pursuing B.S. degree programs."

The AE offices originally were located in rooms 101 and 101a of the Transportation Building. Paul Klevatt ('47) remembers that his first glimpse of the location of the AE department was somewhat disappointing: "Remnants of the 'old days' abounded". He recalls being "amazed by the number of railroad car axles, wheels, sections of track, etc., piled outside the department's classroom areas."

Curriculum

The initial business of the department was the development of a curriculum leading to a Bachelor of Science degree in Aeronautical Engineering. An initial eight-semester curriculum of 136 semester hours was put into operation at the start of the winter 1944 semester. During that semester, only one AE course, AE 1 (aerodynamics), was offered. Professor McCloy writes: "There was only one student taking aeronautical courses, however. Stillwell would spend a week lecturing and making assignments for Wayne Ziegelmiller, and he would spend the next three weeks as a consultant to the Applied Physics Laboratory of Johns Hopkins University."

It was originally planned to offer AE 1, AE 2 (aircraft materials and processes) and AE 22 (aircraft structures) during the spring 1945 semester. Due to research activities and lack of staff, AE 1 and AE 22 were not offered. Instead, ME offered two aeronautical engineering courses, ME 33 (aerodynamics) and ME 34 (aircraft structures), as substitutes for AE 1 and AE 22. Ormsbee notes that ME 33 "was quite primitive by today's standards, containing absolutely no mathematics beyond the standard formulas for lift and drag. He also says that "the first aeronautical course students took was GED5, Aircraft Drafting and Lofting, a sophomore course taught by Professor Randy Hoelscher, a faculty member of what was then the General Engineering Drawing Department".

That initial curriculum included 10 required courses offered by the AE department and 10 new service courses given by other departments. Klevatt's transcript shows that he took the following required courses during his junior and senior years (1945-47):
- Aerodynamics [two courses]
- Aircraft Structures [three courses]
- Airplane Design [three courses]
- Aircraft Drafting and Lofting [one course]
- Aircraft Materials [one course]
- Aircraft Power Plants [two courses]
- Thermodynamics [one course]

- Math (Differential Equations) [one course]
- AC/DC Machinery and Circuits [two courses]
- Fluid Mechanics [two courses]
- Resistance of Materials [two courses]
- Analytical Mechanics and Dynamics of Machinery [one course]
- Machine Tool Lab [one course spent making a one cylinder IC engine]

In the spring 1948 semester, a revised curriculum was placed in operation. This curriculum retained a sustaining core of required courses in fundamental engineering subjects and basic study in the several divisions of aeronautical engineering. However, instead of the technical electives available in the original curriculum, the new curriculum provided for 21 hours of electives: six hours of non-technical and 15 hours of so-called approved electives. This made it possible for a student to select up to 21 hours of non-technical subjects if that was deemed appropriate for the student's program. A focus at that time was the tailoring of the program of study to the interests and needs of the student.

Klevatt remembers that: "The U of I most certainly was not a 'diploma mill'; many students could not cut the tough curriculum. It was a four-seasons grind with only a few weeks between semesters. Most classes ran from 7:00 a.m. to 5:00 p.m., five days a week; a few labs were held on Saturdays and were packed. Homework was all consuming and everyone, who could, carried the maximum number of hours per semester so as to graduate on time."

"As expected, my professors were excellent, hard driving and hard-nosed; they all wanted us to succeed. It was up to us to carry it out; not all could. Standards were high and one had to work for his grades. I still recall Professors McCloy and 'Ziggy' Krzywoblocki, whose patience and skills made learning a pleasurable experience. Professor McCloy taught several courses and also conducted several research projects. I seem to recall that he was working on a ram jet design project at the time and needed undergrads to help him reduce test data via electric Frieden calculating machines. I spent many hours after class and during long weekends churning out results, all at the pay of 70 cents per hour. He raised the pay to 77 cents per hour in my senior year!"

Faculty

After Professor Stillwell was hired to head the department in 1944, the next addition to the faculty was Professor Robert W. (Bob) McCloy in July, 1945. In 1946, Professor M. Zbigniew von Krzywoblocki joined the faculty. Also in 1946, Eldon. J. Seagrist and Allen Ormsbee (after his graduation in September) joined the faculty as Instructors.

World War II ended with the Japanese surrender in August of 1945. One of the results of this was the collapse of the aircraft industry. Ormsbee remembers that: "Many highly qualified engineers, especially aeronautical engineers, were looking for employment. Three who joined the AE faculty were Franz Steinbacher (in 1947) from North American, who taught the structures courses, Jaque Houser (in 1947), who was an experimental aerodynamicist from Boeing, and Art Ogness (in 1948). The buildup for the Korean War and the budding Cold War revitalized the aircraft industry and all three of these professors left for greener pastures after brief service with the department." Others who joined the faculty during these years were Dr. John Coan (1948), Dr. Paul Torda (1949) and Harry Hilton (1949).

With resignations of Houser, Johnson, Ogness, and Steinbacher, the faculty numbered six professors and two instructors at the end of 1949. Figure 1 presents the history of professors on the faculty.

The limited number of faculty, the large undergraduate enrollments, and the added tasks associated with curriculum development had serious effects on the academic staff. The faculty was barely able to keep up with the student enrollment. Teaching loads were high and the time available for research was limited.

Undergraduate Students

At the end of the war in 1945, the GI Bill, which was passed in 1944, generated the means for thousands upon thousands of veterans to attend college, increasing undergraduate enrollment sharply. After the initial AE enrollment of 24 in the winter 1944 semester, the annual enrollment exploded to a maximum of 246 (the combined total for all four undergraduate classes) in 1946. Following that, the enrollment declined to 181 in 1949. This variation in undergraduate enrollment was due first to the veterans returning to school and then due to the collapse of the aircraft industry. The history of undergraduate enrollment is presented in Fig. 2

Beginning in the fall 1948 semester, a significant number of freshmen and sophomores were enrolled at Chicago's Navy Pier branch of the University of Illinois. These students later transferred to the Urbana campus to complete their undergraduate studies.

The first student to receive a Bachelor of Science degree in Aeronautical Engineering was Wayne. E. Zieglemiller who graduated in June 1946. In the time period from 1946 through 1949, a total of 114 students were graduated with B.S. degrees; 53 of those were in 1949. Figure 3 presents the history of B.S. graduations.

In spite of the collapse of the aircraft industry after the war, the opportunities for the employment of AE graduates was strong. This peaked at the beginning of 1949. Even though the demand decreased during 1949, the opportunities remained good. It was also observed that, in the late 1940s, the demand for students with graduate training was increasing.

Graduate Program
The program for the Master of Science degree in Aeronautical Engineering began in the fall 1948 semester with three students enrolled (this number had grown to 16 by the end of 1949). At that time, three new 400-level courses (for graduate students) were added to the 10 existing 300-level courses (for graduate and advanced undergraduate students). In addition, many existing courses in other departments were available to the M.S. students.

Laboratories

In 1945, the Locomotive Laboratory (where full-size steam locomotives were tested) was assigned to the AE department and renamed the Aeronautical Laboratory (later to become Aeronautical Laboratory A [Aero Lab A]). Ormsbee remembers that "The laboratory had access to the old Illinois Terminal Railway tracks that used to cross the campus ... just north of the (Locomotive Laboratory). ... Providing a switch and a spur track to the (lab) explains (its) location and its odd orientation."

Aero Lab A housed the department's machine shop and the first laboratory facility, a low-speed wind tunnel. The initial complement of machine tools, for the shop, was obtained from surplus property. The wind tunnel was designed and built by the faculty, students, and two laboratory mechanics (Martin O'Connor and Paul Tabaka, Navy veterans) during the years from 1945 through 1948. Picture 2 shows O'Conner, Tabaka, and Professor Stillwell in Aero Lab A, inspecting the construction of the tunnel. (Pictures 2 through 11 show the department's faculty, students, and staff throughout the department's history.) The tunnel had an atmospheric pressure, 48- by 30-inch, octagonal test section with a maximum speed of about 100 mph. The tunnel was driven by a modified, four-bladed, airplane propeller powered by a 50 hp. DC motor that received its power from an AC/DC motor generator set. The entire drive system was surplus equipment. Except for the test section, most of the tunnel shell was made of wood.

During the mid forties, the University was authorized, by the State of Illinois, to build the University of Illinois Willard Airport and to establish the Institute of Aviation. The AE department was an active participant in this development. In 1948, as part of this project, the department acquired two Quonset huts at the airport. One was built to house the department's propulsion and flow laboratories and the other was acquired for storage space. Professor McCloy was in charge of developing the laboratories and was an ardent experimenter. Ormsbee remembers that "… consideration of safety matters was not always uppermost in his mind. Minor fires were not rare and students learned early on to be very careful while in that lab. Had OSHA existed at the time, the lab would probably have been put out of business. Fortunately no major calamities occurred."

Ormsbee also recalls that "shortly after the end of the war, the University received a call stating that the Republic Aviation plant in Evansville, Indiana, was to be shut down and if the University wanted any of the stuff in it they would be welcome. The department visited the plant, resulting in the acquisition of several new machine tools, gobs and gobs of miscellaneous hardware, and two complete, never flown P-47 fighters (the wings were removed for transit). It also acquired a (Fisher Body) XP-75 fighter, powered by two Allison V12 engines. These aircraft were cannibalized. The engines from the XP-75 were used to drive the turbochargers from the P-47s to provide an air supply for Professor McCloy's airport lab." This air supply was the beginning of the Air Flow Laboratory that was developed at the airport during the 1950's.

It also should be noted that at about the same time, the department obtained two North American P-51's and one jet-powered Bell P-59.
In the summer of 1949, the Machine Tool Laboratory Building was assigned to the department and renamed Aeronautical Laboratory B (Aero Lab B). This building housed the structures laboratory and would become the home of the combustion and shock tube laboratory.

Research

During the late 1940s, the intense teaching schedule limited the research carried out by the academic staff. Sponsored research in the department was confined to two projects:
1. Supersonic Propeller Investigation for the Aeroproducts Division of the General Motors Corporation.
2. Ejector Ram Jet Investigation for the Office of Naval Research
The supersonic propeller project was entirely analytical and was completed in early 1948. The analytical work for the ejector ram jet project was completed by the middle of 1948, but the experimental work was delayed until late in 1949 because of problems in developing the necessary test facilities. Both of these projects provided an opportunity for a number of undergraduate and graduate students to gain experience in research.

2. The Decades of Growth; 1950-1969

The decades of the 1950s and 1960s were a time of growth, expansion, and change for the AE department. The end of WW II and the GI Bill had driven up enrollment in the mid-1940s. The sharp undergraduate enrollment decrease in the late 1940s, brought on by the collapse of the aircraft industry, was reversed in the 1950s (see Fig. 2). The undergraduate enrollment and the size of the faculty were affected by the Korean War (1950-1953), the beginning of the Cold War, the launch of Sputnik (October, 1957), and the project to land on the Moon in the 1960s. The launch of Sputnik was a symbolic beginning of the space age. Satellites, missiles, launch vehicles, and space flight began to drive both education and research. In the *1960-1961 AE Annual Report*, Professor Stillwell wrote: "Major changes and advances are taking place in the aircraft industry. During the past five years, at least from an engineering standpoint, these have been of greater significance than the important developments of jet propulsion and supersonic aircraft that occurred during the ten-year period immediately following World War II. Space exploration has greatly increased the scope and complexity of engineering problems, and military and civilian requirements continue to challenge aeronautical engineers." And at that time, the Apollo project and the Moon landing mission had not begun.

Influenced by the changes that were occurring, in 1961 the department's name was changed to the Department of Aeronautical and Astronautical Engineering.

Undergraduate Students

As shown in Fig. 2, the steep decline in the undergraduate enrollment of the late 1940s was followed by an equally steep increase in enrollment, beginning in about 1951. This was directly due to the start of the Korean War and the beginning of the cold war. Demand for aeronautical engineers (with both undergraduate and advanced degrees) skyrocketed. Enrollments of undergraduate students followed that trend, reaching a peak of over 600 students near the end of the 1960s.

Gene Hill ('57) has some interesting comments about the program in the mid-1950s: "My aeronautical engineering undergraduate work at the University of Illinois occurred during the 1953-1957 period. During the first two years, I attended the University of Illinois Chicago campus at Navy Pier. My recollection of the two years at the Chicago campus was that the scholastic environment was exceptionally competitive, with more students than space. This was due, in part, to veterans returning from Korea and using their GI Bill educational benefits and the large number of Chicago area students wanting to minimize their college costs by first attending the University of Illinois near home. ...This period was difficult and busy, working a significant number of hours at a job and spending much time in transit to and from school. There was little or no contact with the professors outside of class. The positive was that most of the students felt that class work at the Champaign-Urbana campus would be much easier."

"To my chagrin, the expectations of the Aeronautical Engineering students at the main campus were more stringent. The Aeronautical and Electrical Engineering curriculums were considered the most difficult on campus. Although the class sizes were small, the class content and faculty expectations were exceptionally high. Access to the professors and teaching assistants was good and the lack of outside activities allowed total focus on class work. Little time was available for social activities. Many sunrises were observed while writing laboratory reports or studying for examinations. The Aeronautical Engineering curriculum was heavy on theory, with laboratory classes in structures, propulsion, and a class in the shock tube laboratory. Mathematics continued to be critical in the theory-laden classes. ... Courses offered by the other engineering (departments) were also required, such as metallurgy and thermodynamics courses. The intensity of these courses and grading pressures were less than that for the courses in Aeronautical Engineering."

Graduate Students

In the early 1950s, the demand for B.S. graduates was so great that enrollments for advanced degrees initially were affected adversely. The result was stagnation in the number of graduate students for the first half of the 1950s. This situation finally corrected itself during the last half of the 1950s. Graduate student enrollment grew steadily through the rest of the two decades (see Fig. 4). Graduation data (see Table 1) show that the corresponding graduate degree increases during the 1950s and 1960s are representative of the steady increase throughout the history of the department.. (Although enrollment data for 1958 to 1966 could not be located, graduation data suggest a steady growth in enrollment for that period.)

Faculty

This period began with a faculty of six professors and two instructors. In the late 1940s, expansion of the faculty had been constrained by the sharp decrease in enrollments. With booming undergraduate enrollment, the continuing development of both the undergraduate and graduate curriculums, the need to develop research laboratories and a robust research program, and the difficulty of attracting new staff members (due to competition from the aircraft industry and government laboratories and to limited University budgets), the 1950s were difficult years for the faculty. At the end of the 1950s, the departmental budget allowed for only 10 full-time-equivalent faculty members; at that time, the departmental policy was that all teaching positions be funded with 'hard' money; i.e., State of Illinois money. This situation improved during the 1960s as budgets increased and, the number of professors on the faculty doubled (see Fig. 1). A chronological list of professors, who joined the faculty, is presented in Tables 2 and 3.

Curriculum

The undergraduate curriculum underwent two complete revisions during this period: in 1953 and 1967. The 1953 revisions were guided by the following considerations.
- A revision by the College of Engineering of the common freshman year.
- Insufficient emphasis was being placed on study in the area of social science and humanities. As a result, a list of recommended non-technical electives was established and a required schedule of elective selections was instituted.
- General undergraduate training in aeronautical engineering was considered more desirable than specialization in any of the sub-divisions. Therefore, three additional hours of credit in aerodynamics, propulsion, and structures were required.
- A revision in the introductory thermodynamics course (taught by Mechanical Engineering) was made to improve the background for courses in aerodynamics, aerothermodynamics, and propulsion.

Following the major changes in 1953, frequent additional changes were made to fine-tune the curriculum. Jerry Lundry ('59) contributed this about one of those curriculum changes. "One curriculum change occurred in the fall of 1956. Aircraft Drafting and Lofting was replaced by Advanced Calculus. This was a popular change with many students who had experienced summer employment in the aeronautical industry. That experience typically showed that industry employed draftsmen to make drawings, working under the supervision of graduate engineers. The mathematics class was thus thought to be the more beneficial."

Professor Hilton has this to say about the early curriculum. "The initial curriculum was quite different from the current one. It contained 11 credit hours of engineering drawing including aircraft drawing and lofting. …Aircraft lofting was dropped in the late 1950s and drawing credit was reduced to six hours. The aerodynamic sequence was considerably more modern than the structure/solid (mechanics) offerings. No fluid dynamics course was required, while strength of materials was required.

In 1967, there was a complete revision of the undergraduate curriculum. Courses more closely related to astronautics were introduced and the educational focus was redirected to the sciences of the field. The curriculum included 26 credit hours in technical electives and 18 in the social sciences and humanities. In the junior year, a revised one-year sequence in aerodynamics, structures, and aerospace systems developed a foundation for senior year electives. The drawing requirement was reduced to three credit hours. The strength of materials course was dropped and the required structures courses included elasticity as a beginning topic. New senior laboratory courses were designed. Most of the 300-level courses were re-developed. It represented a major change in the program.

At the end of 1954, the department's Ph. D. program was approved. The department had 13 courses (plus research and special problems) applicable to this program. Other departments in the College offered approximately 35 courses of direct interest to doctoral students. At the time the program was approved, there were six students in the program, two of which had passed the preliminary examinations. The first three Ph. D. degrees in AE were awarded in 1956. Hassan Hassan, the first to receive a Ph.D. in Aeronautical Engineering at the University of Illinois, received his degree in February 1956. He remembers: "Because I was the first Ph.D. candidate, I was kind of a guinea pig in the sense that there was no established procedure to handle such a thing. Before I was allowed to take the Prelims, I was put in a room and all the Aero faculty came and asked me questions on subjects of their choice. I must have done OK because I was allowed to proceed."

"When the (doctoral) committee was formed, I went around and asked what was expected of me, and I was told to prepare for many areas that were not taught in the department at the time. When the time came for the exam, I found out that none of the committee members bothered to ask any questions on the new areas (in which) I was asked to prepare."

Space

Development of the laboratories and the research program was a major part of the department's growth during these decades. These developed in spite of inadequate space and, at least initially, limited funds. By the second half of the 1950s and throughout the 1960's, the lack of adequate classroom, laboratory and office space was a serious problem. The larger class sizes (due to increased enrollments and limited faculty) made finding adequately sized classrooms and laboratory space a problem. The department was inefficiently housed in four buildings on campus (the Transportation Building, Aero Lab A, Aero Lab B, and, during the 1960s, the Wood Shop) and two Quonset buildings at the University airport. By the end of the period, the need for a dedicated AE building had become obvious (at least to the department).

Laboratories

During the early 1950s, a major addition was the Shock Tube Laboratory located in Aero Lab B. The initial development of this facility was under the direction of Professor Walker Bleakney (on sabbatical from Princeton University for the 1950-51 academic year) and his student, Dr.

Charles Fletcher, who joined the department in the summer of 1950. This tube had a 4- by 15-inch cross section, a 10-foot driver section, and a 20-foot driven section. It was capable of generating shock Mach numbers up to 2.5.

The shock tube laboratory was continuously upgraded and improved during both decades. For example, in 1960, a six-foot combustion driver was added to the 4- by 15-inch shock tube. This changed the tube into a double-diaphragm type capable of generating shock Mach numbers up to 20. In 1964, a converging channel was added for use in research on detonation initiation.

Professors Barthel and Strehlow and their students carried out shock wave and combustion research in this facility throughout the period from the 1950s through the 1970s.

A continuing activity of the 1950s was the development of the Air Flow Laboratory at the University airport. Due to the lack of sufficient electrical power, surplus aircraft engines were used to drive compressors for the air supply. This system (with a 1000 cubic foot storage tank added in the mid 1950s) provided 120 psi. dry air for various propulsion wind tunnels, a transonic wind tunnel, and an intermittent supersonic wind tunnel. The laboratory was in operation through the mid-1960s when all flow work was moved to facilities in Aero Lab B and the Mechanical Engineering Laboratory (MEL).

Lundry remembers that: "Professor McCloy had a propulsion laboratory at the (University) Airport. This facility was used for the testing of a valveless pulsejet. A large aircraft engine was used to provide (the air) flow. Ear protection was provided because the valveless pulsejet was extremely noisy. I can remember standing perhaps 10-20 feet from it while it was operating, and feeling a cooling sensation on my exposed face. This was attributed at the time to the intensity of the acoustic pressure fluctuations."

The AE and ME departments developed a close relationship in sharing air flow facilities at the airport during the early 1950s. (In 1954, the Department of Mechanical Engineering changed its name to the Department of Mechanical and Industrial Engineering [M&IE].) In 1956, M&IE installed a new air supply system in MEL. This system included air storage tanks located adjacent to MEL. The air was piped to Aero Lab B (a substantial distance) so that the AE department would have an air supply available for flow studies and student laboratories. Several small transonic and supersonic wind tunnels were built in Aero Lab B to use this air supply. In 1958, additional storage tanks, located adjacent to Aero Lab B, were added to the system. These increased the system's storage capacity by more than 50%, substantially increasing the testing time of the intermittent wind tunnel facilities.

This system was enhanced in 1963 by the addition of nine additional air storage tanks to the Aero Lab B "tank farm". These were connected so that they could also be used as vacuum tanks to increase the performance, of one of the wind tunnels in Aero Lab B, to a test Mach number of 5. In 1964, a heat exchanger type air heater (heating 300 psi air to 1200 deg. F.) and a pebble-bed heater were added to the system.

In 1964, the development of a magnetohydrodynamics laboratory was begun by Professor Bond. This laboratory was located in the Wood Shop Building. It had a 1.5-megawatt power supply that used lead-acid batteries for energy storage.

Research

The research program of the department was enlarged and strengthened during the 1950s and 1960s. This was accomplished in spite of the pressure of large undergraduate enrollments, limited budgets, and (at least for the first 10 to 15 years) limited staff. There is insufficient space here to detail the various research activities during these two decades. In the introduction to the *1964-65 AAE Summary of Engineering Research*, Professor Stillwell summarized the breadth of the research program as: "... investigations ... (have been) undertaken in supersonic, hypersonic and hypervelocity aerodynamics, rarefied gas flows, viscous flows, wave structure and characteristics, magnetohydrodynamics, high-lift devices, detonation, combustion, ejector staging, rocket exhaust plumes, structural behavior at elevated temperatures, aeroinelasticity, viscoelasticity, stochastic structural dynamics, space systems, and space vehicle dynamics."

Additional Anecdotal Comments

Alan Kehlet ('51) remembers how the employment environment was changing in the early 1950s. "In the fall of 1950, many of the class of 1950 were suggesting to us that we should transfer to Mechanical Engineering for better job prospects. In spite of the suggestion, I believe no one transferred. In the spring of 1950, the job opportunities were so few that the 1950 graduates were forced to take jobs on the level of driving milk delivery trucks. The Korean War changed the environment entirely for my class of 1951. By the spring of 1951, we were invaded by government and industry recruiters. ... By graduation time, we had the opportunity to pick and choose."

"Kehlet also has a memory about Professor McCloy. "Professor McCloy, ran the jet engine/wind tunnel operation at the airport and every time we ran the jet engine we had a fire to put out. Putting out fires was not the only physical activity we had. On the last class period on the last day of class just before graduation, McCloy made us sweep the hanger floor and stack the fire extinguishers so we could say we were hands-on engineers."

Earl H. Dowell ('59) has this to say about his experiences as a student. "Of course I have many fond memories of the department and the faculty and the students (now alumni) I had the privilege of knowing there. ... It was and remains my impression that the AAE faculty provided an excellent education for those of us fortunate enough to major in the department and any subsequent success that I may have had has been firmly rooted in that experience."

Professor (M&IE) Clark Bullard (B.S. '64, M.S. '68, Ph.D. '71) provided a story about Professor McCloy. "... one of my first aero courses ... (was a propulsion course) from Prof. Robert McCloy. Like all elementary combustion courses, the numerical calculations were a bit more intensive than my earlier courses, and Professor McCloy was a firm believer in giving us a proper amount of exercise with our slide rules. He was a tough grader, giving little partial credit if the numerical answer was wrong. I recall vividly waiting in line after class on the day our first exam was returned, probably waiting to argue about the number of points taken off. I overheard one of the students ahead of me say: "But Professor McCloy, it was only a slide rule error. I had all the equations right." Professor McCloy looked him straight in the eye and said: "Is that what you plan to say to the wife of a dead test pilot?" The student walked away, and, as I recall, I followed him. To this day, I tell the story to my students, impressing on them the importance of checking their own work and asking: "does this answer make sense"?"

Larry Howell (B.S. '66, M.S. '68, Ph.D. '70) provides some additional comments about the curriculum and the times of the 1960s. "I think most of us who entered the program (in the early 1960s) were excited by the prospects of working in the space program or with advanced aircraft systems. It was an exciting time in the industry; the Space Shuttle concept phase was just beginning, as were the early developments of the F-111, the Boeing 747, and a supersonic transport. ... I tell people I had a great career: an exciting and challenging beginning in aerospace engineering and a finish in the automotive industry during a time of tremendous globalization and technology advancement."

"The curriculum covered every technology necessary to develop a flight vehicle system: aerodynamics, structures, control systems, electrical systems, and propulsion systems, but with focus on the basics of fluid dynamics, solid mechanics, thermodynamics and heat transfer, and so on. More than anything else, I learned how to think logically and analytically in solving complex technical problems."

"The AAE Department was small enough that I always felt part of a family. I felt that all members of the faculty were easily approachable, supportive, and looking out for my best interests. As I became a graduate student, and an Instructor, I became one of "them." I think people like Stillwell, Lin, Hilton, Ormsbee, Strehlow and others were great role models for me "

"Computational methods were in their infancy while I was in school. (But) perhaps there was an advantage to being in school prior to the acceleration of computing technology. By solving problems using analytically-based approximation techniques (e.g., the Hardy-Cross moment distribution method for structural systems), we had to really understand the underlying theory. I have fond memories of computing jet engine efficiency coefficients on the slide rule."

The 1960s were the time of the Vietnam War. Howell remembers about the situation on campus. "Students were marching and rioting on campus. I recall student protestors advancing on the Electrical Engineering Building and throwing rocks at the windows because the EEs were working on defense contracts. ... One evening as I left my basement office in the Transportation Building, I walked outside (into) the middle of a standoff between students and a long line of State Police carrying large batons. After that, I took my thesis material home with me every night, out of fear that the building might be destroyed or burned."

Steve Nagel ('69) has these memories about the AAE experience: "One thing about the curriculum that strikes me is that even though it was called Aeronautical and Astronautical Engineering, I did not take one space related course. (Although Nagel graduated in 1969, he took what essentially was the pre-1967 curriculum. He was in the transition group.)

"Another recollection is that although I enjoyed the courses, many of them were not easy for me -- in fact, getting that degree in aero was the hardest thing I ever did. I remember really enjoying the courses that were a little less theoretical and more practical. Professor McCloy comes to mind with his propulsion course; he was a wonderful man. I also enjoyed Professor Ormsbee in the aircraft design seminar. ... Overall it was a good curriculum and a tough one and I believe that it has been considerably improved since I went through it. The aero degree has served me well, first as a test pilot, later as an astronaut, and even now as a pilot for NASA where I still find myself doing an occasional flight test project." (Note that in the fall of 1969, Professor Prussing joined the faculty and began offering space-related courses in orbital mechanics, optimal control theory, and optimal spacecraft trajectories.)

3. 1970 to the Present

This paper has focused on the first 25 years of the department's history; the period when it was established, grew, and expanded. It is appropriate to finish by including some information on how the department has developed and changed since the end of the 1970s.

Faculty

Driven by the large undergraduate enrollment (see Fig. 2), five new professors were added in 1969. In 1974, because of the plummeting enrollment, four non-tenured professors were terminated. Of the original 1969 additions, only Professors Prussing and Krier remained. The department became insolvent, since it was not generating sufficient instructional units to cover the salaries of the remaining tenured faculty and staff. To relieve this situation, some AAE professors taught courses outside of the department for a number of semesters. No new professors were hired from 1970 through 1979.

Since 1979, although having substantial yearly fluctuations, the faculty has grown from 13 to the present 19 professors. This is the same number as in 1971. However, the breadth of technical fields that are represented has changed substantially (see Tables 2 and 3).

Department Head History

Following Professor Stillwell's death in 1976, Professor Harry Hilton became department head in 1977. Professor Hilton continued as Head until 1985. After he stepped down, Professor Shee-Mang Yen became Acting Department Head until 1988 when Professor Wayne C. Solomon assumed that position. Professor Solomon stepped down in 1999 and was replaced by Prof. Michael B. Bragg (see Table 5).

Undergraduate Students

In the three decades following 1970, undergraduate enrollment has followed a roller-coaster path (see Fig. 1). When the Moon landing mission ended, the space industry began to shrink. At the same time there was an intensified interest in non-aerospace-related activities. The result was a precipitous plunge in undergraduate enrollments in aerospace engineering. Figure 1 shows that the undergraduate enrollment in the AAE department dropped from over 600 in 1968 to almost 150 in 1974.

This plunge was followed by a period of increased enrollments that peaked in the mid-1980s. Part of the steep decrease that followed that peak was generated internally. The student-faculty ratio was simply too high. To alleviate that, enrollment in AAE was restricted by raising the admission requirements. That policy was relaxed in the mid-1990s and the enrollment began to rise again.

The current objective is to stabilize the undergraduate enrollment at about 400 students. In the fall of 2002, 349 undergraduate students were enrolled. By the end of 2002, a total of 3080 B.S. degrees had been awarded by the department.

Graduate Students

The enrollment of graduate students decreased in the 1970s (see Fig. 4). It reached a minimum in 1979 and then began an overall steady rise to today's level.

The peak in graduate student enrollments, in the late 1980s and early 1990s, was an exception. In 1984, Professor Hilton successfully negotiated an agreement, with McDonnell-Douglas, for remote delivery of graduate courses to the St. Louis plant. The graduate courses were identical to courses offered on campus. Courses were delivered via an electronic blackboard system operating over telephone lines. There was two-way audio contact between Urbana and St.

Louis, but no visual contact between professor and students. These were the first stages of remote instruction and a far cry from today's audio-visual-computer teaching systems.

At one time, there were 176 McDonnell-Douglas students enrolled and the number of MS degrees awarded for one year exceeded those awarded to on-campus AAE students. The program terminated in 1992, when the employment situation in the aerospace industry weakened and an insufficient number of new employees were available in St. Louis for a viable program.

In spite of the year-to-year fluctuations, the per-decade graduation rate (see Table 1), has maintained a steady growth throughout the department's history. By the end of 2002, a total of 612 M.S. degrees and 125 Ph.D. degrees had been awarded by the department.

The Curriculum

Over the years, many small changes were made in the undergraduate curriculum that was instituted in 1967. However in 1993, a completely new curriculum was instituted. An introductory course for freshmen and two new sophomore courses were added. New subjects were introduced (e.g., computational methods), the undergraduate laboratory courses were revised, and a second semester of aerospace system design was added to the senior year. The total number of hours in the curriculum now is 134, only two less than the 136 in the original 1944 curriculum. Those two hours disappeared about 1970, when the University dropped the requirement of two hours of physical education.

Space

The possibility of an AAE Building evaporated, in the early 1970s, due to the plummeting enrollments and decreased funds from the State of Illinois. The problem of adequate laboratory space was exacerbated by the demolition of Aero Lab B and the Wood Shop in 1992. This was done to make way for the Grainger Engineering Library. The situation was improved in 1993, when the department moved from the Transportation Building to newly remodeled quarters in Talbot Laboratory. This allowed the AAE faculty, staff, and graduate students to be housed in one building and provided more laboratory space.

Laboratories

There have been many changes in the student and research laboratories during this period. The low-speed wind tunnel, built originally in Aero Lab A, was moved to MEL and in the process upgraded. Then recently, it was removed and replaced by a modern, smaller low-speed wind tunnel for student instruction.

In the early 1970s, under the direction of Professor Sivier, a two-dimensional wind tunnel was built in Aero Lab B. The tunnel had a 5 ft. x 5 ft. x 10-inch test section and could provide speeds up to about 170 mph. Because it was a model for a wind tunnel that the McDonnell-Douglas Corporation was thinking of building, it had an unusual configuration with the fan upstream and the inlet very close to a wall. It was also designed to disassemble into many individual sections, including sections for honeycombs and screens so that the details of the configuration could be studied. In the early-to-mid 1970s, Professor Ormsbee and his students used the tunnel for high-lift airfoil studies.

In 1990, directed by Professor Bragg, a new low-turbulence wind tunnel was built and is located in a remodeled part of Warehouse 1 (located behind the Frederick Seitz Materials Research Laboratory). This tunnel has a 3 x 4 x 8 ft. test section and can provide speeds up to 165 mph. It is used intensively for studies of unsteady aerodynamics, low-speed airfoils, and icing effects on airfoil performance.

In 1983, Professor Hilton obtained the first departmental and College of Engineering computer laboratory through an equipment grant for $2,000,000 from IBM. It consisted of an IBM mainframe, workstations and PCs. It was used by AAE, GE (General Engineering) and ME&I faculty and students for design applications, using software such as CATIA. Then in 1987, a student computer lab was initiated through a grant from AT&T for 30 UNIX-based PCs.

In 1994, Professor Sivier directed the acquisition of a six-degrees-of-freedom Aerospace Flight Simulation Laboratory. It was funded jointly by a National Science Foundation grant, the College of Engineering, and cost sharing by the builder, Frasca International. It has been used regularly for flying demonstrations and for flight test experiments in the airplane dynamics courses. In 2002, under the direction of its current director Professor Selig, the laboratory was extensively upgraded with new software and equipment.

Currently, research and instructional laboratories occupy almost 16,000 sq. ft. of space. Of that, 8,000 sq. ft. is in Talbot Laboratory and the rest is distributed over five other locations. Table 4 presents a complete listing of the current laboratories in the department.

Research

Since the beginning of the 1970s, the department's research program has undergone a tremendous expansion. A large part of this growth has occurred since 1990. An indication of this is the growth in research funding. Data show the yearly research expenditures have grown steadily from about $1,000,00 in the 1989-90 academic year to about $3,800,000 in 2001-02.

The breadth of this research is indicated by the following excerpts from Department Head Bragg's introduction to the *2002 AAE Summary of Engineering Research*. "Active research programs include applied aerodynamics, composites, aircraft icing research, structural dynamics, dynamic fracture, aeroelasticity, stochastic dynamics, combustion, computational fluid dynamics, chemical propulsion, electric propulsion, chemical lasers, optimal orbit analysis, optimal spacecraft trajectories, two-phase flow, systems and control, and wind energy. ... Supercomputer access, departmental workstations, and high-speed networking provide new opportunities for computational research activities in various areas, including fluid dynamics, structural analyses, vehicle performance simulation, space mission analyses, and optimization of high-energy lasers. Current major research initiatives include smart meso flaps for aeroelastic transpiration, self-healing composites, and research for the Center for Simulation of Advanced Rockets in the areas of fracture problems, crack propagation, and the combustion processes of a solid propellant rocket system."

Summary

Although aeronautics education at the University of Illinois began during World War I, the Department of Aeronautical Engineering was established in the fall of 1944. At that time it had 24 undergraduate students, one professor (Associate Professor and Department Head Henry S. Stillwell) and no laboratories. The Department has seen wildly fluctuating undergraduate enrollments and times with too few faculty members, too little funding, insufficient classroom, laboratory, and office space, and difficulties in attracting graduate students. But it has survived, prospered, and grown to its present status of vigor and world-class reputation. There are now 18 laboratories for analytical, computational, and experimental instruction and research. Funded research has grown to about $3,800,000 in the 2001-02 academic year.

From the beginning, the character of the department and its faculty has been conducive to education and research of the highest quality. This is attested to by its reputation and by comments and stories from alumni.

Acknowledgements

The editors have drawn freely from the Department Heads' Annual Reports and introductions to the annual *AAE Summaries of Engineering Research* and from the *Aerospace Alumni Association News*. We extend grateful thanks to Professor Emeritus Alan Ormsbee, who provided much of the history for the first five years of the department, and the present Department Head, Professor Michael Bragg, who provided information about the current state of the department. We thank the alumni who shared their thoughts and experiences. We also thank Professors Bragg and Prussing and Emeritus Professors Hilton and Sentman for reading and critiquing the paper. Finally we thank Artha Chamberlain, Editor Emerita of the *Aerospace Alumni Association News*, for help in finding information about the department's history.

(This information about Professor Henry S. 'Shel' Stillwell's early career is taken f
article written, by Professor McCloy, at the time of Professor Stillwell's death in 197€

Shel became interested in flying while he was in high school. He earned the money for his flying lessons as a draftsman for Tipton Aircraft in 1934-35.

Shel attended the University of Minnesota. He graduated from Minnesota with highest distinction. Before obtaining his bachelors degree, he was getting valuable practical experience by working as an engineer for Porterfield Aircraft Corporation in 1937. The following year he became Chief Engineer.

In 1940 Shel earned his masters degree at Minnesota in nine months. The degree was with highest honors. He then completed the course work for his Ph.D. while he was an instructor of Aeronautical Engineering and Assistant Coordinator of Flight Training for the University.

In 1942 Professor Stillwell was offered the position of Head of the Aeronautical Engineering Department at the University of Kansas. In the two years at Kansas, the department taught aeronautical engineering to Navy V-12 students and civilians. A flight training program was established with the department teaching the ground school program for Army, Navy, and Marine flight training.

During the two years Professor Stillwell was at the University of Kansas, the department made a profit of $100,000 while winning a citation for having an outstanding ground school program.

Early in 1944 President Willard (of) the University of Illinois appointed a committee of educators and outstanding experts in the aviation field to consider the advisability of establishing an Aeronautical Engineering School at Illinois, and the advisability of establishing an Institute of Aviation. As one of the consultants, Professor Stillwell recommended the establishment of the Institute of Aviation separate from aeronautical engineering. The University then asked Professor Stillwell to head the new department. Shel Stillwell was 27 years old when he came to the University of Illinois to establish the Department of Aeronautical Engineering.

Table 1. Degrees Conferred by Decade

Decade	Degrees Conferred		
	B.S.	M.S.	Ph.D.
1944-49	114	0	0
1950-59	445	62	6
1960-69	585	90	23
1970-79	443	94	24
1980-89	853	140	31
1990-99	518	186	33

Table 2. Professors Joining the AAE Faculty; 1944-1990

Name	Period	Area of Specialization
Stillwell, Henry S.	1944-76	flight vehicle design
McCloy, Robert W.	1945-88	propulsion
von Krzywoblocki, M. Zbigniew	1946-60	gas dynamics
Coan, John	1946-60	plate stability
Steinbacher, Franz	1947	structures
Houser, Jaque	1947-48	aerodynamics
Ormsbee, Allen I.	1951-92	aerodynamics
Hilton, Harry H.	1949-90	viscoelasticity, numerical analysis, reliability
Torda, Paul T.	1949-55	propulsion
Fletcher, Charles	1951-54	shock tubes
Yen, Shee-Mang	1956-90	rarefied gas dynamics, naval hydrodynamics
Hicks, Bruce	1957-70	Monte Carlo methods
Barthel, Harold O.	1957-91	gas dynamics
Strehlow, Roger A.	1960-90	combustion, detonation
Lin, Y. K. (Michael)	1960-84	stochastic structural mechanics
Loth, John L.	1962-67	propulsion
Knoebel, Howard W.	1963-74	controls
Bond, Charles E.	1964-99	aerodynamics, renewable energy
Zak, Adam R.	1964-93	solid mechanics, viscoelasticity
Chen, T. C.	1965-67	structural mechanics
Sentman, Lee H.	1965-02	chemical lasers, reactive flows
Regl, Robert R.	1967-74	structural dynamics
Sivier, Kenneth R.	1967-93	flight mechanics, simulation, aerodynamics
Karr, Gerald R.	1969-72	rarefied gas dynamics
Bennett, James A.	1969-74	solid mechanics
Krier, Herman	1969-81	two-phase flows, combustion
Van Tassel, William F.	1969-74	aerodynamics
Prussing, John E.	1969-present	orbital mechanics, optimal trajectories
Wang, Su Su	1980-91	composites
Hale, Arthur L.	1980-84	structural mechanics
Conway, Bruce A.	1980-present	orbital mechanics, optimization
Buckmaster, John D.	1983-present	combustion
Bergman, Lawrence A.	1984-present	structural dynamics, random vibration
Dwyer, Thomas A. W.	1985-90	controls
Lee, Ki D.	1985-present	computational fluid dynamics
Namachchivaya, N. Sri	1985-present	nonlinear and stochastic dynamical systems
Beddini, Robert A.	1985-present	computational aerothermochemistry, propulsion
Solomon, Wayne C.	1988-01	propulsion, systems
Burton, Rodney L.	1989-present	electric propulsion, combustion
Winkler, Karl-Heinz A.	1989-90	high-speed computing

Table 3. Professors Joining the AAE Faculty; 1990-2003

Name	Period	Area of Specialization
Bragg, Michael B.	1990-present	aerodynamics, aircraft icing
Loth, Eric	1990-present	two-phase flow
White, Scott R.	1990-present	composites manufacturing, smart materials
Voulgaris, Petros G.	1991-present	system control
Coverstone, Victoria L.	1992-present	spacecraft control
Selig, Michael S.	1992-present	aerodynamics, design methodologies, turbines
Geubelle, Philippe H.	1995-present	solid mechanics, failure mechanics
Lambros, John	2000-present	dynamic material behavior, experimental fracture mechanics
Frazzoli, Emilio	2001-present	autonomous aerospace systems
Neogi, Natasha A.	2002-present	safety & performance of distributed real-time systems
Elliott, Gregory S.	2003-	experimental fluid mechanics, gas dynamics, laser diagnostics

Table 4. Department Head History

Professor	Period
Henry S. Stillwell	1944-1976
Harry H. Hilton	1977-1985
Shee-Mang Yen (Acting)	1985-1988
Wayne C. Solomon	1988-1999
Michael B. Bragg	1999-present

Table 5. AAE Research Laboratories

Laboratory	Director
Aerospace Flight Simulation	M. S. Selig
Aerothermal Simulations	R. A. Beddini
Applied Computational Aerodynamics	M. S. Selig
Autonomous Networked Vehicle	E. Frazzoli and F. Bullo (G.E.*)
Chemical Laser	L. H. Sentman
Chemical Oxygen-Iodine Laser	W. C. Solomon
Composites Manufacturing (3 labs)	S. R. White
Computational Astrodynamic Research	V. L. Coverstone
Computational Fluid Dynamics	K. D. Lee
Computational Solid Mechanics	P. H. Geubelle
Electric Propulsion	R. L. Burton
High Strain Rate Mechanics	J. Lambros
High Speed Turbulence	E. Loth and J. C. Dutton (M&IE [†])
Linear and Nonlinear Dynamics & Vibrations	L.A. Bergman and A.F. Vakakis (M&IE [†])
Low Speed Turbulence	E. Loth
Nonlinear Systems	N. S. Namachchivaya
Shock Tube and High Pressure Combustion	R. L. Burton and H. Krier (M&IE [†])
Subsonic Aerodynamics	M. B. Bragg and M. S. Selig

* General Engineering
[†] Mechanical and Industrial Engineering

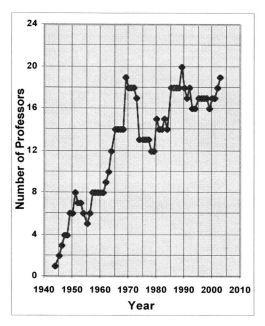

Figure 1. History of Professors on the Faculty

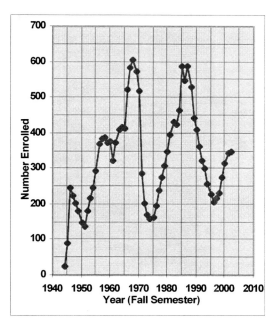

Figure 2. Undergraduate Student Enrollment

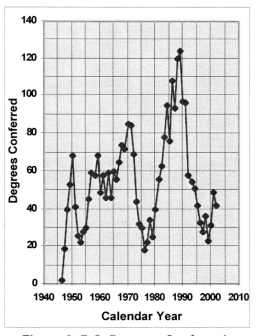

Figure 3. B.S. Degrees Conferred

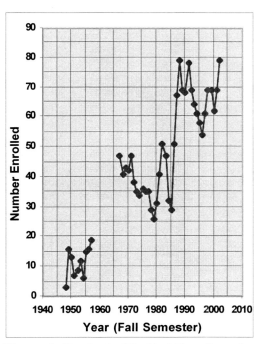

Figure 4. Graduate Student Enrollment

Picture 1.
Professor H.S. Stillwell, Founder and
1st Head of the Department of
Aeronautical and Astronautical
Engineering, 1944 – 1976

Picture 2. Paul Tabaka (left), Professor
H.S. Stillwell (center), and Martin
O'Connor inspect the construction of the
low-speed wind tunnel in Aero Lab A.
(1945)

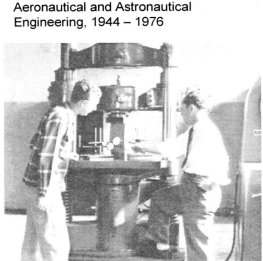

Picture 4. Professors McCloy
(left) and Yen testing a Tesla
turbine model at the University's
Willard Airport. (1958)

Picture 3. Professor H. Hilton (right)
and teaching assistant Bill Simon ('56),
checking out an AE 264 (Aircraft
Structures Laboratory) experiment.
(1956)

Picture 5. Professors A.I. Ormsbee and K.R. Sivier, senior lab mechanic R.W. Fiscus, and graduate student S. Siddiqi (left to right) check out a propeller thrust and torque measuring system. (1975)

Picture 6. Professor J.E. Prussing discusses research in minimum-fuel orbit transfer with graduate students Jo Ann Hoenninger and Kamal Majied II. (1977)

Picture 8. Research assistant Jim Goodding applies an impact hammer to an assemblage of two simple frames as Professors L.A. Bergman and A.L. Hale and research assistant Larry Downey (left to right) observe. (1984)

Picture 7. Professor R.A. Strehlow observes as graduate students Kurt Noe inserts an igniter into a combustion experiment apparatus. (1984)

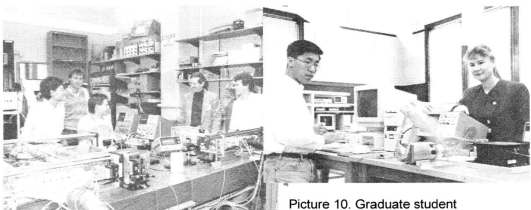

Picture 9. Graduate assistants T. Nguyen, P Theodoropoulos, and R. Waldo, Professor L.H. Sentman, and graduate assistant D. Carroll (left to right) study the performance characteristics of subsonic and supersonic cw HF chemical lasers. (1989)

Picture 10. Graduate student Kangsik Lee demonstrates the capabilities of his nonlinear robust control algorithm to Professor V. Coverstone by controlling an underactuated two-link robot. (1998)

Picture 11. In the Composites Manufacturing Laboratory, graduate students Jennifer Hommema and Jin Li discuss the operation of a filament winder with Professor S. White.(1999)

Chapter 38

Aerospace Education and Research at Princeton University 1942--1975

A. J. Smits[1] and C. D. Perkins[2]
Department of Mechanical and Aerospace Engineering
Princeton University, Princeton, New Jersey 08544-0710

Abstract

A brief account is given of the history of aerospace education and research at Princeton University for the period 1942 to 1975. This period covers the initial establishment of the Aeronautical Engineering in 1942, its move to Forrestal Campus, and the eventual return to Main Campus.

Introduction

Engineering began at Princeton with the establishment of a Civil Engineering program in 1875 [1]. Fourteen years later, Cyrus Fogg Brackett organized what was apparently the nation's first program in Electrical Engineering. It was not until 1921, however, when Arthur Maurice Green Jr. was asked to organize the School of Engineering as an educational unit to be administered by a Dean of Engineering. In his plan for the new School, Dean Greene inaugurated undergraduate programs in Chemical, Geological and Mechanical Engineering. A program in Basic Engineering was added in 1935.

The classes in ME "...were held in the old School of Science, with a makeshift laboratory in a boiler house across Washington Road, until the John C. Green Engineering Building was constructed in 1928. Starting with only two young assistants in mechanical engineering, Dean Greene taught over half the courses in the department in addition to performing his administrative duties. Louis F. Rahm and Alfred E. Sorenson joined the slowly expanding department in 1926, Lewis F. Moody as professor of fluid mechanics and machine design arrived in 1930.

During the Depression, graduating seniors, unable to obtain employment, returned for graduate study, spurring the development of both the engine and hydraulics laboratories. By 1941, when Dean Greene retired and was succeeded as dean and chairman by Kenneth H. Condit, the Department of Mechanical Engineering was both well staffed and equipped, permitting it to acquit itself well during the hectic war years when year-round teaching of military and civilian students was the order of the day" [2].

Dean Condit and other members of the Princeton engineering community expressed an interest in introducing aeronautical engineering into the School of Engineering curricula and started a search for some outside expert who could advise whether or how this might be done [3]. Gordon Rentschler, a Princeton Trustee and an important executive in the aeronautical industry,

[1] Fellow, AIAA
[2] Fellow, AIAA

exerted some pressure on Condit to persevere. He discussed this idea with a former colleague at McGraw-Hill, S. Paul Johnston, then Editor of Aviation magazine. He then decided to invite another former colleague, Daniel Clemens Sayre (figure 1), at that time Associate Administrator of the Civil Aeronautics Board, to make a two month survey. An airplane buff and an MIT graduate, Sayre was a man of many talents. He accepted this assignment and thus started the remarkable relationship between Sayre and Princeton that profoundly influenced the Engineering School in the following years.

Founding of the Aeronautics Department

Dan Sayre made his study in the summer of 1942, and recommended to the Dean that Princeton create a new, free-standing Department of Aeronautical Engineering, independent of the Department of Mechanical Engineering. He so impressed the University and the faculty of the Engineering School that his recommendation was accepted with enthusiasm. The new Department was established with Sayre as a Full Professor and the first Chairman. At the outset he was the entire faculty, and the Department [Sayre] had a desk on the balcony of the Mechanical Engineering Laboratory. This situation didn't last long. Dan wanted a building of his own, but he had no resources or space to build one.

Sayre sought funds to build some cinder block buildings on the lower campus just above Lake Carnegie and in back of the Observatory where small buildings like this were permitted. He approached local wealthy friends of the University and eventually was able to interest Mrs. Edgar Palmer and her sister Helen Hayes Watson who donated $50,000. Two small cinder block buildings were built very rapidly providing Sayre with office space for his staff and some area for research operations. He then set about acquiring a faculty and a small staff to man the Department and also signed up some undergraduates and a few graduate students to open the franchise.

Sayre had his eyes on two new faculty appointments for which slots had been approved in the initial agreement. He had heard through the MIT grapevine of a remarkable engineer at Sikorsky Aircraft who was interested in an academic career. Alexander Alexandrovitch Nikolsky, a Russian aristocrat had escaped the revolution while a naval cadet serving on a ship in Vladivostok harbor. He and a few other cadets made their way to Japan, then Cairo, and finally Paris. In Paris, Nikolsky enrolled in the Sorbonne where he received a diploma in engineering. He then took a job as a seaman on a freighter bound for New York. On arrival he jumped ship and continued on to Boston. The Russian emigree community came to his aid and helped to enroll him in the Aeronautical Engineering Department of MIT. On graduating with an M.S. degree, Igor Sikorsky, another Russian émigré who was looking after his young countrymen, hired him. Nikolsky rose rapidly in Sikorsky's engineering department and was Chief of the Structures Department when Dan Sayre approached him.

Nikolsky came to Princeton in 1943 as the second appointment of the new Department. He taught courses in aircraft structures at the undergraduate level and both taught and did research with his graduate students on predicting loads on rotor blades at the graduate level. The dozen or so students could choose between six undergraduate and eight graduate courses covering the principles of aerodynamics and aircraft structure and design [4]. Nikolsky became nationally known as an expert on helicopters and wrote a famous book on the subject entitled *Helicopter Analysis*. A slight, wiry, friendly man, he was the life of any party and was loved by his many students." [3]

518

The War Years

When the School of Engineering was founded in 1921, the trustees set a limit of 400 for enrollment of undergraduates working toward the degree of Bachelor of Science in Engineering. There were fewer than 100 student engineers at that time, but by the time Dean Greene retired in 1940, the School had reached an enrollment of 394 [1].

Then came the war, and enrollment fluctuated all the way from 85 to 600. During part of that period, the faculty of the School conducted training courses for Army Signal Corps technicians, for Army Specialized Training Program engineers, for various individuals enrolled under the Engineering, Science and Management War Training Program, for Navy V-12 students, and for some 1000 Naval officers being prepared for radar operation and supervision. The teaching load went through cycles of overload and underload. Recruiting of staff in the fields of science and engineering became difficult, impossible, and then to all intents, illegal. Materials disappeared, technical equipment had to be improvised or simply done without [1]. Aeronautical Engineering, with a pocketful of plans for a staff and for laboratories, struggled along as a two-man faculty to teach aeronautical courses as were required.

The Post-War Period

The department came into the post-war era with a small nucleus and some useful experience, and because of the wartime restrictions on hiring, with the majority of its basic faculty positions still open for young people who had themselves helped bring about new developments in jet propulsion and supersonic flight that had revolutionized aviation.

Fig. 1: Daniel Sayre and Courtland Perkins.

When Sayre set out in 1943 to hire the third faculty member of the Department, he approached Courtland Perkins, a flight test engineer who was then at the Army Air Force's Aircraft Laboratory at Wright Field, where he directed research in stability and control. Perkins (figure 1) felt it was not the time to leave Wright Field where the war was generating enormous growth in aeronautical technology, but in early 1945 he accepted the University's offer of an Associate Professorship with tenure. Immediately after V-J Day, Perkins resigned his civil service position in the Army Air Corps, and moved to Princeton, arriving in October 1945.

At this time, almost all of the students were returning veterans with widely varying backgrounds, but with an overwhelming desire to graduate in the very shortest time. Special programs were developed to satisfy their desires and the University requirements. One sophomore who stayed was David C. Hazen who, after he received his B.S. and M.S. degrees became first a Research Assistant, then an Assistant Professor, and ultimately a Full Professor in the department.

Hazen was one of the people helping Perkins to build an instructional wind tunnel, using a large open pit for a return arm of a wind tunnel, leaving the test section at floor level. A coal mine exhaust fan, designed to absorb 150 horsepower, was used to power the wind tunnel. With a 3'x5' throat, a speed at the test section of about 120 mph was possible. Together with a surplus motor and a control system, and a six component balance it turned out to be a very useful and important tool.

The Aeronautical Engineering Department at this time was housed in Sayre's cinder block buildings on the lower campus. All the undergraduate courses were given in the main Engineering Building, then located on Washington Road, about a mile away. There were several sponsored research programs being undertaken by the Department, including one in helicopter rotor de-icing under Sayre, and a theoretical program on rotor loads by Nikolsky. Perkins argued strongly for developing two important fields, one in the theoretical aerodynamics and another in propulsion, with emphasis on Jet Propulsion. Sayre and Nikolsky agreed readily and, with help from President Dodds, had the two slots approved by the University and the Engineering School.

For the Theoretical Aerodynamics position, Perkins approached a friend from MIT and Wright Field days, Lester Lees, then on the staff of the NACA at Langley Field. According to Perkins [3], Lees was an introvert but everyone recognized him as a genius obviously on his way to be becoming a top-flight theoretician. His major research interest was in theoretical aerodynamics with emphasis on the character of laminar and turbulent boundary layers. In particular, Lees had became one of the country's authorities on compressible flow, and Perkins offered him a position as an Assistant Professorship. The work going on in fluid dynamics in the Princeton Physics Department, which included Walker Bleakney doing experimental work with his shock tube, and the work being done by Dr. Ladenberg on interferometry also interested him. Lees joined the faculty in 1946. He brought with him several of his colleagues from NACA, including Abe Kahane and Jerry Schaffer. Importantly, he also brought with him an ONR contract for research in high speed fluid dynamics. He helped start the experimental work in supersonic flow that later was the heart the Princeton Gas Dynamics Laboratory. He was promoted to the rank of Associate Professor in short order, but some years later, when his next promotion was held up at Princeton, CalTech offered him a Full Professorship. Lees took it, and the department lost one of its most important faculty members in 1954.

While these negotiations were underway, a search was started for a candidate to fill a position in the field of Jet Propulsion. At that time the major academic program in Propulsion, and in particular Jet Propulsion, was at CalTech under the leadership of von Kármán. Perkins called von Kármán who suggested Joseph V. Charyk, a Canadian who had received his B.S. degree from the University of Alberta and then went to CalTech for his graduate program. Joe received his Ph.D. in the spring of 1946, and he had a strong interest in propulsion.

Joe came to Princeton to meet Dan Sayre, Nikolsky, and Dean Condit, and he agreed to an appointment as an Assistant Professor and moved to Princeton in the fall. He rose rapidly to Associate Professor, became a first class teacher-scholar, helped stabilize the new Ph.D. program (started in 1946 [5]), and established projects in propulsion that led ultimately to the organization of the Guggenheim Propulsion Laboratories.

In the early 1950s, under intensifying Cold War pressure, and the demands of the Korean War effort, the U.S. greatly accelerated its missile and space programs and industry started actively recruiting staff for new programs. University faculties were a prime target. In 1955, Charyk finally succumbed to a very significant offer made by Lockheed through its powerful Board member, Lt. Gen. (Ret.) Pete Quesada. This was another great loss to the Department. Not long after this Charyk, with others, broke away from Lockheed and formed a company which after several evolutions became the Aeroneutronics Division of Ford. Several years later he left Ford to become the U.S. Air Force Chief Scientist. In less than a year he rose to the position of Assistant Secretary of the Air Force for R & D, and then ultimately Under Secretary. While in this last job he helped define a new private/public organization for handling space communications, called COMSAT. Charyk became its first President and eventually its Chairman.

The arrival of Lees and Charyk had brought the faculty to five by the end of 1946. Interestingly enough, of the five only one, Charyk, had a Ph.D. The rest had attended MIT and had M.S. degrees. Only a short time later appointment to the Princeton faculty without a Ph.D. became almost impossible.

With the appointment of Lees and Charyk the department had the major elements for growth. The faculty was spread widely, if thinly, across the various modern technologies of aeronautics. All five faculty members were eager to build the new Department into an aeronautical engineering program that would attract national attention. All agreed that Princeton University was a fertile environment to achieve this objective. The department was building up undergraduate demand, attracting graduate students to M.S. and Ph.D. programs, and conducting sponsored research, even if in rather make-shift facilities.[3] All of this became possible through the slow development of a sponsored research program which provided the resources for current activities and future growth, which came faster than anticipated.

In the four years from 1947 to 1950, four windfalls accelerated the Department's growth:

- The award by the Air Force, Navy, and Army of a graduate program in Applied Aeronautical Engineering with a principal focus on flight research.
- The development of the Navy's Project Squid with Princeton named prime contractor.
- An award to Princeton of a program for editing the twelve volume series on high-speed aerodynamics and jet propulsion.
- A joint award to Princeton and CalTech to be the Guggenheim Jet Propulsion Centers.

The first, the decision of the Armed services to send their top young officers to Princeton for a special M.S. program in Applied Aeronautical Engineering with a principal focus on flight research, was instigated by the Air Force. This service wanted to build up a graduate program in Applied Aeronautical Engineering to balance those that it already had in Theoretical Aerodynamics and Rocket Propulsion at CalTech, and in Instrumentation, Guidance, and Control at MIT. While at Wright Field during the war, Perkins had recommended such a graduate program to the Air Force, and now he had the chance to set it up at Princeton. First the Navy

---

[3] "As I waited at the dinky station in Princeton, a slick black Olds coupe drives up and the inhabitant introduces himself as Joe Charyk and he takes me to a bunch of Quonset huts and concrete buildings that protruded underneath the stands of the football stadium and then a whole new world opened up to me." (Glassman, on a job visit in 1950 [10].}

asked to be included, and then the Army joined in, and it developed into one of the department's most important graduate activities.

The impact of this large group of excellent students, many of whom later became illustrious graduates, was very great. A number of civilian students in a similar program contributed to the significance of the group.[4] Their substantial thesis undertakings led to extensive sponsored research and ultimately to the Princeton Flight Research Laboratory, a unique activity that added to the Department's reputation.

The second windfall, code named "Project Squid," originated at the end of World War II when the Navy became interested in the possibilities of pulse jets and ram jets. Eager to influence education and research in these areas, the Navy under the guidance of Dr. Frank Parker, then a Department Head of The Bureau of Aeronautics, decided to set up a research program under contract to one university cooperating with four or five others. A committee made up of the partner universities would run the program with the Navy playing an approval and support role.

The Navy decided in 1946 to award the program to Princeton, much to the surprise of many. Joe Charyk's close connection with the CalTech group who were advising the Navy at the same time may have been part of the reason. Project Squid was duly organized at Princeton and housed in the department's cinder block office. Frank Parker was hired to come to Princeton as executive head of the program. At first the other Universities involved were Purdue, RPI, NYU and Cornell.

As Project Squid got under way, a large budget slowed approval and contracting of the program. Parker and the Navy were most anxious to get some money spent before the end of the first fiscal year. They therefore expanded Project Squid with a task order to include the ram rocket project under Charyk and the development of supersonic blowdown wind tunnels under Lees. It was difficult to spread the umbrella of Project Squid over a study of shockwave boundary layer interaction, but the project allowed considerable flexibility to do many new things. It also gave the Department important visibility in the propulsion community, and built up the research facilities and the support for graduate students and research staff. Lees' program developed ultimately into the Princeton Gas Dynamics Laboratory, and Charyk's program, combined with others, later emerged under the banner of the Guggenheim Jet Propulsion Laboratory. Project Squid was a very useful program for Princeton in its early years.

One of the most noteworthy participants in the later stages of Project Squid was John Fenn. Fenn had received a B.A. in chemistry from Berea College in 1937 and a Ph.D. from Yale in 1940. In 1959, after a dozen years in industry he came to Princeton as the Director of Project Squid, and as a Professor of Aerospace and Mechanical Sciences. At Princeton, he studied mass spectroscopy using supersonic jets with skimmers. He returned to Yale as Professor of Chemical Engineering and Chemistry in 1967, remaining there as a Research Scientist after becoming Emeritus in 1987. In 1993 he moved to Virginia Commonwealth University as Research Professor. In 2002, Fenn was honored with the Nobel Prize in Chemistry for his invention of a pioneering technique that allows researchers to "weigh" large biological molecules such as proteins with unprecedented

---

[4] A brief list includes Glen Edwards*47, Pete Conrad*53, Norm Augustine'57, *59, and Phil Condit*65. Edwards died in 1948 evaluating the Northrop YB-49, the all-jet version of the flying wing bomber, and gave his name to Edwards Air Force Base. Pete Conrad was the third man to walk on the Moon. Augustine became CEO of Lockheed-Martin, and Condit is currently CEO and Chairman of Boeing.

accuracy. The technique is a variation of his earlier spectroscopic work called Electrospray Mass Spectrometry whereby species are ionized and then accelerated by strong electric fields. The technique is used in chemistry laboratories around the world to rapidly and simply reveal what proteins a sample contains, contributing to the development of new pharmaceuticals.

Another outgrowth of Project Squid was the slow development of the concept of publishing a series of volumes on modern developments in high speed aerodynamics and jet propulsion. The idea took shape through discussions between Lees and Charyk with von Kármán at CalTech, and Hugh Dryden, the Administrator of NACA. It finally came to a head in 1948 when Princeton received a contract from the Air Force and the Navy to develop a twelve volume series on these topics. It was to be an attempt to update modern aeronautical technology along the lines of the Durand Series of prewar fame. The Princeton University Press assumed responsibility for the contract for publishing the series under the leadership of Herbert Bailey'42, one of the Press's most able and vigorous administrators. Princeton organized an editorial board composed of von Kármán, Hugh Dryden and H. S. Taylor, and hired Dr. Martin Summerfield from CalTech's JPL to be the Editor. Joe Charyk became Associate Editor.

Development of the series turned out to be a long and agonizing effort, but the twelve volumes were finally printed successfully. It took several changes in editors and contracting arrangements, and much sweat and tears before it was finished. It did bring notice to the Department during this early period, but, more important, Martin Summerfield's broad interest in propulsion and his own expertise in solid propellant rockets led to his appointment as a Full Professor in 1950.[5] Charyk took over as Editor of the Princeton Series" [10]. This dynamic and brilliant man played a large role in the development of the Department's Propulsion Laboratories.

The Guggenheim Foundation

In 1948 the new Aeronautical Engineering Department at Princeton initiated discussions on the future of rocketry and jet propulsion with Harry F. Guggenheim, the head of the Daniel and Florence Guggenheim Foundation of New York. This Foundation had a history of interest in the various fields of aviation. It supported the development of aeronautical engineering education and research facilities at several leading schools including MIT, CalTech, NYU and Georgia Tech. It also supported work in aviation safety, helped Charles Lindbergh in many ways, and became involved with the work of Robert H. Goddard, the famous rocket pioneer.

After World War II, Harry Guggenheim began a study of how a relatively small institution like the Daniel and Florence Guggenheim Foundation might advance the new fields of rocketry and jet propulsion. He was helped by his friend and advisor on rocketry, G. Edward Pendray. At first they considered commissioning studies at several universities to design rocket ships for commercial missions like carrying the mail from New York to London. It did not take long to discover that the state of the art of rocketry was still far too primitive for such schemes. A return to fundamentals was needed, including acceleration of research in rocketry, but even more important, a major improvement in graduate education programs and motivation for bright young students to participate in the burgeoning new technology.

---

[5] According to Glassman, "he refused to kowtow to Senator Joe McCarthy and was told he could no longer work for ONR. Princeton and the Aero Department showed their colors by giving him a professorship.

Guggenheim decided to create two centers for improving the technologies of rocketry and jet propulsion: one at the California Institute of Technology and the other at Princeton University. It surprised no one that he picked CalTech as it was already involved with rocket experiments. Princeton's Aeronautical Engineering Department also had a strong interest in propulsion and it was a young and hungry Department, well oriented to working in new fields. Princeton advised the Guggenheim Foundation that their resources were needed for educational purposes, faculty and fellowships, and not for funding actual research, which would be more appropriately conducted under contract with concerned government agencies.

On December 14, 1948 the two grants were announced, each for $240,000. The programs were initially referred to as Daniel and Florence Guggenheim Jet Propulsion Centers and included faculty and fellowship support for an original term of five years. The major figures involved with this program are shown in figure 2.

Fig. 2: Rear two rows, from left to right: Arthur Kovitz, John Scott, Edward Kepler, George Sutherland, Richard John, Irvin Glasmman, Gary Mallard, Andrew Hammitt, Sydney Reiter, Daniel Bershader, David Harrje, Walter Warren, Bud Marshall. Second row from front, left to right: Unknown, Joseph Charyk, Lester Lees, Ronald Probstein, Seymour Bogdonoff, Arnold Brooks, Edward Potlein. Front row, left to right: Daniel Sayre, Luigi Crocco, Theodore von Kármán, Dean Kenneth Condit, Walter Davis, Martin Summerfield. Circa 1950.

The Guggenheim grant provided each school with resources to fund a Chair in jet propulsion. To honor the great rocket pioneer that the Guggenheim Foundation had supported so strongly, these were referred to as the Robert H. Goddard Chairs. At Princeton, a worldwide search was started for an eminent scholar for the position, culminating in the appointment of Dr. Luigi Crocco of the University of Rome. Crocco was the son of a renowned aeronautical scientist, General Gaetano Arturo Crocco, who was internationally known for his work in theoretical aerodynamics. Crocco came to Princeton in 1948, becoming the Departments sixth faculty member. A former student of Crocco, Thomas Brzustowski*63, currently President of Canada's National Science and Engineering Research Council, described him as "a patrician Italian, white-haired, distantly related to Pope Pius XII, and with a fabulous accent he cultivated."

Fig. 3. Left: Princeton Cessna 140, used as a flight laboratory [7]. Right: Modified North American Navion A, used to study flight stability (Courtesy Rob Stengel).

The Guggenheim Center not only provided the Department with a world class professor, but helped focus research programs in many aspects of the fields of jet and rocket propulsion. It also led directly to the establishment of our Guggenheim Jet Propulsion Laboratories some years later.

By 1950 the Department had grown rapidly, although it was still housed in the cinder block buildings that Dan Sayre had developed on the lower campus. Research facilities had been expanded into areas of the campus that permitted makeshift construction, primarily temporary structures just behind Palmer Stadium, neighbors to Walker Bleakney's Shock Tube Laboratory and, a little farther away, George Reynold's Cosmic Ray Laboratory, both part of the Physics Department.

During this period the department maintained a small airplane, a Cessna 140, for undergraduate flying laboratory experiments and also for flight research (figure 3). At first it was flown out of the Nassau Airpark, then on Route 1, but later it moved to the Robbinsville Airport, and finally to the Princeton Airport on Route 206. The department eventually acquired a Stinson L-5, and still later a Navion (figure 3). For a short time the Air Force lent a B-25 that the department's Air Force officers operated from McGuire Air Force Base.

Fig. 4: Chairman Daniel Sayre and Lester Lees look on as Seymour Bogdonoff adjusts a supersonic wind tunnel [9].

As the programs grew, they brought in many highly competent research staff, financed through the various sponsored research programs. In the beginning many were part-time graduate students with a rank of Research Assistant. As they completed their graduate program they would be shifted to full-time Research Associates. In some cases they were appointed to the faculty as Assistant Professors. One of the first new research appointments was Seymour Bogdonoff who had been a colleague of Lees at the NACA. He came to Princeton in 1946 with Lees to become his Assistant in Research and to work out a Master's degree. Bogdonoff was soon heavily engaged in setting up the blowdown tunnel facility (figure 4). Bogdonoff got his M.S.

degree in 1949 and was immediately appointed an Assistant Professor. Another of the outstanding graduate students at that time was Edward Seckel. A former Princeton undergraduate, he had returned after the war to work on an M.S. degree, becoming an Assistant in Research to Nikolsky. After getting his degree he took a position on the technical staff of the Cornell Aeronautical Laboratory. Perkins enticed him back to Princeton and made him an Assistant Professor in 1950.

By the end of this period the major foci of the Department's activities emerged more clearly. First were the flight activities run by Perkins, heavily supported by our large military officer program.

The second emerging focus was a deep interest in high speed aerodynamics, led by Lester Lees and Seymour Bogdonoff. They taught both graduate and undergraduate courses in this area, but spent a great deal of effort in developing one of the world's first supersonic blowdown wind tunnels. They were studying, both theoretically and experimentally, the problems of boundary layers and their interaction with shock waves. Many graduate students were interested in this area and one of them received one of the department's first Ph.D.s in 1949.

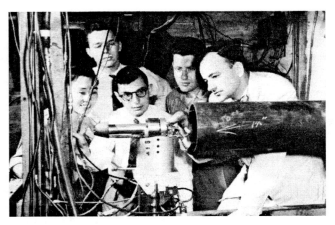

Fig. 5: Left to right: Graduate students Sin-I Cheng*52, John Scott, Jr.* 54, *59, Clarence B. Cohen*52, *54, David Ross*52, and Professor Joseph V. Charyk. Circa 1952.

The third area of concentration was in aspects of propulsion and combustion processes, strongly supported by the Guggenheim Jet Propulsion Center award with its Goddard Professor, Luigi Crocco, and his chief assistant David Harrje*53. Joe Charyk was doing research in ram jets, later assisted by Martin Summerfield whose research interest was in solid propellant rocket motors (see figure 5).

A fourth area of interest centered on Nikolsky and his work on helicopters and the aerodynamics of rotors. Ed Seckel, and later another graduate assistant Pat Curtiss, helped him in this work. This group started to develop unique experimental facilities for investigating rotor problems of helicopters, VSTOLs and others.

A fifth area of faculty interest emerged at the end of this era, the aerodynamics of low speed flight. It grew from the development of undergraduate experimental facilities. It had originally been an interest of Dan Sayre's, aided strongly by Dave Hazen. They performed many experiments in flow visualization, including smoke tunnels and water tables. Most of this work was sponsored by the Navy.

The faculty, then, in the fall of 1950 was as follows: Dan Sayre (Chairman), Nikolsky, Perkins, Lees, Charyk, Crocco, Bogdonoff, Summerfield and Seckel. Some of these people are shown in figures 6--8. Six out of these nine were subsequently elected to The National Academy of Engineering after it was formed in 1964. The Department was growing its professional

Fig. 6: Left to right: Professors Jerry Grey, Luigi Crocco, Robert G. Jahn'51, *55 and Martin Summerfield; Senior Research Associate J. Preston Layton, Professor Irvin Glassman. June 1962.

research staff including graduate students, Research Assistants and Research Associates. Some years later an even higher research rank was established, that of Senior Research Associate. Many of the research appointments carried the additional statement "With rank of Associate or Full Professor."

Besides a faculty of nine in 1950, there were 40 graduate students, twelve of them in the Ph.D. program. Nineteen were civilians, 14 were Air Force, 4 were Navy, and 3 were Army. Total staff of all ranks amounted to 70. The value of the sponsored research program totaled about $550,000. At that time only the Physics Department had a larger program; the rest of the Engineering School was not building up equivalent programs.

Fig. 7: Graduate student W. A. Sirignano *64 prepares equipment to study rapid transients in liquid rocket instability. June 1962.

Establishment of Forrestal Campus

In 1949 and early 1950, Dan Sayre and Court Perkins started to study seriously how they could better house the sprawling Department. Every cubic foot in the original buildings was filled, and the distance from the rest of the University was workable but inconvenient. The cinder block or wooden construction hardly conformed to Princeton architecture and was barely adequate for the department's needs. "Charyk's work was both noisy and noisome, producing malodorous fumes that did even less to endear the department to the astronomers in the nearby FitzRandolph

Observatory. Adding to the cacophony of roaring test engines was the caterwaul of the so-called `blowdown' wind tunnel used by Lees and Bogdonoff." [4][6].

Sayre and Perkins made a drawing of a possible new building, between the Stadium and Prospect Street, where the University built its Computer Center some years later. Lawrence Rockefeller, at that time a member of the Departmental Visiting Committee, helped with the study and made suggestions for raising the million dollars it was estimated to cost. Within a few weeks a remarkable event forced a solution to these housing problems that dominated the life of the Department for many years to come.

In the words of Court Perkins [3]: "During the fall term of 1950, I was teaching

Fig. 8: Left to right: Lester Lees (project head), Chairman Daniel Sayre, Rear Admiral Thorvald A. Solberg (chief of the Office of Naval Research), President Harold W. Dodds, and Seymour Bogdonoff (project engineer) [6].

Fig. 9: Aerial view of the Forrestal Research Center (Princeton University Archives).

an undergraduate course in the Engineering Building when the door opened and Dan Sayre put his head in and told me to dismiss the class and follow him. We got in his car and drove out Route 1 to the grounds of a Division of the New York based Rockefeller Institute for Medical Research. The Princeton Division concentrated on plant and animal pathology. This large property extended from Carnegie Lake nearly to the Pennsylvania Railroad, an area of 825 acres, with some twenty buildings on it (figure 9}. Dan told me that The Rockefeller Institute was going to close down this whole Division

---

[6] The tunnel was built out of war surplus material "to hold down the costs so that it could be used ... even if outside support ... should ever be withdrawn." At the wind tunnel's dedication (figure 11), President Dodds commended the idea of recycling military supplies, saying, "a valuable research weapon has been forged from an especially warlike set of materials. The compressors and red flasks outside this room were originally designed for ... naval torpedoes ...The sides of the tunnel were intended for armor plating on the USS Illinois. That material designed for wartime use can be converted to high scientific purpose should encourage us in our hope that scientific purpose can also serve the cause of peace..." [6], as quoted by [5].

and return parts of it to New York City. The land, buildings, and improvements were being put up for sale, and our alumni, John D. III, and Lawrence Rockefeller were trying to interest the University in buying it. We were being asked whether or not the Aeronautical Engineering Department would move into this area if the University acquired it. If we did, it would solve our space problems for all time, but we would be four miles from the main campus instead of one, and we had found one inconvenient. The advantage to the University of acquiring the property, almost doubling the University campus, was so obvious that we felt we should agree to move in, if it was purchased, and try to solve the distance problem."

Fig. 10: Portrait of James Forrestal by Raymond P. R. Neilson. "In his early thirties, he looked not unlike the actor Jimmy Cagney playing the role of a champion welterweight or a well-groomed, almost respectable gangster." (Hoopes & Brinkley, quoted by [4]).

With considerable help from the Rockefellers, the University purchased the property for $1,500,000, a magnificent price for this large new area. Another $500,000 would be needed for the renovation and conversion of facilities [4]. The whole Aeronautical Department was moved into this area and it was also used to house expanding research interests of other departments. The deal was finally consummated and the complicated move to the new area began. Many Princeton alumni contributed to this new acquisition and the whole area was named The James Forrestal Research Center after the United States' first Secretary of Defense, who served from 1947 to 1949. Forrestal (figure 10) was a Princeton graduate, Class of 1915, and a former Charter Trustee. During his term as Secretary of Defense he was responsible for founding the Office of Naval Research, he initiated work that led to the introduction of nuclear powered submarines, and he authorized research on a satellite space platform [7].

According to Perkins [3]: "The decision to move ended what I like to refer to as the 'cinder block' era of the Department. No longer were we almost sitting on top of each other in makeshift facilities. Nevertheless, we had all been having a grand time. We had built up a nationally ranked Department of Aeronautical Engineering, and at little cost to the University. We all felt euphoric about our situation. We had a very friendly group of about 75 in 1950, including faculty, research staff, technicians, graduate students, mechanics, and secretaries. As a majority were married, it was a large group of people when we all got together, which we did quite often. The Department became famous for its Christmas parties, and many of our friends outside the Department were delighted to be invited to these festive occasions. These activities were usually led by our super party man, Dan Sayre, but he was ably supported by the rest of us, including Nikolsky, our vigorous Russian, and our most enthusiastic graduate students." "..... this was a highly motivated, friendly and successful group. We were delighted with what we were accomplishing and enjoyed being together."

There was considerable debate within the University as to how to utilize this new facility that nearly doubled the size of the campus. It was decided the new Center should be used for expanding research programs of various university departments, that it should house the Aeronautical Engineering Department and, very importantly, that it should be free of normal University rules on architecture. It was also decided to put the activities of the Center under a new administrative office which would supervise grounds and buildings, planning for new occupants, financing and working with the University administration. The Administrative Office of The James Forrestal Research Center was established with Dan Sayre in charge. The responsibility was complicated, and Sayre found that it took him full time. President Dodds then asked Court Perkins to replace Dan Sayre as Chairman of the Department, and Perkins took over this position on February 19, 1951.

After the move, the Aeronautical Engineering Department started a relatively rapid expansion of their facilities in this new area. Other programs in the University also started to move in. One of the first was a new research program in fusion power under the leadership of Dr. Lyman Spitzer, then Chairman of the Astronomy Department. A doughnut-shaped torus for magnetic containment of a plasma of tritium and deuterium was conceived, and the AEC funded a program at Princeton with the first experiments carried out in several modified buildings at Forrestal. Because of the secrecy involved, a whole new set of buildings was constructed. At the outset the program was called "Project Matterhorn," and the main building became known as the Matterhorn Building. When it became evident very rapidly that the problem was very complex and that the physics of high temperature plasmas was not understood, the classification of the program was removed and the program took on the more appropriate title of Plasma Physics Laboratory.

Within a few years the development of a proton synchrotron was undertaken --- again under the aegis of the AEC. This project was led by Dr. Milton White, an eminent experimental physicist who had helped design many high energy physics research facilities. The AEC arranged to have the University of Pennsylvania associated with this project, and the new accelerator became the Princeton-Penn Proton Synchrotron. No sooner had it been completed and the first experiments run, when it was closed by the AEC. The machine was obsolete, and the whole project cancelled. This, of course, was an academic trauma for the Physics Department and a financial one for the University.

The two large physics experiments plus the multiple small facilities of the Aeronautical Department were the heart of The Forrestal Research Center. The Forrestal Administrative group occupied the first floor of the main office building (later named Sayre Hall) and the Aeronautical Engineering Departmental headquarters was on the second along with several members of the faculty. A conference room and later a fine library, were also established on the first floor.

After Perkins became Chairman a set of departmental offices were established on the second floor of Sayre Hall. Then, all of the department's research operations were moved into this new area. This required the modification of many buildings for housing of these facilities and for the faculty, graduate students, research staff, mechanics and secretaries.

The undergraduate laboratory facilities with the 3'x5' wind tunnel were among the first to move in. Behind Sayre Hall was an elaborate series of greenhouses that had been used by the Rockefeller Group for plant disease studies. In the middle was a fine small building that had been used as a potting shed. This was converted to house the wind tunnel. In a greenhouse next to the shed another wind tunnel facility for instructional purposes was built. Designed by Dan Sayre, it had a central diffuser and power section feeding two separate test arms. One circuit had a 3'x2' test section and the other had a 6'x1' test section. An auditorium to house an audience of about 150 was built, and soon became an important part of the life of the Department in its new location. At the back of the greenhouse area a utility building was converted to house Nikolsky's research with an addition for his experiments on helicopter rotor dynamics.

Moving the high speed aerodynamics work of Lees and Bogdonoff into one of the long shed-like animal research laboratories required a great deal of modification to house not only the supersonic throats, but also the reciprocating compressors and the high pressure bottle farm. This was a most difficult transition, particularly as the sponsors of this program were nervous about continuing their support despite long delays.

The work in propulsion and combustion under Charyk, Crocco and Summerfield was another difficult transition. At the outset they set up shop in another of these animal research units, but they didn't fit in very well. Planning proper housing for this expanding operation, which later became one of our largest research activities, had to begin at once.[7] Flight research was improved immeasurably by acquisition of a very large field suitable for an airport right in the middle of the Forrestal area. The very soft field gave trouble at first, but eventually a macadam runway and proper hangars were built, and it became a very fine and unique operation.

The Department during 1951 to 1953 was heavily involved with this move, but left it with almost unlimited room for expansion. The principal disadvantage was the distance from main campus. There was little inconvenience to the graduate students who had cars, and could have offices in the new buildings of the Forrestal Research Center (FRC). However, although the faculty gave their undergraduate courses in the Engineering Building on main campus, they started giving some graduate courses in the FRC. Undergraduate laboratories were also in the new facilities, and the undergraduates had to be transported by bus. Nevertheless, most of the undergraduates enjoyed coming out to the FRC where they could see first line research in progress, and in some cases they were able to get involved in the various programs. They liked this new arrangement, but they would have been happier with the Department offices and the faculty at closer range. The distance problem continued to nag the Department for many years, but Forrestal campus provided a remarkable area for expanding departmental operations, that now included five Laboratories. These Laboratories were entitled: the Flight Research Laboratory; the Gas Dynamics Laboratory; the Guggenheim Propulsion Laboratory; the Low Speed Aerodynamics Laboratory; and the Rotor Dynamics Laboratory.

---

[7] ."Every so often I would hear the very loud noise of Kahane's intermittent ramjet experiment, but I would jump every time there was an explosion which I learned was due to rocket blowing up due to a delayed ignition....The challenge we had was to have the rocket running as soon as possible which meant we had to resolve the delayed ignition problem....We developed a count down procedure, siren on, hit the ignition switch and [count to 30]....Once we had the rocket running without the delayed ignition we were free for a good run. One time we thought we were free and all of the sudden the rocket burned through its side wall. In determining the cause we found somebody flushed the one toilet we had in our fancy complex. In my hurry to get the rocket running I didn't realize that those who preceded me had simply connected our cooling water inlet to the rocket to the quarter inch water line that fed the complex. John and I quickly designed a pressurized tank water system. While waiting for the delivery of the tank our count down procedure now had an item which read `Put sign on toilet door - Please do not flush toilet while the siren is running'" [10].

Aeronautical Engineering at Forrestal

During the decade between 1953 and 1963, the Department had its greatest growth and reached considerable strength as an Aeronautical Engineering Department. The launch of Sputnik in 1957, and the failure of the Vanguard rocket in the same year gave an enormous boost to all U. S. research efforts in aerospace, and the departments activities in propulsion research benefited greatly.

The first faculty appointment made after moving into the FRC was that of David Hazen who was promoted from Research Staff to Assistant Professor in 1953. He was engaged in setting up research programs for the Navy in flow visualization, and is perhaps best known for his film on the subject as part of the NSF Films in Fluid Mechanics, distributed by the Encyclopedia Britannica. However, the biggest changes in the Department faculty were not additions, but two serious resignations. In 1953 Lester Lees left to accept a Full Professorship at CalTech. Soon thereafter Joe Charyk left to accept his research appointment with Lockheed.

When Lees left, he was the Head of the rapidly expanding program in high speed aerodynamics, but he had a dynamic and able associate, Seymour Bogdonoff who took over running this program. Since Bogdonoff was basically an experimentalist, a search was started to find a theoretical person to fill the shoes of Lees. This brought Dr. Wallace Hayes to Princeton in 1954. Hayes had achieved considerable fame for his theoretical work in transonic flows, and laid down some of the theoretical basis for the experimental work done by Whitcomb of the NACA, that brought the Area Rule into prominence. Hayes had received his Ph.D. from CalTech and then worked as a theorist, first for Lockheed, and later for the U.S. Navy. He was later to write a most influential book *Hypersonic Flows Theory* with Ronald Probstein*52, as well as another on *Gasdynamic Discontinuities*.

Some excellent research talent was attracted to the staff of this Laboratory, many of them ending up on the faculty. Among these were Sin I. Cheng, Harvey S. H. Lam, George Bienkowski, Enoch Durbin, and Jerry Smith. Two very able Senior Research Associates, Andrew Hammitt and Irvin Vas, also played an important role in the development of this Laboratory. Bogdonoff rapidly had this group working smoothly with a very sophisticated program in super and hypersonics, both theoretical and experimental, and the activity was called the "Gas Dynamics Laboratory." The combination of experimental and theoretical work proved a powerful combination (figure 11}. By the end of the decade this group was full of graduate students and running a renowned program in high speed theoretical and experimental aerodynamics.

Fig. 11: Schlieren photograph of a blunted flat plate in helium at Mach 11.6. The boundary layer is visible over the top of the plate. The shock shape is close to a power-law shape, with a measured exponent of 0.66 ± 0.01. This may be compared to the theoretical value given by Hayes & Probstein of 0.667. Photo courtesy of the Gas Dynamics Laboratory, Princeton University.

Charyk was a great loss, but Crocco and Summerfield remained, together with Dr. Irvin Glassman, a Research Associate under Charyk. Glassman was appointed to the staff in 1954 and was made an Assistant Professor in 1955. Drs. Jerry Grey and Robert Jahn were added to the faculty and Preston Layton was appointed as a Research Associate to handle the complicated propulsion installations. This group organized themselves around the Guggenheim grants, and they became the "Guggenheim Laboratory for the Jet Propulsion Sciences." Crocco and Cheng made major contributions to the understanding of rocket motor instabilities, and in 1956 coauthored a highly influential monograph on *Theory of Combustion Instability in Liquid Rocket Motors*. Another important contribution was Martin Summerfield's *Burning Rate Theory of Solid Propellant* in 1958. In 1962, Bob Jahn founded the Electric Propulsion and Plasma Dynamics Laboratory, and in 1968 wrote the book *Physics of Electric Propulsion* which was instrumental in founding the field of electric propulsion for space applications.

The Rotor Dynamics Laboratory, under Nikolsky, was strengthened by Pat Curtiss, first as a Ph.D. student, then as Research Staff, and then as an Assistant Professor after he completed his Ph.D. in 1965. Two very able research staffers who kept the whole program afloat were Joe Traybar and William Putman.

In other areas, Dunstan Graham'43, was appointed to the faculty in 1958. Tom Sweeney, became a Research Associate in 1958, and Barry Nixon, a fine test pilot and engineer, joined the staff in 1959. Also, Woldemar von Jaskowsky joined Bob Jahn's growing effort in electric propulsion in 1962.

During this decade the department suffered two further grievous losses. Dan Sayre died of cancer in 1956, and Alexander Nikolsky died of a sudden heart attack in 1963. These were the department's first two faculty members and they had played a large role in getting the Department started. Perkins was now the only member of the first five still present in 1963.

At the end of 1963 the department had 70 graduate students and an equivalent number of undergraduates. The sponsored research funds totaled just over $1,000,000, and the total staff came to 250. At this time a national survey of Departments of Aeronautical Engineering rated Princeton the #2 school in the country.

A new office-type building was built to house the Gas Dynamics Laboratory, as well as a few ancillary buildings for new blowdown wind tunnels and facilities to recover the working fluid when it was helium. A grant from the NASA was used to build the Guggenheim Laboratories. Rocket pits, instrumentation rooms, molecular beam rooms and others were added to this fine building. The flying field was improved by building a 3,200 ft. runway, including taxiways and ramps. A hangar was added with rudimentary rooms for some staff and graduate students. Some of the greenhouses were replaced with a new building to house the Low Speed Aerodynamics Laboratory and finally the research facilities for the Rotor Dynamics Laboratory were expanded. This expansion was completed with the building of a unique facility called "The Long Track" designed to test helicopters, VSTOL designs and ground effect machines. This facility was basically a long shed that housed a 450 ft track (later lengthened to 750 ft) for flying aircraft in still air at low speeds, especially useful for studying problems encountered in hover (figure 12). The Princeton group, led by Curtiss, became a center for helicopter studies, especially for understanding problems in stability and control.

Fig. 12: Left: Aerial view of the Long Track Laboratory used in studying helicopter and VSTOL aircraft aerodynamics, stability and control. Right: Interior view of the Long Track Laboratory (courtesy Rob Stengel).

All of these expansion efforts cost some three to four million dollars, so fund-raising programs were essential. A major resource for these purposes was the "Corporate Associate Program" where industry was asked to support the Department by making annual contributions to this special fund. In return, they received all of the research reports, invitations to the Department for special programs, better access to the graduate students for recruiting, and access to the faculty for special discussions. For this service they would make an annual payment ranging from five to twenty thousand dollars. This program grew to about $200,000 per year and it was referred to as "White Money" because it was fully credited to the Department. The University charged no overhead against the expenditures from this fund and the Departmental Chairman could approve all outlays. This resource was used to build facilities, and occasionally for other Departmental purposes such as add-ons to fellowship awards, special research projects, and so on. This money gave the department great freedom of decision and some freedom from normal University constraints. When the rapid Departmental growth was questioned, President Dodds would turn to critics and say, "But they cut their own feed" [3].

During this decade, Perkins was on leave without pay from Princeton on two occasions to take assignments from the government. The first was an appointment as U.S. Air Force Chief Scientist, a rotating position advisory to the Chief of Staff of the Air Force, at that time General Nate Twining, serving in the Pentagon from June 1956 through September 1957. The second leave, from February 1960 through January 1961, was a Presidential appointment as Assistant Secretary of the Air Force for R & D. In each case, Nikolsky was made Acting Chairman.

Aerospace and Mechanical Sciences

This was the state of affairs at the start of the 1963-64 academic year. At this time the Mechanical Engineering Department was falling on difficult times. Most of the new high technology fields that might have been taken over by the ME Department were swallowed rapidly by the AE Department, leaving ME few new objectives to grow on. They were more accustomed to the normal Princeton Engineering scene and viewed the startling growth of AE with some concern. ME had taken on a program in Solid State Physics that had largely been abandoned by the Physics Department, but besides that there was little excitement in this Department, and there had been a drop in student demand for ME. The Department was becoming largely moribund [3].

Fig. 13: Aerial view of the Engineering Quadrangle, circa 1971 [8].

The Engineering School made a study of ME and what might be done. They seemed to have three choices. They could try to revitalize the Department at the risk of running head on into the programs of the AE Department. They could abolish the Department altogether, but this would be anathema to many of the powerful Princeton alumni and to the strong professional society, the ASME, a politically difficult choice that was soon abandoned. The third possibility was to merge ME with AE, in the hope that the sum of the parts would be greater than the sum of each. The steady development of and growth of the engineering science approach had brought the curricula, course offerings, and research interests of the two departments closer and closer together. On the decision of Professor Drake, Chairman of the ME Department, to resign from the University in June 1963, it therefore seemed reasonable to consider the merger of the two departments [8]. This made good sense, and the merger was approved by the University and the Engineering School and announced on May 31, 1964. Perkins was asked by President Goheen to Chair this new Department, and after much study and argument the newly merged Department was named the Department of Aerospace and Mechanical Sciences (AMS). This new department combined the full activity of the old AE Department based at Forrestal with the activities of the old ME based in the Engineering Quadrangle, a new building completed in 1962 to house the entire School of Engineering and Applied Science (figure 13).

At first Perkins maintained the Departmental offices in Sayre Hall in the FRC, and set up another office in the old ME Chairman's office in the Quadrangle. Slowly but surely, however, the Departmental office in the Engineering Building became headquarters and office space was

535

provided for AE faculty members in the Engineering Quadrangle. This arrangement went a long way to alleviate the problem of distance from the department's students and the rest of the University faculty. At the same time it started moving attention from the FRC and back to the main campus. This was destined to have a damping effect on the growth of the Department. Inevitably the tightly knit group that had formed in the old AE Department was starting to break up, and the department was never were quite the same again.

A study was carried out how to operate this newly merged Department and its much larger faculty and student body during the academic year 1963--64. The faculty of the AMS Department was now one of the largest in the University with a record number of graduate students, and had one of the largest sponsored research programs in the University and a very complex group of research facilities. It was a complicated organization.

From 1963 to 1966 the AMS Department grew to be one of the largest in the University, with 29 regular faculty members and a large research staff, 130 graduate students and a sponsored research program of over $3,000,000, most of which was based on the activities at Forrestal.

During the next few years the department lost quite a few of the original ME faculty. Robert Drake, who was Chairman of the ME Department at the time the merger was being discussed, had left to join the Engineering Department at the University of Kentucky. A year later he took Roger Eichhorn with him. John Fenn left to go to Yale. Jerry Grey of the old AE Department left to form his own company, Greyrad. One of the ME faculty that stayed on, Barrie Royce, continued to pursue his interests in spectroscopy and solid state physics. A student of his, Eli Harari*73, went on to found SanDisk Corporation, and currently serves as President and CEO. George Mellor's contributions evolved from his original interests in turbomachinery, to modelling turbulence, to modeling geophysical flows. He was instrumental in bringing the Bureau of Commerce's Geophysical Fluid Dynamics Laboratory to Forrestal Campus in 1968 from its original home in Washington, D.C.

There were some additions also. Cemal Eringen came from Purdue to help develop a Solid Mechanics Laboratory to parallel the fluid mechanics activities. Bill Sirignano completed his thesis in January 1964 (one of the first two AMS PhD's), and in the same spring was hired at Princeton as a Research Staff Member and Lecturer, and was appointed as an Assistant Professor, along with Jerry Smith and Arnold Kelly from CalTech, in 1967. Richard Grot joined the department from Purdue, and Francis Moon came from Cornell. Mark Knowlton'51 was appointed a Lecturer in 1962, serving for some 30 years in that capacity. Francis Hama joined the department as Senior Research Scientist in 1969, and remained at Princeton for about 20 years. Sirignano left Princeton in September, 1979 to assume the ME Department Head position at Carnegie-Mellon, on his way to become the Dean of Engineering at UC Irvine in 1985.

The 60's saw the development of Princeton as a center for fundamental studies in engineering and applied science. It became particularly noted for its efforts in developing the fundamental understanding of fluid mechanics and combustion processes. In fluid mechanics, Hama's contributions to the understanding of laminar-to-turbulent transition over the entire speed range from hypersonic to incompressible flows are still widely recognized for their importance. The work in hypersonic and supersonic flows continued unabated under the direction of Bogdonoff, Cheng, Hayes, and Bienkowski. The combustion group, led by Martin Summerfield, Irv Glassman and Bill Sirignano, included Frediano Bracco*70 and Fred Dryer*72 who were later to join the faculty. Glassman's book *Combustion* appeared in 1977 (currently in its third edition), and founded the Combustion Science and Technology Journal. In dynamics and control, Rob

Stengel*68 designed the Lunar Module manual control system, and Earl Dowell and Pat Curtiss authored the book *Modern Course in Aeroelasticity* in 1978.

In 1974, after a record 23 years, Court Perkins stepped down as Chairman, and was succeeded by Seymour Bogdonoff. According to Glassman "One of the most important elements [of the department] was the ambience created by Court Perkins --- it was simply an exciting and fun place [to be]...." [10]. Perkins stayed on as Associate Dean for one year under the new Dean of the Engineering School, Bob Jahn, another AE faculty member, and on July 1, 1975 Perkins went on a three-year leave without pay to become President of the National Academy of Engineering in Washington, D.C. In 1978 he was given a four-year extension of his term at the Academy, and became Professor Emeritus. Apart from many other honors he has received through the years, he was awarded an Honorary Doctorate from Princeton University in 2001.

Accreditation

Accreditation was important to the department, but not all accreditation visits went smoothly. According to Perkins [3]: "In the early 1960s we had a visit from the ECPD (Engineering Council for Professional Development) Accreditation Committee that operated in conjunction with the Engineering Professional Societies. We had to undergo these visits every sixth year. The Accreditation Committee this year was led by my old friend Dr. Raymond Bisplinghoff, then on the MIT faculty in Aeronautical Engineering. I had first met Ray at Wright Field during World War II, when he was serving in the Dynamics Branch of the Aircraft Laboratory. His major interest at that time was wing flutter, and he eventually wrote a book on Aeroelasticity with his colleague Holt Ashley. Now, at that time, we were not very strong in those areas of aeronautical technology that dealt with structures. Our only faculty member who was competent in these areas was Nikolsky, and his interests had shifted to other fields. We had to shotgun some of our younger faculty into teaching undergraduate courses in this field and there was essentially no graduate program at all.

As was inevitable, Ray Bisplinghoff complained of this situation and asked us how we expected to be a first class Department of Aeronautical Engineering without a strong program in the structural fields that would match our powerful programs in Flight, Fluid Dynamics, Propulsion and Combustion, Rotor Dynamics and Low Speed Aerodynamics. We certainly recognized this problem, so asked Ray to recommend someone to us from his Sc.D. candidates at MIT. He recommended to us Eugene Brunelle, and after a visit to us Brunelle was appointed to the faculty as an Assistant Professor in 1962, but shortly left to accept a higher ranked position at RPI.

We then went back to Ray Bisplinghoff for a suggestion of a more acceptable candidate for us. This time he recommended one of his top Sc.D. students, Earl Dowell, for this position. We found Earl excellent and appointed him as an Assistant Professor in 1963. Earl satisfied our needs for a faculty member in structures and structural dynamics, and in short order proved himself to be a top flight teacher-scholar with a growing sponsored research program and many graduate students. Perhaps he was too good, for some twenty years in 1982 Earl was made Dean of Engineering at Duke University, and the Department was back where we started. In 1983, Dr. Jim Marr, now MIT's expert in structural dynamics visited the Department as a member of its Visiting Committee. He remarked, 'How can Princeton have a first class program in Aeronautical Engineering without a faculty competence in structures and structural dynamics?' "

More Recent Times

The department continued to prosper under Bogdonoff, who served as Chairman from 1975 to 1983. He was followed in that position by Harvey Sau-Hai Lam (1983-1989), Irvin Glassman (1989-1990), and Garry L. Brown (1990-1998). The present Chairman is Alexander J. Smits.

The period from 1975 to the present day deserves a separate treatment, as the department saw through some momentous changes, not the least of which were the impact of co-education that began in 1969, the end of the space race following the successful landings on the Moon, the Arab oil embargo of 1973 and the ensuing energy crisis, the aftermath of the Vietnam War that ended in 1975, and the strong anti-war sentiment that grew during the protest era.

One particularly noteworthy development was the establishment in 1970 of the Center for Energy and Environmental Studies (CEES). Irvin Glassman became its director in 1972, and "He, Enoch Durbin, and other combustion scientists turned their attention from rocket and jet engines to improving fuel efficiency in automobile engines. David Harrje, who had worked with Luigi Crocco on liquid-fuel rockets, applied his technical expertise to problems of home-energy conservation, using the nearby housing development of Twin Rivers, New Jersey, as his laboratory" [4]. Robert Socolow joined the department in 1974, and on his arrival took over the leadership of CEES. In 2000, the Center was subsumed in the Princeton Environmental Institute (PEI).

In another development, the AMS Department changed its name in 1978 to the Mechanical and Aerospace Engineering Department, in alignment with national trends to gain better name recognition from prospective students.

Four of the department's students became astronauts: Pete Conrad'53, Gerry Carr*62, Jim Adamson*77, and Greg Linteris*90.

The department currently has 24 Faculty (16 Full, 4 Associate, and 4 Assistant Professors), 4 Senior Research Scientists, 22 Research Staff, 16 Research Associates and Assistants, 11 Professional Technical Staff, 4 Senior Technical Support Staff, and 13 Administrative Support Staff. It is loosely organized into five groups, based on the faculty research interests: Materials (Professors Royce, Soboyejo, Srolovitz, Suo); Applied Physics (Professors Choueiri, Jahn, Miles, Suckewer); Fluid Mechanics (Professors Brown, Martinelli, Smits, Martin) Dynamics and Controls (Professors Holmes, Kasdin, Leonard, Littman, Rowley, Stengel) Combustion (Professors Bracco, Dryer, Law, Ju). In addition, Professor Nosenchuck is in Design and Manufacturing, and Professor Socolow continues his work on energy assessment through PEI.

The enrollment statistics for 2002--2003 show a student body of 34 sophomores, 39 juniors, and 32 seniors (105 total). Of the 32 seniors, 10 intend to graduate as mechanical engineers, 19 as combined aerospace and mechanical engineers, and 3 as aerospace engineers). The graduate student body for 2002--2003 totals 89, with 79 Ph.D., 4 MSE, and 6 ME candidates. The research income for 2001-2002 is reported at $11,600,000. In the most recent NRC Report by Goldberg & Meher (1995), the Aerospace Program was ranked #4 in the country, and the Mechanical Engineering Program was ranked #6.

Acknowledgments

Our thanks go to Ms. Maureen Hickey, for rescuing and archiving many of the documents relating to the history of the department, and to Barrie Royce, for his comments on an earlier draft.

References

[1] Condit, K. H. (1949) Engineering at Princeton University. Princeton Alumni Weekly, February 18, 1949.

[2] Hazen, D. C. In Alexander Leitch, "A Princeton Companion." Princeton University Press (1978).

[3] Perkins, C. D. (1985) "Recollections," **2**, Chapter 6: Princeton University. Privately published.

[4] Merritt, J. I. (2002) "Princeton's James Forrestal Campus: Fifty Years of Sponsored Research," Princeton University.

[5] Bix, A. S. (1993) "Backing into Sponsored Research" Physics and Engineering at Princeton University. (unpublished paper).

[6] "Supersonic Wind Tunnel Dedicated." Princeton Alumni Weekly, February 10, 1950.

[7] Press Release, Department of Public Information, Princeton University, October 12, 1958.

[8] Elgin, J. C. (1974) "An Account of The School of Engineering and Applied Science of Princeton University," Princeton University.

[9] "Engineering at Princeton University." Princeton Alumni Weekly\/}, February 18, 1949.

[10] "The Early Days of MAE." http://www.princeton.edu/\~{}mae/SHL/glassman.html.

Chapter 39

Aerospace Engineering at the University of Southern California

Richard E. Kaplan*, University of Southern California

Introduction

There have been distinct eras of engineering education at the University of Southern California (USC) that mirror the stages of its evolution as a recognized research university. In a part of the nation noted for its excellent public education, USC, together with a handful of small private universities, most notably Stanford and Caltech, tried to find a niche in the educational spectrum. For most of the past century, USC had a reputation of more of an athletic powerhouse than of a top ranked academic institution. Trojan Football provided a model for the western athletic university, most often discussed following the Rose Bowl Game on New Year's Day. Caltech, one of USC's geographically closest neighbor institutions, had already achieved eminence as a leader in science and engineering education and research, and was an influential role model for USC to follow.

Perhaps the most surprising fact about USC today is summarized in the following table that compares USC's enrollment to California best known public research universities:

Enrollment	UCB	UCLA	USC
Undergraduate	22,705	24,668	15,883
Graduate	8,632	10,007	13,049
Total	31,337	34,675	28,932
% Graduate	28%	29%	45%

First, USC is large – the second largest *private* university in the nation (after NYU) – and second, it is predominantly a graduate and professional education institution[1]. Approximately 37 percent of the university's total enrollment is composed of American minorities; a further 16 percent are international students. USC is ranked third among all American universities in international enrollment.

Engineering had an early start as an academic program at USC. For a complete history of the early years of the USC School of Engineering, please see **The USC Engineering Story**[2], the memoir of Robert E. Vivian, longtime Dean of the USC School of Engineering.

---

* Associate Fellow
[1] These data are for the latest years available on the web (UC 1999, USC 2001). A table comparing selected *private* research universities is shown near the end of the paper.
[2] USC Press, Los Angeles, 1975

The Earliest Years

The first classes in engineering disciplines were offered in 1906. Those early classes were offered by the College of Liberal Arts - through its first engineering departments: civil engineering, mechanical engineering and electrical engineering. The tradition of separate engineering departments in a broadly based college was established from the inception, and was a model for subsequent USC history.[3]

The first engineering professor, John B. Johnson, was hired in 1908, the same year that the college awarded its first engineering degree in Civil Engineering. The first USC electrical engineer graduated in 1911. USC awarded its first mechanical engineering degree and its first master's, in chemical engineering, in 1921; its first degree in architectural engineering in 1926; and its first petroleum engineering degree in 1927. By 1927, USC had awarded 254 engineering degrees, mostly B.S. degrees in civil and electrical engineering.

1927, two decades after USC offered its first engineering course, a separate College of Engineering was established, with five departments: chemical, civil, electrical, mechanical and petroleum engineering. In many regards, the USC departmental mix at the inception of the school reflected the industry in Southern California – petroleum and its refining were to remain major activities in the Los Angeles area for years to come. The aircraft industry was growing, but a school as small as Caltech, home of GalCIT, the Guggenheim Aeronautical lab under the direction of Dr. Theodore von Kármán, was able to provide a stable source of well trained engineers who would become industry leaders.

Near the start of the Second World War, the USC College of Engineering had 10 full-time faculty, one secretary and 230 students. "In spite of the lack of equipment and small budgets," Vivian wrote, "there was optimism, initiative, cheerfulness and a willingness to work hard on the part of faculty and students. These are the elements which make progress possible." With the awarding of its first PhD. In 1939, USC entered the war years. The war expanded on every aspect of the character of the college.

In1940, Vivian became acting dean - an appointment that was made permanent in 1942 - and presided over nearly two decades of solid growth, starting an upward curve that is still accelerating. When Vivian took office, USC Engineering was devoted almost exclusively to teaching: "You should leave the research to Caltech," a trustee bluntly told Vivian at one point. World War II changed that.

The War Years – Research as well as Teaching

At first, the school's growth remained in teaching: almost immediately after Pearl Harbor, USC Engineering became a major center for the Engineering Science and Management War Training Program, to fill the burgeoning need for technically trained management and supervisory personnel for war industry. Some 50,000 students swarmed through the program, the largest single-campus effort of its kind in the country. Simultaneously, the college gave naval officers crash courses in various engineering disciplines, with teachers putting in 12- to 14-hour days.

---

[3] Much of this history is taken from the USC web site (the author paraphrased Dean Vivian's book)

Another aspect of the USC location during this period was to afford the opportunity to many local Caltech graduates to carry on high-risk, pre-industrial war related research in an academic setting. USC established its Engineering Center, with a $10,000 contract from Lockheed, on spot-welding aluminum alloy sheets in 1944. The success of this first effort resulted in applications for more and varied contracts in subsequent years for the National Advisory Committee on Aeronautics (NACA), the Office of Naval Research (ONR) and the Air Force. USC staff designed and operated a missile test facility at the Point Mugu Naval Station near Oxnard California.

It was in this environment that Roy Marquardt, who had been working at Northrop, moved his ram-jet research activities to USC. Several large test cells were built at the Kaiser Steel Works in Fontana, east of Los Angeles. The rapid growth in this areas caused Marquardt to leave USC to form his own company. Dr. Marquardt retained his interest in USC, and later was a member of the department's advisory committee in the mid 70's.

One other notable accomplishment of this era was the design and construction of the world's first low density wind tunnel by Dr. Raymond Chuan. This facility was built on campus, and originally stood on a site now occupied by a newer engineering building. So advanced was this 1950's design that the essential component of the tunnel, its cryogenic pump, was moved to a consolidated Aerospace Laboratory in the early 1980's.

After the war, the college's proficiency and efficiency in teaching advanced-level engineering led to another notable industrial collaboration that continues to this day. Hughes Aircraft Co. contracted with USC to offer master's-level instruction to its engineers, under what was called the Cooperative Engineering Program. Hughes employees received tuition, books and flexible work hours, enabling them to get advanced degrees.

Collaboration with Industry – Teaching Working Engineers

This collaboration with industry was started with the assistance of Simon Ramo during his years at Hughes. Ramo, a Caltech graduate in Electrical Engineering, came to the West Coast following the war – leaving the General Electric Company and joining Hughes to produce missiles, a new and exciting technology in the late 40's. Ramo saw that the engineers needed to push the knowledge envelope would have to be trained and retrained – a model more complex than graduate education of that day, because the knowledge of the state of the art practice was changing daily. Working engineers would have to have the skill continually upgraded, with greater emphasis on mathematical analysis than most engineering schools had required in the past.

As Ramo later described it, Caltech was too small to fulfill this requirement – but its model of graduate education was overwhelmingly successful. The UC system, particularly UCLA, would have been able to meet this goal, and ultimately was the major provider of advanced technical continuing education in the region, if not in the nation. However, in those days more than now, most decision making power resided in Northern California, and the UC system had its plate full with the expanding National (Energy) Laboratories. USC was centrally located, could make decisions promptly, and had the facilities from its wartime experiences and GI Bill enrollments.

Many of the instructors in these programs were the best in the world in their field – they worked for the companies and taught in the evening. USC sat at the core of an educational experiment for working engineers from all of the companies in the area: Hughes, North American (later Rockwell), Lockheed, Northrop, Douglas (later McDonnell-Douglas), Rocketdyne, *et. al.* These engineers were daily involved in the leading technical problems of their day, and at night sat in class with colleagues from every other company in the LA Basin. They would make those important human contacts that could not have happened in in-plant programs, or even in traditional graduate programs, where an engineer would take a year or two to get a degree, often changing employers when he (more often than she in those days) received his MS or PhD.

Continued defense spending on high-technology, plus a flood of veterans using GI Bill benefits to build careers in engineering, served to transform the school into a major center for the development of new engineering science and technology.

By 1948, the College of Engineering had 100 graduate students. A decade later, the count had quadrupled to 400 in what was renamed the "School of Engineering" in 1950. For USC's Diamond Jubilee 75th anniversary celebration in 1955, Vivian was able to boast that the school had "graduated 5,000 engineers, 4,200 since 1940. They work for more than 600 companies. Ninety-four are presidents, vice presidents or chief engineers of companies in this country."

Since a large proportion of the graduate students were employed in the local aircraft industry, a demand arose for an Aeronautical emphasis degree. Mechanical Engineering established a graduate program in Aeronautics and Guided Missiles in 1947, with graduates in 1949 and 1950. The program was renamed Aeronautical Engineering, and graduated more than 200 M.S. students between 1953 and the establishment of a free standing department in 1964. The early 60's saw a growth in all engineering activities, including an expansion of classes to Edwards Air Force Base to reach a new group of students involved in flight test of the most modern aircraft of that day. One of these students was our most famous graduate, Neil Armstrong. Actually, Armstrong joined the NASA Astronaut trainee program in 1962 before he had achieved the 27 units for his MS degree. Later, he was awarded 3 units of AE 590; Special Projects, for his moon walk on July 20, 1969, and was awarded an *earned* MSAE following his oral report on the project in January 1970.

During the mid-60s USC had an active program in Graduate AE education and low density, high energy research. However, the USC educational model had produced a graduate program that was composed mainly of full time engineers, and in addition had a well regarded research activity that could have used graduate students working on theses and dissertations as a research work force. Under the leadership of a new Dean, Alfred Ingersoll, a Caltech Civil Engineer, and with the energy of Zohrab Kaplrielian, from Caltech and Berkeley, USC set out to develop full-time graduate programs to complement the existing program. Part of this development was the creative use of Educational Television to broadcast graduate classes directly to local industry – classes could be taken during the day as well as the evening, and the LA freeways were starting their inevitable grid lock that continues to this day.

Establishment of the Aerospace Engineering Department

The other innovation was the establishment of a free standing Aerospace Department in 1964 under the leadership of John Laufer (1921-1983). Laufer was then Chief of the Gas Dynamics section at JPL, having previously been at the National Bureau of Standards (with Hugh Dryden and Galen Schubauer) and Caltech (with Hans Liepman and Theodore von Kármán). Laufer came to USC in January 1964 to form Graduate Department of Aerospace Studies, soon changed to Aerospace Engineering when the Air Force ROTC changed the name of its university programs to Aerospace Studies. (Laufer was elected to the National Academy of Engineering [NAE] in 1977, and was a fellow of AIAA and the American Physical Society [APS].)

Laufer was joined by two colleagues appointed in Mechanical Engineering: Richard H. Edwards (d. 1993) a mathematician from Illinois Institute of Technology (IIT) who had come to USC from Hughes Aircraft, and H.T. Yang from Caltech GalCIT. Laufer hired three new PhD graduates, Scott Hickman from UC Berkeley ME (left for UC Santa Barbara in 1968), William B. Bush from Caltech GalCIT (left USC in 1973) and Richard E. Kaplan from MIT A&A (ret. 2000). Additionally, H.K. Cheng from Cornell and Cornell Aeronautics Lab joined the faculty as a visiting Professor and became full time in 1965. Cheng was elected to NAE in 1988 and is a fellow of AIAA and APS. Although designated Emeritus in 1996, Cheng still comes to the university daily! Beyond these founding members, the faculty history of the department is summarized in the table on *full time faculty*.

A quick reading of the table shows that a large number of faculty hired throughout the history of the AE department followed experimental research interests. Laufer assembled what was perhaps the largest group of experimentalists in fluid mechanics at any university in the US or abroad. While this is no longer the case, much of the early history of the department is colored by this fact.

Undergraduate Program

Although the department was founded for graduate education, it seemed clear there was a ne[4]ed for an undergraduate program in Los Angeles. The Caltech program was small and national in scope, and UC did not offer a separate AE degree at that time. Additionally, the fact that USC was host to all ROTC programs lent a university interest to an undergraduate AE program. However, USC had too small an engineering school to establish an inclusive department, like those at MIT, Georgia Tech or Illinois, where there would be expertise in the department for all of the components of the AE discipline – aerodynamics, flight mechanics and controls, structures, and propulsion. Control was well established in the Electrical Engineering at USC, structures were dominant in Civil Engineering, and Mechanical Engineering had a strong thermo-sciences area in existence. It was decided to use joint appointments with these other departments to ensure that course offerings in these other departments offered materials appropriate for an accredited undergraduate program. Recognizing the importance of a sound basis in mathematics for the program, several joint appointments with the Mathematics departments were also completed.

However, the scaling up of the Viet Nam War caused a reduction in national AE enrollments, and the USC program was almost stillborn. I recall a meeting at the University of Kansas in the early 70's about the future of independent AE departments. USC decided to stay the course, although many students at that time felt that having an AE designation on their diploma might prove a detriment to getting a job in a tough market.

The Boom of the 80's

Division of disciplines at USC meant that many research areas included in the Aerospace Engineering departments at other institutions were housed in other departments at USC. In fact, as USC Engineering joined the leading research universities, the local Aerospace Industry sent more graduate students to sister departments than to Aerospace Engineering. However, undergraduate enrollment grew during this period, until there were more undergraduate AE majors than majors in any other department.[5]

---

[4] However large the total undergraduate *enrollment* in AE became, the number of graduates -- those who stayed to received degrees – grew much more slowly.

Following John Laufer's unexpected passing in 1983, Kaplan, and later Muntz and Redekopp as co-chairs, directed the department's progress. A major effort to strengthen the Astronautics activities both on the graduate and undergraduate levels saw the addition of several new colleagues who had distinctly different disciplinary interests than the core members of the department. Muntz took the initiative in this area, and tried to grow the experimental program in rarefied and energetic gas dynamics. These academic initiatives, and the success of the collaborative research programs, in particular Redekopp's leadership of the Turbulence Control URI, were the major characteristic of this period. Soon however, Redekopp was asked by Len Silverman, Dean of Engineering, to replace Tony Maxworthy as chair of Mechanical Engineering, so Muntz became the sole chair of Aerospace.

During this period, the Astronautics program received greater emphasis and drew a larger share of enrollments. AE joined the USC Electrical Engineering and Computer Science programs as major attractions on the Interactive Instructional Television network, which broadcast regular graduate engineering classes to working engineering at their companies. This continued the tradition started under the impetus of Si Ramo some 50 years earlier.

Merger with Mechanical

The end of the cold war produced lower enrollments in the AE program at USC as well as many other institutions. USC decided to force a merger of the AE department with Mechanical Engineering in 1998, with an aim of strengthening both departments, which had complementary degree programs and research programs. The new department, innovatively called Aerospace and Mechanical Engineering (AME) maintained separate degree programs, but is attempting to achieve efficiencies in staffing and class offerings. The merged department has the distinction of receiving more institutional teaching awards than any other department in the school. As of this writing, a search for a new chair is in progress.

Support Staff

No history of an academic program would be complete without a mention of the support staff. They do the most to cope with the red tape of the institution and shield the faculty from the realities of life. Originally, Elizabeth Harris and Gail Wamsley were the key support members in the office, with Don Kingsbury and Casey De Vries heading the lab staff. Later, Gail Dwinell, Alice Vaughan and Virginia Wright headed the office staff, finally followed by Elsie Reyes and Marrietta Penoliar, with Thane De Witt, Mark Trojanowski, and Dennis Plocher in the Labs.

Disclaimer

I have tried to recall many people and events from the past forty years, but although I have consulted with my colleagues, I bear the onus for forgetting important people and events from this era. I hope that the reader will forgive my lapses in memory, and will write me, either c/o the department or at my e-mail **kaplan@usc.edu**. Please help me correct subsequent revisions of this paper for the permanent departmental history. In particular, there are alumni who have been influential at other universities or research centers, grants that were equally important, and Adjuncts whom I forgot. For these omissions, I apologize in advance. Please send me corrections so that I can set the record straight.

Full Time Faculty with Major Appointment in AE[6]				
Name	Acad. Yr (joint)	Ph.D. (prev. empl.)	Initial Area of Specialty	Comments Honors
Ron Blackwelder	1970	Johns Hopkins	Exp Turbulent Boundary Layers	AIAA, APS
Frederick K. Browand	1967 (CE)	MIT	Exp Fluid mechanics	APS, Ch APS-DFD
Richard S. Bucy	1966 (Math)	UC Berkeley (U Colorado)	Th. Math, Control Theory	Ret 1999 AMS,IEEE
William B. Bush	1965	Caltech	Th. Hypersonic Matched expansions	Left 1973
Hsien K. Cheng	1965	Cornell (CAL)	Th. Hypersonics unsteady Aero	NAE 1988 emeritus 1999 AIAA, APS
Wing T. Chu	1969	Toronto	Exp. Jet Noise	Left 1975 Canadian Natl. Res.Council
Julian A. Domaradzki	1989	U. Warsaw (Flow Corp)	Comp. Fluid Mech.	
Richard H Edwards	1964	UCLA, (Hughes)	App. Math	d. 1993
Dan Erwin	1985	USC	Exp. Laser	
Mike Gruntman	1995	Moscow Phys. Tech Inst	Sci. Instruments – Space Sciences	
Roy Scott Hickman	1965	UC Berkeley	Exp. Rarefied gas dynamics	To UCSB 1968
Chih Ming Ho	1975	Johns Hopkins	Exp Turbulent shear flows	Left 1993 UCLA NAE 1997 AIAA, APS, ASA
Patrick Huerre	1978	Stanford	Th. Nonlinear waves	Left 1989 École Poly (France) Fr. Acad. des Sciences 2002 Carnot Prize 2002
Richard E. Kaplan	1965	MIT	Exp Turbulence	Ret 2000
John Kellam	1966	Johns Hopkins	Exp. Fl Mech	To CSULB 1969
Joseph Kunc	1985	Warsaw	Th. Plasma, Space Sc	APS
John Laufer	1964	Caltech (JPL)	Exp. Turbulence Aeroacoustics	NAE 1977 AIAA, APS d.1983
Tony Maxworthy	1967 ME	Harvard (JPL)	Exp. Aerodynamics Geophysical Fl. Dyn.	NAE 1991 APS, Chr APS-DFD Amer.Acad.Arts&Sci Otto Laport Award, G.I. Taylor Med

[6] Data are per USC records, and are incomplete or missing in some cases.

		Stanford (Brown)	Comp. Fluid Dyn	PYI 1990 Left 2000 UCSB
Eckhart Meiberg	1990	Stanford (Brown)	Comp. Fluid Dyn	PYI 1990 Left 2000 UCSB
Michael Merritt	1967	USC (EE)	Computers-Simulation	Left 1968
E. Philip Muntz	1968	Toronto (GE)	Rarefied Gas Dynamics spectroscopy	NAE 1993 AIAA, APS
Paul K. Newton	1987	Brown	Th. Fluid Mech —waves, stability	
Larry G. Redekopp	1970	UCLA	Th. Non-linear waves	APS, Chr APS-DFD
Paul Scott	1967	MIT	Exp. Gas/Surf Interaction	Left 1971
Donald Shemansky	1996	(U Arizona)	Space Sciences	
Geoff Spedding	1996	U. Bristol (Eng)	Exp Fluid Mech – Bio.	
S. S. Sritharan	1983	Arizona	Comp. Fluid Mech	Left 1989 Naval WSC, San Diego
Bob Suzuki	1967	Caltech	Exp. Shock Tubes	Left 1971 U. Mass President. Cal Poly Pomona 1991-2003
B. Andraes Troesch	1966 Math	ETH (Aerospace)	App. Math	Ret. 1996
Hsun-Tiao. Yang	1964	Caltech	Th. Supersonic	Ret. 1987

Joint Appointments with Other USC Departments (before Merger)				
Fokion Egolfopou-los	1992 (ME)	UC Davis (Princeton)	combustion propulsion	NSF-ERI Silver Medal of Combustion
Henryk Flashner	1983 (ME)	UCB (TRW)	Control	
C. Roger Freberg	1957	(Borg Warner)	Mechanical eng.	Ret 1973
Rick Miller	1975 (CE)	MIT (UCSB)	Structures	Left 1992 U. Iowa President, Olin College 1999-
Paul Ronney	1993 (ME)	MIT (Princeton)	Propulsion combustion	NASA Astronaut PYI, FIP
Costas Synolakis	1983 (CE)	Caltech	Fluid Dynamics	PYI

Mechanical Engineering Faculty added in Merger

Charles S. Campbell	1983	Caltech (Hughes)	Multi-phase flow	PYI
Marijan Dravinski		Caltech	Elastic wave propagation	
Yan Jin	1996	Tokyo (Stanford)	Design	NSF Career Award
Terence G. Langdon	1971 (MatSc)	Imperial College (UBC)	Materials	ASM; ACeramS; FREng, FInstP, FIM
Stephen C-Y Lu	1995 (ISE, CS)	UIUC (UIUC)	collaborative eng-design & manufact.	PYI
Stephen R. Nutt	1994 (MatSc)	U Virginia (Brown)	Composites	ACeramS
Satwindar Sadhal		Caltech	Thermo sciences	PYI
Geoffrey R Shiflet	1979	UCB	Mechanical Systems	
Firdaus Udwadia	CE, Math, Business	Caltech	Theo Mechanics	
Bingen Yang		UCB	Dynamics & Control	

Selected Adjunct Faculty

Malcom J. Abzug	1970		Dynamics & Control	Left 1974 - AIAA
William Rodden	1968	UCLA	Aeroelasticity	Left 1971 AIAA
Paul Arthur	1961	Caltech	Space Systems	Left 1964
Robert Liebeck	1982	UIUC (Mac Doug)	Aircraft Design	Left 2000 to UCI NAE 1992- AIAA
Robert F. Brodsky	1986	NYU	Satellite Design	AIAA, IAE
Peter Lissaman	1990	Caltech	Design	AAAS
Gerald Hintz	1980		Orbital Mechanics	IAE
Jerry Salvatore	1984	USC	Orbit Determination	Hughes
M. Oussama Safadi	1998	UCLA	Structural Dynamics	

Selected Ph. D. Graduates and Post-Docs

Jose R. Canabal – to ESRO	Alex Liang – Aerospace Corp
T. Y. Chen – to Cal Poly Pomona	David Lim – to Titan Research
Russell M Cummings – Cal Poly SLO	James Lo – to U Maryland BC
Adam Fincham – to Grenoble	Margaret Lyle – to West Virginia
Eckhart Freund – to	Don Miller – to Arizona
Christine P. Ge – to Oregon State	Mellissa E. Orme – to UC Irvine
Ashok Gupta – to IIT Kanpur	Robert Peterson – to U Arizona
M. Hafazi – to CSU Long Beach	David Rappaport – to Technion
Mohammed Hafez – to UC Davis	Robert Rosen – to NASA HQ
Joe Haritonides – to MIT	Robert Schlinker – to UTRL
Jean Hertzberg – to Colorado	Pat Weidman -- to U. Colorado
Gabriel Karpuzian – to USNA	James Williams, III – to UNC, Auburn
Lawrence Keefe – to Nielson Engineering	Clint Winant – to UCSD – Scripps
Gary Koop – to TRW Labs	

Major Experimental Research Facilities (More than 1 dissertation)

Facility Name	Year	Role
Low Density Wind Tunnel	1960-	Rarefied Gas Dynamics
Low Turbulence Tunnel	1965-72	Turbulent Boundary Layer
Anechoic Chamber	1970-89	Supersonic Jet Noise
Low Turbulence Tunnel	1972-	Low Turbulence (Closed Circuit)
Dryden Tunnel	1977-	Low Turbulence
Görtler Tunnel	1980	Curved Boundary layers
Mixing Layer Tunnel	1972-95	Turbulent Shear Flows
Blue Water Channel	1984-	Low Turbulence Water Flow

Major Multi-Investigator or Facility Grants

Agency	Period	Purpose
NSF	1964	Low Turbulence Wind Tunnel
NSF	1965	Faculty Development
DOT	1968-75	Supersonic Jet Noise & Facility
ONR	1985-90	URI – Turbulent Flow Control
NASA	1985-88	Hypersonics Training & Research
AFOSR	1990-	URI – Low Density Flows
NSF	1998-2000	KDI- Collaborative Network Engineering Over the Internet

Abbreviations, Acronyms and Jargon

NOTE: Societies are listed when the person was elected to Fellow or equivalent

AAAS – American Academy of Arts & Sciences

AIAA - American Institute of Aeronautics and Astronautics

ACeramS - American Ceramic Society

AMS - American Math Society

APS - American Physical Society

ASM - American Society for Materials

Chr APS-DFD – Past Chairman of APS Division of Fluid Dynamics

FIM - Fellow of the Institute of Materials

FInstP - Fellow of the Institute of Physics

FREng - Fellow of the Royal Academy of Engineering

IAE - Institute for Advancement of Engineering

Inst. Mater - Institute of Materials

Inst. Phys - Institute of Physics

KDI – Knowledge and Distributed Intelligence

NAE - National Academy of Engineering

NSF – National Science Foundation

PYI – Presidential Young Investigator

URI – University Research Initiative

Campus-wide Enrollment at Selected Private Research Universities (latest years)

Enrollment	USC	Chicago	Columbia	Cornell	Harvard	NYU	NWU	Penn	Stanford	Syracuse
Undergraduate	15,883	3,996	6,950	13,658	6,649	18,628	7,669	9,863	6,731	12,464
Graduate	13,049	8,459	13,721	5,601	11,879	18,522	5,791	8,187	7,608	5,608
Total	28,932	12,455	20,671	19,259	18,528	37,150	13,460	18,050	14,339	18,072
% Graduate	45%	68%	66%	29%	64%	50%	43%	45%	53%	31%

USC AE Graduation Statistics

Degrees Conferred

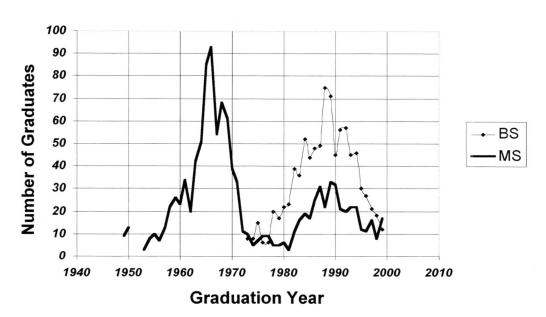

Degrees Conferred

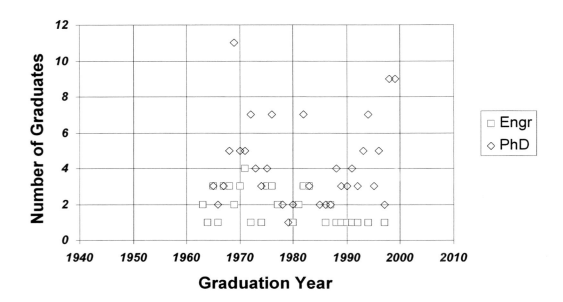

Revision 8/19/03

Chapter 40

Aeronautical Science and Engineering
at the University of California
Davis, CA

Ronald A. Hess
Professor and Vice- Chair
Department of Mechanical and Aeronautical Engineering

Birth of the Program

Aeronautical Engineering became an accredited program at the University of California, Davis in 1984. It was the first accredited Aeronautical/Aerospace Engineering program in the University of California system. While a 20 year period of accreditation can be viewed as brief in comparison to some of the storied programs in the nation, one must bear in mind that the UC Davis College of Engineering, itself, is barely 40 years of age. Although the early aeronautical engineering program at Davis was small in size and enrollment, the department takes pride in the fact that one of its early graduates (1978) was Steve Robinson, currently a NASA Payload Specialist for the Space Shuttle Orbiter and a veteran of two shuttle missions.

Doctors Paul Moller and Bruce White formed the core of the pre-accredited program in the late 1970's. Shortly after Dr. Moller left the faculty for a position in industry, Dr. Paul Migliore joined the fledgling "aero" group within the department. In 1982, this author also joined the faculty. It was with these three faculty members that the Aeronautical Science and Engineering program at Davis was accredited and a formal Division of Aeronautical Science and Engineering was formed within the department. At its inception in 1984, the aero program was administered by the Department of Mechanical, Aeronautical and Materials Engineering (MAME). In early 1990s, the Material Sciences group joined what was then the Dept. of Chemical Engineering and MAME became the Department of Mechanical and Aeronautical Engineering (MAE).

The College of Engineering at UC Davis is the second largest undergraduate engineering college in the UC system, with only UC Berkeley enrolling larger numbers of engineering undergraduates. The MAE department is the third largest of the seven engineering departments at Davis. Undergraduate enrollment in MAE is currently over 650 students, with about one-third of these being Aeronautical Science and Engineering majors.

A Growing Faculty

The Division of Aeronautical Science and Engineering program has grown from three members in 1984 to nine members at this writing. The current members of the Division are listed below in alphabetical order below, along with a very brief description of the research areas of expertise.

Jean-Jacques Chattot, Professor, Computational Fluid Dynamics
Roger Davis, Professor, Propulsion
Mohammed Hafez, Professor, Computational Fluid Dynamics
Ronald Hess, Professor, Aircraft Dynamics and Control

Sanjay Joshi, Assistant Professor, Space Robotics
Lawrence Rehfield, Professor, Composite Structures
Nesrin Sarigul-Klijn, Professor, Aircraft Structures and Aeroelasticity
Case van Dam, Professor, Applied Aerodynamics
Bruce White, Professor, Environmental Aerodynamics

In addition to the faculty above, Dr. Joseph Steger joined the Division in 1989, with expertise in the area of computational fluid dynamics. Although he passed away in 1992, Prof. Steger's contributions to the aero program at Davis and to the entire field of computational fluid dynamics will not be forgotten.

A visit to the department's website http://www-mae.engr.ucdavis.edu/ will provide much more detail concerning the activities and research interests of the faculty listed above.

A Growing Program

As the title of our program suggests, we have traditionally emphasized "aeronautical" as opposed to a "aerospace" engineering. This emphasis has been deliberate. The formative years of the program occurred in the mid-to-late 1970s, and at that time the department faculty agreed that aeronautics best suited the department's research and teaching strengths. This emphasis continued through the 1980s and 1990s. The aero faculty's recognition of the importance of introducing each and every aero student to the practical aspects of flight led the department's purchase of a Cessna 182 in the early 1990s. The fact that the university owns and operates an airport located on campus made the use of this aircraft in undergraduate course work both attractive and efficient. The integration of real flight experience into the classroom was a major benefit to both the aeronautical and mechanical engineering programs. Unfortunately, a substantial increase in insurance premiums forced the department to sell the aircraft in 1997.

The MAE department Cessna 182 with Prof. James Baughn (center) and aero students

The loss of the department aircraft was somewhat compensated by the addition of a new experimental wind-tunnel facility in 1997. The tunnel, with a 3 ft x 4 ft x 12 ft test section and a maximum velocity of 150 mph, is used both for instructional and research activities and operates under the direction of Prof. Case van Dam.

An aircraft model in the MAE wind tunnel

This facility serves to complement the department's large environmental wind tunnel that emulates the earth's boundary layer. This facility, built "in-house" in 1980, is used to study the effects of winds on buildings and other structures. Under Prof. Bruce White's direction, this facility was employed to optimize the orientation of Pac Bell Park, the new home of the San Francisco Giants baseball team.

Prof. White in the MAE environmental wind tunnel with a model of Pac Bell Park

The Unique "Double Major"

The benefits of being part of a joint department are exemplified by the fact that aero students at Davis have the opportunity to pursue a double major in Aeronautical *and* Mechanical Engineering. As opposed to a joint major, the students pursuing the aeronautical/mechanical double major must satisfy the requirements of both degrees. The double major grew out of a request that an aero student made early in the program's history. By carefully selecting elective courses, the student found that the requirements for both the Aeronautical Science and Engineering and Mechanical Engineering degrees could be met. The student petitioned the college to allow the official transcript to indicate this and the "double major" was born. To this writer's knowledge this option is unique in aero programs nationwide. Indeed, the great majority of aero students at UC Davis are double majors. This feature optimizes the employment opportunities of the graduates in that they are equipped to weather the well-documented cyclical nature of aerospace hiring.

Student Projects for Aeronautical Science and Engineering Students

Since its inception, the Aeronautical Science and Engineering program has encouraged student projects as a means of bridging the gap between theory into practice. An example of this project-oriented education is the annual "Aerobrick" design project, initiated in 1994 by Prof. Nesrin Sarigul-Klijn with recent participation by Prof. Jean-Jacques Chattot. This yearlong project allows students to obtain hands-on experience in the design of a radio-controlled aircraft whose purpose is to lift a maximum payload. The students are required to conceive, design, fabricate and test the aircraft. This project integrates all the major sub-disciplines in aeronautical engineering including, aerodynamics, structures, propulsion, and stability and control. After completing the project the students typically compete in an international competition sponsored by the Society of Automotive Engineers.

The 1998 "Aerobrick" team and aircraft. Prof. Sarigul-Klijn is standing in the first row and Prof. Chattot is third from the left in the second row.

The Graduate Program

The graduate program in Aeronautical Science and Engineering has grown apace with that of the undergraduate program. In our combined department, there is considerably less distinction made between the aeronautical and mechanical engineering graduate programs as compared to the undergraduate programs. Currently there is no separate listing for aeronautical and mechanical engineering courses at the graduate level, all now bearing the MAE prefix. The nature of the student's course work and the thesis/dissertation clearly define the interests and capabilities of the graduate. Students completing graduate degrees are awarded an MS or Ph.D. in Mechanical and Aeronautical Engineering.

The historic emphasis on aeronautics in the UC Davis program is giving way to a curriculum that includes topics on aero*space* engineering. This evolution was prompted by the initiation of a *SpaceED Program* within the department in 2001. This program was made possible by a grant from the California Competitive Space Grants Program and builds upon existing courses at the graduate level in the department. An executive committee for the SpaceED program has been formed consisting of Profs. Sarigul-Klijn , Bruce White and Fidelis Eke. Professor Eke is from the mechanical engineering side of the department. Members of the Aeronautical Science and Engineering Division complete the list of participants in this program.

Enrollment Trends

Historically, undergraduate enrollments in Aeronautical Science and Engineering at UC Davis have paralleled those of other aero programs nationwide. Once again the advantage of being part of a joint department comes into play here as transient reductions in aero enrollments are accommodated by increases in the number of students pursuing mechanical engineering degrees. Conversely, surging aero enrollments have offset falling numbers of mechanical engineering enrollees in the past. For the past two years, freshman enrollments have been nearly identical for the Aeronautical Science and Engineering and Mechanical Engineering programs, with enrollments of 60-70 students for each.

The Future

Engineering Colleges in California are roughly half the size of their counterparts in the rest of the country. This means that California continues to be a substantial net importer of engineers. This fact bodes well for the growth of these institutions within the state. The College of Engineering at UC Davis stands poised to take a major step forward in the coming decade to earn a place among the leading engineering institutions in the state and nation. The Division of Aeronautical Science and Engineering within the Department of Mechanical and Aeronautical Engineering is eager to participate in this projected growth in size and reputation.

Chapter 41
Aerospace Engineering in Buffalo -- The X-factor

A. Patra[1] , D. Taulbee[2] , and W. J. Rae[3]
Department of Mechanical and Aerospace Engineering
University at Buffalo, Buffalo, NY 14260

Aerospace and Buffalo have had a long association. This history and tradition provides a rich fountain of inspiration and resources for the Aerospace engineering program. Hence, it is appropriate to start this article with a description of this history. Beyond, just providing nice stories it provides us a long list of distinguished pioneers who go back and forth from industry to the class room. This is our unique strength – the X- factor in our program.

I. Aerospace and Buffalo -- a long history

From the earliest days of flight, the Buffalo-Niagara region of Western New York has been at the center of aerospace research, development, and production. This focus began with names like Glenn H. Curtiss and flourished during World War II in the form of companies such as the Bell Aircraft Company and the Curtiss-Wright Airplane Company. In the late 1800s, a group of aviation enthusiasts in Buffalo began what became, in 1906, the Aero Club of Buffalo. It was formally incorporated in 1910, as the first in the USA.

In 1915 Curtiss moved his company (which had begun in Hammondsport, NY) to Buffalo as the Curtiss Aeroplane Company, where he developed the R-model airplane which was the forerunner of the famous Curtiss JN-1 ("the Jenny"), and at the close of World War I built the four NC-4 flying boats which were eventually successful in circling the globe. Flying boats were also designed and built by Consolidated Aircraft, which had come to Buffalo in 1924. The last of these aircraft was the PBY of World War II fame.

In 1940, the Curtiss-Wright Airplane Division built a very large factory on property adjacent to the airport, and during the war years designed and built thousands of aircraft, including the P-40 fighter and the C-46 cargo plane. The Curtiss-Wright Airplane Company was unique in establishing its own research department, directed by Dr. Clifford C. Furnas. At the end of the war this department was incorporated as the independent, not-for-profit Cornell Aeronautical laboratory, which carried on major research and development efforts in such areas as Wind-Tunnel technology, Flight Controls, Radar, Electronic Warfare, Dynamics and Crash testing of Road Vehicles, Hypersonics, and numerous other fields. It also gave birth to some 50 Western New York high-tech companies, beginning with Moog Servocontrols. Its founder, Dr. Furnas, later became the founding president of the University at Buffalo, State University of New York.

Consolidated Aircraft, headed by Major Rubin Fleet, came to Buffalo in 1924. In March 1928, Major Fleet persuaded Lawrence D. Bell to join his operation, and seven years later Bell left to found the Bell Aircraft Corporation on July 10, 1935. During World War II that company grew to over 30,000 employees, who produced thousands of aircraft, including the P-39 Airacobra, the

---

[1] Associate Professor; Program Director, Aerospace Engineering
[2] Professor and Chair
[3] Distinguished Teaching Professor Emeritus

P-63 King Cobra, and the first jet-powered fighter in the USA, the P-59 Airacomet. The company continued in the post-war years to design and build a remarkable series of innovative aerospace vehicles, beginning with the rocket-powered X-1, the first airplane to break the speed of sound. This was soon followed by the X-1A and X-1B, as well as the X-2 and X-5 supersonic research airplanes.

In addition to theses supersonic aircraft, Bell also pioneered development of Vertical Takeoff and Landing (VTOL) aircraft, such as the Air Test Vehicle, the XV-3 VTOL aircraft, the twin-jet X-14, and the X-22, whose use of four ducted fans has provided decades of research on the flight of VTOL aircraft. The post-war years also saw Bell's expansion into the fields of helicopter design and production

Figure 1 Bell X1 manufactured at Buffalo's Bell   Aerospace

Bell's activities in rocket propulsion led to their involvement with the Rascal, Meteor, and Nike missile programs, and during the 1960s to a number of space-propulsion systems that played key roles in the US Space Program. The two most important of these were the Agena engine, whose outstanding reliability record made possible the development of orbital rendezvous techniques, and to the Lunar Ascent engine - the one component of the Apollo program which *had* to be successful every time - and was.

Many other important aerospace developments were produced by Bell, including all-weather automatic landing systems, air-cushion technology for surface vehicles and for landing systems of airborne vehicles, the Lunar-Landing Training Vehicles, the Rocket-Belt back-pack flight system, the use of flexible bladders internal to rocket-propellant tanks to assure restart capability, and all of these were accompanied by advances in instrumentation, data acquisition, and computer technology.

The litany of aerospace achievements in Western New York contains many more items of equal importance. Typical of these are Atlantic Research Corporation, where many of the Bell rocket programs have continued, Scott Aviation, Irving Airchute Co, which produced thefirst successful descent in a free-type parachute, the Argonaut Aircraft Co, Gwinn Aircar Co, Wendt Aircraft Corp, Aviation division of Carborundum Co, and the Buffalo Forge Company, whose Chief Engineer Willis H. Carrier is recognized as the founder of Heating, Ventilation, and Air-Conditioning science. And not least of local accomplishments is Buffalo's Burgard Vocational High School, the first federally certified Aviation Mechanics program in a public school in the USA.

II. Aerospace Engineering at University at Buffalo

III.1 History and Background

Given the aerospace roots of Dr. Furnas (the first President of the school), it is not surprising that Aerospace Engineering has long been a key discipline in the School of Engineering and Applied Science at the University at Buffalo. The School of Engineering began its life in 1946 as part of the University *of* Buffalo, the private university founded in 1846 by Millard Fillmore, who later served as president of the United States. When the State University of New York was founded in 1962, the University of Buffalo became one of its four University centers, designated as the University *at* Buffalo.

In its early days, a number of Programs in the School of Engineering were clustered in what was known as the Division of Interdisciplinary Studies and Research, headed by Dr. Irving Shames. Aerospace Engineering was included in this grouping, along with Engineering Science, Nuclear Engineering, and Bioengineering. As interest in aerospace grew, a decision was made in 1983 to merge it with Mechanical Engineering, thus forming the present department of Mechanical and Aerospace Engineering (MAE). In addition the to B.S. degree, the MAE Department offers M.S. and Ph.D. degrees in Mechanical and Aerospace Engineering.

Aerospace enrollment has varied over the years, reflecting the large variations that have occurred in the aerospace industry, and graduating classes have ranged in number from several dozen to seventy or more. However, as the only Aerospace program in the State University of New York System we have a special role. We are often the only affordable Aerospace program for a large number of talented students in the North-eastern United States. Following the trough in enrollments in the mid 1990's enrollment has gone up and been stable. Figure 2 shows the total enrollments of Aerospace Engineers over the years 1994-2002.

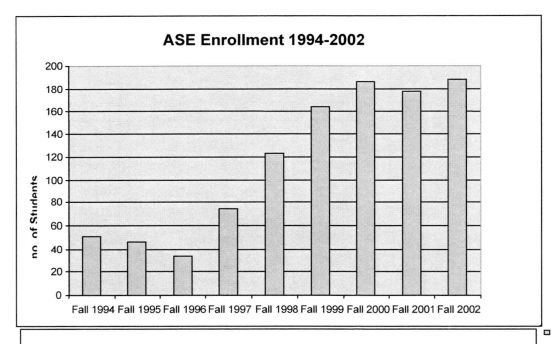

Figure 1 Aerospace Enrollments at the University at Buffalo

II. 2 Curriculum

Our program emphasizes a mix of aeronautical and astronautical topics with the focus on aeronautical aspects. Since 2001 we have also added to our offerings in astronautics related topics with additional offerings in Guidance and Control etc. using recently hired faculty. Table 2 lists the classes with the standard characterization required by the Accreditation Board for Engineering (ABET).

The AIAA has provided two criteria that aerospace engineering curricula need to satisfy. We list below the criteria and our courses that meet these requirements.

1. *"demonstrate that graduates have a knowledge of aerodynamics, aerospace materials, structures, propulsion, flight mechanics and stability and control"*. Aerodynamics is covered in MAE424 Aerodynamics, aerospace materials are taught in the sequence of MAE381/MAE382 Materials class and laboratory sequence. MAE415 Structures I and MAE416 Structures II provide ample coverage of aerospace-structure- related topics. MAE422 Gas Dynamics, MAE 423 Propulsion are used to introduce students to propulsion related issues while MAE 435 Flight Dynamics provides coverage of flight mechanics and control. A required technical elective is used by most graduates to enhance knowledge of one of the above topics.

2.*"demonstrate that graduates have design competence which includes integration of aeronautical or aerospace topics"*. We have integrated a sequence of classes formally dealing with design starting with MAE277 Intro to MAE Practice in the sophomore year followed by MAE377 Product Design in the Junior Year and the Senior level MAE451 Design Processes and MAE434 Aircraft Design. The MAE 434 Aircraft Design class is the formal capstone class that promotes integration of all the aeronautical and astronautical topics. Furthermore, most junior/senior level classes (e.g. MAE 415 Structures incorporate significant design component through the use of class projects) .

II. 3 Laboratories and Facilities
a. General Facilities

The University at Buffalo occupies two campuses in the Buffalo area. The South Campus is located at the northeast corner of the City of Buffalo. It is the original campus of the University and now serves largely as a Health Sciences Campus, containing the Medical School, Dental School, etc. The North Campus is now the heart of the University (particularly for undergraduates) and is located in the suburban Town of Amherst. Construction on the North Campus began in the 1970s and the Department of Mechanical and Aerospace Engineering began to occupy its space in 1978. The department is now primarily contained in the ten-story Furnas Hall and the adjacent Jarvis Hall (named after alumnus and Challenger astronaut Gregory B. Jarvis). The campus itself is now essentially completed although additional space for student housing is still being added.

Because of the relatively new North Campus, the overall environment, classroom and lecture hall space across campus is reasonably modern. The recent addition of computer stations to many classrooms and lecture halls provides the ability to use the most modern visual aids and to perform computer demonstrations when necessary. Yahoo recently ranked The University at Buffalo the tenth most wired university. The Mechanical and Aerospace Engineering Department space for offices, conference and seminar rooms and laboratory facilities is similarly up-to-date and provides a pleasant atmosphere conducive to learning.

561

b. Department Laboratories

The departmental teaching laboratories are briefly described here. They consist of the Materials Laboratory, the Systems and Instrumentation Laboratory, the Transport Laboratory and the PC Laboratory. Modern instrumentation and data acquisition equipment are commonly used, especially where necessary to illustrate principles and capabilities. The descriptions below provide a basic sketch of the nature of the lab and the equipment status for each of our laboratory facilities:

i)   *Materials Laboratory*:

This laboratory supports the required laboratory course MAE 385 (Engineering Materials Laboratory) which follows and supplements the lecture course MAE 381 (Engineering Materials). It includes tensile testing, hardness measurements, the microstructure of steel, the relationship between microstructure and properties, and the character of aluminum alloys. Students are exposed to the study of material characteristics and properties and develop skills in sample preparation and testing. Four formal reports are required. Major equipment includes tensile and hardness testing machines, polishers for sample preparation and mounting, and furnaces for heat treating. The equipment in this laboratory is well maintained and effective.

ii) Systems and Instrumentation Laboratory:

This laboratory supports the required courses MAE 334 (Instrumentation and Computers) and MAE 340 (Systems Analysis) both of which contain laboratory sessions integrated into the course. It also supports our elective controls courses MAE 443 and MAE 444.
The MAE 334 course has two lecture meetings and one laboratory session per week. The course introduces instrumentation and data acquisition. Formal written reports are required and graded for both language and technical content. Thermocouples, pressure transducers, motion transducers, function generators, analog/digital converters and PC data acquisition are the key components of the laboratory.

The MAE 340 course has three lecture meetings and one laboratory session per week. The laboratory material is coordinated with the course lecture material and studies the characteristics of dynamic system components including springs, dampers, fluid capacitors, etc. The characteristics of DC motors are studied along with motor transient behavior. Hydraulic supplies, function generators, servovalves, electronic power amplifiers, pressure and motion transducers, DC motors and PC data acquisition are the important laboratory components.

ii)   *The Transport Laboratory*:

This laboratory services our required course MAE 338 (Fluid and Heat Transfer Laboratory) which follows and supplements the lecture courses MAE 335 (Fluid Mechanics) and MAE 336 (Heat Transfer). Student groups each complete a series of experiments from the nine experiments available in the laboratory and including a mix of fluid mechanics and heat transfer topics. Preparation quizzes must be passed before labs may begin. A lab notebook is required. After data collection and analysis, the lab notebook is submitted and a checkout quiz is completed. The experiments are mostly custom-made and use relatively simple instruments including manometers, flowmeters, stop watches and pitot tubes. Hot film velocity probes used in a turbulent jet experiment are the most sophisticated instruments in the lab. This lab functions well, using experiments and instrumentation emphasizing "first principles".

iii) *The Aerodynamics Laboratory*:

This laboratory services the course MAE 424 Aerodynamics. The course consists of three hours of lecture per week and associated laboratory experiments. Most of the experiments involve a wind tunnel out fitted with a six component force balance for doing airfoil and airplane model studies. An extended test section is being developed for wake and boundary layer experiments. Automated data acquisition is accomplished with a PC and Labview software.

v) *The Personal Computer Laboratory*

: In addition to the required laboratories above, the MAE Department shares an instructional PC Laboratory with the Chemical Engineering Department. This lab contains thirty well-configured Dell PCs and supplements the SEAS-wide computing facilities which are based on Unix workstations. It also contains a demonstration computer with projector that can be used directly for instructional purposes. Our PC Lab became operational in Fall 2000 and provides platforms for the most popular PC engineering software. The lab is formally used in the instruction of MAE 377 (Product Design in a CAD Environment) where it operates like a design studio. And in MAE 477 (CAD – Applications) and MAE 473 (CAD – Graphics) using ProEngineer software, ProMechanica and Visual C++. The laboratory is also used informally to support word processing, web browsing, spreadsheeting and other computational needs for a variety of courses. The Science and Engineering Node Computing Services Group provides support for the facility.

II.   Faculty

Our faculty are diverse and well qualified. Most are active in cutting edge research in aerospace related fields. A common set of faculty serve both the Aerospace and Mechanical Engineering programs – thus enabling us to have more faculty than an isolated Aerospace program would ordinarily be able to support.

Table 1 Faculty and their expertise

Faculty	Expertise
Bloebaum, Christina	Design, Optimization, Visualization
Chopra, Harsh	Material Science
Chung, Deborah	Material Science, Smart Materials
Crassidis, J	Guidance and Control
Desjardin, Paul	Propulsion, Computational Fluid Dynamics
Felske, James	Thermal Sciences
Krovi, Venkat	Mechatronics, Robotics
Kesavadas, T	Manufacturing, Visualization and VR
Lewis, Kemper	Design, Optimization
Liu, Ching	Applied Mathematics
Madnia, Cyrus	Propulsion, CFD
Mayne, Roger	Design, CAD, Control Systems
Medige, John	Biomechanics
Meng, Hui	Fluid Mechanics
Mollendorf, Joseph	Thermal Sciences, Aircraft Design
Mook, D. Joseph	Systems and Controls

Patra, Abani	Structures, Computational Mechanics
Singh, Tarunraj	Control Systems,
Soom, Andres	Design, Tribology
Yu, C.P.	Fluid and Aerosal Mechanics
Taulbee, Dale	Fluid Mechanics, Turbulence
Wetherhold, R.	Structures, Mechanics of Composites

II.   5 AIAA Student-Chapter

We have an unusually active AIAA student chapter on campus. For most aerospace engineering majors it provides an opportunity to learn and have fun while doing so. Each year the chapter participates in several major competitions like the Design/Build/Fly and the Microgravity (see Figure 3) contests. They also undertake several field trips every year to places of Aerospace interest. Of course the ready accessibility of Veridian Flight Research, the Niagara Fall Air Force Base, Moog Aerospace , the Niagara Aerospace Museum, and others provides many local opportunities for fascinating field trips.

III Final Thoughts

After more than 50 years of aerospace-related education, UB aerospace still has the X-factor of a glorious tradition inspiring a new generation of graduates to go out and conquer air and space. Our graduates work all over the country from the Lockheed Skunk works out west, to right here at Moog Aerospace, and to Pratt and Whitney on the East Coast. Over the last few years we have replaced almost half the faculty, our enrollments have stabilized at high levels and we have revamped the curriculum. With many new investments in laboratories and faculty, the best, as they say, is yet to come.

Figure 2 AIAA student chapter activities -- Microgravity project and Design/Build/Fly project participants

Table 2. Basic-Level Curriculum
Aerospace Engineering (For students entering Fall 2002)

Year/ Semester	Course and Number	Math & Basic Sciences	Engineering Topics	General Education	Other
1st year 1st semester	EAS 140 Engineering Solutions		3		1
	CHE 107 General Chemistry I	4			
	MTH 141 Calculus I	4			
	ENG 101 Writing				3
	General Education (1)			3	
1st year 2nd semester	MAE 177 Intro to CAD*				2
	EAS 230 Higher Level Language		1		2
	PHY 107 Physics I	4			
	MTH 142 Calculus II	4			
	ENG 201 Adv. Writing				3
2nd year 1st semester	MAE 277 Introduction to MAE Practice		3		
	EAS 204 Thermodynamics		3		
	EAS 207 Statics		3		
	PHY 108 Physics II	3			
	PHY 158 Physics II Lab	1			
	MTH 241 Calculus III	4			
2nd year 2nd semester	EAS 200 EE Concepts		3		
	EAS 208 Dynamics		3		
	EAS 209 Mechanics of Solids		3		
	Science Elective**	4			
	MTH 306 Differential Equations	4			

3rd year 1st semester	MAE 334 Instruments and Computers		3		
	MAE 335 Fluid Mechanics		3		
	MAE 376 Applied Math for Mech & Aero Eng	3			
	MAE 377 Product Design w CAD		3		
	MAE 381 Engineering Materials		3		
	General Education (2)***			3	
3rd year 2nd semester	MAE 422 Gas Dynamics		3		
	MAE 336 Heat Transfer		3		
	MAE 340 Systems Analysis		4		
	Applied Math Elective****	3			
	MAE 385 Engineering Materials Lab		1		
	General Education (3)***			3	
4th year 1st semester	MAE 338 Fluid/Heat Transfer Lab		1		
	MAE 451 Design Process & Methods		3		
	MAE 423 Propulsion		3		
	MAE 424 Aerodynamics		4		
	MAE 436 Flight Dynamics		3		
	MAE 415 Structures I		3		

4th year 2nd semester	MAE 434 Aircraft Design		3(3)		
	MAE 416 Structures II		3		
	Technical Elective		3		
	General Education (4)***			3	
	General Education (5)***			3	
TOTALS - BASIC-LEVEL REQUIREMENTS		38	63	15	14
OVERALL TOTAL FOR DEGREE	130 hrs				
PERCENT OF TOTAL		29 %	48 %	12 %	11 %
Total must satisfy one set	Minimum semester credit hours	32 hrs	48 hrs		
	Minimum percentage	25 %	37.5 %		

Chapter 42

FIFTY PLUS YEARS
OF ENGINEERING EXCELLENCE:
DEPARTMENT OF AEROSPACE ENGINEERING
AT THE UNIVERSITY OF MARYLAND

DARRYLL PINES
ALFRED GESSOW
WILLIAM FOURNEY
JOHN ANDERSON
JEWEL BARLOW
MARK LEWIS
DAVE AKIN
INDERJIT CHOPRA
ROBERT SANNER
NORMAN WERELEY
CHRISTOPHER CADOU
BEN SHAPIRO

**Department of Aerospace
Engineering
3181 Glenn L. Martin Hall
University of Maryland
College Park, Maryland**

"The way to build an aircraft or to do anything else worthwhile is to think out quietly every detail, analyze every situation that may possibly occur, and, when you have it all worked out in a practical sequence in your mind, raise heaven and earth and never stop until you have produced the thing you have started to make."

Glenn L. Martin (1886–1955)

INTRODUCTION

Advancing flight …
Exploring space …
Designing new smart material systems,
Vehicles and devices …

These are the abiding interests of faculty and students in the Department of Aerospace Engineering at the University of Maryland. Since the department's founding in 1949, we have been at the forefront of aerospace science—conducting important research, developing new technologies and educating new generations of talented engineers to study and improve the vast world of flight. In all of our endeavors, we continue to reach beyond traditional engineering research and education. The following pages explore the department's history, growth and significant milestones during the last half-century.

PRE-FLIGHT CHECK
The Years Before 1949

The end of World War II brought both challenge and opportunity to the field of aeronautical engineering in the United States. The emerging Cold War put America's engineers on short notice to develop prevailing military aircraft to counter those of the Soviet Union. Also, advanced research and development in the new area of long-range intercontinental missiles would become paramount. Military matters aside, America's postwar public thirsted for new technological advances in commercial flight—better, faster, farther, higher could be heard resonating across the nation.

Glenn L. Martin, one of the nation's first aviation pioneers, saw early on the value of research and education in the aeronautical sciences. Martin himself lacked a formal technical background, yet the company he founded in 1912 would grow to become the leading airplane

Glenn L. Martin seated in 1ˢᵗ Generation Aircraft of the Martin Aircraft Corp.

manufacturer in the nation. From the beginning, Martin continuously hired skilled engineers to design his planes and talented managers to run his factories. The Martin Aircraft Company provided training and experience to a remarkable number of future aviation manufacturers: William Boeing, Donald Douglas, Lawrence Bell and James S. McDonnell all worked for Martin before forming their own companies.

Martin was pivotal in developing an aerospace engineering program at the University of Maryland. In 1944- the 50th Anniversary of engineering at the University of Maryland- Martin made a gift of $1.7 million to the university to establish instruction and research in the

Bomber On Engineering Field in Front of Martin Hall in 1950

Photo of Engineering Classroom Building in 1949

aeronautical sciences. A second gift of $800,000, named in honor of Martin's mother, Minta Martin, was made the following year. Today this endowment is a major source of research funding for the A. James Clark School of Engineering.

The State of Maryland, in support of Martin's efforts and foresight, appropriated funds in the late 1940s to construct four new buildings at the university including a state of the art wind tunnel. In recognition of his philanthropic gifts and pioneering spirit in the field of aeronautical engineering, the University of Maryland in 1949 designated the College of Engineering as the Glenn L. Martin College of Engineering and Aeronautical Sciences. That same year, a newly formed Department of Aeronautical Engineering began providing full-time instruction in the aeronautical sciences.

THE DEPARTMENT TAKES FLIGHT
The Sherwood Era: 1949–1968

Before 1949, aeronautics research and instruction at the University of Maryland were limited in scope and were carried out by but a few dedicated individuals—most notably Professor John Younger of the mechanical

569

engineering department. Younger shared a similar vision to that of Glenn L. Martin—that is, that commercial passenger flight on a large scale would soon become a reality.

Prof. Wiley Sherwood standing inside the Glenn L. Martin Wind Tunnel

In fall of 1949, the aeronautical sciences option was separated as a discipline from the mechanical engineering department. A. Wiley Sherwood was chosen as chair of the new department, beginning what is commonly referred to within the department as the Sherwood Era. During these formative years, most of the department's activity was focused on developing a quality undergraduate curriculum in aeronautical engineering.

With the launch of Russia's Sputnik in 1957, the department, along with the rest of the nation, took a keen interest in aerospace flight. The name of the department was changed from aeronautical engineering to aerospace engineering that same year.

To support instruction in the course work necessary for an aerospace engineering degree, Sherwood used lecturers and faculty from the nearby Naval Ordnance Laboratory and the Institute for Fluid Dynamics.

During those years, the emphasis was on teaching. Professor Gerald Corning taught leading edge courses in aircraft design and was the author of a number of nationally used texts in subsonic and supersonic aircraft design. The recently constructed and self-supported Glenn L. Martin Wind Tunnel, under the direction of Donald Gross, was widely used in service testing by the automotive and aviation industries.

One of the most significant accomplishments during this time was the establishment of the master's and doctoral degree programs in aerospace engineering. This led to the department's first master's degree graduate, Dale Scott '50 (deceased). After graduation, Scott went on to a successful engineering career at the Martin Aircraft Co., NASA Goddard Space Flight Center and later as a private consultant. The first doctoral degrees in aerospace engineering were awarded in 1963 to Irvin Pollin and John Nutant.

TURBULENCE AHEAD
The 1970s: Low Enrollments Amid a Decline in the Aerospace Industry

The 1970s proved a difficult time for the aerospace engineering community nationwide. A national decline began in 1969 after Neil Armstrong became the first man to set foot on the moon. In much of the public's eye—and also within the political sphere that funded NASA—America had won the space race and further space exploration was not needed. Consequently, NASA's manned space flight program shrunk precipitously.

In addition, the U.S. Air Force had reached a plateau in its deployment of intercontinental ballistic missiles and began to cut back on the testing and production of ICBMs. As if this were not enough to diminish research and production in the aerospace field, the commercial transport business had also peaked. Boeing, the nation's largest airplane manufacturer, laid off a large number of professional employees. This was the period of the famous billboard in Seattle—where Boeing maintained its headquarters and a large work force—that read: "Will the last person leaving Seattle please turn out the lights."

As a result, the early 1970s saw the bottom drop out of the job market in aerospace engineering. This led to an almost catastrophic drop in university enrollments in the field. From 1970–73, the aerospace engineering program at Maryland had a 70 percent drop in enrollment—very similar to the enrollment decline in aerospace studies at other universities across the nation.

This was not the best environment for Professor John Anderson to assume his position as new department chair in May of 1973. Among his most serious concerns were a low faculty morale; poor "weighted credit hour" statistics associated with low enrollment—with the consequent danger of the department being eliminated by the university; and the virtual non-existence of funded research programs in the department. From 1973–80 the department made progress to reverse these adverse trends in enrollment. It began by making the university community more aware of the nature and importance of aerospace engineering as a separate department. Also, the importance and value of funded faculty research was encouraged. On a sad note, the department lost Gerald Corning due to an untimely death. Corning was a well-known airplane design professor and author of two books on vehicle design.

Robert Rivello, a well-respected professor of aircraft structures who had served as acting chair of the department, retired to go full time with the Johns Hopkins Applied Physics Laboratory. Former department chair A. Wiley Sherwood also retired during this period. With these changes, the "old guard," who had been so instrumental in the early success of the department, were gone. The department continued to build research programs—predominantly in the area of aerodynamics and propulsion—and by the end of the 1970s had established a program of hypersonic aerodynamic research through support of the NASA Langley Research Center.

Subsequently, a major general aviation program with Langley was initiated, resulting in both computational and wind tunnel research in that area. Finally, a significant administrative change was made. The Glenn L. Martin Wind Tunnel, which had previously reported directly to the dean of engineering, was now made part of the Department of Aerospace Engineering. Jewel Barlow was made director of the facility, which greatly augmented the research funding in the department. The wind tunnel expanded its mission to include academic research in addition to its usual corporate and government-sponsored investigation. In addition to jet aircraft research, the list of the more than 1,500 projects to date includes helicopter, submarine, automobile and sailboat design testing and development.

NEW IDEAS AND A NEW ERA
The 1980s: Emergence of Rotorcraft Research and Education

With the appointment of Professor Alfred Gessow as chair of the department in 1980, the next decade saw the department experience tremendous growth in undergraduate and graduate enrollments and degrees awarded; in research expenditures and scholarly presentations

The Alfred Gessow Center for Rotorcraft Education and Research, established in 1982 as a U.S. Army Center of Excellence in Helicopter Technology, is arguably the leading such rotorcraft center in the world. The center now conducts research in rotorcraft acoustics, CFD, aerodynamics, dynamics, flight mechanics and smart structures applications to rotorcraft.

and publications; and in national and international recognition of its programs. The surge in undergraduate students reflected a national trend, but a major source of growth at Maryland was when Gessow, as new chair of the department, made a commitment to expand the department's research activities and establish new rotorcraft graduate education and research programs in aerodynamics, dynamics, flight mechanics and composite structures.

Gessow came to the university after a long and distinguished career as a researcher and administrator. As one of the world's leading helicopter experts, he provided new theoretical approaches to expand knowledge of helicopter aerodynamics and flight dynamics to improved designs. His textbook on the subject is still in use today, 50 years after its first printing.

During his eight-year tenure as department chair, Gessow's research and contacts with government and university researchers across the nation greatly expanded the department's research and educational activities in rotorcraft, composite structures, hypersonics and space systems. The undergraduate and graduate aerospace programs also achieved national rankings during this period. In 1981, Gessow founded, and for the next 12 years directed, the Center for Rotorcraft Education and Research. The center is now an internationally recognized research center in rotorcraft science and technology with leading-edge research conducted by professors James Baeder, Roberto Celi, Inderjit Chopra, J. Gordon Leishman, Darryll Pines, Frederic Schmitz, Norman Wereley and Anthony Vizzini. Professors Jewel Barlow and Sung Lee were earlier contributors to the center, and Professor Chopra is the current director of the center.

The rotorcraft center boasts one of the most comprehensive graduate programs in the country, and graduates hold high level positions with NASA, the Federal Aviation Administration, the U.S. Army and Navy, and in private industry (Sikorsky, Boeing and Bell) as well as other universities nationwide. The center has awarded more than 200 master of science and 45 doctoral degrees to date.

Funded mainly by extensive NASA, military and industry grants to the rotorcraft center, the department's yearly research expenditures rose during the 1980s from several hundred thousand dollars to $2.3 million. This research income provided much needed operational and teaching expenses for the department, as well as funding for the construction of a number of specialized research facilities. Rotorcraft funds provided for an upgrade to the department's Glenn L. Martin Wind Tunnel; complex articulated and bearingless rotor systems for testing in

An endowed chair named for Alfred Gessow (far left) was established in 1998 through an endowment given by Gessow's son, Jody (left).

that tunnel; two hover test facilities; an invacuo 10 ft. diameter rotor test facility; wide-field shadowgraphy and schlieren flow visualization facilities; and a 3-D laser Doppler system. Spawned by the activities of the Rotorcraft Center and the ONR-URI contract awarded to Sung W. Lee, the Composites Research Laboratory was founded in 1987 by Anthony J. Vizzini. The laboratory has developed extensive capabilities to build and test helicopter blades, perform leading-edge research in composite structures and manufacturing, and interact with local government laboratories and industry.

One indication of how far the department advanced in the 1980s was its rankings compared to other departments in the college and other departments nationwide. Aerospace engineering outperformed all other departments in the college by most measures of productivity per faculty member: undergraduate students; research dollars awarded; and refereed journal articles.

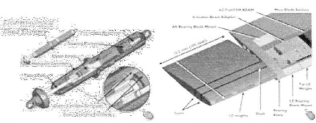

Also during the 1980s, the hypersonics program under John Anderson contributed significantly to the stature of the department. During much of the decade, the department almost single-handedly provided the United States with young professionals with specialized training in hypersonics. Its graduates have all been recruited vigorously by industry and government. Mark J. Lewis joined the faculty in 1988 and greatly expanded the program and steered it in new directions and toward new heights.

Current research in the department includes development and testing of ER and MR dampers (above) by Professor Norman Wereley and his graduate students. Other research involves development of an active blade tip (above) designed by Professor Inderjit Chopra with assistance from his graduate students.

Graduate Student Mahendra Baghwat makes Laser Velocimetry Measurements of Rotor Blade Tip Vortices (Research of Prof. J. Gordon Leishman)

HIGH FLIGHT
The 1990s: Space—The Final Frontier

The department entered the 1990s in an excellent position to take on new challenges, particularly in its research and graduate programs. The already highly successful rotorcraft team, with the further addition of dedicated and talented faculty, expanded its efforts and funding in its core programs and branched out in the area of smart structures research and applications to rotorcraft.

In 1992, the Army Research Office selected the University of Maryland as one of three University Research Initiatives on multidisciplinary activities in smart structures. Under the leadership of Professor Inderjit Chopra, faculty from throughout the A. James Clark School of Engineering—including those from aerospace, mechanical, electrical and materials engineering—worked together on research projects.

In response to ever-increasing importance of composite materials in aerospace structures, the department has developed a strong program in composite materials. The Composites Research Laboratory is comprised of modern equipment including a computer-controlled autoclave and a vacuum hot press which permit the manufacture, inspection, testing and analysis of composite materials and structures.

In 1995, the Department of Defense awarded Chopra and colleagues at affiliate universities a five-year grant for research on the reduction of rotorcraft vibration and noise through the use of active control and smart structures technology. These two competitively won awards boosted the international reputation of the rotorcraft center to its highest level.

The Composites Research Laboratory expanded its relationship with NASA Goddard in the 1990s, working on programs such as FUSE, MAP, TOPHAT and AIMS. The laboratory currently is designing and manufacturing a composite skirt for the Mars Micro Mission.

The department continued to make other significant strides, including a more comprehensive expansion into high-speed research. This was motivated by a national interest in the X-30 high-speed research vehicle, more commonly known as the National Aerospace Plane (NASP). The promise of NASP was a vehicle that could take off and land like a conventional aircraft, but also accelerate to speeds beyond Mach 25 using air breathing propulsion. NASA realized that to design such a vehicle,

Star Body Configuration for low drag

it was necessary to train engineers in the field of hypersonics, including aerodynamics, propulsion, design, structures, materials and flight dynamics and control. This lead to the creation of the NASA-sponsored Center of Excellence in Hypersonics at the university. Now in its ninth year and under the direction of Professor Mark Lewis, the center's ongoing projects include hypersonic missiles, transatmospheric cruisers, accelerators, SSTO, TSTO and reentry vehicles. Other research activities—focused on propulsion, fluid dynamics, inverse design and vehicle optimization, flight dynamics and control and structures—are conducted in conjunction with industrial partners and government laboratories. An experimental high speed combustion and propulsion capability was added when Prof. Ken Yu joined the department from the Naval Surface Weapons Center in China Lake, California. This "star body" (insert figure) is used in tests for low drag configurations.

While activities in hypersonics extended the research capability of the department to the uppermost region of the earth's atmosphere, the department was still lacking in the area of astronautics research and education. Prior to 1990 there were a limited number of courses devoted to space science and engineering. This changed with the addition of Professor David Akin to the faculty.

The Space Systems Laboratory (SSL), previously located at MIT, moved to the University of Maryland in 1990. The SSL has built eight robots, including the Multimode Proximity Operations Device (MPOD), an orbital maneuvering vehicle or "space tugboat;" the Secondary Camera and Maneuvering Platform (SCAMP), a "floating eyeball;" and Ranger, a robot designed to be capable of replicating a Hubble Space Telescope servicing mission. A modified version of Ranger, Ranger TSX, is awaiting a launch on the Space Shuttle.

The Space Systems Laboratory faculty, staff, and students

The Space Systems Laboratory is a nationally recognized leader in the area of astronautics. Much of the research is centered around a 50-foot diameter, 25-foot deep water tank used to simulate the microgravity environment of space. Maryland's neutral buoyancy tank is the only such facility in the nation that is located at a university. There are currently five robots being tested, including Ranger, a four-armed satellite repair robot. Ranger and its predecessor robot were both constructed in the Space Systems Lab.

In 1992, the Neutral Buoyancy Research Facility (NBRF) was completed with substantial funding from NASA. It is the only neutral buoyancy facility located at a university in the United States, and gives the SSL world-class research facilities. The SSL continues to work cooperatively with other research facilities at the university, including the Human-Computer Interaction Lab and the Autonomous Mobile Robotics Lab. With the addition of Professor Robert Sanner, the SSL began research into advanced spacecraft control systems.

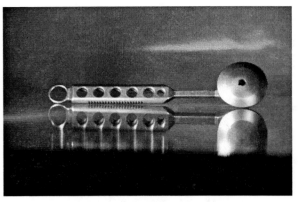

Brian Roberts, a graduate student in the aerospace engineering program, received a grant from NASA to develop a ratchetless wrench for use in the weightless environment of space. The wrench, designed for ease of use while wearing bulky gloves, flew aboard the Discovery shuttle mission that carried Sen. John Glenn into orbit in fall of 1998.

In 1997, NASA Goddard transferred operation of the Solar, Anomalous and Magnetospheric Particle Explorer (SAMPEX) to the Flight Dynamics and Control Laboratory (FDCL) housed within the department. SAMPEX gathers information on high-velocity radiation arriving at Earth from the sun and interstellar space. In collaboration with Mr. Tom Stengle of Goddard Space Flight Center and the Computer Science Corp., the FDCL has developed the capability to provide real-time mission support of SAMPEX as well as other NASA spacecraft. Aerospace undergraduate and graduate students perform orbit determination, orbit prediction, scheduling of ground-station access times, ground tracking, attitude determination and monitoring of the satellite's sensors. Students learn about mission control in the classroom and then put this knowledge to practice on the SAMPEX project.

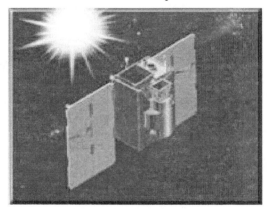

Through collaboration with NASA Goddard, aerospace students provide real-time mission support for several NASA spacecraft, including the SAMPEX satellite.

The SSL is also developing control systems that work well with humans, and can adapt to changing conditions that affect the spacecraft's motion, such as temperature extremes, hardware failures and decreasing mass due to fuel loss. The addition of the SSL and FDCL to the department has had a significant impact on the expansion of the department's astronautics curriculum.

A New Millennium in Aerospace Research and Education
2nd Century of Powered Flight

Entering the 2nd century of powered flight, the department as well as the college has embarked on several bold new research initiatives including the creation of a Small Smart Systems Center, a NASA Space Launch Institute and partnership in the first ever National Institute of Aerospace.

In support of the new research area of small smart systems as well as ongoing activities in space systems, four new faculty members joined the department including Ella Atkins, Christopher Cadou, Benjamin Shapiro and Alison Flatau. With the addition of Professor Atkins, the department's activities in autonomous systems with applications to air and space vehicles has grown considerably. The addition of Cadou and Shapiro has permitted research activities in the department to expand into two exciting research thrusts involving micro-fluidics and micro-combustion. Prof. Flatau has added strength and depth in the area of novel actuators development for a variety of applications..

Complementing these emerging research activities in small smart systems, faculty members in the AGRC have spearheaded research activities involving the design, analysis and development of autonomous "micro" hovering air vehicles. These activities have lead to the development of many unique testing and manufacturing facilities for small scale systems, as well as the development of the first micro-hovering prototype displayed below.

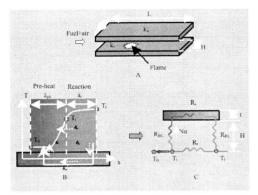

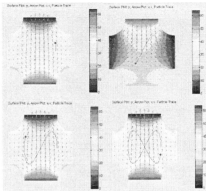

Professor Cadou and his students are working on developing models of flame stabilization in micro-combustors. A flame is stabilized in a microchannel (A) by heat from the reaction zone that conducts upstream thereby pre-heating the incoming reactants to the ignition temperature as illustrated in figure B. Note that conduction upstream through the structure in addition to conduction through the gas is included. Calculations are simplified by representing these pathways using an equivalent thermal circuit as pictured in C.

Prof. Shapiro and his students are working on path control of a single particle in a micro fluidic system at low Reynolds numbers. A micro fluidic cross channel is shown from above. The channel is actuated by 4 pressure actuators (for larger channels) or equivalently by four electrodes that cause electroosmotic flow (for smaller channels). Arrows denote the flow field and the color denotes the pressure or lines of constant voltage. A single uncharged particle such as a cell (the blue dot) is floating in the liquid. The actuators are turned on in such a way that the fluid motion causes the cell to execute arbitrary motion.

In 2001 NASA selected a multi-university team led by Professor Mark Lewis as one of seven University Research, Education and Technology Institutes-URETI. The Institute is working to achieve NASA Aerospace Technology program objectives for access to space, which requires major improvements as reusable launch vehicles. Maryland's URETI research and education goals are focused on the fundamental technologies, analysis, design, integration and optimization of third-generational launch vehicles that can utilize the potential advantages of air-breathing engines on the way to orbit. By treating this class of vehicle as a fully coupled system, the institute will undertake the most challenging, and ultimately, most beneficial problems in launch vehicle design and integration.

"What this accomplishes is not evolutionary, but revolutionary in how we achieve access into space," says Dr. Lewis.

Following the success of becoming one of seven NASA URETI Institutes in 2001, the department became a founding partner in the creation of a National Institute of Aerospace-NIA in 2002 along with the AIAA Foundation, the campuses of Virginia Polytechnic Institute and State University, University of Virginia, North Carolina State University, North Carolina Agricultural and Technical State University, and Georgia Institute of Technology. This institute will be located in close proximity to the NASA Langley Research Center.

The specific goal of the NIA is to become a premier aerospace research and education institute. The NIA will link all six participating universities through the creation of a graduate curriculum. Six prestigious tenured Langley professorships, one at each member university, will be created to foster research into earth system science, advanced aerospace systems architecture, high confidence computational systems, quantum/molecular materials design for sensors, multifunctional design, smart, adaptive aerospace vehicle technology and concept development.

i

Micro-Coaxial Rotorcraft (MICOR) developed in the Alfred Gessow Rotorcraft Center by Professor Pines and students in the rotorcraft center.

As the department's research programs continued to grow, the department and rotorcraft community suffered an enormous loss when Professor Alfred Gessow passed away in the Spring of 2002. As the founder and first director of the rotorcraft center of excellence at Maryland, Prof. Gessow played a large part in improving the research and educational activities of the department. Because of Prof. Gessow's vision and Prof. Chopra's leadership, as well as the committed effort of rotorcraft faculty and students, the center was presented the Grover Bell Award for its research in smart structures technology in 2002 at the annual American Helicopter Forum held in Montreal, Canada.

Aerospace Engineering Undergraduate Program

The Department of Aerospace Engineering housed within the A. James Clark School of Engineering is among the best undergraduate aerospace programs in the nation. With a strong base in physics and mathematics, the curriculum divides in the third year into specializations in aerospace or aeronautics. Both areas are strongly supported by faculty research and unique laboratories. The aerospace undergraduate program has a strong advising system. Each faculty member is assigned 12 to 16 students consisting of a mix of freshmen, sophomores, juniors and seniors who stay with that advisor throughout their four years and meet on a regular basis to talk about curriculum, student activities, research opportunities and career possibilities.

Aerospace Engineering Curriculum

The curriculum to obtain an undergraduate degree in the field of Aerospace Engineering has undergone careful refinement over the last 20 years. Probably the most significant changes have been the reduction in overall credit hours from 130 to 124, the creation of two specialty tracks in the junior/senior years, and the creation of an Honors Program in 1998 to challenge the most academically talented students as well as expose these students to research opportunities. The current curriculum consists of a common set of courses taken by all students and an additional set of courses that provides more specialized training in one of two areas of specialty. Students interested in vehicles that remain in Earth's atmosphere pursue the 'Aero-Track' while students interested in vehicles that leave Earth's atmosphere will pursue the "Astro-Track". The table below summarizes the coursework required for an undergraduate degree in Aerospace Engineering. The sequence begins with training in basic math and science and moves on to more specialized topics by the middle of the junior year. Students begin the sequence of courses in their specialty area in the second semester of their junior year and continue on with these courses in their senior year. In addition, to the technical coursework, students are required to fulfill the University's general education requirements referred to as the 'CORE Liberal Arts and Science Studies Program'. The program is designed to ensure that all students receive a balanced education while they pursue their particular area of interest.

Aerospace Engineering Honors Program

In 1998, under the direction of Prof. Norman Wereley, the department initiated a formal Aerospace Engineering Honors program to provide a rigorous and comprehensive education for a career in technical leadership and scientific or engineering research. The current director is Prof. Ken Yu. Honors course work encompasses the required curriculum for all University of Maryland Aerospace Engineering students, but at an advanced level. Students also have the opportunity to engage in a number of independent research projects. The curriculum is designed to prepare our students for the world of engineering and emphasizes design and hands-on experience. The honors curriculum consist of the following classes:

ENAE 283H: Introduction to Aerospace Systems
ENAE 311H: Aerodynamics I
ENAE 423H: Vibration and Aeroelasticity
ENAE 398H: Honors Research Project

Requirements for Undergraduate Degree

Academic Year	Semester	Course Number	Course Name	Cr
Freshman 30 credits	Fall	ENES 100	Introduction to Engr Design	3
		ENAE 100	The Aerospace Engineering Profession	1
	*ENGL 101 should be taken during the first year	CHEM 135	General Chemistry for Engineers	3
		MATH 140	Calculus 1	4
		CORE *		3
	Spring	ENES 102	Statics	3
		ENAE 202	Aerospace Computing	3
		MATH 141	Calculus 2	4
		PHYS 161	General Physics: Mech. And Part. Dyn.	3
		CORE *		3
Sophomore 33 credits	Fall	ENES 220	Mechanics of Materials	3
		ENAE 283	Introduction to Aerospace Systems	3
		MATH 241	Calculus III	4
		PHYS 262 or 260/261	General Physics Vib., Wvs., Heat, Elc., & Mag	4
		CORE		3
	Spring	ENES 221	Dynamics	3
		ENAE 201	Aerospace Analysis and Computation	3
		MATH 246	Differential Equations for Scientists & Engrs	3
		PHYS 263 or 270/271	General Physics: Elc dyn, Light, Rel, & Mod	4
		CORE		3
Junior 31 credits	Fall	ENAE 311	Aerodynamics I	3
		ENME 232	Thermodynamics	3
		ENAE 301	Dynamics of Aerospace Systems	3
		ENAE 362	Aerospace Instrumentation & Experimentation	3
		CORE		3
	Spring	ENAE 324	Aerospace Structures	3
		ENAE 432	Control of Aerospace Systems	3
		ENGL 393	Technical Writing	3
		CORE		1

	AERO TRACK:	ENAE 414	Aerodynamics II	3
	ASTRO TRACK:	ENAE 404	Space Flight Dynamics	3
Senior 30 credits	Fall	ENAE 423	Vibration and Aeroelasticity	3
		**CORE or Elective		3
**Dependent upon coursework completed in CORE	AERO TRACK:	ENAE 403	Aircraft Flight Dynamics	3
		ENAE 455	Aircraft Propulsion and Power	3
		ENAE 461	Principles of Aircraft Design	3
	ASTRO TRACK:	ENAE 441	Space Navigation and Guidance	3
		ENAE 457	Space Propulsion and Power	3
		ENAE 463	Principles of Space Systems Design	3
	Spring	ENAE 464	Aerospace Engineering Lab	3
		**CORE or Elective		3
		Aerospace Elective		3
		Technical Elective		3
	AERO TRACK: (Capstone)	ENAE 482	Aeronautical Systems Design	3
	ASTRO TRACK: (Capstone)	ENAE 484	Space Systems Design	3

Total credits must be ≥124

DEPARTMENT OF AEROSPACE ENGINEERING TIMELINE:

1939 Professor John Younger of Mechanical Engineering Receives Aeronautic Award

1939 Flying Course Here Seen As Possibility by Dean Steinberg

1939 Maryland Gets CAA Air Course

1939 Forty-One Take Aviation Course Offered By CAA

1939 CAA Aviation Students Start Flight lessons

1940 Engineers Conduct Study of Airplanes

1940 Professor John Younger Predicts Great Aviation Future

1940 Professor John Younger Begins Aero Experiments

1941 Professor Younger Receives ASME "Spirit of St. Louis" Gold Medal for Great Service to Aviation

1947 Glenn L. Martin Engineering Building Plans Revealed to Public

1948 Glenn L. Martin Engineering School Building Plans Go To Board

1949 Four Engineering Buildings Cost $2,700,279.48, New Figures Show

1949 Wind Tunnel Opening Slated For April

1949 Martin College Occupied

1949 Wind Tunnel Starts Operation

1949 Aerospace Engineers Now In New Home\

1950 Pines, Oaks, Shrubs Planted on Martin College Mud Flats

1950 Glenn L. Martin Institute of Technology Expands Into $8,000,000 Building Project

1952 Engineering College Conducts Tests on Modern Aircraft

1963 Outer Space expands Aerospace Engineering

1966 Wind Tunnel Assists Experimentation

1977–78 First Full Year of Operation of Glenn L. Martin Wind Tunnel as part of the Department of Aerospace Engineering

1982 Center for Rotorcraft Education and Research is founded

1987 Composites Research Laboratory is founded

1990 Space Systems Laboratory s founded

1992 Composites Research Laboratory moves to a new 5000-square-foot facility in the Manufacturing Building

1993 Aerospace Curriculum undergoes major changes to create two-track system

1994 NASA Center for Excellence in Hypersonics created

1994 Ranger, a four-armed NASA satellite repair robot is rolled out for display

1995 Smart Structures Laboratory created in the J.M. Patterson Building

1996 Rotorcraft Center and Hypersonics Center moved to J.M. Patterson Building

1997 Flight Dynamics and Control Laboratory founded in the GLM Wind Tunnel Bldg.

1997 Undergraduate students control NASA Goddard's SAMPEX satellite orbit and attitude determination

1998 Anechoic chamber built for rotorcraft acoustics research

1998 Department's graduate programs are ranked 10th in the nation by U.S. News & World Report

1999 Meyers Building undergoes renovation to house aerospace research activities

1999 Autonomous Vehicle Laboratory founded in Meyers Bldg.

2000 College creates Small Smart Systems Center

2000 Alfred Gessow Rotorcraft Center is renewed for another 5-year term.

2001 Combustion/Propulsion Facility established in the AeroLab building.

2002 NASA Awards URETI to University of Maryland and its partners (Univ. of Michigan, Univ. of Washington, NCAT and APL/Johns Hopkins University)

2002 University of Maryland teamed with 5 other universities win prestigious NASA grant to create a National Institute of Aerospace (NIA)

2003 Maryland ranked in the top 10 in undergraduate and graduate programs by U.S. News and World Report

Vital Statistics of Department: (2002–2003)

Full-time tenured or tenure track Faculty   20
Number of undergraduate students   320
Number of students enrolled in M.S. program   62
Number of students enrolled in Ph.D. program   70
Average SAT scores of incoming freshmen   1346
Average GPA of incoming freshmen 3.97
Average GRE scores of incoming graduate students (Q+A) 1463
Average GPA of incoming graduate students   3.7
Department Research Expenditures $10.64 million
Department Rankings:
- Undergraduate Program (7th as listed by most recent Gorman Report)
- Undergraduate Program (10th as listed by 2003 U.S. News & World Report)
- Graduate Program (10th as listed by 2003 U.S. News & World Report)

Department Faculty and Staff

Faculty Emeritus (University where highest degree earned in parentheses)	Current and Past Department Staff
John Anderson (Ohio State)	Pat Baker
Everett Jones (USC)	Terry Clark
Alfred Gessow (NY University) (Deceased)	Deborah Chandler
	Brian Cugle
Current Tenured and Tenure-Track Faculty	Sue Cunningham
David Akin (MIT)	Matt Fox
Ella Atkins (University of Michigan)	Janet Giles
James Baeder (Stanford)	Michael Green
Jewel Barlow (University of Toronto)	Chris Fuller
Christopher Cadou (UCLA)	Bryan Hill
Roberto Celi (UCLA)	Elton Howard
Inderjit Chopra (MIT)	Julia Johns
Alison Flatau (Utah)	Bernard LaFrance
William Fourney (University of Illinois, Urbana)	Dawn Leavell
Sung Lee (MIT)	Kevin Lewy
J. Gordon Leishman (Glasgow)	Maureen Meyer
Mark Lewis (MIT)	Janet Murphy
Darryll Pines (MIT)	Mark O'Connor
Robert Sanner (MIT)	Vjekaslav Pavlin
Frederic Schmitz (Princeton)	Mike Perna
Benjamin Shapiro (CalTech)	Carol Pironto
Anthony Vizzini (MIT)	Nicole Roop
Norman Wereley (MIT)	Rebecca Sarni
Allen Winkelman (University of Maryland)	Pat Salvotore
Ken Yu (UC Berkeley)	Jennifer Widdis
	Rita Woodell

Department Chairs

Time Period	Name
1997-2003	William Fourney
1995-1997	William Fourney (Interim Chair)
1993-1994	David Schmidt
1991-1993	Sung Lee (Acting Chair)
1990-1991	Bryan Hunt
1988-1990	Inderjit Chopra (Acting Chair)
1980-1988	Alfred Gessow
1973-1980	John Anderson
1970-1973	Robert Rivello (Acting Chair)
1968-1970	Richard Thomas
1949-1968	A. Wiley Sherwood

Advisory Board Members	Academy of Distinguished Alumni
In 1995, the department established an advisory board to guide and support the graduate and undergraduate programs. The board, comprised of leading experts in the field of aerospace engineering, meets once each semester.	In the fall of 1999, the department established an Academy of Distinguished Alumni to acknowledge the accomplishments of students who have made significant contributions to the field of Aerospace Engineering.
Current Board Members	**1999-Class**
John Langford, Aurora Flight Sciences Corporation	Glenn L. Martin (1886-1955)
Leo Dadone, Boeing Corporation	Kevin G. Bowcutt,, B.S. '82, M.S. '84, Ph.D. '86
Antonio Elias, Orbital Sciences Corporation	Gary L. Curtain, B.S. '65
Richard Freeman, The Aerospace Corporation	Michael D. Griffin, Ph.D. '77
Michael Griffin, In-Q-Tel (formerly with Orbital)	Bastian 'Buz' Hello, B.S. '48
Barnes McCormick Professor Emeritus, Pennsylvania State Univ.	**2003-Class**
Vincent Pisacane, US Naval Academy	William Bissell, B.S. '52
Mike Miller, Sikorsky Aircraft Corp.	George Orton, B.S. '69
Alan Sherman (chair), Consultant (formerly Lockheed-Martin Corp.)	Norris Krone, B.S. '55, Ph.D. '74
John Clark, Navy Patuxent River	Michael Torok, M.S. '86, Ph.D. '89
Daniel Held, Northrop Gruman	
Michael Ryschkewitch, NASA Goddard	
Douglas Dwoyer, Observer (NASA Langley)	

Chapter 43

Some History and Recollections
of the Aero Program at Rensselaer

by John A. Tichy
Chair
Department of Mechanical, Aerospace, and Nuclear Engineering
Rensselaer Polytechnic Institute
Troy NY 12180-3590 USA

Aeronautical and aerospace engineering have a noble and storied history at Rensselaer Polytechnic Institute, since the founding of the Aeronautical Engineering Department in 1935, to the present Department of Mechanical, Aerospace, and Nuclear Engineering. Only a few historical documents could be unearthed in preparing for this project. Therefore, my recollections and consultation with colleagues comprise the primary source of information for this article. It is frightening to reflect that I have been at RPI since 1976, that is, for 27 of the 68 years of the aero program's existence, so I imagine that makes me somewhat of an authority.

According to an article in our *Alumni News,* June 1943, the first graduating class of 1937 was 13 strong. The author of this article was the first chair of the fledgling department, Dr. Paul E. Hemke (1890-1974), in whose honor our current Hemke Lecture Series is named, as is our Paul E. Hemke Award, going to our most outstanding Bachelor's Degree recipient in Aeronautical Engineering. Dr. Hemke had met Ludwig Prandtl and Theodore von Kármán and authored a widely used textbook entitled *Elementary Applied Aerodynamics*, published in 1946. He remained as department chair until 1950, when he became Dean of the Faculty.

The student numbers steadily increased through wartime and the Hemke era. Rensselaer graduates contributed significantly to the war effort and many went through the program in two years to become pilots. In 1943, nearly 100 women studied Aeronautical Engineering in a streamlined 10-month course, and they typically worked in wartime plants on propeller manufacturing and the like. A 4 x 6 ft test section wind tunnel designed by Hemke still exists in our Fluid Dynamics Laboratory of the then new Ricketts Building.

Dr. Russell P. Harrington became head of the Department of Aeronautical Engineering in 1950, following a nine-year stint as department head at Polytechnic Institute of Brooklyn. Unfortunately, I could find no historical records from 1950 until the 1970s. I have filled in the blanks with some oral history and personal recollections. As was fashionable at the time, in 1970, RPI converted their academic departments to a so-called division structure. The aeronautical faculty members were part of the Thermal, Fluids and Systems Engineering Division (one of four such divisions). One can imagine a loss of focus and identity during this period. This "division" experiment was widely considered to be a disaster and the school returned to conventional academic departments in 1974.

The Department of Mechanical Engineering, Aeronautical Engineering and Mechanics was born at that time, and the name continued until 2001. The department chair was Dr. Frederick F. Ling, a mechanics-oriented mechanical engineer. The aeronautical engineering faculty, perhaps 12-15 of 35-40 total faculty members held a separate identity, although part of the larger department. The aero group in the 1970s and 1980s was informally headed by Professor Robert E. Duffy. When I first came, I recall some minor overtones of unhappiness among the aero faculty concerning their status within the larger department, but these problems seemed to dissipate with time and we were proud (as we are to this day) of having a most collegial group. If our aero alumni of the 1975-1990 era have a primary recollection of their education, it is that the program was *tough*. As was typical, many more credits were crammed into the four-year curriculum than at present, but in addition, the courses of Professors Brower, Bursik, Duffy and Hagerup (who is still on the faculty) *et al*. were fearsome undertakings. I recall trying to help students with their homework and thinking how glad I was *they* were the students, not I.

From 1979 to the late 1990s, numerous students worked on three generations of a composite materials sailplane glider. The program, which was originally sponsored by NASA and the AFOSR, aimed at research and development of the then revolutionary composite materials. As part of the project, students worked directly on the design and fabrication of the glider. On September 16, 1980, the inaugural flight of the RP-1, a 19 foot long, 38 foot wingspan glider, weighing just 137 pounds began its maiden flight. A larger and more complex glider, the RP-2, took off in 1986. This second generation featured an enclosed cockpit, more streamlined design, more tapered wings, and split flaps to slow the craft during landing. The far more ambitious RP-3 featured a two person side-by-side cockpit, which offered better communication between instructor and pilot, a single set of instruments and other advantages. The cockpit could be reconfigured for a single occupant. Several unique manufacturing processes were used, such as a moldless construction technique for the wings. The RP-3 was designed in the late 1980s, mostly manufactured in the early 1990s and after a series of ground and wind tunnel experiments and re-examination of the design, finally flew in 1999, see the figure below. An RP-4 bush plane is currently in the conceptual design stage.

The aeronautical engineering program had a period of considerable glory within Rensselaer in the mid-seventies. The 14th President of the Institute from 1976 to 1984 was George M. Low, an alumnus of the class of '48, Deputy Administrator of NASA, and head of the Apollo Program that landed Americans on the moon in 1969. In the same period, Dr. Robert G. Loewy (class of '47) joined RPI as Provost in 1974. In 1978, Dr. Loewy stepped down as the Institute's second in command, and assumed the position of Institute Professor in the department. He later founded the USARO Rotorcraft Technology Center in 1982 and served as its director until 1993. Dr. Loewy is currently Chair of the School of Aerospace Engineering at Georgia Institute of Technology. As part of the activities of the Center, new undergraduate courses were added to the curriculum, including a "capstone" rotorcraft design course. The strong interest in rotorcraft science and technology continues to this day.

We took on our present configuration as the Department of Mechanical, Aerospace, and Nuclear Engineering in 2001, when nuclear engineering was merged with mechanical and aeronautical engineering. We decided to give in to the reality that we had an aerospace curriculum, and replaced "aeronautical" with "aerospace." The degrees still read "aeronautical" and our ABET accreditation is still under this rubric, but we are undertaking the formalities to change all these consistently to "aerospace." Presently, aero has a more distinct identity than over the past decades, as Professor Zvi Rusak was named the Director of the Aerospace Engineering Program.

Currently, Rensselaer has a robust program in research and education. Research areas include theoretical and computational aerodynamics, structures and aeroelasticity, transatmospheric beamed energy propulsion, multidisciplinary design optimization, smart rotor design, micro air vehicles, CFD and turbulence modeling, flow control, spacecraft dynamics, multibody dynamics, interfacial hydrodynamics. The faculty is widely known for its contributions to aerospace science and to AIAA and AHS. Our undergraduate program is noted for educating top quality engineers for industry, national laboratories or further academic study in fixed wing, rotorcraft and spacecraft studies. Traditionally, the number of our graduates is strong, but volatile, varying between 15 and 65 B.S. degrees per year over the last 20 years.

The RP-3 takes flight in 1999

Chapter 44

The Aerospace Program at Boston University
—and the origins of the College of Engineering

J. Baillieul[*]
Dept. of Aerospace and Mechanical Engineering
Boston University

During the first half of the twentieth century, as it became increasingly clear that aviation was an economically and strategically important technology, many of the American universities which offered instruction in engineering launched programs in aeronautics---usually beginning as a set of courses within mechanical engineering. Boston University's early involvement with aeronautics followed a rather different path, and contrary to what happened at other universities, what began as a program of instruction in aviation technology led eventually to the founding of Boston University's College of Engineering in 1963. While this article is intended to provide a short history of the aerospace program at Boston University, the people and events that are recalled are a significant part of the overall history of the College of Engineering. To keep the article of manageable size (and meet the deadlines given for the writing), I have chosen to focus exclusively on the people and events that were significant in establishing the College and Department. Because of a sense of urgency to record events that are only available from personal memories, the emphasis is on the early history of the program. Recent events and a recording of the intellectual and technical contributions made by a number of the people mentioned below will have to be presented on a different occasion.

Boston University was founded in 1839 by delegates of the Methodist Episcopal Church as a school for the improvement of theological training. Over nearly a century, it grew into a large liberal arts university under the stewardship of a succession of presidents who were also Methodist ministers. By the early 1930's, Boston University was suffering a great deal from the effects of the Depression. In 1931, University President Daniel L. Marsh asked all employees to take a "voluntary" five per cent pay cut. The following year, there was a ten per cent cut which was not voluntary, and the year after that an additional cut, bringing the total depression-era salary

Copyright © 2004 by J. Baillieul. Published by the American Institute of Aeronautics and Astronautics, Inc., with permission.

[*] **Acknowledgment** This article reports historical research which received no financial support whatever--- except for the the author's salary that was unwittingly contributed by his University. Nevertheless, there is a great debt of gratitude to a number of senior Boston University faculty whose efforts in earlier years made our current aerospace program possible and whose memories of these efforts have made this brief article possible. In particular, I would like to cite James Bethune, Dick Vidale, and Dan Udelson, and a very special word of thanks goes to Merrill L.\ Ebner and Arthur T.\ Thompson, who were enormously generous with their time and who, along with Dick Vidale, played key roles at a critical time in the College's history. It is also appropriate to acknowledge Kara Peterson, who is by no means an old timer. Finally, a word of thanks to Conrad Newberry of the Naval Postgraduate School for inviting the piece and then persistently but not unpleasantly pestering the author until the job was done.

reduction to 19%[1]. In this context, the operative University policy was that there would be no expansion, and in particular there would be no new academic departments. Not having foreseen the economic strain of the times, however, in the late 1920's, President Marsh hired Hilding N. Carlson and Arthur G.B. Metcalf to start an Aeronautical Engineering Department.

In 1928, Carlson, who Marsh had appointed Professor of Science and Mathematics in the B.U. College of Business Administration, began conducting evening classes in aeronautical engineering at Boston's Logan Airport. ([4]) According to Kathleen Kilgore's account ([1]):

> "Carlson and Metcalf, both M.I.T. trained engineers and pilots, began an ambitious four-year program that included design, shop in a hangar at [the] airport, meteorology and navigation, and even salesmanship. In addition to a full teaching load, Metcalf was working on two of his own projects. The first was a new type
>
> of fighter plane that would attain what *The Boston Globe* called the "amazing speed" of 350 miles per hour.
>
> The second was the 'Aeromobile,' a plane whose operation would require the same amount of training and skill as would an automobile..." "...Depression economics prevailed over the dream of creating an aeronautical engineering school, and in 1937, Marsh was forced to drop the aeronautics program. Metcalf had, in the meantime, founded his own Engineering Research Company on Newbury Street. [Soon afterward] Carlson organized his own private New England Aircraft School."

Figure 1: Dean Sheley (seated at the right) with visitors at the Logan Airport facility.

In 1940, Carlson and a colleague, Professor Walter J. Goggin, left B.U. to found the New England Aircraft School in a building he acquired at Logan Airport.

With World War II creating great demand for personnel with knowledge of aviation technology, Carlson's school prospered. He was contracted by the U.S. Army Air Corps to provide ground training for a large body of Air Corps personnel. According to Kilgore's account, the enrollment in the school was about three hundred by 1950. The facilities included an airplane hangar, machine shop, welding shop, metal-working shop, and wood-working facilities. In 1943, after the death of his business partner Walter Goggin, Carlson approached B.U. President Marsh "with the proposition that Boston University absorb his school as a separate department, appoint him its dean, and appoint him once again as professor in the College of Business Administration. In exchange, the building at Logan Airport would be leased to Boston University and all the equipment donated... The trustees finally approved the arrangement in December 1950."

The newly acquired school was called the College of Industrial Technology. Carlson served as the founding Dean, and the old Aircraft School's Registrar, Mr. B. LeRoy Sheley, was appointed Assistant Dean. The College's course of study appeared for the first time in the University Bulletin (Course Catalogue) of the academic year 1951-52. The opening paragraph read: "The College offers a variety of courses, both day and evening, which are adapted to the skills and abilities of all types of young men. The Aeronautical Engineering Courses train young men for engineering and design positions with aircraft factories and airlines, as well as with the manufacturers of aircraft accessories. The Aircraft Maintenance Engineering Courses are for those interested in maintenance, overhaul, and repair of aircraft and aircraft engines." Reflecting its vocational school roots, the College opened with three programs: "a four-year program leading to the Bachelor of Science in Aeronautical Engineering, a two and one-half year program in Aircraft Maintenance Engineering leading to the degree of Associate in Science, and a two-year Certificate Course in Aircraft Maintenance."

Within a few years the College had changed its offerings to include the following three programs: (1) a four year program in Engineering Administration leading to a B.S. degree, (2) a four year program in Aeronautical Engineering, also leading to a B.S. degree, and (3) a two and one-half year program in Maintenance Engineering leading to a Civil Aeronautics Authority Federal Certificate. This certificate program was notable for its requirement that students take a semester of social science and humanities classes on the main campus. In 1954, the College of Industrial Technology awarded its first degree in aeronautical engineering. While the programs in the 1951 course catalogue may have been "adapted to the skills and abilities of ... young men," such gender bias soon disappeared, and program publications referred to opportunities for both men and women. In the early '50's, the College admitted its first woman student, Anne Everest (BS '56), who went on to also become the College's first woman graduate. Also in 1956, Sheley was promoted to Dean of the College---a position he held until his death in 1962.

Figure 2: An archived photograph of one of the many posterboard instructional aids at the airport facility of the College of Industrial Technology. Also, what was an up-to-date jet engine of the period. While there was significant technical content in the early program of instruction, it retained many of the vocational aspects of the predecessor curriculum at the New England Aircraft School.

During the period 1951 through 1963, the College of Industrial Technology divided its operation between space allocated on the main (Charles River) campus and facilities at Logan Airport. In 1958, the on-campus portion moved from the top floor of 680 Commonwealth Avenue (now the College of Communications) to the basement of 775 Commonwealth Avenue (now the School of Fine Arts).

Arthur T. Thompson was appointed Dean of the College of Industrial Technology in July, 1963. The University Administration under President Harold C. Case had recruited him from Penn State, where he was an Associate Dean of the College of Engineering and Architecture. Thompson immediately began a vigorous effort to broaden and deepen the technical scope of the College. He petitioned Case and the Administration to change the name from the College of Industrial Technology to the College of Engineering. The petition was granted and the name changed in 1964. During a sabbatical leave from Penn State, Thompson had worked in manufacturing with the Crane Company while getting an MBA at the University of Chicago. Based on this experience, he envisioned a new engineering discipline in manufacturing, and he set out accordingly to redesign Boston University's B.S. program in Engineering Management (the successor of Engineering Administration). The program was renamed Manufacturing Engineering, and in the fall of 1964, it was placed under the leadership of the newly appointed Assistant Professor Merrill Ebner. Ebner was one of two assistant professors appointed that year, the other being Richard F. Vidale, who would go on to be in charge of the Systems Engineering program.

Figure 3: The opening of the College of Engineering Building at 110 Cummington Street in January, 1964. Left to right are, Merritt A. Williamson, Dean of the College of Engineering and Architecture at Penn State, B.U. President Harold C. Case, and B.U. College of Engineering Dean Arthur T. Thompson. Williamson gave the Convocation Address at the inaugural ceremony of the College. The building had been a stable for carriage horses at the beginning if the twentieth century.

In 1965, new B.S. degree programs in Systems Engineering and Information Engineering were started, and Aeronautical Engineering was renamed Aerospace Engineering. As conceived in 1965, Aerospace Engineering was to be a nine-semester program. This was scaled back a few years later to a more standard eight semesters in response to criticism from an accreditation team that visited the College on March 10,11, 1969. Although the formation of a curriculum in Information Engineering appears to have been forward looking from a contemporary vantage point, the program attracted few students and was terminated a few years later (partly in response to criticisms from the 1969 accreditation visit). In 1966, a B.S. program in Bioengineering was initiated, and the year after that, two M.S. programs---Manufacturing Engineering and Systems Engineering---were started. Continuing the programatic expansion, in 1968, the College created a joint Masters degree program in Manufacturing Engineering and Business Administration, and in 1969, Bioengineering was renamed Biomedical Engineering.

Thompson's vision of engineering education was that there should be a universal "liberal engineering" degree for all undergraduates with specialization provided through several focused topical programs. Conversations with people responsible for accrediting undergraduate programs in engineering, however, persuaded Dean Thompson that separate discipline-specific departments were needed. The programs were duly placed within distinct departments in 1969, and the College underwent an accreditation review in 1971. The B.S. programs in Aerospace, Manufacturing, and Systems Engineering were awarded accreditation in 1972, following the 1971 review. The Manufacturing Engineering program was the first to be accredited in the U.S., and it is a tribute to the ingenuity of Thompson and Ebner that they were able to persuade the Accreditation Committee of the Engineers Council for Professional Development that manufacturing was such an important "golden disciplinary" core of Industrial Engineering that it should be able to stand on its own as an accredited program.

It was not until 1974 that the College started B.S. programs in Mechanical Engineering and Computer Engineering. The Mechanical engineering program was introduced within the Aerospace department, giving rise to a renamed Aerospace and Mechanical Engineering Department. Computer Engineering and Systems Engineering were joined with a newly approved Electrical Engineering program in 1976, creating a Department of Electrical, Computer, and Systems Engineering (ECS). Electrical Engineering was the final undergraduate program for which approval was sought, and with its addition, the Boston University College of Engineering assumed its present form. One footnote to be added, however, is that Systems Engineering was eliminated as an undergraduate major in the academic year 1995-96, and the parent department was renamed Electrical and Computer Engineering.

As the decade of the 1960's ended and the 1970's began, the University found itself in the midst of social unrest and once again felt economic distress. In 1967, Arland F. Christ-Janer became the sixth president of Boston University. It was a tumultuous time. In the University records of past presidents we read that "The week of his inauguration, the Students for a Democratic Society declared a Stop the Draft Week. Soon after, an African-American student organization issued a list of demands and staged a non-violent sit-in in the President's Office. President Christ-Janer agreed to all their demands, but campus demonstrations and radical student actions continued. After serving for three years, President Christ-Janer resigned in July 1970. Commencement that year had been canceled because of the threat of violent protests." The University operated in the red for several years, and a search was begun for a new president who could bring the operating budget back into balance. John R. Silber, a Kant scholar and former Dean of Arts and Sciences from the University of Texas, was appointed president in 1971. That year, the budget showed a deficit of almost $2.5M, which while only a small fraction of the operating budget, could not be sustained for more than a very few years. Silber knew that anything he hoped to accomplish at Boston University would depend on his being able to run the organization within a balanced budget.

Like all business and academic units in the University, the College of Engineering came under budgetary scrutiny, and Dean Thompson was asked to make significant cuts. With his three department chairmen, (Ming Chen in Aerospace, Merrill Ebner in Manufacturing, and Dick Vidale in Systems), Thompson drafted a budget that would go as far as possible to meeting Silber's directive. During one of President Silber's 8:00am Sunday morning budget hearings, the Dean presented his budget to an assemblage of University Administrators. There is no written record of the discussions that ensued, but the severe cost-cutting measures of the Thompson budget were deemed to be adequate by the University Administration. The College was kept open---although it became a challenge to keep it running on the meager budget that had been agreed to. Apparently President Silber liked Thompson's management style well enough that he asked him to serve concurrently as the Dean *ad interim* of the College of Business Administration. He served as Dean of both units from 1972 through 1973 when he agreed to take charge of the Boston University Overseas Program in Heidleberg, Germany. At that point, Professor Ebner became

Dean *ad interim* of the College of Engineering, and a search was launched to find a new permanent Dean.[2]

In 1975, President Silber and the Board of Trustees selected Professor Louis N. Padulo from Stanford University as the next Dean of Engineering. Padulo had prior administrative experience as a department chairman at Morehouse College and had been named Associate Dean of the Stanford Graduate School when recruited to B.U. by Silber. He was charming, self assured, and interested in the job. He had recently published a well regarded control theory text with the computer scientist Michael Arbib ([2]), and he was known and respected by academic colleagues at leading universities. Dean Padulo's mandate was to grow the small College of Engineering into a visible and respected institution. Padulo was Dean from 1975 until late 1986. He was a brilliant recruiter of faculty and students, and the College enjoyed a phenomenal period of growth during these years. There had been 163 students (graduate and undergraduate) registered in 1973. By 1987, the year after Padulo stepped down, there were 1,933 full-time students (mostly undergraduates) and 548 part time students (mostly masters degree students) with the first doctoral students fleshing out the new graduate and research programs. Padulo also increased the size and quality of the faculty. Ming Chen, who was Chairman Aerospace and Mechanical Engineering

Figure 4: Professor Merrill Ebner, who with Arthur Thompson created the first Manufacturing Engineering Program in the country---which later evolved into the first accredited Manufacturing Engineering Department---served as Dean *ad interim* from 1974 to 1975.

--

[2] Arthur T. Thompson remained active in academic administration for many years. From 1974 through 1978, he served as Boston University Associate Vice-president for Overseas Programs and expanded the University's operation in Europe from 15 to 44 educational centers. He later became Provost and member of the Board of Trustees at the Wentworth Institute of Technology, from which he retired in 1988.

when Padulo arrived, had been the first faculty member in the College who had been hired with a PhD. Although others had followed, Padulo began a drive to recruit faculty who could be expected to become distinguished researchers as well as good teachers. Some idea of the change that occurred in the faculty size and quality may be found in comparing the record of 1968-69 (when the College's first accreditation visit took place under Dean Thompson) with the corresponding data in the Fall of 1986, the final year in which Lou Padulo served as Dean. In 1968-69, there were fewer that 20 full time faculty, and only six held a doctoral degree of some kind. By 1986-87, there were 67 full time engineering faculty, all but four of whom held doctoral degrees.

Figure 5: From left to right: an unnamed visitor, Dean Louis Padulo, and Ming Chen, Chairman of the Department of Aerospace and Mechanical Engineering.

Figure 6: Professor Daniel Udelson, Chairman of the Department of Aerospace and Mechanical Engineering from 1981 to 1991, with the supersonic wind tunnel.

From 1987 through 1989, Carlo De Luca, a Professor of Biomedical Engineering served as Dean *ad interim*. Taking note of Padulo's accomplishments, De Luca wrote: "We have reached an appropriate plateau in undergraduate enrollments. . . . Our most immediate challenge is to develop our growing graduate/research program so as to bring the College to national and international prominence." After a lengthy national search, Charles Delisi was recruited from Mount Sinai School of Medicine to be the next Dean, starting in 1990. Trained as a physicist, but also having once been an undergraduate history major, Delisi was widely regarded as an intellectual visionary. He was credited for his work several years earlier, at the U.S. Department of Energy, to launch the government's efforts to map the human genome. Just as Padulo had greatly increased the size and quality of the undergraduate student population in the College, Delisi recruited a star studded array of new faculty---including a number of members of national academies and other senior faculty with large and highly visible research programs. During the decade in which Delisi served as Dean, the level of sponsored research in the College increased by a factor of four. In August, 2000, David K. Campbell, former Director of the Center for Nonlinear Systems at Los Alamos National Lab and Chairman of the Physics Department at the University of Illinois, assumed the position of Dean of the College. He continues in this post.

The records of those serving on the faculty and administration of the Department of Aeronautical Engineering during the earliest days of the College of Industrial Technology are difficult to trace. During the first part of Dean Thompson's term, Charles L.D. Chin was the Chairman. He served until the academic year 1967-68, when Ming Chen was appointed Chairman. Daniel Udelson succeeded Chen in 1981, and served for a decade until 1991. The author joined the Department during this time, in 1985. After this, a number of important faculty appointments were made by

Udelson. In 1988, he recruited Tom Bifano, who is presently Chairman of Manufacturing Engineering and an internationally recognized authority on micro-electro-mechanical systems (MEMS). He also hired Don Wroblewski who currently serves as the Department's Associate Chair for the Undergraduate Program in Aerospace. Toward the end of his tenure, in the Fall of 1991, he recruited Michael S. Howe, a distinguished and prolific authority on the acoustics of fluids, and the late Charles Speziale, who was one of the world's leading authorities on turbulence modeling. It should be added that this appointment established turbulence modeling as an area of excellence that would be sustained going forward with the appointment of Victor Yakhot and later the more junior Assad Oberai.

Figure 7: Professor Allan D. Pierce, a major figure in acoustics, was chairman from September, 1993, through June,1999. Pierce recruited a number of very bright junior faculty members, and during his term, two members of the National Academy of Engineering joined the Department.

Following Udelson's term, the department had two interim chairs: Jeffrey Forbes, from Fall 1991 through Spring of 1992, and the author from 1992 through July, 1993. On August 1, 1993, Allan D. Pierce became Chairman. A distinguished acoustician, Pierce, was recruited from Penn State, as Dean Arthur Thompson had been thirty years earlier. During his term as Chairman, a number of people of prominence in acoustics were recruited. It was also at this time that National Academy of Engineering members Donald C. Fraser and Leopold B. Felsen joined the department.

As I write this short history in the summer of 2003, a fair summary statement seems to be that the aerospace program has undergone a phenomenal transformation from its origins at Logan Airport in the1930's to its present state of being part of a highly regarded department in a major research university. As it has flourished, the department's needs for space have grown. In 1999, it was finally able to claim the entire building at 110 Cummington Street on the (main) Charles River

Campus of the University, the same building that housed the entire College of Engineering in 1964. (See Figure 3.) The building has undergone multiple major renovations (including a recent investment of approximately $2M to upgrade faculty office space and undergraduate teaching labs), but its space is already too limited to house all department operations. The department now uses classroom facilities in the adjoining Photonics Center (an $80M building completed in 1997) and the Arthur G.B. Metcalf Center for Science and Engineering (Figure 9) which was dedicated in December, 1985.

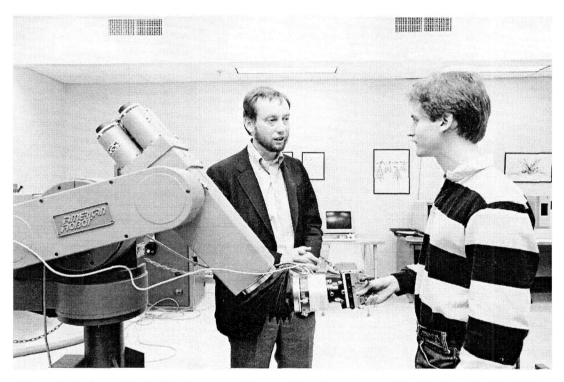

Figure 8: Professor John Baillieul, current chairman of the department was hired by Dean Lou Padulo in 1985. Padulo wanted him to create a robotics laboratory, and Baillieul is shown in this 1987 photo with a graduate student and one of several industrial robots that were used in early laboratory experiments. (The student, Richard Goldenberg, received a masters degree and became a patent attorney.)

While the department maintains a healthy respect for the kind of entrepreneurial abilities displayed by its Depression era forebears Carlson and Metcalf, the tendency in recent years has been to recruit faculty based on their technical creativity and intellectual stature. Research active faculty members are no longer expected to teach the three courses a term that were the norm during the terms of Deans Thompson and Padulo. Among the current 24 full time faculty members there are two members of the National Academy of Engineering, present and past editors-in-chief of leading technical journals in various subdisciplines (the *Journal of the Acoustical Society of America*, the *IEEE Transactions on Oceanic Engineering*, and the *IEEE Transactions on Automatic Control*), and a solid corps of active researchers who in some cases average more than $1M per year in funded research projects. It is widely recognized that contemporary engineering has become an interdisciplinary endeavor, and most of our current faculty are not uniquely specialized in either

Figure 9: The entrance (left) to the Arthur G.B. Metcalf Center for Science and Engineering, which houses the department's Laboratory for Nanometer Scale Engineering (right).

aerospace or mechanical engineering. Both research and instruction supporting both disciplines are focused on a wide range of enabling technologies which make fundamental use of electronics, information technologies, and modern physics. The research thrust areas[3] which the department has targeted in its current five year plan are on the one hand nontraditional while at the same time being similar to the focus areas found in a number of forward looking aerospace and mechanical engineering programs at leading universities.

The growth and advancement have been a part of the wider success and growth of the College, which for the first time this year was listed among the top 50 graduate schools in engineering by U.S. News. There has been a long period of sustained investment in the department by both the College and University. With continuing support, a number of superb new faculty appointments have been possible during the past few years. The University now seeks to appoint its ninth president, and it is notable that the leading candidate holds a degree in Mechanical Engineering and has held a very public position in aerospace. Encouraged by the remarkable institutional transformation that has continued for many decades, there is a sense of optimism that the Aerospace program will continue its prominence within the larger Aerospace and Mechanical Engineering Department, which in turn will be increasingly recognized as a technical leader in the thrust areas mentioned above.

Chronology of Department Chairmen From 1964 to the present						
Charles L.D. Chin	Ming Chen	Daniel G. Udelson	Jeffrey Forbes	John Baillieul	Alan D. Pierce	John Baillieul
Until AY '66-'67	1966-1981	1981-1991	*ad interim* AY '91-'92	*ad interim* AY '91-'92	1993-1999	1999 - Present

---

[3] Without going into detail, we list the five current departmental thrust areas: 1. Nanotechnology and ultra-small structures, 2. Dynamics, Control, and Mechatronics, 3. Acoustics, 4. Fluid mechanics, and 5. Biological and medical applications of mechanical engineering. Details of faculty involvement, recent recruiting, etc. are available in [6].

References

[1] K. Kilgore, 1991. *Transformations: A History of Boston University*, Boston University, 479 p. (Available in the Boston University Library.)

[2] L. Padulo & M.A. Arbib, 1974. *System Theory; a Unified Sate-Space Approach to Continuous and Discrete Systems*, Philadelphia, Saunders, xvii, 779 p.

[3] Kara J. Peterson, 2003. "Timeline: The History of the Boston University College of Engineering," *BU College of Engineering Magazine*, Spring, 2003.

[4] E. Ray Speare, 1957. *Interesting Happenings in Boston University's History, 1839 to 1951*, Boston University, 204 p. (Available in the Boston University Library.)

[5] A.T. Thompson, 1971. "Engineering Accreditation: A Saga," Report to the Engineers Council for Professional Development, Documentation prepared for the Second Boston University College of Engineering Accreditation Visit.

[6] Boston University Department of Aerospace and Mechanical Engineering Five Year Plan, Fall, 2002. Available from Department of Aerospace and Mechanical Engineering, 110 Cummington Street, Boston University, Boston, MA, 02215.

[7] All pictures are thanks to the Boston University Office of Photo Services.

Chapter 45

Aerospace Engineering and Its Place in the History Of The University of Arizona

John J. McGrath
Department Head and Professor

Compiled by

Dianne Smith
Administrative Assistant for the Department Head

The Formative Years

Tucson is Arizona's oldest city with a unique blend of Indian, Spanish, Mexican and Anglo Heritages. It was founded in 1775 as a walled presidio or military garrison to protect settlers from Apache raids. Some of the early University of Arizona students walked to classes from homes adjacent to this wall.

Theodore B. Comstock, became the first president of the University of Arizona in 1892. By this time the University consisted of four colleges: Agriculture, Mines, Natural Science, and Letters. Included under the College of Mines were the school of mines, school of civil engineering, and school of mechanical engineering.

In 1895 the Reverend Howard Billman became the second president of the University and developed a strong emphasis on a military academy type of organization. Within a year the prep students wore grey uniforms trimmed with black braid like the dress uniform of the West Point cadets, and the day was organized by bugle calls.

By February 17, 1910, Charles E. Hamilton and Glenn H. Curtis had flown an eight-cylinder Curtis biplane to Phoenix and then took it by train to Tucson where it reached an altitude of 900 feet and flew one mile in one minute and twelve seconds. These were heady times that caused men literally and figuratively to turn their eyes toward the heavens.

The aircraft industry received an infusion of capital as a result of the war. In 1919 the first continental airport in the United States was organized in Tucson on south Sixth Street, and on November 20 the first plane landed on the 2,000 square foot field.

The War Years

On April 2, 1917, the Congress of the United States declared war on Germany. The University of Arizona became deeply involved. The Dean of the College of Mines and Engineering, Gurdon Montague Butler, administered three war-training contracts. Although promising students were eligible for deferment, most of them entered the service. Dean Butler expected to retain only two seniors and six juniors in the spring term of 1918.

At the same time the dormitories were turned over to the military for housing during the summer until temporary barracks could be built by fall. Butler received a letter of high commendation from the War Department for his work.

Martin Lynn Thornburg played a pivotal role in the future of the Mechanical Engineering Department, at that time paving the way for the future Aerospace program. Thornburg, a 1915 graduate of Purdue University, joined the Mechanical Engineering faculty in 1924 and remained as head of the department until 1957. In 1917-18 he was captain in the Army Air Corps and taught in the Army School of Military Aeronautics at the University of Illinois.

In the fall of 1926 Charles Lindbergh flew into Tucson in the *Spirit of St. Louis*. He toured the city and University of Arizona in an open car, and had lunch at the University Commons. Shortly after, Professor Thornburg helped students acquire a plane and form a Wildcat Aero Club.

Albert Hamilton (ME '33) soloed after only two and one half hours of instruction and also made a parachute drop. Another pilot, however, crashed the plane and that ended the glamorous part of the club. In 1939-40 the Civil Aeronautics Authority (CAA) had introduced a primary flying course, and 25 students received their private pilot licenses.

On Sunday, December 7, 1941, the carrier based aircraft of the Japanese fleet struck Pearl Harbor in a surprise attack. The next day Congress passed a declaration of war against Japan, and when Germany and Italy declared war on the United States a few days later Congress passed another declaration of war against them. This action did not catch the University of Arizona by surprise. Dean Butler was already administering three training contracts for the government, and by June 1941 he could count 183 people as having completed the programs.

Members of the faculty left for the military or for war related industries. Captain Harold Jimmerson (ME '37) was captured along with General MacArthur's entire forces in the Philippines and was forced into the Bataan death march. Bernard Harvey (ME '49) recalled the times vividly. "The only permanent University male population seemed to be in the Navy V-5 program that totally occupied Old Main." The U of A had a great ROTC department that may have saved lives later because all healthy male students were compelled to take ROTC. "We were a Calvary Unit and were trained in basic military science, which gave us an advantage in survival when the real war grabbed us," remembers Harvey. The formations of B-24s originating from Davis-Monthan continually flew overhead as a constant reminder of the impending interruption (which for many turned out to be the end) of their University careers.

Unlike the First World War, the Second World War, brought a tremendous amount of activity and dollars into Arizona. In the Phoenix area, Williams and Luke Air Force bases had 10,000 servicemen. Around Tucson, Davis-Monthan was expanded to 7,000 acres and trained 3,000 men in B-17, B-24, and B-29 airplanes of the First Bombardment Wing. By February 1942, Consolidated Vultee Aircraft Corporation opened a Tucson division to modify the B-24s. They built three large hangers and had 6,000 employees when in full operation. Professor John Park was hired for the summer as office engineer at Marana Airbase, and he brought Theda Plumb Adams (CE '43), John Brown (CE '43), and several other civils to help in the design and computations of the auxiliary fields.

"A station wagon picked us up from our various living quarters in Tucson each day," Mrs. Adams reported. By the end of the war, Marana had trained 10,000 men and was the largest pilot training center in the world.

Dean Butler's 1939 contract with the CAA to provide civilian pilot training was extended in 1941, and nine instructors gave the courses for credit. Thornburgs' World War I experience with the Army Air Corps and his teaching experience with the Army School of Military Aeronautics gave him the background needed to teach aeronautics. The program had one complete plane and two engines. By the summer of 1942 the CAA program was closed, and flight school was restricted to the military. By this time the ground training included spherical trigonometry and military astronomy in addition to the mechanics and navigation studies. Altogether, 331 students were awarded private flying licenses.

The Navy Aviation School was started in May 1943 under a six-month contract to train flight instructors, only to have it changed in July to produce combat pilots. By August 1944, when the program was terminated, 591 students had enrolled.

In June of 1945 University President Alfred Atkinson reported to the Board of Regents that, in the three and a half years since war had been declared, the University of Arizona had trained 13,666 people in defense and war related work. Among them the Navy Indoctrination School had produced 9,841 officers. The course in aviation was now given each semester, and there was rapid growth of interest in aeronautical engineering.

June 6, 1944 was D-Day, and General Dwight Eisenhower launched the invasion of Europe. May 8, 1945 was V-E Day with the Germans accepting unconditional surrender, and most of the servicemen began returning to their homes and civilian life. Finally, on August 6, a B-29 bomber dropped the first atomic bomb on Hiroshima and a second on Nagasaki three days later. Japan surrendered August 14, 1945 and World War II was ended. Yet less than a hundred years earlier a handful of Spaniards and a few thousand Indians occupied an Arizona where even the smallest change took centuries.

The Space Age Arrives

The Aeronautical Engineering Building was finished in 1947 and contained some drafting rooms, fluid mechanics lab, a small wind tunnel, and at long last a hydraulics facility. The mechanical engineers developed aeronautics, and in 1958 the Department of Aerospace and Mechanical Engineering came to be. Technology moved at a much more rapid pace in the 1950s that earlier. Transportation no longer meant horse and foot, or even steamship and locomotive, but jet propulsion and, for military payloads, missiles.

Harvey D. Christensen became head of the newly formed Department in 1958. Christensen was a head with the mission to change the directions of the department to reflect the just arrived space age. He had taught at Oregon State University for 11 years and received his PhD from Stanford ('60). Christensen headed the department until 1970.

The new program in aerospace engineering had been approved for the 1959-60 catalog, and the first undergraduate class of nine students received degrees in June 1960. The department also enrolled 20 students in the MS and four in the PhD programs.

Manfred Ronald Bottaccini arrived on campus with Christensen. He had been teaching at the State University of Iowa since 1954, and in 1958 he received his PhD there in fluid mechanics. He soon designed and built a low velocity wind tunnel at Arizona. This was a major instructional facility for several reworked courses.

Edwin K. Parks (1960) came from teaching at the University of Kansas with his PhD in aerophysics from the University of Toronto ('52). He had been an instructor in radar and air navigation with the Royal Canadian Air Force and now was interested principally in Aerodynamics. He installed a supersonic wind tunnel, a plasma tunnel, and worked on a flight simulator. His main interest during the latter part of his career was airplane stability and control.

Tom Vincent (AE '63) was the first PhD to graduate under the Aerospace program; he became a faculty member who recently retired. He is still active as an Emeritus with a research project, nicknamed "Shadow," is an underwater device that sits on the ocean floor and photographs an octopus when it becomes active. "Shadow III" will help scientists track giant pacific octopi in their natural habitat. 1963 was the year Helen Wong was the first woman to receive a BS in Aerospace Engineering.

William R. Sears (1974) arrived in Tucson following a distinguished career as professor of aeronautical engineering at Cornell. He brought great prestige to the department. After Sears retired from the University he received the Daniel Guggenheim Medal for his contribution to the world of aviation. Sears' contributions to the field of aeronautics came from research focused on the general field of "unsteady aerodynamics." Before coming to the University Sears worked for Northrop Aircraft as the chief of aerodynamics and head of flight-testing and developed the "flying wing" bomber which was the model for the Stealth Bomber's design.

One of the most notable alumni was Francis R. Scobee (AE '65). Scobee was Pilot and second in command on the Challenger STS, 41-C in April 1984. The mission saw the successful retrieval, on-board repair and redeployment of a damaged Solar Maximum Satellite, as well as flight testing of a Manned Maneuvering Unit. His second mission was as Commander of the fatal January 1986 flight of the Challenger 51-L launched from Kennedy Space Center in Florida.

Richard Seebass joined the faculty from Cornell in 1975 as a professor. The whole field of Aerospace engineering was transformed through his contributions. The focus was on major research activities and graduate education. After leaving the University of Arizona he became the dean of the University of Colorado at Boulder's College of Engineering and Applied Science.

Fernando "Frank" Caldeiro (ME '84) is the only Hispanic astronaut in the country. President Bush appointed Caldeiro last year to the Commission on Educational Excellence for Hispanic Americans. In 1996 he was selected as an astronaut candidate and reported to the Johnson Space Center in Houston. Caldeiro is in charge of shuttle software testing at the Shuttle Avionics Integration Laboratory. Before the explosion of the space shuttle Columbia in January, he had been testing software for the International Space Station. Since the accident, he has been assigned to software verification. Caldeiro says he hopes to go into space in the next two years.

Lawrence B. Scott was department head from 1970-80. He received his PhD in aeronautical engineering from Stanford University. He had a special interest in wind and solar energy, and in wind tunnel testing for aircraft design.

(See appendix for listing of Department Heads and their tenure from 1980 to 2003)

The recent $28 million AME building is a world-class facility with over 90,000 square feet of space. It consists of two buildings connected by elevated walkways and a ground-level courtyard. It houses classrooms and other educational spaces, state-of-the-art teaching and research laboratories, conference spaces, and office space for graduate students, faculty and staff.

The AME Department offers ABET accredited undergraduate programs in aerospace engineering and mechanical engineering. AME also offers graduate programs leading to the MS and PhD degrees in aerospace engineering and mechanical engineering. The Department has a strong presence in the Applied Mathematics and Bio-medical Engineering interdisciplinary programs.

Research activities are concentrated in fluid mechanics and aerodynamics, multi-body dynamics and control, heat transfer, solid mechanics and composite materials, space technology, biomedical engineering and reliability. Some of the emerging areas of concentration include micro-electrical-mechanical systems (MEMS), nano-technology and opto-mechanics.

The Aero program places special emphasis on the design of aircraft, rockets, satellites and spacecraft. For the aerospace senior design project, each team of students generates and realizes the design of an aircraft. Individual students are responsible for particular aspects of the design, such as lift and control surfaces, propulsion, control, and fuselage design.

The graduate student body is composed of 100 students, about half are MS students and half are PhD. Ninety-four MS degrees and 24 PhD degrees have been awarded since the Aerospace degree was first offered in 1960.

Teaching facilities include: classrooms, large auditorium and lecture hall (all equipped with multi-media projection equipment), undergraduate and graduate computer laboratories with modern workstations, high-bay laboratories housing a modern wind tunnel and other large equipment, controls laboratory, aerospace structures laboratory, instrumentation laboratory, fluid-thermo sciences laboratory, large design studio and machine shop.

The Department is active in both fundamental and applied research. Students at all levels (including undergraduates) are involved in research projects, which are coordinated by a faculty advisor. Research activities encompass the following broad areas: Aerospace, Thermal/Fluid Science and Solids, Dynamics, Controls and Reliability.

Department Leadership 1980-2003

Tony Chen	1980 - 1989
Steve Crow	1989 - 1991
Parviz Nikravesh	1991 - 1993
Joseph "Pepe" Humphrey	1995 - 1997
Tom Balsa	1998 - 2002
John J. McGrath	2002 -

Current Aerospace faculty and some their research interests.

Thomas Balsa - Fluid mechanics; boundary layers, jet noise, nonlinear waves in shear flows
Weinong Chen - Experimental solid mechanics, high strain rate response of materials, fatigue of solids
Herman Fasel - Computational fluid dynamics, hydrodynamic stability, turbulent flows, aerodynamics
Jeff Jacobs - Experimental fluid mechanics, hydrodynamic instabilities, turbulent mixing, optical flow
Ed Kerschen - Fluid mechanics, applied mathematics, unsteady flows, aeroacoustics, aerodynamics
Erdogan Madenci - Composite structures, thermo-mechanical fatigue, micro material characterization
Kumar Ramohalli - Space technologies, propulsion, combustion, acoustics, solid propellant rockets
Sergey Shkarayev - Fracture mechanics, structural analysis, software development
Israel Wygnanski - Aerodynamics (related to fixed and rotary craft); high lift devices, drag reduction

A few notable alumni

Helen Wong ('63) Engineer, Los Alamos
Harlan Goudy ('64) President, Goudy Engineering, Inc.
Francis S. Scobee ('65) Commander Challenger STS 51L
Capt. Alan Gemmill ('68) US Fighter Pilot - served on USS Theodore Roosevelt
Gene Cliff ('70) Professor Virginia Polytechnic Institute, notable achievement in Central Design
Ron Cottrell ('70) Master Thesis ultimately furthered Advanced Guidance Systems used today at Raytheon
Peter Schwartz, Jr. ('71) Swartz Engineering, Inc.
Lt. Col. John Sherfesee ('71) Associate Dean USAF Academy, Colorado Springs, CO
Peter Amundsen ('72) Engineering Development Sr., Honeywell International
Robert Hanson ('73) Director of Pima Air Museum in Tucson, AZ
Co. Sherwood Spring ('74) Astronaut Space Shuttle "Atlantis"
Jeffrey Glover ('77) Aeronautical Engineer, Northrop Grumman Corporation
Col. Francis Shelley ('79) Test Pilot U.S. Air Force
Michael Langan ('80) National Aeronautics & Space Association
Michael Jenkins ('82) Design Software / training simulator for Apache Helicopter
John Redington ('85) Experimental Test Pilot, US Army
John T. Ahearn ('85) Engineering Manager, Lone Star Energy Company

Acknowledgements

Information for this paper was compiled from a book on the history of the College of Engineering and Mines, *"Rah For The Engineers! A Century of Change."* Also, thanks go to members of the Department who helped with gathering and proofing the information, Professors, Skip Perkins, Tom Vincent, and Tom Balsa; Graduate Program Administrator, Barbara Heefner; and Student helper, Ryan Reese. This project has been fun and exciting, but the real challenge was keeping it to only five pages.

Chapter 46

AEROSPACE ENGINEERING AT THE UNIVERSITY OF TENNESSEE

Mancil W. Milligan, Professor of Mechanical and Aerospace Engineering
The University of Tennessee, Knoxville, Tennessee
Roy J. Schulz, Professor and Chair, Mechanical and Aerospace Engineering
University of Tennessee Space Institute, Tullahoma, Tennessee

Aerospace Engineering at Tennessee, as is true for many AE Programs, had its roots in Mechanical Engineering. An interest in aerodynamics existed from the early 1930's. As early as July 31, 1937, notes from the minutes of UT President Hoskins' Selective Council indicate that Colonel Harry S. Berry, then State Administrator, expressed a need for men trained "in the field of aeronautical engineering and that he thought this constituted a marvelous opportunity for the University." The notes further state, "[Colonel Berry] knew of one aeroplane factory which was proposing to locate in the eastern portion of Tennessee and that he thought there would be other plants locating in the State." These facts, combined with the completion of an airport in Knoxville led the council to consider funding options for beginning aeronautic classes (1). Some courses in Aerodynamics and Airplane Design appeared in the Mechanics and Materials or Civil Engineering portions of the catalog until the late 1950's. In the Fall of 1959 a young man named Mancil Milligan joined the Mechanical Engineering faculty. He had just come from the Boeing Company in Seattle and had a great interest in Aerospace Engineering. Under his leadership all of the aero related courses and activities were consolidated into the Mechanical Engineering Department. In 1960, the name of the department became Mechanical Engineering-Including Aerodynamics. After offering elective courses in low and high speed aerodynamics, orbital mechanical, and propulsion for three years an Aero Option was officially offered in Mechanical Engineering and in 1964 the name of the Department was changed to Mechanical and Aerospace Engineering.

In 1965, Dean Charles H. Weaver, then Dean of the College of Engineering, came to Milligan and stated that the college needed an Aerospace Engineering degree program. His instructions were that it could be a new stand-alone department or continue in the Mechanical and Aerospace Engineering Department. Discussions between Milligan and Joel Bailey, who was

Head of the Mechanical and Aerospace Engineering Department (MAE), resulted in the decision to keep the AE programs in the MAE Department.

Milligan succeeded Bailey as Head of the MAE Department in 1973 and served in that capacity until 1982. By that time the ME and AE programs were fully integrated, with many common courses such as thermodynamics, heat transfer, dynamics, controls, instrumentation, etc. The administrative arrangement with both AE and ME degree programs being in one department has existed over the almost forty year period and continues today. Degree programs were established at the BS, MS, and PhD levels with the first BSAE degrees being granted in 1968. The BSAE degree program has been ABET accredited since its initial eligibility. Through the years many faculty members made major contributions to the AE undergraduate program with Mancil Milligan and Harvey J. Wilkerson being the major continuing contributors. Wilkerson retired in 1999. Additional information on current faculty, curriculum, and research can be viewed at www.engr.utk.edu/mabe/.

In the late 1950's an off-campus graduate program in ME was established at the USAF Arnold Engineering Development Center in Tullahoma, Tennessee. As would be expected much of the interest there was AE related and the program developed into The University of Tennessee Space Institute (UTSI). A large portion of the graduate activities in AE over the years have been centered at UTSI with a much smaller graduate program activity on the Knoxville campus. Academically the graduate programs at both locations have been fully integrated since the inception. A presentation of the UTSI history is given below.

The BSAE degree program is relatively small with 15 to 20 graduates per year. It is the only aerospace engineering degree program in the state and also serves surrounding states that do not have aero degree programs. BSAE degree recipients have populated the full spectrum of professional activities. We have graduates on the faculty of major Universities such as Georgia Tech and Penn State and serving as Dean of Engineering at David Lipscome University. Our graduates have progressed to very responsible positions in industry including Vice-Presidents at Chrysler and Lockheed-Martin and Plant Manager at IBM. More recently, BSAE degree recipients have been recognized as outstanding students. In 2000 Randy Warren was selected by Sigma Gamma Tau as the most outstanding senior in the Southeast Region and in 2002 Yvonne Jones was selected by Sigma Gamma Tau as the most outstanding AE senior in the United States.

University of Tennessee Space Institute History

The University of Tennessee Space Institute (UTSI) is a graduate education and research institute of the University of Tennessee, Knoxville. UTSI offers the same M.S. and Ph.D. degree programs in Aerospace Engineering that have been established at the University of Tennessee, Knoxville. The history of graduate aerospace engineering programs at UTSI begins with the history of UTSI, which is tied in turn to the history of the United States Air Force Arnold Engineering Development Center (AEDC), at Arnold Air Force Base, reference 2. AEDC is one of the world's premier aerospace systems test centers, with test cells that simulate air and space environments.

The stimulation for the creation of AEDC, was a report authored by a committee under the chairmanship of Dr. Theodore Von Karman, and called "Toward New Horizons," reference 3 and 4, and a memo (the so-called "Transaltantic Memo") written by Dr. Frank L. Waltendorf to Brigadier General Franklin O. Carroll, recommending a new Air Force research and development center with a wide range of testing and simulation capabilities. The essence of the documents that lead to the creation of AEDC was the impression that German aeronautical research and development was stimulated by close interaction and integration of military, academic, and corporate organizations. It was felt that an advanced aeronautical research center should have available to its technical staff both academic and research resources, specifically, the opportunities for staff members to work on and obtain advanced academic degrees, as well as the research services of well-trained faculty. The history of the conception and creation of UTSI was documented by Payne, reference 5, and will not be discussed in detail here. In any case, UTSI as a separate campus of the University of Tennessee in Knoxville grew from an on-base graduate courses program to an off base but near by academic institute and campus in the years roughly from 1956 to 1966.

The aerospace engineering program at UTSI and UT has always been integrated with the mechanical engineering program as described above by Milligan. Graduate courses in turbomachinery, rocket propulsion, and some of the fluids related courses are taught under the ME designation. A representative description of the propulsion related courses at UTSI was presented by Crawford and Schulz that is indicative of the approach of melding of the aerospace and mechanical engineering curricula, reference 6. This approach has met with great success at UT and UTSI since it fosters an interdisciplinary interaction between faculty, students, and research staff. UTSI has had a significant number of military graduates of its AE and ME programs, including US Air Force officers, some of which were Air Force Institute of Technology (AFIT) students, as well as National Aeronautics and Space Administration (NASA) employees. UTSI has as members of its alumni, eight astronauts who have graduated from the programs. NASA official and Astronaut Henry (Hank) Hartsfield (MS '71) was one of the first to complete UTSI's academic program. Many other UTSI graduates are distinguished engineers, corporate leaders, and faculty at other US and overseas universities.

B. H. Goethert, the first Director and Dean of UTSI, and R. L. Young the first Associate Dean provided most of the on-site leadership for the development of UTSI. Goethert had established a name in aeronautical research and development, first in Germany and then later in the United States. He was a major proponent of the concept of UTSI and played a central role in its creation (reference 5). Above all Goethert was a futurist, and his advanced thinking shaped much of what UTSI and its AE/ME programs were to become. Under Goethert's leadership, UTSI developed a relationship with the Technical University (the RWTH, or Rheinisch-Westfalische Technische Hochschule) of Aachen Germany, which lead to sharing of students and faculty. The cooperation between UTSI and RWTH was created through the joint effort of Dr. Goethert of UTSI, and Professor-Doctor August W. Quick of RWTH. An annual Quick-Goethert Lecture program shared by UTSI and the RWTH still marks the international connection between UTSI and the RWTH. Payne (reference 5) also describes the beginning of the UTSI-RWTH relationship in some detail.

Robert L. (Bob) Young was a young but rapidly becoming well-known professor in the early 1950's, having co-written a textbook on thermodynamics and heat transfer with Professor Edward F. Obert in 1949. Bob Young was recruited to head up the then on-base AEDC graduate courses program being run by the University of Tennessee. With the physical establishment of the separate UTSI campus, Young was made Associate Dean, principally for Academic Affairs, and he worked closely with the UT Engineering School to implement academic programs offered on campus at UTSI. Young played a major role in defining, establishing, and maintaining academic quality in all of UTSI's graduate programs, and especially in the AE/ME program.

In addition to the normal academic programs in engineering, physics, and others, at UTSI, Dr. Goethert also defined the educational mission of the Space Institute to include, from UTSI's inception, a now internationally-recognized continuing education program based primarily on technical (and occasionally non-technical) short courses. In the Aerospace Engineering discipline, UTSI short courses such as the B. H. Goethert Short Course on Aeropropulsion, the Ground Test Facilities Short Course, the Solid Rocket Motor Fundamentals Short Course, the Stability and Control Flight Testing Short Course, and many others have served thousands of engineers, scientists, aviators, program managers, and workers in the aerospace and aeronautical engineering field. These continuing education programs have been offered and conducted by UTSI faculty with great success for the last 40 years, and continue to provide a valuable resource in aerospace engineering education.

UTSI programs, including the Mechanical and Aerospace Engineering and Engineering Science Programs, as well as the Short Course Programs, can be viewed at www.utsi.edu.

References

1. President's Selective Council, Volume 16, p 292.
2. www.arnold.af.mil
3. Von Karman, T., et al., "Toward New Horizons," a report to General of the Army H. H. Arnold, submitted on the behalf of the A. A. F. Scientific Advisory Group, December 1945.
4. http://www.arnold.af.mil/aedc/baseguide/zhistory.pdf
5. Payne, W., *Web to the Stars*, Kendall/Hunt Publishing Company, 2460 Kerper Boulevard, P. O. Box 539, Dubuque, Iowa, 52004-0539, 1992. ISBN 0-8403-7088-1.
6. Crawford, R. A. and Schulz, R. J., "Graduate Propulsion Education and Research at the University of Tennessee Space Institute," AIAA paper no. 94-3116, presented at the 30th AIAA/ASME/ASEE Joint Propulsion Conference, June 27-29, 1994, Indianapolis, Indiana.

Chapter 47

A Brief History of Aerospace Engineering at Syracuse University

Thong Dang & John LaGraff
Department of Mechanical, Aerospace and Manufacturing Engineering
Syracuse University

The L.C. Smith College of Engineering and Computer Science at Syracuse University was established in 1901, the fifth of the 11 schools and colleges within Syracuse University today. An added option in aeronautical engineering was introduced into the BME (Bachelor Mechanical Engineering) program in the Mechanical Engineering department in 1949, but it was not until 1964 that the department offered a separate BS degree in Aerospace Engineering. Over the past 40 years, the graduating class in Aerospace Engineering has varied in size, between 10 and 40 students. Today, the BS degree in Aerospace Engineering is offered by the department of Mechanical, Aerospace and Manufacturing Engineering, with an entering freshman class among the largest in the college. The department also offers an MS degree in Aerospace Engineering, and a joint PhD degree in Mechanical & Aerospace Engineering. The mission of the Aerospace Engineering program at Syracuse University is to educate and promote learning and discovery in aerospace engineering and to prepare students for a career in technical excellence, professional growth, and eventual leadership in a complex and competitive technological environment.

The founders of Syracuse University Aerospace Engineering Program

The development of aeronautical engineering content in the curriculum at Syracuse University began within the existing mechanical engineering program after World War II. At that time, Professor Martin Barzelay joined the department from Republic Aviation Corporation in Farmingdale, L.I., New York and provided the leadership to introduce an option in Aeronautical Engineering within the BSME degree. Professor Barzelay was also instrumental in developing an impressive laboratory facility for undergraduate "aero" labs utilizing the surplus General Electric jet engines from the manufacturing and testing facility in nearby East Syracuse.

It was not until the mid-fifties that key faculty appointments were made in the department. These faculty appointments provided the critical mass and breath of expertise to not only greatly strengthen the undergraduate aeronautical engineering option within the BSME degree, but more importantly to establish a graduate program of study with world-class research programs in the field of aeronautical engineering. The contributions of these new faculty members also included the development of new facilities related to aeronautical engineering, many of which have been modernized and are still in use today for both undergraduate teaching and research. Together with Professor Barzelay, this group included: Dr. Ross Evan-Iwanowski, an expert in structural vibrations and control, who joined the department from Cornell University in 1954; Dr. Salamon Eskinazi, a turbulence expert, who joined the department in 1955 from Johns Hopkins University; and Dr. Darshan Dosanjh, a compressible flow and aero acoustics expert, who joined the department in 1956 and also from Johns Hopkins University. Together, these four faculty members are credited with the establishment of first-rate teaching and research programs in aeronautics at Syracuse University. The group members not only quickly established individual international research reputations, they collectively provided the enthusiasm and support for a quality undergraduate aeronautical option.

Professor Ross M. Evan-Iwanowski, known affectionately to his colleagues as E-I, joined the faculty as an Assistant Professor in the summer of 1954 after receiving his PhD in Applied Mechanics from Cornell University. He was promoted to the rank of Associate Professor in 1958 and to Professor in 1962. His area of expertise was in the linear and nonlinear vibrations of discrete and continuous mechanical systems, an area that he pursued vigorously both experimentally and theoretically until his retirement from Syracuse University in the late 1980's. The Applied Mechanics Laboratory that he established was a state-of-the-art facility used for graduate and undergraduate teaching and research. In the early 1990's the laboratory was renamed the Composite Materials Laboratory and its main focus shifted from the exploration of oscillatory phenomena to investigations related to moderate and elevated temperature delamination of composite materials.

Professor Dosanjh is an expert in compressible flows, and his research work resulted in many fundamental contributions in the fields of jet flows and jet noise. Professor Dosanjh built numerous test facilities during his 40-year tenure in the department (1956-1996). These facilities include a fully instrumented shock tube facility with optical experimental capabilities (built in late 50's), a jet flow facility (built in 60's), a Mach 4 supersonic wind tunnel facility (built in 60's), a hypersonic-flow facility (late 70's-80's), and jet noise facilities (60's - 80's). The latter facilities included an anechoic chamber, reverberation room with a heated multi-jet facility. The shock tube facility was built and used for shock wave research in the late 50's, which was considered to be a pioneering field of research in US academic institutions. Over the years, many of these facilities have been modernized and are still being used for both teaching and research. For example, our aerospace engineering senior lab course still uses the supersonic wind tunnel and the shock tube facilities to validate the theories learned in the first-course in compressible flows. Finally, it is worth noting that Prof. Dosanjh had support from NSF for undergraduate participation in research for 11 consecutive years, spanning from the 60's to the 70's. This effort resulted in many of our undergraduates winning many national competition awards. Professor Dosanjh retired from the faculty in 1992, but taught part-time until 1996.

Professor Salamon Eskinazi came to Syracuse University in 1955 as an Assistant Professor and rose through the ranks to become chairman of the Mechanical Engineering Department in the mid 1960's. During his tenure as chairman, the shift from an aeronautical engineering option in the BSME degree program to the emerging field of aerospace engineering took place. While providing leadership during the development of the undergraduate and graduate programs in Aerospace Engineering, Prof. Eskinazi continued to teach undergraduate fluid mechanics and had many undergraduates involved in his research laboratories. He established a research program in fluid mechanics and became internationally recognized for his work in turbulence. A low-speed wind tunnel he designed and built was still used for undergraduate laboratory until recently. His contributions to engineering education also included the publication of three popular undergraduate textbooks in fluid mechanics.

The addition of Dr. Charles Libove (PhD'62, Syracuse University) to the faculty in the early 1960's was an important step towards creating a separate BS in Aerospace Engineering. Professor Charles Libove had already established a research career at NASA Langley prior to joining the department, and he quickly became the "heart" of the undergraduate program in Aerospace Engineering. He put his considerable teaching talents to work in all aspects of the program. For example, he established specific requirements for the BS Aerospace Engineering curriculum, including the creation of aero-specific courses such as aerodynamics, gas dynamics, aero structures, flight performance, and design. Many of these newly-developed courses were taught by Professor Libove himself, including aero structures, flight performance, and aero design.

In 1970, Dr. Fred Lyman's and Dr. John LaGraff's appointments rounded out this initial phase of assembling a core group of aeronautically focused faculty. Dr. Fred Lyman (PhD'61, RPI) joined the faculty after doing research in propulsion at NASA Lewis Research Center and serving as a faculty member at Case Western Reserve University (his MS degree from Syracuse University was supervised by Prof. Eskinazi). Dr. John LaGraff, currently Vice President for Education for AIAA, also joined the department in 1970 after receiving his PhD from Oxford University. His teaching supported the Aerospace Engineering Program, and his research area was in the field of hypersonics.

Today's Aerospace Engineering program at Syracuse University

In the 1980's, another phase of faculty hiring in aerospace engineering took place at about the time when Prof. John LaGraff became the first separate director of the Aerospace Engineering Program. Partly in response to the retirements of professors Barzelay, Eskinazi, Evan-Iwanowski, and a strong increase in student enrollment, an active faculty recruitment effort was undertaken under the leadership of Prof. John LaGraff. During this period, 5 new aerospace faculty members joined the department. Dr. Vadrevu Murthy (PhD'74, Georgia Tech) came to the department in 1983 after spending 9 years at various industrial and government research laboratories. Dr. Eric Spina joined the department after receiving his PhD from Princeton University in 1988. In 1989, Dr. Hiroshi Higuchi (PhD'77, Cal Tech) joined the department with an established research career at NASA Ames Research Center and at the University of Minnesota. Dr. Thong Dang (PhD'85, MIT) joined the department in 1989 after spending four years in industry, and Dr. Barry Davidson (PhD'88, Texas A&M University) joined the department in 1990 after spending 2 years at the Jet Propulsion Laboratory. Dr. Mark Glauser (PhD'87, SUNY Buffalo) joined the department in 2001 after spending 14 years as a faculty member in the Department of Mechanical & Aerospace Engineering at Clarkson University.

Today, this core group of 7 faculty members is engaged in a variety of teaching and research activities connected to the Aerospace Engineering program. Research areas related to aerospace engineering include bluff-body aerodynamics (Higuchi), composite materials (Davidson), flow control (Glauser, Higuchi), helicopter dynamics (Murthy), jet noise (Glauser, Spina), personal environmental quality control (Dang, Glauser, Higuchi), supersonic/hypersonic flows (LaGraff, Spina), and turbomachine aerodynamics and heat transfer (Dang, LaGraff). Several new lab facilities have been built for both research and undergraduate teaching. In the thermo-fluids area, a low-speed wind tunnel, two water tunnels, a Mach 6 test facility, a new high-speed heated anechoic jet noise facility, and a fluid transport chamber for environmental quality study have been added. These facilities are equipped with substantial instrumentations, including two PIV stereo systems, a time-resolved PIV system, and a fiber optic based LDA/PDA measurement system. In the structures/materials area, a composite materials lab has equipment for the fabrication, mechanical testing and inspection of a variety of materials, including 10 and 200 KIP MTS servohydraulic mechanical testing machines with digital controllers, 20 KIP INSTRON servohydraulic mechanical testing machines with a digital controller, advanced ultrasonic inspection system with high-speed computer data acquisition and custom imaging software, and a 300psi/800°F computer controlled autoclave. Several computational labs are also available for teaching and research, including a recently acquired Beowulf parallel computer cluster equipped with 60 64-bit AMD Opteron 242 processors, 60GB of memory, and a Gigabit Ethernet network.

Over the years, many undergraduates participated in research projects with these faculty members, and their work have resulted in many prestigious awards from AIAA. Major student accomplishments in recent years included:

1. 1st place undergraduate paper, National 2000 AIAA paper competition (Anthony Vinciquerra).
2. 2nd place undergraduate paper, Northeast region 2000 AIAA paper competition (Arun Chawan).
3. 2nd place team, AIAA Design/Build/Fly 1998 competition.
4. 1st place paper, Zerem Award in Astronautics, 1997 (Paul Cataldi).
5. 38th AIAA/ASME/ASCE/AHS/ASC Structures, Structural Dynamics and Materials Conference, Jefferson Goblet Student Paper Award, 1997 (Simon Gharibian).

At present, the department is also engaged in many exciting initiatives specifically aimed at improving undergraduate education. One example is the Advanced Interactive Discovery Environment (AIDE) for Engineering Education project, which focuses on developing a virtual environment that integrates and advances the best features and best-practices of virtual, collaborative engineering environments, state-of-the-art simulation tools, and advanced learning management systems. The AIDE itself is a virtual environment that contains application-specific content, application-appropriate simulation and software packages, distributed learning modules, expert systems, knowledge bases, and synchronous and asynchronous communication tools, including message boards, instant messaging, chat, and multi-point audio and video. The AIDE addresses many of the current needs of NASA, which must use engineers and scientists at geographically distributed sites to design complex vehicles and missions. Funded by NASA, NY State, and AT&T Foundation, the AIDE project is currently the senior capstone design projects for aerospace engineering students.

So our good fortune was to have a new generation of aerospace engineering enthusiasts in place to build on the strong foundation started by Marty Barzelay!

Chapter 48

The early history of aeronautics at Stanford University and the founding of the Department of Aeronautics and Astronautics

Brian Cantwell

In 1957 Stanford University began a new initiative that would culminate two years later in the creation of the department of Aeronautics and Astronautics. It all started with Fred Terman's recruitment of Nicholas Hoff from the Brooklyn Polytechnic Institute in New York. But the history of aeronautics at Stanford started much earlier and is almost as old as the university itself. It began with the appointment of William F. Durand in 1904 only a year after the first flight of Wilbur and Orville Wright.

Stanford University opened its doors on October 1, 1891. The first student body consisted of 559 men and women. There were 15 members of the faculty seven of whom came from Cornell. After the death of Leland Stanford in 1893 the university went through a prolonged period of financial uncertainty until the release of his estate from probate in 1898. Only the determination of his wife Jane Lathrop Stanford and the skill of the first president David Starr Jordan kept the

First Classman Durand, United States Naval Academy, 1880

dream of a university alive during that uncertain period. In 1903 Mrs. Stanford finally transferred full control of university affairs to the board of trustees and the university has continued to grow and thrive ever since. In contrast to the great educational institutions of the East, Stanford was co-educational from the outset, non-denominational and above all practical: designed to provide an education that would produce cultured *and useful* citizens.

After 1903 one of the first departments created was a department of mechanical engineering and among the first faculty was a former naval officer and Annapolis graduate William F. Durand. Durand was recruited like so many others from Cornell University. He joined several of his former Cornell colleagues as Chairman of ME at Stanford and served in that capacity for the next twenty years. At Cornell, Durand had been a marine engineer and acting director of the Sibley College. At Stanford he became essentially a hydraulic engineer and was heavily involved

Durand with some early workers in Aeronautics at Stanford in 1941. From left to right: Professor Elliott G. Reid, Aerodynamics; Professor Alfred S. Niles, Structures; Professor William F. Durand, Professor Everett P. Lesley, Aerodynamics; Henry Jessen, Aerodynamics, Ralph Huntsberger, Aerodynamics

in the design of dams and the development of the precious water supply system of the West. He served as a consultant on the Hoover, Grand Coulee and Shasta dams. But the compelling developments in aeronautics that occurred during the period leading up to World War I eventually demanded more and more of his attention and in 1915 at the age of 56 Durand established the second course in aeronautics offered at any institution of higher learning in the United States; the first being at MIT.

The National Advisory Committee for Aeronautics (NACA) the forerunner of NASA, was created in 1915. Durand was appointed a member and shortly thereafter became its first civilian chairman. As such he helped to plan the committee's famous laboratory at Langley field. In 1917 he served as scientific attaché to the American embassy in Paris where he helped organize the post-war Inter-allied Inventions Committee. During this assignment he made the acquaintance of a young Naval Lieutenant and aviation enthusiast named Harry Guggenheim. This chance friendship was eventually to have a profound impact on the development of aeronautics research and education in the United States.

Durand's world-wide eminence continued to grow and in 1918 he was the first American to deliver the annual Wilbur Wright lecture to the Royal Aeronautical Society of Great Britain.

Durand retired from academic life in 1924 only to begin what was to become one of the most important chapters in his extraordinary career. In 1925 he became a member of the President's Aircraft Board. This famous "Morrow Board" led to the appointments of assistant secretaries for Air in the War and Navy Departments and fostered passage of the basic Civil Aeronautics Act by Congress. In 1926 he became a trustee of the Daniel Guggenheim Fund for the promotion of aeronautics. Much of the enthusiasm with which Harry Guggenheim prevailed on his father to establish this fund came from his association with Durand during the war. This fund was used to establish and support new aeronautics departments throughout the country. This early and timely endowment has served the country extraordinarily well and provided many of the trained engineers that enabled the United States to reach a position of pre-eminence in aeronautics.

Another great engineer at Stanford at the time was Everett Parker ("Bill") Lesley. Lesley received his bachelor's degree from Stanford in 1897, his master's degree from Cornell in 1905 and joined the Stanford faculty in 1907. This was shortly after the great earthquake and Lesley spent a good deal of his time on the reconstruction of the university.

Durand and Lesley worked together on the design of propellers, a subject made vitally important by the events of World War I. In 1915 they built the first of Stanford's three wind tunnels to carry out a project on propellers for the National Aeronautical Commission. They tested 125 propellers designed in families that varied by blade twist and several aerodynamic shape characteristics. Fifty of these propellers are now on display in the Engineering Library housed in the Terman Engineering Center. The result of their research was a large catalog of propeller data used by aircraft engineers for more than twenty years.

Durand and Lesley's tests were finished by 1926 after which Stanford's aeronautical engineering program moved ahead with a substantial grant from the Guggenheim Foundation. Two appointments were made using Guggenheim support. Elliott Grey Reid came to Stanford from the NACA Langley laboratory in Virginia in 1927 as the youngest full professor at the university. Reid is remembered as a superb experimentalist with uncompromising standards and one who, according to the memorial written after his death, "insisted on a precision of understanding and expression beyond that required by most teachers". He is best known for his book, *Applied Wing Theory*, published in 1932. The second faculty member brought in with Guggenheim support was Alfred Salem Niles, who had received his degrees from Johns Hopkins and MIT. Niles was the head of the structures unit at Hope Field, Ohio, near Wright-Patterson Air Force Base. While Reid specialized in aerodynamics, Niles did teaching and research in aircraft structures. Niles is best remembered for his book *Aircraft Structures* co-authored with Joseph Newell of MIT. This book remained the bible of aircraft structural design well into the 1950's.

In 1929 Durand began work as editor of the six-volume work *Aerodynamic Theory* under the sponsorship of the Guggenheim fund. The set was completed in 1936. In 1933 Durand resigned from the NACA and in 1935 he became chairman of the special committee on airships that was to recommend future design practices after the loss of the airship Akron off the New Jersey coast and the Macon lost off the California coast. Durand was a skilled diplomat and an eloquent speaker and was much in demand for national and international committees. He was elected president of the World Power Congress and at the opening meeting in Washington in 1936 he addressed the delegates in English, then in French, German and Spanish, all languages he had mastered. Today a memorial in Stanford's Durand Building reads: "His first professional assignment in 1880 was on the USS Tennessee, a full rigged wooden ship with auxiliary steam power. His last, 1942-46, was as chairman of the National Aeronautical Commission for the development of jet propulsion for aircraft." A true aeronautical pioneer whose career spanned the era from steam-power to jet-flight, Durand died in 1958 at the age of 99.

By 1939 there were 42 Stanford aeronautics graduates active in the airframe and airline industry, in the military and in government research. The area next to Moffett Field was designated the Ames Research Center by the NACA in 1939 and became active in a range of aeronautical research activities. But after World War II, activity in aeronautics slowed greatly at Stanford just as it did in industry during the postwar slump. The Guggenheim fund ran out of money in 1939 and by the late 1950's with the retirement of Reid and Niles approaching and with the number of students down to a trickle of four or five a year, Dean of Engineering Frederick Terman seriously considered discontinuing what was then the aeronautics division of mechanical engineering.

When graduates heard this they offered to raise money from the aircraft industry to save the program. A committee led by John Buckwalter ('24, Engr. 32) of Douglas Aircraft and Philip Coleman ('34) of Lockheed formed a committee and asked each major western aircraft company to contribute $5,000 per year for five years to help re-invigorate aeronautics at Stanford and get it back to a position where it could attract students and research support. Douglas, Boeing, Convair, Northrup, North American, Hughes, and Lockheed all agreed to provide support. This committee was the predecessor of the Aero/Astro affiliates program that has remained the department's primary connection to industry for the last forty-five years.

One of Alfred Niles' PhD graduates was Nicholas Hoff (PhD '42) who was then head of the aeronautical engineering department at the Brooklyn Polytechnic Institute. Hoff had gained a reputation as a superb researcher and administrator at Brooklyn and was one of the rising stars in aeronautics. He was also one of the top consultants for the Lockheed Company. Willis Hawkins, head of Lockheed's missile program, offered Hoff a full time job and when Hoff declined Hawkins asked if he would consider going to Stanford as a professor. Lockheed would cover half of his salary for five years and provide him with consulting opportunities.

The success of this fundraising effort and the leadership that Hoff had demonstrated at Brooklyn greatly impressed Terman. This was a time of resurgence for aeronautics and emergence of the new field of astronautics. Sputnik, the first satellite, had launched the Space Age in 1957, and soon afterward the importance of high-speed flight and space flight would be widely recognized. In 1957 Terman brought Hoff to Stanford with the understanding that he would head a new department. A Centennial Conference in honor of Durand in August of 1959 pointed to the growth of the department with a substantial student body and research contracts and gave Terman the clear evidence he needed to determine that the new department could stand on its own. In Sept. 1959, the Department of Aeronautics and Astronautics was formally established as a separate, graduate-only department with Nicholas Hoff as its first chairman. Housed in the building named after Durand, the department is a continuing, vibrant tribute to the individual whose life meant so much to the American aeronautical enterprise.

The close connection between the newly founded department and local industry as well as the nearby NACA Ames Research Center would have a profound impact on the shape of the department for many decades. Hoff brought to Stanford Lockheed scientist Dan Bershader, an expert in high temperature gas dynamics, to teach part-time for free. Bershader's appointment was eventually converted to a full-time tenured one as were several others at the time. In addition, Hoff hired Walter Vincenti in 1957 from Ames Research Center whom he had known since his student days. Vincenti had become well known for his work in transonic and supersonic aerodynamics. At Ames he had supervised research in supersonic wing design and had begun to move into the emerging area of aerothermodynamics which involved flight at extreme speeds where air begins to undergo chemical reactions. Vincenti's research moved the department for the first time into space and re-entry applications and he worked with Ronald Smelt of Lockheed to develop a hot-shot wind tunnel for studying hypersonic flow. Soon afterward Vincenti recruited his Ames colleague Milton Van Dyke who, together with his thesis supervisor at Caltech Paco Lagerstrom, had pioneered methods for analyzing the complex flow field about aerodynamic shapes in subsonic and supersonic flow. At Stanford Van Dyke developed an active program in hypersonic flow theory.

More appointments would soon follow. These included: Krishnamurty Karamcheti, a recent Caltech PhD and expert on flow generated noise, Howard Seifert an authority on propulsion and in 1960 President of the American Rocket Society, Donald Baganoff, a fresh PhD from Caltech and an expert in gasdynamics, Jean Mayers a renowned expert in aircraft structures and I-Dee Chang a superb theoretician and another student of Lagerstrom. Using his persuasive charm and the lure of a California lifestyle, Hoff continued to recruit highly visible faculty from the East coast and in 1965 he brought from MIT Holt Ashley, one of the nation's foremost researchers in high speed flight and aeroelasticity.

Hoff also moved the department into the increasingly important field of guidance and control and hired Robert Cannon from MIT. It was Cannon's inertial guidance system design that had been chosen to navigate the *USS Nautilus* in its famous voyage under the North Pole in 1958. In the early 1960's Cannon used Air Force sponsorship to rapidly build up a world class research group. Hoff and Cannon recruited more faculty from Lockheed including Benjamin Lange, who worked on the so-called drag-free satellite and Dan DeBra, a superb experimentalist and designer who applied the idea to a family of Navy navigation satellites. They also brought to Stanford from Lockheed John Breakwell a world-class theorist in orbital mechanics. Eventually they would attact Arthur Bryson from Harvard who was recognized as one of the most brilliant and creative researchers in the field of automatic control. In 1966 this group awarded a PhD to a young Naval Academy graduate by the name of Bradford Parkinson. Six years later Parkinson would become the director of the department of defense's new GPS program office where he would play a key role in utilizing the best features of several competing designs to create the basic GPS architecture in use today. For this work and his subsequent research in industry and later as a faculty member in the Aero/Astro department, the National Academy of Engineering would honor Brad Parkinson and Ivan Getting of the Aerospace Corporation with the 2003 Charles Stark Draper Award; the most prestigious award in engineering.

Aeronautics and Astronautics Faculty in January 1968 together with faculty jointly appointed from Mechanical Engineering. Top row, left to right: Daniel DeBra, Jean Mayers, Charles Steele, Erastus Lee, Holt Ashley. Middle row, left to right: Donald Baganoff, John Breakwell, Irmgard Flugge-Lotz, Walter Vincenti, Nicholas Hoff, Daniel Bershader, Bryan Noton, Benjamin Lange. Bottom row: I-Dee Chang, Milton Van Dyke, Howard Seifert, R. M. Carlson, Chi-Chang Chao, Krishnamurty Karemcheti, Max Anliker. Inset, upper right: Robert Cannon, inset upper left: Alfred Niles, just below right: Elliott Reid.

This remarkable series of appointments under Hoff's leadership solidified Stanford's position at the forefront among research universities. The department continued to grow throughout the 1960's and by 1970, with the new Durand building in place, more than a million dollars yearly in research, and over two hundred graduate students, Stanford surpassed MIT as the nations largest producer of PhD graduates in Aeronautics and Astronautics.

In commemoration of the 100th anniversary of the Wright Brothers flight the AIAA Journal recently published a special issue (Volume 41, number 7A, July 2003) devoted to reprinted articles of special significance that have appeared in the journal and its predecessors going back to 1934. Four of the thirty-six landmark papers selected for this issue were authored by current and former faculty members of the department including, Holt Ashley, Dean Chapman, Milton Van Dyke and Bob MacCormack.

Today with fourteen faculty and approximately two hundred and twenty graduate students, Stanford continues to rank in the top tier of departments of Aeronautics and Astronautics in the nation.

Acknowledgement:

I would like to thank Walter Vincenti and Donald Baganoff for sharing their personal recollections of the department and for their help with names and dates. Any errors are mine alone.

References:

1)Leslie, Stuart W. 1993 *The Cold War and American Science*, Columbia University Press

2) Reid, Elliott G. 1944 William Frederick Durand a biographical sketch, in *Selected Papers of William Fredrick Durand* reprinted in commemoration of the 85th anniversary of his birth with preface by Theodore von Karman, California Institute of Technology.

Chapter 49

The Evolution of Aerospace Engineering Education at California State Polytechnic University, Pomona

Conrad F. Newberry, Professor Emeritus; Ali R. Ahmadi, Professor and Interim Chairman; Paul A. Lord, Professor Emeritus; and George R. Graves, Professor Emeritus

Abstract

This paper describes the evolution of the aerospace engineering program at California State Polytechnic University, Pomona. Undergraduate education is the primary mission of Cal Poly, Pomona and its Aerospace Engineering Department. The Cal Poly, Pomona aerospace undergraduate curriculum at its inception in 1958, middle age in 1980, and maturity in 2002 is contrasted in terms of theoretical foundation, design, and engineering practice. A brief description is given of novel undergraduate program features, undergraduate enrollment and graduation histories, capstone design, faculty, and student successes. Commentary is also presented on interdisciplinary learning venues, the flexibility of the senior project classes, some innovative teaching paradigms, and student performance in regional and national competitions. Some discussion is provided regarding student and faculty research. A successful, almost unique, external aeronautical Master's degree program is described in some detail

Introduction

California State Polytechnic University, Pomona (CSPUP) traces its origin to a California vocational high school located just outside of San Luis Obispo, midway between Los Angeles and San Francisco on a campus which now encompasses some 2,850 acres in the foothills of the Santa Lucia mountains, some twelve miles from the Pacific Ocean. In 1927, the educational level of the institution was elevated to that of a junior college. Shortly thereafter, in 1933, Cal Poly became a two-year and three-year technical college. 1936 saw the addition of a degree transfer program. The California State Board of Education authorized Cal Poly to grant the Bachelor of Science degree in 1940; the first baccalaureate degrees were awarded in 1942 [1].

Southern Campus

In 1938, Charles B. Voorhis of Pasadena and his son, former Congressman Jerry Voorhis, deeded a completely equipped school and farm of 157 acres to Cal Poly. The Voorhis school was located near San Dimas in Los Angeles County, some 36 miles east of Los Angeles. The Voorhis school became the southern campus of Cal Poly. This southern campus was a near ideal location for technical instruction in citriculture, deciduous fruit production, agricultural inspection, and landscape gardening. Due to the exigencies of World War II, this campus was closed in 1942 and reopened in 1945 [1].

During World War II, the U. S. Army used the Kellogg Ranch just west of Pomona as a remount station. In 1949, the W. K. Kellogg Foundation of Battle Creek, Michigan gave the ranch to the State of California i.e., California State Polytechnic College. The ranch consisted of approximately 816 acres of ranchland located some 33 miles from Los Angeles and near the Voorhis campus. The deed of this gift required that Cal Poly maintain an Arabian horse breeding program (and Sunday horse shows) that had already been established on the Kellogg ranch. With the construction of a Science Building in 1956, instruction on the southern campus was shifted from the Voorhis campus to the Kellogg campus [1]. The Voorhis campus was retained as a continuing education facility. The Voorhis campus was sold in 1978.

Initially, Cal Poly was a coeducational institution. In 1929 the enrollment of women was discontinued. The enrollment of women was resumed in 1956 on the San Luis Obispo campus and in 1961 on the Kellogg-Voorhis (southern) campus [1].

In 1966, the Trustees of the California State Colleges elevated the Kellogg-Voorhis (southern) campus of the California State Polytechnic College to a separate college in its own right; California State Polytechnic College, Pomona. The San Luis Obispo school became California Polytechnic State College, San Luis Obispo; both schools refer to themselves as Cal Poly - to the continuous, constant consternation of many people. The two schools continue to cooperate on a number of projects, e.g., a two-piece (often prize winning) float in the annual Rose Parade. On June 1, 1972, California State Polytechnic College, Pomona became California State Polytechnic University, Pomona, primarily an undergraduate institution. Cal Poly, Pomona has grown from an institution having six undergraduate programs enrolling approximately 550 men in 1956 to one in 2003 with 71 undergraduate and graduate programs enrolling some 17,800 men and women [1,2]. Although situated in a very large metropolitan area, the large university campus provides a quiet, unhurried, almost ranchland environment for learning (once you get past the parking lot).

Engineering Programs

The southern (Kellogg) campus undergraduate aeronautical engineering program, along with programs in civil, electronic, mechanical, and industrial engineering, was established in 1956, when instruction was moved from the Voorhis campus to the Kellogg campus (a distance of some seven miles). *Learn by Doing* is the educational paradigm followed by both Cal Poly schools. All Cal Poly, Pomona engineering programs were created to produce graduates well versed in design, planning, product development, production, operations, management, service, and technical sales [1]. This practical feature is seen in the initial curricula for all Cal Poly, Pomona engineering programs. Thus, each engineering program provided a theoretical basis enriched with production methods and advanced mathematics. Graduates are expected to contribute in their professional position from their first day of employment. Alternatively, graduates are well prepared for graduate studies.

Aerospace Engineering Program

The initial Cal Poly, Pomona faculty consisted of the Acting Department Head Wallace E. Nally. Cal Poly, Pomona classes have always been given in a quarter system. In keeping with the *Learn by Doing* paradigm, the aerospace engineering curriculum has generally had at least one aerospace engineering course in each quarter of instruction throughout the twelve-quarter program. Cal Poly, Pomona students are expected to declare a major when they enroll, if they expect to graduate in four years. Each aerospace engineering student is required to complete a two-quarter project during their senior year; essentially this project is a senior thesis. Generally, this project is expected to combine theoretical and experimental work. The undergraduate program has been accredited since 1970 by the Accreditation Board for Engineering and Technology (ABET).

In general, Cal Poly, Pomona graduates are prepared, upon graduation, to enter graduate school, be productive upon their entry into industry, and conduct personal lifelong learning as the professional aerospace engineering community grows. Few engineers perform research (something like less than 3-4%), most assignments are related to design, production and sales. Cal Poly, Pomona has generally had a design, production oriented faculty. At one time, the average faculty member in the Aerospace Engineering Department had spent approximately ten years in industry, in a wide variety of assignments.

The Department obtained a Charter for an Institute of the Aeronautical Sciences (IAS) Student Branch in 1959. George R. Graves was the first Faculty Advisor, serving for five years. The IAS Student Branch became the American Institute of Aeronautics and Astronautics (AIAA) Student Branch, when the IAS and the American Rocket Society merged in 1963. Professor Newberry followed Professor Graves as the Student Branch Faculty Advisor and also served in this position for five years. Over time, almost all Departmental faculty members have served as the Student Branch Faculty Advisor.

The Cal Poly, Pomona AIAA Student Branch has always been an active group. It has sponsored car rallies, conducted field trips to nearby aerospace production facilities (for its members), and hosted the AIAA Region VI Student Conference at least three times (1969, 1978, and 1998). Students from the Branch have nearly always participated in the annual AIAA Region VI Student Conference and in the annual AIAA vehicle and engine design competitions, winning many awards in both sets of competitions. For example, in the undergraduate portion of the of the Region VI Student Conference, Cal Poly, Pomona students have placed (overall) third (Gary Clark) in 1966, second (Motoaki Ashizowa) in 1967, first (Dennis R. Roark) and third (James R. Knepshield) in 1970, first (Wayne E. Hall) and third (Richard C. Barbiere) in 1972, and first (Marco A. Luque) in 2001 and a tie for third.(Rebecca L. Carter and Jose G. Armentra) in 2003. Placements in the design competitions are discussed in a later section of this paper.

Since 1993, an industry advisory committee known as the Aerospace Action Council was formed to provide guidance to the Aerospace Engineering Department, via quarterly meetings with the faculty. The current, approximately twenty member, advisory committee has representatives from Loral-EOS, Hoh Aeronautics, Northrop Grumman, Lockheed Martin, Boeing, JPL, and Aerojet,

Cal Poly, Pomona graduate education in aerospace engineering began in 1972. With few exceptions, the Cal Poly, Pomona campus graduate engineering student will receive the Master of Science in Engineering degree upon the completion of 45 quarter units of coursework, of which at least 24 units must be in 500 or 600 level courses.

Undergraduate Program

Students were first accepted into the Cal Poly, Pomona Aeronautical Engineering Department in the fall of 1957, although the facilities for this program were not scheduled for completion until 1958. The initial laboratories consisted of a low-speed wind tunnel, a smoke tunnel, a cold flow supersonic "blow-down" wind tunnel, and a structural strong wall. The Department name was changed to the Department of Aerospace Engineering in 1963.

Circa 1958. Initially, Wallace E. Nally was the Acting Department Head and its only faculty member. The initial aeronautical engineering curriculum is shown below. It will be noted that the curriculum is very much a "hands-on" course of study, in keeping with the *Learn by Doing* paradigm. The curriculum was designed to make its graduates productive from their first day of industrial employment, while providing sufficient theoretical content to enable graduates to learn on their own after graduation and/or pursue graduate study at any major research university.

In this initial curriculum, it will be noted that considerable attention was devoted to manufacturing and production processes; something now thought to be important. Considerable attention was also given to design across the curriculum; again something now considered important by accreditation organizations.

The teaching load for faculty was typically twelve quarter units (minimum) per academic quarter. The academic quarter was ten weeks in length with an eleventh week for final examinations. A faculty member's nine-month salary was distributed over twelve months.

The curriculum required 210 quarter units for graduation. Most similar programs required some 186 quarter units. The approximate 24-unit difference was largely devoted to the practical manufacturing and production aspects of aerospace engineering.

By 1962 Rodney D. Sutherland was Acting Department Head. Sutherland was made Department Head in 1965 and Department Chairman in 1968. There were two additional faculty members: George R. Graves (1958) and Conway H. Roberts (1961). Horatio O. Morgan [Col. USAF (ret.)] was hired in 1963 to help with the laboratory courses. Colonel Morgan could spend the day in a laboratory with lathes, milling machines, presses and welding machines operating full tilt and never have a drop of oil or anything else despoil his blue suit - it wouldn't dare. Albert D. Sanford and Conrad F. Newberry were hired in 1964.

Circa 1980. The Cal Poly, Pomona Aerospace Engineering Department had changed considerably since 1958. There was less emphasis on drafting, production, and manufacturing processes, compared to the 1958 curriculum. Also, there was less emphasis on design. However, as a result of these changes, the program was accredited some years before (1970) by the Accreditation Board for Engineering and Technology (ABET).

Aeronautical Engineering Curriculum 1958					
Aeronautical Engineering Courses			**Support and Elective Courses**		
Aircraft Machinery & Structures	Aero 124	(3)	Engineering Drafting	ME 121	(2)
Aircraft Machinery & Structures	Aero 125	(3)	Engineering Drafting	ME 122	(2)
Aircraft Machinery & Structures	Aero 126	(3)	Descriptive Geometry	ME 125	(3)
Aeronautical Laboratory	Aero 144	(1)	Machine Shop	MS 144	(2)
Aeronautical Laboratory	Aero 145	(1)	Machine Shop	MS 145	(2)
Aeronautical Laboratory	Aero 146	(1)	Sheet Metal Laboratory	MS 155	(1)
Aircraft Drafting	Aero 247	(2)	Sheet Metal Laboratory	MS 156	(1)
Aircraft Drafting	Aero 248	(2)	Welding	Weld 144	(1)
Aerodynamics	Aero 301	(3)	Welding	Weld 145	(1)
Aerodynamics	Aero 302	(3)	Mathematics for Engineers	Math 117	(5)
Aircraft Stress Analysis	Aero 327	(3)	Mathematics for Engineers	Math 118	(5)
Aircraft Stress Analysis	Aero 328	(3)	Calculus	Math 201	(3)
Aircraft Detail Design	Aero 344	(2)	Calculus	Math 202	(3)
Aircraft Detail Design	Aero 345	(2)	Calculus	Math 203	(3)
Aircraft Detail Design	Aero 346	(2)	Differential Equations	Math 316	(3)
Aircraft Design Layout	Aero 444	(3)	General Physics (Mechanics)	Phys 131	(4)
Aircraft Design Layout	Aero 445	(3)	General Physics (Thermofluid/Optic)	Phys 132	(4)
Aircraft Design Layout	Aero 446	(3)	General Physics (Electric/Magnet)	Phys 133	(4)
Aerodynamics	Aero 404	(3)	Engineering Statics	ME 201	(3)
Aerodynamics	Aero 405	(3)	Engineering Dynamics	ME 204	(3)
Aeronautical Laboratory	Aero 457	(2)	Strength of Materials	ME 202	(3)
Aeronautical Laboratory	Aero 458	(2)	Strength of Materials	ME 203	(3)
Aircraft Propulsion Systems	Aero 421	(4)	Strength of Materials Lab	ME 249	(1)
Aircraft Propulsion Systems	Aero 422	(3)	Kinematics	ME 223	(3)
Senior Project	Aero 461	(2)	Thermodynamics	ME 301	(3)
Senior Project	Aero 462	(2)	Heat Transfer	ME 313	(3)
Undergraduate Seminar	Aero 463	(2)	Electrical Engineering	EE 231	(3)
Advisor Approved Electives		(12)	Electrical Engineering	EE 232	(3)
			Electronic Engineering	EE 222	(3)
			Electronic Engineering	EE 223	(3)
			Chemistry	Chem 321	(4)
			Chemistry	Chem 322	(4)
			Industrial Relations	EC 412	(3)
			General Education Courses		(38)
Total		(78)	Total		(132)

Table 1
CSPUP Aeronautical Engineering Curriculum 1958

Although space topics were introduced in a number of courses, the curriculum leaned a bit to the side of aeronautics. Mathematically, the students were fluent with matrices, differential equations, finite differences, vector analysis, complex variables, transforms, and tensor analysis. Most courses contained computer applications in some form, either the use of an existing code or the writing of new code by the individual student. All students could write FORTRAN code and most could write BASIC code as well. A variety of software was used by students in a number of classes.

As mentioned above, the Department sponsors an active Student Branch of the American Institute of Aeronautics and Astronautics (AIAA). Qualified students had the opportunity to join the Tau Beta Pi (engineering) and/or Sigma Gamma Tau (aerospace engineering) honor societies. Although the coursework was, arguably, more theoretical than in 1958, an attempt was made to retain the practical aspects of aerospace engineering. The curriculum was still designed to prepare the student to be (1) productive from his first day in industry, (2) capable of learning on his or her own, or (3) proceed directly to graduate school.

The teaching load was still typically twelve quarter units (minimum) per academic quarter. Rodney D. Sutherland continued to serve as Department Chairman. The other full time faculty members were Donald C. Curran, George R. Graves, Conrad F. Newberry, and Paul A. Lord. The 1980 Aerospace Engineering Department curriculum is shown below.

Aerospace Engineering Curriculum 1980					
Aerospace Engineering Courses			**Support and Elective Courses**		
Aero Engineering Fundamentals	AR0 124	(2)	American Civilization	AMC 201	(4)
Aero Engineering Fundamentals	ARO 125	(2)	American Civilization	AMC 202	(4)
Aero Engineering Fundamentals	ARO 126	(2)	Life Science	Bio 110	(3)
Introduction to Systems Engineering	ARO 244	(2)	General Chemistry	Chem 111	(3)
Systems Engineering	ARO 245	(2)	General Chemistry	Chem 112	(3)
Experimentation Systems	ARO 246	(2)	General Chemistry Laboratory	Chem 151	(1)
Vector Fluid Dynamics	ARO 301	(4)	General Chemistry Laboratory	Chem 152	(1)
Subsonic Aerodynamics	ARO 305	(4)	Principles of Economics	EC 201	(4)
Gas Dynamics	ARO 311	(4)	Elements of Electrical Engineering	ECE 231	(4)
Propulsion Systems	ARO 312	(4)	Electronic Instrumentation & Control	ECE 333	(4)
Introduction to Structural Mechanics	ARO 326	(4)	Analog Computers Laboratory	ECE 343	(1)
Aerospace Structural Mechanics	ARO 327	(4)	Freshman Composition	ENG 104	(4)
Energy, Mass & Momentum Transfer	ARO 401	(4)	Descriptive Geometry [or Computer Aided Drafting MFE 210 (2)]	MFE 125	(2)
Supersonic Aerodynamics	ARO 404	(4)	Analytical Geometry and Calculus	MAT 114	(3)
Dynamics of Aerospace Systems	ARO 406	(4)	Analytical Geometry and Calculus	MAT 115	(3)
Advanced Aerospace Analysis	ARO 444	(3)	Analytical Geometry and Calculus	MAT 116	(4)
Advanced Aerospace Analysis	ARO 445	(3)	Calculus of Several Variables	MAT 214	(4)

Advanced Aerospace Design Project	ARO 446	(2)	Calculus of Several Variables	MAT 215	(4)
Senior Project	ARO 461	(2)	Differential Equations	MAT 216	(4)
Senior Project	ARO 462	(2)	Laplace Transforms & Fourier Series	MAT 317	(3)
Undergraduate Seminar	ARO 463	(2)	Mathematical Analysis (Vector Analysis) of Engineering Problems	MAT 318	(3)
Advisor Approved Electives		(16)	Vector Statics	ME 214	(3)
			Vector Dynamics	ME 215	(4)
			Thermodynamics	ME 301	(4)
			Material Science	MTE 207	(3)
			General Physics (Mechanics)	PHY 131	(4)
			General Physics (Thermofluid/Optic)	PHY 132	(4)
			General Physics (Electric/Magnet)	PHY 133	(4)
			Basic Science Elective		(3)
			Social Science-Humanities		(15)
			Electives		(6)
Total		(82)	Total		(120)

Table 2
CSPUP Aerospace Engineering Curriculum 1980

Circa 2002. In 2002 the Aerospace Engineering Department was forty-five years old. Its undergraduate program was generally recognized as one of the better ones in the nation. Its students were still being educated to be either productive from their first day of employment or successful in graduate school. It had an on-campus as well as a nationally recognized off-campus external Master's degree program operating in El Segundo, California (Northrop Grumman) and in Palmdale, California (Lockheed Martin).

Rodney D. Sutherland, George R. Graves, Donald C. Curran, Robert F. Davey, Paul A. Lord, and Conrad F. Newberry had retired by this date. During the decade of the 1990s Professors Davey (1989-1991) and Lord (1991-1997) also served as Department Chairmen. William E. Mortensen was the Department Chairman (1997-2003) through June of 2003, when he retired. The Department is now somewhat smaller, reflecting the downsized aerospace industry. The current Departmental faculty members are Ali R. Ahmadi, Gabriel G. Georgiades, and Donald L. Edberg. Dr. Ahmadi is currently the interim Department Chairman. Professors Graves, Lord, and Davey still occasionally teach a course or so when needed.

Aerospace Engineering Curriculum 2002					
Aerospace Engineering Courses			**Support and Elective Courses**		
Intro Aerospace Engineering I	ARO 101A	(1)	CME Thermodynamics I	CHE 302	(4)
Intro Aerospace Engineering II	ARO 102A	(1)	Analytical Geometry & Calculus II	MAT 115	(4)
Intro Aerospace Engineering III	ARO 103A	(1)	Analytical Geometry & Calculus III	MAT 116	(4)
Aero Engr Compute Graphic/Lab	ARO 127/L	(2)	Calculus of Several Variables	MAT 214	(3)
Fundamentals of Systems Engr	ARO 201L	(1)	Calculus of Several Variables	MAT 215	(3)
Fundamentals of Aeronautics	ARO 202L	(1)	Differential Equations	MAT 216	(4)
Fundamentals of Astronautics	ARO 203L	(1)	Laplace Transforms & Fourier Series	MAT 317	(3)
Fluid Mechanics	ARO 301	(4)	Mathematical Analysis (Vector Analysis) of Engineering Problems	MAT 318	(3)
Subsonic Aerodynamics	ARO 305	(4)	Materials Science	MTE 207	(3)
Astronautics	ARO 309	(3)	General Physics (Mechanics)	PHY 131	(3)
Gas Dynamics	ARO 311	(3)	General Physics (Thermofluid/Optic)	PHY 132/L	(4)
Aerospace Propulsion Systems	ARO 312	(4)	General Physics (Electric/Magnet)	PHY 133/L	(4)
Aerospace Feedback Control Sys	ARO 322/L	(4)	Elements of Electrical Engineering	ECE 231	(3)
Introduction to Structural Mech	ARO 326	(4)	Elements of Electrical Engr Lab	ECE 251L	(1)
Aerospace Structural Mechanics	ARO 327	(3)	General Education Area A		(12)
Aero Structural Mech & Design	ARO 329	(3)	General Education Area B		(16)
Fluid Mech/Heat Transfer Lab	ARO 351L	(1)	General Education Area C		(16)
Aerodynamics & Propulsion Lab	ARO 352L	(1)	General Education Area D		(20)
Aerospace Structures Laboratory	ARO 357L	(1)	General Education Area E		(4)
Heat, Mass & Momentum Transfer	ARO 401	(4)	Vector Statics	ME 214	(3)
High-Speed Aerodynamics	ARO 404	(3)	Vector Dynamics	ME 215	(4)
Aerovehicle Stability & Control	ARO 405	(4)			
Dynamics of Aerospace Systems	ARO 406	(4)			
Senior Project	ARO 461	(2)			
Senior Project	ARO 462	(2)			
Aerospace Concepts Integration	ARO 490L	(1)			
Introduction to Vehicle Design	ARO 491L	(2)			
Vehicle Design I Laboratory	ARO 492L	(2)			
Vehicle Design II Laboratory	ARO 493L	(2)			
Advisor Approved Electives		(12)			
Total		(81)	Total		(121)

Table 3
CSPUP Aerospace Engineering Curriculum 2002

The balance between aeronautics and astronautics is better in this curriculum than in prior curricula. The teaching load remains at twelve-quarter units (minimum) per academic quarter. Generally this means teaching three courses (minimum) per academic quarter. Compared to the 1980 curriculum, the 2002 curriculum features more design. Computer programming or usage is a component in most courses; C++ seems to be the coding language of choice. The AIAA Student Branch remains quite active. The 2002 aerospace engineering curriculum, requiring 202-quarter units, is shown above.

Some, mathematics [such as Analytical Geometry and Calculus MAT 114 (4)], chemistry, and physics courses are contained within the General Education requirements. There are five areas within the General Education requirements: Area A, Communication and Critical Thinking; Area B, Mathematics and Natural Sciences; Area C, Humanities; Area D, Social Sciences; and Area E, Lifelong Understanding and Self-development. Although twelve units in these general education areas are elective, most of the courses are specified.

Student Data. Aerospace engineering student enrollment and graduation data for Cal Poly, Pomona reflect the oscillating character of the aerospace industry - out of phase by three or four years. Figure 1 shows the undergraduate enrollment data from 1981 through the present.

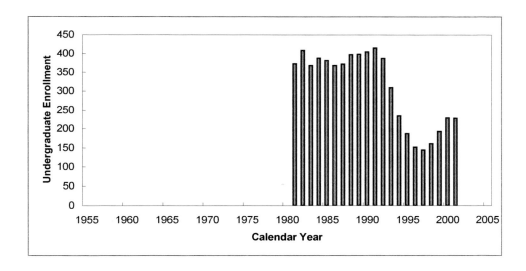

Figure 1
Aerospace Engineering Undergraduate Enrollment

Prior to 1983, the data are not as reliable as the authors would like and are not shown for that reason. From the 1960s to the present, the Department undergraduate student enrollment (freshman, sophomore, junior, and senior years) has roughly varied between the low-70s to just a little over 400. Figure 2 shows student graduation numbers over time since 1983. Again, some of the early numbers are not considered reliable and are, therefore, not shown.

The cyclic nature of the data is apparent. Data shown prior to 1981 are good faith estimates. Data are not shown where these estimates do not exist. As noted above, aerospace engineering undergraduate enrollment tends, by some three or four years, to be out of phase with aerospace industry employment cycles.

Post-Undergraduate Studies. The Cal Poly, Pomona Aerospace Engineering Department has sent numerous students to graduate schools across the nation. For example, a number of students have completed graduate work (master's and/or doctoral work) at the University of Southern California, UCLA, University of Kansas, University of Washington, Virginia Polytechnic Institute and State University, MIT, Stanford, CSU Fresno, California Institute of Technology, George Washington University, the Naval Postgraduate School, West Coast University, and the State University of New York (SUNY) at Buffalo. At least two students have used the undergraduate aerospace engineering program as the stepping stone to law school.

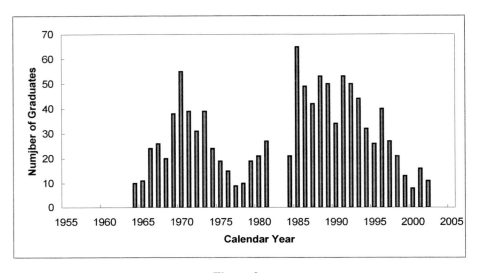

Figure 2
Aerospace Engineering Timewise Graduates

Graduate Program

The Cal Poly, Pomona graduate aerospace engineering program has consisted of two separate components: on-campus and external. Initially, the Master of Engineering degree was awarded to students successfully completing either program. Currently, the Master of Science in Engineering with emphasis in Aerospace Engineering is awarded to students successfully completing on-campus graduation requirements. The Master of Engineering with emphasis in Aeronautical Engineering is awarded to students successfully completing the external degree program.

Although the two programs are similar, the on-campus program is perhaps a bit more theoretical, and the student may elect to do a thesis. With few exceptions, graduates of the external degree program (at the request of industry) are required to take a comprehensive examination covering his or her graduate coursework. Only a few courses are actually common to the two programs. The external degree program coursework is specifically tailored to the needs of aircraft manufacturers. A minimum of 45 quarter units (roughly eleven courses) must be completed in either program.

Campus Degree Program. Each Cal Poly, Pomona graduate engineering program of study consists of at least fifteen quarter units of breadth courses, at least fifteen quarter units of elective technical specialty courses, and either a thesis or a comprehensive examination. Thus, the on-campus graduate aerospace engineering student has considerable flexibility in structuring his or her individual Master's degree program. For example, an on-campus graduate aerospace engineering student may want breadth in his or her technical specialty courses by taking advanced courses in aerodynamics, structures, propulsion, and controls or, alternatively, the student can specialize in one particular disciplinary area such as, say, flight controls.

The on-campus student may elect to either complete a research thesis or take a comprehensive examination over his or her graduate coursework. Four-to-six quarter units can be allocated to the thesis. Two units may be allocated to the comprehensive examination.

External Degree Program. During the mid-1980s, the Northrop Corporation wanted to give their engineering employees the option of obtaining an in-plant advanced aeronautical engineering degree. In response to a Los Angeles, California metropolitan area-wide request-for-proposal, Northrop selected the external degree program proposed by Professor Newberry at Cal Poly, Pomona. This external Master of Engineering with emphasis in Aeronautical Engineering degree provided the student with more breadth than was required by the campus program; it also permitted the student to select a specialty from among the disciplinary areas of aerodynamics, airbreathing propulsion, flight controls, or structures [3,4]. The faculty for this external degree included members of the Cal Poly, Pomona campus faculty as well as qualified members of the Northrop Corporation technical staff. The first graduates from this program completed their studies in 1986.

External Aeronautical Master's Degree Curriculum Specialties					
Aerodynamics & Propulsion		**Flight Controls**		**Structures**	
Five Breadth Courses		*Five Breadth Courses*		*Five Breadth Courses*	
Methods of Engr Analysis	ARO 501	Methods of Engr Analysis	ARO 501	Methods of Engr Analysis	ARO 501
Diff Eqs & Transforms	ARO 502	Diff Eqs & Transforms	ARO 502	Diff Eqs & Transforms	ARO 502
Airbreathing Propulsion	ARO 510	Airbreathing Propulsion	ARO 510	Airbreathing Propulsion	ARO 510
Aircraft Structures	ARO 506	Aircraft Structures	ARO 506	Advanced Aerodynamics	ARO 515
Aircraft Stability	ARO 578	Advanced Aerodynamics	ARO 515	Aircraft Stability	ARO 578
Four Emphasis Courses		*Four Emphasis Courses*		*Four Emphasis Courses*	
Advanced Aerodynamics	ARO 515	Aircraft Stability	ARO 578	Mechanics of Composites	ARO 504
Aerodyn of Wings/ Bodies	ARO 516	Modern Control Theory	ARO 535	Aircraft Structures	ARO 506
Comput Fluid Dynamics	ARO 518	Aircraft Flying Qualities	ARO 545	Structural Dynamics	ARO 521
Flight Sciences	ARO 598	Flight Sciences	ARO 598	Flight Sciences	ARO 598
Two Elective Courses		*Two Elective Courses*		*Elective Courses*	
Structural Dynamics	ARO 521	Aerodyn of Wings/Bodies	ARO 516	Finite Ele Structures II	ARO 508
Numerical Analysis	ARO 503	Numerical Analysis	ARO 503	Numerical Analysis	ARO 503
Aircraft Design	ARO 614	Aircraft Design	ARO 614	Aircraft Design	ARO 614
Or any other elective course		Or any other elective course		Or any other elective course	

Table 4
CSPUP External Aeronautical Master's Degree Curriculum Options 2002

The external degree program has an oversight committee comprised of Cal Poly, Pomona Aerospace Engineering Department and College of Engineering personnel as well as Northrop Grumman (or Lockheed Martin) technical staff and administrative personnel. From the beginning, qualified graduate students from any industrial or governmental organization in the Los Angeles metropolitan area were welcomed into the program (for financial viability). Students from the Antelope Valley area were admitted to the program in 1997. Students employed by Northrop Grumman, Boeing, Lockheed Martin, Raytheon, Edwards Air Force Base, McDonnell Douglas, Systems Technology, Hughes Aircraft, NASA Dryden, North American Rockwell, and the Computer Science Corporation have participated in this external degree program. Instructional sites (often corporate facilities) include locations in Hawthorne, Pico Rivera, El Segundo, Palmdale, and Long Beach, California. The external degree graduate curriculum is shown in Table 4.

The curriculum shown in Table 4 is circa 2002. This curriculum has changed somewhat from its initial form, and features only three specialty areas rather than the original four. Figure 3 shows the number of external degree program graduates over time.

In the past seventeen years, several hundred engineers have participated in, and some 100 plus individuals have graduated from, this external degree program. It should be noted that the external aeronautical engineering graduate degree program has been much larger than the on-campus graduate aerospace engineering program. However, the current downsizing of the aerospace industry is having a negative impact on the continued viability of the external degree program. No course offerings are being made in the external degree program at this time.

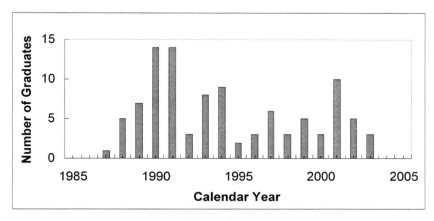

Figure 3
CSPUP External Aeronautical Master's Degree Timewise Graduates

Curriculum Innovations

The Aerospace Engineering Department faculty at Cal Poly, Pomona has supported a number of innovative concepts in undergraduate education. Some of these concepts are discussed below.

Self-Paced Instruction. Self-paced instruction enables the student to learn at his or her own pace rather than at the pace established by the instructor or the class average. It enables the student to master the course material in a manner rarely possible in the conventional lecture mode of instruction. During 1974, Professor Newberry taught the (spring) Energy, Mass and Momentum course, ARO 304 (4), and the (fall) Vector Fluid Mechanics course, ARO 301 (4), using his own self-paced course materials; student assistants required for the course offerings were supported by the Department.

The student learning experience appeared to be enhanced by the self-paced approach. However, compared to the conventional lecture method of instruction, the instructional labor intensive nature of the self-paced method and the (at least perceived) increased student learning effort of the self-paced method to achieve mastery did not seem to warrant continued use beyond its initial trial offerings.

Senior Project. The senior project (senior thesis) has been a part of the Cal Poly, Pomona engineering instructional paradigm since its inception in 1957. The typical senior project is a two-course sequence during the senior year of study. Each course in the sequence is allocated two-quarter units and treated as a laboratory. Thus, each student is expected to spend in the neighborhood of some sixty hours of effort in each of the two courses. Ideal projects include (equally weighted) both experimental and theoretical components. The project must be approved by and closely supervised by at least one of the faculty members.

By the time students reach their senior year of study, each student has had each Department faculty member for at least two classes, generally. Thus, each rising senior knows each faculty member fairly well (the reverse is also true) as well as the research interests of the faculty member. The approved senior project is usually associated with a shared interest by both student and faculty member.

Typical senior projects have covered such diverse topics as conceptual design weight estimation methods for missiles, ring-wing performance, construction and test of a vorticity meter, and the velocity mapping of a subsonic wind tunnel test section. The senior project course sequence has also been used for airbreathing propulsion system design, wherein each member of the design team is responsible for the design of a significant component of the system [4]. Senior project students have also been involved in faculty research and school-wide interdisciplinary projects (see EIC below).

Rose Float. The annual New Year's Day Pasadena Rose Parade provides an opportunity for students from Cal Poly, Pomona and Cal Poly, San Luis Obispo to share a non-academic educational experience. The two universities annually share the design and construction experience associated with a Rose Parade float entry - with minimum help from and supervision by a variety of faculty members from both universities. The experience is one of shared teamwork and expertise - requiring more time than that available during the student's Christmas vacation. Engineering students are usually involved in the design and construction of the float frame and its guidance and automation mechanisms and controls. Aerospace engineering students have often contributed their talents to this enterprise. It is believed that a Cal Poly float was the first Rose Parade float to feature an automated moving float element. These Cal Poly Rose Parade floats have won numerous awards over the years.

EIC. The Engineering Interdisciplinary Clinic (EIC) has performed fixed-price contract applied research for external agencies and corporations. This research is performed by interdisciplinary teams of students and faculty members. A student participating in these studies is selected by a faculty committee and is expected to take three consecutive three-quarter unit classes devoted to a particular project. Often the student receives senior project and/or senior seminar credit for these activities (in lieu of actually taking these courses).

Aerospace engineering students have often participated in this activity; they are prized for their ability to work in teams, their systems engineering savvy, and their analytical capabilities. Aerospace engineering students have worked on Pacific Gas & Electric sponsored environmental projects concerned with reduced emissions due to improved burner characteristics, a Northrop Grumman sponsored project concerned with solar panel generated electric current transmission, a General Dynamics sponsored project concerned with the performance of a metal-organic chemical vaporization process unit, and an Allied Signal sponsored project concerned with the impact of the thermal environment on material properties during material testing. The Clinic has not been active for the past couple of years.

Aerospace Construction Laboratory. Until 1972, the Aerospace Engineering Department required sophomores to take an Aerospace Construction Laboratory, ARO 252 (2). This two-unit (six contact hours) laboratory gave the students the opportunity to utilize their production and manufacturing expertise. Prerequisite for this class were courses in production and manufacturing, e.g., Manufacturing Processes, MPE 221 (2) and Weldment Engineering, WE 238 (2).

Individually, and/or in teams, students completed a number of unique and interesting projects. For example, one group of students designed, constructed, and tested a circular hydrofoil test basin (the sophomore "swimming pool") some fifteen feet in diameter. A vertical sting was suspended from the end of a circular rotating arm with a radius of some five feet; the test specimen (e.g., a hydrofoil) was attached to the bottom of the sting, some eight inches below the water surface. The test specimen could be moved at rotational speeds up to 15-20 fps. A special student-built balance permitted the measurement of lift and drag forces.

Over a period of several consecutive years (the task was too big for one class), the sophomore construction students built a full-scale, flyable Baby Ace (kit) airplane. Each major subassembly was FAA inspected as it was completed. In some instances, junior level students performed a stress analysis of some portion of the aircraft. Upon completion, the aircraft was donated to an Air Scout squadron sponsored by a National Guard unit based at the Ontario, California airport.

A circular, open-channel Aqua Ring testing device was designed, built and tested as another unique project. Rotation of the open-channel ring was in the vertical plane. Water was held in the rotating channel ring by centrifugal force. The test specimen, such as a small hydrofoil, was held by a non-moving strut anchored at the ring center. The Aqua Ring was built in two sizes; two were built with a diameter of some five feet, the second size (only one was built) was some 10-12 feet in diameter. One was not advised to stand too close to the ring on the spin-down to a stop, unless one wanted to get very wet.

A thirty-some foot long open-circuit environmental wind tunnel was built by one or more of the sophomore construction classes. The wind tunnel had a ten-foot long, four-foot square test section. Seniors (via Senior Projects) provided some assistance with this project. The atmospheric boundary layer could be simulated in this tunnel. The tunnel provided students with the opportunity to measure forces on and air circulation movement around a variety buildings and towers. All of these projects required design, production, and manufacturing capability, as well as a lot of teamwork.

Design. The Cal Poly, Pomona Aerospace Engineering Department has always had a strong design component. This strength is believed to be due to the *Learn by Doing* educational paradigm of the University and the industrial experience typically brought to the classroom by the faculty.

Although the design emphasis is on fixed-wing aircraft, some attention is given to rotary-wing aircraft, spacecraft, and missile design. Rotary-wing topics are covered in some of the required aerodynamics courses. There are required courses as well as electives in astronautics to support spacecraft design. A missile performance/design course has been offered occasionally as an elective, either as ARO 412 or as ARO 499 [6].

Capstone Design Program. Aircraft design has been a strong component of the Cal Poly, Pomona Aerospace Engineering Department curriculum since its inception, with more success in some years than others, as the curriculum evolved. There has been a continuous effort to provide aircraft design across the curriculum. The attempt is to provide assurance that any design attribute needed in the capstone design sequence has been considered in at least one course prior to the student taking the design course sequence.

In general, few lectures have been given in the design course sequence. Each design team provides a response (usually a 100-page design report) to a Request-for-Proposal (RFP) issued by a Department faculty member, the AIAA (annual design competition), or developed in conjunction with the Universities Space Research Association (USRA). The instructor in the capstone design sequence courses acts more as an advisor than a lecturer. Usually, lectures are given only when topics arise which have not been addressed in prior coursework.

The strong Cal Poly, Pomona Aerospace Engineering Department curricular emphasis on engineering design in general is carried over to the capstone design sequence. The capstone design course sequence has grown from a single senior course in the 1970s to a two-course then a three-course sequence [7,8,]. Two measures of the quality of the Departmental design program are suggested by (1) the number of student awards received in national design competitions, and (2) selection as one of some twelve universities to participate in the NASA/Universities Space Research Association (USRA) Advanced Design Program in Aeronautics.

At the end of the design course sequence, students are required to present the results of their design effort to a panel of industrial and government designers. The panel critiques the student effort(s) after the presentations. These student designs usually incorporate a number of current or new technologies. The senior seminar class has therefore been used as one of the design sequence courses for a number of years. The design course sequence is currently a three-course sequence spanning the three-quarter senior year of study.

Cal Poly, Pomona Aerospace Engineering Department students have won a number of national aircraft, spacecraft, and (airbreathing) engine design competitions sponsored by the AIAA and industry. For example, in the past two decades, the students of Professors Newberry and Lord have won (or placed) in several of these competitions, as indicated in Table 5.

CSPUP Aerospace Engineering Department AIAA Design Awards			
Year	Competition	Faculty Advisor	Placement
1981-1982	Individual Aircraft Design	Lord	Hon Mention
1985-1986	Individual Aircraft Design	Newberry	2nd
1986-1987	Team Spacecraft Design	Newberry	2nd
1986-1987	Individual Aircraft Design	Newberry	1st
1986-1987	Team Aircraft Design	Newberry	Hon Mention
1988-1989	Team Airbreathing Engine Design	Newberry	1st
1989-1990	Individual Aircraft Design	Newberry/Lord	1st , 3rd
1989-1990	Team Aircraft Design	Newberry/Lord	Hon Mention
1989-1990	Team Airbreathing Engine Design	Newberry/Lord	3rd
1996-1997	Individual Aircraft Design	Lord	2nd

Table 5
CSPUP Aerospace Engineering Undergraduate Design Awards

The1989-90 individual and team entries were co-advised by Professors Newberry and Lord. There were years in the decade of the 1990s that Cal Poly, Pomona design teams worked on projects other than the national competitions.

During the 1987-1992 time frame, Cal Poly, Pomona was selected by NASA/USRA to participate in their Advanced Design in Aeronautics Program. This participation brought with it a funding level of some $ 25,000 per annum, plus the opportunity for the Design Fellow (a graduate teaching assistant) to spend the preceding summer at a NASA Center (NASA Ames for Cal Poly, Pomona students). The funding provided a stipend for the Design Fellow during the academic year, travel, teaching allowance, and supplies. Participation in this program greatly enhanced the undergraduate design program at Cal Poly, Pomona. Ms. Renee Yadzi, and Messrs. Steven H. Cass, and David Polladian served as Design Fellows (at different times) in the Cal Poly, Pomona NASA/USRA program.

The results of the NASA/USRA studies were reported at the annual spring student conference sponsored by NASA and USRA [9]. These conferences provided useful interaction between students at all of the participating universities. These conferences were usually held at or very near a NASA Center, thereby providing outstanding field trips (during the conference) for the students.

Facilities

A number of facilities exist to support the Aerospace Engineering Department curriculum. The facilities are updated aperiodically. The Department had an altitude chamber and a high-speed centrifuge for several years; maintenance costs and lack of use led to the decision to survey these facilities.

Subsonic Wind Tunnel.

The low-turbulence closed-circuit subsonic Aerolab wind tunnel has a 36 x 42-inch rectangular cross section with a three-point balance. Speeds of up to 250 mph can be achieved in the test section. An electronic read-out system provides real-time lift, drag, and pitching moment values during test operation. The tunnel supports coursework, senior projects, research, and design. Figure 4 shows students operating the tunnel.

Figure 4
CSPUP Subsonic Wind Tunnel Operation

Supersonic Wind Tunnels.

The Department had two supersonic wind tunnels of the "blow-down" type. One of the tunnels had a pebble heater and could be used for thermal testing. A common air supply supported both tunnels and provided some two minutes of test time. A schlieren system supported flow field visualization in both tunnels. The thermal tunnel has been surveyed. The cold-flow tunnel is still operational and has nozzle blocks that allow operation at roughly transonic, M = 2, M = 3, and M = 4 speeds. Early in the program, a sting mount was damaged beyond repair by excessive starting loads. This event put a point on the need for students to follow proper laboratory procedure. The tunnel supports coursework, senior projects, research and design. Figure 5 shows students working with the cold-flow tunnel.

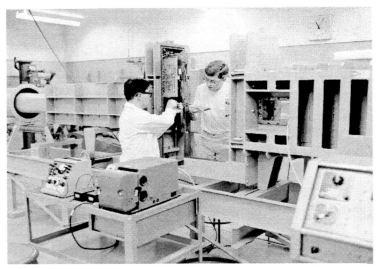

Figure 5
CSPUP Cold Flow (Blow-Down) Supersonic Wind Tunnel
Test Section

The transonic section has been used sparingly. One of the first times it was used was as a class demonstration. With flow initiation, the test-section glass panel, on the side of the tunnel away from the students, blew-out. A concrete wall was some eight feet away - nothing was between the glass panel and the concrete wall. To this day one can find small glass fragments in the wall. The contractor had installed the panel incorrectly. This experience puts a point on the importance of laboratory safety.

Flight Structures Laboratory. The test jig (20' x 10' x 35') and strongback are adequate for any testing needs. The MTS materials and testing system has the capacity for any foreseeable dynamics needs. The MTS system has an electronic data acquisition system. This facility supports coursework, senior projects, and research. Figure 6 shows some students operating the equipment.

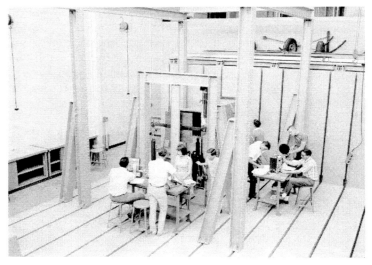

Figure 6
CSPUP Flight Structures Laboratory

Fabrication Laboratory. The fabrication laboratory was used for the sophomore constriction work described above. The facility is equipped with lathes, milling machines, welding equipment, and presses for the fabrication of any needed models and/or fixtures. This laboratory is used extensively by senior project students.

Design Laboratory. The design laboratory has some twenty-five work stations. Computer access permits the students to use a variety of software when working on their design projects.

Faculty

The Department has been fortunate to have had, over the years, an outstanding teaching (and research, or engineering practice) faculty. At one time the average faculty member had almost ten years of engineering practice in the aerospace industry. Faculty members often spent their summers in industry. Several had a license to fly. The value of these attributes should not be underestimated - as is the case at some research institutions. In Table 6, the faculty members are listed in the order in which they joined the Department; the institution where they received their highest degree is also listed.

Dr. Newberry was the recipient of the prestigious national John Leland Atwood Award (Outstanding Aerospace Engineering Educator) in 1986/1987. This was a joint award by the American Institute of Aeronautics and Astronautics (AIAA) and the American Society for Engineering Education (ASEE). Professor Newberry was selected as a Fellow of the AIAA in 1989. In 1997 he was honored by the ASEE with the Fred Merryfield Design Award for excellence in teaching engineering design.

CSPUP Aerospace Engineering Department Faculty		
Name	**Highest Degree/University**	**Tenure**
Wallace E. Nally		1958-1962
George R. Graves	MS; CSU, Long Beach	1958-1992
Rodney D. Sutherland	MS; University of California , Los Angeles	1960-1991
Conway H. Roberts		1961-1963
Horatio O. Morgan	BS; Pomona College	1963-1973
Albert D. Sanford	BS; University of Colorado	1964-1970
Conrad F. Newberry	D.Env.; University of California, Los Angeles	1964-1990
Donald C. Curran	AE; California Institute of Technology	1966-1986
Bruce D. Shriver	MS; West Coast University	1967-1971
Arthur G. Powell	Ph.D.; University of Southern California	1969-1979
Joseph W. McKinley	Ph.D.; University of Arizona	1969-1980
Paul A. Lord	MS; University of Colorado	1980-1997
Robert F. Davey	Ph.D.; California Institute of Technology	1981-2000
William E. Mortensen	MS; University of Southern California	1982-2002
Ali R. Ahmadi	Ph.D.; Massachusetts Institute of Technology	1985-present
Gabriel G. Georgiades	MS; Pennsylvania State University	1985-present
Donald L. Edberg	Ph.D.; Stanford University	2000-present

Table 6
CSPUP Aerospace Engineering Department Faculty

Over time, Cal Poly, Pomona Aerospace Engineering Department faculty members have been active in several professional engineering societies. For example, Professors Davey and

Georgiades, at different times, served as Chairman of the San Gabriel Valley Section of the American Institute of Aeronautics and Astronautics (AIAA). Professor Newberry has served as Chairman of the AIAA Los Angeles Section. Several faculty members have served as AIAA deputy regional directors, members of technical committees, and members of standing committees such as the Academic Affairs Committee. Professor Shriver did his undergraduate work at Cal Poly, Pomona, graduating in 1964. Also, Dr. Ahmadi did his undergraduate work at Cal Poly, Pomona, graduating in 1971.

Research

Cal Poly, Pomona is primarily a teaching institution. Teaching loads are heavy. However, faculty members are still able to conduct some research. As noted above, the senior project classes can support faculty research project. Two typical research projects are discussed below.

Over a period of five years (1982-1987), NASA Ames awarded an Educational Research Grant to Professor Paul A. Lord to investigate STOVAL ground effect aerodynamics. The Grant was valued at approximately $100,000. Thirty-six senior projects related to STOVAL ground effects were completed to support this Grant. A number of these projects involved subsonic wind tunnel testing.

Professor Robert F. Davey obtained a $ 100,000 grant from the General Dynamics Corporation to develop, during the 1990-1992 time period, a unique undergraduate controls laboratory. Tutorial software developed under the grant was used to lead the student through each experiment. The tutorial software and companion workbooks allowed students to conduct the experiments with minimal assistance, thereby freeing the instructor to help students analyze the results and explain differences between them and the theoretical predictions [10].

Most Department Professors perform consulting work, particularly during the summer months, to maintain and hone their engineering skills. For example, Professor Ahmadi has consulted with Bolt, Beranek & Newman on rotor aerodynamics. Professor Georgiades has been a consultant to the Naval Weapons Center at China Lake, California. Professor Graves was a (consulting) member of the Jet Propulsion Laboratory technical staff working on reliability problems. Professor Mortensen was a consultant to Leighton Corporation for a number of years. At various times, Professor Newberry was a member of the technical staffs of Northrop Aircraft, Lockheed Aircraft Service, Celesco, and Rockwell Corporation working on a variety of aerothermodynamic problems.

Annual Banquet

Since May of 1959, the Department and the Cal Poly, Pomona Student Branch of the American Institute of Aeronautics and Astronautics, have sponsored an annual spring banquet. This banquet provided the faculty with the opportunity to recognize outstanding freshman, sophomore, junior, and senior students. Student awards often consisted of a textbook(s) the student would use the following year. Some twelve students and faculty members attended the first banquet. Since 1995 the banquet has been named the Outstanding Alumni and Student Awards Banquet.

The banquet also gives the students the opportunity to say farewell to the faculty members that have guided (sometimes herded?) them for the preceding four or five years. These farewells took many forms. One year Professor Newberry was given a device (battery powered) whereby he could write on a chalkboard; an eraser, some twelve inches behind the chalk, would erase what had been written. The following year Professor Newberry distributed copies of his lectures to his students. That spring he was awarded a nice small plaque to which had been affixed a plaster hand painted gold - "The Golden Handout." The students accused Professor Newberry (erroneously, unfortunately) of owning stock in a paper company.

Outstanding Alumni

The Cal Poly, Pomona Aerospace Engineering Department undergraduate program has had a considerable number of outstanding graduates. Since 1995 an outstanding graduate has been honored at the annual spring Outstanding Alumni and Student Awards Banquet. Table 7 indicates the graduates who have been so honored.

Outstanding CSPUP Aerospace Engineering Department Graduates		
Name	**Class**	**Position**
Roger H. Hoh	1964	President Hoh Aeronautics
Dr. Lloyd H. Smith	1964	Director Naval Air Warfare Center, China Lake
Dr. Randall L. Peeters	1967	Vice President of Engineering GenCorp Aerojet
Stephen R. Cavanaugh	1968	Vice President & Program Director Orbiter Program at Boeing North American Space Systems Division
Richard C. Barbiere	1972	Senior Manager, Engineering Projects Raytheon
Michael P. Garland	1976	Director, Flight Test Research Center Edwards, AFB
Gregory B. Mutch	1978	Director of Airline Analysis Bombardier Aerospace
Richard S. Christiansen	1979	Director of the Aerospace Research Division NASA Ames Research Center
Michael T. Huggins	1985	Chief, Space & Missile Propulsion Division U.S. Air Force Research Center, Edwards, AFB

Table 7
CSPUP Aerospace Engineering Department Outstanding Alumni

The spectrum of the positions and organizations listed in Table 7 is both wide and varied. The success of these graduates may be one measure of the quality of the Cal Poly, Pomona Aerospace Engineering Department program, to say nothing of the talent of the individuals themselves. Since 1995 the Banquet has been completely sponsored by the companies represented on the Aerospace Action Council. Many other names could be (and will be in the future) added to the list given in Table 7.

References

1. _____, 1962-63 California State Polytechnic College Bulletin, California State Polytechnic College, San Luis Obispo, California, July, 1962, pp. 15-16, 289-294.

2. _____, 2001-03 Cal Poly, Pomona Catalog, California State Polytechnic University, Pomona, July, 2001, p. 15.

3. Newberry, C. F. and B. L. Hunt, "An External Degree Program in Aeronautical Engineering That Meets the Requirements of Both Industry and Academia," Preprint AIAA-86-2753, AIAA/AHS/ASEE Aircraft Systems, Design & Technology Meeting, Dayton, Ohio, October 20-22, 1986.

4. Newberry, C. F., A. R. Ahmadi, J. M. Schoenung, A. A. LeBel, D. J. McNally, and V. H. Garner, "The Development of a Mature External Master's Degree Program in Aeronautical Engineering: A University/Industry Partnership," Preprint AIAA-92-4256, AIAA Aircraft Design Systems Meeting, Hilton Head, South Carolina, August 24-26, 1992.

5. Newberry, C. F., "Using Senior Projects to Teach Airbreathing Propulsion Engine Design," Preprint AIAA-88-2976, AIAA/ASME/SAE/ASEE 24th Joint Propulsion Conference and Exhibit, Boston, Massachusetts, July 11-13, 1988.

6. Newberry, C. F., "The Conceptual Design of Missiles," The International Journal of Engineering Education, volume 4, number 1, 1988, pp. 35-40. Also, 94th Annual American Society for Engineering Education (ASEE) Conference Proceedings, Cincinnati, Ohio, June 22-26, 1986.

7. Newberry, C. F. and Paul A. Lord, "Aerospace Vehicle Design at California State Polytechnic University, Pomona," Preprint AIAA-86-2637, AIAA/AHS/ASEE Aircraft Systems, Design & Technology, Dayton, Ohio, October 20-22, 1986.

8. Newberry, C. F., "The Undergraduate Education of the Configurator," Preprint AIAA-87-2890, AIAA/AHS/ASEE Aircraft Design, Systems and Operations Meeting, St. Louis, Missouri, September 14-16, 1987.

9. Cass, Steven H. and Christopher M. Ball, "Platform Effects on High Speed Civil Transport Design," Preprint AIAA-88-4487, AIAA/AHS/ASEE Aircraft Design, Systems and Operations Meeting, Atlanta, Georgia, September 7-9, 1988.

10. Davey, Robert F., "Computer-Controlled Experiments for an Aerospace Controls Laboratory," Preprint AIAA-93-0425, AIAA 31st Aerospace Sciences Meeting & Exhibit, Reno, Nevada, January 11-14, 1993.

Chapter 50

Aerospace Engineering at SDSU – A Brief History

Joseph Katz
Dept. of Aerospace Engineering and Engineering Mechanics
San Diego State University

Preface

Forty years may not appear to be a 'long time' when discussing 'history'; however, during the information gathering process for this article, it became evident that the SDSU Aerospace Engineering (AE) story is worth telling. The department has seen years of tremendous growth, only to be stopped by unforeseen, sudden disasters, leading to today's moderate recovery. The significant part of all this, however, is the process itself which educated several thousand students; most of them still driving the national economy. Some of those students are now in key positions, but due to the short notice couldn't participate in preparing this report. Our hope is that with time, more information on our student body will surface and the next version of this report will be less one sided (e.g. documenting mostly faculty history). In the meantime here is a brief description of the events and of the people involved in keeping the wheels of the AE department rolling.

The Program

San Diego, the home of the Pacific fleet during World War II, was always one of the major aeronautical and later space related industrial centers. Companies such as Convair, Ryan Aeronautical, and Rohr were involved in many high Profile airplane and space propulsion projects. Therefore, it was only natural for the educational system to support this large-scale engineering effort by training the young and providing advanced degrees to the mature engineering workforce (as early as 1922). Although SDSU was founded in 1897 with an initial role of educating teachers, it rapidly grew and by 1964 the College of Engineering was formed with Aerospace Engineering being one of the first four departments in the new College. Its first Chair, Bill Shutts, was an aeronautical expert and masterminded the wind tunnel laboratory long before the official establishment of the department. Both the closed circuit low-speed wind tunnel and the blow-down supersonic wind tunnel were planned in the late 1950's, with help from the Rohr Company, which eventually became one of the first major customers of the facility. With the design of these important laboratories being complete, the Engineering building was designed and the building itself was inaugurated in 1960 (the year that the California State College System was formed). It is interesting to point out that engineering courses appear in the catalog as early as 1922 and by 1961 the programs were fully functional within the school of Engineering, with graduate courses appearing in that year's catalog. Three years later, in 1964, four official 'Engineering' departments were formed but were officially listed only in the 1968 catalog. During the 1965-66 academic year the graduate program leading to the MS degree in Aerospace engineering was started. By 1971 the whole university had grown dramatically, and the word 'University' was added to establish the current name of 'San Diego State University.' Consequently, the AE undergraduate program was first accredited in 1974 and since the department also offered engineering mechanics courses, it was renamed in 1978 as the 'Dept of Aerospace Engineering and Engineering Mechanics.' Lastly, the joint PhD program with UCSD was started in 1989/90 with Steven Yon, the first graduate of the program, defending his dissertation in 1994.

The Early Days (of the AE Dept.)

Aviation in the 20th century was a new discipline, which was started with the Wright brothers and peaked during World War II. After a short decline, jet engine aircraft begin to crowd the skies in the 50's and the space age really started with the first Sputnik (1957) and gained momentum in the 60's eventually leading to the dramatic moon walk (1969). With those amazing developments in the background, the new field of Aero/Space evolved and many potential students vied to work in this field. Responding to needs of the local industry, the Aerospace Engineering program was officially started (as noted earlier) in 1964 with the three founding Professors. Professor Bill Shutts (1958-77), active in the field of aerodynamics and wind tunnel testing, became the first Dept. chair and served between 1964-68. Professor John Conly (1962-2003), an aerodynamicist and wind tunnel experimentalist, and Prof. Nadar Dharmarajan (1960-97 – 'Rajan' in short) the first expert in light structures completed the list of the opening trio. In terms of supporting laboratories, the two (subsonic and supersonic) wind tunnels were quite exceptional facilities available to educate the students of the young department. At this point it's worthwhile to pause for a moment and recall that SDSU is part of a large, state supported university system, where budget and other important decisions are made based on student enrollment. Therefore, the very existence or survival of a program, for that matter, depends on the number of student-enrolled. In the spirit of these words the AE department enrollment diagram is presented in Fig. 1. As will be discussed later, this figure is far more than just a statistical account; it possesses a strong correlation with the events dominating the history of the department. Also there is no data prior to 1974, but the 'old and knowledgeable' say that enrollment was less than 100 (total) during the initial years and serious growth began only around 1976.

Now that the student enrollment versus time data is graphed, we can continue with the historical account. On the research side the wind tunnel facilities were busy, primarily with the Rohr Company, and Shutts was able to buy a three-component balance for the low-speed wind tunnel. The increased usage of the facility by Rohr convinced them within a few years to upgrade the load measuring system by acquiring a new six-component balance system. With the increased activity and sizeable student body the department asked for more faculty and new Professors were quickly hired. Harry Spencer (1963-66) was an expert in the then popular subject of turbulence, but left the department to pursue other career avenues. The next hiring wave was quite large and the flight mechanics discipline was started by newcomer Bob McGhie (as in the old Janis Joplin song). Bob came in 1967, the year that Howard Chang (1967 –73) was also hired. Howard was a fluid dynamics expert, earning his fame in civil engineering applications in addition to his AE background. The following year Balbir Narang (1968-) was hired to help with the high-speed aerodynamics area and Andy Crooker (1968- 73) with the fluid dynamics labs. During the early seventies, David Eggleston (1971-75, Flight mechanics), Lou Fontenot (1980-83, aerodynamics) and Dale Moses (1980-83 experimental aerodyanmics) came and went (as Conly recalls this). Although in the early 70s aerospace activity was at a high level, popularity of defense oriented programs wasn't high (due to the Vietnam war), and student enrollment didn't increase as expected. It appears that because of this lack of growth some of the above mentioned Professors left the department. Also for the same reason, Howard Chang left the department for Civil Eng. in 1973, toward the end of the 60s courses and activities in the department crystallized and due to the increased activity in the labs, Bill Fleck (1968-84), an 'old navy guy' was hired as the first department technician. One year later, a department secretary, Geraldine Keck 1969-80 was hired to help with the growing number of students and lab oriented engineering activities. Also it was agreed that the department chair position would rotate and Rajan became the second Dept. chair from 1968 to 1971. John Conly was next, and chaired the dept between 1971 – 74, and was replaced later by Bob McGhie (1974-77).

Sudden Growth

Continuing development in both aeronautics and space in the form of newer supersonic airplanes and the lunar missions made aerospace engineering studies highly desirable for the younger students. It became even more popular after the end of the Vietnam war (1972) when defense related issues became less graphic. At the other hand the 'cold war' was still in its prime so that the prospective job market looked quite promising. These developments set the scene for the rapid increase in enrollment as seen in Fig. 1. In less than 10 years student population increased several folds, reaching the magic number of over 500 enrolled students. At this peak the number of graduate students was well over 40. As a direct result of this rapid growth a new group of faculty was hired. Prof George Faulkner (1978-92), a retired Navy pilot became one of the most popular 'full time' part-time faculty. He was teaching design, flight mechanics, and capstone project related courses and systematically topped the title of 'most influential Professor.' The students always referred to him as Captain Faulkner, indicating that the 'Prof' title is much lower on their food chain. David Chou (1979-80), with main interest in plasma physics, and Mauro Pierucci (1979-) in the field of mechanical vibration and acoustics were hired later, and Chou left at the end of the year to New Mexico. As noted, Prof Pierucci's expertise was in structural mechanics and acoustics but later he became instrumental in computer aided teaching and web-based education. The following year, a well-established fluid dynamicist, K. C. Wang (1980-99) was hired from the Martin RIAS company to boost local expertise in computational fluid dynamics.

As growth continued, McGhie was replaced by Conly as the next dept chair. The three year rotation of the position cooled off and Conly stayed in this position from 1977 until 1984. During this time the department secretary Gerry (Geraldine) retired in 1976 and was replaced by the 'French lady' Terry Caron (1976-82). After her tenure, Gerry's was replaced by the first legendary secretary; Helen Benzing (1982 - 96). Helen established the tradition in the department; that the secretary's word is first and then Dept chair and so on. She knew more about the program than any of us and was liked by all, including students and faculty. Also by 1984 Bill the Lab technician retired and Dave Seltenrich (1984 - 87) was hired for that position.

The next wave of new faculty was started by the arrival of Nagy Nosseir, a fluid dynamicist, coming to us from USC. Over the years Prof. Nosseir taught numerous fluid dynamic related courses and most importantly, the capstone design project. With the rapid growth of the department, management related instability followed and the Associate Dean, Nihad Hussain was acting chair for one year (1984-85), after Prof. Conly's tenure. During Hussain's reign, a search was conducted to find the next department chair and Prof. Allen Plotkin (1985-), an established fluid dynamicist from the University of Maryland was selected for the 'job'. Plotkin was immediately charged with meeting the demands of the growing department by hiring several new faculty. During his first year, he hired Joe Katz (1986 -), an aerodynamicist working in the huge wind tunnel at NASA Ames, and during the next year, Costas Lyrintzis (1987 – 96) for the structural sciences position. Also at the same time (1986) Prof. Krishnamoorthy Govindarajalu ('Krish' in short), whose field is engineering mechanics, joined the department from the Civil engineering department. Krish joined the Civil Eng. Dept. in 1968, so he was already a veteran in the College.

During the 80s space shuttle programs became 'routine' and in general the aerospace industry matured (e.g., no new major inventions). At the same time, civil transportation continued to grow and the job market was still steady and the AE department enjoyed prosperity. The only other personnel change not mentioned before was in the position of the department technical support. Dave left in 1987 after a short tenure and Dept chair, Plotkin made one of the most important hiring decisions by bringing in Stig Johansson (1987-). Stig proved to be one of those irreplaceable resources who can design and build anything mechanical, electronic, or computer driven (I guess nothing else is left). Also during 1987 a large equipment grant from NSF allowed Katz to upgrade the wind tunnel lab which now had both a 6 component scale and a 6 component balance. With the same grant the first dept. computer was purchased – the so called MicroVAX station which was the most popular compact computer of that era. To illustrate the capability of that machine one must remember that it was used for wind tunnel data acquisition from 1988 till 2002 when it was finally replaced due to steam damage (one of the steam pipes exploded – resulting in humidity destroying all delicate instruments nearby).

On the research side a computational code PMARC was developed (1986-93) in close cooperation with NASA Ames and a consortium program was established. Students who participated in the program spent up to one year at Ames Research Center and participated in numerous research programs. Years later, this effort was recognized by NASA and the joint group of AMES/SDSU received the 1997 NASA Space Act Award for creative development (for the development of the code PMARC). Also in the 80s, race-car activity was quite intensive in the wind tunnel and with the support of the Mazda racing group numerous race car projects were conducted. Among the most successful were the 91 GTO and the 92 prototype cars (the former winning the IMSA title that year). In view of this successful research program, the early 90s were filled with lots of optimism. The newly approved joint PhD program with UCSD was started with AE graduates being the first students enrolling into the new program. Research supported by NASA and other resources resulted in numerous publications and the AE department was nationally visible. At about this time Prof. Nagy Nosseir replaced Allen Plotkin as Dept. chair for the next three years (1990-93). As expected, optimism was high and nobody could predict what was coming next, and placing Fig. 2 at this point seems appropriate for documenting the pinnacle of the 'good years'. This picture was taken circa 1991 with the supersonic wind-tunnel in the background. We all look like a happy family and the names of all are listed in the figure captions.

Bad politics and then tragedy

The world was rapidly changing during the late 80s and early 90s. The cold war ended, the Berlin wall collapsed (1989) and the defense industry begun to shrink. With this trend many aerospace companies left San Diego and employment opportunities in town shrank as well. On top of all this, in 1992 the state of California saw a rapid decline in revenues (from taxes). Since most of the university budget is based on CA State support the declined revenue had a direct impact on the university. Faced with expected budget cuts (of about 8%) the President of the University invented his most brilliant slogan (one can imagine how brilliant he really was): 'Deep and Narrow' cuts. His model postulated that instead of equal budget reduction across the board – he will eliminate certain programs entirely, but those that remain will not lose any support and may even see increased budgets. With some magic wand each College was to delete one department and in engineering the ax fell on the Aerospace program. During the last week of the 1992 fall semester, the Dean called the AE faculty and staff one by one and gave them the 'you're fired' routine. The plan called for closing the department and transferring all students to the remaining departments (mainly to ME). Of course new students were not admitted for the following year. The AE faculty and staff received their lay-off notices, to be effective at the end of the academic year. The only two escaping the ax were Profs. McGhie and Krish who had a previous appointment in the Civil Eng. Dept. Therefore, they were transferred back to that department and were spared from the firing ceremony. Also, Prof. Faulkner retired from official teaching, just to continue to lead the local AIAA student chapter on a voluntary basis. The void in teaching the flight mechanic courses was filled out by part time Professor Jeff Butler (1988 -) of General Dynamics.

The decision was devastating for the AE program. Although the layoff-call was rescinded by the end of 1992, the spirit was lost and new students were not admitted to the program until fall 1994. A glance at Fig. 1 now explains the dramatic drop in enrollment after 91. This was the beginning of the bad years. At the end the 'smart' university president was let go (probably with a sizeable pension plan) but area colleges remained under the 'Closed department' impression for quite a while. Well, who wants to go to a program that may disappear shortly? Thus, at the beginning of the 1992 fall semester, no new students were admitted but the program was left intact, so that the remaining students could graduate. The verdict was slowly removed entirely and the program was reinstated in fall 1994, but spirit was low. Dept chair Nosseir was devastated and at the end of his 3 years, in 1993 was replaced by Allen Plotkin as the next Dept Chair.

The mention of 'recovery' was not an issue and the closing of the department was continuously voiced by administrators – now on the basis of diminishing student body and the mass exodus of the aerospace industry from San Diego. With this tone the activities continued to fade and as Fig. 1 shows student population kept on shrinking. This trend continued until late 1996 due to pre 1994 students graduating while only a small number of new freshman showing up in the department gates. In 1996 Joe Katz replaced Allen Plotkin as the new chair. Also, Joan Rollins joined the department (coming from Civil Eng) as the department secretary, replacing Helen who left for the Exercise and Nutrition Science department. Joan was warmly received and continued the legend of 'super secretary' in the department. At the same time the new Dean of the college (appointed one year earlier) promised Katz help in turning the department's bad fortunes around. But it was too early for optimism:

It was the end of summer 1996, August 15, 2:00 pm, to be exact. Fred Davidson, a graduate student in the ME dept. was preparing his thesis defense. Fred was an AE undergraduate but due to recent events thought that a Master's degree in ME would improve his chances in the job market. Professor Costas Lyrintzis was on his thesis committee along with two others Professors from the ME department. For some reason Fred believed that his Professors were trying to spoil his chances of being hired by a certain company and in his sick mind he planned to get even with them. Without any notice, during the first minutes of the thesis presentation, he walked to a first aid box on the wall and pulled a handgun placed there earlier by him. Without any further ado he pointed the gun at the surprised committee and killed instantly all three Professors. The event was devastating for the whole University and the loss of Costas Lyrintzis was a major blow to the already wounded AE department. The main question after all this was: Can the AE program recover at all?

The recovery

In the wake of the University administration's ill-treatment of the 1992 budget crisis and the infamous 'Thesis defense shooting', the healing process was left to the department itself. The magnitude of the task can be illustrated via another glance at Fig. .1. Clearly, a full recovery to the pre 1991 era is quite unlikely within several years. However the slope of the enrollment curve (and corresponding funding) slowly turned around and a continuous growth is maintained. Another look at the data in Fig. 1 may suggest that a 'typical recovery curve' may take as long as ten years..

One of the important parameters helping the recovery was the cooperation with the Ryan Aeronautical group involved in developing the 'Global Hawk', a large UAV (unmanned aerial vehicle). One of the department alumni, Hermann Altman was in charge of the project and orchestrated the cooperation. This cooperation not only resulted in excellent student training and the generation of valuable research results, but also familiarized the students with the Ryan project (so when hired by the company they could be useful from day one). The project was a great success and Hermann won the 99 prestigious AIAA Aircraft Design Award for the UAV design. During this process, the wind tunnel, which was used for the project, was updated. The model's (pitch, yaw, etc) are now controlled by stepper motors, and flow visualizations are aided by several computer controlled video cameras. Data acquisition is based on virtual instrumentation (e.g., Labview) with full real-time graphic presentation of results. A new transonic section was designed and built for the existing supersonic tunnel, which is now controlled by a new digital actuator and computer.

At the same time a CFD (computational fluid dynamic) lab was formed with a whole range of computational tools. Visualization workstations were added and by the end of the millennium, potential flow based panel codes, airfoil design codes, and full-Navier-Stokes solvers were available and were applied by both graduate and undergraduate students. Also, the old fluid-lab experiments were all replaced by modern computer controlled test stations using modern sensors. To recognize the recovery of the AE program at SDSU, by the end of 2001,the department was finally granted a new faculty position. With great enthusiasm, Prof. Venkataraman (2002-), an expert in Structural optimization (grad of Univ. of Florida) was hired.

Laboratory modernization was not limited to the graduate program only and undergraduate students had access to all experimental hardware and CFD tools, particularly in their design related courses. Project topics span a large variety of disciplines and included studies such as: the design of solid-sail land yachts, aerodynamic development of golf club shafts, wind turbines, antennas for phones and satellites, sail boat masts and sails, roof of single a family home, flapping birds, and even the a study on optimizing the motion of a swimmer hand during the swim-stroke (to support the athletic department).

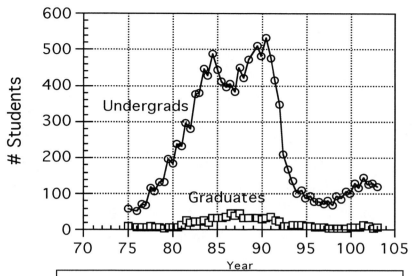

Fig 1. Student enrollment in the AE department. Note the sudden increase between 1975 –85 and then the dramatic drop after 1991.

The program now is focused on 'hands on' and the use of 'modern tools' so that after graduation the students can be productive engineers from day one (on the job). This approach resulted in continuous success of our students in national competitions. For example, Chad Berman used his senior project report as the basis of an AIAA paper and won first place in the nation in the AIAA Undergraduate Paper Competition (AIAA –2003-0113). His study described the testing and design of a modified propeller blade, with the design portion utilizing three-dimensional CFD tools validated by full-scale experiments. Another success followed by a group of undergraduates that used extensive numerical simulation to design a UAV for the annual AIAA competition. In April 2003 they won first place in the international AIAA Student Design/Build/Fly Competition.

Fig. 2 The AE department in the midst of the 'good years', circa 91. From left to right, sitting: Prof. Faulkner, Prof. Dharmarajan, Prof. Pierucci, Prof. Nosseir, Prof. Lyrintzis, Prof. Narang. Standing, from left: Prof. Katz, Prof. Conly, Prof. Wang, Prof. Plotkin, secretary Helen, Prof. Krishnamoorthy, Prof. McGhie, and technical director Johansson.

APPENDIX

List of faculty in chronological order

Bill Shutts (58-77)   Wind tunnel technology
Nadar Dharmarajan (60-97)   Structural engineering
John Conly (62 -03)   Aerodynamics
Harry Spencer (63-66)   Turbulence
Howard Chang (67 –73)   Fluid mechanics
Robert McGhie (67- 92)   Flight Mechanics
Balbir Narang, (68 -)   High-speed aerodynamics
Andy Crooker (68-73)   Aerodynamics
David Eggleston (71-75)   Flight Mechanics, Structures
Lou Fontenot (80-83)   Experimental Aerodynamics
Dale Moses (80-83)   Aerodynamics
George Faulkner* (78-92)   Design, Flight mechanics
Mauro Pierucci (79-)   Acoustics, mechanical vibration
David Chou (79-80)   Plasma Physics
K. C. Wang (80-99)   Fluid mechanics
Nagy Nosseir (83 -)   Fluid Mechanics,

Allen Plotkin (85 -)	Classical Aerodynamics
Krishnamoorthy Govindarajalu (86 -92)	Structures
Joe Katz (86 -)	Fluid dynamics
Costas Lyrintzis (87 - 96)	Structural Mechanics
Geoff Butler * (88 -)	Flight Mechanics
Sachi Venkataraman (02-)	Structural optimization

* Part time faculty that served over 5 years.

Technicians

Bill Fleck (68-84)
Dave Seltenrich (84 - 87)
Stig Johansson (87)

Secretaries

Geraldine Keck 69-76
Terry Caron 76-82
Helen Benzing (82 - 96)
Joan Rollins (96 -)

Chapter 51

History of the Aerospace Engineering Program at Howard University

Peter M. Bainum, Distinguished Professor Emeritus of Aerospace Engineering,

The Aerospace Engineering Program at Howard University has been a part of the undergraduate curriculum as far back as the late 1950's, and possibly before that. One of Prof. Wiley Sherwood's low speed wind tunnels, complete with manometer banks still exists in the Departmental Aerospace Laboratory. Dr. Abdul K. Azad was hired in the late 1950's to introduce courses in Aerodynamic Theory and Experimental Aerodynamics using the low speed wind tunnel. In the early 1960's he receive a contract together with faculty at Catholic University from NASA to further develop and improve instruction in Aerodynamics and Aerospace. Current undergraduate Aerospace Engineering courses include: Introduction to Flight, Fluid Dynamics, Aerodynamics, Aerospace Structures, Aerospace Propulsion, and a two-semester Capstone Senior Design sequence which has had a number of participating aerospace industry companies including: Hamilton Sundstrand, and Boeing Helicopters. These companies have provided motivation for and the definition of the senior design topic, as well as continuous guidance under the leadership of their engineers and scientists, both on-site at Howard as well as during visits to their home facilities on the part of our senior class students. The Capstone Senior Design sequence was initiated by the current Department Chairman, Dr. Lewis Thigpen. The Aerospace Engineering Option has been supported over the years by a maximum of three full-time equivalent (FTE) Aerospace faculty. Other Departmental faculty have taught related courses such as: Thermodynamics, Heat Transfer, Systems Dynamics, Automatic Controls, Vibrations, Instrumentation, and Experimentation.

At one time during the late 1970's and early 1980's, the Department gave serious consideration to changing it name to: The Department of Mechanical and Aerospace Engineering, but, unfortunately, this did not materialize. The Department has sent a representative to the Aerospace Department Chairmen's Association (ADCA) for at least 15 years and one of our faculty served as the Chair of ADCA for two years.

In response to the strong interest in Aerospace Engineering, the support of our then Department Chair, Dr. Charles B. Watkins, Jr., and the then AIAA Chair of Student Programs. Mr. Jeff Irons, the Howard Student Chapter of the AIAA was chartered in 1976 and continues to the present time. The first Co-Faculty Advisors were Dr. Azad and Dr. Bainum; since Dr. Azad's retirement he has been replaced by Dr. Sonya T. Smith. This chapter has been cited three times as the Best Student Chapter in the Mid-Atlantic Region for the 1976/77, 1978/79, and 1999/2000 academic years. The success of such a student chapter depends to a large degree on the enthusiasm and efforts of the student officers (especially the Chapter President), and also on the competition from other student organizations (such as ASME) in a combined Department. During the 1983/84 academic year the Howard University AIAA Student Chapter organized and hosted the Mid-Atlantic Regional Student Conference for the first time.

Graduate programs in the Department were initiated at the Master's level in 1967. Initial programs were offered in Aerospace Engineering (Aerodynamics and Dynamics and Control) and also in Fluid/Thermal Sciences. For the planning of the new program in Aerospace Engineering, Dr. Gabriel D. Boehler from Catholic University was employed as a consultant since Catholic University had a vast experience in this type of curriculum dating back to the pioneering work of Dr. Max M. Monk in the early 1900's.

The first graduate of the program was in the Aerospace Engineering option in 1969. Since that time 23 Masters degrees in Aerospace Engineering have been awarded, approximately 50 per cent of the total Masters graduates from the Department.

During the early to mid 1970's the Department prepared its proposal to inaugurate Ph.D level programs in two areas: Aerospace Engineering (Aerodynamics and Spacecraft Dynamics and Control) and also in Fluid/Thermal Sciences). After several years of revisions to the original proposal and various levels of internal and external approvals, these Ph.D. level programs were initiated in 1976. Up to the present time a total of 12 Aerospace Engineering Ph.D. degrees have been granted, again, representing approximately 50 percent of the total Ph.D. yield of the Department. The total FTE faculty of the complete Department has averaged between 9 and 12 during this period, with 2.5-3 FTE Aerospace faculty. In other words, roughly 25% of the total FTE of the Departmental faculty have produced approximately 50% of the total graduates.

Virtually all of the Ph.D. and Masters graduates are working in the Aerospace industry (although two who were trained in space robotics are now working in the area of automotive robotics). Aerospace Engineering graduates from the Department are/were working at: NASA Goddard, NASA Langley, JPL, Computer Sciences Corp, Jackson & Tull, Hughes STX, ITT Optics, Booz-Allen-Hamilton, Motorola, Chrysler Research Lab., Cadillac Research Division, Lockheed Martin, The Indonesian Space and Aeronautics Institute (LAPAN), the Brazilian Space Research Institute (INPE), and Grambling State University. Three students have returned to their home institutes in Indonesia and Brazil, while all the others are working in the USA. Only one of the Aerospace Ph.D. graduates has returned to full-time academia, although a few are teaching part-time.

Dedicated FT Aerospace Engineering faculty have served as Principal Investigators on more than 40 research grants/contracts with a total value of more than $5 million, not including their contributions as Co-PI's on large long-term institute (block) grants, estimated at $2-3 million additional .

Aerospace Engineering faculty and students have presented their research results at a number of national and international conference sponsored by the AIAA, AAS, ASME, American Physical Society, International Astronautical Federation, Institute of Automatic Control, Pan American Conference of Applied Mechanics, VPI & SU, NASA, ESA, etc. All of the Masters graduates have produced at least one conference paper and approximately half have also published in refereed journals. All Ph.D. graduates have published extensively, both in conference proceedings and also in refereed archival journals.

Aerospace Engineering faculty have also hosted eleven post-doctoral level visiting scientists and faculty from institutions in the US, China, Brazil, and Korea. These highly motivated visitors have, on average, produced two jointly authored publications for each year in residence. One visiting scientist, Dr. Ijar M. Fonseca from Brazil, returned, after six years, for a second assignment at Howard.

CHAPTER 52

AEROSPACE ENGINEERING AT THE UNIVERSITY
OF MISSOURI-ROLLA:
THE FIRST THIRTY FIVE YEARS

Fathi Finaish[1], Harry J. Sauer[2],
Department of Mechanical, Aerospace Engineering
And Engineering Mechanics

Diana L. Ahmad [3]
Department of History and Political Science

University of Missouri-Rolla
Rolla, Missouri, 65401, USA

1. UMR Launches a New Program

In 1870, the University of Missouri in Columbia expanded its programs to Rolla, Missouri, a small town one hundred miles south of the main campus. The new campus would focus on technological studies. Known as the Missouri School of Mines and Metallurgy (MSM), the new school offered mechanical engineering courses as early as 1871[3]. As the first technological institution west of the Mississippi River, MSM offered scientific and practical education during the post-Civil War industrial boom. Daniel Read, the President of the University of Missouri, stated, "This school [MSM] is to be a school both of science and of its applications; its purpose is to teach knowledge and art—first to know and then to do, and to do in the best manner." The mission of the Mechanical Engineering Department vision grew and expanded just as President Read hoped.

After many successful years as the Missouri School of Mines, the school decided to expand its mission by enhancing its curriculum and increasing its degree offerings. As such, in 1964, MSM became the University of Missouri – Rolla (UMR). Only three years later, in 1967, the Department of Mechanical Engineering also expanded its curriculum to include an aerospace program. Dr. Aaron J. Miles, an aviation enthusiast[1] and better known as "Doc," launched the first two aerospace engineering courses in the Department of Mechanical Engineering. Doc's excitement with flight began in 1927 when he served as a flying cadet in the Army Air Corps. Dr. Miles was hired by UMR as an Assistant Professor of Mechanical Engineering in 1937. By 1942, he became department chair, and in 1965, Miles became the first dean of UMR's School of Engineering. Figure (1) shows a photo[4] of Dr. Miles in 1971 holding a medallion created to celebrate the 100th birthday of the University of Missouri Rolla.

---

[1] Professor of Aerospace Engineering, Associate Fellow AIAA
[2] Professor of Mechanical and Aerospace Engineering
[3] Assistant Professor of History and Political Science and Archivist

Because of his enthusiasm for flying, Doc convinced his colleagues to add two new "aero" courses, Aerodynamics and Aircraft Structures, to the mechanical engineering offerings. By the 1950s, the department continued to expand its curriculum by adding an aeronautical engineering emphasis within the traditional mechanical engineering program. The new emphasis area expanded in 1962 to include aircraft and space vehicle propulsion, as well as aerospace mechanics courses. Finally, in 1964, with more and more students showing excitement over the new course plan, the department decided to include an aerospace engineering preference program. Once again, meeting with success, the new chair of the department, Dr. Thomas R. Faucett with support from Chancellor Merl Baker, proposed[2] to the now Dean Miles a new program that would lead to a degree in Aerospace Engineering. The purpose for the new degree arose from the need to serve society and provide industry with the best trained engineers possible for the growing demands of civilian and military aircraft and space vehicles. The new degree would be separate from the mechanical engineering degree, although the Mechanical Engineering Department would continue to house the new program. Dr. Faucett not only sought support for the Bachelor's degree, but also for a Master's degree in Aerospace Engineering.

In May 1967, the University of Missouri Board of Curators approved the curriculum for the new bachelor and master's degrees in Aerospace Engineering. The new B.S. degree required 143 credit hours (Figure 2) and included seven new required courses and five new elective classes that needed to be developed prior to the inauguration of the program. Further, to better reflect the changing degree offerings, the mechanical engineers voted to change the name of their department to the Department of Mechanical and Aerospace Engineering. In 1968, the department hired its first full time aerospace faculty [Professors H. F. Nelson and B.P. Selberg] and granted its first Bachelor of Science in Aerospace Engineering in 1969. Professor Selberg led efforts to develop a supersonic tunnel capable of producing flows up to Mach 3. Professors Bruce Selberg and Ronald Howell cooperated on the development of this facility and completed the project by 1970. The tunnel was utilized to produce the first Master's degree in aerospace engineering awarded by the program. During the same year [1970], Chancellor Merl Baker provided funding to develop a compressible flow laboratory on the west side of campus. Dr. Robert B. Oetting acquired a closed loop subsonic wind tunnel from Case Western University and led efforts to develop a driving system for the tunnel. The tunnel was assembled, integrated into the laboratory, and was operational by 1973. By 1973, 55 bachelor and 6 master's degrees had been awarded. Despite the large number of graduates, the department remained strong with 68 majors continuing to pursue their degrees in fall 1973.

In 1973, the Engineering Council for Professional Development (EPCD) accredited the new expanded department for the maximum 6-year period. The following year, with efforts led by Professor Bruce Selberg, the department presented a 5-year development plan for the aerospace program. The undergraduate curriculum included only 132 credit hours, or a reduction of 11 hours. The development plan included a statement of objectives for the aerospace program. It included:

> The UMR Aerospace Engineering program objectives[5] are: provide the students with a solid background in the laws of physics and mathematics, 2) Provide the student with a broad basic understanding of aerospace principles and application of these principles to aerospace systems, 3) provide the student with technical and humanities electives so that the student may mold his education to meet his total career objectives. Areas which will be stressed in addition to the basic aerospace fundamentals are: Engineering Design, Systems Approach, STOL-

VTOL Systems, Environmental Systems, Energy Systems, Transportation Systems.

Figure (1) This photo shows Dr. Aaron J. Miles in 1971 holding medallion created to celebrate the 100th birthday of University of Missouri-Rolla [Taken from Reference 4]

Freshman Year

Chemistry 1 or 2 – General Chemistry	5	Physics 21 – General Physics	4
English 1 – Rhetoric and Comp	3	Econ 100 – Principles of Econ 1	3
Hist 60 – American Civilization[3]	3	ET 10 – Engineering Drawing	3
Math 8 – Cal with Anal. Geom I	5	Math 21 – Cal with Anal. Geom II	5
AE 1 – Intro to Aerospace Engineering	0	ME 53 – Intro to Mfg Proc	3
M 10 – Military Fundamentals (if elected)*		M 20 – Military Fundamentals (if elected)*	
PE 1 – Physical Education	0		
	16		18

Sophomore Year

Phy 25 – General Physics	4	E ME 51 – Eng Mech-Statics	3
Phy 22 – General Physics Lab	1	Engl 60 – Exposition[2]	3
ET 25 – Graphical Design Lab	2	Math 201 – Differential Equations	3
Engl 105 – American Literature I	3	ME 110 – Kinematics	2
Math 22 – Cal with Anal. Geom III	4	ME 121 – Thermodynamics	3
Mgmt 201 – Engineering Economics	3	M 40 – Military Fundamentals (if elected)*	
C SC 73 – Intro to Comput Tech	1	Elective – Humanities or Social Studies[1]	3
M 30 – Military Fundamentals (if elected)*			
	18		17

Junior Year

EE 171 – Electrical Circuits	3	EE 173 – Electronic Devices	3
E ME 102 – Heat Transfer	2	EE 174 – Electronics Lab	1
E ME 103 – Mechanics of Materials	3	AE 229 – Aerospace Mechanics	3
E ME 104 – Materials Testing Lab	1	AE 251 – Aerothermodynamics	4
AE 245 – Materials for Aerospace Engineering	3	AE 281 – Aerodynamics I	3
Phy 107 – Atomic and Nuc Phy	3	AE 285 – Aerospace Structures I	3
Elective[4]	3	Elective[4]	3
	19		19

Senior Year

ME 120 – Mech Instrumentation Lab	1		
ME 201 – Heat Transfer	3	AE 295 – Flight Dynamics and Control	3
AE 243 – Intro to Aerothermochemistry	3	AE 296 – Aerospace Systems Design	4
AE 271 – Aircraft and Space Vehicle Propulsion	3	AE 298 – Experimental Methods in Aero Eng	2
AE 287 – Aerodynamics II	3	Electives[4]	8
AE 293 – Aerospace Structures II	3		
Elective[4]	3		
	19		17

*Basic Military (M 10, 20, 30, 40) may be elected in the freshman and sophomore years, but are not creditable toward a degree. See page 205 on
constructive credit for previous service, and Advanced Military.
[1]See pages 85, 196, and 210 on requirements in Humanities and Social Studies.
[2]Or Engl 61.
[3]Or Pol. Sci. 90.
[4]Electives: Nine hours must be in the humanities or social studies, and the remaining hours may be technical courses.

Figure (2) 1967 UMR AEROSPACE ENGINEERING CURRICULUM (143CREDIT HOURS)

2. Enrollment Trends, 1968-2002

Following the national trends, UMR's aerospace enrollment experienced two peaks; the first in 1969 and the second in 1985 (Figure 3). In 1970, the department contained 188 students reflecting the nation's interest in the Apollo projects and NASA's attempts to land a man on the moon. After Neil Armstrong's successful moon landing in July 1969, the enthusiasm for the space program dropped off, as is reflected in the nearly immediate decline in aerospace engineering students. During these years, the military aircraft industry suffered under reduced military spending by the federal government. Furthermore, the oil embargo by Middle Eastern oil-producing countries produced higher
fuel prices that led to limited airline traffic and subsequently, resulted in a drop in orders for civilian transport aircraft. As such, fewer jobs were available for aerospace engineers. By 1975, the department only contained 59 majors. The downward trend stopped that year and the department experienced steady growth until 1985. Between 1975 and 1985, the department grew from 59 to 351 students. However, beginning in 1986, the department experienced another downward trend reflecting the downsizing of the aerospace industry during the Reagan years. This time, the enrollment fell to 66 students, but once again, enrollment grew steadily, and the department expects over 120 majors in the fall 2003 semester.

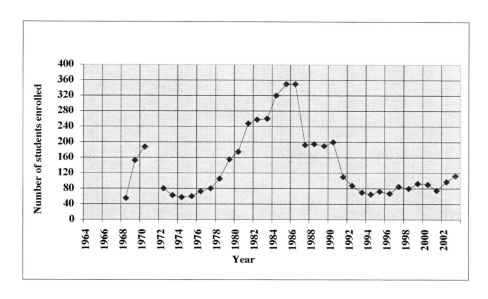

Figure (3) UMR Undergraduate Aerospace Engineering Enrollment

3. Curriculum Development and Revisions

As noted in Section 1, the Department of Mechanical Engineering started an aerospace program because of the excitement of one man, Doc Miles. The program blossomed because of the space race and an affluent society's demands for greater exploration of earth's nearest neighbor, the moon. The earliest program was simply an extension of the Mechanical Engineering program. Soon, however, more students developed an interest in AE itself, and the faculty decided to establish a separate degree plan emphasizing aerospace engineering as a unique field separated from the mechanical engineering degree.

Between 1974 and the late 1980s-early 1990s, the curriculum required the students to take 132 or 133 semester hours for their degree; however, a number of significant changes took place that strengthened the program. Now, instead of using professors trained in fields that did not focus on aerospace engineering, the department hired young faculty members with advanced degrees in aerospace engineering. In addition, in 1987, the department formed a group, the Aerospace Committee, who addressed and monitored matters related to the aerospace engineering program. In October 1987, the aerospace committee began discussions on how to improve the aerospace engineering curriculum. The committee desired more laboratory and control courses for its majors. The faculty also suggested changes to the existing courses and the addition of several new classes. Out of these discussions, two new classes resulted, aircraft stability and control and advanced and composite structures. Effective in Spring 1988, the laboratory course sequence also changed from Mechanical Instrumentation (ME 240) and Experimental Methods in Aerospace Engineering (AE 280) to Stability and Control of Aerospace Vehicles. (AE 361). A new aerospace laboratory course, AE 283 Experimental Methods in Aerospace Engineering II, resulted from the discussions as well. The Committee also approved the development of the flight simulation laboratory as promoted by Professor Robert Oetting.

In 1988, the Department of Engineering Mechanics merged with the Department of Mechanical and Aerospace Engineering creating the Department of Mechanical and Aerospace Engineering and Engineering Mechanics (MAEEM). Now, three of the nine members of the engineering mechanics faculty strictly focused on Aerospace Engineering (AE). Because of the expertise of the engineering mechanics faculty, the department incorporated several elective courses into the Aerospace Engineering curriculum, including Finite Element, Fracture Mechanics, Fatigue Analysis, and Mechanics of Composite Materials. Seniors and aerospace engineering students could avail themselves of these classes.

The department experienced more changes in 1990, including the incorporation of the Computer Aided Design course, "AE 321 Aerodynamics CAD Design." This course became part of the regular curriculum by the late 1990s. More changes to the curriculum led to the dropping of ME 225 Heat Transfer and replacing it with a technical elective. A required course in modern physics (Physics 107) replaced the old basic science elective. Then, in 1992, the department added a one-hour course, AE 210 Senior Seminar, to the curriculum. The idea of the senior seminar was intended to develop a student's sense of moral, ethical, and professional obligations. Speakers from industry, members of the UMR Academy of Mechanical and Aerospace Engineers, and faculty members from the UMR campus gave lectures to the seniors about various topics involving AE.

In 1991, the Aerospace Committee decided to improve the design content of the curriculum. The result of the discussion led to the most significant development in the program's curriculum in the 1990s. Prior to the change, in the 1970s and 1980s, the two required senior capstone design courses (AE 280 Aerospace Systems Design I and AE 281 Aerospace Systems Design II) were purely theoretical courses. In 1992, Curators' Professor Walter Eversman suggested that the design courses become practical applications of everything the students had learned in their years at UMR. The students would build and fly remotely piloted aircraft designed to perform specific missions as an indication of their understanding of the field of aerospace engineering. The course emphasized team building and communication skills. The project demanded that the students deal with a specified set of constraints to design a remotely piloted vehicle (RPV) that operated at optimum capacity. The course professor, industry representatives, and members of the UMR Academy of Mechanical and Aerospace Engineers judge the students' performance. If the RPV entered a national competition, such as the Society of Automotive Engineers Heavy Lift Competition, then the results of the competition also weighed in on the students' performance in class. Professor Eversman's vision and dedication to redevelop and teach the senior capstone design courses during the last decade led to a significant enhancement in the design content of the AE curriculum and generated a great deal of excitement among AE students and faculty. Each year, UMR students continue to do well in national competitions. For instance, in April 1999, UMR aerospace engineering students [Figure (4)] won second place at the national Society of Automotive Engineers Heavy Lift Competition, and in May 2003, the UMR Advanced Aero Group (AAVG) won first place in the national Open Class Heavy Lift Competition.

Figure (4) UMR team with their winning aircraft at the **Society of Automotive Engineers Heavy Lift Competition [DeLland, Florida, April 1999]**

Another change to the design content of the program included a new course, AE 180 Introduction to Aerospace Design. Ready for the sophomore students in 1994, the course required the students to design, construct, and test two small aircraft models. They must also analyze and solve open-ended design problems, build models that illustrate the analysis phase, test and experiment with different design concepts, work in design teams, and develop technical reports. More recently, two courses in the area of spacecraft design, "AE 380: Spacecraft Design I" & "AE 382: Spacecraft Design II," have been incorporated into the curriculum. In these courses, seniors and first-year graduate students design and develop space vehicles.

More recently, the UMR School of Engineering strongly suggested that engineering departments reduce the number of required hours for the B.S. degree to 128 hours. Partly due to a national trend in fewer students enrolling as engineering majors, the lower requirements will allow students to graduate in four years. It is further hoped that the students will pursue an M.S. degree in engineering that will allow them more in-dept training in their areas. In response to the request, the AE faculty lowered their requirements to 128 hours by dropping one course, AE 233 "Aerothermochemistry." Further, the manufacturing elective and one humanities/social science course were changed to free elective courses. The new curriculum, depicted in Figure (3), was approved by the campus in February of 2003.

The master's degree program also experienced major changes over the years, including revisions to existing courses and the development of new classes in the areas of stability and control, propulsion, hypersonic flows, aerodynamics, structures, spacecraft dynamics and design, and aeroelasticity. Currently, there are approximately 50 graduate level courses that master's and doctoral students can select from to craft their program of study. The research of Aerospace Engineering faculty and graduate students is externally funded research by government agencies and aerospace industries.

In the early 1980s, with efforts led by Professor Selberg, a proposal that aimed at improving the aerospace program was developed and submitted to McDonnell Douglas Foundation. In 1986, the Foundation funded the proposal and gave the UMR aerospace program a \$1.2 million grant. With matching funds from the State of Missouri, the department hired several new aerospace engineering faculty and enhanced the aerospace laboratories and the undergraduate/graduate curricula. In 1989, an outgrowth of the enhancement funds was the development and approval of a Ph.D[6]. in aerospace engineering at UMR. The doctoral program provides the students with advanced courses and opportunities to conduct externally funded research with the AE faculty on a wide range of topics. Currently, the AE faculty and graduate students are involved in research efforts on topics pertaining to control of aerospace and mechanical systems, flight and orbital mechanics, spacecraft mission design, satellite attitude optimization identification and estimation, composite materials, smart structures, fracture mechanics, fatigue and failure analysis, noise control, acoustics, vibrations, aircraft aeroelasticity, aerodynamics, flight simulation, aero-structure interaction and control, radiative transfer, laser-gas and/or surface interaction, computational fluid dynamics, and hypersonic propulsion systems. Various government agencies and companies in the aerospace industry fund these projects.

4. Program Administration and Faculty

Since the beginning, Drs. Miles and Faucett established a tradition of strong leadership for the Aerospace Engineering program. Those who followed in the chair's position, including Dr. Walter Eversman, Dr. Bassem Armaly, Dr. H. Dean Keith, and currently Dr. Ashok Midha, remained committed to a solid aerospace engineering program. In the 1960s, a number of faculty members were instrumental in developing the aerospace program. Dr. Harry "Hank" Sauer came to UMR after holding positions in aerophysics testing and propulsion at Convair and Boeing. He initiated the new propulsion and aerospace mechanics courses. Dr. Robert Oetting and Professor Bruce Selberg, the first faculty who possessed aeronautical/aerospace engineering degrees at MSM/UMR, served on the committee to develop the original undergraduate curriculum for aerospace engineering at UMR. From 1972 until his retirement in 1999, Professor Selberg served as the Professor-in-Charge and then, as Associate Chairman for the Aerospace Engineering Program. Dr. Fathi Finaish, assumed the Associate Chairman's position upon Selberg's retirement.

Currently, the UMR Aerospace Engineering program includes 12 faculty members (1 Curators' Professor, 10 Professors, and 1 Associate Professor). Of the 12 men, 5 focus exclusively on AE, while the remaining 7 hold joint appointments with Mechanical Engineering or Engineering Mechanics. Each member of the faculty advises and teaches undergraduate and graduate students, actively participates in research, and possesses service commitments to the university and aerospace communities. In the mid 1990s, the department took over NASA's Missouri Space Grant Consortium, a federal program designed to encourage high school, undergraduate, and graduate students from around the state of Missouri to pursue science, technology, and engineering careers. Professor Selberg served as consortium director until his retirement in 1999, and was succeeded in this position by Dr. Fathi Finaish. To date the consortium has provide support for over 400 undergraduate, graduate, and high school students and has reached thousands of kids and teachers throng a wide range of public education and outreach programs. The students supported by the consortium participate in aerospace engineering research and at the end of the session, present papers about their findings. Further, the program includes grants for high school teachers that give them aerospace experiences to take back to their classes.

Table (1) provides a list of departmental faculty who currently have a major role in the teaching of the undergraduate aerospace engineering program. The table indicates their teaching and research interests.

5. Today's Program

In May 2002, the Department of Mechanical and Aerospace Engineering and Engineering Mechanics contained 114 Aerospace Engineering students working toward a B.S., a M.S., or a Ph.D. in aerospace engineering. Between the program's inception in 1967 and the May 2003 graduation, UMR has conferred 747 bachelor, 91 master, and 8 doctorate degrees in aerospace engineering.

Name	Areas of Teaching Responsibility	Research Interests
Balakrishnan, S. N.	Stability and control.	Control of aerospace and mechanical systems, flight and orbital mechanics, optimization, numerical methods and, and neural networks.
Chandrashekhara, K.	Structures, composite materials, and finite element.	Composite materials, smart structures, structural dynamics, finite element analysis, composite manufacturing and experimental characterization.
Dharani, Lokesh	Structures, composite materials, fracture mechanics, and fatigue analysis.	Micromechanics of bi-material interfaces, composite materials, fracture mechanics, fatigue and failure analysis of welded structures, wear and friction in composites, fracture and failure of laminated glass.
Eversman, Walter	Design, stability and control, vibrations, aeroelasticity , and acoustics.	Noise control, acoustics, vibrations, aircraft structural dynamics and aeroelasticity, systems and control.
Finaish, Fathi	Aerodynamics, fluid dynamics, design, and experimental methods.	Aerodynamic testing, unsteady flows, vortex dynamics in separated flows, physical and numerical flow visualizations, and flow control.
Isaac, K.M.	Aerodynamics, gas dynamics, CFD, experimental methods, and propulsion.	Aero-structure interaction and control, active flow control, wave-riders, MEMS, multiphase flow, emissions from combustion and evaporative systems, active combustion control, atomization and sprays, particle image velocimetry (PIV) and CFD
Mac Sithigh, Gearoid	Structures and orbital mechanics	Finite elasticity, viscoelasticity, liquid crystal hydrodynamics, solid and continuum mechanics.
Nelson, H. Fred	Aerodynamics, aerothermochemistry, Hypersonics, stability and control.	Radiative transfer, laser-gas and/or surface interaction, aerothermochemistry, combustion, applied aerodynamics.
Pernicka, Hank	Astrodynamics, orbital mechanics, spacecraft design.	Astrodynamics, orbital mechanics, spacecraft design, spacecraft mission design, satellite attitude dynamics, nonlinear analysis, dynamics and control, optimization.
Riggins, David	Propulsion, aerodynamics, fluid dynamics, computational fluid dynamics.	Fluid dynamics, computational fluid dynamics, hyper/sonic propulsion systems, computational analysis of jet mixing, flow losses and mixing enhancement in combustors, aircraft gas turbine ramjet propulsion systems, and scramjet performance.

Table(1) Current Faculty with Major AE Teaching Responsibility and Research Interest

Driven by the new Accreditation Board for Engineering and Technology (ABET) vision and the needs of industry, the aerospace engineering faculty developed a list of goals aimed at producing quality aerospace engineers at UMR. Working with the department's Industrial Advisory Committee and its Academy of Mechanical and Aerospace Engineers, the department finalized its goals in November 1999. The goals are:

1. A solid foundation of principles of science and engineering with a strong background in mathematics and physics to serve as a foundation for life-long learning.
2. A solid technical knowledge in the areas of aerodynamics, materials, structures, stability and control, propulsion, and aerothermochemistry, including cross-linkage among the areas.
3. The ability to apply engineering knowledge and skills to engineering analysis, solve open-ended problems, design projects, and develop useful products and processes.
4. The ability to work in a team environment, create group synergy in pursing a given goal, and communicate technical information in written, oral, visual and graphical formats.
5. An awareness and understanding of their moral, ethical, and professional obligations to protect human health and the environment.

The above educational objectives focus on developing students into aerospace engineers who are broad-based in technical knowledge, experienced with reality, and prepared to interact with other engineers, as well as society. These goals remain consistent with UMR's mission to meet Missouri's needs for engineering education, while emphasizing leadership skills. Leadership in an engineering environment is not acquired in class, but instead by preparation and experience learned through the students' curriculum and in student organizations. Consistent with the Department mission, the objectives of the degree program emphasize the application [Figure (5)] of learned skills while embracing an understanding of theory and emphasizing a technical practicality that is functional in reality. To achieve these goals, the aerospace engineering curriculum includes basic science and engineering principles, the mentoring of students through undergraduate research projects, by individual advising, and by encouraging students to gain practical experience through student aerospace organizations.

Figure (5) This photo shows AE Students building and testing a radio-controlled aircraft model that they are required to design, build, and test during the sophomore year. Daniel Read, the President of the University of Missouri, stated in 1870, "This school is to be a school both of science and of its applications; its purpose is to teach knowledge and art—first to know and then to do, and to do in the best manner." The mission of the UMR Aerospace Engineering program vision grew and expanded just as President Read hoped.

The current undergraduate curriculum requires 128 semester credit hours for the B.S. degree. As shown in Figure (6), the 128 semester credit hours include 34 hours of mathematics & basic science, 58 hours of engineering topics, 21 hours of general education, and 15 hours of other topics, including 6 hours of free electives. The basic science and engineering courses include mathematics, physics, chemistry, statics, and dynamics. Students gain technical knowledge in aerospace engineering disciplines in classes that study aerodynamics, materials, structures, stability and control, propulsion, and aircraft and spacecraft design.

5.1 Student Activity and Organizations

Today, aerospace students are active in extracurricular activities, particularly in campus organizations such as the UMR Chapter of the American Institute of Aeronautics and Astronautics (AIAA), an organization that represents the students in the Aerospace Engineering program. The UMR Student Chapter of AIAA began in 1959 when it was chartered as the American Rocket Society. In 1963, it merged with the Institute of Aeronautical Sciences to form the AIAA. This chapter continues to be an outstanding student organization on campus and keeps students actively involved in the aerospace engineering field. The student chapter hosts industrial speakers who provide opportunities for students to learn about technical problems in the aerospace industry. Also, the chapter sponsors an annual program designed to help the Boy Scouts of America earn the Space Exploration Merit Badge. Further, the group participates in the annual AIAA Region V Student Paper Meeting where outstanding graduate and undergraduate research papers are presented. Most recently, nine students presented papers at the 2003 AIAA Region V Student meeting held in Boulder, Colorado. UMR students won the Regional Best Chapter Award and secured first place in the undergraduate and graduate categories.

Students are also active in other organizations such as Sigma Gamma Tau, the national Aerospace Engineering Honorary Society, the advanced Aero-Vehicle Group that competes in design and flight test competitions, and the Society of Flight Test Engineers Chapter. In Fall 1999, three graduate aerospace students proposed a new student organization that would allow students to compete with their design projects. By Spring 2002, MAEEM and the UMR School of Engineering agreed to sponsor the new organization. Since its inception, this student group has competed yearly in the Society of Automotive Engineering International Heavy Lift competition. In May 2003, the group's design, a 35 lb. aircraft (empty) lifted over 70 lbs. of payload. Their efforts won first place in the competition held in Dayton, Ohio. Following the successful beginning of the group, students organized a chapter of the Society of Flight Test Engineers. Officially sanctioned in January 2003, the group won the approval of the SFTE Board of Directors, making the UMR Chapter the first official student-run chapter in the United States. The undergraduate aerospace students actively participate in the undergraduate research programs on campus. Nearly 10% of the undergraduate students are involved in research with the aerospace faculty. These research projects are funded mainly by NASA, through the NASA Missouri Space Grant Program, and UMR through the Opportunities for Undergraduate Research Experience (OURE) program. In recent years, the undergraduates won numerous paper presentations awards at national and regional student meetings.

5.2 Laboratories

During the last 30 years, instructional laboratories at UMR have made significant progress. Currently, the department possesses seven main aerospace laboratories that support undergraduate and graduate students, as well as faculty teaching and research activities. A brief description of these facilities is listed below:

Aerospace Flow Laboratory: The aerospace flow laboratory, located on the west side of the campus, contains 3,000 square feet. This facility supports various instructional and research activities in areas pertaining to steady and unsteady aerodynamic testing, flow visualizations, and flow instrumentation. Test facilities housed in this laboratory include three subsonic wind tunnels and a wide range of equipment and instrumentation for airflow measurements and control. The largest wind tunnel is a closed circuit tunnel equipped with a 32"X48" test section and is capable of producing flow speeds of up to 200 mph. The laboratory also contains a second tunnel designed for steady and unsteady flow testing and is an open return type capable of producing flow speeds of up to 90 mph in an 18"X18" square test section. The third tunnel housed in the laboratory is a variable density wind tunnel used for flow testing under controlled conditions of flow temperature, pressure, and humidity.

Freshman Year

Basic Engineering 10	1	Basic Engineering 20		3
Chemistry 5[1]	5	Math 15[4]		4
English 20	3	Physics 23[4]		4
Math 14[4]	4	H/SS Economics elective[3]		3
H/SS History elective[2]	3			
Semester Hours	16	Semester Hours		14

Sophomore Year

Comp Sci 73-Basic Scientific Programming	2	AE 180-Intro to Aerospace Design		2
Comp Sci 77-Computer Programming Lab	1	EMech 160[5]-Eng Mechanics-Dynamics		3
Bas Eng 50 or 51-Eng Mech-Statics	3	ME 219[4,5]-Thermodynamics		3
Math 22[4]-Calculus/Analytic Geometry III1	4	Math 204-Elementray Differential Equations		3
Physics 24-Engineering Physics II	4	Bas Eng 110-Mechanics of Materials		3
AE 161-Aerospace Vehicle Performance	3	Elective/Literature		3
Semester Hours	17	Semester Hours		17

Junior Year

AE 213[4]-Aerospace Mechanics	3	AE 251[4]-Aerospace Structures I		3
AE 231[4,5]-Aerodynamics I	3	AE 261-Flight Dynamics and Control		3
AE 377-Principles of Engineering Materials	3	AE 271-Aerodynamics II		3
EE 281-Electrical Circuits	3	AE 282-Experimental Methods in AE I		2
Elective/Advanced Math/Computer Science[6]	3	Elective/Free[10]	3	
		Elective/Communications[8]		3
Semester Hours	15	Semester Hours		17

Senior Year

AE 210-Seminar	1	AE 233-Intro to Aerothermochemistry		3
AE 235-Aircraft & Space Vehicle Propulsion	3	AE 281-Aerospace Systems Design II		3
AE 253-Aerospace Structures II	3	Elective/Technical[7]		3
AE 280-Aerospace Systems Design I	2	Elective/Technical[7]		3
AE 283-Experimental Methods in AE II	2	Elective/Free[10]		3
Elective/Technical[7]	3	Elective/Humanities/Social Sciences[9]		3
Elective/Humanities/Social Sciences[9]	3			
Semester Hours	17	Semester Hours		15

List of Notes:

1 Chemistry 1 and 2 or Chemistry 5, depending on placement and Chemistry 4 or an equivalent training program approved by UMR.

2 Must be one of the following: Political Science 90, History 112, History 175, or History 176.

3 Must be one of the following: Economics 121 or Economics 122.

4 A grade of "C" or better in Math 14, 15, 22, and Physics 23 is required both for enrollment in ME 219, AE 213, AE 231, or AE 251 and for graduation.

5 A grade of "C" or better in EM 160 and ME 219 is required both for enrollment in any courses which require either EM 160 or ME 219 as prerequisites and for graduation.

6 Must be one of the following: Comp Sc 228, Math 203, Math 208, or any 300-level math or computer science course approved by the student's advisor.

7 Electives must be approved by the student's advisor. Nine hours of technical electives must be in the Mechanical and Aerospace Engineering and Engineering Mechanics department. Three hours of departmental technical electives must be at the 300-level. Honors students have special requirements for technical electives.

8 This course can be selected from English 60, 160, SP&MS 85, or the complete four-course sequence in Advanced ROTC (Mil Sc 105, 106, 107, and 108 or Aerospace Studies 350, 351, 380, and 381).

9 All electives must be approved by the student's advisor. Students must comply with the School of Engineering general education requirements with respect to selection and depth of study. These requirements are specified in the current catalog.

10 Each student is required to take six hours of free electives in consultation with his/her academic advisor. Credits which do not count towards
   this requirement are deficiency courses (such as algebra and trigonometry), and extra credits in required courses. Any courses outside of
   Engineering and Science must be at least three credit hours.

NOTE: All Aerospace Engineering students must take the Fundamentals of Engineering Examination prior to graduation. A passing grade on this examination is not required to earn a B.S. degree, however, it is the first step toward becoming a registered professional engineer. This requirement is part of the UMR assessment process as described in Assessment Requirements found elsewhere in the catalog. Students must sign a release form giving the University access to their Fundamentals of Engineering Examination score.

Figure (6) 2003 AEROSPACE ENGINEERING CURRICULUM (128CREDIT HOURS)

Sophomore Design Laboratory: Also located on the west side of the campus, the Sophomore Design Laboratory supports hands-on activities, including construction and testing, that are taught in AE 180 Introduction to Aerospace Design. Offered in the second semester of the sophomore year, this course requires students to design and build two aircraft models. The equipment in this laboratory includes a system for launching small aircraft models, a test stand for measuring engine performance, a weight and balance arrangement, construction work stations, computer stations, tools, and supplies.

Senior Design Laboratory: The laboratory in the Mechanical Engineering building provides space and equipment for manufacturing remotely piloted aircraft designed by student teams in their senior design class. The laboratory includes several workstations for the construction and assembly of aircraft models. Several hardware arrangements including an engine test stand for measuring the static thrust of model aircraft engines, a weight and balance setup to determine the total weight and center of gravity location, and a number of standard woodworking power tools are available in the laboratory. In addition, the laboratory provides an assortment of specialized tools for student use.

Composite Materials Manufacturing and Characterization Laboratory: This laboratory possesses facilities for manufacturing and characterization in the area of composite materials. A laboratory scale pultrusion machine is available to manufacture polymer matrix composites. Hot presses are also available for student use in the laboratory. Ceramic processing facilities in the laboratory include equipment for synthesizing and forming ceramics and ceramic matrix composites. An MTS material test system with a 22 kip fatigue rated load frame, tensile test module, cyclic function generator, and data acquisition system are also available. A Fatigue Dynamics, Inc. fatigue machine, in which direct stress and combined tests can be performed, is part of this facility. High air temperature testing of ceramics and ceramic composites up to 1400°C are done using an Instron 4204 microprocessor-controlled 11 kip test system. For creep testing, an ATS 2430 Lever Arm Tester located in the MRC is available. A gas gun consisting of a high pressure chamber with a 3/8-inch diameter and a 6-foot long tube is available for shooting low velocity projectiles directed towards a composite target. A drop-test tower is also available. The resistance to wear and the friction stability of brake lining composites can be characterized using a Chase machine that is also provided in the laboratory.

Flight Simulation and Training Laboratory: This laboratory contains flexible flight models that simulate the flight characteristics of fixed-wing aircraft. Aircraft aerodynamics and propulsion can be tested in this laboratory for best performance. This laboratory provides a learning environment that complements the theoretical presentation students receive in their classes, such as Stability and Control, Propulsion, Aerodynamics, Performance, and, Aircraft or Subsystem Design, and Human

Factors Engineering. Flight training software and hardware arrangements for introducing students to flight and aircraft operations are also available in the laboratory. Also, the laboratory possesses an FAA certified Personal Computer Airplane Training simulator as general aviation training is employed for teaching juniors in the aerospace engineering program {Figure 7)].

Figure (7)Photo shows an example of today's laboratories
[flight simulation laboratory]

Fluid Dynamics and Combustion Laboratory: Experimental facilities in this lab include a supersonic wind tunnel, a shock tube, a combustion rack, a detonation apparatus, a supersonic atomization apparatus, a 48"x 30" test section, an active flow control rack, and a high-temperature, high-pressure, open jet combusting flow apparatus. The laboratory is equipped with state-of-the-art instrumentation for flow velocity and turbulence measurements, liquid and solid particle size measurement in two-phase flow, and chemical species such as NOx and CO in combustion systems. Two powerful pulse lasers form the core of the instrumentation. These are used for particle image velocimetry (PIV), flow visualization, and laser-induced fluorescence. A laser diffraction particle analyzer is available for instantaneous measurement of solid and liquid particle size. A powerful pulse xenon light source is used for short duration Schlieren studies and UV diagnostics. A hot wire anemometer is available for velocity and turbulence measurements.

Space Systems Engineering Laboratory: This laboratory supports the overall astronautics curriculum through individual experiments that focus on one or more disciplines of spaceflight, including structures, thermal, attitude, power, and orbit determination/tracking. Also, the laboratory is employed in activities that are related to the design and fabrication of small spacecraft. The students using this lab are primarily graduate students and those in the senior design class who choose to focus on spacecraft design, instead of aircraft design. Several arrangements including a clean room hood, a ground station to control and command spacecraft, and various testing equipment are available in the laboratory. Images of the Missouri area, including visible and infrared spectra, can be down linked from various NOAA spacecraft in the laboratory.

6. Summary

This paper summarizes the development of the Aerospace Engineering program on the University of Missouri-Rolla campus. Although the program was established in 1967, courses and studies in aerospace engineering began in the 1950s by an aviation enthusiast, Dr. Aaron J. Miles, within the Department of Mechanical Engineering. Today, the UMR aerospace engineering program provides a curriculum that permits students to gain theoretical knowledge and skill to apply such knowledge to generate original solutions to practical and hands-on problems. This is accomplished by involving students in research and by providing numerous opportunities for graduate and undergraduate students to participate in hands-on design projects and in competitions that allow them to develop as engineers, as leaders, and as individuals. The program possesses a dedicated faculty who teach, mentor, and advise students and are committed to producing new knowledge by researching, discovering, and publishing. More importantly, the program enjoys focused dynamic students who are excited about future possibilities and challenges. The UMR aerospace program can be proud of awarding over 800 undergraduate and graduate degrees since its inception. These graduates are making significant contributions in the design and development of aerospace and related systems in the aerospace industry and in various government agencies.

Acknowledgment

The authors would like to acknowledge colleagues in the Department of Mechanical and Aerospace Engineering and Engineering Mechanics and on the UMR campus who freely gave of their time and expertise to provide information for this manuscript. In particular, we would like thank Professors Robert Oetting, Bruce Selberg, Fred Nelson, and Walter Eversman for their insights into the birth and growth of the UMR aerospace program.

References

1. Our Flying Dean Emeritus, Interface, University of Missouri-Rolla office of public information, June 1969.

2. Recommendation for Program Leading to Degree in Aerospace Engineering, Thomas R. Faucett, Internal
    Memorandum ,Department of Mechanical Engineering, June 1967.

3. Building Dedication, Harry J. Sauer, Internal Report, Department of Mechanical and Aerospace Engineering and
    Engineering Mechanics, September 1969.

4. MSM Alumnus, MSM-UMR Alumni Association, University of Missouri - Rolla, October 1971

5. Aerospace Engineering Five Year Development Plan, Internal report, Department of Mechanical and Aerospace
    Engineering, 1974.

6. A proposal for A Doctor of Philosophy Degree in Aerospace Engineering, Internal report, Department of Mechanical
    and Aerospace Engineering and Engineering Mechanics, December 1989.

Chapter 53

Aerospace Engineering at The University of Texas at Arlington

Donald R. Wilson
Professor and Chair
Department of Mechanical and Aerospace Engineering

Introduction

Aerospace engineering at UTA developed and grew during the last half of the first century of powered flight, with formal inauguration of the AE program in 1964. Located in the center of one of the largest aerospace industrial complexes in the nation, the development of the program has been shaped by the nature of the local industry, which consists primarily of aircraft, missile and rotorcraft companies. The academic program has benefited immeasurably by interaction with the local industry, and hopefully has contributed significantly to development of that industry. The following paragraphs trace the historical development of the program through four rather distinct periods, highlight the initiation of major education and research initiatives, and describe significant accomplishments of faculty, students and alumni.

Historical Development of the AE Program

The Early Days (1964-1973)

The Engineering School at The University of Texas at Arlington began in 1959, when then Arlington State College was elevated to a four-year senior college status to serve the educational needs of the rapidly expanding Dallas-Fort Worth metroplex. Because of its unique location at the center of a large industrial and business complex, the development of engineering and business programs received a high priority and many of the initial faculty members were recruited from the local aerospace and electronics industry (i.e. General Dynamics, LTV Aerospace, Bell Helicopter, and Texas Instruments). This resulted in a particularly well qualified faculty with a depth of experience in engineering research, development, design, and production. Initially, Arlington State College was part of the Texas A & M System, but in 1965, the legislature transferred ASC to the University of Texas System, and the name was changed to The University of Texas at Arlington.

Because of the presence of a major aerospace industry in the metroplex, Aerospace and Mechanical Engineering Department Chair Carl Files saw a need to initiate a degree program specifically in aerospace engineering. Thus, in 1964, he recruited Dr. Jack Fairchild (Ph.D., Oklahoma), a former employee of both Bell Helicopter and Chance Vought Aircraft, and who had recently completed his Ph. D. degree to join Arlington State College. Dr. Fairchild was charged by Dean of Engineering Wendell Nedderman "with prime responsibility to develop an aerospace engineering program". Dipl. Engr. Paul Buetiker from the Swiss Federal Institute and he were the faculty for the first year of the AE program. With the cooperation of the Aeronautical Technology Department, Dr. Buetiker and a team of students designed and built a high quality low speed wind tunnel for use in undergraduate instruction and research. Dr. Fairchild organized the AIAA Student Branch, which was chartered in 1965. Additional faculty were soon recruited from the local industry to form the core of the aerospace program; aerodynamics specialist Dr. Don Seath (Ph.D., Iowa State) from General Dynamics in 1965, propulsion specialist Don Wilson (M.S., UTSI) from LTV Aerospace and Defense Company in 1968, and viscous flow specialist Dr. Fred Payne (Ph.D., Penn State) from General Dynamics in 1969. These four, along with three engineering mechanics specialists already on the engineering faculty; Drs. Wynn Dalley (Ph.D., UT-Austin), J. H. Gaines (Ph.D., UT-Austin), and Joseph Stanovsky (Ph.D., Penn State), rounded out the core faculty for the aerospace program.

The program developed rapidly with the consistent support of Prof. Files. The initial ECPD (now ABET) accreditation was approved in a record time of three years in 1968 after awarding its first degree in 1965. The first female graduates from the program were Ms. Margaret Kantz and Mrs. Penny Carlisle. Both were members of the 1967 graduating class, and were instrumental in founding the Society of Women Engineers at UTA. Full six-year ECPD accreditations were granted in 1971 and again in 1977. Dr. Payne founded the UTA Chapter of Sigma Gamma Tau, AE national honor society in 1970. The M.S. Graduate Program was petitioned for, approved, and initiated, with the first M.S. degrees awarded in 1970. An undifferentiated Ph. D. degree was added in 1972, with the first Ph. D. awarded in 1973. Prof. Don Wilson was one of the early Ph.D. graduates, finishing in December 1973.

For complex reasons, Dean Andy Salis decided to remove the AE program from the Aerospace and Mechanical Engineering Department and form a separate Department. This was accomplished informally in 1971 and formally in 1972—a particularly inopportune time due to a major weakness in the aerospace industry resulting from winding down of the Vietnam War and completion of the Apollo program and subsequent massive lay-offs in the aerospace industry. However, the fledgling new Aerospace Engineering Department was able to adapt to the low undergraduate enrollments by expanding its graduate programs to attract the many out-of-work aerospace engineers in the local area. Dr. Fairchild served as the first AE Department Acting Chair until Dr. Seath took over as Chair in 1973.

The undergraduate degree program was unique for its time in requiring "core" courses in viscous flow, astronautics, flight dynamics and control, propulsion, and heat transfer as well as traditional ones in aerodynamics and structural analysis. The program culminated in a student-initiated senior laboratory project and a "capstone" vehicle design project which integrated all of the disciplines in the design of an airplane to satisfy a given RFP requirement. The "capstone" design course later became an ABET requirement of all aerospace programs in the U.S. and the AIAA initiated its student design competitions. In order to serve the local aerospace industry, the program was oriented toward atmospheric air-breathing aircraft analysis, design, and research.

Growth of the Graduate Program (1973-1984)

The graduate aerospace engineering program experienced significant growth during the period from 1973-1984 under Dr. Seath's leadership. Dr. Payne took over as graduate advisor in 1973, and immediately established the graduate seminar to improve oral defenses of theses and dissertations, and to prepare students for presentation of their research at national and international conferences. During this period eight Ph.D. and 35 M.S. degrees were awarded, and a number of these degrees were earned by part-time students employed full time in the local aerospace industry. Graduate student research covered a wide range of topics in the traditional areas of aerodynamics, turbulence, propulsion, structures and flight dynamics, as well as non-traditional areas including mass transportation, plasma dynamics and MHD power generation.

A number of notable events occurred during this period. Dr. Fairchild, together with members of the AIAA North Texas Section held the first AIAA North Texas Mini-Symposium in 1973. This was done to give graduate students and young engineers working in the local aerospace industry an opportunity to meet and exchange ideas without the expense of traveling to national conferences, and established a tradition that lasted well into the late 80s. Dr. Fairchild also introduced an innovative program in composite aircraft construction in 1977 to give students some hands on experience in building and testing composite aircraft materials. Although the project envisioned using the airplane ultimately in the Flight Test Engineering Course, legal and liability issues raised by the UT System Administration made this goal unattainable. Dr. Stanovsky was invited to be a Visiting Professor at the University of Petroleum and Minerals, Saudi Arabia from 1974 to 1976 and helped established its engineering programs. Dr. Fairchild was the 1975 recipient of the ASEE/AIAA Aerospace Division Educational Achievement Award. Dudley Smith (MSAE, 1976) became the first UTA student to win the AIAA Graduate Student Paper Competition, with his paper on laser doppler anemometry. Dr. Payne, together with Drs. A. Haji-Sheik (ME), Constantin Corduneanu (Math) and Leo Huang (CE); began planning for a series of international conferences on integral methods in 1983. The first two IMSE (Integral Methods in Science and Engineering) conferences were chaired by Dr. Payne at UTA in 1985 and 1990. Subsequent meetings were held in Japan (1993), Finland (1996), Michigan State (1998), Banff (2000) and France (2002). The next meeting in Orlando, Florida is scheduled for 2004.

Expansion of Funded Research Programs (1984-1992)

Dr. Seath announced his decision to step down as Department Chair in 1983 and after an abortive search for a replacement, Dr. Bill Jiles, Professor of Electrical Engineering and a controls specialist was appointed Interim Chair in September 1984. That same year, Dr. Dale Anderson (Ph.D., Iowa State), Director of the highly successful NASA CFD Center at Iowa State, and Dr. Rich Hindman (Ph.D., Iowa State) were recruited to establish a new program in computational fluid dynamics. This was a new direction for the AE Department, and was in part the result of strong support from Dr. Dick Bradley of General Dynamics, who was a member of the AE Department Advisory Board and largely responsible for establishment of CFD activities at General Dynamics. Other areas of the department also experienced growth. Early in 1985, the AE Department established a plan for growth that would continue to build upon its existing strength in fluid dynamics, and then focus in turn on building research capabilities in structures and flight mechanics. Three new CFD faculty members were added during the next few years; Dr. Ijaz Parpia (Ph.D., Purdue) in 1986, and Drs. Steve Kennon (Ph.D., UT-Austin) and Dave Thompson (Ph.D., Iowa State) in 1987. The experimental gas dynamics program was also strengthened by the addition of Dr. Frank Lu (Ph.D., Penn State) in 1987, and Dr. Shiv Joshi (Ph.D., Purdue) was recruited from Arizona State in 1988 to establish a research program in composite and smart structures. In 1991, Dr. Dan Tuckness (Ph.D., UT-Austin) was recruited from Lockheed Engineering and Sciences to develop a graduate program in astronautics.

Funded research programs grew significantly during this period, particularly in the computational and experimental fluid dynamics and composite structures areas, and the AE faculty numbered 13 at its peak in 1988. Unfortunately, changes in the world geopolitical scene led to an abrupt downturn in the aerospace and defense industries, and consequently AE enrollment entered a period of significant decline. AE enrollment dropped from a peak of 468 in

1988 to 191 in 1992. This drop in enrollment derailed further planned expansion of the AE faculty, and faculty leaving for other opportunities or retiring during this period were not replaced. As a result the active AE faculty dropped from 13 in 1988 to 11 in 1992.

Several notable events occurred during this period. Dr. Fairchild was invited to be a Distinguished Visiting Professor at the U.S. Military Academy, West Point, NY during the 85-86 academic year. Dr. Anderson received the College of Engineering 1988 Haliburton Award for Excellence in Research and the 1990 Professional Achievement Citation in Engineering from Iowa State University. Dr. Thompson was the 1990 recipient of Haliburton Award for Excellence in Teaching, and Dr. Parpia received the same award in 1991. Dr. Wilson together with Clark White from the NASA Ames Research Center developed a unique graduate course in powered lift concepts in 1986 that was taught by a team of instructors from NASA Ames, Langley and Lewis Research Centers, with the closing lecture given by Dr. Hans Mark, Chancellor of the UT System and former Director of the NASA Ames Research Center. Dean John McElroy provided budget support to cover travel expenses of the NASA instructors, commenting that this was undoubtedly the most expensive graduate course offered by UTA. The course was a great success, and was attended by a large number of industry personnel in addition to our full time students.

Recent History (1992-Present)

In September 1992, as a result of declining aerospace engineering enrollments and the retirement of Dr. Bill Jiles as AE Chair, Dean John McElroy merged the Aerospace Engineering and Mechanical Engineering Departments to form the Department of Mechanical and Aerospace Engineering. Dr. Kent Lawrence (ME) was appointed Chair, and Dr. Wilson (AE) was appointed AE Program Director. The combined department had 27 faculty and 618 students. Aerospace engineering enrollment continued to dwindle, particularly at the undergraduate level. However the program survived, due in part to a continued strong graduate program, substantial research funding brought in by AE faculty, and a supportive administration. The merger actually strengthened the AE program by adding the considerable expertise in structures and materials residing in the ME faculty with the fluid dynamics expertise that resided primarily in the AE faculty. This combination of faculty strength gave us the ability to put together a winning proposal for one of three NASA Hypersonic Research Center awards in a national competition held in 1993.

Dr. Lawrence stepped down as Chair in 1994 and was replaced by Dr. J. Ronald Bailey (Ph.D., NC State), a senior manager with IBM Corp. Dr. Dora Musielak (Ph.D., University of Alabama-Huntsville) joined the faculty as Assistant Professor of Mechanical Engineering, although she also supported the AE program by taking over the graduate propulsion course. Dr. Bailey was recruited to be Dean of Engineering at UTA in July 1996, and appointed Dr. Roger Goolsby (ME) as Acting Chair of MAE. After a national search Dr. Wilson was appointed as Chair of the MAE Department in July 1997, and Dr. Tuckness was appointed Director of the Aerospace Engineering Program. In 1998, Dr. Anderson was made Vice-President for Research and Dean of the Graduate School, and Dr. Tom Lund (Ph.D., Stanford) from the Stanford Center for Turbulence Research joined the faculty as Associate Professor of Aerospace Engineering. Dr. Parpia resigned in 1997 to form his own company, and after an extensive search, Dr. George Dulikravich (Ph.D., UT-Austin) was recruited from Penn State in 1999. Dr. George Emanuel (Ph.D., Stanford) retired in 1999 from the University of Oklahoma, but not being ready to leave the engineering field, joined UTA as a Research Professor. He is currently working on several projects, including ignition enhancement of supersonic fuel-air mixtures with corona discharges and a novel concept for improving the performance of COIL lasers.

The aerospace engineering enrollment decline finally bottomed out at 71 in 1998, and entered a period of impressive growth. By the fall of 2003, undergraduate enrollment had quadrupled. Graduate enrollment during this same period experienced a small but steady growth, but is anticipated to grow much more rapidly as a result of increasing numbers of new engineers recruited by Lockheed Martin following award of the JSF contract, and anticipated hiring of new engineers in the remaining North Texas aerospace companies as their aging engineering workforce begins to retire. Thus as we start into the next 100 years of powered flight, the future growth of the aerospace engineering program at UTA certainly looks promising.

Several notable events occurred during this period. Dr. Anderson received the 1993 Haliburton Award for Excellence in Teaching. Dr. Tuckness was the 1996 recipient of the General Dynamics Robert Q. Lee Award for Excellence in Teaching. A UTA team directed by Dr. Lu won first place in the 1997 AIAA Graduate Missile Design Competition. Dr. Lu also served as Chair of the North Texas Section of the AIAA during 1997 and 1998. Dr. Wilson was awarded the 2001 AIAA Ground Testing Award, Dr. Emanuel the 2001 AIAA Plasmadynamics Award, and Dr. Dulikravich was the recipient of the 2001 Eli Carafoli Award and Commemorative Medal from Politechnica in Bucharest, Romania. The AE faculty also became active in hosting national and international meetings on campus. Dr. Wilson hosted the 79th Supersonic Tunnel Association International Meeting in April 1993. Dr. Joshi hosted the First ARO Smart Structures Workshop in September 1993, The MAE Department and Bell Helicopter co-hosted Rotorcraft Industry Technology Association (RITA) Meetings in 1999 and 2002, bringing over 150 representatives from NASA, Army, FHA, industry and academia to campus at each meeting. They also co-hosted the AHS Specialist Meeting on Powered Lift Technology in 2001. Dr. Lu hosted the 23rd International Symposium on Shock Waves in 2001 which brought 210 delegates from 21 countries to the UTA campus.

Major Research Initiatives

Early Research Programs

Early graduate research in the Department involved experimental studies of high-angle-of-attack unsteady aerodynamics utilizing the newly completed low-speed wind tunnel and funded by the National Science Foundation; an experimental, analytical, and computer simulation study to determine how the fuel efficiency of Dallas transit buses could be improved during the "oil crunch" of the early 1970's funded by the Urban Mass Transportation Administration (DOT); investigations of fixed and rotary wing tip vortex structures; development of a new wing loading theory; experimental investigation of propulsive wing concepts; and studies sponsored by the Bell Helicopter to develop a new anti-torque system for use on helicopters. This multi-year helicopter study involved theoretical, experimental, and design parameter analysis of ducted tail rotors and provided thesis work for four graduate students. One of these, then Army Capt. Robert Stewart, ultimately became an astronaut on the Space Shuttle Challenger and along with Bruce McCandless accomplished the first untethered flights with the manned maneuvering unit (MMU). These tail rotor studies led to the development of a full-scale tail rotor system, dubbed a "ring-tail" rotor by Bell Helicopter, which was slated to be used on the new Bell Model 400 helicopter after successful flight tests until the project was cancelled for marketing and economic reasons.

Turbulence Modeling

Dr. Payne initiated research in turbulence with a grant from NASA Ames Research Center in 1975 to investigate the large eddy structure of turbulent boundary layers. Further turbulence research included development of large eddy interaction models for turbulent wakes, instability modes in turbulent boundary layers, microburst simulations via vortex ring and turbulent jet models, self-preservation in turbulent free jets, and computational modeling of turbulent boundary layers via a novel direct formal integration method.

Dr. Lund initiated a new line of turbulence modeling research when he joined the faculty in 1997 from the Stanford University Center for Turbulent Research. He has developed new dynamic subgrid-scale models for large eddy simulation of complex flows, and hybrid RANS/LES models for rotorcraft applications (sponsored by Bell Helicopter). Dr. Lund also was involved in the NASA-sponsored High Lift Flight Tunnel (HiLiFT) Feasibility Study. This was a cryogenic, pressurized semi-free flight tunnel concept designed to simulate full scale Reynolds numbers for large transport aircraft.

Experimental Gas Dynamics

Current research in experimental gas dynamics is housed in the UTA Aerodynamics Research Center, whose origins can be traced to an effort initiated in 1976 by graduate student, J.P. Angelone to develop a shock tube for experimental research in gas dynamics. The shock tube was fired for the first time on May 25, 1978. Also in 1976, Department Chair Don Seath and Dr. Don Wilson visited both AEDC and the NASA Marshall Space Flight Center to view their transonic Ludwieg tube facilities, with the idea of building a much smaller version. Upon arrival at AEDC, they discovered that the Ludwieg tube, which was the prototype of the Air Force concept for the National Transonic Facility, had just been decommissioned when the decision was made to build the NTF at NASA-Langley Research Center. With tremendous help from Rogers Starr and Dave Stallings, both who had been deeply involved in the Ludwieg tube development project, the AEDC Ludwieg tube showed up at the rail head behind the Arlington library on November 11, 1978. The shipment required three rail cars, one an oversized 88 foot flat bed that held the 112 foot charge tube along with a smaller 90 foot tube. The remaining equipment was mounted to two standard 60 foot flatbed cars located at each end of the longer car so that the curves along the way could be negotiated with the oversized tube. Unfortunately, the AEDC donation did not include the air compressor, hydraulic, pneumatic and computer control and data acquisition systems required to run the facility. It took five years and nearly $50,000 to assemble the required support systems, and on January 12, 1984 the first firing of the Ludwieg tube by a team of students headed by James Sergeant, now Flight Test Manager of the JSF Integrated Test Force at Lockheed Martin, was successfully conducted. The first funded research project to be conducted in the transonic Ludwieg tube was an experimental investigation of the helicopter blade vortex interaction (BVI). The project was supported by an ARO grant and managed by Dr. Henry Jones of the Army Aeroflightdynamics Laboratory at NASA Ames Research Center, and one of the first graduate students from the AE program.

With the initiation of funded research programs in the transonic facility, Drs. Seath and Wilson were able to convince Dean John Rouse to approve a new research lab to house the transonic tunnel. The donation of valuable test equipment; which included nozzles and test sections from supersonic and hypersonic wind tunnels from Jim Cooksey of the LTV Gas Dynamics Lab; a 5-stage Clark air compressor rated at 3000 psi and 2000 cfm from Dr. Henry Jones of the Army Aeroflightdynamics Lab and a Thermal Dynamics 2MW DC arc heater from Dr. Mike High of AEDC, enabled us to make a case for expanding the scope of the lab so that we could provide test capability spanning the range from low to hypersonic speeds. Construction started on the new facility, located on the southeast corner of the campus, in March 1986 and the official open house celebration was held on December 3, 1986. This event was attended by over

140 people, and gave our graduate students an opportunity to display their research activities to a wide audience. The original low speed tunnel was moved to the new lab, and Dr. Seath acquired two smaller low speed wind tunnels from Edwards Air Force Base that were installed in the Engineering Lab Building on the main part of the campus to support the undergraduate lab program.

Continued development of the ARC facility has occurred since 1986. Initial operation of the Hypersonic Shock Tunnel occurred in August of 1989, due in large part to the efforts of graduate students Ragu Murtugudde and Scott Stuessy, and the first funded research grant was obtained from NASA-Langley Research Center for research on hypersonic shock boundary layer interactions. Dr. Frank Lu, recently recruited from Penn State University, took over the operation of the shock tunnel, and was principal investigator for this program. Award of a $300,000 grant from the UT-System Permanent University Fund in June 1989 enabled the development of an Aeropropulsion Laboratory to house the AEDC arc heater. After a lengthy development program, this facility became fully operational in November 1996, again with the principal development effort carried out by graduate students, William Jacqmein, Chris Roseberry and Zak Boonjue. This facility is currently being used to investigate fuel reformation of methane into hydrogen and carbon to be used as fuels in supersonic combustors. UTA was invited to join the Supersonic Tunnel Association, which required us to present a briefing on the facility capabilities and research programs in place at the ARC to the STA membership at a meeting at Princeton University in October 1989. UTA subsequently hosted the 79th STA meeting in March of 1993, which brought scientists and engineers from laboratories around the world to the UTA campus for a two-day meeting. Award of a $49,500 grant from Dean John McElroy provided the funds necessary to complete the supersonic wind tunnel, which became operational in 2001. The development of this facility, which is currently being used to investigate performance of an integrated supersonic inlet/ejector-augmented pulse detonation rocket, was conducted by Joji Matsumoto, one of our first Japanese graduate students.

The facilities at the ARC are unique. To our knowledge, no university in the U.S. has the capability of simulating atmospheric flight conditions spanning the complete spectrum from take-off to Mach 16. A proposal reviewer stated that our transonic wind tunnel "is unique, there are no comparable facilities anywhere in this country." A complete description of our hypersonic shock tunnel was included in the recent AIAA publication *Advanced Hypersonic Facilities*, and Dennis Bushnell, Chief Scientist of the NASA Langley Research Center, has stated publicly that the ARC is "a jewel, unequalled among universities in the U.S." Since development of the ARC was initiated in 1984, its unique facilities have resulted in $3.15M in research funding and $735K in equipment grants and donations. Furthermore, two of the facilities donated by the government to establish the ARC had a stated replacement value of $2.25M. Ten Ph.D., 64 M.S., 6 M.Eng. and 32 B.S. students have completed research projects at the ARC, resulting in 24 journal publications, 112 conference papers, 28 technical reports, and issue of 2 provisional patents. . We have hosted two international conferences, and scientists and engineers from 14 countries and numerous aerospace companies and government labs have toured our facilities. Currently, 2 Ph.D., 6 M.S. and 2 B.S. students are conducting research projects in the ARC

Computational Fluid Dynamics

Computational fluid dynamics emerged as a critical technology in the aerospace industry in the late 70's and early 80's. This trend was recognized by the Industrial Advisory Board of the Aerospace Engineering Department and led to a recommendation to expand faculty expertise in this area in 1982. The department made a substantial commitment based on this recommendation

683

The Metroplex is home to a large contingent of aerospace companies including Lockheed Martin Aeronautics, Lockheed Martin Missiles and Fire Control, Bell Helicopter and Vought Aircraft Industries. At that time, Lockheed Martin Aeronautics (then General Dynamics) and Lockheed Martin Missiles and Fire Control (then LTV Missiles and Space Division) provided additional support to the fledging CFD program by providing consulting opportunities for new CFD faculty, and a ready source of young engineers who entered the graduate program to receive education in CFD.

Dr. Dale Anderson, former Director of the NASA-sponsored CFD Center at Iowa State, and colleague D. Richard Hindman were recruited from Iowa State to formally establish the CFD Program at UTA in 1984. Three additional young faculty members, Drs. Ijaz Parpia, Steve Kennon and Dave Thompson were hired over the following two years. This provided a complement of five faculty members in this critical area. In the ensuing years, the CFD program at UTA provided substantial support for the industry and played an essential part in the development of CFD expertise in local industries. Substantial sponsored research in CFD flowed to UTA, funded by local companies, NASA, DOD, NSF and numerous other organizations. The UTA CFD group was particularly well-known for their code development work involving finite difference and finite volume formulations of compressible Euler and Navier Stokes equations, adaptive grid generation, unsteady aerodynamics, and non-equilibrium chemical kinetics. The support for graduate students provided through this research effort was essential for program health and a large number of students have completed graduate programs with a CFD emphasis since the inception of the CFD program at UTA. The faculty size has dwindled in recent years as a result of several faculty departures not being replaced because of declining enrollment in aerospace engineering, and the current faculty in this area includes Dr. Anderson, Dr. Tom Lund (Ph.D., Stanford) and Dr. George Dulikravich (Ph.D., UT-Austin). Dr. Dulikravich also initiated a major new program in Multidisciplinary Inverse Analysis and Design Optimization, in addition to his participation in CFD.

Composite and Smart Structures

Dr. Shiv Joshi was recruited from The University of Arizona to initiate a composite structures program in the Aerospace Engineering Department in 1989. He joined Dr. Wen Chan of the Mechanical Engineering Department, who had been recruited from Bell Helicopter one year earlier, and together they formed the nucleus of our composite structures program. Actually, both had been trained by Dr. C.T. Sun at Purdue University, one of the leading experts in composite structures in the world. Research in composite structures has been funded by National Science Foundation, DARPA, Army Research Office, NASA, Army Aviation Technology Lab, Northrop-Grumman, Lockheed Martin Aeronautics, Bell Helicopter, Boeing, American Eurocopter, and a number of smaller companies. Specific research projects have included development of analysis and design models for composite structures, damage tolerance, durability, structural health monitoring, delamination analysis, fatigue and fracture mechanics. Later research has evolved into investigations of smart structures, development of nano-films for high resolution pressure sensing, and manufacturing of bio-MEMS devices. The Center for Composite Materials, direct by Dr. Chan, was formally established in 1994 and includes participating faculty from mechanical, aerospace and materials science and engineering. Dr. Joshi is currently working with NextGen Aeronautics, a Northrop-Grumman spin-off company, funded by DARPA to investigate structural morphing of unmanned aircraft to optimize performance over the full extent of the flight envelope. Dr. Chan will assume the position of Interim Chair of the Department later this summer, replacing Dr. Wilson, who is returning to full-time teaching and research.

NASA Hypersonic Research Center

UTA, together with The University of Maryland and Syracuse University, successfully competed in 1993 for the award of three NASA-sponsored Centers for Hypersonic Research. Each university was awarded a three-year $600,000 grant, starting in September 1993. NASA funding for the UTA Center was supplemented by support from Lockheed Martin Tactical Aircraft Systems, NASA-Langley Research Center, the State of Texas Advanced Technology Program, MSE, Inc., and Cray Research. The principal objectives of the Hypersonic Center program were to develop modern educational programs in hypersonics, and provide graduate students with opportunities to conduct state-of-the-art research in technical disciplines that were critical to the development of hypersonic flight vehicles. The UTA Center included faculty and students from Aerospace, Mechanical and Materials Science and Engineering programs. The NASA Center was co-directed by Dr. Wilson (AE), whose technical specialty was gas dynamics and propulsion, and Dr. Anderson (AE), who specialized in computational fluid dynamics. Other faculty members and their technical areas of expertise include Dr. Pranish Aswath (MSE), high-temperature materials; Dr. Wen Chan (ME), composite materials; Dr. Dora Musielak (ME), propulsion; Dr. A. Haji-Sheikh (ME), heat transfer; Dr. Shiv Josh (AE), composite structures; Dr. Kent Lawrence (ME), structural dynamics and multi-disciplinary design optimization; Dr. Frank Lu (AE), experimental fluid dynamics; Dr. Ijaz Parpia (AE), computational fluid dynamics; and Dr. B.P. Wang (ME), structures and design optimization.

Specialized courses were developed in hypersonic aerodynamics, aerothermodynamics, propulsion, structures, high-temperature materials, flight dynamics and hypersonic vehicle design. Complementary research programs focused on the high-temperature aspects of sustained flight at hypersonic speeds. Lockheed Martin personnel provided invaluable support in developing and teaching a new graduate course in Hypersonic Vehicle Design, offered for the first time in the fall of 1995. The Lockheed participants in the design course included Dr. J.V. Clifton, flight vehicle design and aerothermodynamics and Lockheed coordinator for the course; Dr. Armand Chaput, program management; Mr. Don Couch, propulsion; Mr. Dennis Finley, aerodynamic design; Mr. Mike Henson, structures; Dr. Bill Garver, materials; Mr. Scott Hames, flight vehicle performance; and Mr. Grant Castleberry, subsystems design. The course culminated with a student team design of a single stage to orbit hypersonic flight vehicle.

The NASA program was terminated in 1997 due to funding cutbacks in hypersonic research at NASA. However, during the four years of operation, 11 faculty members and 22 graduate students (eight Ph.D. and 14 M.S.) were involved in the Center program. Virtually all of the students graduating from the program have taken positions in the aerospace industry. The Center also supported 14 undergraduate students, four of whom went on to become involved in the Hypersonic Program at the graduate level. New courses specializing in hypersonics were developed, and a total of 19 new research projects were initiated. The $700,000 provided in NASA funding was supplemented by external funding of $1,016,000. A total of 55 journal articles and conference papers were published by faculty and students. Research in hypersonics, primarily in the areas of hypersonic aerodynamics and propulsion has continued since the termination of the program, but at a lower level of intensity.

Pulse Detonation Engine Research

Research in pulse detonation engines was initiated with an award of an $180,000 grant from the Texas Higher Education Coordinating Board in January, 1993. Industry collaborators in the project included Lockheed Fort Worth Company and the Rocketdyne Division of Rockwell International. Initial research focused on investigation of detonation ignition and propagation in a detonation tube, using mixtures of oxygen with hydrogen, propane or methane. The mixtures were ignited with a high-energy arc discharge from a capacitor bank. Chapman-Jouguet detonations with supersonic wave speeds approaching 2500 m/sec were achieved for all three mixtures. The effects of combustor geometry, equivalence ratio, initial pressure level, ignition system energy level and location, and various deflagration-to-detonation transition devices were investigated. Demonstration of a pre-ignition concept in which a detonation wave initiated in a hydrogen-oxygen mixture was used to detonate a hydrogen-air mixture was also accomplished. A rectangular detonation wave chamber was developed for optical flow visualization of detonation waves. Wave phenomena associated with the initiation and development of detonation waves will be visualized via an existing schlieren system. Furthermore, planar laser induced fluorescence (PLIF) will be used to determine the mole fraction of selected chemical species in the vicinity of the detonation wave. The objective of the investigation is to determine the role of the transverse wave structure on the energy release mechanisms of detonation waves. The PLIF study will utilize a new excimer laser and intensified charge coupled device camera purchased with funding support from the Office of Naval Research DURIP Program. UTA faculty involved in the PDE program included Drs. Wilson, Lu and Gonzalez, an ME faculty member now with TRW.

Following completion of initial detonation tube experiments, a repetitive pulse detonation engine was designed and fabricated. An ignition system capable of delivering 25 Joule discharges at 200 Hz was developed as part of a NASA-Langley grant, however the current rotary valve fuel injection system limits the engine operation frequency to 30 Hz. Development of solenoid valves capable of 100 Hz operation with flow rate levels compatible with the current engine size are in progress. The effect of Shchelkin spirals and aft nozzles is currently being evaluated in the PDE. In a related series of experiments, the feasibility of electrical power generation via magnetohydrodynamic energy conversion was investigated by measuring the electrical conductivity of the exhaust from the PDE. The reactants were near-stoichiometric mixtures of hydrogen or propane with oxygen or air, and were seeded with cesium and potassium carbonate. The study was supported by MSE, Inc. as part of a Phase 1 STTR funded by NASA-Langley Research Center.

A novel multi-mode, single flow path, pulse detonation propulsion system concept with applications to both hypersonic cruise and access to space is currently being investigated. The basic concept includes the following four modes of operation: an ejector-augmented pulse detonation rocket (PDR) for take off to moderate supersonic Mach numbers, a pulsed normal detonation wave engine (NDWE) mode for operation at flight Mach numbers from approximately 3 to 7, which corresponds to combustion chamber Mach numbers less than the Chapman-Jouguet Mach number, an oblique detonation wave engine (ODWE) mode of operation for flight Mach numbers that result in detonation chamber Mach numbers greater than the Chapman-Jouguet Mach number, and pure PDR mode of operation at very high Mach numbers and altitudes. The program was initiated as part of an Air Force Phase 1 SBIR funded to a team of HyPerComp, UTA and Lockheed Martin. Computational simulations indicated that thrust and specific impulse were comparable or superior to existing RBCC designs over most of the flight trajectory. The ejector-augmented pulse detonation rocket mode is currently being evaluated experimentally in UTA's supersonic wind tunnel.

Atmospheric/Space Flight Mechanics

Research in atmospheric flight mechanics was conducted principally by Dr. Fairchild until his retirement in 1989. This research focused on a variety of both fixed and rotary wing topics. Typical studies included topics such as the response of large transport aircraft to severe turbulence and wind shear, lateral and directional stability of small civil aircraft, stability augmentation for statically unstable fighter aircraft, and open-loop dynamics of helicopters. When Dr. Tuckness joined the faculty, research emphasis shifted to high-speed missile and spacecraft flight mechanics, guidance, navigation and control.

Multidisciplinary teams consisting of students from aerospace, mechanical, electrical and computer science won either first, second or third place four times during the first five years of the Unmanned Vehicle Systems International Arial Robotics Competition. The program was temporarily discontinued following the departure of the team advisor Dr. Cliff Black (ME), who left to take a position in industry. However, Dr. Atilla Dogan (Ph.D., Michigan), who joined the AE faculty in 2002, has revitalized the program together with Prof. Arthur Reyes of Computer Science and Engineering, and they have established a new design lab, are assembling a team of students, and plan to participate in the next national competition. Dr. Dogan is also initiating major research programs in flight dynamics and control of unmanned aerial vehicles, and is establishing a digital flight controls lab. We have also hired Dr. Kamesh Subbarao (Ph.D., Texas A & M) to replace Dr. Tuckness, who left for a faculty position at the University of Oklahoma. Dr. Subbarao is charged with rebuilding a strong graduate program in space flight mechanics.

Distance Education

The graduate program expanded in the early 1970's to serve the local aerospace industry through The Association for Graduated Education and Research (TAGER), a closed-circuit TV network between UTA, other universities in the North Texas area, and local aerospace and electronics companies. Talk-back provisions were included so that interaction between professor and student was possible. On-campus students provided a live audience to provide a more realistic environment for the courses. Examinations were initially given on campus to assure quality control, but later given on-site and monitored by appointed company coordinators from the various human resource offices. The program grew in popularity, and many M.S. Degrees were granted through it. The M.S. program always required a thesis and many thesis topics were generated by industry students on topics of direct application to their work. At the peak of graduate enrollment in 1986, over half of the AE graduate students were full-time working engineers that took their courses over the TAGER system.

Unfortunately, the aerospace and electronics industry downturn in the metroplex starting in the late 80s caused a significant decline in part-time enrollment. Consequently, many of the industries terminated their membership in the TAGER system. The decline in enrollment coupled with increasing obsolescence of the original TV equipment caused Dean Ron Bailey to look at alternative means of providing distance education. As an interim approach, the TAGER studios were used to videotape lectures which were mailed to students on a weekly basis. Once the technology improved to a satisfactory level, the College of Engineering started broadcasting courses on the internet using the video streaming process. Moreover, most instructors now routinely post their course notes on departmental web sites. The MAE Department bought into the program, and now offers masters degrees in both aerospace and mechanical engineering via the internet.

Notable Alumni Accomplishments

Robert L. Stewart

Capt. Robert L. Stewart was sent to UTA by the U.S. Army to earn an M.S. degree in aerospace engineering in 1970. During his stay at UTA, he initiated research in ducted fans that ultimately led to Bell's Ducted Tail Rotor concept. After graduation in 1972, Capt. Stewart served for two years in Korea with the 309th Aviation Battalion (Combat) as a battalion operations officer and battalion executive officer. He then attended the U.S. Naval Test Pilot School at Patuxent River, Maryland, completing the Rotary Wing Test Pilot Course in 1974, and was then assigned as an experimental test pilot to the U.S. Army Aviation Engineering Flight Activity at Edwards Air Force Base, California. He joined the NASA Astronaut Program in August 1979, and flew as a mission specialist on STS-41B in 1984 and STS-51J in 1985, logging a total of 289 hours in space, including approximately 12 hours of EVA operations. During the STS-41B Challenger mission, Stewart and Bruce McCandless participated in two extravehicular activities (EVA's) to conduct first flight evaluations of the Manned Maneuvering Units (MMU's). These EVA's represented man's first untethered operations from a spacecraft in flight.

In 1986, while in training for his scheduled third flight, Col. Stewart was selected by the Army for promotion to Brigadier General. Upon accepting this promotion Gen. Stewart was reassigned from NASA to be the Deputy Commanding General, US Army Strategic Defense Command, in Huntsville, Alabama. In this capacity Gen. Stewart managed research efforts in developing ballistic missile defense technology. In 1989, he was reassigned as the Director of Plans, US Space Command, Colorado Springs, CO. Gen. Stewart retired from the Army in 1992, and became Director of Advanced Programs for the Nichols Research Corporation, Colorado Springs. He is the recipient of numerous service medals, as well as the AHS Feinberg Memorial Award and AIAA Oberth Award.

Kalpana Chawla

Kalpana Chawla, or KC as she preferred to be known as, entered the masters program at UTA in 1982, after graduating with a B.S. degree in aeronautical engineering from Punjab Engineering College, India. While at UTA, she participated in an experimental research investigation of the cross flow fan, a novel propulsive wing concept originated at LTV. Following graduation in 1984, she moved to Colorado where she earned a Ph.D. degree in aerospace engineering from University of Colorado in 1988. She then joined NASA at its Ames Research Center, working in the area of powered-lift computational fluid dynamics. Her research concentrated on simulation of complex air flows encountered around aircraft such as the Harrier in "ground-effect." Following completion of this project she supported research in mapping of flow solvers to parallel computers, and testing of these solvers by carrying out powered lift computations. In 1993 KC joined Overset Methods Inc., Los Altos, California, as Vice President, where she was responsible for development and implementation of efficient techniques to perform aerodynamic optimization.

KC was selected by NASA as an astronaut candidate in December 1994. After completing an intensive year of training, she was assigned as crew representative to work technical issues for the Astronaut Office EVA/Robotics and Computer Branches. Her assignments included work on development of Robotic Situational Awareness Displays and testing space shuttle control software in the Shuttle Avionics Integration Laboratory. In November, 1996, KC was assigned as mission specialist and prime robotic arm operator on STS-87, a microgravity research mission that flew in

November 1987. In January 1998, she was assigned as crew representative for shuttle and station flight crew equipment, and subsequently served as lead for Astronaut Office's Crew Systems and Habitability section. Her second Shuttle mission was on STS-107, another microgravity research mission. KC logged 30 days, 14 hours and 54 minutes in space, but tragically, the STS-107 mission ended abruptly on February 1, 2003 when Space Shuttle Columbia and her crew perished during entry, 16 minutes prior to scheduled landing. KC was inducted into the College of Engineering Hall of Achievement in 1988, and in honor of her accomplishments, UTA established an endowed scholarship in her name, and will name its new residence hall Kalpana Chawla Hall.

James Royce Lummus Jr.

Royce Lummus graduated in 1972 with a M.S. degree in Aerospace Engineering. He returned to earn a Ph.D. in Aerospace Engineering in 1980. His Ph.D. research involved an experimental investigation of the fountain effect on VSTOL aircraft configurations. He also received a master's degree in Biomedical Engineering from Southern Methodist University in 1975 and a master's degree from the University of Dallas in 1986. Dr. Lummus joined Lockheed Martin Tactical Aircraft Systems in 1977 as an aerodynamics specialist and project engineer in Advanced Programs. He helped create the design programs that led to the F-22 and A-12 aircraft configurations. In 1984, Dr. Lummus was appointed deputy program manager and chief engineer for the Indigenous Defensive Fighter Program, a cooperative effort with the industry and government of Taiwan that he managed from initial concept to the successful first flight. In 1992, Dr. Lummus was appointed program director of the KTX-2, a cooperative trainer aircraft development program with the Republic of Korea.

Dr. Lummus is currently Director of Cross-Product Integration Operations and Advanced Development Programs at Lockheed Martin, where he is responsible for operational management of product improvements and derivatives programs for the company's aircraft systems product lines. He supervises a staff of 600-800 at three locations. He is also President and Partner of Consulting Services International, based in Fort Worth, and is an adjunct professor of engineering at both UTA and the University of Dallas. He was inducted into the UTA College of Engineering Hall of Achievement in 1999, and was named outstanding alumnus of the College of Engineering in 2001.

B. Robert Mullins

Dr. Bob Mullins dream of being an aerospace engineer began as a boy watching Convair B-36 bombers from Wright-Patterson Air Force Base rumble over his home in West Virginia. After earning a B.S. degree in Aerospace Engineering from West Virginia University he joined General Dynamics as an Associate Engineer working on flight test instrumentation for the F-111 program. He progressed rapidly up through the ranks to the position of Senior Engineer in charge of various electronic warfare system simulations, earning a M.S. degree in Engineering from Southern Methodist University along the way. He also had experience with Texas Instruments working on the Hellfire Missile Infrared Imaging Seeker project, United Airlines developing flight training simulators, Rediffusion Simulation developing real-time simulation software and Bell Helicopter where he directed systems engineering and modeling efforts on the Advanced Rotorcraft Integration (ARTI) Program. While working with Bell, he started the Ph.D. degree program at UTA, where he conducted research of the flight of large transport aircraft through intense microbursts and wind shear. Following graduation in 1985, he remained at UTA as Visiting Associate Professor until 1987, when he rejoined General Dynamics as a Senior Engineering Specialist in charge of integrated flight controls for the National Aerospace Plane. In 1988, he joined Texas Computing and Simulation as Senior Partner, performing consulting with BEI Defense Industries, Bell Helicopter, Ishida Corporation, Airline Pilots Association and a various legal firms involved in aircraft accident investigations. He continued to teach at UTA on a part time basis, and ran several research projects as an Adjunct Research Professor, including an FAA Regional Vertiport Study. Dr. Mullins rejoined Bell Helicopter in 1996, and has now risen to the position of Chief, Research and Technology Division, where he is responsible for Bell's IRAD and CRAD Programs, and Chair of the RITA Technology Advisory Committee. Bob also serves on the AE Department Industry Advisory Board and chaired the Board during the 99 academic year.

Lawrence (Larry) W. Stephens

Larry Stephens holds B.S. (1972) and M.S. (1979) degrees in Aerospace Engineering from UTA. While working on his undergraduate degree, Larry was Chairman of the AIAA Student Branch and President of the Sigma Gamma Tau Chapter. His M.S. thesis research was a sensitivity study of mass transit bus performance. Following graduation, Larry joined then LTV Missiles and Space Division, beginning his career as an aerodynamics engineer. He quickly advanced in responsibility as a designer and analyst. Later he served as chief engineer or project engineer for advanced technologies, space systems, anti-armor and air defense weapons programs. Larry has accumulated over 25 years of experience in aircraft, space and missile systems development in engineering and project management positions. Before moving into functional engineering management, he held positions of increasing responsibility in product development where he successfully led the technical activities of several aerospace systems programs, from concept studies to flight demonstrations.

Larry is currently Director of Systems Engineering for Lockheed Martin Missiles and Fire Control where he directs a department of 300 technical specialists engaged in flight technologies, system analysis and design, integration and test activities. He is responsible for technical excellence in these disciplines through application of new technologies, tools and methodologies development, technical support to programs, and staff development. Prior to becoming Director, he served as Senior Manager of the Flight Technologies Section including aerodynamics, guidance, navigation and controls, performance and trajectory analysis, and modeling and simulations. In addition to his management responsibilities, he also teaches engineering and management topics in the Lockheed Martin Leadership Development Program and the Professional Development Program.

An Associate Fellow of the American Institute of Aeronautics and Astronautics, Larry received the Distinguished Service Award for his contributions to the Institute. He is a member of several professional societies including the International Council on Systems Engineering and the Project Management Institute, and is a past or present member of many professional and university advisory boards and committees. He chaired the MAE Department Advisory Board for three years, and continues to serve as a member of the Board.

The Next 100 Years

It is extremely difficult to predict the future in such a volatile and changing industry as the aerospace industry, but the UTA AE Program seems to be well-positioned to continue to thrive as a productive and viable academic program. Many of the original faculty have retired or soon will in the near future. They are being replaced by bright new faculty with impressive teaching and research potential. Hopefully, interaction with the local aerospace industry, long a hall-mark of the

UTA program, will continue to grow, and will be supplemented by expansion to include increased interaction with industry on the national scale. The AE Program at UTA has long had a reputation of turning out excellent graduates that were particularly well-suited to join multidisciplinary industry design teams. Our goal is to continue this tradition, while at the same time, growing in reputation as a major graduate research program.

Chapter 54
Aerospace Education at the University of Central Florida

J.D. McBrayer[1]
Department of Mechanical, Materials, and Aerospace Engineering
University of Central Florida, Orlando, Florida 32816-2450

Abstract

A brief account is given of the history of aerospace[2] education and research at the University of Central Florida. The paper covers the time period from the initial establishment of the university in 1963, to the present.

The Founding Years

The need for a university in the central part of the state of Florida was first felt in the late 1950's when forecasts predicted tremendous growth in college enrollment. In 1963 the University of Central Florida became the seventh of the eleven state universities to be established.

"The University of Central Florida" was one of the first names suggested for the new university, but the Orlando community preferred the name "Florida Technological University," which reflected the space coast area's need for programs in engineering and the physical sciences. Florida Technological University (FTU) officially began on June 10, 1963, when the State Legislature passed Bill No. 125 and Governor Farris Bryant signed it into law. An act authorizing the state board of education to establish a state university or a branch of an existing state university in the east central part of Florida took effect July 1, 1963. The University was originally authorized to open with schools of Business Administration, Education, Arts and Sciences, and General Education. In 1966 the Board of Regents approved the addition of a College of Engineering. To more accurately reflect its mission as a university with a wide range of academic programs, on December 6, 1978, the Florida Legislature approved a change of the name of the university. It has been called the "University of Central Florida" (UCF) since that time.

Construction on the university began in January 1967, and the first classes were offered on October 7, 1968, with an enrollment of 1,948 students. Among the first degrees offered at FTU was the Bachelor of Science in Engineering (BSE). Along with Dr. Robert D. Kersten as the founding dean of the College of Engineering, there were eight faculty and two staff for the college when it opened. The original departments within the college were: Civil Engineering and Environmental Sciences, Electrical Engineering and Communication Sciences, Engineering Materials Science, Engineering Mathematics and Computer Systems, Industrial Engineering and Management Systems, and the Mechanical Engineering and Aerospace Sciences (MEAS). Dr. Ronald D. Evans was the initial chair of the Department of Mechanical Engineering and Aerospace Sciences.

[1] Associate Fellow, AIAA

The aerospace oriented courses stated in the university bulletin for 1968-1969 included:

- Aerodynamics
- Stability and Control
- Space Mechanics
- Flight Vehicle Structures
- Propulsion Systems

These courses were, obviously, not all available during the inaugural year. While all engineering majors were eligible to take these classes, they were intended primarily as electives for mechanical engineering majors. These courses continue to be listed in the university bulletin through 1973-1974. In support of the aerospace courses, a supersonic wind tunnel was installed in the early 1970's.

M.S.E. programs began in 1971 with options in each department.

Figure 1 – Supersonic Wind Tunnel

In the MEAS department the only M.S.E. option initially was Mechanical Engineering. In 1973 the department shifted its undergraduate emphasis to other areas, and the Stability and Control course was deleted from the offerings. In 1976 the baccalaureate degree offered by the College was changed to the B.S.E. with majors, instead of the previous concentrations. The majors available were: Environmental Engineering, Electrical Engineering, Engineering Mathematics and Computer Systems, Engineering Mechanics & Materials Sciences, Industrial Engineering, and Mechanical Engineering. The aerospace oriented courses were further reduced in 1976,

Fig. 2. Mechanical/Aerospace Laboratory

with the elimination of Space Mechanics and Flight Vehicle Structures. This minimal availability of aerospace oriented courses continued until 1985.

The doctoral level engineering programs were initiated in 1982, with degrees in Environmental Engineering, Electrical Engineering, Industrial Engineering, and Mechanical Engineering.

The Maturation of the Aerospace Engineering Program

Under the leadership of the department chair, Dr. Stephen L. Rice, the decision to initiate a BSE with a major in Aerospace Engineering was approved in 1985. This program was developed with the intent of seeking professional accreditation. In the fall of 1986 the B.S.E. in Aerospace Engineering was introduced, with three "hard core" aerospace faculty members.

The 1986 degree requirements included:

- Aerodynamics I
- Propulsion Systems
- Engineering Design
- Aerodynamics II
- Heat Transfer
- Intermediate Fluid Mechanics
- Flight Structures
- Vibration Analysis

This Aerospace Engineering program was initially reviewed by ECPD (now ABET) in 1987 and has been continually accredited since that time. In 1989 Gary E. Whitehouse became the

Fig. 3. 1991 SAE Aero Design Team

second dean of the college and Dr. David W. Nicholson was named MEAS department chair in 1990. Also in 1990 the designation for the masters' degree was changed from options within the M.S.E. degree to a major designation, in the case of the MEAS department to the M.S.M.E. degree with a major in Mechanical Engineering

The Aerospace Engineering program success was becoming apparent as illustrated by the enrollment of its students and graduation rates. In 1991 the students from the student branch of AIAA continued their successful quest for excellence in the SAE Aero Design Competition, taking first and second place at Dayton. Also in 1991, the name of the department was changed to the Department of Mechanical and Aerospace Engineering (MAE).

The Aerospace Engineering program experienced typical curricular revisions throughout the next few years in order to maintain professional accreditation standards. By 1991, the degree requirements included:

- Fundamentals of Aerospace Flight
- Mechanics of Materials
- Junior Aero Lab I
- Flight Mechanics
- High Speed Aerodynamics
- Aerospace Design I

- Aerodynamics
- Feedback Control Design
- Junior Aero Lab II
- Flight Structures
- Propulsion Systems
- Aerospace Design II

Dr. Martin W. Wanielista was named Dean of the College in 1993. In 1994 a formal Aerospace Systems option was added to the Master of Science in Mechanical Engineering. Dr. Louis C. Chow was named chair of the department in 1996. That same year the name of the department was changed to Mechanical, Materials, & Aerospace Engineering (MMAE) to illustrate the growing importance of material science education in the department.

The student branch of AIAA has been quite successful in its professional activities during the past decade. Between 1995 and 2003, five UCF Aerospace Engineering students have represented the Southeastern region at the student paper competition held in Reno each year. In 1999 the branch received first prize in the national student membership recruitment drive. It also received the outstanding student branch award for the Southeastern region in 1999, 2002, and 2003. Their faculty advisor, Dr. James D. McBrayer, received the national faculty advisor award at the Reno meeting in 2003.

The Master of Science in Aerospace Engineering (M.S.A.E.) was approved in 1998 and introduced in 1999. Two tracks are available: 1) Space Systems Design and Engineering, which includes the fields of controls and dynamics, space environment, instrumentation and communications, structures and materials, thermal analysis and design; 2) Thermofluid Aerodynamic Systems, which includes the fields of controls and dynamics, aerodynamics, propulsion, thermal analysis and design. Each track contains required courses and several option courses.

In 2000 the Department of Computer Science, which had been in the College of Arts and Sciences, was transferred to the College of Engineering. Consequently the Departments of Electrical and Computer Engineering and Computer Science were merged into the new School of Electrical Engineering and Computer Science, and the college was renamed the College of Engineering and Computer Science. Dr. Ranganathan Kumar was named chair of the MMAE department effective January 2003. A major effort is being placed on turbomachinery research, in cooperation with the Siemens Westinghouse Corporation. The process for the implementation of the Ph.D. program in Aerospace Engineering is in place, with anticipated degree granting status in 2005.

The 2002-2003 Aerospace Engineering degree requirements included:

- Fundamentals of Aerospace Flight
- Structures and Properties of Aerospace Materials
- Solid Mechanics
- Aerospace Engineering Measurements

- Flight Mechanics
- Propulsion Systems

- Aerospace Design I
- Space Elective

- Aerodynamics
- High Speed Aerodynamics
- Feedback Control
- Design of Aerospace Experiments
- Flight Structures
- Discrete Control of Aerospace Vehicles
- Aerospace Design II

Laboratory Facilities – Recent Changes

In 2003 a number of new or improved laboratory facilities have been added or were in the process of implementation. These facilities include:

- Wind Tunnel and Flow Laboratory
  This facility became available for use at the beginning of the Spring 2002 semester. It provides direct support to the course a number of undergraduate and graduate courses as well as improved research capabilities. The primary undergraduate teaching tools in this laboratory include the Engineering Lab Design 1ft.x1ft. test-section, low-speed (100 mph maximum speed) wind tunnel, and the 4"x4" supersonic wind tunnel (nominally 2<M<4). The low-speed tunnel has two Engineering Lab Design force balances, a number of airfoil, cylinder, etc. models, and the appropriate additional measuring equipment.

  The supersonic tunnel has a remotely controlled sting, allowing changes in physical location (vertical) and angle-of-attack during a test run. A Schlieren system is the primary flow visualization tool for undergraduate experimentation. Pressure distributions along the tunnel wall and on the various models can be determined with either Bourdon tube-type gauges or by electronic pressure transducers, which can then be fed to a data acquisition system. Both of these tunnels have been relocated to the new, larger, facility.

  A third, transonic, wind tunnel is under construction in 2003. Initially, as a consequence of the funding mechanisms, this tunnel will only be available for research activities. This tunnel is being built through the joint effort of the Siemens Westinghouse Corporation and the Florida High Tech Corridor. This project is being funded on a 50%-50% cost-sharing basis over a five-year period. This tunnel is a four-passage cascade facility, designed to study the heat transfer in cooled turbine blades and vanes. It is a

695

transonic tunnel (M~0.8) and has a 0.5 m x 0.5 m x1.5m long test section. One exciting aspect of this effort is that, in addition to the funding noted above, Siemens Westinghouse has established three Ph.D. scholarships, each paying $29,000 per year for three years. This should assist in attracting our high quality aerospace engineering graduates to remain at UCF for graduate work.

- Measurements and Fluid Mechanics Laboratory
  An Eidetics Corporation University Desktop Water Tunnel, Model 0710 was put into service in 2002. This included the electronic speed control, both a sting-mounted model support and a tunnel wall-mounted model support, an upstream dye wand, and the complete experiment package including the airfoil experiment, the cylinder experiment, the forebody/projectile experiment and the delta wing experiment.

- Computational Mechanics Laboratory/Satellite Ground Control Station
  This laboratory is primarily a research laboratory used by the graduate and doctoral students in Computational Mechanics and Satellite Ground Control Station (SGCS) group. The SGCS group has two Gateway workstations, two SGI workstations, two SGI OCTANE SE (IRIX workstation), One ULTRASPARC Sun UNIX workstation, and hardware components for the ground control station antenna.

- Energy Systems Laboratory
  This laboratory contains a number of small, low speed wind tunnels, with various different test section configurations. Two of these tunnels were obtained from NASA Langley, where they were used to verify designed modifications to large wind tunnels there. Equipment in this lab includes TSI two channel hot wire anemometer, TSI aerosol generator, TSI laser Doppler anemometer, hot-wire probe traversing mechanism, Photo-kinetics high speed film camera (10,000 pps).

Current and Recent Research Activities

These research projects include:

- Design of an Assured Crew Recovery Vehicle;
- Corrosion inhibition in launch platforms at the Kennedy Space Center.
- Serpentine robot for inspection of the Space Shuttle;
- Robot for tending crops in space vehicles on long flights;
- Experimental modal analysis for damage detection in Orbiter tiles;
- Design of space structures to withstand impact with orbital debris;
- Development of a new thruster for deep space applications;
- Supersonic combustion;
- Supersonic mixing;
- Building a transonic cascade wind tunnel for turbomachinery heat transfer studies;
- Development of fault-tolerant launch monitoring and control system;
- Studies on an rail gun for rocket launching;
- Sounding rocket payload development;
- Photovoltaics for space applications;

Research Affiliations

The Aerospace Engineering program has a long affiliation with a number of the centers and institutes at UCF. These include the: Advanced Materials and Processing Analysis Center, Center for Research and Education in Optics and Lasers, Florida Solar Energy Center, Florida Space Grant Consortium, Florida Space Institute, and the Institute of Simulation and Training. Approximately twelve joint faculty appointments are held with several of these centers and institutes. A description of the activities of several of these centers and institutes is given below, to illustrate the obvious avenues for research collaboration.

Heritage of the Florida Space Institute

In 1987 and 1988 the Governor's Commission on Space studied how the State of Florida should become involved in the space industry. In 1988 the State Legislature passed the Mandate to Astronaut and Space Research. In the fall of 1989, UCF convened a Space Task Force composed of members from both the faculty and administration, to study how UCF could respond to this mandate. The UCF Space Task Force filed its report on November 22, 1989 entitled "Structural Recommendations for the University's Involvement in Education and Research in Areas Related to Space". As a result of this report, in 1990 UCF founded the Space Education and Research Center (SERC) as a Type 2 SUS institute of faculty and academic departments within UCF. Encouraged by the Spaceport Florida Authority, in 1996 UCF changed the name of SERC to the Florida Space Institute (FSI), and expanded the nature of the institute. FSI was reshaped as a consortium of state universities, community colleges, and private schools, and expanded its research and teaching activities to the space center. An agreement was signed by the presidents of UCF, Florida Institute of Technology (FIT), Brevard Community College (BCC) and the director of Spaceport Florida Authority to support Florida's space industry and to expand Florida's space industry beyond the space launch activity. To do this FSI has set its objectives and mission to support the goals of Kennedy Space Center, the Eastern Range, and Florida's space industry. Since then the consortium has been expanded. The current academic members are: FIT, Embry Riddle Aeronautical University (ERAU), UCF, Florida A & M University (FAMU), U of Miami, Florida Atlantic University (FAU), University of South Florida (USF), BCC and Broward Community College, and through NASA's Florida Space Grant Consortium some 16 additional universities and colleges throughout Florida. In November 2000 the Florida Board of Regents voted to redesignate FSI as a Type-1 institute, with a concomitant statewide charter as a Center-of-Excellence for space research and education.

The objectives of the Florida Space Institute were set to meet the goals, to respond to the external forces, and to fulfill the recommendations of the Strategic Plan of the University of Central Florida. The objectives of the Florida Space Institute are to:

- Support the space industry in Florida;
- Enhance and expand the diversity of space-related industries within Florida;
- Consolidate and focus Florida's space research and educational efforts;
- Provide academic support to government and commercial space programs;
- Enable Florida's universities and community colleges to establish an internationally recognized research and education institute;
- Foster cooperation among government, industry, and education organizations.

Heritage of the Institute of Simulation and Training

The Institute of Simulation and Training (IST) is an internationally recognized research institute that focuses on advancing modeling and simulation technology and increasing the understanding of simulation's role in training and education. Founded in 1982 as a research unit of the University of Central Florida and reporting directly to the Vice President for Research, the institute provides a wide range of research and information services for the modeling, simulation and training community. The wide range of expertise provides capabilities in virtually all key areas of simulation research and development.

The Goals of IST are to get the job done, research teams include personnel from IST and UCF and, when required, from other universities and schools in Florida and nationwide.

- With University of Central Florida academic colleges, serve as a research and development focal point for development and expansion of the simulation and training community;
- Develop and conduct basic and applied sponsored research programs and academic services;
- Identify future directions for simulation and training technology;
- Review the vital aspects of challenges to the simulation community;
- Generate ideas and demonstrate opportunities for extending simulation and training technologies into new capabilities and application areas;
- With the University of Central Florida be a link between industry and others and create networks for transferring and diffusing technology;
- Partner with industry and others to develop useful services and products based on simulation and training technologies;
- Provide an environment for students and faculty in the simulation and modeling discipline;
- Provide continuing education services for the simulation community.

Heritage of the Center for Research and Education in Optics and Lasers

The School of Optics/CREOL is a graduate school for optical science and engineering education and research. Its mission is to:

- Provide the highest quality education in optical science and engineering
- Enhance optics education at all levels
- Conduct scholarly, fundamental and applied research
- Aid in the development of Florida's and the nation's technology based industries.

The School of Optics is recognized as one of the top three independent optics academic departments in the nation. The School offers interdisciplinary graduate programs leading to MS and Ph.D. degrees in Optics. CREOL--The Center for Research and Education in Optics and Lasers--is integrated in the school as its research arm.

The School of Optics/CREOL has grown to an internationally recognized institute with 37 faculty members and associated faculty, 39 research scientists and 139 graduate students with research activities covering all aspects of optics, photonics, and lasers. It is housed in a state-of-the-art 82,000 sq. ft. building dedicated to optics research and education.

The School of Optics is one of the world's leading graduate institutions in optics and photonics education and research. The School offers a comprehensive interdisciplinary graduate program covering all aspects of optics, photonics, and lasers leading to Master's and Doctoral degrees in Optics. The School has twenty-four full time faculty members and more than one hundred graduate students.

Heritage of the Advanced Materials and Processing Analysis Center

The Advanced Materials and Processing Analysis Center (AMPAC) strives to excel in the development, processing and characterization of advanced materials and achieve national prominence in targeted research areas that include energy, microelectronics and laser materials technology by providing facility, faculty and student support to College of Engineering and Computer Science and College of Arts and Sciences and related disciplines. AMPAC has six full time faculty members.

The educational goals of AMPAC are to:

- Prepare students through education, training, and mentoring for challenging careers in vital technology areas;
- Provide industrial internships, government fellowships and co-op opportunities to students;
- Enhance and expand the graduate and undergraduate program offerings in accordance with the current technological needs of industries.

Chapter 55
The Unusual Evolution of Aerospace Engineering
At The George Washington University

John L. Whitesides and Michael K. Myers
The George Washington University
NASA Langley Research Center

Background

Since its charter by the Congress of the United States in 1821, long after the company whose shares were bequeathed by George Washington to endow a national university in the District of Columbia had gone bankrupt, The George Washington University (GW) has had many and varied interactions with the Federal Government[1]. This non-sectarian, private University has benefited in many ways from its location in the heart of the District and from its programs with Federal Government Agencies.

The Corcoran Scientific School of Columbian University was organized in 1884 to offer evening courses in science and technology leading to the degrees of Bachelor of Science, Civil Engineer, Mechanical Engineer and Mining Engineer. The curriculum emphasized both theoretical and practical knowledge stressing creativity rather than conformity. In 1962, the school became the School of Engineering and Applied Science and continues the practice of emphasizing not only current technology, but also the underlying theory supporting the technology. The engineering school has been noted for its service to the nation such as developing seven special courses in 1940 to prepare the country for the stresses that war would put on U.S. productive capacity and inaugurating a program for the degree of Master of Engineering Administration in 1954 to prepare industry and government executives to manage complex, technologically demanding programs. Aerospace engineering itself had not yet developed at GW although many government, particularly defense, projects were undertaken at the school. Until 1968, the School of Engineering and Applied Science was still essentially a night school for part-time students in engineering[2].

Beginning of Aerospace Engineering at GW

All that changed with the appointment of Harold Liebowitz, formerly of ONR, as Dean of the School of Engineering and Applied Science in January of 1968. Shortly after his appointment that month, at an AIAA meeting, Liebowitz, by chance, met with John Duberg, Associate Director, NASA Langley Research Center (LaRC), whom he knew from the structures and materials research community. Their conversation centered around LaRC's need for a graduate education program at LaRC in Hampton, Virginia some 170 miles south of D.C. Duberg indicated that LaRC had maintained some sort of graduate program for their employees on LaRC for many years, but that the participating universities always required some type of residency on campus prior to completion of degree programs. LaRC would like to see that residency requirement waived for M.S. and D.Sc. degrees. Liebowitz was, of course, interested in the possibility of GW being involved with one of the foremost research labs in the country which had outstanding researchers and facilities.

Waiving the residency requirement would be a bold academic departure and required serious consideration by the faculty at GW. Many meetings were held during the spring semester of 1968 between senior members of GW and LaRC. The University found that many LaRC researchers were eminently qualified to teach graduate courses; indeed, many had already taught for other universities. In addition, appropriate facilities were available for classrooms on-site. Liebowitz knew that SEAS had a long history of working with government employees as part-time students, but did not offer aerospace engineering. During this period of discussions between GW and LaRC, Liebowitz persuaded S. W. Yuan, who had been one of Liebowitz's professors at Brooklyn Polytechnic, to join the GW faculty to help develop the GW-NASA LaRC interaction, particularly in the aerospace area. Following many meetings and much discussion, the GW faculty approved offering a resident-credit MS program at NASA LaRC – by one vote – Liebowitz! The program at LaRC would have to be an integral part of SEAS, not the standard off-campus program.

Meetings between senior GW faculty and LaRC researchers continued through the spring and into the summer of 1968. Electrical Engineering and Engineering Mechanics Departments were the principals at GW while senior members of the entire LaRC research and engineering staff contributed input to the meetings. Finally it was agreed by all parties that a full resident credit Master of Science program would be offered at LaRC by GW with majors in aeronautics, astronautics, applied mechanics, mechanical engineering, electrical engineering and computer science beginning in the Fall 1968. Many LaRC senior researchers were approved as part-time faculty by GW, and an entire graduate degree area, aerospace engineering, was approved for the new GWU-NASA LaRC Graduate Program even though it was not offered on the Washington, D.C. campus. The University appointed J. L. Whitesides to coordinate and develop this off-campus program at the Langley Research Center.

Thus aerospace engineering at GW began, not in downtown Washington, but in Hampton, Virginia at the NASA LaRC in the Fall semester of 1968. Twelve graduate courses were offered by GW and 114 part-time students registered for degree or non-degree resident credit courses. The GWU-NASA LaRC Graduate Program had begun.

Research and Education Program

During subsequent years this graduate study program expanded substantially in both participants and scope. The curriculum was constantly improved as the needs and interests of Langley Research Center and The George Washington University changed. After gaining confidence with this new off-campus graduate program, the two institutions felt that even greater benefits could be derived from increased interaction in both research and education. The needs and capabilities of both GW and LaRC were then considered, and a plan was devised to satisfy the needs of each institution by using the capabilities of both to their maximum advantage.

The LaRC had extensive facilities and a number of outstanding researchers who enjoyed teaching part-time. Much of the research at LaRC, of necessity, had become project oriented and, in general, less basic. On the other hand, University faculty and graduate students usually conducted very basic research but lacked the extensive facilities such as those at LaRC[3].

Consideration of the factors cited above led to the establishment in the fall of 1970 of a pilot program, supported by a LaRC cooperative agreement with GW, of research and education in aeroacoustics located at LaRC. Since neither GW nor LaRC had a wealth of aeroacousticians, they enlisted the help of the premier acoustics research organization, The Institute of Sound and Vibration Research (ISVR), Southampton, England, to establish a research team consisting of one full-time George Washington University research faculty member and four Graduate Research Scholar Assistants (GRSA) to join the GW faculty already at the Center to form the nucleus for an increased and more responsive interaction with Langley Research Center. The research team worked directly with the NASA aeroacoustics researchers and participated in the graduate study program already at Langley. To initiate this program, ISVR sent a Visiting Professor to GW to assist in supervising the GRSA and teach in the graduate program at LaRC. The Visiting Professors from ISVR generally spent one year at LaRC and then were replaced by someone with a different speciality. This rotation brought fresh ideas to the program in both academics and research. The ISVR Visiting Professor program continued until 1975 during which time GW was able to appoint two outstanding faculty, J. C. Yu and M. K. Myers, full-time to the aeroacoustics program. The aeroacoustics research mirrored that of LaRC except that, in general, the University faculty and graduate students studied more fundamental aspects of each problem. Generation, propagation and alleviation of noise from ducts, jets, rotors, and other sources were all studied by the faculty and GRSA.

Each year the Graduate Research Scholar Assistants were recruited nationally and were given support for up to two years to complete their M.S. degrees. Each of the GRSA was given a NASA mentor and a GW advisor to insure that the research conducted satisfied both NASA's needs and the University's academic requirements. At least once a year, the full-time GW faculty and the part-time NASA faculty met to determine course needs for the next academic year. In fact, this pilot program was organized such that both NASA Langley Research Center and The George Washington University had inputs to all phases of its operation.

Joint Institute for Advancement of Flight Sciences

A similar program of research and education in Flight Sciences was established in 1971 by a separate cooperative agreement between LaRC and GW. This program which emphasized aerodynamics, performance, stability and control required additional research faculty and GRSA to be brought to LaRC by GW. Many full-time faculty have been associated with this research and education activity particularly J. P. Campbell and J. L. Whitesides. Research in the Flight Sciences program focused on fluid mechanics and dynamics. Projects studying the aerodynamics of portions of aircraft, conventional and V/STOL aircraft, and helicopters were undertaken for all speed ranges. These were conducted in LaRC's wind tunnels, in flight, both full scale and models, and with computational fluid dynamics.

There were a sufficient number of full-time GW faculty at LaRC for the School to approve offering the D.Sc., with no residency requirement, at LaRC by late 1971. GW activity at LaRC was an integral part of the School of Engineering and Applied Science even though separated by 170 miles.

It became clear that many potential cooperative programs between LaRC and GW were possible, and that a formalization between the organizations would be useful. A Memorandum of Understanding was developed and signed by the Center Director and The University President establishing the Joint Institute for Acoustics and Flight Sciences (JIAFS) in late 1971. JIAFS would then encompass the GW-NASA LaRC Graduate Program, the Aeroacoustics Program, the Flight Sciences Program and all future interactions between the two organizations. The Institute was established as a cooperative effort between NASA LaRC and GW to serve the national interests in scientific research, engineering technology and engineering education. Its purpose was to increase the nation's research and engineering capabilities by bringing together researchers and scholars from government, industries and universities for the exchange of technical information, by enabling researchers to use the extensive equipment and facilities, and by educating graduate engineering students and professional engineers. J. Duberg was named Director and H. Liebowitz the Co-Director of the new Institute with faculty at LaRC (Whitesides) and in Washington, D.C. (Yuan) delegated the operation of JIAFS.

Early in 1972, the University proposed another research and education program in Computer Aided Structural Dynamics (later changed to Structures and Dynamics) to LaRC. The proposal was funded by the Center and an outstanding researcher, A. K. Noor, was appointed by the University to lead the new research group. This program concentrated on developing new finite elements and techniques for solving structures and dynamics problems using the most advanced computational facilities available.

During 1973 a different type of joint program was established in JIAFS. The Visiting Members program was established to allow outstanding researchers to work with GW faculty and LaRC scientists and engineers on current projects. Over 150 individuals conducted research in aerospace engineering in JIAFS through this program as Visiting Scientists or Engineers, Staff Scientists or Engineers, and as Postdoctoral Research Associates. Visiting Members in JIAFS were associated with all of the research divisions at LaRC, and thus, their research was in a multitude of areas.

Since LaRC was becoming more heavily involved in atmospheric sciences in 1974, a joint GW-NASA LaRC proposal was approved by NASA Headquarters to establish a research and education team in Environmental Modeling at LaRC similar to the current programs in JIAFS. B. R. Barkstrom was appointed by the University to lead the program and additional GRSA were recruited. Research was conducted in a number of areas in atmospheric science including modeling of circulation and radiation budgets particularly in the presence of clouds.

The following year, in an effort to understand better the macroscopic properties of composites and other aerospace materials starting from the microscopic fundamentals, another research and education program in Materials Science was initiated in JIAFS. The University recruited D. M. Esterling to lead this program.

It became clear that the name originally given to the Institute was not reflecting the broad interactions between GW and LaRC. Thus, when the Memorandum of Understanding was renewed in 1976, the name was changed to Joint Institute for Advancement of Flight Sciences to more accurately reflect the extensive activities of the Institute.

A program in Flight Dynamics was established in JIAFS in 1978 similar to the previous research and education activities. This research was primarily in the area of parameter identification methods applied to aircraft flight dynamics and has been led by V. Klein since its inception.

In 1987, A. D. Cutler was appointed to the GW faculty in JIAFS to lead a new research and education program in Experimental Fluid Mechanics. As with the other research and education activities in JIAFS, new GRSA were recruited for this program. This research activity was primarily focused on solving the fundamental physics of reacting supersonic jets for application to hypersonic vehicles.

An Astronautics research and education program was initiated in 1987. G. D. Walberg was chosen to lead the program, but left for a position in North Carolina State University after two years. R.

H. Tolson was then appointed to the faculty and has built this activity substantially. Research in this program has been in many areas such as aerobraking for Mars missions, including award-winning real time participation in the Mars Global Surveyor Mission, many projects associated with the International Space Station, design of spacecraft for various missions and mission planning.

Since many of the premier aircraft designers had retired from LaRC, a research and education program in Aircraft Design was begun in 1995. An outstanding aircraft designer from industry, R. R. Sandusky, was appointed as the faculty lead of this program. Faculty and graduate students in this program have been associated with almost all of the systems studies and proposed aircraft designs at LaRC since its inception.

All of the activities discussed above were either in aeronautics, astronautics or both. Although each of the programs possessed many similarities such as: faculty and GRSA, responsiveness to LaRC needs, separate funding for each; each program operated somewhat differently due to the research operations and personalities of the different LaRC researchers and GW faculty involved. The freedom to operate independently and the separate funding mechanisms insured that each program satisfied the needs of all parties, i.e., LaRC researchers, GW faculty and GRSA. The size of each of the research and education programs varied each year due to changes in NASA's budget or priorities and availability of GRSA. The Environmental Modeling Program closed in 1979 and the Materials Science activity ended in 1983. The other programs are currently completing their last GRSA.

Faculty

In any graduate educational endeavor, the faculty are the key. The interactions between GW and LaRC blossomed due to the progressive-thinking faculty at GW and the outstanding researchers at LaRC who enjoyed teaching and were talented instructors.

The University initially appointed full-time faculty at LaRC as Research Professors. In 1976, following a number of successful years in JIAFS, most of the faculty were given regular tenure track positions. Each of the full-time faculty was selected specifically for the research and education programs in JIAFS after a nationwide search by the University. Since it was a Joint Institute, many of the University's search committees included a senior LaRC researcher. There have been 30 different full-time GW faculty in JIAFS, see Figure 1. Each contributed to the success of the Institute. Several faculty left GW to become full-time at NASA LaRC, but, in general, continued interacting with the graduate program. In addition several LaRC employees who had been part-time faculty joined the University as full-time faculty following retirement from LaRC. All but two of the full-time faculty in JIAFS were associated with aerospace engineering. In addition to the full-time GW faculty in JIAFS, over one hundred and fifty different researchers, primarily from LaRC, taught in the Institute, see Figure 2. Each part-time instructor went through the departmental approval process at GW just as other part-time faculty in Washington. Some of the part-time faculty taught only once or twice, but a number of them taught each academic year. Generally, the part-time faculty taught their research specialty while the full-time faculty taught the required courses which benefited the graduate students immensely.

Graduate Students

The GW-NASA LaRC Graduate Program initially offered majors in aeronautics, astronautics, applied mechanics, mechanical engineering, electrical engineering and computer science to part-time graduate students employed by NASA LaRC. One of the primary reasons for the success of GW LaRC interactions in graduate engineering education is the flexibility of a private university in modifying, adding, or deleting courses and majors. Since there are no state councils from which to seek approval or other schools to battle over turf, it becomes relatively easy to modify curricula. This capability was used by GW at the beginning (aeronautics, astronautics) and during the establishment of the research and education programs. New curricula were developed for aeroacoustics and environmental modeling (atmospheric science), and extensive modification to reflect aerospace engineering was accomplished in all major areas. Some of these graduate programs had relatively short lives, e.g., Environmental Modeling 1974-84, Materials Science 1975-84, while others such as aerospace engineering were vital during the entire period of the GW-NASA LaRC Graduate Program. In addition, many Special Topics courses were approved by the School of Engineering and Applied Science and offered at LaRC.

The number of graduate students in JIAFS is shown in Figure 3 which also shows aerospace engineering students by year. In the early years of the graduate program, there were many electrical engineering and computer science students. Due to lack of LaRC hiring, the majors in electrical engineering and computer science were eliminated in 1984. A graduate program in engineering management was initiated in 1976 and continued in JIAFS until 1997 when it was moved to GW's Hampton Roads Center in Newport News, Virginia. There was a great deal of interest in the engineering management degree program particularly from non-NASA local residents. After 1997 the remaining students were all in aerospace engineering either aeronautics or astronautics.

The composition of the student body was changing due to the research and education programs. As shown in Figure 4, the mix of aerospace engineering students began to change when the Graduate Research Scholar Assistants were appointed in JIAFS. From an entirely part-time student body in 1968 until the present, the number of full-time students has been increasing. Most of these full-time students are GRSA appointed to the research and education programs, but some are also self paying students who relocated to the area to go to graduate school with GW at the NASA Langley Research Center. The GRSA were very competitive and the selected students had an average gpa of 3.7. From just over 40, the number of aerospace engineering graduate students grew to over 110 and is still 77. These numbers are fairly substantial, but become rather large when it is noted that, due to LaRC restrictions, they were all U.S. citizens!

With a growing number of full-time students, GRSA, the University established an AIAA Student Branch at LaRC in 1974. The GW-Langley Student Branch has been quite active. It has been the local organization for students without a classical campus. In addition the members of the Student Branch have been quite successful in AIAA's graduate student paper competitions having been awarded first place 8 times in the last 16 regional paper competitions. Since 1998, students have also received 8 of 15 AIAA Graduate Student Aircraft Design awards in national competition.

Institute Funding

Obviously with many different types of programs between LaRC and GW there were also many varied types of funding arrangements in JIAFS. Initially LaRC contracted with GW for each graduate course at a much lower tuition rate than the campus rate since LaRC provided offices and computing facilities at no charge to the University. Within a few years there were enough non-NASA students in the graduate program to charge tuition directly to each student. LaRC received a tuition rate of less than half the campus rate and non-NASA students paid the standard off-campus tuition. This formed the basis for JIAFS' academic budget.

Each of the research and education programs were supported by one year renewable cooperative agreements from LaRC to GW. The amount of these cooperative agreements varied dependent on the number of faculty and GRSA involved in the program.A task contract was negotiated between LaRC and GW to allow specific, generally short-term visits by outside researchers to undertake projects with NASA scientists and engineers, and GW faculty and students. In 1980 a small core fund was established to allow GW to bring new activities to LaRC and to provide flexibility in appointing GRSA. In addition many of the faculty obtained research grants from other agencies such as NASA Headquarters, NSF, ARO, ONR, AFOSR and JPL to support on-going projects in JIAFS.

During the 35-year history of JIAFS over 70 different grants and contracts were obtained. A number of these, particularly the research and education programs, lasted for many years.

Results of JIAFS Activities

Since 1968, some 2,831 different individuals have taken classes with GW at the NASA Langley Research Center. Many of these were individuals taking graduate courses as non-degree students to better their research capabilities. Others took courses when their career changed directions. Of course, many were seeking advanced degrees to move into better positions. Still others merely wanted to continue the process of lifelong learning.

The primary output of graduate education and research activities are the degrees awarded and the publications and presentations of the faculty, visiting members, and graduate students. These numbers are shown in Figure 5 for JIAFS. Over 600 of the degrees are in aerospace engineering and the number will increase as the final group of GRSA graduate at which time there will have been over 1,000 degrees awarded by GW through JIAFS.

Mechanical and Aerospace Engineering Department

From 1968 until about 1985, while the Aerospace Engineering curriculum was developed and refined at JIAFS, the courses comprising the curriculum were listed in the GW Bulletin with an asterisk denoting the fact that they were given only at NASA Langley Research Center. The full-time faculty at JIAFS became regular members of the Department of Civil, Mechanical, and Environmental Engineering but, in fact, there was virtually no aerospace instruction on the DC campus. Aerospace research there was confined to work by S. W. Yuan on vortex alleviation by tangential blowing that he continued after moving to GW from the University of Texas. However, opportunities that arose in the late 1980's for new activities in research and education with NASA Goddard Research Center and an increasing demand for aerospace curricula from part-time students in the DC area led the GW Mechanical Engineering faculty to begin offering a number of the JIAFS-designed courses on the DC campus. From that time until the present, the GW Mechanical Engineering graduate program in DC has included a strong complement of aerospace offerings taught by qualified part-time faculty in DC and also given in the past few years in person and via television by JIAFS faculty themselves. At the time of the reorganization of the departmental structure of the GW School of Engineering and Applied Science in 1999, the University formally recognized the vital role played by the aerospace component of the mechanical engineering graduate program by the creation of its new Department of Mechanical and Aerospace Engineering. In addition to the fact that the first chairman of this department, M. K. Myers, is a long-time JIAFS faculty member, the strong influence of the Institute on the programs of the home department is reflected in the introduction, in 2000, of an undergraduate Aerospace

Option in Mechanical Engineering designed primarily by the JIAFS faculty. Currently, this program is a rapidly growing component of the on-campus educational activity.

Since the GW-NASA LaRC Joint Institute for Advancement of Flight Sciences officially closed January 1, 2003 and is currently completing the research and graduate programs of the final Graduate Research Scholar Assistants, it is only appropriate that the legacy of this dynamic educational experiment will be, not only its graduates and research output, but also the establishment of a vibrant aerospace engineering program at The George Washington University in the nation's capital.

Faculty Member	Period	Notes
John L. Whitesides	1968-present	Professor and Associate Director, JIAFS
Selwyn Wright	1970-1971	Visiting Professor, ISVR
Gary Koopmann	1971	Visiting Professor, ISVR
Ahmed K. Noor	1972-1990	Structures and Dynamics, joined UVA
James C. Yu	1972-1977	Aeroacoustics, joined LaRC
Alvin M. Bloom	1972-1975	Aeronautics, went to GW in DC
Christopher Rice	1972	Visiting Professor, ISVR
Maurice Petyt	1972	Visiting Professor, ISVR
George Kuhn	1972	Visiting Professor, ISVR
Michael K. Myers	1973-present	Professor and Chair, MAE
Parmanand Mungur	1973-1975	Visiting Professor, ISVR
Paul L. Coe	1973-1974	Aeronautics, joined LaRC
Michael J. Griffin	1973	Visiting Professor, ISVR
John P. Campbell	1974-1988	Aeronautics, retired LaRC, joined GW, retired
Bruce R. Barkstrom	1974-1979	Environmental Modeling, joined LaRC
Richard S. Brice	1974-1978	Structures and Dynamics, joined another University
Donald M. Esterling	1975-1983	Materials Science, went to GW in DC
Harry L. Runyan	1975-1977	Aeronautics, retired LaRC, joined GW, retired
Vladislav Klein	1978-present	Professor Emeritus, Flight Dynamics
Fereidoun Farassat	1978-1979	Aeroacoustics, joined LaRC
Ivatury S. Raju	1979-1983	Structures and Dynamics, joined LaRC
Mohamed M. Hafez	1980-1982	Aeronautics, joined UC/Davis
Robert J. Kurzeja	1980-1984	Environmental Modeling, went to industry
John E. Duberg	1981-1985	Engr Management, retired LaRC, joined GW, retired
Andrew D. Cutler	1989-present	Associate Professor, Experimental Fluid Mechanics
George R. Brier	1989-1997	Engr Management, joined GW's HRC
Gerald D. Walberg	1989-1991	Astronautics, retired LaRC, joined GW, joined NCSU
Robert H. Tolson	1991-present	Professor, Astronautics, retired LaRC, joined GW
Robert R. Sandusky	1996-present	Professor, Aircraft Design
Paul A. Cooper	2001-present	Research Professor, Structures & Dynamics, retired LaRC, joined GW

Figure 1. Full-Time Faculty in JIAFS at LaRC

Flight Sciences-Aerospace Engineering Faculty

M. S. Anderson
R. Barnwell
H. L. Beach
I. E. Beckwith
R. C. Blanchard
R.L. Bowles
D. A. Brewer
J. D. Buckley
R. D. Buehrle
J. J. Buglia
A. Caglayan
G. Canavos
M. F. Card
J. Carter
R. Chambers
J. B. Davidson
P. Dechaumphai
S. S. Dodbele
P. Donely
K. E. Dutton
D.L. Dwoyer
J. W. Edwards
W. D. Erickson
M. J. Ferebee
W. B. Fichter
R. Foye
R. W. Fralich
R. E. Fulton
L. B. Garrett
I. E. Garrick
C. H. Gerhold
P.A. Gnoffo
W. L. Grose

W. Gunn
M. Gunzburger
A. A. M. Halim
J.C. Hardin
L.C. Hartung
L. G. Horta
M. Y. Hussaini
C. J. Jachimowski
A. R. Johnson
G. S. Jones
J. Juang
C. E. Kirby
E. Kruszewski
J. E. Lamar
D. Lansing
A. L. Leybold
K. B. Lim
C-H. Liu
L. K. Loftin
E. R. Long
J. M. Luckring
R. J. Margason
G.L. McAninch
M. M. Mikulas
R. C. Montgomery
E.A. Morelli
M. K. Morin
J. N. Moss
V Mukhopadhyay
T.K. O`Brien
W. Olstad
J. H. Park
S. K. Park

L. D. Pinson
E. B. Pritchard
M. J. Queijo
H. G., Reichle, Jr.
T. W. Roberts
J. B. Robertson
A. Saunders
J. Scheiman
J. M. Seiner
J. D. Shaughnessy
J. N. Shoosmith
J. Shuart
W. A. Silva
G. L. Smith
J. Sobieski
B. Stein
M. Stein
T. Straeter
K. Sutton
R. C. Swanson
T. Talay
D. R. Tenney
A. Tessler
J. L. Thomas
L. W. Townsend
J. Tulinius
J. Unnam
R. P. Weston
A. W. Wilhite
J. Williams
E. C. Yates
T. A. Zang
W. E. Zorumski

Figure 2.a. Part-Time Faculty in JIAFS at LaRC

**Engineering
Administration Faculty**

F. Alario
M.P. Clark
W. R. DeLoach
D. E. Eckhardt
T. L. Griswold
J. E. Harris
D. P. Hearth
D. E. Henderson
W.P. Henderson
R.D. Hofler
G. T. Malley
J. Mathews
S. Mauldin
J. F. McNulty
S.F. Pauls
L. E. Putnam
R. B. Reynolds
R. P. Rhode
R.E. Snyder
D. Tabak
W.L. Williams

**Electrical Engineering and
Computer Science Faculty**

E.A. Armstrong
C. A. Balanis
W. Batte
M. Burce
W. Chang
J. I. Cleveland
W. Croswell
E. C. Foudriat
C. Fricke
G. W. Haigler
J. D. Harris
J. E. Hogge
F. Hohl
L. Jones
S.M. Joshi
D. Loendorf
D. D. Moerder

P. Mumola
R. Noonan
J. Painter
D. Reaugh
J. L. Rogers
J. M. Russell, III
J. R. Scheiss
L. D. Staton
R. L. Stermer
O. O. Storaasli
C. Swift
A. Trehan
J. S. Tripp
J. H. Tucker
T. Walsh
C. Wang

Figure 2b. Part-Time Faculty in JIAFS at LaRC

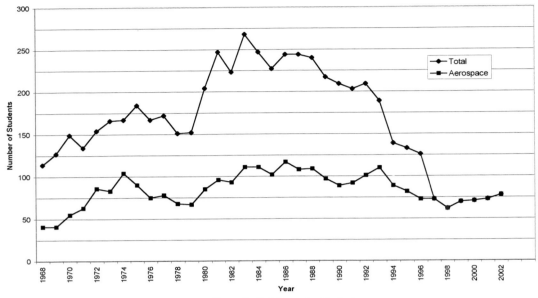

Figure 3. Fall Semester Students

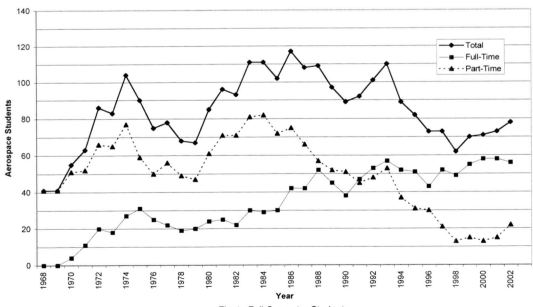

Fig 4. Fall Semester Students

710

Degrees Awarded Since 1968

	M.E.M.*	M.S.	Eng.	D.Sc.	Total
NASA Employees	27	181	11	15	234
GRSA -		361	9	18	388
Local Residents	270	48	2	2	322
Total	297	590	22	35	944

Research Output Since 1968

Publications and Presentations	-	1392
Major Conferences Sponsored	-	17
Seminars	-	421

*Masters in Engineering Management

Figure 5. Results of JIAFS Activities

References

1. The George Washington University Bulletin, 2002-2003, Washington, D.C.

2. From Strength to Strength, "A Pictorial History of The George Washington University 1821-1996," George Washington University, 1996.

3. Whitesides, J. L. and Yuan, S. W., "JIAFS – A Pattern for Graduate Engineering Education of the Future?," #2560, American Society for Engineering Education Annual Conference, June 16-19, 1975, Fort Collins, Colorado.

Chapter 56

AEROSPACE ENGINEERING
at
Old Dominion University

E. James Cross, Professor of Aerospace Engineering

INTRODUCTION[*]

In 1930 Old Dominion University (ODU) opened as a one-building branch of the College of William and Mary, the nation's second oldest institution of higher education, at Norfolk in the Tidewater region of southeast Virginia. Early classes at the College's Norfolk Division included a

two-year program for teachers, and freshman and sophomore engineering classes that prepared students for Virginia Polytechnic Institute. The two-year school evolved into a four-year branch, then gained full independence as a state-supported college in 1962, taking on the name Old Dominion College. Soon the college was greatly expanding its research facilities and preparing to offer doctoral degrees, and in 1969 the Board of Visitors authorized that the name of the institution be changed to Old Dominion University. Old Dominion has been offering master's degree programs since 1964 and doctor of philosophy degrees since 1971. Students at Old Dominion currently choose from 66 baccalaureate programs, 67 master's programs, two education specialist programs, and 23 doctoral programs

Engineering and Technology

[*]The Norfolk College administration was well aware of a significant need in the Tidewater region for a four-year engineering degree program. And, in 1960 engineering societies in the Hampton Roads area appealed to President Lewis Webb to seek state approval for such a program and he subsequently petitioned the State Council of Higher Education (SCHEV) for approval. The

[*] Excerpts from the Old Dominion University Catalog 2002-2004
- Sweeney, J.R. "Old Dominion University: A Half Century of Service" ODU Office of Printing and Publications, 1980

University of Virginia and the Virginia Polytechnic Institute had engineering programs of their own and vigorously opposed the Hampton Roads proposal in order to avoid sharing limited state funding with a third competing program. The SCHEV appointed a committee to determine whether the Norfolk area needed a four–year engineering degree program and in 1961 the Council voted in favor of recommending the establishment of an engineering program at the Norfolk College of William and Mary. In 1962 SCHEV authorized full independence as a state-supported college and a name change to Old Dominion College. That same year the Virginia General Assembly authorized the college to offer the Bachelor of Science degree in engineering, and Dr. J. Harold Lampe was appointed as Dean of the new School of Engineering. Lampe had previously served as Dean of the College of Engineering at the North Carolina State University for seventeen years, and prior to that he was the founding Dean of engineering at the University of Connecticut. That experience was especially valuable during the formative years of the emerging engineering program. A faculty was formed of a mixture of good people from a variety of backgrounds, and, programs of instruction and an engineering matriculation were established by 1964. Four interdisciplinary groups were formed in addition to engineering technology. These groups were Thermal Engineering Group, Mechanics and Materials Group, Electrical and Electronics Group, and the Civil Engineering Design Group. The first engineering baccalaureate degrees were granted by 1966 and in 1967 the engineering degree was fully accredited. During this same period SCHEV granted authority to establish a Master of Science degree in Engineering. Lampe retired in 1966 and was followed by Dr. Ralph Rotty as Dean for the period 1966-1972. Subsequently Dr. Donald Ousterhout (1972-1974), and Dr. John Weese (1974-1980) served in the critical Dean position. Dean Weese established the College of Engineering in1975 and formed four new departments – Civil Engineering, Electrical Engineering, Mechanical Engineering and Mechanics, and Engineering Technology. In 1982 Dr. Weese accepted an appointment to the National Science Foundation and Dr. Bob Ash was designated Interim Dean. A national search concluded with the appointment of Dr. Jim Cross as permanent Dean in 1984. He served in that position for 13 years and generated substantial change. The College name was changed to the College of Engineering and Technology, a new department (of Engineering Management) was formed, and several research enterprise centers were established, including an engineering clinic, a center for modeling and simulation, the Virginia Spaceport Authority, the Applied Research Center, the Langley Full Scale (Wind) Tunnel, and a center for advanced ship repair and maintenance. And, in 1991 the Aerospace Engineering Department was formally established. Dr. Bill Swart followed as the Dean in 1998 and left in2002. Dr Oktay Baysal, Associate Dean, has served as Interim Dean since 2002.

Currently, the College of Engineering and Technology at ODU offers degrees in engineering and in engineering technology. The course of study that leads to engineering degrees is characterized by a solid foundation in the theoretical basis of engineering found in mathematics and physics. Graduates are well prepared to pursue graduate education, or professional registration, and directly enter the engineering professions. The curriculum that leads to engineering technology degrees is characterized by strong laboratory experiences that will prepare the graduate to hit the ground running as a technical partner of the engineer who can implement advanced design and development concepts. The engineering technology degree is considered to be a terminal degree and graduates are not expected to pursue graduate degrees or professional registration, although they are not excluded from doing so.

Aerospace Engineering

Historically, the Mechanical Engineering and Mechanics (MEM) Department, and the antecedent Thermal Sciences Group, sought opportunities to partner with engineers and scientists at the nearby NASA Langley Research Center in aeronautical research projects that were of mutual interest. These partnerships produced research revenue for support of the studies of many student-engineers that led to engineers trained in the aeronautical sciences with Masters and Doctoral degrees. The teaming of faculty, graduate students, and NASA researchers produced a synergism that resulted in improved research productivity, and education. Courses and degree programs were constructed to satisfy the needs of the NASA community and the many small, engineering-intensive businesses within the greater region. Most of these ODU activities were traditional aeronautical or aerospace and characterized the interests and effort of many of the faculty people in the MEM department. At this time, the MEM department consisted of 21 well-qualified and experienced people, all with PhD degrees. It was decided to split the department along lines of technical disciplines to form two new departments. Fourteen faculty people elected to form a new Mechanical Engineering Department that would concentrate on mechanical design, structures and materials, control systems, and fluid mechanics at the undergraduate and graduate level. The bachelor of science degree in mechanical engineering was offered, as well as the master of engineering, master of science, and doctor of philosophy degrees in mechanical engineering and engineering mechanics. At the same time, seven faculty people with strong credentials in aerospace engineering, research, and graduate education elected to form a separate aerospace engineering department.

The Aerospace Engineering Department was formed in 1991 by a faculty of seven highly motivated and exceptionally well-qualified individuals. The new department was enthusiastically endorsed and encouraged by the NASA Langley Research Center director, at the time, Dr. Paul Holloway. Each of the faculty persons had substantial research experience and a long record of external funding, and publications, and they were totally dedicated to graduate-level education. Courses, and laboratories and diverse curricula were developed to support the instructional needs of several degree tracks spanning the research interests of the faculty. New degrees were authorized for the Master of Engineering and Master of Science degrees, as well as the PhD in aerospace engineering and in aerospace engineering mechanics. The following are those of the initial and defining faculty:

Dr. Thomas Alberts; PhD, Georgia Tech; Control Systems
Dr. Robert Ash; PhD,Tulane U; Space Systems/Space Travel
Dr. Oktay Baysal; PhD, LSU; Computational Aerodynamics
Dr. Colin Britcher; PhD, U of Southampton; Applied Aerodynamics
Dr. Osama Kandil; PhD, Virginia Tech; Computational Aerodynamics
Dr. Norm Knight; SD, George Washington U; Computational Structures
Dr. Chuh Mei; PhD, Cornel U; Structural Analysis/Non-linear Vibrations

Dr. Osama Kandil was elected by the faculty to serve as Chairman of the faculty and administrative head of the department.

During the next several years the program and the department grew rapidly in size and influence. At this same time Federal budgets and the Virginia-state budgets were showing large surpluses and the overall economy was vigorous and confident. This happy situation was mirrored by increased budget allocations at the university and a prevailing optimism that allowed additional fully funded faculty positions to accommodate growth. During 1996 the faculty size was at 12 tenure track positions. The additional people are as follows:

Dr. Brett Newman; PhD, Purdue U, Orbital Mechanics/Trajectory Analysis
Dr. M. Aminpour; PhD, U of Washington; Structures

Dr. Gary Gibbs; PhD, Virginia Tech; Acoustics/ Structures
Dr. J.J. Ro; PhD, Catholic U; Experimental Structures
Dr. P. Balakumar; PhD, MIT; Computational Aerodynamics
Dr. Don Kunz; PhD, Georgia Tech; Rotary Wing
Dr. Drew Landman; PhD, ODU; Applied Aerodynamics
Dr. Ahmed Noor; PhD, U of Illinois; Structures

There was a concentration of resources, faculty positions and laboratories, in several aerospace sub-disciplines that best served the perceived opportunities and needs of the Langley Research Center and the regional industrial base. These are:

Applied Aerodynamics – Britcher, Landman
Computational Aerodynamics – Kandil, Baysal, Balakumar
Control Systems – Alberts, Gibbs
Space Systems – Ash, Newman
Structures and Vibrations – Aminpour, Gibbs, Mei, Noor, Ro
Rotary Wing – Kunz

Not all of these people were at ODU at the same time.

By the end of the 90's decade the national economy was in a dreadful decline, tax revenues were much less than optimistic forecasts and university (ODU) budgets were subject to devastating recalls that amounted to nearly 20%. Engineering struggled to meet the need for reductions in spending and a major restructure and redistribution of resources was required. Subsequently, normal attrition and scavenging of faculty- position funding for the Aerospace department resulted in a severe reduction in force and restructure of effort. The faculty was reduced to a minimum possible cadre of six. By the year 2001 the faculty consisted of these seven:

Dr. Thomas Alberts – Control Systems
Dr. Colin Britcher – Applied Aerodynamics
Dr. Osama Kandil – Computational Aerodynamics
Dr Drew Landman – Applied Aerodynamics
Dr. Chuh Mei – Structures and Vibrations
Dr. Brett Newman – Orbital Mechanics and Trajectories
Dr. Ahmed Noor - Structures

It was no longer possible to cover all of the (6) sub-disciplines, and it was decided to concentrate on three, i.e.

Aerodynamics
Control Systems
Structures and Vibrations

In addition, there are two minors available for undergraduate enrollment. They are Aerospace Engineering and Motorsports Engineering. Each minor requires completion of a four-course (12credit hour) sequence. These courses are available for enrollment during the students Junior and senior years

At the same time, 1991, Dr. Kandil completed ten years of service as the Chair and he returned to full-time teaching and research as a university Eminent Scholar. University policy limits Chair appointments to ten years. At the start of academic year 2002 Dr. Jim Cross was designated Interim Chair for a two-year term, pending a national search for qualified persons for permanent appointment. At this point the faculty is fully tenured and consists of three Eminent Scholars, three professors, and two associate professors.

As the size of the aerospace industries decline and many of the historical airplane companies disappear, and the national defense needs shift dramatically it seems likely that the national requirement for aeronautical research and facilities will decline and the need for engineers and technicians trained in the aeronautical sciences will also decline. Clearly, NASA has seen a substantial decline in the need for wind tunnels and other high-operational-cost experimental facilities and has responded by closing many facilities, agency wide. The 30x60 wind tunnel and others at Langley have been closed and the 80x120 at Ames has been closed and others, including the 40x80, will soon be closed and removed from service by abandonment. These events have triggered a review of possibilities at ODU within the aero department to consider alternative applications of aerospace technologies and possible non-traditional opportunities for aerospace employment. Accordingly, in 1995, ODU ASE made a bold move to acquire access and operations authority for the very large NASA 30x60 wind tunnel at the LaRC.

Langley Full Scale Tunnel

The NASA 30x60 wind tunnel, (NACA Full Scale Tunnel), was constructed in 1930 and was for many years the premier (and the largest) low speed wind tunnel in the world. It was used extensively during World War II for full-scale, power-on drag reduction tests of nearly every fighter produced by the US war effort. Following the war years, the tunnel continued to be used for a variety of aerodynamic and acoustic testing until its shutdown in 1995. During the tight budgeting cycle of 1995 it was declared surplus to the Agency's needs and abandoned. The College of Engineering and Technology (CoET) with the aero department set about determining if operation of the tunnel could be commercially viable and if NASA (specifically the 1995 LaRC management team) would consider the possibility of transferring operational authority to the CoET at Old Dominion University. It was established that privatization was possible by the authority of the 1958 Space Act and, also, that a sufficient market existed for non-traditional use of the tunnel to justify the business development effort. The tunnel would become the primary facility in an enterprise center to support traditional faculty research, student projects, undergraduate co-op students and a cadre of graduate students.

A formal agreement was established in August 1997 for a ten-year period with automatic renewals in three-year increments that allowed operation of the facility as a commercial enterprise, but it disallowed use of the tunnel for tests of any aerospace vehicle, and, indeed, disallowed any NASA support. So, the CoET set about to development a customer base and develop the tunnel for nontraditional applications. The motorsports industries were determined to be the best possible market opportunity, but major changes to the tunnel and support component development would be required for tests of racecars. A proper balance system was designed and fabricated in-house that measured the aerodynamic forces on the car in drag and downforce at each wheel. A state-of-the-art data acquisition system was developed that used LabView as the primary software, sensors and a car lift system were also developed, and necessary modifications to the tunnel circuit and groundboard were completed. The first racecar tests were completed in January 1998. Since that time tunnel utilization has been on a single shift operation and varies between 800 to 1200 hours per year. The operation is fully self-supported and currently employs approximately 15 students, 1 full-time faculty person, several other part-time (50%) faculty people and a full-time professional staff of 4 people. The marketing effort has targeted aerodynamic improvement of racecar performance and heavy truck fuel efficiency improvements. The constraint on tests of aerospace vehicles has been relieved and now approximately 35% of the tunnel utilization is airplane development. The tunnel is currently the largest university-operated wind tunnel in the world, and the largest operational wind tunnel in the hemisphere.

The following photos illustrate the size and racecar test utility of the tunnel.

External View – Langley Full Scale Tunnel

Mutual Interference Effects – Cars in Passing Penske Racing – South

Conclusion

The possibilities for application of aerospace technologies to other than air and space systems are enormous and mostly self-evident. The aerospace engineering department at ODU is focusing on aerospace technologies to transportation systems for applications to class 8 tractor-trailers (heavy trucks), high-speed trains, motorsports (mostly racecars), and small aircraft transportation systems. The relevant technologies are aerodynamics (experimental and computational), lightweight high strength materials and structures, and automatic control systems. Job opportunities for new-graduates in aerospace engineering are substantially improved with specific course content in non-traditional applications. The Aerospace Engineering Department at Old Dominion University is planning for the future by increasing the context of curricula to include innovative applications of traditional degree content.

Chapter 57
Aerospace Engineering at UCLA

H. Thomas Hahn, Professor and Chair, Mechanical & Aerospace Engineering

Introduction

The College of Engineering at UCLA was created in 1948. Its first Dean, L. M. K. Boelter, strongly believed that an engineering education should be interdisciplinary in nature. During the early years of the new college the curricula reflected the Dean's philosophy, and accredited engineering degrees were only offered in a single curriculum called Unified Engineering. The unified curriculum remained in place until the early 1980's at which time ABET accreditation was obtained in several professional categories including Mechanical Engineering and Aerospace Engineering. The move toward traditional engineering curricula was accompanied by a corresponding reorganization of the departmental structure.

Although the early educational framework emphasized the unified engineering point of view, many of the course offerings and much of the research was closely linked to aerospace engineering. UCLA was (and still is) located in close proximity to several major aerospace companies, and a large percentage of our graduates seek employment in the aerospace industry. During the 50's and 60's the UCLA faculty included such noted aerospace researchers as John. W. Miles and Nicholas Rott in the fluids area, and William T. Thomson, Robert E. Roberson, Samuel Herrick, and Peter W. Likins in the astrodynamics/attitude dynamics area. Other faculty working on aerospace related research topics included Robert E. Kelly (fluid mechanics), Francis R. Shanley (structures), Richard B. Nelson (structures), Lucien A. Sch[1]mit (structural optimization), William Meecham (acoustics), Kurt Forster (astrodynamics), Walter C. Hurty (structures) and D. Lewis Mingori (attitude dynamics and control).

During the 1970's, one of our most distinguished faculty members, Professor Julian D. Cole (fluid mechanics and perturbation methods) urged the school and the department to increase its commitment to aerospace engineering by forming an aerospace engineering department with a focused aerospace engineering curriculum. New faculty were recruited in the areas of propulsion (Ann Karagozian), dynamics and control (J. Steven Gibson) experimental fluids (Peter Monkewitz) and rotary wing aircraft/structural dynamics (Peretz P. Friedmann). The idea that aerospace engineering should be strengthened was shared by many, and while a stand alone aerospace engineering department was never formed, many of the ideas in Professor Cole's proposal were employed when the Mechanical, Aerospace and Nuclear Engineering department was created in 1984. This new department offered accredited undergraduate degrees in both mechanical engineering and aerospace engineering. Graduate degrees were offered in all three of the areas listed in the name of the department.

Perhaps one of the strongest advocates of aerospace engineering at UCLA was Professor Peretz Friedmann. During his term as department chair, additional faculty including Jason S. Speyer (control systems) and Oddvar O. Bendiksen (aeroelasticity) were added to strengthen the program. Subsequent years saw the addition of Chih-Ming Ho (experimental fluids), John Kim (numerical fluids), Xiaolin Zhong (numerical fluids), Greg Carman (smart materials/structures), H.

Thomas Hahn (composite materials/structures), and Satya Atluri (structures/solid mechanics). As the department evolved, activity in the area of nuclear engineering was reduced due to several factors including retirement, departure of some faculty to other universities, and a change of research emphasis on the part of other faculty. Ultimately, this de-emphasis on nuclear engineering resulted in a decision to change the name of the department to Mechanical and Aerospace Engineering (MAE) in 1996.

Enrollment

Today the department maintains strong programs of instruction and research in both mechanical and aerospace engineering. The department has about 530 undergraduate students enrolled, of which about 200 are in aerospace engineering, Fig. 1. The graduate students number about 230 in both disciplines with 100 enrolled in the Ph.D. program. The enrollment trends over the past 5 years are also shown in Fig. 1.

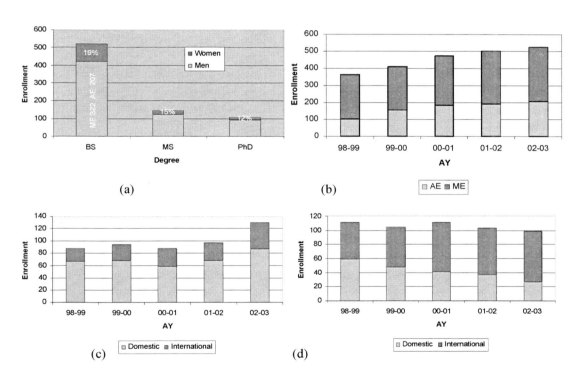

Figure 1. Enrollment trends over the past years: (a) AY 202-2003, (b) B.S. students, (c) M.S. students, (d) Ph.D. students

Faculty

The size of the ladder faculty has remained fairly stable at 30 or so since the beginning of the MANE Department, Figure 2. Complementing the ladder faculty are the lecturers from industry who bring industrial experience into the classroom. As aerospace engineering is intimately intertwined with mechanical engineering, grouping of faculty into these two disciplines is rather blurred. Rather, the faculty expertise is represented by the following seven major fields: dynamics, fluid mechanics, heat and mass transfer, manufacturing and design, nanoelectromechanical and microelectromechanical systems, structural and solid mechanics, systems and control. Based on the courses taught, the ratio between AE and ME is roughly 1 to 2.

The faculty is active in professional societies: they collectively boast 20 fellow grades, 2 members of foreign national academies, and two members of the National Academy of Engineering. Several of them serve as editors of technical journals including Fusion Engineering and Design, IEEE/ASME Journal of Microelectromechanical Systems, International Journal of Fluid Mechanics Research, Journal of Composite Materials, Journal of Heat Transfer, and Physics of Fluids. In addition, most of our faculty are recipients of various awards.

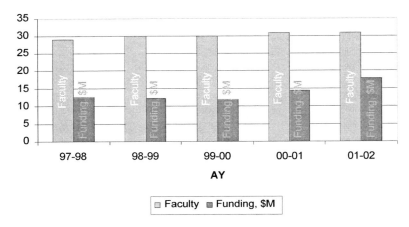

Figure 2. Ladder faculty and extramural research funding over the past 5 years.

Research

Research funding in the department is at record levels: the total extramural funding for AY 2001-2002 was $18M, which is almost $600k per faculty, Fig. 2. Through the Industrial Advisory Board and research collaborations, the department maintains close ties with its industrial neighbors such as BEI Technologies, Boeing, Capstone Turbine Corp., Honeywell, HRL Laboratories, Lockheed-Martin, Northrop-Grumman, Rand Corp., Raytheon, Rand Corp., Rockwell Scientific, TechFinity, the Aerospace Corporation, NASA Dryden Flight Research Center, and NASA Jet Propulsion Laboratory, all major players in the U.S. aerospace program.

The department is the home to a number of interdisciplinary programs especially in the areas of nano and micro engineering, and systems and control. The department faculty is providing leadership roles to the NASA Institute for Cell Mimetic Space Exploration, the California NanoSystems Institute, and a number of MURI programs funded by AFOSR, DARPA, and ONR. The other active areas of research include aerodynamics, propulsion, and smart materials/structures.

Alumni

Most of our graduates tend to stay in California, finding employment in aerospace industry. In AY 2002-2003, about 40% of graduating seniors were planning to go to graduate school with the rest being interested in industrial and government jobs.

Curriculum Reform

The department is currently going through a curriculum reform to better prepare our students for new developments in bio, info and nano technologies as well as for responsible citizenship. We continue to strive to provide an environment which would instill a love of life-long learning, a high degree of professional integrity, and a sense of leadership in our students.

Chapter 56

CALIFORNIA STATE UNIVERSITY, LONG BEACH

Aerospace Engineering program at California State University, Long Beach

Hamid Hefazi, Chair, Department of Mechanical and Aerospace Engineering
California State University, Long Beach

Background

California State University, Long Beach (CSULB) is one of the 23 campuses of the California State University (CSU) System. CSU is a statewide group of comprehensive and polytechnic universities and the California Maritime Academy, that were brought together as a system by the Donahoe Higher Education Act of 1960. It offers more than 1,800 bachelor and master's degrees in more than 240 subject areas, employs more than 22,000 faculty, and serves over 400,000 students. The system serves as a gateway institution for the great majority of students seeking a baccalaureate education in California, and for those who seek professional training as teachers, nurses, social workers, business managers and engineers. It awards more than half of the bachelor's degrees and 30 percent of the master's degrees granted in California. The CSU has awarded some 2 million bachelors and masters degrees since 1961.

Located in the southern part of Los Angeles County, and bordering northern Orange County, CSULB serves a large diverse urban population. Established in 1949, it has grown into a comprehensive University, with a student population of over 30,000. It offers 194 baccalaureate programs through 70 degrees, 88 masters programs through 61 degrees, and a joint doctoral degree. Students come from 58 California countries, 47 states and U.S. Territories, the District of Columbia, and 104 foreign countries. Since its founding, CSULB has awarded 151,374 undergraduate degrees and 30,310 graduate degrees. The University's physical plant includes 322 acres, with 80 buildings consisting of 3,061,429 square feet, inclusive of State-funded and non-State-funded buildings. The campus is beautifully landscaped. A central feature of the landscape design is more than 3200 flowering peach trees donated by the citizens of Long Beach.

As CSULB moves toward the next century, it does so with Vision 2001, a strategic plan jointly composed by representatives from across the campus as well as the community. The goal established under Vision 2001 is that CSULB will be the campus of choice for Southern California students. Its acclaim will be the quality of baccalaureate and master's programs, commitment to excellence, and preparation of students to function effectively in a global society. As part of this vision, CSULB has attracted more than 250 California high school valedictorians and National Merit Finalists among its current undergraduate population.

One of the seven academic colleges, the college of engineering (COE) consists of five departments: Civil, Chemical, Electrical, Computer Engineering and Computer Science, and Mechanical and Aerospace Engineering. All engineering programs are accredited by ABET. The Aerospace Engineering program is housed in the Department of Mechanical and Aerospace Engineering (MAE), and consists of the Bachelor of Science in Aerospace Engineering (BSAE) degree and the Master of Science in Aerospace Engineering (MSAE) degree. In addition, the Department offers a certificate program in Aerospace Manufacturing, and through the College of Engineering offers a Ph.D. in Engineering and Industrial Applied Mathematics jointly with Claremont Graduate University.

Bachelor of Science in Aerospace Engineering (BSAE) Program

The Bachelor of Science in Aerospace Engineering is a four-year degree, with the possibility for students from California Community Colleges to transfer to the program through the articulation agreements existing between the University and individual Community Colleges. The BSAE was initiated in 1994. Following rigorous University, CSU, and Statewide approval processes, the program was approved in 1996. The first students to obtain the BS in Aerospace Engineering graduated in the Spring of 1999. The program went through its first ABET accreditation visit in Fall 2000.

Despite the challenges of recruiting qualified students in the late 90's, due to the combined actions of the people involved in diffusing information and recruiting, the number of full time students enrolled in the BSAE program increased from 2 in 1995-96 to 53 in 1998. The enrollment has continuously increased to the current 119 (F 03) as shown in Table 1. The breakdown of current student levels is as follows:

FALL 2003		Semester	Number of Majors
Class Level	Number of Majors	Fall 1998	53
Freshman	28	Fall 1999	76
Sophomore	34	Fall 2000	83
Junior	22	Fall 2001	117
Senior	35	Fall 2002	120
Total:	119	Fall 2003	119

Table 1: Recent Enrollment History and Distribution by "Class" – BSAE Program

Program educational objectives

The goal of the BSAE program is to produce well-rounded engineers who possess the skills required for a successful career in aerospace engineering. This goal is reached by:

1. Providing the students with a comprehensive education in:
   1.1. non-technical areas, particularly communication in the English language and critical thinking, physical universe, humanities and arts, social and behavioral sciences and history, and self-integration;
   1.2. mathematics and basic sciences;
   1.3. general engineering topics and computer and software fundamentals, and
   1.4. aerospace engineering topics (aerodynamics; aerospace materials and structures; propulsion; space environment and space systems; communications and avionics systems; orbital and flight mechanics; and stability & control).
2. Preparing students for careers in aerospace engineering by emphasizing:
   1.5. analysis and problem solving;
   1.6. preliminary design, design for manufacturing and systems engineering;
   1.7. teamwork, professionalism, economic fundamentals, commercial viability of projects and business cases, and communication skills.
3. Providing student projects, internship and research opportunities to expose students to the professional environment, to stimulate their creativity, and to foster lifelong learning.

The achievement of the program educational objectives is mainly ensured through stated program outcomes and certain strategies used to assure outcomes. They are primarily articulated around the curriculum and the support provided by the department and the university.

All aerospace engineering classes are small (less than 15-20 students), thus allowing for an excellent interaction between students and faculty. The numerous class projects also offer an added vehicle for faculty-student interaction. The program has a special Advising and Mentoring System which assigns students to various faculty who serve as their mentors thus enhancing the interaction between faculty and students. The current four year program is shown in Table 2.

Master of Science in Aerospace Engineering (MSAE) Program

The (MSAE) program has four different emphases:
- Aircraft Systems Engineering
- Space Systems Engineering
- Aerodynamics and CFD
- Aerospace Structures and Materials

There are two paths available to students. The "thesis" path requires 24 units of approved courses and a thesis for 6 units. The "project" path requires the completion of 30 units approved course work and 6 units of directed research (project). The definitions of thesis and project are as follows.

Thesis

An engineering thesis is the written product of the systematic study of a significant problem. It identifies the problem, states the major assumptions, explains the significance of the undertaking, sets forth the sources for and methods of gathering information, analyzes the data, and offers conclusions and recommendations. The finished production evidences scientific originality which makes contributions to the field of study, critical and independent thinking, and appropriate organization and format, and thorough documentation. In general, the investigation results in a refereed technical publication or a presentation at a professional meeting.

Project

An engineering project is a significant undertaking appropriate to professional fields. It evidences originality and independent thinking, appropriate form and organization, and a rationale. It is described and summarized in a written report that includes an abstract and the project's significance, objectives, methodology, and a conclusion and recommendation.

The majority of MSAE students are engineers working in the local aerospace industry. The enrollment in the MSAE program has substantially declined since its peak of 90 students in the early 1990s. This is a reflection of Aerospace Industry employment in Southern California. Various anecdotal indicators (such as inquiries) point to improved enrollments in the MSAE program. Recent enrollments in the MSAE program are given in Table 3.

MSAE Enrollments	
Fall 1998	10
Fall 1999	16
Fall 2000	13
Fall 2001	11
Fall 2002	7
Fall 2003	18 (est)

Table 3. Recent MSAE Enrollments

California State University, Long Beach
UNDERGRADUATE CURRICULUM GUIDE
Bachelor of Science in Aerospace Engineering
2003-2004 Catalog (Code: 3-4310)

FRESHMAN YEAR

FIRST SEMESTER			SECOND SEMESTER		
COURSE	TITLE	UNITS	COURSE	TITLE	UNITS
MAE 101A	Intro to Aerospace Engineering	1	MAE 172	Engineering Graphics	3
UNIV 100	Intro to the University	1	PHYS 151	Mech. & Heat (B1b)	4
MATH 122	Calculus 1 (B2)	4	MATH 123	Calculus II (B2)	4
GEN ED	General Education (A2)	3	GEN ED	General Education (A3)	3
GEN ED	General Education (D1b)	3	GEN ED	General Education (B1a	3
GEN ED	General Education (A1)	3		or C3 or E)	
		15			17

SOPHOMORE YEAR

GEN ED	General Education (C1)	3	EE 211	Electric Circuits I	3
PHYS 152	Elec. & Magnetism (B1b)	4	EE 211L	Electric Circuits Lab	1
MATH 224	Anal. Geom. & Calc. III	4	MATH 370A	Applied Math I	3
MAE 205	Computer Methods in ME	2	MAE 371	Anal. Mech. II (Dynamics)	3
CE 205	Anal Mech I (Statics)	3	GEN ED	General Education (D2a)	3
			CHEM 111A	General Chemistry (B1b)	5
		16			18

JUNIOR YEAR

MAE 333	Engr. Fluid Mechanics	3	MAE 453	Stability&Control AV	3
MAE 373	Mech. of Deformable Bodies	3	MAE 334	Aerodynamics I	3
MAE 350	Flight Mechanics	3	MAE 365	Aerospace Structures I	3
GEN ED	General Education (D1a)	3	MAE 381	Spacecraft Systems Eng	3
MAE 330	Engr. Thermodynamics I	3	ECON 300	Fund. of Economics (D2b)	3
GEN ED	General Education (C2a,b, or c)	3	MAE 390	AE Seminar & Com Skills	1
		18			16

SENIOR YEAR

MAE 465	Aerospace Structures II	3	MAE 452	Propulsion	3
GEN ED*	General Education (C2a,b,or c)	3	MAE 486	Avionics Systems	3
MAE 434	Aerodynamics II	3	MAE 479	Aerospace System Des II	3
MAE 408	Systems Engr. & Integration	3	MAE XXX	Elective	3
MAE 483	Space Flight & Orbital Mech.	3	MAE 374	Mech.Prop. Matls Lab	1
MAE 478	Aerospace System Design I	3	MAE 440	Aerodynamics Lab	1
		18			14

*General Education: A minimum of nine (9) units of Upper-Division (UD) are required. Six (6) units must be Upper-Division Interdisciplinary (UDI) in one particular THEME (please see the College of Engineering section in the University Catalog). A minimum of nine (9) units of Capstone courses (_), with three (3) units of Global (_) and three (3) units of Human Diversity (_), are required. Please consult a General Education advisor. The above 132 units minimum program allows for a six (6) unit General Education waiver (3 units in category D2, and 3 in B1a or C3 or E). All IC and category A classes must be taken for a letter grade. Students are advised to plan their programs with care in order to meet all departmental and General Education requirements within 132 units.

Table 2. BSAE Four-Year Program

Joint PhD program in Engineering and Applied Mathematics with Claremont Graduate University

This college-based program is offered jointly by CSULB and the Department of Applied Mathematics at Claremont Graduate University. Under this program candidates are admitted to both institutions, finish a total of 76 units of approved graduate work (38 units in each institution, including 12 units of thesis and directed research), pass preliminary and final exams given by a joint faculty committee, and conduct their thesis in a relevant subject under the supervision of a joint faculty committee. The research subject is expected to be in an area encompassing Applied Mathematics and Engineering sciences. The Aerospace Engineering program is actively involved in the joint Ph.D. program. Two students from the AE program have graduated, and three more are at various stages in the program. Some examples of dissertations include: *Prediction of High Lift Flows* (Eric Besnard), *Constructive Neural Networks for Function Approximation and Their Application to Computational Fluid Dynamics Optimization* (Adeline Schmitz) and "Delamination detection in composite laminates using neural networks and genetic algorithm optimization"(Hie Le)

Students

A variety of approaches are taken to enhance the education of students. For example, it is the ABET stated goal of the program that 2/3's (67%) of undergraduate students will have participated in an internship/research program and/or an industrial project. To achieve this ambitious goal, the program actively employs students on various department research projects, sponsors and develops research and development programs through AIAA student chapters, and constantly works with the local Aerospace Industry to create internship opportunities. Some highlights are as follows:

Assistantships

During the summer of 2002 and Academic Year 2002-2003, six (6) undergraduate students and two (2) graduate students were working on externally funded research programs in the department and laboratories.

Student projects

The California Launch Vehicle Education Initiative (CALVEIN), a.k.a., CSULB Rocket Project, was established in 2001 using departmental resources with the objectives of: (1) providing engineering students with hands-on experience on a flight vehicle; (2) providing senior students with design projects, which go from requirement definition all the way to the realization and test – in flight – of a manufactured component; and (3) providing graduate students and faculty with opportunities for the research, development and testing of new technologies. This program is unique in the nation because, unlike many other academic institutions launching rockets which use consumer rocket motors (solids mostly), students design and build the entire propulsion system, including the rocket engines. Propellants are Liquid Oxygen (LOX) and ethanol.

The outcome of the program was the successful development, launch, and recovery of the Prospector-1 vehicle (June 2001). The program was extended with a grant from the California

Technology Trade and Commerce Agency ($110,000) in June 2001. As part of this grant, Prospector-2 (P-2) was successfully flown in February 2002, and Prospector 3, a more advanced vehicle which incorporated the first version of a student-developed thrust vector control system, was flown and recovered in February 2003.

Figure 3. Prospector-3 team and P-3 at take-off

Another noteworthy achievement of the program was the successful development, by students, of an aerospike engine. Unlike conventional bell nozzles, which are optimized for operating at one particular altitude, aerospike nozzles provide altitude compensation, and thus offer higher thrust coefficient below the design altitude. The second iteration of the design was successfully static fire tested in June 2003, and the first launch of an aerospike engine using liquid propellants after more than forty years of research, is scheduled to take place early Fall 2003.

Figure 4. Aerospike static fire test, June 2003   Figure 5. Aerospike engine mounted to P-2
(P-3 seen hanging, above)

The program also gives opportunities for CSULB students to work on projects with students from other universities. For example, Stanford students flew a payload onboard P-2 and USC students flew 2 payloads onboard P-3, including Microelectromechanical System (MEMS) samples made by JPL. Such interactions also help students to learn to work with individuals and teams from other organizations, skills critical for a successful career.

In addition to its educational benefits, the program also serves as a vehicle for developing new low cost technologies, like the aerospike engine or thrust vector control system (flight tested on P-3), for future small launch vehicles. The university is also involved with Garvey Spacecraft Corporation on the development of its Nanosatellite Launch Vehicle (NLV), slated to deploy payloads up to 10 kg in LEO.

Internships

In February 02, the MAE department signed a Memorandum of Agreement with Honeywell Space and Systems (Formerly Allied Signal) in Torrance CA to establish a novel internship program. Under this program, the department screens applicants and hires students (primarily juniors and seniors) to fill internship positions that are available at Honeywell. Interns also go through a rigorous interview process at Honeywell. If selected, they work under the supervision of Honeywell program managers on various projects at the Honeywell site in Torrance, CA.

This program gives students valuable work experience and gives Honeywell the opportunity to recruit qualified engineers when they graduate. Since its inception, the program has been very successful. Fourteen students have been placed. Three who graduated have been hired by Honeywell into full-time engineering positions. During the Summer of 2003, ten (10) interns worked at Honeywell as part of this program. In addition to this program, a number of students find internship opportunities on their own in the local aerospace companies.

Professional Organizations

AIAA student chapter

CSULB has an active student chapter of AIAA. The organization is involved in the rocket project described above, as well as several other activities, such as organizing regular guest lectures from industry who discuss issues of professional development or on-going development projects.

Student Accomplishments

Both graduate and undergraduate students are encouraged to participate in conferences, competitions and professional events to learn life long learning skills. Prominent among recent accomplishments are:

Graduate student Stanley Baksi won 2nd Place at the AIAA Student Conference, in the Graduate Section, for his paper on the thrust vector control system development. Stanley also presented a more complete version of this paper at the 2003 Joint Propulsion Conference in Huntsville, Alabama, in July 2003. Additionally, Stanley was the recipient of the Department Outstanding Thesis Award at the 2003 Commencement Ceremony for his thesis titled "*Dynamic Stress*

Analysis of an Epicyclic Gear Pump". Starting Fall 03, Stanley will be pursuing his PhD at UCLA.

Graduate Tadashi Murayami won the second place award in 2002's State-wide CSU research competition for his thesis entitled "Computational Analysis and Optimization of vertical Axis Wind Turbine Blades".

Graduate student Micah Abelson won the MSAE Outstanding Thesis Award at the 2003 Commencement Ceremony. His thesis title was: *Multidisciplinary Optimization of a Simplified Linear Aerospike Engine.*

Undergraduate student Kelly Carter won the BSAE Outstanding Undergraduate Award at the 2003 Commencement Ceremony.

International Exchanges

As part of the California State University System, CSULB has access to a large number of opportunities for students to study, for a semester or a year, abroad. Several of the partner universities offer aerospace classes which can be transferred to CSULB for credit towards the bachelor's degree.

The MAE Department also regularly receives Visiting Research Scholars and Students, who come to conduct research projects at the university.

Faculty

The faculty primarily responsible for the Aerospace Engineering program consists of four tenured/tenure-track professors, one full-time lecturer, six part-time lecturers from industry, and a full time research associate. A number of Mechanical Engineering faculty also contribute to the program by teaching some classes and supervising students. Typically, most of the theoretical topics of the curriculum are taught by tenured faculty who all hold a Ph.D. degree and who are actively involved in research and development in their respective areas. The more applied areas of the curriculum, such as systems engineering or manufacturing, are commonly taught by instructors from industry who bring their extensive knowledge and experience to the classroom. CSULB is strategically located near many aerospace companies such as Boeing, Northrop Grumman, Raytheon, the Aerospace Corporation, etc., and has access to much well-qualified engineering talent to teach these courses. Many executives from the local aerospace industry teach in our AE program. For example, two (former) vice presidents of the Boeing Company currently teach classes in the program. This combination of tenured faculty who are up-to-date with the latest developments in their respective specialties, and of practicing senior engineers and executives, offers the required coverage and balance of all the theoretical and applied areas of aerospace engineering.

Research

As previously noted, faculty are heavily involved in research and development with industry and government agencies through grants, contracts, and independent research. All tenured faculty members are active in professional societies and regularly publish and present papers at national and international symposia. The total research expenditure from external funding during the past ten years has consistently been over $1 million/year. Additional substantial research support such as assigned time, student payroll, and mini-grants, also comes from departmental and University internal resources. Both graduate and undergraduate students are encouraged and given opportunities to work on these projects. General areas of research include: Aerodynamics, Hydrodynamics, Computational Fluid Dynamics (CFD), Structures, Design Analysis and Testing of Composites, Neural Networks and Optimization Methods. Highlights of some of these research programs are described below.

Building on its decades of experience in the aerospace industry, CSULB has been focusing on fast ship research since 1997. More specifically, in 1997-98, CSULB was selected by the Office of Naval Research to conduct a CFD study of hydrofoil design and optimization for fast ship concepts. [1]

Since 2000, following the ONR project, the same team of faculty, research associates and students has been continuously working on a multi-year program on fast ship research funded by the Center for Commercial Deployment of Transportation Technologies (CCDoTT). This Center (http://www.ccdott.org/)) is a chartered university center functioning as a partnership of academic institutions, government, and commercial corporations. The United States Transportation Command (USTRANSCOM), and the Department of Transportation (DOT), through the Maritime Administration Office (MARAD), have oversight responsibility for the CCDoTT projects. The three year CCDoTT program (00, 01, and 02) has been conducted in partnership with Pacific Marine (Navetek Ltd.) of Hawaii.

The year 00 program entitled "CFD Design tool Development and Validation for Fast ship Applications" focused on the development of an accurate "free surface" method for wave drag calculations, as well as design method development and application to underwater hull configuration optimization. The objective of the latter part was to optimize the shape of one of the proprietary Pacific Marine's HYSWAC hull forms with under water lifting bodies. [2]

During the 01 program, the CSULB-optimized configuration was tested in a joint "At Sea Trial" program with Pacific Marine [3]. The test confirmed the optimization and also provided test data for the validation of the CFD methods being developed. Also, during this program the CFD methods were improved and validated. The development of new CFD methods, based on Reynolds Averaged Navier Stokes (RANS), was also initiated [4]. The second important element of the year 01 program was the development and use of advanced design/optimization methods using Neural Networks. These tools were applied to the shape optimization of a second proprietary Pacific Marine hull form with under water lifting bodies. [4]

(a)   (b)

Figure 1a-b: (a) Scaled model of Pacific Marine's WB optimized by CSULB on a test craft
(b) CSULB/PM Joint At Sea Trial Program

The current year program, which is ending in August 03, has finalized the development of our free surface RANS method [5]. The main focus of the 02 program, however, has been on optimization methods based on Neural Networks. The major accomplishment of this work has been the successful application of Neural Networks to shape optimization of the BWB body using a training set based on only 1000 CFD runs. The traditional optimization approach, in which the CFD method is coupled with the optimizer and executed during the optimization process, was run and compared with that using the Neural Network based on a 1000-point training set. Because of the rather larger CPU requirements of the CFD method, the number of optimization iterations of the Genetic Algorithm was limited to 2000 for the 28 design variables. In contrast, the optimization using the Neural Network was allowed to conduct a much more extensive design space exploration and the net result was an optimum which was 5% better (verified by the CFD tool post optimization) with the Neural Network than with the traditional approach, in roughly half the CPU time. These developments have culminated in the design and construction of Pacific Marine's HDV-100 technology demonstrator/Patrol boat/fast ferry. HDV-100 a 100 foot boat designed to go at speed in excess of 47 knots. It is expected to go to sea around January 04.

Other research activities conducted by Aerospace Engineering faculty include:

The prediction of ice accretion on airfoils, multielement airfoils, and wings: a computer code, LEWICE, which can be used to predict the ice shapes on airfoils and wings, was previously developed at NASA Lewis. The LEWICE code was modified by CSULB to extend calculations to three-dimensional flows and include the viscous effects in the flow field. The modified LEWICE code was used to predict the ice shapes on a NACA 0012 airfoil for a wide range of icing conditions and on a MS-317 swept wing.

Boundary layer transition and stability analysis: Several computer codes have been developed to predict the onset of transition on airfoils and wings in low speed, transonic and supersonic flows. These codes were used as analytical tools in applications of Laminar Flow Control.

Aeroelastic analysis of wings, wing/fuselage configurations and high-lift systems: An interface method was developed to couple aerodynamics and structures for aeroelastic analysis. The method was used to evaluate the aeroelastic effects of an advanced transport wing at cruise and critical conditions. The method was also applied to more complicated configurations, including the MD90 wing/fuselage configuration, a simple three-element high lift system, and a rather complicated high lift system of an advanced high-wing transport.

Development of drilling simulation tools: The objective of this ongoing research is to develop analytical tools and a database to design optimal drilling processes and tools for aircraft manufacturing and other precision manufacturing industries. A finite element model of drilling/burr formation has been developed to simulate the multi-layered drilling process. Besides, a comprehensive drilling database will be developed using laboratory data and results from finite element analysis. Students in the Department are working with Faculty and Boeing engineers at the Boeing Manufacturing Laboratory, located at CSULB, to conduct laboratory tests on coupons simulating the C-17 parts. Measurements of burr height, hole diameter, etc., will be analyzed to optimize the drilling process.

Students conducted laboratory tests using the electroimpact drill test cell
at the Boeing Manufacturing Laboratory

AE faculty and graduate students have also been involved in research and development activities associated with aerospace vehicle composite structures. From 1991 to 1994, CSULB was funded by the McDonnell Douglas Corporation (Currently the Boeing Company) to develop new engineering and design techniques and conduct finite element analysis for the Innovative Composite Aircraft Primary Structures (ICAPS) program. This program was conducted at Boeing under the contract from NASA to develop aircraft primary structural components including wing and fuselage using composite materials.

CSULB was also funded by Boeing at Long Beach to develop technology of microcracking analysis for long-term high-temperature durable composite materials relevant to High Speed Civil Transport (HSCT) project. Under this task a theoretical model that was able to accurately simulate the microcracking characteristics of materials, including fracture toughness and strain energy release rate subjected to applied tensile loads was established.

In 2002, the Army Research Office funded a CSULB project to develop an efficient and accurate technique for characterizing and detecting delaminations in composite structures using built-in piezoelectric sensors. In this project, a hierarchical identification technique will be developed to quantify delaminations using a combination. This program is conducted jointly with Arizona State University (ASU).

Facilities

A number of research and educational facilities and laboratories serve the Aerospace Engineering Program. Prominent among them are:

Composite Research Laboratory (http://www.csulb.edu/colleges/coe/ae/composite/)

The Aerospace Engineering Program at CSULB has a large capacity to fabricate and test composite materials. The University has been supporting the composite materials research and development effort for Boeing-Long Beach engineers for many years. The facility contains a composite manufacturing center that includes a 3`x5` high temperature autoclave (800°F and 415 psi pressure).

Figure 6. High Pressure Autoclave & Instron test Frames in Composite Research Lab.

The other portion of the facility is a large-scale long-term durability testing facility. The test facility has twelve (12) Instron servo-hydraulic testing frames with computer controlled thermal chambers capable of temperatures ranging from –120°F to 700°F and ranging from 22 Kip to 110 Kip.

Past programs conducted at the Composite Research Laboratory have included:

1. High Speed Civil Transportation (Boeing/NASA)
2. X-37 Development Research (Boeing)
3. ACT (Advanced Composite Testing) Program (Boeing)

The HSCT program involved composite development work for 300 seat High Speed Civil Transportation (HSCT) design at temperature up to 350°F, speed up to Mach number 2.4 and altitude about 60,000 ft. The Advanced Composite Technology (ACT) program, involves building a highly durable light weight _ scale MD-90 wing of stitched composites.

The facility is also used for teaching and training. Two training programs in Advanced Composite Materials Manufacturing, funded by the State Department of Trade and Commerce and from the National Science Foundation, have so far been conducted in this facility. These 6-week summer programs were designed to offer trainees the skills useful for local manufacturing industries.

Figure 8. Students Training in Advanced Composite Manufacturing

Computational Fluid Mechanics (CFD)

This facility is used by the student assistants and graduate students who work on projects. The computers in this facility include a SGI Origin 3200 parallel machine with eight processors, two SGI Octane work stations, as well as many other PCs and SGI Unix-based work stations. In addition to many in-house specially developed codes for CFD and hydrodynamics, such as panel and RANS codes for free surface problems, CSULB has extensive experience with the following NASA-developed CFD codes: INS2D and 3D, CFL3D, Over Flow and Pegasus (for Chimera Scheme). Other relevant programs available in the CSULB CFD lab include: Pro-Engineer, CAD software, the ICEMCFD mesh-generation program, Ensight and FieldView flow visualization, as well as iSIGHT numerical optimization.

CAD/CAM FEA

The department has excellent facilities and faculty expertise in CAD/CAM and associated tools for analysis and manufacturing. These associated tools include finite element analysis (FEA) for design, for large deflection analysis, for plastic mold design, for heat transfer, for fluid mechanics, and for mechanisms. They also include associated tools for manufacturing, including laser scanning, rapid prototyping, and CNC machining.

The primary CAD/CAM system being used for upper division, some lower division, and all graduate classes, is IDEAS, from EDS PLM Solutions (formerly SDRC). Some lower division drafting classes are using AutoCAD. Also available are ProEngineer and CATIA. The primary

737

analysis software is the finite element analyzer associated with IDEAS. This includes static linear analysis, modal analysis, dynamic analysis, nonlinear analysis, heat transfer analysis, and mechanism analysis. Other programs used extensively include MSC NASTRAN, ANSYS, DYTRAN, and LSDYNA for large deflection and fluid/structure interaction.

Low Speed Wind Tunnel

Used by the McDonnell Douglas Company for decades, the Douglass Long Beach Wind Tunnel (DLBWT) as it is called was donated to the University in 1993. The closed circuit wind tunnel has a 96.5 X 137 X 305 cm working area. The tunnel has a 10:1 contraction with a maximum free stream speed of 90 m/sec in the working area. The settling chamber has a honeycomb and six screens for flow conditioning. For speeds less than 50 m/sec, the free stream turbulence is less than 0.2 percent and the non-uniformity in the flow across the test section is less than 0.1 percent per 5 cm.

.

Partnerships

The Aerospace Engineering program at CSULB has very close partnerships and links with many local Aerospace Industries, such as Boeing, Northrop Grumman (formerly TRW) and Honeywell (formerly Allied Signal). Prominent among these companies is the Boeing Company.

In addition to being a "Boeing Focus" school, in March 2000, following a cash donation of $1.15 million from the Boeing Company to the University, the Center for Advanced Technology Support for Aerospace Industry (CATSAI) was established in the College of Engineering. CATSAI is a strategic partnership between CSULB and the Boeing Company. The center currently occupies about 10,000 square feet of laboratory space in two buildings and six offices which houses 15 Boeing engineers and scientists from the Boeing C-17 and Phantom Works divisions.

CATSAI's vision is the establishment of an innovative center for manufacturing technology research & development, integration of near term & long term manufacturing technology needs of the Aerospace Industry & its supply chain, improve the overall capability of California Workforce, promote R & D opportunities for faculty and students and bring external resources to the University. CATSAI's primary objectives are to advance workforce education in all disciplines, prioritize, focus and coordinate technology development, accelerate technology implementation, train and work with Boeing subcontractors & regional manufacturers and serve as a conduit for technology exchange and transfer. A number of joint CSULB/Boeing projects have been conducted at CATSAI.

Figure 9: Center for Advanced Technology Support for Aerospace Industry (CATSAI)

One of the educational activities associated with CATSAI is a Certificate Program in Aerospace Manufacturing (http://www.csulb.edu/colleges/coe/ae/boeing/). This program which is developed in collaboration with the Boeing Company and with funding from the Society of Manufacturing Engineering (SME) Foundation and the Boeing Company started in 1997. It is designed to provide students with skills required for an Industrial Engineer or Manufacturing Engineer. Individual courses could also be taken by other professionals to improve understanding in a certain area. As of Spring '03-semester more than 160 students were participating in the program and 79 have completed the program. In addition to Boeing employees, CSULB students and students from other smaller Aerospace Manufacturers participate in this program. Since 2002, Parker Hannifin Aerospace has joined this program. More than 50 Parker Hannifin employees have been taking classes on-site in Irvine, CA.

References:

[1]   E. Besnard, A. Schmitz, K. Kaups, G. Tzong, H. Hefazi, H.H. Chen, O. Kural and T. Cebeci, "Hydrofoil
        Design and Optimization for Fast Ships", presented at the 1998 ASME International Congress and   Exhibition, Anaheim, CA, Nov. 1998.

[2]   "CFD Design Tool Development and Validation", Center for the Commercial Deployment of   Transportation Technology (CCDoTT) report Task 2.8, H. Hefazi, Task Manager, FY 00. Available
        online at www.ccdot.org.

[3]   "CFD Design Tool Development and Validation Part 2 At sea trials", Center for the Commercial
        Deployment of Transportation Technology (CCDoTT) report Task 2.14.1, H. Hefazi, Task Manager,
        FY 01. Available online at www.ccdot.org.

[4]   "CFD Design Tool Development and Validation Part I", Center for the Commercial Deployment of
        Transportation Technology (CCDoTT) report Task 2.14, H. Hefazi, Task Manager, FY 01. Available
        Online at www.ccdot.org.

[5]   "CFD Design Tool Development and Validation with Accompanying Technology Demonstrator",
        Center for the Commercial Deployment of Transportation Technology (CCDoTT), Task 2.2 report,
        H. Hefazi, Task Manager FY 02. Available August 03.

Part V

Military School Programs in Aerospace Engineering

Chapter 59

THE WINGS OF WEST POINT: CONTRIBUTIONS OF THE UNITED STATES MILITARY ACADEMY TO THE DEVELOPMENT OF AERONAUTICS

Captain Steven R. Braddom
Colonel Kip P. Nygren
United States Military Academy

Few institutions are as closely tied to the genesis and development of aviation and Aerospace Engineering education as the United States Military Academy (USMA) at West Point. Since the dawn of the era of powered flight, West Point graduates have been involved in and made lasting contributions to every aspect of aviation from the Wright Flyer to the International Space Station. The United States Military Academy was established in 1802 as the Nation's first Engineering School, and its graduates have been at the forefront of the growth of several pioneering technologies – military armaments and fortifications, the expansion of railroads, the construction of the Panama Canal, and the creation of the first atomic bomb to name a few.

Frank P. Lahm. *USMA 1901 Class Album*   **Thomas E. Selfridge,** *USMA 1903 Class Album*

The Dawn of Flight

As the potential of heavier than air flight became apparent, it was the U.S. Army that purchased the first military airplane from the Wright Brothers in 1909. Two West Point graduates, First Lieutenants Frank Lahm and Thomas Selfridge, both recent Military Academy faculty members, were the Army experts supervising the purchase of the Wright Flyer and the earlier acquisition of the Army's first dirigible. In the fifty year explosion of aeronautics growth that followed, the Army Air Service, the Army Air Corps, the Army Air Force, and the U.S. Air Force, led for the most part by Military Academy alumni, were the primary motivating forces in the pursuit of higher, faster and larger aircraft.

On August 1, 1907, Brigadier General James Allen, USMA 1886, established an Aeronautical Division within the Signal Corps of the U.S. Army. This embryonic organization was the forerunner of the U.S. Air Force, and within six months, a dirigible was ordered for this aeronautical component of the Army. Interest would continue in dirigibles for about 25 years, but the great success of heavier-than-air powered aircraft would spell the end of dirigibles for the Army.[1]

U.S. Army Dirigible Over the West Point Parade Field (1920s), *USMA Photo Collection*

Thomas E. Selfridge, almost certainly the least know major contributor to the early development of the airplane, graduated from West Point in 1903 in the same class as the future

Five-Star General, Douglas MacArthur. At the time Selfridge was a cadet, it can be seen in Table 1, that Mathematics, Science and Engineering constituted 68% of the four year curriculum.

After graduation, Selfridge was assigned to the Artillery and transferred to the Presidio of San Francisco, California, where he continued his great interest in learning by studying a wide variety of subjects. Appointed to teach Ordnance and Gunnery at West Point in 1906, Thomas spent some of his first year on the faculty researching many of the various new technological developments occurring at that time in order to decide where to concentrate his efforts and direct his future career. Once he learned of the work of the Wright Brothers, Selfridge was infected with the aviation bug and wrote the Wrights in the spring of 1907 to offer his services as an associate in their work on powered aircraft during his summer break from West Point. Rebuffed by the always suspicious Wrights, he turned his attention to the experiments that Alexander Graham Bell was conducting with large kites that had attracted the attention of the Army.[2]

Armed with a letter of introduction from the West Point librarian to the famous telephone inventor, Selfridge set off to Bell's summer home at Baddeck, Nova Scotia in June of 1907. Bell was so impressed by Selfridge that he wrote a letter to President Theodore Roosevelt asking that Selfridge be assigned to work with him and the Aerial Experiments Association (AEA) for the next year. Selfridge became the Secretary of the AEA in a partnership created with F. W. "Casey" Baldwin and J. A. D. McCurdy, two young Canadian engineers in Bell's employ, Glenn H. Curtiss, who had built an engine for one of Bell's large tetrahedral kites, and Bell, who was the AEA President. The purpose of the association was simple, "Get into the Air," and Selfridge was selected to design the first of four aircraft named the Red Wing, probably based on his engineering education and teaching experience at the Military Academy.

Academic Subject	Curriculum Weight
Civil Engineering with some Mechanical and Electrical Engineering	3.0
Mathematics *Algebra, geometry, trigonometry and the calculus*	4.0
Physics/Mechanics	3.0
Chemistry	2.25
Ordnance	1.5
Drawing	1.25
English *History, Ethics, Geography, Law, Grammar, Rhetoric*	3.0
French	1.5
Spanish	0.85

Table 1. Academic Subjects for a Cadet in 1903.[3]

To build the Red Wing, the AEA shifted location to the Curtiss shop in Hammondsport, New York in the winter of 1907 to 1908. The new aircraft was to fly from the ice of Lake Keuka, but the inevitable delays set the first flight back to March of 1908, and before Selfridge could test it he was recalled to duty in Washington.

The Aerial Experiments Association in 1907 - Glenn Curtiss, John McCurdy, Alexander Graham Bell, Frederick Baldwin, Thomas Selfridge, *Jack Carpenter* *(GlennHCurtiss.com)*

As a result, the honor of making the first flight on 12 March 1908 fell to Baldwin. "After a run of 200 feet over the ice, the plane rose to a height of between six and ten feet and flew 318 feet, 11 inches. It then went into a stall, the tail structure collapsed, and the machine slid down on one wing. Five days later during another trial, the plane was completely wrecked."[1] Selfridge got his opportunity to fly the next prototype designed by Baldwin named the White Wing. In his first time at the controls of an airplane on 19 May 1908, Selfridge guided the aircraft to a height of 30 feet and covered a distance of 237 feet, making him the first U.S. Army officer to pilot a mechanically powered flying machine. This flight was followed by others, and subsequently Selfridge flew the AEA June Bug, which won the Scientific American Trophy for a public flight of over one kilometer.[1] In July 1908, Thomas Selfridge was reassigned to the Signal Corps Aeronautical Division and tasked to test the first Dirigible the Army was contracting to purchase. He was taught to fly the lighter-than-air craft and assisted with the final testing prior to the Army's purchase of Dirigible No. 1. Also during that summer, the Aeronautical Division was responsible for testing the aircraft that the Wright Brothers had proffered to the Army in fulfillment of a recent Army request. Orville Wright began the preparation of the Military Flyer at Fort Myer, Virginia in August and the first flight test occurred on September 3, 1908. After several flights, it was Selfridge's turn to fly as an observer with Orville, and they took off late in the day on September 17. After a couple of turns around the field at Fort Myer, a propeller apparently failed and the airplane crashed. Orville was badly injured, but Selfridge sustained a skull fracture and died at the Fort Myer Army hospital within a couple of hours. It would be a few months before Orville was able to leave his hospital bed and he would walk with a limp from the accident for the rest of his life.[4]

Thomas Selfridge at the Controls of the White Wing in May, 1908, *Library of Congress*

It his hard to imagine a more exhilarating fifteen months than Thomas Selfridge experienced from June of 1907 to September of 1908. He was near the center of the development of a revolutionary new technology, and was the Army's foremost expert on powered aircraft design and operation. The average life expectancy of airplane pilots would not be very long during the next thirty years of airplane development, but had Thomas Selfridge lived, his contributions might have rivaled the career and reputation of Carl Spaatz or even Hap Arnold.

Thomas Selfridge was not the only aviation pioneer from the engineering education program at West Point. Frank P. Lahm, who graduated from West Point two years before Selfridge, began his career in aviation with free balloon ascents. After graduation in 1901, Lahm joined the Cavalry and was assigned to duty in the Philippines until 1903, when he returned to West Point to teach French. On a summer tour in France in 1904, Lahm was initiated into ballooning by his father, who happened to be a member of the French Aero Club and who also owned a balloon. Lieutenant Lahm earned his license as a balloon pilot from the Federation Aeronautique Internationale the following summer. Then in 1906, after only two years of experience, Lahm entered and won the inaugural Gordon Bennett Trophy Cup by placing first in an international balloon race from Paris to England.

With his unique aviation experience, the Army assigned First Lieutenant Lahm in 1907 to the new Aeronautical Division of the office of the Chief of the Signal Corps in Washington, D.C. At Fort Myer, Virginia, Lahm and a detachment of Signal Corps soldiers constructed a hydrogen generating plant and practiced observation work with captive balloons. Along with Lieutenants Thomas Selfridge and Benny Foulois, Frank Lahm assisted in the testing of the first dirigible purchased by the Army during the summer of 1908. The three lieutenants were also the official Army evaluators of the performance of the Wright Military Flyer. After the death of Selfridge, the injuries to Orville, and the destruction of the Military Flyer, it would not be until the next summer that the Military Flyer trials could continue. Orville flew the duration trials with Lahm and the speed trials with Foulois, which were successfully completed on 30 July 1909. The first military aircraft was officially accepted by the Army on 1 August 1909.[5]

Once the Army owned the airplane, it needed pilots. First Lieutenants Frank Lahm and Frederic E. Humphreys were trained by Wilbur Wright as the first two military rated airplane pilots, achieving their qualifications on 26 October 1909. Lahm and Humphreys then established the first military flight school at College Park, Maryland. Although Frank Lahm had to return to the Cavalry after four years of detached duty, he would return to make several more contributions to aviation during World War I and up to the start of World War II.

The Start of Aerospace Education at West Point

Although the roots of Mechanical Engineering at the United States Military Academy date back to its inception, the appearance and development of Aeronautical Engineering in the curriculum closely followed the ascendance of aviation as a combat arms branch of the U.S. Army. The impetus which forced the inclusion and expansion of Aeronautical Engineering education was the desire of the Air Service to have officers who understood and could utilize aeronautical technology in the field.

Providing a course in Aerodynamics was first considered at West Point in the fall of 1917, when Lieutenant Colonel C.C. Carter became the Head of the Department of Natural & Experimental Philosophy. This early interest resulted in the inclusion of some elements of the "Aero Dynamics of Mechanical flight" into the Academy's existing course in Mechanics. But, the first written mention of creating a new course in Aeronautical Engineering was a memorandum sent from the Director of Military Aeronautics to the Superintendent of West Point in September of 1918. The memorandum was written by Colonel Thurman H. Bane of the Technical Section of the Division of Military Aeronautics in Dayton, Ohio. At the time, the only Aeronautical Engineering education for military officers was provided to a select few who attended the Massachusetts Institute of Technology later in their careers. Colonel Bane felt that this approach was insufficient in terms of both the number of officers who were trained and the gap between undergraduate and post-graduate education due to military service. Colonel Bane's point in the memorandum was clear: "With the great and constantly increasing importance of aviation, it is essential that there should be a permanent group of officers who are aeronautical engineers. The Navy has Naval Constructors, and the Army should have airplane designers."[6]

The Department of Natural & Experimental Philosophy supported the request for the creation of a new course which could be coordinated with the Massachusetts Institute of Technology to better prepare up to twenty-five percent of cadets for post-graduate study. C.C. Carter had recently visited other technical institutions, flying fields and aircraft factories in order to devise a relevant course. But, West Point rejected the proposal since the curriculum was compressed to 3 years at the time due to World War I. The Academy noted that with the 3 year curriculum it was not feasible to teach Aerodynamics because there was insufficient time to teach the cadets all of the required pre-requisite subjects in sufficient depth.

When legislation was introduced into Congress to restore the full 4 year curriculum for the 1920-1921 academic year, the Air Service again requested to create a course in Aerodynamics. The Air Service became a combat arm of the U.S. Army under the Army Reorganization Act in June of 1920. In a memorandum to the Superintendent of West Point, Chas T. Menoher, the Chief of the Air Service and a West Point graduate, wrote: "In view of the fact that the Air Service is now a coordinate and combatant branch of the Army, and that it will play a major part in warfare, it is desired to renew at this time, most urgently, that special consideration be given to the inclusion of Air Service training and special subjects pertaining thereto, in the curriculum of the United States Military Academy."[7] This time the Academy relented, and in March of 1921 the Superintendent, Brigadier General Douglas McArthur, approved the request from the Department of Natural & Experimental Philosophy to teach 18 lessons of Aerodynamics. The Aerodynamics course was taught for the first time to the entire Second Class from April 15, 1921 to June 4, 1921.[8]

A Department of Natural & Experimental Philosophy Lecture (1900), *USMA Photo Collection*

The goal of the Aerodynamics course, according to C.C. Carter, was to meet the Air Service's requirements for officers who were familiar with aviation technology. In a memorandum to the Superintendent he wrote: "This instruction is fundamental and is intended to cover only so much of the Theory of Flight and Types of Service Planes as should be a part of the education of every officer, whether or not he be assigned to the Air Service."[9] Douglas McArthur later noted in a memorandum to the Air Service that the "...instruction is made as practical as possible by using charts, graphs, models, results of wind tunnel tests, lantern slides, films, lectures and actual demonstrations of an airplane."[10] This focus on practical application remains a hallmark of the Aerospace Engineering curriculum at West Point to this day.

The Aerodynamics course was added to the end of the current Second Class (third year) curriculum, which was divided into The Slide Rule, Precision Measurement and Graphical Methods, Elementary Mechanics and the Properties of Matter, Wave Motion, Sound, Light, Technical Mechanics, Hydraulics, and General Astronomy. The 18 lessons for the course were: Lift (although at the time lift was referred to as "Sustentation") (2 lessons), Relations in Flight, Resistance, Thrust Required, Power Required, Power Available for the Air Propeller and the Air Plane Engine (2 lessons), Relation Between Power Required and Power Available, Climbing and Gliding, Airplane Performance at Different Altitudes, Single and Multiple Planes, Longitudinal Stability (2 lessons), Lateral Stability, Directional Stability, and two Guest Lectures. Each lesson consisted of a class period of 75 minutes with approximately two hours of out of class study.[11]

In order to teach the first iteration of the course effectively, West Point requested visual aids, specifically aircraft models, from the Air Service in March of 1921. Although "not made up with exactness of detail," a number of small scale aircraft models intended for an Air Service exhibit at the Hotel des Invalides in Paris, France, which were already prepared for shipment, were diverted to West Point.[12] The models included a Martin Model TT, a Thomas Morse S-4-C, a Curtiss JN-4-D, and a Loening Monoplane M-H. At the suggestion of the Air Service, a Messenger Plane was also flown to West Point and displayed for the latter portion of the course while the pilot was made available to present lectures to the cadets. In 1925, a complete MB-3A Boeing Pursuit Airplane was sent to West Point as a display. Despite attempts by the Quartermaster to put it in the corner of the Riding Hall, the airplane was placed on display in the only academic building with sufficient space, the Mineralogical Museum.

The Aerodynamics course taught by the Department of Natural & Experimental Philosophy was not the only contact cadets had with aviation technology. At the same time the Aerodynamics course was introduced, cadets were receiving training from many of the other departments on a variety of subjects related to aviation. In terms of academics, the Department of Drawing taught map reading, map making, and one lesson on aerial photography; the Department of Chemistry and Electricity provided 15 lessons on internal combustion engines and 16 lessons in Radio-telegraphy; and the Department of Ordnance taught such topics as machine guns, synchronizing gears, bombs, and bomb sights. In terms of military training, cadets were schooled in the tactics and logistics particular to the Air Service as well as exposed to actual aircraft during their summer military training. Starting in the summer of 1920, the Air Service made several airplanes and a captive balloon available to the cadets at Camp Dix, New Jersey and later at Mitchell Field on Long Island. During summer training, cadets were required to make at least one flight in an airplane and one ascent in a balloon.[10]

The first attempt to provide actual pilot training to cadets (as opposed to familiarization flights) was a 1921 proposal by the Air Service to train up to 30 cadets each year during their summer leave at Carlstrom Field in Arcadia, Florida. However, the War Department quickly denied that request and all of the requests which followed until 1941. In 1927, an Aviation Detachment of 3 officers and 3 enlisted men was established at West Point to allow aviators on the staff and faculty of the Military Academy to maintain their aviation proficiency. Since there was no room for an active runway on West Point, the Detachment used a series of amphibious aircraft which were launched from the Hudson River. The first aviation accident at West Point occurred in the fall of 1933 when a BT-2B airplane capsized in the Newburgh Bay after a rough landing.[13] Fortunately, no one was injured in the accident.

In early 1941, the prospect that the United States might become involved in another World War loomed large on the horizon. General Henry "Hap" Arnold, the Chief of the Army Air Corps, and the Superintendent of West Point, Brigadier General Robert Eichelberger, decided to investigate ways that aviation training at West Point could be improved. They decided that since the majority of officers who ultimately failed to complete flight training could be identified after only fifteen hours of instruction, providing the first segment of flight training to cadets at West Point would minimize the loss of time and money resulting from officers who failed and had to be retrained for another branch of the Army. On December 3, 1941, just four days before the Japanese attack on Pearl Harbor, the plan to provide a short flight training course to First Class cadets (seniors) who volunteered for the Air Service was approved.[14] After the United States entered World War II, General Arnold and General Eichelberger decided to provide full flight instruction to all interested cadets. The flight instruction would allow cadets to graduate with their pilot wings, proficient in flying advanced multi-engine aircraft. The plan, which was approved by the Secretary of War, required dividing the Corps of Cadets into "Air Cadets" and "Ground Cadets." The number of Air Cadets was established by the annual needs of the Air Service, not to exceed sixty percent of each cadet class. The first class of 256 Air Cadets began training in the fall of 1942. Flight training and other aviation subjects were added to the hectic cadet schedule by removing 800 hours of tactics training in such topics as dismounted drill, field artillery, military engineering, military intelligence, movement by motor/rail/water, and inspections.[15]

Significant problems began in October of 1942 when the Academy's curriculum was again reduced from four years to only three. Under the condensed three year program, the differences between the education provided to the Air Cadets and the Ground Cadets were significant. Air Cadets were excused from hundreds of hours of military and academic training, resulting in many officers questioning the benefits of such training and whether the Air Cadets were receiving the breadth of education required of a military officer. In August of 1945, a USMA Curriculum Committee recommended that the Air Cadet program be eliminated on the basis that the training interfered with the academic curriculum and was too specific to be consistent with the mission of the Academy.[15] The problem with providing specialized flight training at West Point was clearly stated by Brigadier General S.E. Tillman, a previous Superintendent of West Point, who wrote in 1919 that "...any attempt to make the Institution both a school of application and an Academy for the proper mental education and training of officers is sure to result in failure in both directions."[16] Even General Arnold, faced with strong opposition to the program after the conclusion of World War II, recommended that it be canceled. The last West Point class to be awarded pilot's wings graduated on June 4, 1946. Cadets received their wings in a ceremony on the day prior to graduation. In all, from 1943 to 1946 West Point graduated 1,031 cadets with wings.[15,17]

Air Cadet Training at Stewart Field, *USMA Photo Collection*

Air Cadets in Training at Stewart Field, *USMA Photo Collection*

Creation of the United States Air Force

The men with the knowledge, foresight and persistence to create the world's preeminent military air forces during the 35 years from the purchase of the first Wright Military Flyer to the strategic bombing campaigns of 1944 were for the most part products of the West Point engineering education and military development programs. Who were these exceptional leaders who also possessed the luck and skill not to perish in an airplane crash as so many of their colleagues did? A representative few of these dedicated innovators from West Point are portrayed below.

Brigadier General James Allen, USMA 1886, was the Chief of the Army Signal Corps from 1906 to 1913. General Allen ordered the establishment of the Signal Corps Aeronautical Division on 1 August 1907 and authorized the procurement of the first military airplanes and dirigibles.[18]

Colonel Samuel Reber, USMA 1886, became the first aviation commander of a U.S. military aviation unit when he was selected in September 1913 to command the U.S. Army Signal Corps Aviation Section.[18]

Major General Mason T. Patrick, USMA 1886, was Chief of the Army Air Service from 1921 to 1926 when it became the Army Air Corps, and he continued as Chief of the Army Air Corps until 1927. Patrick graduated second in his class and taught Engineering at West Point twice during the periods 1892-1895 and 1903-1906.[18]

Major General Frank P. Lahm, USMA 1901, continued to serve in the forefront of aviation and organized the Balloon Service in World War I. He commanded the first Army Air Service Training Center, established at San Antonio, Texas in August, 1920. The center included the primary and advanced flying schools and the School of Aviation Medicine. Lahm was appointed assistant to the Chief of the Army Air Corps with the rank of Brigadier General in July 1930. He received the Legion of Merit for his contribution to the development of Army Aviation during its important formative period and retired from the Army in November 1941.[5]

Major General Oscar Westover, USMA 1906, was a member of the West Point faculty twice from 1911-1913 and from 1916-1917, and Chief of the Army Air Corps from 1935 until his death in air accident in September, 1938. In 1920, Captain Westover transferred to the Air Service and rapidly became a highly skilled aviator through service as both a student and as an instructor at a series of schools: the Air Service Balloon School in 1921, the Air Service Airship School in 1922, the Air Service Primary Flying and Advanced Flying Schools in 1923, and the Air Corps Tactical School in 1927. Westover earned aeronautical ratings as a balloon observer, airship pilot, airplane pilot, and airplane observer. After serving as the assistant to the Chief of the Army Air Corps for four years, Westover was named Chief of the Air Corps and promoted to Major General on December 22, 1935. He used his position to travel the country advocating increased aircraft production and expanded pilot training for the war he knew would come. On September 21, 1938, General Westover lost his life in an airplane accident near the Lockheed plant at Burbank, California, when his plane burst into flames on landing.[19]

General Henry "Hap" Arnold. USMA 1907; Brigadier General, Assistant Chief of the Army Air Corps, 1936-38; Chief Army Air Corps 1938-41; Commanding General, Army Air Forces, 1941-46; Five-Star General of the Air Force, 1949. Hap Arnold's life paralleled the growth of American air power and Arnold personally contributed to a majority of the key milestones in air power growth throughout a military career that lasted 39 years. In 1949, the Congress appointed him to the permanent rank of Five-Star General of the Air Force. He is the first and only Air Force officer to hold that rank. He learned to fly at the Wright Brothers Flying School in Dayton, Ohio, earning his airplane pilot certificate on July 6, 1911. In October of 1912, Lieutenant Arnold was awarded the first Mackay Trophy for the most outstanding military flight

Henry H. Arnold as Five-Star General of the Air Force (Left) and as an Army Lieutenant in a Wright B Airplane in 1911 (Right), *U.S. Air Force Public Affairs*

of the year. He had flown an observation flight between Ft. Myer, Virginia; College Park, Maryland; and Washington, D.C. to pinpoint a troop concentration. He also set altitude records and accomplished the first successful spin recovery in an airplane that year. During World War I, Arnold was the Executive Officer for the Air Division of the War Department and spent most of the war overseeing aircraft production and development in the U.S. He contributed to many of the advances at that time such as oxygen masks, air-to-ground communications, automatic cameras, armored pilot seats, increased firepower and improved aeromedical research. During the interwar period, Arnold took advantage of every opportunity to promote the Air Service all across the country. Hap also continued to make contacts in the civilian aircraft industry that would prove very useful to him during the future buildup for World War II, and while assigned to the Materiel Division at Wright Field, he became acquainted with the Army Aviation research and development structure. While serving as the commanding officer of March Field in California, Hap initiated a lifelong relationship with Professor Theodore von Karman, the director of the Aeronautical Laboratory at the California Institute of Technology. Hap earned a second Mackay Trophy for his command of a roundtrip flight of ten new B-10 bombers from Washington, D.C. to Fairbanks, Alaska to demonstrate strategic bombing capability. Hap Arnold became Chief of the Army Air Corps upon the death of Major General Oscar Westover in September 1938. During World War II, Hap Arnold was the architect of all air activities against both Germany and Japan. He had triumphantly supervised the expansion of the Army Air Forces from 22,000 men and 3,900 planes to 2.5 million men and 75,000 aircraft during his time in command. Throughout his long career, Arnold remained an unrelenting advocate for better equipment, improved performance, and increased aircraft safety. During the third, fourth and the early fifth decades of the twentieth century, Hap Arnold knew the important aeronautics educators, industry leaders, and military leaders, and he was a crucial integrator of the Army's efforts to expedite the development of aviation. It is clear that no one was more closely tied to the development of the airplane and aeronautics in the United States and throughout the world than Hap Arnold.[20]

Carl Spaatz, USMA 1914; Commanding General 8th Air Force, 1942; Commanding General, Strategic Air Force European Theater, 1944-45; Commanding General, Army Air Forces, 1946-47; First Chief of Staff, U.S. Air Force, 1947-48. Carl Spaatz became a military pilot in 1916 and commanded the 31st Aero Squadron in France from 1917-18, where he shot down three German Aircraft. After serving in a wide range of aviation command and staff positions during the interwar years, Spaatz became Chief of the Air Staff, U.S. Army Air Corps under Hap Arnold in 1941. Recognizing his talents, Arnold sent Spaatz to command the 8th Air Force in England and Spaatz later directed the entire strategic bombing campaign against Germany as commander of the Strategic Air Force in Europe. In March of 1946, Carl Spaatz was selected as Chief of Army Air Forces, and in September 1947 he presided over the founding of the U.S. Air Force and served as the first Air Force Chief of Staff until his retirement in 1948.[21]

For each of the more well known Military Academy leaders in the development of aviation in this country, there were numerous lesser known officers whose contributions were also pivotal to the success of the U.S. military air forces. A review of the graduates of the Class of 1928 who branched into the Air Corps revealed that fourteen out of 261 graduates perished in aircraft accidents or in aerial warfare. Flying was clearly risky in the first four decades of flight and many graduates of the Academy paid the ultimate price to advance the cause of U.S. airpower. The crucial roles played by West Point graduates in the development of the U.S. Air Force are clear from the listings below of the graduates who served as either the Chief of Staff of the Air Force, the Superintendent of the U.S. Air Force Academy, or the Dean of the U.S. Air Force Academy.

Chief of Staff, United States Air Force

Carl Spaatz	USMA 1914	1947 – 1948
Hoyt S. Vandenberg	USMA 1923	1948 – 1953
Nathan F. Twining	USMA 1919	1953 – 1957
Thomas D. White	USMA 1920	1957 – 1962
John P. McConnell	USMA 1932	1965 – 1969
John D. Ryan	USMA 1938	1969 – 1973
George S. Brown	USMA 1941	1973 – 1974
(Chairman of the Joint Chiefs of Staff, 1974 -1978)		
Lew Allen, Jr.	USMA 1946	1978 – 1982
Charles A. Gabriel	USMA 1950	1982 – 1986
Michael J. Dugan	USMA 1958	1990

Superintendents of the United States Air Force Academy

Hubert P. Harmon	USMA 1915	1954 – 1956
James E. Briggs	USMA 1928	1956 – 1959
William S. Stone	USMA 1934	1959 – 1962
Robert H. Warren	USMA 1940	1962 – 1965
Thomas S. Moorman	USMA 1933	1965 – 1970
Albert P. Clark	USMA 1936	1970 – 1974
James R. Allen	USMA 1948	1974 – 1977
Kenneth L. Tallman	USMA 1946	1977 – 1981
Winfield W. Scott	USMA 1950	1983 – 1987
Charles R. Hamm	USMA 1956	1987 – 1991

Deans of the United States Air Force Academy

Don Z. Zimmerman	USMA 1929	1953 – 1955
Robert F. McDermott	USMA Jan 1943	1956 – 1968
William A. Orth	USMA 1954	1978 – 1983

Development of Army Aviation

After the creation of the Air Force, graduates of the United States Military Academy continued to shape and develop the role of Army Aviation. Three graduates in particular, James Gavin (USMA 1929), Hamilton H. Howze (USMA 1930), and Harry W.O. Kinnard (USMA 1939), saw the potential for aviation to revolutionize land warfare and led the way in transforming Army Aviation into the flexible, mobile, and lethal force it has been in modern conflicts such as Operations Desert Storm, Enduring Freedom and Iraqi Freedom.

The Army obtained its first helicopters, thirteen Bell YR-13s, in 1946. However, actual helicopter units were not organized, trained and fielded until 1952.[22] Despite the presence of helicopters during the Korean War, their use was limited and few officers grasped their true potential. One of these officers was James Gavin, an avid student of military history and the youngest Brigadier General since George Armstrong Custer. During World War II, Gavin led elements of the 82nd Airborne Division in a daring airborne drop behind enemy lines at Normandy on the eve of the D-day invasion and later became the commander of the 82nd Airborne Division. Gavin realized that the Army required a cavalry force, a force which could exploit a "mobility differential" between itself and other land forces to provide security and reconnaissance. Such a mobile force, like the horse-mounted cavalry of the American Civil War, did not exist in 1954. Gavin wrote "...with the motorization of the land forces, and the consequent removal of the mobility differential, the cavalry has ceased to exist in our Army except in name."[23] Gavin saw the mobility provided by aviation as the key to a new cavalry. Writing about the lack of cavalry and the poor performance of the U.S. Army in the Korean War, he wrote "If ever in the history of our armed forces there was a need for the cavalry arm--air-lifted in light planes, helicopters, and assault-type aircraft--this was it."[23]

Another officer who realized the full potential of aviation was General Hamilton H. Howze. General Howze had been a horse cavalryman as well as an airplane and helicopter pilot. As the Director of Army Aviation in 1955, he wrote that the "...army hadn't grasped at all, from its experience in Korea, the real utility of the light aircraft and what could be done by really integrating them into the tactics and combat support of the army."[24] In 1962, Secretary of Defense Robert McNamara ordered the creation of the U.S. Army Tactical Mobility Requirements Board to "...plan for implementing fresh and perhaps unorthodox concepts which will give us a significant increase in mobility."[24] Unhappy with the conservative approach being taken by the Army at the time, McNamara appointed General Howze and other supporters of air mobility to the Board. The Board reported its findings in August of 1962, directly resulting in the formation of the 11th Air Assault Division (Test).

The 11th Air Assault Division, commanded by Harry W.O. Kinnard, was tested at Fort Benning for three years. In a pun of Churchill's famous quote, Kinnard remarked that "Never have so few been observed by many so often."[24] As a result of Kinnard's drive and innovation, the "Skysoldiers" of the 11th Air Assault Division "...conclusively demonstrated that its elements could seek out the enemy over a wide area despite unfavorable weather conditions, find him, and then rapidly bring together the necessary firepower and troops to destroy him."[24] As a result of the success of the 11th Air Assault Division, the 1st Cavalry Division (Airmobile) was created in 1965 under Major General Kinnard and sent to Vietnam. In Vietnam, the 1st Cavalry Division demonstrated the tremendous capability of aviation to provide mobility in difficult terrain and to rapidly mass troops and firepower. During the Ia Drang Valley campaign in Vietnam in 1965, the 1st Cavalry Division conducted twenty-two infantry battalion and sixty-six artillery battery movements over distances as great as seventy-five miles.[24]

Thanks to the foresight of such pioneers as Gavin, Howze and Kinnard, Army Aviation developed into a modern cavalry, the "...arm of shock and firepower...the screen of time and information."[23]

Contributions to the U.S. Space Program

At 4:17 Eastern Daylight Time on July 20th, 1969, the entire world watched as the crew of Apollo 11 successfully completed the first manned lunar landing, one of the greatest achievements in human history. Of the three crew members of Apollo 11, two were graduates of the United States Military Academy. From the beginning of the Gemini program to the construction and operation of the International Space Station, the long gray line of USMA graduates has played an important role in the development and the success of the U.S. Space Program. Since the dawn of manned space flight, USMA graduates have logged a combined total of more than 5,450 hours in space.[25]

One of the clearest ways to see the great contributions made by West Point graduates to the development of the U.S. Space Program is to look at the history of the Apollo project. Edward H. White II, Class of 1952 and the first American to walk in space as part of Gemini 4, was killed on January 27, 1967 along with Gus Grissom and Roger Chaffee when a fire swept through their command module during a routine pre-flight test on the launch pad at the Kennedy Space Center. The mission of White, Grissom and Chaffe, later designated Apollo 1, was to be the first manned Apollo mission. The remains of Lieutenant Colonel White are interred at West Point. Frank Borman, Class of 1950 and a veteran of Gemini 7 and 8, commanded Apollo 8 and went on to become to the President of Eastern Air Lines, piloting the company through the deregulation of the airline industry. Edwin "Buzz" Aldrin, Class of 1951, became the second man to walk on the moon as part of Apollo 11 as Michael Collins, Class of 1952, piloted the lunar command module orbiting overhead. A veteran of Gemini 10 and the third American to walk in space, Michael Collins went on to become the first Director of the Smithsonian's National Air & Space Museum. Apollo 15, like Apollo 11, had two West Point graduates on the crew. David R. Scott, Class of 1954 and a veteran of Gemini 8 and Apollo 9, commanded the mission and walked on the moon while Alfred M. Worden, Class of 1955, piloted the command module. Donald H. Peterson, Class of 1955, served on the support crew for Apollo 16 and went on to fly aboard STS-6, the maiden flight of the Space Shuttle Challenger.[25]

However, the contribution of West Point graduates as astronauts did not stop with the Apollo project. The graduates who have served as NASA astronauts since the conclusion of Apollo are: Sherwood "Woody" Spring, Class of 1967; Richard M. Mullane, Class of 1967; James C. Adamson, Class of 1969; William S. McArthur Jr., Class of 1973; Richard Clifford, Class of 1974; Charles D. Gemar, Class of 1979; Patrick G. Forrester, Class of 1979; Jeffrey N. Williams, Class of 1980; Douglas H. Wheelock, Class of 1983; and Timothy L. Kopra, Class of 1985.

Four of these astronauts not only graduated from the USMA engineering program, but also returned to West Point to teach and to further develop the Aerospace Engineering curriculum. Frank Borman, James Adamson, William McArthur Jr. and Richard Clifford all served as Assistant Professors in the Department of Mechanics before being selected as astronauts, teaching such courses as Fluid Mechanics; Thermodynamics; Aerodynamics; and Aircraft Performance, Stability and Control.

Current Aerospace Engineering Program

Although West Point does not offer a major in Aerospace Engineering, it does offer an ABET accredited Mechanical Engineering major with a concentration in Aerospace Engineering. This program provides a solid foundation for graduates to study for a Master of Science or Ph.D. in Aerospace Engineering, as several of the current faculty members of the Department of Civil & Mechanical Engineering have earned. Many cadets are still attracted to the allure of aviation and Aerospace Engineering. Thirty to forty-five cadets major in the Aerospace track of Mechanical Engineering each year and more than 100 cadets or ten percent of every graduating class will become Army helicopter pilots within a year of graduation. Currently, USMA provides about thirty-five percent of the Army's new aviation officers each year. It is the belief at USMA that a broad undergraduate engineering educational experience in the multi-disciplinary aspects of Mechanical Engineering is best attuned to the needs of future Army leaders.

The four year educational experience is depicted in Table 2 below.

Academic Subjects	Credit hours
Mathematics	19.0
Chemistry	6.0
Physics	7.0
Engineering	54.5
Information Technology	3.0
Literature and Writing	9.0
Physical Geography	3.0
Philosophy and Law	6.5
History	12.0
Foreign Languages	7.0
Psychology and Leadership	6.0
Political Science and International Relations	7.0
Economics	3.5
Total	143.5

Table 2. Academic Subjects for a ME Major in 2003.

The quality of our engineering program has been recognized not only by our accrediting agencies, to include ABET, but also by *U.S. News & World Report*, which ranks the entire

Military Academy engineering program and the specific Mechanical Engineering program fourth and fifth respectively in the nation for programs not granting Ph.D.s. Our students have competed successfully in many different types of intercollegiate engineering design competitions, from the paper design of innovative aircraft in AIAA and American Helicopter Society (AHS) competitions to the design and actual manufacture of both piston and electrically powered aircraft for intercollegiate contests.

A unique aspect of our Aerospace Engineering program is the inclusion of flight laboratories in the curriculum. Since 1970, Army aviators assigned to the Department faculty have conducted several different flight laboratories with cadets in four Aerospace Engineering courses: Introduction to Aerodynamics, Aircraft Performance and Static Stability, Helicopter Aeronautics, and Aircraft Dynamics and Control. Currently, with four flying positions and two Cessna 182 aircraft assigned to the Department in addition to the use of UH-1 helicopters, cadets will fly a minimum of three two-hour flight laboratories to measure aircraft performance and stability parameters for comparison with theory.

These laboratories also function as orientation flights for cadets to get a real "seat of the pants" feel for the subjects discussed in the classroom. The aviators on the faculty attend a top graduate school for two years to earn a Master of Science in Aerospace Engineering prior to their West Point assignment. Their flight laboratory experience stimulates many of these aviators to become experimental test pilots. Over the last 33 years, 18 members of the Department of Civil & Mechanical Engineering faculty have attended the elite U.S. Naval Test Pilot School.

T-41 Aircraft Used for Flight Laboratories (1970 - 1989), *Department of Civil & Mechanical Engineering*

By all measures, the Military Academy has an exceptional engineering program to prepare cadets for careers in aviation as pilots, engineers or educators. As engineering and technology continue to rapidly advance, and as multidisciplinary education becomes the key to the integrated design of a multitude of disparate technologies, the United States Military Academy will continue to grow and to improve the education of its graduates so they can lead the Army and the Nation into an exciting future.

[1] Wright Patterson Air Force Base Museum, 2003.

[2] Annual Reunion of the Association of Graduates of the U.S. Military Academy, USMA Association of Graduates, 1909.

[3] Centennial History of the U.S. Military Academy, United States Military Academy, 1902.

[4] Tobin, J., To Conquer the Air: The Wright Brothers and the Great Race for Flight, Free Press, New York, 2003.

[5] Lahm, F. P., Biography of Frank Purdy Lahm, undated.

[6] Bane, Thurman H., memorandum to the Superintendent of the United States Military Academy, September 9, 1918, File Number 351.051, Aerodynamics Course (1918-1925), United States Military Academy Permanent Records.

[7] Menoher, Chas T., memorandum to the Superintendent of the United States Military Academy, August 30, 1920, File Number 351.051, Aerodynamics Course (1918-1925), United States Military Academy Permanent Records.

[8] McArthur, Douglas, memorandum, March 22, 1921, File Number 351.051, Aerodynamics Course (1918-1925), United States Military Academy Permanent Records.

[9] Carter, C. C., memorandum to the Superintendent of the United States Military Academy, March 11, 1925, File Number 351.051, Aerodynamics Course (1918-1925), United States Military Academy Permanent Records.

[10] McArthur, Douglas, memorandum to the Chief of the Army Air Service, April 19, 1921, File Number 351.051, Aerodynamics Course (1918-1925), United States Military Academy Permanent Records.

[11] Carter, C. C., memorandum to the Superintendent of the United States Military Academy, April 25, 1921, File Number 351.051, Aerodynamics Course (1918-1925), United States Military Academy Permanent Records.

[12] McArthur, Douglas, memorandum to the Chief of the Army Air Service, March 28, 1921, File Number 351.051, Aerodynamics Course (1918-1925), United States Military Academy Permanent Records.

[13] History of the Air Corps at West Point (1936-1941), United States Military Academy Archives, West Point, New York, undated.

[14] Benson, James T., Crash and Burn: The Failure of Flight Training at the United States Military Academy, United States Military Tactical Officer Education Program, 1999.

[15] Isabell, D., Johnson, J., Cate, A, Flying High: The Evolution of Flight Training and the Air Cadet Program at West Point During World War II, Department of History, United States Military Academy, West Point, New York 1990.

[16] Tillman, S. E., letter to the Chief of the Army Air Service, April 16, 1919, File Number 351.051, Aerodynamics Course (1918-1925), United States Military Academy Permanent Records.

[17] Memorandum from the Commandant of Cadets to the United States Military Academy Adjutant General, January 2, 1945, Air Corps Tactical Instruction (1944-1945), File Number 351.051, United States Military Academy Permanent Records.

[18] West Point Leaders of Flight – Mahan Hall, United States Military Academy Association of Graduates, 1988.

[19] U.S. Air Force Major General Oscar Westover, United States Air Force Biographies, 2003.

[20] U.S. Air Force General of the Air Force Henry H. Arnold, United States Air Force Biographies, 2003.

[21] U.S. Air Force General Carl A. Spaatz, United States Air Force Biographies, 2003.

[22] Weinert, Richard P., History of Army Aviation 1950-1962, Historical Office, Office of the Deputy Chief of Staff for Military Operations and Reserve Forces, U.S. Continental Army Command, 1971.

[23] Gavin, James M., *Cavalry, and I Don't Mean Horses*, Armor, Volume LXIII, May-June 1954, pgs 18-22.

[24] Stanton, Shelby L., Anatomy of a Division, The 1st Cav in Vietnam, Presidio Press, 1987.

[25] Astronaut Biographies, Astronaut Office, Flight Crew Operations Directorate, NASA, Lyndon B. Johnson Space Center.

Chapter 60

Genesis of the United States Naval Academy Aerospace Engineering Department

By

David F. Rogers
Aerospace Engineering Department
United States Naval Academy
Annapolis, Maryland 21402

Associate Fellow

Background

The official beginnings of the Aerospace Engineering Department occurred in 1964. However, initial efforts at changing the curriculum at the Academy began as early as the 1957–1958 academic year (Ref. 1). Prior to the late 1950s, the curriculum at the Academy was the same for all midshipmen. The only choice within the curriculum was what foreign language an individual midshipman took. A typical semester course load was 20 credit hours.

During the 1958–1959 academic year some additional flexibility was introduced in the form of electives (Ref. 1). For example, the 1961–1962 academic year catalog (Ref. 2) lists 11 electives in engineering, including two in aerodynamics — E803 Aerodynamics I and E804 Aerodynamics II. The first aerodynamics course covered topics such as lift and pressure distribution, boundary layer effects, subsonic compressibility and shock waves, as well as wind tunnel work. The second course added thin airfoil theory, sweepback and transonic effects, propeller analysis and airplane performance, as well as static stability. The text used was Dommasch, Sherby and Connolly (Ref. 3). These elective courses were either taken in lieu of other courses in the basic curriculum or over and above the required basic curriculum, i.e., as overloads.

In May 1959, an external curriculum review board chaired by Dr. Richard G. Folsom, President of Rensselaer Polytechnic Institute, was established to examine the curriculum changes already in place and to make recommendations for additional changes (Ref.4). In June 1959 the Faculty was reorganized into three departments — Naval Science, Social Sciences & Humanities and Science and Engineering. Interestingly enough, it is a little known fact that the Naval Academy teaching faculty has always been composed of approximately equal numbers of military officers and civilian professors. Also beginning with the 1959– 1960 academic year, entering midshipman were allowed to validate courses based on previous academic work and an examination.

The Beginnings

In 1961, LCDR Doc G. Faulkner, with an MS in AE from Princeton, along with a young civilian, Tom York, with an MS in AE from Penn State and LT Keith Nelson, with a BS in AE from the Naval Postgraduate School, joined the faculty and began to further develop and teach the aerodynamic courses. Tom York left in 1963 to pursue a PhD, while both Doc Faulkner and Keith Nelson remained through 1964.

Doc Faulkner was a naval aviator with a background in carrier aviation. Arriving at the Naval Academy, he was surprised to find an engineering department comprised of committees rather than a college of engineering with departments. The Second Class Committee taught the Second Classmen (juniors), the First Class Committee taught the First Classmen (seniors), etc. The major surprise, though, was that there was no Aeronautical Engineering Committee. Even though carrier aviation was critical to the Navy, aeronautical engineering was not being offered.

Initially Doc taught fluid mechanics and thermodynamics as a member of the Second Class Committee. However, he was really uncomfortable with the absence of a curriculum in aeronautical engineering. In due course, this led him to approach the Head of the Engineering Department, Capt Wayne Hoof, to make the case for an aeronautical engineering curriculum. Capt Hoof was very receptive and eager to bring engineering education at the Naval Academy up to date. From that point, Capt Hoof took the ball and ran with it. He sold the concept of a separate aeronautical engineering curriculum to the new Civilian Academic Dean, Dr. A. Bernard Drought, who had joined the Faculty in 1963, and to Rear Admiral Charles C. Kirkpatrick, the Superintendent of the Academy.

The timing of Doc Faulkner's request and Capt Hoof's efforts was auspicious, because in 1963 a major change in educational philosophy that had been percolating for some time surfaced. At that time, driven by the Folsom Report (Ref. 4) and internal reviews, the fundamental decision to change the thrust of midshipman education from traditional professional naval-oriented subjects to a more mainstream university style education was made. One could say that the Trade School on the Severn, as the Academy was sometimes known, was to become the Navy's undergraduate university. Understandably, the change began in engineering and in particular with aerospace engineering.

By the 1963–1964 academic year the number of electives in engineering had expanded to 15, with two additional aeronautical engineering courses in gas power propulsion. At that time the decision was made to form a separate Aerospace Engineering Committee as an initial step to establishing a separate curriculum for aerospace engineering. Capt Hoof now became busy with finding and hiring civilian faculty to support this new effort.

Throughout late 1963 and early 1964 Capt Hoof had been talking to Dr. Richard D. Mathieu, a Penn State PhD then working for General Electric, about joining the Naval Academy Faculty as Academic Chairman of the Aerospace Engineering Committee. In March/April of 1964 the author was shown an advertisement for an academic position at the Academy. At that time the author had completed all course work and was working on his doctoral dissertation in hypersonic aerodynamics at Rensselaer Polytechnic Institute under Dr. Henry Nagamatsu, one of Von Karman's students. The advertisement looked interesting, but where was Annapolis? Mailing a response to the advertisement on a Monday, the author was surprised to receive a telephone call from Capt Hoof the following Friday! Could he come to Annapolis for an interview — right away?

Arriving in Annapolis one afternoon about two weeks later, the author followed military precise instructions for the walk from his hotel to the home of Capt Hoof on the Academy grounds to attend a cocktail party that evening. Upon arriving at the destination, and knowing almost nothing about the military, the author found a large structure. Unfortunately, the only door appeared to be at the top of a rather long flight of stairs. Oh well, up the stairs and knock on the door — and knock on the door. Just as he was about the give up, a lovely lady opened the door; the author identified himself and said he was looking for Capt Hoof. At that point Mrs. Mary Hoof exclaimed "Oh, you came to the back door". You might say the author joined the Naval Academy faculty through the back door! The rest of the interview, both in the Hoof's quite impressive quarters and the next day in the engineering building, went quite well.

The author joined the newly formed Aerospace Engineering Committee on 1 August 1964 as an Assistant Professor. He was the first civilian faculty member specifically hired for the Aerospace Engineering Committee. Doc Faulkner extended his tour for a fourth year in order to work through the first academic year of the new committee. He served as the Military Chairman of the Committee. Assistant Professor Charles O. Heller, who had joined the engineering department the previous year, came over from the Third Class Committee to teach structures. Heller had both BS and MS degrees from Oklahoma State University. LT Keith Nelson, who had been teaching the aerodynamics courses, also joined the Committee.

Later in August Capt Vadym V. Utgoff joined the Committee as an Associate Professor. Vad Utgoff graduated from the Academy in 1939. He was commander of a squadron of PBY Black Cats in the South Pacific during WW II and a squadron of PBM Mariner seaplanes in the Korean conflict. He ended a distinguished naval career as the Commanding Officer of the Naval Air Facility at Sigonella, Sicily. Deciding that he wanted to teach, he had spent the previous year at MIT refreshing his 1948 BAE degree from the Naval Postgraduate School and his 1949 MIT masters degree. He is also the only man I know who sunk two PBYs on succeeding days, one during an attempted rescue of an American pilot and the second by hitting a log in Manila harbor on take-off.

Unfortunately, Dick Mathieu was unable to join the Committee as Academic Chairman until January 1965. For the remainder of the 1964–1965 academic year, Dick and Doc Faulkner jointly oversaw development of the major.

Perhaps a few words about the early administration organization are appropriate. Prior to and for the first few years after establishment of the Aerospace Engineering Committee, a dual administration structure was maintained. Each Committee/Department had both a civilian academic chairman and a military chairman. Each of the three Departments/Divisions had both a military head and a civilian senior professor who acted as an advisor to the military head on academic matters. Although it sounds cumbersome, it actually worked quite well. This structure was changed with the 1970–1971 academic year such that the chairmen of the now academic departments were directly responsible for all academic matters and reported directly to the Academic Dean. The various Divisions were headed by a Navy Captain, who was responsible for allocation of material resources. He also reported to the Academic Dean.

The Initial Curriculum

About a week after the author joined the Faculty, Doc Faulkner walked into the office and said "We need a curriculum. Can you have one next week?" Of course, the answer was yes. Of course, there were some ground rules. Specifically, the major courses were to be taken in addition to or as alternates to courses in the basic curriculum. To be certified as completing a major, a midshipman must have a cumulative QPR of at least 2.3 in the major and have no grade less than a C in any elective courses counted towards the major (Ref. 5). At that time the basic curriculum consisted of 160 credit hours, of which approximately 145 were academic credit hours (Ref. 6). A minimum of 18 additional credit hours was required for a major.

The basic curriculum contained one humanities course each of the eight semesters, e.g., composition, history, economics and government. Four semesters of a foreign language were required. A professional subject was also required in each semester, e.g., piloting and navigation, leadership, meteorology and naval operations. Mathematical preparation included three semesters of calculus and analytic geometry and a semester of differential equations. Engineering and science subjects in the basic curriculum included two semesters of physics (taken in the sophomore year), two semesters of chemistry, a semester each of engineering drawing and descriptive geometry, strength of materials, statics and engineering materials. In addition, four semesters of electrical science, two semesters of fluid mechanics and two semesters of thermodynamics were required. Thus, on paper, an acceptable base on which to build a major in aerospace engineering was in place.

Being somewhat 'wet behind the ears', the author needed a model for the curriculum. As a Rensselaer aeronautical engineering graduate, the Rensselaer catalog was an obvious choice. Hence, the first Naval Academy aerospace major is based on Rensselaer's. The result is shown in Table 1. Doc Faulkner and the author convinced Capt Hoof that revised Aerodynamics I and II courses could be substituted for Fluid Mechanics I and II in the basic curriculum and that a revised Gas Power Propulsion I course could substitute for Thermodynamics I in the basic curriculum. This bought 10 credit hours. An expanded statics course absorbed one of those credit hours. All together the major required 20 additional credit hours over and above the basic curriculum. The graduating class of 1967 was the first to fully benefit from the aerospace engineering major.

The Introduction to Aerodynamics course was developed and initially taught by the author and Doc Faulkner. The course was mathematically and physically based and set the level and tone for subsequent courses. The course was based on Principles of Aerodynamics by James H. Dwinnell of the Boeing Airplane Company (Ref. 7). The course covered Chapters 1–8, 10 and most of Chapter 14. Specifically, Newton's laws as applied to air, perfect fluids including Euler's and Bernoulli's equation, aeronautical nomenclature, experimental facilities, airfoil characteristics, finite wing effects including aspect ratio and planform effects, viscous phenomena, compressible phenomena, auxiliary lift devices and static aircraft performance were covered.

When the author first taught the revised Aerodynamics I course, it was based on Kuethe & Schetzer (Ref. 8) and included the kinematics and dynamics of fluid fields, the flow about a body, i.e., sources, sinks and vortices in a uniform parallel flow, the Kutta condition and Kutta–Joukowski airfoils, thin airfoil and finite wing theory and an introduction to viscous flow.

Aerodynamics II, taught from the author's notes, was revised to include more advanced material, including derivation of the Navier–Stokes equations, the exact solutions of the Navier–Stokes equations, derivation of the boundary layer approximation, solutions of the incompressible and compressible similar boundary layers, an introduction to turbulent flows, supersonic wing theory and hypersonic and high temperature flows.

Dr. Andrew Pouring, with BAE and MSAE degrees from Rensselaer and a D.E. from Yale, initially joined the faculty of the Second Class Committee in the Fall of 1964 and subsequently transferred to the Aerospace Engineering Committee in the summer of 1965. Andy taught the Gas Power Propulsion I and II courses. The first course covered the fundamentals of classical internal and external gasdynamics, including one-dimensional compressible subsonic and supersonic flows, flow in ducts including viscous and thermal effects, nozzle and diffuser theory, shock waves, Prandtl-Meyer flow and characteristic theory in nozzle design. However, the second course, based on Professor Joe Foa's book (Ref. 9), was rather unique in that, at an undergraduate level, it covered detonation and deflagration theory; one-dimensional nonsteady flows; characteristics and waves in nonsteady flows; shock tube theory; pressure exchange and combustion in nonsteady flows; and steady and nonsteady thrust generators.

The stability and control course, taught by Vad Utgoff and the author from Etkin's Dynamics of Flight book (Ref. 10), included both static and dynamic stability including the solution of the characteristic equation. The Structures I course, developed and taught by Charlie Heller, included an introduction to bending of thin flat plates and membrane stresses in pressure vessels, as well as an analysis of skin-stringer structural systems.

The Minors Program

During this early period a minors program in Aerospace Engineering was also offered. Basically, the program consisted of 4 or 5 specific courses in Aerospace Engineering, e.g., Introduction to Aerospace Engineering, Aerospace Performance, Aerospace Structures, Stability & Control and an elective. Students in the minors program took two semesters of fluid dynamics from the basic curriculum instead of the Aerodynamics I & II (see, for example, Ref. 11). The minors program was discontinued with the Class of 1970. After the Class of 1970, all midshipmen had to complete a major either in engineering, science or the humanities; some 15 different majors were offered (Ref. 12).

Accreditation

As Chairman of the Aerospace Engineering Committee, Dick Mathieu provided outstanding leadership and reasoned council both to the committee faculty and to the Academy administration. Among his many contributions is his early championing of efforts to achieve accreditation of the program by the Engineering Council for Professional Development (ECPD). This effort started in 1968. The first accreditation visit occurred on 9 and 10 March 1970. A full six year accreditation for the Aerospace Engineering curriculum was granted in August 1970, applicable to the Class of 1969. Prior to this time all graduates of the United States Naval Academy received a simple Bachelor of Science degree no matter what their field of study. However, with the Class of 1969 designated degrees in engineering where established and awarded, including aerospace engineering, mechanical engineering, electrical engineering and systems engineering. Full accreditation has been received on each succeeding visit. In 1969 Dick went on to become Senior Professor in the Engineering Department and eventually Director of Research and then Vice Academic Dean.

The Trident Scholar Program

Recognizing that faculty involvement in research and consulting was fundamental to a vibrant educational program, faculty were both encouraged and expected to develop areas of research and expertise within their disciplines. As part of this effort and to encourage a limited number of outstanding midshipmen to stretch themselves by conducting independent research, a Trident Scholars Program was started in the Spring of 1963. Midshipmen from among the top 10 per cent of their class were invited to participate. Each Trident Scholar applicant was required to select a faculty advisor and in consultation with their faculty advisor to submit a detailed proposal at the end of their Second Class (junior) year. Proposed research was expected to be at a masters degree level and in an area of expertise of the faculty advisor. The research and the required thesis were expected to constitute the major portion of their academic program during their First Class year. Consequently, Trident Scholars carried a reduced course load during their First Class (senior) year. Competition to be selected as a Trident Scholar was fierce. During that initial year, six midshipmen were selected as Trident Scholars. Over the years the Aerospace Engineering Department faculty have attracted and supported numerous Trident Scholars (see Table 2). The Trident Scholar thesis topics are indicative of the breadth and range of research interests of the department faculty.

Experimental Facilities

Initially aerospace engineering experimental facilities were minimal. When the program started in 1964, the principle available facility was a subsonic recirculating wind tunnel with a 30 x 42 inch test section (see Fig. 2). The tunnel, equipped with a pyramidal balance, was capable of speeds up to 225 mph. Located on the ground floor of Isherwood Hall in a 77 seat amphitheater, the tunnel was used principally for demonstrations. The initial design of the tunnel was started in 1957–1958 by LT Bruce Johnson in consultation with Wiley Sherwood of Aerolabs. Wiley designed and provided the balance and the propulsion system, and Bruce did the remainder of the design. The main portion of the tunnel was constructed in house by the engineering machine and wood shops. Initially this tunnel served the experimental needs of the aerodynamic courses. Experiments included tunnel calibration, including the classical turbulence sphere experiment, lift and drag on airfoils, the effect of auxiliary lift devices on finite wings, etc.

The lack of experimental facilities to support gasdynamics and supersonic flows was addressed by acquiring a 1 inch supersonic blow down wind tunnel from Wiley Sherwood at Aerolabs. The tunnel, acquired in the 1967–1968 time frame, was equipped with a Schleiren visualization system and was capable of pressure measurements on simple bodies, e.g., wedges. The tunnel was equipped with a variable nozzle which allowed Mach numbers from about 1.5 to 4. The tunnel, which cost about $6000, was self-contained and could be rolled into a classroom.

By the 1969–1970 academic year additional experimental facilities included a second subsonic wind tunnel with a test section of 38 x 54 inches, which was acquired by Dr. Bernard (Bud) H. Carson from government surplus when Fairchild Aviation decided to cease wind tunnel operation. Bud, with a PhD from Penn State, joined the faculty in 1967 and was to be a mainstay of the faculty for more than three decades. He found the tunnel on some government surplus equipment list, filled out the appropriate paperwork and submitted it. Nothing was heard for quite some time, and he had basically forgotten about it, i.e., until one day a small fleet of flatbed trucks arrived at the Academy gate wanting to know where he wanted to put the tunnel! A quick decision was made to temporarily store the tunnel in a field across the Severn River. In due course the tunnel and the six component balance were refurbished, with money provided by Howard Law, the Director of Navy Laboratories, and installed on the ground floor of Griffin Hall, one of the three buildings in the Engineering Complex. The tunnel, which was designed and manufactured by Aerolabs of Pasadena California, is still in use today.

A small helium tunnel capable of Mach numbers of 4 to 10 provided for the study of hypersonic flows. The study of nonsteady flows was provided by a 1 x 3 inch shock tube. Both were designed by Andy Pouring. A variable speed water table used for the simulation of low- and high-speed flows was also available. Three small classroom demonstration wind tunnels and a smoke tunnel were also available to illustrate aerodynamic characteristics.

Computers

When the author arrived at the Academy in 1964, the principal computer available to faculty was a centrally located IBM 1620. Of course, it read cards. It did, however, have a Fortran compiler! By the 1967–1968 academic year the Academy was renting timesharing services from a Washington, DC based company. Every academic department had access to at least one timesharing terminal.

In teaching viscous flow, the author found students had a very difficult time grasping the concept of guessing and iterating on the initial condition for the shear stress in order to satisfy the asymptotic free stream boundary condition for the similar boundary layer solutions. A BASIC computer program was quickly written to perform the solutions on the timeshare computer. Students were then brought to the teletype terminal located in a separate room in an adjacent building and the program run while students estimated the new initial conditions.

Well, this resulted in about three students really participating in the discussion while the other 15 skylarked. The teletype needed to be moved into the classroom and the results made visible to the entire class. However, that required stringing a telephone cable to the classroom in the adjacent building. Being somewhat impatient, the author came in one Saturday morning and simply strung the cable. On Monday, the administration was somewhat unenthusiastic about the author's actions. They were somewhat more concerned that he might have fallen two stories off a roof. The author, however, was concerned about breaking the adjacent window if he missed the open one with the bow and arrow he used to shoot the cable messenger across the open space. This was the first remote terminal classroom in the Isherwood complex.

With the cable run, the teletype was mounted on wheels and rolled to the classroom. A borrowed video camera focused on the teletype paper and a monitor allowed all the students to see the results and participate in the discussion (see Fig. 1). It also saved class time (see Ref. 13). Once the correct initial condition was found, an overhead projector was used to display graphs of the boundary layer velocity profiles, etc. Almost immediately the students challenged the author to have the computer plot the results. That took a bit of time but in due course was accomplished. This simple beginning in computer graphics was to be the genesis of the Engineering Departments efforts in computer aided design and interactive graphics (CAD/IG) as well as computer aided manufacturing. Within a short period the beginnings of a CADIG group in the form of a single programmer and the author as the faculty advisor was established.

As early as 1966 the aerospace engineering faculty were engaged in computer assisted education projects. Dick Mathieu, Don Mathews and Dick Goodspeed, as well as the author, wrote extensive programs for aircraft design, stability and control, and aircraft performance (see Refs 14-19). Programs such as simulation of a wind tunnel, aircraft level flight power required, weight and balance, stick free static stability calculations, etc. were developed. LCDR Barry Gastrock developed programs for experimental wind tunnel data reduction that were used as the experiment progressed (see Fig. 3 and Ref. 16). Robert W. Werlwas, who joined the faculty in 1967, developed programs to support an orbital mechanics course. Charlie Heller developed programs to support structural analysis, e.g., determination of the wall thickness of a rocket booster. Although in the context of programs that are available today, at the time these programs were on the cutting edge of computer utilization in college level education.

The 1970s

Reorganization

The 1970s were both busy and exciting, beginning with a further reorganization of the academic side of the Academy. Five divisions, which roughly correspond to schools in a typical university, were created: Engineering and Weapons, English and History, Mathematics and Science, Naval Command and Management, and United States and International Studies, each headed by a Navy Captain with the title of Director. Within each Division were academic departments, each typically chaired by a civilian member of the faculty (Ref. 20). Within the Division of Engineering and Weapons were the Departments of Aerospace, Electrical, Mechanical, Naval Systems (with naval architecture, ocean and marine engineering majors) and Weapons and Systems Engineering Departments. It was at this time that the formal dual military and civilian chain of command was discontinued within the academic program.

Although the original reorganization indicated that the department chairman was to be a civilian, in fact, this was modified very shortly to allow the faculty within a department to choose the chairman subject to confirmation by the Academic Dean. As Table 3 shows, the Aerospace Engineering Department most frequently chose a military chairman. The naval officer faculty have always contributed a wealth of administrative ability along with solid academic credentials in the form of at least a masters degree in aerospace engineering. The department, with a small civilian faculty, has always been very fortunate to have outstanding naval officers willing to step up to the plate and perform the many administrative tasks that devolve onto a department chair while giving due consideration to the civilian faculty's advice on academic matters. Graduates of the department have returned and accepted the department chair position: specifically, Paul Schlein, Bill McCracken and Ken Wallace.

During the 1970–1971 academic year, the Aerospace Engineering Department had a total of 17 faculty — 6 civilians and 11 military. A total of 73 midshipmen graduated with an accredited degree in aerospace engineering (see Table 4).

Flight Test Course

On June 9 1970 CDR Chuck Gerhan and Professors Bud Carson and Vad Utgoff journeyed to Wichita, Kansas to pick up a new 1970 F33A Beech Bonanza that the department was to lease from Beech Aircraft to support an elective course in flight test engineering (see Fig. 2). The flight test course was run for 15 years, supporting up to 24 midshipmen a year. It was discontinued at the end of the lease in 1985. Many of the students in the course were inspired to consider flight test as a career choice and were successful in garnering a position at the Naval Test Pilot School after becoming Naval Aviators and completing a couple of squadron tours. One of those students is now the chief test pilot at a major business aircraft manufacturer. In addition to supporting the flight test engineering course, every student in the introductory course enjoyed a demonstration flight. Furthermore, every student in the Stability and Control course received a demonstration flight illustrating the concepts of static and dynamic stability, including the phugoid and dutch roll modes. Finally, the course was reinstituted in January 1998 upon the urging of Professor Rogers and CDR Wallace. A rental A-36 Beechcraft Bonanza (see Fig. 3) is used as the flight test aircraft (Ref. 21). Currently, to contain costs, the use of the aircraft is limited to 12 students in the flight test course. Professor Rogers, who holds an ATP pilot certificate, is currently the chief pilot for the course. Capt Niewoehner, the Navy's former lead test pilot on the F/A-18E/F, shares the flying responsibilities.

769

The Dartmouth Time Sharing System

In December 1970, under the leadership of Admiral James Calvert, the Academy purchased a Honeywell 635 computer to support the growing educational requirement for timesharing computing. At the same time, the Academy entered into an agreement with Dartmouth College to provide the operating system and initial support services. This decision was to prove pivotal in really launching the Academy into modern computing.

The author immediately expanded the use of timesharing in both teaching viscous flow and in computer graphics. Shortly, advanced plotting teletypes (Typographs) and even CRT (Cathode Ray Tube) alphanumeric and storage tube-based terminals were acquired and in use. LCDR Barry Gastrock expanded the work in laboratory data acquisition. Dick Mathieu continued his work in simulating wind tunnel experiments and LCDRs Don Mathews and Dick Goodspeed expanded their previous work in aircraft performance calculations (Ref. 16). All the Aerospace Engineering faculty took advantage of the availability of onsite, immediate access to computer facilities to enhance their teaching and research.

In 1969 the Aerospace Engineering Department introduced the concept of externally funded and sponsored professorships to the Academy when Professor Carson was named as NAVAIR Research Professor. The author was named as ONR Research Professor in 1972. A long line of visiting and research professors has followed in these footsteps (see Table 5).

During that same time period Professor Mathieu spent two years (1971 and 1972) in London, England as an ONR Liaison Scientist, followed immediately by Professor Carson in 1973. During the 1977–1978 academic year the author was attached to ONR London while he spent the that academic year at University College London studying naval architecture with the Royal Corp of Naval Constructors.

The Aerospace Engineering Department also introduced the concept of immediate graduate education to the Academy. In 1967–1968 several graduates of the aerospace engineering program were sent to graduate school at North Carolina State immediately after graduating from the Academy. Today, the Academy has a significant immediate graduation program, including participation in the Rhodes, Guggenheim, Burk, etc. scholarship programs.

The New Building

By 1973 it became obvious that the existing experimental facilities were inadequate. The Aerospace Engineering Faculty, in particular Andy Pouring, Bud Carson and the author, wanted real supersonic and transonic wind tunnels. With the increasing numbers of students (see Table 4), a single subsonic wind tunnel was not adequate; at least two were needed for scheduling. The quality of the tunnels and the balance systems also needed improving. The structures program needed additional laboratory space and equipment. Prof. Vad Utgoff had become very interested in rotorcraft to the extent that he was designing a personal rotorcraft using flexible rotor blades. A propulsion facility that could handle running piston, turbojet and ramjet engines was needed. The author wanted a small motion-based variable stability flight simulator. In addition, Prof. Bruce Johnson, who had rejoined the faculty in the Naval Systems Engineering Department, was talking about a large high speed towing tank (380 feet and 50 feet per second). The computer aided design/interactive graphics effort as well as the computer aided education efforts were expanding. The author was talking about additional equipment and room to house it as well as additional staff.

After a visit in 1972 to Pierre Bezier's laboratory outside Paris, the author was proposing a significant computer aided manufacturing effort to support both the towing tank and the wind tunnel model making efforts. Other departments also wanted additional facilities. Serious consideration was given to modifying and remodeling the existing engineering complex to accommodate these new facilities and the space to house them. In due course it became obvious that the space requirement, especially those for the proposed towing tank and the aerospace engineering laboratories, made the cost of remodeling prohibitive; a new building would actually be cheaper.

Along with a new building went a collateral equipment list. Everything from chairs to supersonic wind tunnels went on the collateral equipment list. The faculty was intimately involved in every aspect of the design of the building — a very busy and intense time that occupied nearly three years. The new building, called Rickover Hall, was occupied in April 1975. There was a new 6 x 6 inch blowdown supersonic wind tunnel (see Fig. 4), a new 6 x 6 inch blowdown transonic wind tunnel; both the Fairchild and original Isherwood subsonic wind tunnels had been refurbished and moved to the new building. A large structures laboratory with a large steel frame was completed; an engine propulsion laboratory and a large rotor test stand with a sliding roof were included. The motion-based flight simulator was purchased from Singer-Link and sent to the naval laboratory at Warminster, PA to be modified for variable stability. The open jet wind tunnel used in the Shapiro fluids films had been replicated and installed in a classroom for use in demonstrations in the introductory course. The 380 foot towing tank and companion 120 foot tank were also installed on the laboratory deck of Rickover Hall.

Computer Aided Design

The collateral equipment list also provided the author an opportunity to expand the computer aided design, interactive graphics and computer aided manufacturing capability. Some 7254 square feet of classroom, office and laboratory space including raised computer flooring and special air conditioning, were included in the building. State of the art computer graphics equipment, including a PDP-11/45 mini-computer, an Evans and Sutherland (E&S) Picture System 1 (serial number 9) refresh calligraphic real time display system and a Xynetics 5 x 8 foot flatbed plotter, provided a research capability (see Fig. 5). These systems were used for some of the first research into real time interactive graphical manipulation of B-spline (NURBS) curves and surfaces (see Refs. 24 and 25). To support computer aided manufacturing, the author specified a large Pratt & Whitney TriMac milling machine for the engineering shop complex. Direct output from the PDP 11/45 and the E&S Picture System was used to generate towing tank and wind tunnel models on the TriMac (see Ref. 26).

While the author was at University College London (UCL) in 1977, he and David Chalmers of the RCNC developed a program to graphically interactively load and analyze the moment, shear and bending on a uniform beam. David Chalmers developed the analysis part of the program, and the author developed the graphical user interface. The program was used for instruction of masters students in the RCNC program at UCL. Upon return to USNA the author and Steve Satterfield from the CADIG Group implemented the graphical user interface on the PS1. Eventually this effort led to a number of interactive real time educational programs on the PS1 and its successors, the E&S PS300s. These included a program to show the motions of both the longitudinal and lateral dynamic stability modes for an aircraft. In addition, in collaboration with Robert Siddon of the physics department, a program was developed to interactively graphically place sources, sinks, vortices, doublets and uniform flows in a fluid medium. By touching a spot on the screen, a streamline (or potential) line was drawn through that spot. Grids of streamlines or potential lines could also be specified. This program was used in teaching aerodynamics and fluid dynamics.

In addition to the PDP 11/45 and the E&S PS1, 50 of the newly available Tektronix 4051 computer systems where acquired. The Tektronix 4051 is one of the first personal computers available for scientists and engineers. It quite literally changed the face of scientific and engineering computing. Nearly 40 of these systems were installed in two classrooms to provide graphical stand-alone and timesharing access for midshipmen. The rest were used in laboratories and faculty offices. Many of the early computer assisted education programs initially developed were moved to these systems and improved and extended. One of the systems was very cleverly interfaced to the CNC (computer numerical control) computer in the TriMac CNC milling machine by Francisco Rodriguez of the CADIG Group to provide direct control of the milling machine when creating towing tank and wind tunnel models. The accompanying program, running on the 4051, was able to take raw three-dimensional coordinates produced by the E&S PS1 and generate the necessary commands to drive the milling machine (see Ref. 26).

One of the interesting specifications that the author included in the collateral equipment list was that every office, classroom and laboratory contain a computer access point. The twisted pair wires for each of these access points were routed to a manual patch panel in a central location within CADIG. Thus, even a limited number of computer access lines could provide computer services to any location within the building. This foresight proved especially valuable when the entire building was recently wired for Ethernet access.

Sir Frank Whittle

Over the years the Aerospace Engineering Department has hosted many visitors. Perhaps the most distinguished was Sir Frank Whittle, the inventor of the turbojet engine. As usual, Sir Frank's visit came about rather serendipitously. Some time during the Spring of 1976, Professor Andy Pouring visited the Naval Air Propulsion Laboratory in New Jersey. During a tour of the facility, one of the staff members happened to mention that Sir Frank Whittle had visited just previously. Andy literally burst out with an exclamation "You mean that he is still alive!" (Later, Frank told Andy that everyone says that!).

Shortly afterwards, Andy contacted Sir Frank with a phone number provided by the British Embassy and invited him to give a distinguished visitor lecture sponsored by the Naval Air Systems Command. The following Fall, while chatting after the lecture, Andy asked Sir Frank what his future plans were. Sir Frank said he planned to marry a United States Navy Nurse (Tommy) that he met many years ago and live in the States. Andy then asked: "What would you do if invited to serve as Naval Air Research Professor in the Aerospace Engineering Department?" He immediately responded "Why, I would accept it of course". Sir Frank's association with the Academy continued for a number of years after the initial one year NAVAIR appointment was concluded.

Andy had the good fortune to be the only one to ever teach with Sir Frank. They jointly taught a course in propulsion, centered mostly on jet propulsion, during the academic year 1977–1978. Andy found it quite a challenge to sit in Sir Frank's class at 9am, amplify his detailed notes prepared just in time for his class and present them himself at 11am. Those notes were later incorporated into a United States Naval Academy Engineering and Weapons report published in June 1979 (Ref. 22). These notes were subsequently used in the propulsion course. The notes were extensively amended and revised and submitted to Pergamon Press and published as "Gas Turbine Aero-Thermodynamics" in 1981 (Ref. 23).

During the Spring of 1978 an historic first meeting between Sir Frank and Dr. Hans J.P. von Ohain (Ref. 27), who nearly simultaneously invented the jet engine, took place at a conference organized by Professor Pouring at the Academy (see Fig. 6). The two gentleman became good friends and freely acknowledged each others contributions.

Sir Frank served as advisor for several research students. It was a pleasure to see his steady hand draw detailed diagrams and conduct precise analytical work in his study of low speed drag reduction. Many of the models he designed and tested are still available and occasionally used in further student research. His friendly smile, cheerful disposition and contribution to the Aerospace Engineering Department are still remembered.

The 1980s and the Astronautics Program

In the early 1980s it became obvious that the Navy was increasingly interested in space. With experience on Saturn V, the Lunar Rover and Skylab, Professor Bill Bagaria was the first within the department to recognize this trend. By the Fall of 1983 he, along with CDR Bill McCracken, then chairman of the department, had convinced the Faculty that the time had come to revise the curriculum to incorporate more astronautics courses (see Ref. 28). In addition to internal discussions, ideas were sought from a number of outside government and university sources. A number of individuals from NASA Goddard Space Flight Center and the Johns Hopkins University Applied Physics Laboratory made significant contributions. Very early in the discussions three fundamental decisions were made: first, that there would be only one degree program with options or tracks in aeronautical engineering and astronautical engineering; second, that as much commonality between the tracks would be maintained as possible; and third, that the objective of the astronautics track was to prepare midshipmen to perform a spacecraft vehicle design just as the objective of the aeronautics track was to prepare midshipmen to perform an aircraft vehicle design.

The development of courses for the astronautics program began during the 1984–1985 academic year. As with the original aeronautical program, the first courses were offered as electives and overloads. Because of the lack of experience among department faculty, the first courses were offered using outside help. On an informal basis, the Applied Physics Laboratory provided staff to initially teach Spacecraft Systems, Space Environment and Spacecraft Design. NASA Goddard Space Flight Center also informally provided a number of visiting professors from its staff. During the 1985–1986 academic year the Naval Space Command created a research professor chair, which they continue to support and fund. Goddard Space Flight Center provided two visiting faculty members during the Spring semester. In May 1986 the Academy administration approved the astronautics track, and ABET accredited the program in July of 1988.

The astronautics track is supported by excellent laboratory facilities, including a satellite ground station serviced by a 10 meter dish acquired from surplus by Professor Bagaria (see Fig. 7), and an orbital mechanics simulation laboratory. The ground station is used to actively communicate with satellites, including a successful launch on 30 September 2001 of the first small satellite (PC-SAT) designed and constructed by midshipmen. After more than a year it is still operational. The ground tracking station is used to track Naval Academy ocean sailing vessels and Yard Patrol Craft during midshipmen summer cruises using GPS and satellite communication. The orbital mechanics laboratory simulates spacecraft rendezvous, interplanetary trajectories and spacecraft ground track problems.

Today the Navy is the biggest user of space, and interest in space within the Navy has continued to increase. Command and control, navigation for aircraft, ships and weapons, communication both tactical and for personnel moral on long deployments, to mention only a few uses, all depend on space. Perhaps the Navy's interest in space is best summed up by a recent remark attributed to Vice Admiral Dick Mayo: "Ain't no fiber off the fantail".

Endowed Chairs

At the end of the Christmas break at the start of the new millennium, we learned that the Aerospace Engineering Department had received a wonderful present —an anonymous donor had endowed The Robert A. Heinlein Chair in Aerospace Engineering. Robert Heinlein was a graduate of the Naval Academy, Class of 1929, and a well known science fiction author. He served in the Navy until 1934, when his career was cut short by tuberculosis. He began his writing career in 1939 with a short story submitted to Astounding Science Fiction. He wrote such classics as Space Cadet, Starship Trooper and Stranger in a Strange Land. We were fortunate to have Professor Vincent L. Pisacane join the Faculty in the Summer of 2002 as the Robert A. Heinlein Chair.

In April of 2000 Kevin W. Sharer, USNA Class of 1970, and a graduate of the Aerospace Engineering Department, endowed the David F. Rogers Chair in Aerospace Engineering. Kevin Sharer is currently Chairman of the Board and Chief Executive Officer of Amgen, one of the nations largest biotechnology companies. After graduation from the Academy, Sharer attended the Naval Postgraduate School and received a masters degree in aeronautical engineering in 1971. Subsequently he served aboard two nuclear attack submarines.

Our Graduates

The measure of the success and contributions of a program are embodied in the graduates of that program. Untold numbers of graduates of the Naval Academy and of the Aerospace Engineering Department have gone on to become Naval Aviators, both pilots and naval flight officers, squadron commanders, test pilots, flight test engineers and even Admirals. As an example, the Naval Academy has produced 54 astronauts, more than any other individual institution in the western world. Of those 54 astronauts, 15 graduated from the Academy before the first aerospace engineering major graduated from the early curriculum. Of the remaining 39 Naval Academy graduates, 16 are graduates of the Aerospace Engineering Department (see Table 6), more than any other department in the country. One of those graduates, Mike Smith, Class of 1967, died in the Challenger accident. Others have gone on to hold high positions within NASA. For example, Bill Readdy, Class of 1974, is now NASA Associate Administrator for Space Flight, and Dave Leestma is now Deputy Director of Engineering at JSC in Houston, to mention only two. In addition, the current Naval Academy Superintendent, Vice Admiral Richard J. McNaughton, Class of 1968, graduated with a minor in aerospace engineering.

Part of the Mission of the United States Naval Academy states ".....to assume the highest responsibilities of command, citizenship and government." In keeping with the mission, our graduates who leave the Naval Service make significant contributions in civilian life. For example, Kevin Sharer, Class of 1970, is Chairman and CEO of Amgen as mentioned above, Mike Bangert, Class of 1970, is a senior pilot for a major airline, as are a number of other graduates; Ken Reightler, Class of 1973, is a Vice President at Lockheed Martin Corp; Joe Sweeney, Class of 1974 is a test pilot with Lockheed Martin Tactical Aircraft in Ft Worth and was Lockheed's Chief Test Pilot on the Joint Strike Fighter competition; Steve Oswald, Class of 1973, is Vice President for the Space Shuttle at Boeing Corporation, to mention only a few. Obviously, the Aerospace Engineering Department has contributed to the fulfillment of the academy's mission.

Acknowledgements

A number of people contributed to this paper including Bill Bagaria, Bud Carson, Doc Faulkner, Charlie Heller, Dick Mathieu, Keith Nelson and Andy Pouring. Richard Davis, the Naval Academy Registrar, was particularly helpful in digging out some of the numbers as was Glenn Gotschalk of institutional research. Admiral Bob McNitt suggested that I look at the Yarbro and Folsom reports for some of the early history. Robin Kirkendahl collected and formatted much of the detailed data. My wife Nancy performed yeoman duty as copyeditor. I have tried to be as accurate as possible. However, I am sure that there are still small inaccuracies. Those are mine.

References

1. Yarbro, J.D., USNA Curriculum Development Report, October 1966, revised January 1974, USNA Nimitz Library Archives.

2. United States Naval Academy 1961–1962 Catalog of Information.

3. Dommasch, Daniel O., Sherby, Sydney S. and Connolly, T., *Airplane Aerodynamics 3/e*, Pitman Publishing Corporation, New York, 1967.

4. Folsom, Richard G., Report of the Curriculum Review Board of the United States Naval Academy, 1959, available from the United States Naval Academy Archives.

5. United States Naval Academy 1964–1965 Catalog of Courses.

6. United States Naval Academy 1963–1964 Catalog of Information.

7. Dwinnell, James H., *Principles of Aerodynamics*, McGraw–Hill Book Company, New York, 1949.

8. Kuethe, A.M. and Schetzer, J.D., *Foundations of Aerodynamics*, John Wiley & Sons, New York, 1950.

9. Foa, J.V., *Elements of Flight Propulsion*, John Wiley & Sons, New York, 1960.

10. Etkin, Bernard, *Dynamics of Flight, Stability and Control*, John Wiley & Sons, New York, 1959.

11. United States Naval Academy Catalog 1965–1966.

12. United States Naval Academy 1969–1970 Catalog of Information.

13. Rogers, D.F. and Adams, J.A., Computer Aided Analysis in Aerospace Engineering, Proceedings of the XXIst International Astronautical Congress, pp.1045–1061, North-Holland, 1971.

14. Mathieu, R.D., Introduction of Computer-Aided Instruction into an Aerospace Engineering Curriculum, Fifth Space Congress, Cocoa Beach, Florida, March 11–14, 1968.

15. United States Naval Academy Feasibility and Appreciation Study of Remote Terminal On-line Computing in Education, Sponsored by the Office of Naval Research, October 1966, Academic Computing Center, United States Naval Academy, Annapolis, MD, 21402.

16. Mathews, D.W. and Goodspeed, R.E., The Introduction of the Remote Teletype Computer Terminal into the Engineering Classroom, United States Naval Academy Engineering Report E-68-12, September 1968.

17. Mathieu, R.D., Computer-Aided Education at the United States Naval Academy, presented at Sixteenth Annual Conference on Readying the Gifted for the Coming World of the Computer, Pennsylvania Association for the Study and Education of the Mentally Gifted, Hershey, PA April 28–29, 1967.

18. Report on Spot Testing of Computer-Assisted Education, Bureau of Naval Personnel TDP 43-03X Education and Training Development Project, Appendix F, Report No. PR0767-3, December 1967.

19. Conord, A.E., Report on The implementation of Computer Assistance in Education, Report No. PR-0169-14, Academic Computing Center, United States Naval Academy, Annapolis, MD 21402, September 1968.

20. United States Naval Academy Catalog 1970–1971

21. Rogers, David F., An Engineering Flight Test Course Emphasizing Flight Mechanics Concepts, AIAA JAC, Vol. 39, No. 1, pp. 79–83, 2002.

22. Whittle, Sir Frank, USNA Engineering and Weapons Report EW-7-79, June 1979.

23. Whittle, Sir Frank, Gas *Turbine Aero-thermodynamics, With Special Reference to Aircraft Propulsion*, Pergamon Press, New York,1981.

24. Rogers, D.F., B-Spline Curves and Surfaces for Ship Hull Design, Proceedings of SNAME, SCAHD 77, First International Symposium on Computer Aided Hull Surface Definition, 26–27 September 1977, Annapolis, MD.

25. Rogers, D.F. and Satterfield, S.G., B-Spline Surfaces for Ship Hull Design, SIGGRAPH '80, 14–18 July 1980, Seattle, WA.

26. Rodriguez, F.A. and Rogers, D.F., Interactive Graphics and the DNC Production of Complex Three Dimensional Shapes, Proceedings of NCGA Conference, 14–18 June, 1981, Baltimore, MD, pp. 153–159.

27. Private communication A.A. Pouring, October 2002. 28. Bagaria, William J., Undergraduate Astronautics At the United States Naval Academy, Engineering Education, pp. 386–388, Vol. 81, No. 3, April 1991.

Chapter 61

Falcons Soaring: USAFA Department of Aeronautics Contributions to Aerospace Power During a Century of Manned Flight

Col D. Neal Barlow*, Scott R. Nowlin[†,] LtCol Dave Bossert[#]

Department Of Aeronautics, United States Air Force Academy

ABSTRACT

The Department of Aeronautics (DFAN), in existence since the organization of the United States Air Force Academy (USAFA) in 1955, has offered an accredited Aeronautical Engineering degree since 1967. DFAN has remained dedicated first and foremost to USAFA's unique and primary mission statement: *To inspire and develop outstanding young men and women to become Air Force officers with knowledge, character, and discipline; motivated to lead our Air Force and nation.* In the course of fulfilling this mission, cadets and faculty serving in DFAN have consistently achieved exemplary results in their military service, academic scholarship, and research during and after their USAFA attendance. This invited paper will present noteworthy contributions made by DFAN to the military and civilian aerospace community during the latter part of the century since the Wright Brother's flight in December 1903, emphasizing the current state of the department.

INTRODUCTION

"Capable of flying at tremendous speeds and executing astonishingly agile maneuvers, the Peregrine Falcon is a true ruler of the air, fully worthy of its fame as a raptor of unequaled predator abilities."[1]

"Noble birds, magnificent hunters, some of the fastest of all fliers – falcons are universally admired... in power diving from great heights to strike prey, the Peregrine may possibly reach 200 mph."[2]

A better mascot couldn't have been chosen for the men and women attending and serving at the United States Air Force Academy. With a sharp vision for integrity, service, and excellence, USAFA faculty and graduates have high achievements in all facets of our nation's military and civilian life. Cadets (officer candidates, formerly civilians or airmen) are challenged and stretched in their knowledge and intellect with an accredited curriculum, an intense leadership laboratory environment, a comprehensive peer-driven character development program, and strenuous physical training. They are exposed to all facets of our nation's world-class air and space force. Many taste the freedom and discipline of flight in various training programs for the first time during their cadet career. Graduates leave USAFA with a Bachelors of Science degree and a commission as a Second Lieutenant. They join a "long blue line" of officers dedicated to leading airmen. They often continue their education and training as a weapons system operator, technical

* Permanent Professor and Head, AIAA Senior Member
† Instructor of Aeronautics and Personnel Officer, AIAA Member
# Associate Professor and Deputy for Plans and Programs, AIAA Member

expert, instructor, and/or specialist. As they've reached the upper levels of leadership in our nation's military service or civilian workforce, they've truly demonstrated the value of the unique USAFA educational experience.

The root of that experience is found the values of the institution, expressed daily by a faculty of over 600 dedicated instructors and military training staff, composed of civilian professors, officers, and enlisted personnel, supported by an exceptional civil service and contract staff. The men and women of the USAFA faculty are role models and mentors for the cadets, sharing the warfighter spirit as well as their technical and leadership expertise.

With respect to the Department of Aeronautics, from the earliest days of the Academy, DFAN's outstanding faculty, facilities, curriculum, and culture have made a substantial contribution towards molding these young people into preeminent aerospace force officers. This paper will highlight DFAN's impact on the overall mission of the academy, on the men and women who pass through the institution, and therefore on our Air Force and nation.

FLEDGLINGS

USAFA traces its institutional roots back to the first days of military aviation. Less than four years after the Wright Brother's first sustained, powered, controlled flight of December 17th, 1903, the United States Army established an Aeronautics Division. In 1907, this forerunner of the now-separate United States Air Force (USAF), with a single officer assigned, began to seek airframes to develop a military aviation capability. By 1912, there were seventeen Army pilots flying later-model Wright Flyers.[3]

From 1916 to 1918, the US Army's 1st Aero Squadron encountered great difficulties operating and maintaining their aircraft in campaigns against Mexico's Pancho Villa and across Europe's WWI battlefields. These challenges motivated early Army leaders such as Gen William "Billy" Mitchell to press for an independent, professional military Air Service, separate from the Army, as early as 1920. In 1926, Congress moved in that direction, establishing the Army Air Corps and helping re-define the mission of the Air Service from combat support to offensive war-fighting. By 1941, with World War II growing in scope and aviation technology and capability exploding on both sides of the war, the Army created the quasi-independent Army Air Corps (AAC) to organize and apply airpower across the globe.[4]

The Allied victory of WWII, due in large part to airpower and its ability to deliver massive destruction on distant targets, led to further pressure for an independent Air Force. This time pressure came from President Truman, fresh from the experience of having an AAC aircraft deliver the first and second atomic weapons to the Japanese mainland, forcing their capitulation and ending the war. Congress, too, recognized the on-going radical evolution of air power, including the development of jet-powered aircraft, the achievement of supersonic flight, the tactics of strategic bombing, and a growing Cold War menace requiring deterrence and containment. On July 26th, 1947, the National Security Act and Executive Order 9877 were both signed, creating the USAF as a separate service and assigning its mission: strategic warfare. [5]

The leadership of the AAC and earlier Army aviation units benefited greatly from the professional military education offered by the US Military Academy at West Point. Early Air Force leaders (many themselves West Point graduates) immediately began studies and plans for a USAF Academy. On April 1st, 1954, Congress authorized the construction of USAFA after six years of Air Force and civilian leadership review of Military and Naval academy programs, including academic and military training curriculum, and a search for an appropriate location. The 18,000 acre site northwest of Colorado Springs was chosen for its relative isolation, good weather, and incentives offered by the local community and state. [6]

With construction underway in 1955, the first class of cadets entered training at a temporary location at Lowry AFB, Denver, CO. In August 1958, the Cadet Wing occupied the Academy itself, and the first class of 207 officers graduated on June 3rd, 1959. The "long blue line" of professional aerospace officers had begun.

STRETCHING WINGS

From the beginning, thanks to the mission of the newly-organized USAF, aeronautics and related fields contributed heavily to the curriculum levied on Academy cadets. Of course, a premium was put on the development of officers and, therefore, the requisite leadership training, rather than pure academic studies. Since the first graduating class, every cadet at the academy has experienced a well-integrated military training program including academics, leadership development, physical fitness, intercollegiate or intramural athletic competition, and flight training.

DFAN has had a profound impact on cadets' academic experience. In addition, its faculty has regularly volunteered its expertise to enhance the other mission elements. Organizationally, DFAN's roots are in separate Departments of Aeronautics and Thermodynamics, both established in 1955. Cadets were required to take classes in both disciplines enroute to a General Academics Bachelor of Science degree. In 1960 the departments were combined into one Department of Aeronautics, which gained accreditation in 1967 from the Engineering Council for Professional Development. That same year, the first class of fourteen Aeronautics majors left DFAN and USAFA for active duty service. Including the Class of 2001, USAFA/DFAN has produced over 1,500 Aeronautics majors from a curriculum now accredited at the highest level possible by the Accreditation Board for Engineering and Technology.[7]

Despite the formulation of an Aeronautics major (currently 49 credit hours) and other specific technical and non-technical majors in 1966, a comprehensive core curriculum (currently composed of 109 credit hours) has remained in place for every cadet, providing a broad range of required courses in engineering, basic sciences, and humanities. DFAN has taught the Aerodynamics and Thermodynamics courses within the core, helping mold every cadet, regardless of academic major, into an officer ready to support all components of the Air Force as a weapon system operator, maintainer, logistician, acquisition officer, or other specialist.

Since 1955, over 280 faculty members have served within DFAN, supervised by 12 Department Heads. Led by the Academy's Dean of Faculty, Department Heads serve as Permanent Professors on extended assignment to USAFA, providing continuity in the curriculum and organization of each department. Colonel D. Neal Barlow, a Distinguished Graduate of the Class of 1978, is the first Aeronautical Engineering major and graduate to return as the DFAN Permanent Professor and Head. Historically, a majority of the faculty members have been active-duty military members, serving several years at the Academy before moving on to operational military assignments, advanced educational opportunities, or civilian life. With a Master of Science or Doctoral degree in Engineering, these military faculty members range in academic rank from instructors to full professors. Above and beyond the technical knowledge shared in the classroom, they provide critical military role modeling for cadets. Their "war stories", involving character, integrity, and leadership issues as well as technical "been there, done that" wisdom, help the cadets internalize the Air Force's core values: *Integrity First, Service Before Self,* and *Excellence in All We Do.*

DFAN's civilian faculty members, either permanent or visiting, provide the backbone of continuity to the department's technical expertise and curriculum organization. With Doctoral degrees in Engineering and a wide range of prior military, academic, and industrial experiences, these professors remind cadets of their customers and benefactors – the citizens of a nation blessed by freedom and prosperity. Currently a quarter of DFAN's 26 faculty members are civilian. This number is expected to grow in the years to come, thanks to demands on the military faculty members to serve in critically-manned operational flying and engineering-oriented units. While some may see this as a threat to the military training environment, DFAN has developed an exceptionally beneficial faculty culture that emphasizes the unity, rather than the potential conflict, of the USAFA mission and the aeronautics curriculum outcomes. Civilian instructors, many with prior military service, utilize operational Air Force air and spacecraft as laboratory demonstrators or design case studies. National and international current events, including USAFA and USAF policies, plans, and activities world-wide, are discussed regularly in the classroom alongside the aeronautics curriculum, keeping the institution's goal always front and center: *develop professional aerospace officers.*

The DFAN faculty is unique in its cultural "family". After an assignment to HQ USAFA/DFAN, instructors, professors, and technicians may move on to other military or civilian pursuits, but always remain a member of the department in a Not Currently Assigned (NCA) status. Members never truly leave the DFAN faculty, and annual events draw attendees from decades past to share the department's heritage with new members. Of special note is DFAN's secretary, Ms Shirley Orlofsky, who has maintained the organizational and administrative excellence of the department, and answered the Department Head's telephone with a warm, welcome greeting, since 1961 – over 40 years of outstanding service!

Beyond the classroom, the annual contributions of the DFAN faculty to the overall mission of the Academy via extracurricular activities are staggering. For example, just from July 2000 to July 2001, the 26 members of the DFAN faculty oversaw over 60 cadet research and development projects in aeronautics and related fields, authored or co-authored over 80 professional papers and presentations, spent well over 100 hours in local schools or hosting student visits to the Academy, providing aerospace education to the local community, and served in over 35 different after-hours military leadership/mentorship positions with the cadet wing, providing academic counseling, ethics training, or supervision of the cadet-to-cadet training program.[8]

LEARNING TO FLY

A critical component of the Aeronautics major's education -- hands-on research and development -- occurs in the DFAN Aeronautics Laboratory. The Laboratory is a 55,000 ft^2 facility housing nine major wind tunnels capable of various test velocities from low speeds to Mach 4.5, a water tunnel, three jet engine test cells including operational J-69 and F-109 cycles, a rocket/internal combustion engine test cell, two Genesis 3000 flight simulators, and a variety of smaller experimental equipment. The laboratory also includes two computer-based design classrooms and a complete machine shop. A robust computational capability added in the last few years has become the overall USAFA High Performance Computing Initiative, used by several departments for interdisciplinary research as well as unsteady, turbulent computational fluid dynamics.

In the labs, cadets benefit from working alongside experienced faculty members and technicians. Working as a team, the cadets and their instructors make significant contributions to real-world research, development, and operational programs sponsored by the USAF, NASA, and other DoD, civilian, and educational agencies. For example, a recent nationally-recognized five-year research effort provided critical performance improvements to the USAF AC-130 Spooky/Spectre gunships. Managed by Dr. Tom Yechout and Mr. Ken Ostasiewski, three dozen cadets designed fairings and other external flowfield improvements that will increase duration by 20 minutes and ceilings by 2,000 ft once implemented.[9] Similar on-going work sponsored by NASA is resulting in recommended improvements to the design of the X-38 International Space Station Crew Return Vehicle.[10]

With respect to the actual discipline of flight, all cadets participate in some form of airmanship activity each of their four years at the academy. For example, during their first summer at USAFA, cadets are introduced to flying through orientation rides in a sailplane and Cessna 172. Freshmen take an aviation fundamentals course that includes classroom instruction, T-37 simulator rides, and basic navigation. Sophomores participate in the "Soar-For-All" program that offers every Academy cadet the opportunity to fly and solo in an unpowered glider. This program has trained over 26,000 cadets and soloed over 16,600. Over 23,000 sorties are flown in the Soar-For-All program annually, making it the world's largest soaring operation.[11]

Many cadets also participate in the Academy's free-fall parachuting training program and earn their military jump wings (over 16,000 cadets since 1967). In addition to being the world's largest free-fall jump program, it's unique in that the first jump is a solo free fall event, not a static line or tandem jump. The Academy's Wings of Blue collegiate parachute competition team has been the collegiate national champions 26 of the last 33 years.[11]

Key to the Academy's success in these airmanship programs is the critical element of cadet-to-cadet soaring and parachuting instruction -- over 90% of course instructors are current Academy cadets! Each year, after a highly competitive selection process and a yearlong course of concentrated instruction, 80 cadet soaring instructors and 50 cadet jumpmasters serve as primary cadet trainers. At no other flight school in the world do students teach other students in such large numbers. The time and energy each cadet dedicates to the flying curriculum, despite the many demands of the academy's overarching military and athletic training programs, illustrates a true motivation for flight. In total, these outstanding programs have provided soaring and parachuting opportunities to more than 35,000 USAFA and ROTC cadets. Skills taught by DFAN to all cadets in the core or aeronautics curricula provide the essential theoretical knowledge that makes these cadets competent and confident to safely execute the cadet-to-cadet training programs. Moreover, Aeronautics majors consistently represent the highest percentage of all academic disciplines among the instructor pilots.

At an even higher level, the Academy annually facilitates training for hundreds of cadets and newly-minted Lieutenants in the Introductory Flight Training (IFT) program. IFT leads to a private pilot's license, and is mandatory for all graduates who will attend the military's various Joint Specialized Undergraduate Pilot Training (JSUPT) programs. In the last two years, this program has trained 247 cadets and 699 Lieutenants, and flown over 42,000 hours.[11]

The impact of these comprehensive flight training programs is far-reaching and widely lauded. For example, 2Lt Tracy Neddleblad, Class of 2001, was recognized as the top collegiate female flyer in the Rocky Mountain region during her senior year. The USAFA Flying Team regularly garners national awards for demonstrated precision cross country and pattern flying skills.

For the active duty Air Force, these flight training programs help develop the physical skills and mental discipline required for success in operational cockpits. For example, at the pinnacle of their profession, hundreds of USAFA graduates – pilots or flight test engineers – have attended the USAF Test Pilot School (TPS) at Edwards Air Force Base, California, having been well-prepared by DFAN. In fact, the DFAN faculty regularly teaches several portions the TPS academic curriculum as visiting instructors, covering topics such as modeling and simulation, propulsion, sub- and supersonic aerodynamics, and flight mechanics, stability, and control.

With respect to TPS, of special note is DFAN's Flight Test Techniques course, a special semester-long immersion in the academic and operational aspects of aircraft performance testing. In this highly-competitive elective course, students practice flight test data collection and analysis methods. At the end of the semester, they visit the actual TPS facilities in California to validate their skills during USAF T-38 Talon supersonic trainer flights. At least ten aeronautics majors have gone on from this unique experience to graduate from an Air Force, Navy, or international TPS program.[12]

LEAVING THE NEST

Since the Class of 1959, USAFA has graduated over 33,000 lieutenants with a Bachelor of Science degree. These men and (since 1980) women have gone on to serve the nation as Air Force and civilian leaders in all aspects of our nation's defense. They've greatly contributed to the development of America's military, commercial, and industrial aerospace might.

For their military service, a majority of Aeronautics majors do enter JSUPT soon after graduation and, after several years of additional training, serve as an expert weapon systems operator and warfighter. The DFAN curriculum, comprehensive in breadth and depth, has served graduates well in preparing them for the rigors of flight school. Lieutenants enter the one-year pilot training program with a fundamental knowledge of fluid dynamics, propulsion, energy systems, structures, stability, and control. They are easily a month ahead of their peers in the academic portion of JSUPT training, leaving more time for the study and practice of actual stick-and-rudder skills. According to 2Lt Nick Rutgers, Class of 2001, the core aeronautics course (part of the mandatory curriculum every cadet experiences) provided him an exceptional foundation in understanding and applying the concepts of turn, climb, and takeoff performance, as well as specific excess power. Majors-level courses provided him a theoretical and working-level knowledge of various stability augmentation systems and dampers utilized on military aircraft, engine components and their performance (including anomalies such as compressor stalls), and subsonic and supersonic aerodynamics. Importantly, the overall stress of Academy academic and military life prepared him well to deal with the stress of UPT. The long days of training and numerous high-profile check flights degrade the performance of many students, especially those who hadn't experienced the USAFA training environment.[13]

Besides pilot training, USAFA/DFAN graduates provide the USAF an especially exceptional resource in the aerospace engineering and aircraft maintenance career fields. The various branches of the Air Force Research Lab employ Aeronautics majors in numerous basic and applied research efforts for aircraft and space systems. Graduates serving as maintainers have a theoretical understanding of their particular weapon system that contributes immensely to the practical service and repair tasks they're responsible for.

SOARING FREE

This paper has had much to say about USAFA's unique undergraduate learning environment, and how DFAN's faculty, culture, facilities, and curriculum benefit every cadet's Academy and active duty career. One might think that an academic program – and an Aeronautics curriculum in particular – inserted in the midst of USAFA's intense military training environment wouldn't fare well when validated externally or internally; that it might be minimized or marginalized in its academic rigor. Just the opposite is true! From a student perspective, Princeton Review's summary of student inputs rated the USAFA undergraduate academic experience number one nationally in 2000, followed by a fourth-place finish in 2001.[12] The Academy's overall engineering program is recognized consistently as one of the nation's top five undergraduate programs.[13] The aerospace curriculum in particular was recently rated as the second-best undergraduate program in the nation.[14] Furthermore, USAFA won the National Aeronautics Association 2001 Cliff Henderson Award for "significant and lasting contributions to the promotion and advancement of aviation and space activity in our United States and around the world."[11]

The achievements of specific graduates provide further evidence of institutional excellence. Since 1959, 310 graduates have become general officers, to include the previous Air Force Chief of Staff (General Michael Ryan, Class of 1965) and the Academy's Superintendent (Lieutenant General John Dallager, Class of 1969 and an Distinguished Graduate in Engineering Mechanics), Commandant of Cadets (Brigadier General S. Taco Gilbert III, Class of 1978 and a Distinguished Graduate in Civil Engineering), Dean of the Faculty (Brigadier General David Wagie, Class of 1972 and a Distinguished Graduate in Astronautical Engineering), and Director of Athletics (Colonel Randall Spetman, Class of 1976). The first USAFA graduate to reach the rank of a four-star general is General (Retired) Hansford Johnson, Class of 1959 and a NCA DFAN faculty member.

160 graduates have made the ultimate sacrifice as combat casualties, including Captain Lance Sijan, Class of 1965. Captain Sijan was posthumously awarded the Medal of Honor for gallantry displayed as a prisoner of war (POW) at the hands of the North Vietnamese. 35 graduates are repatriated POWs. 2 graduates are combat aces.

31 graduates are astronauts, including Aeronautics majors Loren Shriver, Class of 1967; Charles Precourt, Class of 1977; Sidney Gutierrez, Class of 1973; Richard Searfoss, Class of 1978; Susan Helms, Class of 1980, first female graduate in space and a NCA DFAN faculty member; and Gregory Johnson, Class of 1984.

Academy graduates have received national scholarly recognition for excellence across the board: 31 Rhodes Scholars; 92 Guggenheim Fellows, 70 National Science Foundation Fellows, 30 Fulbright Scholars, 35 Hertz Fellows, 4 Marshall Fellows and 7 Truman Scholarship winners. In addition, 52% of all USAFA graduates commissioned as USAF Lieutenants are still on active duty serving the world's greatest aerospace power, testifying to a high level of institutional mission accomplishment.[11]

Numerous other graduates are now serving in the United States Civil Service or civilian sector, working as the engineers, pilots, and corporate leaders maintaining America's premier aerospace industrial base. By way of brief example, Representative Heather Wilson, Class of 1982, serves the citizens of New Mexico in the U.S. Congress, while Harry Pearce, Class of 1964, is a Vice Chairman of General Motors.

CONCLUSION

The United States Air Force Academy has made, and will continue to make, significant and lasting contributions to the promotion and advancement of aviation and space activity in our United States and around the world. USAFA's Department of Aeronautics is an integral part of the institution, providing critical components of the academic curriculum while supporting the professional military officer training program in numerous ways. USAFA graduates have proved the worth of their education in numerous ways in the military and civilian spheres of influence. DFAN is proud of its contributions to aerospace power in these last four decades of a century of manned flight.

ACKNOWLEDGMENTS

The authors acknowledge the sustained contributions of Brigadier General Daniel Daley, USAF (Ret) and Brigadier General Mike Smith, USAF (Ret), to the development and sustainment of the DFAN culture of excellence in faculty, curriculum, facilities, and espirit de corps.

REFERENCES

1. Snyder, Noel & Helen. Raptors: North American Birds of Prey. Voyageur Press, Stillwater, MN, 1997, pg 172.

2. Kaufman, Kenn. Live of North American Birds. Houghton Mifflin Co., New York, NY. 1996. pp 140-141.

3. Contrails: Air Force Academy Cadet Handbook. Volume 34. Capt Michelle S, Atchison, OIC. 1988. pp 38-39.

4. ------. pp 39-64.

5. ------. pp 65-68.

6. ------. pp 13-14.

7. USAFA Department of Aeronautics Department Yearly Activity Reports. USAFA Department of Aeronautics, USAFA, CO 80840. 1955-2001.

8. -----. 2001.

9. Tudor, Jason. "Academy staff, students streamline Gunships." Air Force Print News, Dec 14, 2001.

10. Johnston, Cheryl and Nettleblad, Tracy. X-38 Rudder Control and Speedbrake Performance Analysis for a Mid-Rudder Configuration. USAFA/DFAN Report 00-03. September, 2000.

11. 2001 Cliff Henderson Award Citation. National Aeronautics Association, 1815 N Fort Myer Dr #500, Arlington VA 22209

12. Scott, William B. "Cadets Introduced to Flight Testing," Aviation Week and Space Technology. January 17, 2000.

13. Personal communication from N. Rutgers to T. Yechout, November 27, 2001.

14. The Princeton Review. 2315 Broadway, New York, NY 10024. http://www.review.com/college/ studentsayrankings.cfm

15. "Best Colleges." U.S. World News and World Report. September, 2000 and June, 2001.

16. "Best Colleges." U.S. World News and World Report. June, 2001. http://www.usnews.com/usnews /edu/college/rankings/engineering/nophd/aero

Chapter 62

THE EVOLUTION OF AIR FORCE AEROSPACE EDUCATION AT THE AIR FORCE INSTITUTE OF TECHNOLOGY

By Peter J. Torvik
Fellow, AIAA
Professor Emeritus of Aerospace Engineering and Engineering Mechanics
Air Force Institute of Technology
Wright-Patterson AFB Ohio

Abstract

Education in Aerospace Engineering for the USAF began in 1919 as a yearlong postgraduate program. This activity continued through World War II. After the war, the Army Air Forces Institute of Technology (now AFIT) opened with a two-year program comparable to the last half of undergraduate engineering degree programs. These evolved into BS degree programs in Aeronautical and Electrical Engineering. In 1950, emphasis shifted towards graduate education with graduate-level programs in 1951, degree granting authority in 1954, and ECPD (now ABET) accreditation in 1955. Regional accreditation was earned in 1960 and extended to the doctoral level in 1965. Through 1997, totals of 920 BS, 8198 MS, and 283 Ph.D. degrees were awarded in the disciplines of the School of Engineering. Since the early 70's emphasis has moved increasingly towards research. Originally available only to officers of the Air Force, sister services, and allied nations, AFIT enrollment is now open to the public on a tuition basis through the Dayton Area Graduate Studies Institute, a collaborative program involving several local universities. In addition to enrollment in MS and Ph.D. degree programs, the school also offers post-doctoral research with the resident faculty in areas of Air Force interest.

Introduction

The evolution of engineering education at AFIT may be viewed in three periods, or generations, each being approximately 27 years in length. In the first, from the formation in 1919 through the end of WWII, postgraduate studies in aeronautical engineering in the form of one-year programs with small enrollments were offered. In the second period, from 1946 until 1973, degree programs at BS, MS, and Ph.D. levels were developed and earned full accreditation, both by the regional accrediting agency and, when appropriate, by bodies granting professional accreditation. In the third period, from 1974 to the present, the Institute progressed from being primarily a teaching institution into one that emphasizes graduate education along with a significant production of research in science, engineering and management for the United States Air Force and for the technological infrastructure of the nation.

Postgraduate Programs: 1919-1946

The history of the Air Force Institute of Technology may be traced to 1919, when Colonel Thurman H. Bane, Commanding Officer of McCook Field at Dayton, Ohio, formed an Air School of Application. In his request for authority to begin, he wrote, "No man can efficiently direct work about which he knows nothing." His argument was evidently convincing, as he was ordered to begin the instruction on Nov 10, 1919. Seven officers were enrolled in the first course. The task of organizing and running the new school fell on Lieutenant Edwin E. Aldrin holder of a MS from MIT and (later to become) the father of Col. "Buzz" Aldrin.

At that early date McCook Field was already a large experimental laboratory, with about 400 scientists, engineers and technicians engaged in many research and development projects. Between 1919 and 1922 Air Service engineers had designed and built 27 airplanes of all types. With this base of expertise, it is not surprising that most of the instruction in the new school was provided by engineering specialists assigned to the base. While not originally intended to produce engineers, by 1923 the yearlong course in General Aeronautical Engineering was primarily concerned with the design of aircraft and aircraft engines. Graduates of the Class of 1923 included aviation notables Lt. John Macready, (the only three time winner of the McKay trophy) and Lt. James H. Doolittle.

But changes in the role of the Air Service engineers were taking place. Spokesmen for the infant aircraft industry sought a larger role in aircraft development, seeking contracts for experimental airplanes. Consequently, the role of Air Service engineers changed from that of designing and building new aircraft to that of approving designs submitted to the Air Service, and to providing consultation to outside designers and manufacturers. In 1927 the school (by this time renamed the Air Corps Engineering School) was moved, along with engineering and test activities, to the 4500 acre tract of land donated to the government by the citizens of Dayton and designated as Wright Field.

Changes also occurred in the school. Whereas the original curriculum had featured application, emphasizing individual student projects and minimizing formal lecturing, the curriculum was revised in 1926 to a greater use of the lecture method and the first steps were taken towards the creation of a permanent faculty. Nonetheless, most of the instruction continued to come from the Wright Field staff as it was thought that the branches concerned could only properly accomplish the teaching of highly specialized subject matter. Among these instructors was Bradley Jones (instructor in instruments and navigation, 1923-1929), who later headed the aeronautics program at the University of Cincinnati. S.D Heron (Instructor in fuels and air cooled cylinder design and development from 1921-1934) and C. J. Cleary (instructor in materials) are familiar names in the Wright Laboratories to this day as the Propulsion and Materials laboratories, respectively, honor their outstanding scientists and engineers through awards in their names. And as indicated in the preface of *Airplane Performance Stability and Control*, by Perkins and Hage, the first version was developed while teaching in the Engineering School.

The curriculum was revised to appeal to younger students, all graduates of West Point or civilian technical institutions. The program culminated with a course in performance and flight-testing, with the Commandant assigning an airplane design problem each year. After analyzing the tactical requirements, the students completed a preliminary design which including weight estimate and balance, performance, stability, a stress analysis of various components, and a detailed designed of some components. The class of '34 designed an observation airplane and a reconnaissance airplane for a future Air Force, the class of '35 designed a basic trainer, the class of '36 class was assigned an interceptor pursuit airplane, and the class of '37 designed a slow-speed observation and liaison airplane. A curriculum from this period is given as Table I in the Appendix.

Courses were suspended for the 1939-40 school year as all available personnel were temporarily diverted from research and development to the massive procurement and production effort then being initiated. The Material Division "froze" development on the best of the existing planes and ordered them into huge production, after which research and development of the next generation was resumed. The yearlong course resumed in the fall of 1940 with a small class of six lieutenants, one of these being Lt. Bernard A. Schriever ('41).

By the outbreak of the war, the school had graduated a total of 230 officers. Their achievements before and during the war were notable. Captain George V. Holloman ('35) earned, with Carl J. Crane, the MacKay Trophy for 1937 for the Airplane Automatic Landing System. Graduates played key roles in the development of jet aircraft. Benjamin Chidlaw ('31) was assigned the task of developing a jet aircraft project using the Whittle engine for the AAC; Laurence Craigie ('35) became the first military man to fly a US jet, the XP-59A Aircomet; and Ralph Swofford Jr. ('36) was assigned as project officer for the next generation P-80.

By 1943 it was clear that younger officers must be prepared to replace the rather limited number of engineering officers. When civilian schools could not produce the numbers, nor the most critical specialties the Air Force needed, the AAF Engineering School was reopened on 1 April 1944 with a compressed program. One of the graduates was Captain Dan Daley ('45), who returned to the AFIT faculty and later served (1967-1984) as Head of Aeronautical Engineering at the United States Air Force

Academy. Degree Programs and Accreditation 1946—1973

In 1945 the Army Air Forces (AAF) Scientific Advisory Group, directed by Dr. Theodore von Kármán, completed a study of the role of research and development for the post-war AAF. In the cover letter to the report, von Kármán wrote, "the men in charge of the future Air Forces should always remember that problems never have final or universal solutions, and only a constant inquisitive attitude toward science and a ceaseless and swift adaptation to new developments can maintain the security of this nation through world air supremacy." On 15 December 1945 the AAF Institute of Technology was officially authorized.

The Army Air Corps made the decision to reconstitute the school at Wright-Patterson into a technical institute comparable to the best in the US, beginning as an undergraduate school and granting only undergraduate degrees. A graduate school with courses comparable to similar courses provided by the best civilian technical institutions was to be established as soon as a qualified staff and students could be acquired. The initial restriction to undergraduate studies was a matter of feasibility. By the end of the WW II, the officer corps of the Army Air Force was found to be somewhat lacking in educational attainment, a situation brought about by the fact that a college degree had not been a requirement for earning a commission through pilot training.

The "blueprint" for the new school was created by a committee of civilian educators headed by Dr. John Markham of MIT. Other members were Professors W. H Pickering and E. E. Sechler of Cal Tech and T. H. Troller of the Case School of Applied Science. All except Pickering were aeronautical engineers. The committee provided the detailed design of a two-year curriculum, including descriptions for individual courses. The initial curriculum is in Table II of the Appendix.

Hiring took longer than anticipated, but by the fall of 1946 eight civilians and five officers had been recruited for the faculties of Engineering and Industrial Administration. The AAFIT opened on 30 September 1946 with 189 officers in the student body. Of these, 132 were in the College of Engineering with 100 enrolled in the new two-year program. But the memories of the first class of students had been overestimated. During the first week it was found necessary to stop all courses so that the instructors could teach a six-week review of mathematics. A one-month review in mathematics was built into later programs. Originally, the Institute had no laboratories of its own as it was intended that on-base laboratories would be used. But the availability of those facilities was soon found to be uncertain and the faculty began to develop their own. It is of interest to note that in 1949 the USAFIT student branch of the Institute of the Aeronautical Sciences had 24 members.

But the hiring of faculty, the "rustiness" of student backgrounds, and the development of laboratories were not the only challenges facing the new school. Even after it opened, differences of opinion remained as to the appropriate nature of instruction in the Institute. Opposition to accreditation and degrees emerged within the Air Force and was strengthened by a group of distinguished educators newly constituted to provide advice to the new AF on educational matters. They recommended that the undergraduate work should be highly specialized – limited to courses of study not normally available in civilian institutions. To some AF leaders, it became apparent that the kind of AAFIT seen by the Markham Committee and supported by Air Force would prove to be a duplication of civilian schools and that – as a result – sooner or later the Air Force would lose it. Nonetheless, the decision was made to continue development of a "technical school comparable to the best in the United States," awarding degrees at both undergraduate and graduate levels.

The Air Force needed officers with graduate degrees as well and wished to send the new graduates (who now held diplomas, but not degrees) to civilian institutions. Some of the graduates convinced their prior undergraduate school to accept AFIT credits in transfer, thereby enabling them to earn an undergraduate degree. Others earned admission to graduate schools through performance on the Graduate Record Examination.

By 1951, the curriculum of engineering sciences had been divided into two sections, aero-mechanical and electronics, with a common first year and specialization in the second. And, as the school moved towards earning accreditation, these became formally treated as two separate programs and were so designated from 1956 through the end of the program.

In 1950, the Institute began developing graduate programs. Enrollments of undergraduates were reduced, retaining enough to keep a sound base under the graduate programs. The first students to enter resident graduate studies completed a yearlong Advanced Engineering Management Course in December of 1951. Students graduating in August of the following year (1952), including nine students in Aeronautical, completed the Institute's first graduate programs in engineering. The program was originally one year in length, but was soon expanded into an eighteen-month program. An early curriculum is given in Table III of the Appendix. An Air Ordnance program was begun in the fall of 1953. Later renamed Air Weapons, it continued for 21 years.

The Department of Defense submitted to the Bureau of Budget in 1952 and in 1953 proposed legislation authorizing the award of master's and doctor's degrees to students in the Resident College. But there were difficulties. The regional accrediting agency (North Central) was reluctant to support the granting of undergraduate degrees by institutions that were primarily technical or scientific. It initially (1953) recommended that USAFIT concentrate its efforts on the graduate programs rather than seeking authority to grant undergraduate degrees. The US Office of Education was unconvinced that government-supported schools should grant degrees at all, and there were those in the Office of the Secretary of Defense who did not feel that the Air Force should be conducting 'schools of higher education.' But on August 30 1954, President Dwight Eisenhower signed public Law 733 of the 83rd Congress, giving degree granting authority for programs completed in the Resident College of the United States Air Force Institute of Technology, subject to accreditation by a nationally recognized accreditation association or authority.

Arrangements for an accreditation visit were made, and in October of 1955, ECPD granted accreditation for both curricula. With accreditation obtained, Public Law 733 took effect, and the

Institute had the authority to grant degrees. On March 13 1956 the first 22 MS degrees were granted. Later in that year the first accredited graduate and undergraduate degrees in aerospace engineering were awarded. Students who had completed the program and earned a diploma prior to accreditation were given an opportunity to earn a degree through a process of completing additional courses.

ABET had suggested that emphasis be increased in the area of socio-humanistic studies. Faculty identified the requirements of 12 other programs, and found the range to be from 4 to 45. The average degree requirement was found to be 218 QH, with a non-technical requirement of 40, including 11.5 in English and Speech, and 20 in Socio-Humamistic studies. The (predictable) AFIT response was to replace three electives and one management course with courses taken from a quickly created department of Humanities.

As graduate enrollments increased, the size of the undergraduate classes was further reduced. By 1956, the faculty had grown to a total of 55, with 25 of these being in Aeronautical, Mechanical Engineering, and Mechanics. The ratio between military and civilian faculty was maintained at 50%.

A proposal to eliminate the AFIT School of Engineering and create graduate programs at the newly formed Air Force Academy was given serious consideration in 1959. But the arguments for retaining graduate engineering education at Wright Patterson were found to be compelling. On April 1, 1960, on the basis of programs in the School of Engineering, the Institute was accredited by the North Central Association of Colleges and Secondary Schools to grant degrees through the master's level. A resident doctoral program in Aerospace Engineering was begun in July of 1965, with preliminary accreditation by North Central in August of 1965 and full accreditation in 1972.

The successful launch of Sputnik on 4 October 1957 resulted in an immediate response. Courses in astronautics were added to existing curricula and, on September 8, 1958, the first class began a two-year astronautics program at AFIT. The faculty designed a curriculum built around sequences of courses in mathematics, basic sciences, and engineering sciences. To this foundation were added advanced and highly specialized courses more directly applicable to work with anti-missile systems, reconnaissance satellites, space weapons, or other new weapon systems. The original curriculum was very broad, highly structured, and rather inflexible. The eight-quarter curriculum contained required courses from six different departments and had only two electives. One of the students in that first class, Captain Donn Eisele, would ten years later take part in the first Apollo flight.

A program in Space Facilities introduced in 1962 was of a special interest and uniqueness. It had the objective of providing the educational basis for the engineering competence to design, develop, test and eventually operate and maintain fixed facilities and supporting systems in free space and on the lunar surface. The curricula included such courses as lunar soil mechanics, direct methods of energy conversion, the methodology of engineering systems required for sustained space operations, and studies of the structure of space, moon, and planets. The program was discontinued in the early seventies when it was determined that the Air Force had no manned mission in space.

Ten astronauts have attended AFIT, with eight graduating from the Department of Aeronautics and Astronautics or its predecessor departments: Fabian , Bluford, Mullane, Lindsey, Grissom, Cooper, Eisele, and Brown. Additionally, Ford is in training. Anders and Chaffee attended AFIT, but in other programs.

The doctoral program initiated in 1965 was designated as a program in Aerospace Engineering but was actually implemented as an interdisciplinary program involving all the instructional departments of the School of Engineering. Two years of course work were completed at AFIT, with research for the dissertation to be done in a specially arranged four-year assignment to an AF laboratory.

The Air Force strategy for managing advanced degree education changed in 1969. Under the new concept, specific educational requirements were determined from a billet-by-billet audit, rather than by estimating the requirement to educate towards a broader, or career-long, need. This philosophy contributed to a reduction in the "apparent" need for advanced education in the Air Force. The change in concept was implemented in AFIT programs through the identification for each graduate program of a small core program consisting of mathematics and the appropriate basic sciences and engineering. These provided the background for a more advanced specialty, consisting (typically) of two graduate sequences and the thesis.

A program in Systems Engineering was developed by the faculty and implemented in 1971 with first graduates in 1973. The highly interdisciplinary program contained a core program emphasizing mathematics and aerospace engineering with advanced work in probability and statistics, systems management, and systems modeling. This program was unique for AFIT in that a group design study was used in lieu of a thesis.

With 31 assigned faculty, the combined size of the mechanics, aeronautical engineering and mechanical engineering faculties reached its peak in the late '60s. This represented 40% of the entire faculty of the school of with aspects of aerospace engineering were introduced at AFIT between 1946 and 1974. The numbers of students graduating from each are given in Figure 1.

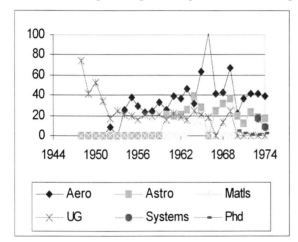

Figure 1 Graduates in Aerospace Engineering

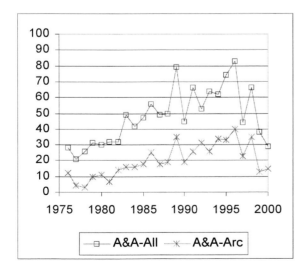

Figure 2 Research Publications, Aero and Astro

An Emphasis on Research, 1974-2001

By the summer of 1969 only graduate studies were offered in Aerospace Engineering. And by the early 1970's research activity by the faculty was well established, forming a base from which significant increases would occur as the School of Engineering focused on graduate education and research. Research productivity measures for the Department of Aeronautics and Astroautics are given in Figure 2.

791

AFIT programs leading to the MS degree are virtually unique in that they do not offer a thesis option. They offer a thesis requirement. Originally awarded seven quarter-hours credit, and later raised to 12, the thesis was expected to be original research, with about half leading to some form of publication. Upon completion, the thesis is graded and included in the computation of the final grade-point average. Because the doctoral program has always been small, the research of MS students is taken seriously, and plays a significant role in the faculty research programs.

The structure of the doctoral program was modified significantly in 1975. Experience gained in the first eight years showed that the expectation that students could complete their research for the dissertation while in laboratory assignments was not being met. Beginning with students entering in 1975, the assignment in residence at AFIT was lengthened to three years, and students were expected to complete all requirements, including the dissertation, while assigned to AFIT. The program was redesignated as a Ph.D. program, (as opposed to a Ph.D. in Aerospace Engineering), with departmental majors. The language requirement was dropped in December of 1974.

Education in aeronautical engineering had begun in 1946 under the two departments of aeronautical engineering and mechanics, with a third department of mechanical engineering being formed in the summer of 1949. These three were reorganized back into two in 1969 (with the renamed aero-mechanical engineering department offering aerodynamics, fluid mechanics and propulsion and the department of mechanics emphasizing solid mechanics, materials, dynamics, and controls). In 1977 these two were combined into one, the present department of aeronautics and astronautics. By 1974, the combined strength of the aero-Mechanical and mechanics faculties had reached 23, but further reductions in the post-Viet Nam era reduced this to 17 by the time the departments were merged in 1977. Faculty size rose again to 23 in the '80s, and has since declined to its present size of 16. The total faculty of the School of Engineering first reached 100 in 1984, and is approximately of that size today. The present core and specialty sequences for the three primary curricula in Aeronautics and Astronautics are given in Table IV of the Appendix.

By 1979 the Air Force was facing a severe shortage of engineering officers, with the deficit estimated at 1200. To meet this need, AFIT began in 1980 a new concept in engineering education, a highly structured, minimum time path by which holders of a baccalaureate degree in a strongly quantitative area could earn a fully accredited baccalaureate degree in engineering. The AFIT faculty determined that carefully screened and well-motivated holders of degrees in such "hard" sciences as physics, chemistry and mathematics should be able to complete all additional work required for a baccalaureate in engineering in six academic quarters.

In the spring and summer of 1980, Air Force recruiters sought students from the target population of math and science majors and offered them an opportunity to enter officer candidate training, earn a commission in the Air Force, and immediately begin a guaranteed assignment at the AFIT School of Engineering. In a few months, over 100 volunteered, with about 50% being declared eligible by the faculty. The entrance requirements for Aeronautical Engineering were: a baccalaureate degree with a strong quantitative background in mathematics and science; differential and integral calculus, and one course in ordinary differential equations; 24 quarter hours of basic sciences, including 12 hours of calculus based physics taught at a level offered for science and engineering majors; and 24 hours of humanities or social sciences.

Courses in the curriculum were as follows:

Aeronautics (Performance, Stability and Design)
Aerodynamics (Fluids and Aerodynamics I & II)
Thermal Sciences (Thermo, Heat Transfer, Propulsion)
Structural Analysis (Strength of Mat, Structures I&II)

Mechanics (Statics, Dynamics I&II, Vibrations)
Electrical Engineering (Circuits, Electronics, Controls)
Support Courses (FORTRAN, Materials, Engineering Economy, and electives)
Mathematics (Applied Mathematics I, II & III)

The program also included separate laboratory courses in circuits, wind tunnel, structures, and vibrations. Students interested in an option in Astronautical Engineering used their electives for this purpose.

The initial class of 30 admitted to aeronautical engineering in September of 1980 had undergraduate majors as follows: mathematics (13), physics (8), chemistry (4), computer science (2), meteorology (2), and one each from several other disciplines. The average undergraduate GPA was 3.19. Between 1982 and 1985, 109 students completed the aeronautical engineering program. A substantial number remained for full careers and many later earned MS and Ph.D. degrees. It was especially gratifying that these students, nearly all of whom would have been highly sought as graduate students in their original disciplines, choose instead to pursue a BS in engineering and a career as an Air Force officer.

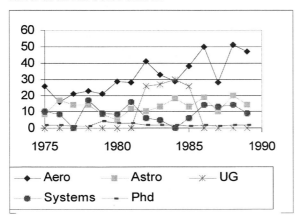

Figure 3 Degrees Granted in Aerospace

After the turbulence and uncertainties of the 70's, the decade of the '80s represented a period of relatively stable growth for the Institute. This is reflected in the level of activity within individual programs of the Department of Aeronautics and Astronautics during this period. The numbers of graduates of the various programs are shown in Figure 3. In several of these years, AFIT ranked first nationally in the total number of graduate degrees granted in Aerospace Engineering.

The decade of the '90s, however, was a period of declining enrollments as the Air Force reduced the size of the officer corps by approximately 30% from 1989 levels. In addition, increased emphasis in other areas brought about a substantial reduction in the number of students graduating in the aerospace disciplines, as may be seen in Figure 4.

As graduate students at AFIT are free of outside responsibilities such teaching or part-time employment, the credit load has traditionally been heavy. Although the degree requirement has always been only 48 hours, including the credit for thesis, a typical student's program of about 90 hours over six quarters resulted from prerequisite work and courses taken to provide additional mate-

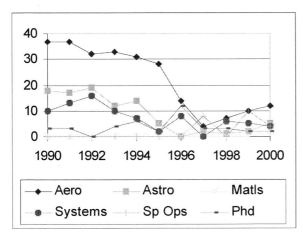

Figure 4 Degrees Granted in Aerospace

rial considered necessary to meet Air Force educational requirements. In 1989, a review of student loads led to a determination that students were overloaded, especially during the latter half of program when the thesis research was being conducted. To give greater emphasis to research, the expectation of the average load carried by a full-time student was lowered from 15 credits per quarter to 12. "Front-loaded" programs were developed, with students typically carrying 14-15 credits for the first three terms, and 9-10 (including thesis) for the last three. Students now typically complete 72 quarter hour programs, with even more substantive theses.

Originally, all students in the School of Engineering were officers in the USAF, along with a few RAF exchange officers and occasional officers from the Navy and the US Coast Guard. The programs were opened to AF civilian employees on full-time, long-term training programs shortly after accreditation was earned. Later, the US Army and other allied nations began to enroll its officers in AFIT programs. In 1973, AFIT courses were made available to AF civilian employees on a part-time basis, and since 1976 MS degrees (but not the Ph.D.) can be earned from AFIT totally through part-time study.

AFIT opened its doors to the public in 1993 by establishing a procedure for the payment of tuition. US citizens or permanent residents from outside the Wright-Patterson community can now enroll on a part or full time basis and earn a degree from AFIT, although a few courses are unavailable due to security considerations. In 1995, a cooperative was formed between three partner schools, the University of Dayton, Wright State University, and AFIT. This agreement enables students at any one to take courses from the other two without payment of tuition. Further, the schools agreed that up to one-half of the credits submitted towards a graduate degree could be earned from one or both of the other two schools. Such courses are not treated as (nor subject to the limitations of) transfer credit, but are treated as if taken in the home school.

Because the primary customer for AFIT research is the USAF, and because salaries of faculty and officer students may not be charged to another Air Force organization, the amount of external funding is not an appropriate measure of the total scope of the AFIT research program. AFIT may and does, however, accept from research sponsors funds for such research activities as equipment, materials, travel, salaries for post-doctoral students, and travel. Since 1990, the annual external funding for the School of Engineering has averaged something over $3M. AFIT does not receive directly from the AF any funding for research and may not accept reimbursement from AF organizations for the time which faculty and funded students spend on AF research problems. These activities do, however, provide AF laboratories and other offices with an annual cost-avoidance that is somewhat greater than the cost to the Air Force of operating AFIT. Nonetheless, as the entire twelve-month costs of faculty salaries along with some research equipment are included in the budget line as a cost of education, the reported cost of education at AFIT is artificially inflated.

When the Air Force prepared its requirements for the FY 98 Budget Submission to Congress in the fall of 1996, provision for graduate education at AFIT was deleted as the Secretary of the Air Force intended to terminate resident graduate programs. But before public announcement of closure could be made, the local congressional delegation took a strong interest in the proposed closing of the schools. They requested the Air Force to provide a thorough study of the cost and benefits of alternative means for obtaining the necessary programs of advanced technical and managerial education for officers.

At the same time, the local business community and the State of Ohio showed a strong interest in retaining specifically defense-related advanced technical education as a critical segment of the scientific and technical infrastructure of the region and the state. Accordingly, the Ohio Board of Regents requested the development of a means for providing all AF required graduate programs—essentially, a plan for the privatization of the Institute. Under the concept developed, a small private corporation - a New AFIT - of about 1/3 the size of the existing AFIT would be created to provide education in areas not offered by any of the other institutions and to serve as the interface with the Air Force. This new entity would be incorporated and chartered by the State of Ohio to grant graduate degrees. One of five member schools of the Dayton Area Graduate Studies Institute (Ohio State, Cincinnati, Wright State, Dayton and the New AFIT) would take primary responsibility for each of the former AFIT programs. Thus, while some degrees would be granted by each of the schools, a high degree of program efficiency would be achieved as the resources (faculty and courses) of all five institutions would be available to each program. All instruction would be on, or near, Wright-Patterson and base facilities would be used in research, when appropriate. Ohio, through the Board of Regents, would provide significant funding for all programs through the normal means of tuition assistance. The only cost to the Air Force would be that of tuition, to be set at a level somewhat between that of state-supported and private universities ($2K/quarter per student) and a long-term research contract of $7M/year.

As the plan for privatization was being developed during the summer of 1997, the Air Force conducted a parallel study of the actual costs that would be entailed if all possible required graduate programs were to be placed in civilian institutions. A number of universities were invited to offer proposals to develop and provide each of the required programs, including provision for required research opportunities for all MS and doctoral students. While not all institutions invited chose to submit proposals, enough did as to enable the Air Force to establish a projected cost of placing the AFIT programs in civilian universities.

With the privatization proposal from DAGSI and the results of the cost study of university-operated programs available, the Air Force compared the cost to the Air Force of each of these with the actual cost of continuing to obtain the required graduate programs through AFIT. The study showed that the proposal from Ohio would be the most economical, the cost of using universities would be somewhat more expensive, and that the cost of continuing the operation of AFIT would not be substantially more than using the bidding universities. The perceived benefit of the increased responsiveness of retaining an Air Force operated institution was found to be sufficient to justify the slightly greater cost, and the decision to terminate the Graduate School of Engineering was reversed by a new Secretary of the Air Force early in May of 1998.

Summary

The tradition of educating Air Force officers in the environment of Air Force research and development is as old as the Air Force. That concept, initiated in 1919 through the Air School of Application, was implemented through post-graduate education in aeronautical engineering up through the Second World War. When the Institute of Technology was formed after the war, the dominant activity was in the engineering sciences supporting aeronautical engineering. While the breadth of Air Force activity in aerospace has expanded significantly since that time, interest in other disciplines of science, engineering, and management have expanded to an even greater degree. As a result, aerospace engineering is no longer the discipline of dominant interest and now constitutes a relatively small fraction of the total Air Force requirements for graduate education. In consequence, the aerospace faculty now constitutes only 16% of the school's total.

Although the decision to close the Institute was reversed, the future of graduate study in science and engineering at AFIT is not secure. A fundamental problem remains. Several recent studies, including those by the National Research Council and by the Air Force Association, have called attention to the decline in Air Force investment in science and technology. But it is the level of Air Force interest in science and technology that drives the need for a strong technical competence within its officer corps. Thus, the future needs for advanced degrees in technical areas will depend on a restoration of Air Force interest in science and technology.

Visionaries in 1919 and 1945 saw that the application of science and technology to the development of airpower would place special demands on the officer corps. Through its evolution into a graduate institution, the Air Force Institute of Technology came to be the 'flagship' organization for this philosophy. Thus, the future, and fate, of AFIT are tied to the extent to which the vision of Thurman Bane and Theodore von Kármán remain the vision of the United States Air Force.

Acknowledgments

What is now termed the discipline of aerospace engineering has been practiced at AFIT by several departments. Aeronautical Engineering, Mechanical Engineering and Mechanics were created, divided and then merged over the years into the single present Department of Aeronautics and Astronautics. Accordingly, I have taken the liberty of grouping the activities of all three into the single category of aerospace engineering. Unless otherwise noted, totals given are for one, two, or three departments, as was appropriate to the era.

The opinions put forth are strictly my own. The colleagues with whom I shared 32 years on the faculty may have different interpretations, or even different recollections, as to what transpired. And finally, I am indebted to those members of AFIT who, through the years, have left a trail of documentation for others to follow. Most notable of these is Sanders A. Laubenthal who, while at AFIT some twenty years ago, and on the occasion of the sixtieth anniversary of AFIT education, wrote the wonderful account, "Yesterday, Today, Tomorrow." Much of what I have written of the first half of the AFIT story is based on her work.

Principal Sources

History of the Air Force Institute of Technology,
 Individual Volumes, 1946-1986.
Integrator, (AFIT yearbook), 1948 to 1956,
 Air Force Institute of Technology
Catalog, Air Force Institute of Technology, 1963-1991
Yesterday, Today, Tomorrow, Capt. Sanders Laubenthal Air Force Institute of Technology, 1979
Institutional Self-Study Report for Reaccreditation Review by North Central Association of Colleges and Schools, Air Force Institute of Technology, 1990, 2000.

Appendix: Aerospace Curricula

Table I
Engineering School – 1935
(One-year long with 29.5-32.5 contact hours per week;
~34 semester hours estimated using 3:1 ratio)

I. Shop work, Factory Prod and Processing
   (~ 5 SH)

Drawing	42 hr
Shop Practice	87.5 hr
Inspection	50 hr
Other Topics	42 hr

II. Strength of Materials and Structures
   (~ 8 SH)

Mechanics	105 hr
Strength	87.5 hr
Structures	52.5 hr
Chemistry, Metallurgy and Test	92.5 hr

III. Thermodynamics and Engine Design
   (~ 7 SH)

Thermodynamics	70
Engine Laboratory	77 hr
Engine Design	133 hr
Other Topics	20 hr

IV. Theoretical Aviation
   (~13 SH)

Lighter-than-air craft	60 hr
Aerodynamics	150 hr
Propeller Design	60 hr
Airplane Design	240 hr
Performance/Flight-Test	60 hr

V. Miscellaneous   (~ 1 SH)
   Armament, Radio, and Instrumentation

Table II
College of Engineering Sciences, 1946-1947
x-y denotes weekly hours of lecture - laboratory
(Year consisted of 4 semesters plus 3 weeks vacation)

Semester I (18 credits)

Mathematics	5-0
Physics	3-3
Engineering Reports	2-0
Drawing Lab	0-6
Logistics	2-0
Mechanics	3-0

Semester II (17 credits)

Mathematics	3-0
Strength of Materials	3-0
Testing Lab	0-6
Materials	3-0
Electrical Engineering	3-3
Business Management	2-0

Semester III (0 credit)
Orientation to WP

Semester IV (16 credits)

Mathematics	3-0
Strength of Materials II	3-0
Electrical Engineering	3-3
Fluids	3-3
Graphical Analysis	2-0
Processing	2-0

Semester V (18-19 credits)

Aerodynamics	3-0
Thermodynamics	3-0
Machine Design	3-3
Controls	3-3
Electives	4-0 or 5-0

Semester VI (18-19 credits)

Aerodynamics	3-0
Thermodynamics	3-0
Machine Design	3-3
Measurements	3-3
Electives	4-0 or 5-0

Table III
School of Engineering- 1954
Aeronautical Engineering – Graduate
(Year consisted of 4 Quarters and 4 week short term)

Reviews in Dynamics and Mathematics

First Quarter

Theoretical Aerodynamics	4
Thermodynamics	4
Aircraft Structures	4
Advanced Engineering Math	4

Second Quarter

Applied Aerodynamics	3
Thermodynamics of Gas Flow	4
Airplane Detail Stress Analysis	4
Advanced Engineering Math II	4
Compressible Flow Lab	1

Third Quarter

High Speed Flight Problems	4
Aeronautical Lab Techniques I	2
Aircraft Propulsion Systems	4
Adv. Engineering Math III	4
Technical Reporting Techniques	3

Fourth Quarter

Aircraft Preliminary Design	5
Seminar	1
Experimental Stress Analysis	4
Aircraft Structural Vibrations	4

Aerodynamics Option
Fifth Quarter

Aeronautical Lab Techniques II	3
Advanced Fluid Dynamics	4
Independent study	3
Research Management	3

Sixth Quarter

Aircraft Stability and Control	4
Seminar	1
Independent Study	4
Elective	3-4

Propulsion Option
Fifth Quarter

Heat Transfer	4
Gas Dynamics I	4

Independent Study	3
Research Management	3

Sixth Quarter

Advanced Propulsion	4
Seminar	1
Independent Study	4
Elective	3-4

Structures Option
Fifth Quarter

Special Methods in Structures	4
Aeroelasticity	4
Independent Study	4
Research Management	3

Sixth Quarter

Design Problems of Structures	4
Seminar	1
Independent Study	4
Elective	3-4

Later modifications expanded options by adding two advanced courses. The independent study was increased to 12 credits.

Table IV
School of Engineering and Management - 2001
Aerospace and Systems Engineering

Aeronautical Engineering
Reviews: Dynamics; Aeronautics; Computers; Mathematics

A. Core Program (7 courses)
Low or High-Speed Aerodynamics; Aircraft Stability;
Solid Mechanics; Propulsion; Materials Seminar;
Two courses in mathematics

B. Two Specialty Sequences
(3 courses each: 11-12 Q credits)

Aerodynamics
Computational Fluid Dynamics
Air Breathing Propulsion
Heat Transfer
Aircraft Stability and Control
Control and Optimization

Structural Analysis
Vibration Damping and Control
Structural Materials
Air Weapons
System Analysis and Design
Aerospace Robotics

C. Independent Study (12)
<u>**Astronautical Engineering**</u>
Reviews: Dynamics; Physics; Computers;
Mathematics

A. Core Program (10 courses)
Intermediate Dynamics; Control and State
Space;
Linear Systems Analysis; Space Flight Dynamics;
Attitude Determination and Control;
Satellite Communications; Space Surveillance;
Space Environment;
Two courses in mathematics

B. Two specialty sequences
(3 courses each 11-12 Q credits)
Mechanics and Control of Space Structures
Aerospace Robotics
Aerospace Navigation
Advanced Astrodynamics
Vibration Damping and Control
Rocket Propulsion
Space Facilities
Structural Analysis
Structural Materials
Estimation Theory
Control and Optimization
Reliability

C. Independent Study (12)

<u>**Systems Engineering**</u>
Reviews: Dynamics; Computers; Mathematics

A. Core Program (7 courses)
Linear Systems; Operations Research;
Introduction to System Design
Systems Optimization OR Spacecraft Systems
OR
Systems Design;
Life-Cycle Cost OR Engineering Economics;
Two courses in probability and statistics

B. Two specialty sequences
(3 courses each: 11-12 Q credits
One of these:
Operations Research
Optimization
Reliability
Simulation
Plus one approved sequence
in any area of engineering

C. Group Design Study (12)

799

CHAPTER 63

Naval Officer Graduate Education in Aerospace Engineering at the Naval Postgraduate School

Max F. Platzer, Distinguished Professor and Chairman
Richard W. Bell, Louis V. Schmidt, Conrad F. Newberry, Professors Emeriti
Department of Aeronautics and Astronautics
Naval Postgraduate School
Monterey, CA 93943

Abstract

Although the Naval Postgraduate School (NPS) was established in 1909 on the campus of the United States Naval Academy, the Department of Aeronautics was not established until 1947. The initial mission of the Department of Aeronautics was to better prepare naval aviation officers for the transition from piston engine powered aircraft to gas turbine powered jet aircraft. In 1987 the Department was expanded to include the field of astronautics. In this paper the educational objectives, programs, and developments in the major areas of concentration are briefly described for the purpose of providing a historical perspective on the Department's development, major accomplishments and current status.

Introduction

Navy and Marine Corps aircraft are designed to operate aboard ships as part of a larger battle group. Challenges normally not considered by aircraft operating from land bases become design constraints for shipboard compatibility. Therefore, the U.S. Navy has found it necessary since the beginning of naval aviation in the 1920s under the leadership of Admiral Moffett to establish its own aircraft design bureau, the Bureau of Aeronautics (now the Naval Air Systems Command), and to involve a sufficient number of naval officers in the design, development, and testing of naval aircraft. This, in turn, necessitated officer education in aeronautical engineering beyond the baccalaureate degree. From this requirement evolved a number of courses in aeronautical engineering which led to the formal establishment of a Department of Aeronautics in the Naval Postgraduate School in 1947. It is the objective of this paper to describe the major developments since that time and to highlight some of the major accomplishments and the present status of aerospace engineering education at the Naval Postgraduate School.

The decisive role played by ship-based aircraft in World War II left no doubt at war's end about the need to fully transition from battle ship to carrier-based naval operations. This, in turn, required the transition from piston engine powered propeller aircraft to gas turbine powered jet aircraft. To support the technical challenges posed by this transition, Dr. Wendell Coates, an engineering professor in the Naval Postgraduate School located on the Naval Academy campus, proposed and obtained approval to establish an aeronautical engineering department in 1947. He

---

served as its chairman until 1962 and, after the School's transfer to Monterey, was instrumental in attracting additional faculty and providing the Department with outstanding laboratories. He was succeeded by Professor Richard Bell (until 1978), Professor Platzer (until 1988), Professor Wood (until 1991), Professor Collins (until 1997), Professor Lindsey (until 1999) and Professor Platzer again since January 2000. Important additions to the Department's curricula occurred in 1979 and 1987, respectively, when the joint Naval Postgraduate School/Naval Test Pilot School and the space systems engineering/ operations curricula were implemented. Both additions reflect the Navy's recognition of the continuing need for flight test expertise and the growing use of satellites for naval operations.

Educational Objectives

The aeronautical engineering curricula have been fully accredited for the past fifty years; the astronautical engineering curriculum became accredited in 1995. Both curricula were reviewed again by ABET (Accreditation Board for Engineering and Technology) in October 2001 and received full six-year ABET re-accreditation in July 2002. In addition to the ABET requirements, the two curriculum sponsors, the Naval Air Systems Command and the Naval Space Systems Division, impose several additional requirements, namely:

• **Cost Effective Education:** In order to minimize the time spent on graduate studies the Department offers year-round instruction and course sequencing such that the officer students may enroll at any time during the year. Also, students who would not attempt a graduate degree because of their undergraduate background or their time away from academia are brought up to the necessary standards.

• **Broad-Based Engineering Education:** The officers need to be provided with a broad-based aeronautical or astronautical engineering education to qualify for future assignments in a variety of jobs rather than in a narrow specialty. Also, they need to be familiarized with modern computing, design, measurement and testing techniques. This calls for the availability and maintenance of modern laboratories.

• **Navy/DOD Relevant Engineering Education:** The officers need to be familiarized with problems in past and current aircraft and weapon systems developments in order to sensitize them to the current state-of-the-art and to the uncertainties involved in a typical development program. This calls for faculty with previous industrial experience and strong involvement in current Navy/DOD projects.

• **Total Systems Design Education:** It is well recognized that the success of any aircraft or spacecraft crucially depends on competent integration of all the disciplines needed for their design and development. Thus, a comprehensive capstone design project has been made a firm requirement in both curricula. Moreover, the aircraft or spacecraft, in turn, is part of a larger weapons system. Therefore, there is a need to expose the students to the Total Systems Design approach. By working together with other students and faculty in the recently established Institute for Defense Systems Engineering and Analysis, Aero/Astro students learn how air and space vehicles become part of a larger combat system that includes all aspects of war fighting.

Educational Programs

The typical graduate from the normal two-year NPS Aeronautical or Astronautical Engineering curricula is a career military officer who will, in the future, encounter many technical challenges currently unknown by both the student and the faculty. As a result, the best education is believed to consist of a program that will address the fundamentals of engineering and scientific principles, including experience in the application of these principles to unique Navy/DOD problems and issues. The NPS student population is composed of officers from the U.S. services and foreign countries. This joint, international class composition provides additional leverage in defining the challenges of the future. The department offers four curricula, which are available to U.S. and foreign military officers and U.S. government civilian employees: Aeronautical Engineering, Aeronautical Engineering (Avionics), NPS/Test Pilot School Program, and Astronautical Engineering.

Aeronautical Engineering
Aeronautical Engineering (Avionics)

Both of the Aeronautical Engineering Programs are designed to meet specific needs of Navy technical managers with a broad-based graduate education in aerodynamics, flight mechanics, propulsion, flight structures, systems integration and/or avionics. Additionally, students receive graduate level instruction in aircraft/missile design and aero-computer science. The programs are divided into preparatory, graduate and advanced graduate phases. During the advanced graduate phase, all students receive in-depth graduate coverage through advanced electives in areas of their choice including flight dynamics, gas dynamics, propulsion, structures, avionics, and aircraft or missile design. Over the past fifty years the number of students completing these programs varied from 20 to 80 per year.

NPS/Test Pilot School Cooperative Program

The NPS/Test Pilot School Cooperative Program combines portions of the Aeronautical Engineering and Aeronautical Engineering (Avionics) curricula with the complete U.S. Naval Test Pilot School syllabus. After completion of the requirements at NPS, students proceed to Patuxent River for the full Test Pilot School Curriculum. This program is very competitive, and students accepted to this program are typically exceptional undergraduate engineering students and aviators who are capable of completing all the graduate education coursework in 5-6 quarters. Graduates receive the Master's degree in Aeronautical Engineering at the completion of Test Pilot School. During the past 14 years ten students per year completed this program.

Astronautical Engineering

The Astronautical Engineering program provides officers with a comprehensive scientific and technical knowledge of military and Navy space systems through graduate education. This curriculum is designed to equip officers with the theoretical and practical skills required to design and integrate military space payloads with other spacecraft subsystems. Graduates will be prepared by their education to design, develop and manage the acquisition of space communications, navigation, surveillance, electronic warfare and environmental sensing systems. Since its inception in 1987 the number of students per year who graduated from this program varied from 15 to 25.

Special Courses

The Department of Aeronautics and Astronautics provides curriculum strength in a number of areas. Some of the strengths are in rare, if not unique, courses and course sequences.

Design. Students are required to participate in a major design project where they perform as members of a design team to an aircraft, avionics, spacecraft or missile design project. At the conclusion of the design project, the student teams present their work to a panel comprised of military, science, and industrial experts. Unless the design project is completed for NAVAIR or some other branch of the Department of Defense, the design teams often elect to compete with other universities in national graduate design competitions sponsored by such organizations as the American Institute of Aeronautics and Astronautics (AIAA) or the American Helicopter Society (AHS). NPS student teams have consistently ranked high in these competitions, lending credence to the quality of their efforts.

Survivability/Lethality. Aircraft Combat Survivability and Air Defense Lethality are two courses believed to be unique to this NPS program. Both courses were developed by Distinguished Professor Robert E. Ball who is the author of the only textbook in this relatively new field of lethality and survivability.

Aircraft combat survivability brings together all of the essential ingredients in a study of the survivability of fixed-wing aircraft, rotary-wing aircraft, and cruise missiles operating in a hostile (non-nuclear) environment. The technology for increasing survivability and the methodology for assessing the probability of survival in such an environment are presented in some detail. Topics covered include current and future threat descriptions; mission/threat analysis; combat analysis of Southeast Asia and Desert Storm losses

The air defense lethality course is concerned with the design and effectiveness of anti-aircraft guns and missiles. Target detection, target signatures, warheads, fuzes, damage mechanisms, and countermeasures are considered. Total system lethality is evaluated by determining the probability of target kill given a single shot and given an encounter.

Missile Option. An elective five-course Missile Option is available for qualified students. However, most of the students who take advantage of this option are in the Combat Systems Sciences and Technology (physics) curriculum. A gas dynamic, thermodynamic, and fluid dynamic background is prerequisite for the option. The five required courses are Missile Aerodynamics, Tactical Missile Propulsion, Missile Flight Analysis, Air Defense Lethality, and Missile Design. Non-aero students taking this option often take additional elective course work in aerodynamics, performance, controls, and structures. Students taking the missile option often qualify for the Master of Science in Engineering Science degree.

The combat systems students bring an academic background in electromagnetic radiation, signal processing, explosives, warheads, combat simulation, detection and engagement elements. Most students also have some academic background in the military acquisition process. The missile option serves as a sequence of service courses for the combat systems curriculum.

Research. Students are required to complete a thesis project in order to receive their degree. The thesis serves as an integral part of the NPS education process by giving students an opportunity to conduct individualized research in a subject of their choosing. At the completion of their thesis project, students present their work to the faculty and students. Often, the students' contribution is part of a larger research project by his/her thesis advisor, and therefore is routinely presented by the student (or his/her advisor if the student has already graduated) at scientific conferences and published in scientific journals. An exception to the thesis requirement is made for the NPS/TPS students, whose final flight test report at TPS serves in lieu of a thesis.

Interdisciplinary Efforts. Interdisciplinary efforts combine faculty and students from across campus. The Department of Aeronautics and Astronautics is an integral partner in interdisciplinary projects which bring together students from across campus to participate in a "total concept" analysis. Students from the engineering disciplines provide the technical design, operations research students provide logistics and analysis, national security affairs students provide political-military perspectives, business students address manning and costs, applied science students tackle the environment. The project for 2001 was called "Crossbow" which originated with the President of the Naval War College who proposed studies to determine the feasibility and operational worth of a small, high-speed aircraft carrier concept. NPS students chose to pursue a high-speed ship design that supports an air wing composed primarily of Unmanned Air Vehicles.

Department Faculty

Brij N. Agrawal, Distinguished Professor, Associate Fellow AIAA
Oscar Biblarz, Professor
Russell W. Duren, Associate Professor
Garth Hobson, Professor, Associate Fellow AIAA
Richard M. Howard, Associate Professor
Isaac I. Kaminer, Associate Professor
David W. Netzer, Distinguished Professor
Max F. Platzer, Distinguished Professor, Fellow AIAA
I. Michael Ross, Associate Professor, Associate Fellow AIAA
Raymond P. Shreeve, Professor, Associate Fellow AIAA
Michael G. Spencer, Assistant Professor
E. Roberts Wood, Professor, Associate Fellow AIAA

Adjunct Teaching Faculty:
S.K. Hebbar, Senior Lecturer, Associate Fellow AIAA
Barry A. Leonard, Visiting Associate Professor

Research Faculty:
Christopher M. Brophy, Assistant Professor
M.S. Chandrasekhara, Professor, Associate Fellow AIAA
Kevin D. Jones, Associate Professor
Ramesh Kolar, Assistant Professor
Jose Sinibaldi, Assistant Professor
Oleg Yakimenko, Associate Professor, Associate Fellow AIAA

Emeritus Faculty:
Robert E. Ball, Distinguished Professor, Fellow AIAA

Richard W. Bell, Professor
Allen E. Fuhs, Distinguished Professor, Fellow AIAA
Charles Kahr, Professor
Donald M. Layton, Professor
Gerald H. Lindsey, Professor
James A. Miller, Associate Professor
Conrad F. Newberry, Professor, Fellow AIAA
Louis V. Schmidt, Professor, Associate Fellow AIAA
Edward M. Wu, Professor
Robert D. Zucker, Associate Professor

Curricula

On December 22, 1951, by order of the Secretary of the Navy, the United States Naval Postgraduate School was officially disestablished at Annapolis, Maryland, and established at Monterey, California. During the period from November 21, 1951 to February 16, 1952, the entire Naval Postgraduate School at Annapolis – faculty, students and equipment, was moved to Monterey. This move, unique in character, involved the transcontinental transportation of approximately five hundred families, civilian and military, their household effects, and some three million pounds of school equipment. What had been the U.S. Naval Postgraduate School, Annapolis, was redesignated the Engineering School of the U.S. Naval Postgraduate School. The aeronautical engineering curricula have evolved over time as indicated by the following discussion.

In the 1950s the M.S. degree in aeronautical engineering required three years for completion. The first two years were taken at the Naval Postgraduate School. Qualified students then were selected to take the third year at a civilian engineering school (Princeton, Michigan, Stanford, CalTech, Stevens Institute of Technology, Cranfield Institute of Technology, etc.). The third year courses at the various civilian institutions were arranged to provide emphasis on such fields as aircraft structural analysis, aircraft propulsion, high-speed aerodynamics, seaplane design, pilotless aircraft, aircraft performance, as well as general aeronautical engineering. Completion only of the two-year curriculum provided the student with what amounted to a Master's degree level of education. However, the Master's degree was not awarded. The first year of the two-year program normally covered the junior and senior years of a civilian university. The second year included airplane design, flight analysis, flight testing, electronics, human engineering, principles of industrial organization etc. When practicable, a summer period was spent at a civilian institution in classes offered in industrial engineering, prior to reporting to a new duty station.

In the 1960s the requirement to take the third year at a civilian institution was dropped and the M.S. in aeronautical engineering was awarded after two years at the Naval Postgraduate School. The aeronautical engineering faculty had grown from nine in 1952 to nineteen in 1968 and the faculty became more research oriented leading to the award of the first Ph.D. degrees.

In 1979 an agreement was signed with the U.S. Naval Test Pilot School (TPS) at Patuxent River, Maryland, to offer a joint program wherein highly qualified students could enroll in a five or six-quarter aeronautical engineering program, followed by the standard one-year TPS program, qualifying the student for the award of the Master's degree in aeronautical engineering at the completion of the TPS program. Over the years this program led to the enrollment of ten students per year in the joint NPS/TPS program.

In 1987 the Department of Aeronautics became the Department of Aeronautics and Astronautics when it started to offer a curriculum in astronautical engineering. The average number of aeronautical and astronautical engineering students in the Department substantially exceeded one hundred, but then started to drop in the 1990s together with a gradual decline in the number of tenure-track faculty to ten at the beginning of 2003.

Students typically receive their orders to report to the Department after completion of their first or second operational tour of duty, and therefore have been away from the academic environment for some four to six years. Therefore, one or two refresher quarters are offered to ease some of these students back into the academic environment. Six-week refresher courses are also available.

The requirements for entry into the aeronautical or astronautical engineering programs consist of a baccalaureate degree, or its equivalent, with an above-average Quality Point Rating, preferably in engineering or the physical sciences; mathematics through differential and integral calculus; and completion of a calculus-based physics sequence.

A typical course sequence leading to the Master of Science in aeronautical engineering is:

Quarter 1: flight structures, digital computation, differential equations, software methodology
Quarter 2: matrix analysis, gas dynamics, aerodynamics, partial differential equations
Quarter 3: flight mechanics I, aircraft navigation, digital avionics I, structural analysis
Quarter 4: propulsion, measurement techniques, aerodynamic analysis, aircraft design tools
Quarter 5: aeromechanics/aeroelasticity, flight mechanics II, flight controls, digital avionics II
Quarter 6: design course, aircraft survivability/reliability/safety engineering, thesis, elective
Quarter 7: aircraft structural fatigue, thesis, thesis, elective
Quarter 8: thesis, elective, elective, elective

The aeronautical engineering/avionics curriculum offers courses in electronics, communications engineering, radar systems, avionics software engineering, digital avionics, and avionics system design.

A typical course sequence leading to the Master of Science in astronautical engineering is:

Quarter 1: digital computation, structures, differential equations, digital logic circuits
Quarter 2: orbital mechanics, space environment, matrix analysis, controls
Quarter 3: spacecraft dynamics, signal analysis, remote sensing, guidance & control
Quarter 4: spacecraft propulsion, communications engineering, systems dynamics, vibrations
Quarter 5: attitude dynamics & control, electromagnetic waves, microprocessors, astrodynamics
Quarter 6: spacecraft design tools, space power, military space applications, thesis
Quarter 7: spacecraft design I, spacecraft communications, thermal control, smart structures
Quarter 8: spacecraft design II, elective, elective, thesis
Quarter 9: systems acquisition, thesis, thesis, elective

Laboratories

Laboratories support instructional and research programs in aerodynamics, flight mechanics, flight controls, avionics, structures and composite materials, scientific computing, aircraft and spacecraft design, gas dynamics, turbopropulsion, rocket and ramjet propulsion, and dynamics and nondestructive evaluation. The major facilities include two low speed wind tunnels with 28-by-45-

inch and 3-by-5-foot test sections, a 5-by-5-foot open circuit flow visualization wind tunnel, a 15-by-20-inch water tunnel, a supersonic blow-down tunnel with a 4-by-4-inch test section, a shock tube, three test cells equipped with diagnostic apparatus for investigating solid, liquid, gaseous and hybrid rockets, solid fuel ramjets, pulse detonation engines, and gas turbine combustors, a 10-by-60-inch test section rectilinear cascade wind tunnel, a large three-stage axial research compressor, two fully instrumented transonic turbine and compressor test cells, a spin-pit for the structural testing of rotors up to 50,000 RPM, a transonic cascade wind tunnel, two flight simulators, an unmanned air vehicle research laboratory, an MTS electro-hydraulic closed-loop fatigue testing machine, a flexible spacecraft simulator, a space robot simulator, a three-axis spacecraft simulator, a Navy communications satellite, a smart structures laboratory, a flight controls laboratory, an avionics laboratory, a computation laboratory, and aircraft and spacecraft design laboratories.

Major Areas of Concentration

The major curriculum areas of concentration include disciplinary, research, and design components. Typically students select only one area of design (from the four areas available in aeronautical engineering) and one area of research.

System Design

Since the inception of the Department, system design has been recognized as an important ingredient for the education of aeronautical engineering duty officers. After graduation, naval officers may be assigned to class desks or program acquisition offices. NAVAIR sponsored research by Professor Ulrich Haupt in the mid-70s addressed the basic needs of aircraft design education in the United States [1,2]. The quality of the NPS curriculum design component can, in part, be measured by student success in national graduate design competitions, contributions to advanced system development at the Naval Air Systems Command (NAVAIR), and national recognition of outstanding system design instruction. Depending upon their particular program, students will be involved in one or more system design projects. These projects may be aircraft, missile, rotary wing, engine, avionics, spacecraft, and/or weapons system design.

Aircraft Design. For more than a decade, Professor Newberry taught aircraft design at the Naval Postgraduate School in a team environment. Each team consists of some 6-10 students functioning as an Integrated Process and Product Development Team (IPPDT). The two-quarter course design effort takes place in a special Aeronautics Design Laboratory containing computers, software, design literature, and work spaces. It should be noted that the same design laboratory is used by aircraft, missile, rotary wing, and end engine design teams (without mutual interference). The aircraft design team effort is in response to a Request-for-Proposal (RFP) generated by a variety of sources, e.g., NAVAIR, national AIAA graduate design team competitions, or sponsored research. Each design team addresses aerodynamic, structural, propulsion, stability & control, performance, total ownership cost, survivability, system effectiveness, weight, armament, deployment, utilization, maintainability, manufacturing, and system parameter trade-off issues impacting the subject aircraft system. At the completion of the project, the students present their design effort to an evaluation panel consisting of representatives from NAVAIR, NASA, industry, and academe; the composition of the panel varies from presentation to presentation. The student design effort is documented in a comprehensive report which, when applicable, is submitted for national competition judging.

One indication of the quality of the design experience is evidenced by NPS team placements in AIAA/McDonnell Douglas national graduate student design competitions. Aircraft design team placements during the 1990s were 2nd and 3rd in 1992-93, 1st in 1993-94, 1st and 2nd in 1994-95, 1st and 2nd in 1995-96, 1st and 2nd in 1996-97 and 3rd in 1997-98.

From 1990 to 2002, the aircraft design teams were directed by Professor Newberry. He was honored with the Fred Merryfield Award of the American Society for Engineering Education (ASEE) for excellence in teaching engineering design in 1997.

However, not every NPS aircraft design team enters a national competition. During 1993 and 1994, for example, an M = 6 waverider configured interceptor was the focus for two student design teams supporting independent NPS waverider research [3].

From 1999 through 2001, NPS aircraft design teams supported NAVAIR advanced UCAV planning studies. This support resulted in a number of NPS graduate student team generated Uninhabited Combat Air Vehicle (UCAV) configurations of interest to NAVAIR and ONR. This UCAV design experience also enabled NPS aircraft design students to support the NPS Systems Engineering CROSSBOW Project discussed below.

Missile Design. Interested students may take an elective five-course sequence in missile systems. Courses in missile aerodynamics, tactical missile propulsion, missile flight analysis, and air defense lethality precede a single course in missile design. Generally, an NPS missile design team will consist of from six to ten students, but due to the elective nature of the missile option sometimes results in a team as few as five members. The team functions as an IPPDT.

The missile design effort is in response to an RFP generated by the naval or military service, NASA, a national academic design competition, or some defense related source. Each design team addresses aerodynamic, structural, propulsion, stability & control, trajectory, performance, cost, storage, weight, system effectiveness, lethality, warhead, deployment, utilization, manufacturing, and systems parameter trade-off issues impacting the subject missile configuration. As in the aircraft design project, the students present their design effort to an evaluation panel and document their work in a comprehensive report. When applicable, the report is submitted for national design competition judging.

From 1990 to 2002, Professor Conrad F. Newberry directed the NPS missile design teams. One measure of the quality of the missile design experience is provided by their placements in the national AIAA/Northrop Grumman missile design competitions, namely 1st and 2nd in 1992-93, 2nd and 3rd in 1993-94, 2nd in 1996-97. Due to decreasing student demand the sequence was offered infrequently during the late 1990s.

Rotary-Wing Design. The procedures for design of helicopters and other types of rotorcraft parallel that of fixed-wing aircraft design. Typically, 6-10 students work on the helicopter design project, which is a response to a Request for Proposal written by one of the three major U.S. helicopter manufacturers (Sikorsky, Boeing, and Bell).The design competition is sponsored, judged and managed by the American Helicopter Society (AHS). Student teams advised by Professor E. Roberts Wood achieved 1st place in the 1995 AHS/NASA Student-Industry Helicopter Design Competition and 2nd place in the 1996, 1997, 1998 and 1999 AHS/NASA Student-Industry Helicopter Design Competitions.

Engine Design. The engine design experience consists of a one quarter course. An engine RFP, either a past military RFP or from a current AIAA competition, is selected for the course. For example, the JAST RFP (which evolved into the Joint Strike Fighter (JSF) program) was chosen as the design problem when the course was concurrently taught by Distance Learning to the Naval Air Warfare Center. Each student performs constraint and mission analyses, selects and sizes an engine, in the first half of the course. A selection is then made from these candidate designs. In the second half of the course, the class, working as a team, carries out the preliminary design of the components. GASTURB is used in the engine selection phase. Codes developed in-house are used for the fan, compressor and turbine designs. The student team advised by Professor Raymond Shreeve placed 2nd in the 1997 AIAA/Rockwell Rocketdyne Engine Design Contest.

Avionics Design. Digital design and hardware/software integration is taught by Associate Professor Russell Duren through a series of small design projects and a more complex final project, such as the development of video controllers or serial communications controllers. PCs are equipped with modern CAD software and instrumentation for digital design. Designs may be entered in any combination of schematics, HDLs (Hardware Description Language) including VHDL and Verilog, or commercially available IP (Intellectual Property) modules. Hardware designs are verified using computer-aided functional and timing simulation tools. Assembly language and C programs are verified using microprocessor simulation programs and commercial software development tools. The designs are then implemented using combinations of FPGAs (Field Programmable Gate Arrays ranging from 10,000 to 1,000,000 gates) and micro-controllers. The designs are then verified using PC-based logic analyzers and digital oscilloscopes.

Spacecraft Design. The spacecraft design is taught in a dedicated laboratory that uses computer-aided design tools, such as GENSAT, Aerospace Conceptual Design Center software, STK, NASTRAN, IDEAS, and MATLAB/ Simulink. A student team advised by Professor Brij Agrawal achieved 2nd place in the 1997 AIAA/Lockheed Martin Spacecraft Design Contest. During summer 2000, the students finished a preliminary design of a Bifocal Relay Mirror Spacecraft under the sponsorship of Air Force Research Laboratory (AFRL).The Bifocal Relay Mirror Spacecraft is composed of two optically coupled telescopes used to redirect the laser light from ground-based, aircraft-based or spacecraft-based lasers to a distant point on the earth or in space. The design effort identified the need to develop new technologies for beam acquisition, tracking and pointing.

Weapons (Total) System Design. During the 2001/2002 academic year, the Department of Aeronautics and Astronautics aircraft design teams worked with the Total Ship System Engineering (TSSE) students (ship design) and the Systems Engineering students (system requirements) to develop the CROSSBOW battle force system. CROSSBOW is essentially a small, fast carrier task force supporting global, littoral warfare scenarios. The Systems Engineering students developed the CROSSBOW ship and aircraft requirements. The TSSE students designed the small, fast carrier, *SEA ARCHER* .The Aeronautics students designed two aircraft capable of operating from the *SEA ARCHER* .SEA ARROW was designed for an armed reconnaissance (UCAV) mission; SEA SPECTRUM was designed for the intelligence, surveillance, reconnaissance and combat (ISRC)UCAV mission. SEA SPECTRUM also has the capability to operate unassisted from an LHA.

Aerodynamics, Aeroelasticity, V/STOL Aircraft.

A better understanding of viscous flow effects throughout the whole Mach number and Reynolds number regimes has been and continues to be a serious challenge for the design and operation of various aerospace vehicles and propulsion systems. Of special importance is the prediction and measurement of the onset of flow separation (stall) on airfoils, three-dimensional wings, and helicopter and jet engine blades. For this reason Distinguished Professor Platzer has developed a computational and experimental research program to investigate steady and unsteady flow problems relevant to naval aircraft and weapons problems.

To this end, he also established a joint program with the NASA Ames Research Center in 1986. Professors Bodapati and Chandrasekhara developed a special wind tunnel, located in the Fluid Mechanics Laboratory of NASA Ames Research Center, which permits the detailed measurement of the dynamic stall flow phenomena on helicopter blades using modern point diffraction interferometry and Laser-Doppler velocimetry. These measurements are complemented by Navier-Stokes computations in the NPS Computation Laboratory (which has 17 Silicon Graphics workstations and a parallel cluster of fifteen PCs running Linux), and on NASA and DOD supercomputers. Experiments are also performed in the NPS Aerodynamics Laboratory which consists of a low-speed flow visualization tunnel with a 5x5 inch test section and a 15x20 inch water tunnel. Laser Doppler velocimetry is available in both tunnels.

Current projects address the computational prediction of abrupt wing stall on F-18 wings using modern Navier-Stokes codes and the development of helicopter blades capable of controlling the onset of dynamic stall. Another major project has been and is directed at the development of a micro-air vehicle which uses flapping wings requiring detailed wind/water tunnel studies and computations of the flow past flapping wings. Also, a new type of lift fan, the cross-flow fan, is being investigated experimentally and computationally to analyze the aerodynamic fan characteristics, optimize thrust and propulsive efficiency, and determine the applicability of cross-flow fans to VTOL aircraft. The current members of the aerodynamics research group are Research Professor M.S. Chandrasekhara and Research Associate Professor K.D. Jones. Past members and visiting researchers were J.M. Simmons, U Queensland, K. Vogeler, TU Aachen, H.H. Korst, U. Illinois, J. Ekaterinaris, KETA Greece, W. Sanz, TU Graz, F. Sisto, Stevens IT, W. Geissler, DLR Goettingen, I. Tuncer, METU Ankara, J. Lai, Australian Defense Force Academy Canberra, M. Nakashima, IT Tokyo, T. Fransson, EPFL Lausanne, S. Weber, TU Aachen, C. Dohring, German Armed Forces University Munich.

Flight Mechanics and Control

Extensive teaching and research work in these fields address real fleet-and-field problems in the areas of unmanned air vehicle performance; flying qualities; guidance navigation and control; precision airdrop of military re-supply; use of Unmanned Air Vehicles (UAVs) for winds extraction and particle sensing for chemical/biological attack response; and integrated plant controller optimization for high speed civil transport aircraft. Current faculty supporting these efforts include Associate Professor R. Howard, Associate Professor I. Kaminer, Research Associate Professor O. Yakimenko, and NRC Associate V. Dobrokhodov. Professor Emeritus L.V. Schmidt, author of the AIAA text "Introduction to Flight Dynamics", continues to contribute his expertise.

Several laboratories support this work. The Unmanned Air Vehicle Flight Research Laboratory (UAV FRL) is used to conduct flight research with scaled radio-controlled and semi-autonomous aircraft to study problems identified with fleet UAVs and to design, implement and test new concepts in flight performance, flying qualities, guidance, navigation and control. Research vehicles include fixed-wing and rotary wing platforms. Telemetry is available for transmission of data, video images, and infrared images. The Flight Controls Laboratory presently consists of four hardware-in-the-loop stations designed to conduct extensive hardware-in-the-loop studies of guidance, navigation and control systems. These stations are supported by a family of Realsim and MATLAB rapid prototyping tools.

Aircraft Structures

Studies of aircraft structures include classical approaches, finite element methods, fatigue life estimations, and design of composite aircraft structures. Coverage in this discipline has been a focal point for the department since its founding, involving Distinguished Professor W. Coates, Professors C. Kahr, L.V. Schmidt, G. Lindsey, and more recently Professor E. Wu, who established a composite materials laboratory. Academic studies have been supported by laboratory experiments using test machines to demonstrate both tensile and shear stress effects upon structures, full-scale wing structures to show properties of multicell thin-wall beams under combined bending and torsion loadings, fatigue testing, and composite material behavior.

Turbo-Propulsion and Gas Dynamics

As already mentioned, the transition from piston engines to jet engine technology presented the U.S. Navy with a special challenge. Therefore, approval was obtained shortly after the Department's establishment in 1947 to construct and equip a "Turbo-propulsion Laboratory" on the new campus in Monterey. Distinguished Professor Michael Vavra was instrumental in establishing the laboratory and directing it until his death in 1975. The laboratory (TPL), now operated together with the Gas Dynamics Laboratory (GDL), comprises three large and unique buildings. The compressed-air power systems in the three buildings supply facilities operating in three different speed regimes. The low-speed building houses a large high Reynolds number cascade wind tunnel, radial cascade wind tunnel, and a large three-stage low-speed research compressor. The high-speed building contains a 1200HP air supply system and two explosion-proof test cells for transonic compressor and turbine testing; an engine-scale vacuum spin pit; a probe calibration and turbocharger test facility; control, data acquisition and computer rooms; three offices and a conference room. The Gas Dynamics Laboratory building, with a compressed air system providing 8000 cubic feet of storage at 20 atmospheres and 2000 scfm continuously, contains a variable Mach number supersonic wind tunnel, a small transonic cascade wind tunnel, two free-jets and a three-inch shock tube. Also, two micro-jet engine test stands are installed; one in a free jet.

While Vavra initially brought young postdoctoral engineers from Europe to work with him (Willi Schlachter from Switzerland and K. Papailiou from Greece), after a research charter was written for TPL in 1978, well recognized visiting professors and research investigators were invited to work with Vavra's successor, Professor R. Shreeve. Examples include Dr. Dan Adler (Technion, Israel), Mr. Roy Peacock (Cranfield, U.K.), Mr. John Erwin (NASA), Dr. H.J. Heinemann (DLR Germany), Professor Charles Hirsch (Vrije U, Belgium), Dr. Greg Walker (U. Tasmania, Australia), Professor John Kentfield (U. Alberta, Canada), Professor Ahmet Ucer (ODTU, Turkey) and Dr. Theo von Blackstrom (U. Stellenbosch, S. Africa). Dr. Atul Mathur (VPI, Virginia) spent three years at TPL. Also, research studies were performed at TPL toward doctoral degrees granted later at their home institutions; examples include Hans Zebner, Dieter Schulz and Thomas Vitting at U. Aachen; Friedrich Neuhoff at GAFU, Munich; and Ian Moyle at U. Tasmania. Postdoctoral NRC Research Associates also included Shmuel Eidelman, David Helman, Upender Kaul and (currently) Anthony Gannon. The laboratory has received recognition particularly for the development of the Dual Probe Digital Sampling (DPDS) Technique for rotor exit flows, for viscous code validation measurements in controlled-diffusion compressor cascades, for reviving interest in wave-rotors and wave engines (leading to a NASA experimental program), and for the first successful operation of an air-breathing detonation engine.

Practical instruction and advanced research in air-breathing propulsion and gas dynamics remain the charter functions of the Turbo-Propulsion Laboratory, an NPS Research Center, and the Gas Dynamics Laboratory under the direction of Professor Raymond Shreeve, working in close association with Professor Garth Hobson. Realistic (engine-scale) experimental studies are enabled by, for a university, unusually high power levels, large scale or high speeds of the test rigs. Exploiting this uniqueness, the emphasis is on developing and applying advanced measurement techniques to obtain data to validate emerging computational (CFD) predictions and new designs. In addition to the application of CFD codes to experimental test geometries, a new geometry package has been developed to optimize the aero-structural design of a compressor or fan rotor. Most recently, that package was used to obtain optimized redesigns of two turbomachinery CFD test cases; namely, the Sanger rotor being tested currently at TPL, and the NASA Rotor 67. It is anticipated that an optimized rotor design will be evaluated next in the transonic compressor rig. Also, under NASA sponsorship, a cross-flow fan investigation, proposed for a V/STOL aircraft application, is currently underway in the turbine test rig. As the highest Navy priority, high-cycle fatigue structural test techniques are being developed for use in Navy spin-pit facilities at the Naval Air Warfare Center, Patuxent River. This program is a joint program between the Navy and the Air Force. Finally, small turbo-engine variants for Unmanned Air Vehicles or missiles are being explored experimentally. Current programs are tied to Navy-critical engine Research & Technology programs, and very close coordination is maintained with the Propulsion and Power Division at the Naval Air Warfare Center, Patuxent River.

Rocket and Ramjet Propulsion

High-speed propulsion systems on both military and commercial platforms require a thorough understanding of the existing gas dynamic and chemical processes within these systems. Propulsion systems commonly investigated include solid propellant and liquid propellant rocket engines, ramjets, and pulse detonation engine systems. The Rocket Propulsion and Combustion Laboratory (RPCL) was established by Dr. Roy Reichenbach. Since joining the Naval Postgraduate School in 1968 it was further developed by his successor, Distinguished Professor David Netzer, who is currently supported by two research assistant professors, Dr. Christopher Brophy and Dr. Jose Sinibaldi. Many of the high-speed propulsion courses in the department involve experience at the laboratory investigating advanced systems and their related technologies. The laboratory has received support from a variety of government agencies, including the Office of Naval Research, Air Force Research Lab, and Naval Air Warfare Center Weapons Division as well as commercial companies such as General Electric Aircraft Engines and Pratt and Whitney.

The laboratory consists of three hot-fire test cells, two cold flow testing areas, and a control room capable of monitoring experiments throughout the lab. The laboratory is capable of testing both solid and liquid rocket engines up to 500-lbs.thrust.Ramjets can be tested with vitiated air heaters which provide airflow rates up to 8 lb per second at 750 K. Gaseous and liquid-fueled pulse detonation engines can be tested up to 100 Hz operation, and comprehensive conventional and optical diagnostics are available to characterize performance and system operation. The hardware and infrastructure of RPCL is complemented by a wide range of diagnostic capabilities required for the investigation of various propulsion systems. Some of the diagnostic capabilities existing at the lab include a Phase Doppler Particle Analyzer (PDPA), Malvern particle analyzers, copper vapor laser system for Particle Image Velocimetry (PIV), Nd:YAG laser, high speed intensified CCD cameras, visible and infrared imaging systems, spectro-radiometers, and a wide range of additional laser systems. PC-based, high-speed data acquisition systems are located throughout the laboratory and are used to monitor the diagnostic systems, thermocouples, and high frequency pressure transducers.

Avionics

Starting in 1996 Associate Professor R. Duren developed a series of avionics courses, including an avionics design course, and he built up an avionics laboratory to study fleet related problems including real-time software design, fusion algorithms, software re-hosting, software engineering methods, open systems, and computer architectures. Schematic and HDL circuit design tools and modern software development tools are hosted on ten Pentium III and IV class PCs for hardware and software development. The Machine Transferable AN/AYK-14 Support Software System (MTASS/M) software development tools are hosted on a Sun SPARCstation 10. Some tools were provided by the Navy's F/A-18 Advanced Weapons Laboratory in China Lake, CA. These tools include assemblers, compilers, and simulators to develop software for the AYK-14 Mission Computer of the F/A-18 aircraft. These resources enable students to engage in research that directly assists the Advanced Weapons Laboratory in the support of the F/A-18 aircraft.

Rotary Wing Aircraft Technology

A helicopter technology course was first taught since 1969 by Professor J.A.J. Bennett, the former head of the Department of Aeronautics of Cranfield Institute of Technology. Following the death of Professor Bennett in 1971, instruction in rotary wing technology continued under Prof. Donald Layton. After the appointment of Professor E. Roberts Wood in 1988 he expanded the offerings in rotary wing technology to three courses (including a separate helicopter design course) and established a close cooperation with the NASA Ames and the Army Flight Dynamics Directorate at Moffett Field for field trips and thesis studies. Professor Wood also developed the Rotorcraft Laboratory which is designed to provide a multi-faceted approach to the problems encountered in flight by rotary wing and Vertical Take-Off and Landing (VTOL) aircraft. The testing portion of the lab consists of flight testing, structural dynamics testing, wind and water tunnel testing, acoustic testing and flight simulation. The jewels of the rotorcraft lab are the two OH-6A helicopters. Through a cooperative agreement with Mississippi State, one helicopter is certified for use in flight testing. Cockpit components of the other helicopter are used as part of a flight simulator developed with Advanced Rotorcraft Technologies in Mountain View, CA. The fuselage of the second helicopter serves as part of the structural dynamics testing at NPS. Making use of additional test facilities at NPS, models have been developed for both the water tunnel and wind tunnel to study circulation control. In conjunction with the Physics Department at NPS, an acoustic test facility has been developed. The modeling and simulation portions of the lab consists of several computers using commercial-off-the-shelf software such as NASTRAN ® DYTRAN ®, MATLAB ®, Simulink ®, Maple ®, and FlightLab ® to study problems in rotor dynamics, acoustics, structural dynamics and flight performance. The Joint Army/Navy Rotorcraft Analysis and Design (JANRAD) computer program was developed at NPS to perform performance, stability and control, and rotor dynamics analysis during preliminary helicopter design efforts.

Aircraft Combat Survivability

An essential aspect in the education of the warfighter is the study of aircraft combat survivability. The core of the survivability discipline was developed during the past twenty years by NPS Distinguished Professor Robert E. Ball who published a book on this topic in 1984 (recently greatly expanded in a second edition). The course on aircraft survivability emphasizes the operational considerations and analytical methodologies necessary to design aircraft, both fixed-wing and rotary-wing, that are survivable in the combat environment. The resources available to educate the war fighter include the Survivability and Lethality Assessment Laboratory, Distance Learning and Short Courses, and a variety of multi-discipline survivability and lethality related graduate courses at NPS.

Spacecraft Systems, Attitude Control and Smart Structures

Starting in 1988 Distinguished Professor Brij Agrawal developed several unique laboratories to provide hands-on experience in the design, analysis, and testing of space systems and subsystems and to enable experimental research on current problems on DOD spacecraft. He also succeeded to attract several postdoctoral researchers, namely, H. Bang (South Korea), G. Song (U of Houston), G. Ramirez (Tennessee TU), H. Chen (Columbia U), and M. Romano (U of Milan). The Spacecraft Attitude Dynamics and Control Laboratory is used to perform research on developing improved control techniques for attitude control of flexible spacecraft and flexible robotic manipulators. The emphasis has been to develop improved control laws for fast slew maneuvers of flexible spacecraft. The laboratory has three simulators to validate the improved control techniques experimentally: Flexible Spacecraft Simulator (FSS), Space Robot Simulator (SRS), and Three-Axis-Spacecraft Simulator (TASS). The FSS simulates attitude motions of the spacecraft in one axis. The SRS consists of a two-link manipulator with rigid and flexible links. The TASS simulates a free floating spacecraft with a platform that incorporates rate gyros, sun sensors, and magnetometers, three reaction wheels and a laptop computer. The platform floats on a spherical air-bearing stand, thus giving the simulator three degrees of freedom for attitude control. The simulator also has an optical payload consisting of a fast steering mirror, jitter control system, and camera for acquisition, tracking and pointing. The integrated system is used as a simulator of a relay mirror spacecraft.

The Smart Structures Laboratory is used to perform research on active vibration control, vibration isolation, and fine pointing by using smart sensors and actuators. This laboratory has three main experiments: Ultra Quiet Platform (UQP), Positioning Hexapod and the NPS Space Truss. The UQP is used for testing control algorithms for vibration isolation of an imaging payload. It has six piezo-ceramic actuators and a geophone sensor. The Position Hexapod is used for testing control algorithms for both vibration isolation of an imaging payload and fine steering. It is based on the arrangement of six self-supporting electromagnetic voice coil actuators with in-line accelerometers and position sensors. The NPS Space Truss is used for testing control algorithms for active structural control and vibration isolation. The overall dimension of the truss is 3.76 m long, 0.35m wide and 0.7 m tall. It has piezo-ceramic struts as actuators and a linear proof mass actuator as source of disturbance. The FLTSATCOM Laboratory consists of a qualification model of the Navy communications satellite, FLT-SATCOM, and ground TT&C system. This laboratory is kept operational in cooperation with Naval Satellite Operational Center, for use by students in classes and by NAVSOC for analyzing on-orbit anomalies. Commands are sent to the satellite for wheel spin-up, firing of thrusters and rotation of solar array drive.

The Satellite Servicing Laboratory is a new laboratory used to develop and operate a servicing spacecraft simulator to conduct research into autonomous rendezvous, docking and control of a small manipulator vehicle. The servicing spacecraft simulator floats on a granite table using air pads to provide a frictionless 2-D simulation of on-orbit operations. A new joint NPS and Air Force Research Laboratory (AFRL) Optical Relay Spacecraft Laboratory was dedicated in June 2002. This laboratory is used for both instruction and research on acquisition, tracking, and pointing of flexible military spacecraft. The test bed consists of a spacecraft attitude simulator, which can simulate spacecraft three-axis motion, and an optical system simulating a space telescope. The simulator has three reaction wheels and thrusters as actuators; rate gyros and sun sensors; on-board processor; batteries; and it is supported on a spherical air bearing. The optical system consists of a laser source, a fast steering mirror, jitter sensor, and a video camera as a tracking sensor.

Spacecraft Guidance, Control and Optimization

Associate Professor M. Ross developed the NPS Astrolab and the Space Technology Battlefield Laboratory to study high-speed precision guidance and control of space vehicles and ballistic missiles. He achieved a significant breakthrough by a revolutionary approach to the design of feedback laws. In this approach, the "laws" are determined on-line with an adaptive nonlinear model instead of the traditional off-line design and implementation. This system can adapt to changing mission objectives while maintaining optimal performance. Two software packages have been developed at NPS in this field. DIDO is the implementation of a pseudo-spectral method invented at Astrolab. It is a one-of-a-kind method to provide automatic "adjoint sensitivities" or co-vector information for complex non-smooth problems. ACAPS is software developed for the Jet Propulsion Laboratory for the preliminary design of interplanetary aero-assisted missions. It has also been used by Raytheon to support JPL missions.

Research

The research projects carried out by Aeronautics and Astronautics faculty mostly are focused on topics of critical importance to military users and, typically, are funded by various Navy and other DOD sponsors. These externally funded projects ensure a continued close interaction between the faculty and the sponsors and thus provide the students with valuable insight into current naval aircraft and weapons development, maintenance, and operational problems during their formal courses and, especially, during their thesis project studies. The total externally funded research amounted to $ 2.86 million in FY2001, where more than 50% of the funds came from various Navy laboratories, 20% from Army laboratories, 12% from the Air Force, the remainder from NASA and other agencies. In FY2002 the externally funded research increased to over $ 3.5 million. Output from these projects is documented in 28 journal publications and 82 conference papers published during the past three years. Additional information is available from the individual professors and from the NPS Research Office.

Cost Effectiveness Considerations

These days most universities struggle with the significantly higher costs involved in maintaining engineering schools compared to, say, business management schools. This "bean counting" approach is widely practiced in evaluating federal government activities. In the NPS context the argument is often used that the naval officers only need to have a thorough management education to oversee the procurement of aircraft and weapons built in the private sector. To counter this argument we present a few specific examples which should help to elucidate the fallacy of this argument.

CDR D. Lott analyzed the previously unrecognized P-3C static aeroelastic wing behavior using a finite element analysis to show the cause of wing leading edge rib section failure

CAPT J. Clifton was the first to accurately model the unsteady motion of the 20,000 foot long trailing antenna wire when towed by the orbiting E-6A aircraft during TACAMO missions in the presence of wind gradients

After his PhD studies, CDR R. Niewoehner served as Navy chief test pilot during part of the F/A-18 E/F flight test program where he was responsible for the Navy's share of the envelope expansion flying, including flutter, flying qualities, and high angle-of-attack/spin testing. This included both discovery and resolution of the Super Hornet's well publicized transonic wing drop. Note that the F/A-18 E/F program was the Navy's largest development program of the 1990's and in 2000 the program received the Collier award for innovative contributions to aeronautics.

Distinguished Graduates

Each year, the NPS Department of Aeronautics and Astronautics and the AIAA Point Lobos Section jointly make an award to the best student in both the aeronautics and astronautics programs. The Admiral William Adger Moffett Award is presented annually to the outstanding aeronautical engineering officer student on the basis of academic excellence, including thesis, and career potential. The Astronaut Michael J. Smith CAPT, USN, Astronautics Award is presented annually to the outstanding astronautical engineering student on the basis of academic excellence, including thesis, and career potential.

Over the years, the Department of Aeronautics and Astronautics has graduated many naval and other officers who reached the rank of Admiral. Many have made important contributions to the development of naval/military aviation.

The Department is especially proud that the following six graduates became APOLLO astronauts:
COL Gerald Carr, USMC, CAPT Edgar D. Mitchell, USN (sixth man on the moon), CAPT Eugene A. Cernan, USN (last man on the moon), CAPT Ronald E. Evans, USN, COL Robert Overmyer, USMC, CAPT Paul J. Weitz, USN (Skylab2).

The following twenty-two astronauts became SPACE SHUTTLE astronauts:
COL Jack R. Lousma, USMC, CAPT Michael J. Smith, USN, CAPT David C. Leestma, USN, COL David C. Hilmers, USMC, CAPT Michael L. Coats, USN, CAPT Winston E. Scott, USN, CAPT Kenneth S. Reightler, USN, CDR Michael J. Foreman, USN, CDR Kent V. Rominger, USN, LCOL Jeffrey N. Williams, USA, CDR Michael E. Lopez-Alegria, USN, CDR Brent W. Jett, Jr., USN, CDR Scott D. Altman, USN, LCDR Robert L. Curbeam, USN, LCDR William C. McCool, USN, LCDR Lisa N. Nowak, USN, LCDR Christoper J. Ferguson, USN, LCDR Stephen N. Frick, USN, LCDR Mark E. Kelly, USN, LCDR John B. Herrington, USN, LCDR Alan G. Poindexter, USN, LCDR Kenneth T. Ham, USN

The following aeronautical engineering graduates received the Admiral Moffett Award:
Peter T. Rodrick USN 1972, Philip R. Elder USN 1973, Marle D. Jewitt USN 1974, Paul B. Schlein USN 1975, E. Fenton Carey USN 1976, Donald L. Finch USN 1977, Joseph A. Strada USN 1978, Michael J. Arnold USN 1979, Thomas R. Darnell USN 1980, Stephen T. Van Brocklin USN 1981, Richard W. Campbell USMC 1982, Daniel T. Kuper USN 1983, David B. Cripps USN 1984, William L. Posnett III USN 1985, Richard W. Cummings USN 1986, Michael J. Foreman USN 1987, Jeffrey N. Williams USA 1988, Thomas M. McKannon USN 1989, Walter L. Rogers USN 1989, George D. Duchak USN 1990, Richard B. Bobbitt USN 1991, William H.

Reuter USN 1992, Kerrin S. Neace USN 1993, Mark A. Couch USN 1994, William Donovan Jr. USN 1995, Grant R. Stephenson USN 1996, Vincent M. Tobin USA 1997, Osa E. Fitch USN 1998, Christopher W. Rice USN 1999, Rendell K.W. Tan Singapore 2000, Breno M. Castro Brazil 2001, Justin M. Verville USN 2002

The following astronautical engineering graduates received the Michael Smith Award:
Scott E. Palmer USN 1985, Jill A. Vaughn USN 1986, Austin W. Boyd USN 1897, Neal R. Miller USN 1988, Charles B. Cameron USN 1989, Carl E. Josefson USN 1990, Robert O. Work USMC 1991, Jeffrey A. Haley USN 1992, Michael A. Hecker USN 1993, James D. Atkinson USN 1993, Lester C. Makepeace III USN 1994, Markham K. Rich USN 1995, Kirk E. Treanor USN 1996, Douglas C. Eskins USN 1997, Karl E. Jensen USN 1998, Stephen B. Zike USN 1999, Stephen G. Edwards USAF 2000, Christopher M. Senenko USN 2001, Vince Chernesky USN 2002

Summary and Outlook

On 17 December 2002 a Memorandum of Understanding was signed by the Secretaries of the Navy and Air Force to transfer the NPS aeronautical engineering programs to the Air Force Institute of Technology effective 1January 2003. The astronautical engineering program remains at NPS for the time being. Hence this decision consolidates the graduate aeronautical engineering education for Air Force and Navy officers in one institution only and thus terminates the unique Navy/Marine Corps oriented aeronautical engineering programs described in this paper. It remains to be seen whether this decision is in the Navy's near- and long-term interest because of the continuing need to offer interdisciplinary systems studies which require aeronautical engineering expertise. Also, it remains to be seen whether the NPS leadership recognizes the need to maintain the unique NPS aeronautical engineering laboratories to support such interdisciplinary systems studies.

Books/Contribution to Books

Vavra, Michael H., "Aero-Thermodynamics and Flow in Turbomachines", John Wiley & Sons, 1960.

Ball, Robert E., "The Fundamentals of Aircraft Combat Survivability Analysis and Design", Second edition, AIAA Education Series, 2003.

Schmidt, Louis V., "Introduction to Aircraft Flight Dynamics", AIAA Education Series, 1998.

Brophy, C.M. Sinibaldi, J.O. and Netzer, D.W., "Detonation of a JP-10/Air Aerosol for Pulse Detonation Applications," High-Speed Deflagration and Detonation, pp.207-222, ELEX-KM Publishers, Moscow, 2000.

Roy G., Frolov, S., Netzer, D.W., and Borisov, A., eds., "Control of Detonation Processes", ELEX-KM Publishers, Moscow, Russia, 2000.

Sutton, G.P. and Biblarz, O., "Rocket Propulsion Elements", 7th Edition, John Wiley & Sons, New York, NY, 2001.

Jones, K.D., Lund, T.C., Platzer, M.F., "Experimental and Computational Investigation of Flapping Wing Propulsion for Micro Air Vehicles", chapter 16, pp. 307-339, Progress in Aeronautics and Astronautics, Volume 195, 2001.

Newberry, Conrad F., ed., "Perspectives in Aerospace Design", American Institute of Aeronautics and Astronautics, Washington, D.C., 1991.

Wimpress, John K. and Newberry, Conrad F., "The YC-14 STOL Prototype: Its Design, Development and Flight Test", American Institute of Aeronautics and Astronautics, Reston, Virginia, 1998.

Zucker, R.D. and Biblarz, O., "Fundamentals of Gas Dynamics", John Wiley & Sons, New York, NY, 2002.

References

1. Haupt, Ulrich, "Needs and Challenges in Education for Aircraft Design," Report NPS-57HP73121A, Naval Postgraduate School, Monterey, California, November 1973.

2. Haupt, Ulrich, "Decision-Making and Optimization in Aircraft Design," Report NPS-67Hp77021A, Naval Postgraduate School, Monterey, California, February 1977.

3. Newberry, Conrad F., "The Conceptual Design of Deck-launched Waverider-Configured Aircraft," Aircraft Design, Vol. 1, No. 3, September 1998, pp. 159-191.

Dedication

This paper is dedicated to the memory of Dr. Wendell M. Coates, Distinguished Professor and founder of the Department of Aeronautics, and to Dr. Michael H. Vavra, Distinguished Professor and founder of the Turbopropulsion Laboratories.

Wendell M. Coates
Ph.D. University of Michigan

Michael H. Vavra
Dr. Tech. Sci., Technical University of Vienna

Part VI

Proprietary School Programs in Aerospace Engineering

Chapter 64

From Parks Air College to Parks College
of Engineering & Aviation:
75 years of Legacy in Aerospace Engineering

K. Ravindra, Professor & Chair, Aerospace & Mechanical Engineering
John Waide, Archive Librarian, Pius Library
Saint Louis University

Abstract

Oliver L. Parks' vision of aviation and aerospace engineering in 1927 spawned the birth of Parks Air College. He began the school with the motto of "learning by doing", which to this day is being practiced at Parks College of Engineering & Aviation. Parks has produced many aerospace engineers from across the United States and around the world. Parks graduates continue to hold key positions and make contributions to the aerospace industry and education. Many accounts exist in literature about the aviation history of Parks. The present manuscript is a focus on the evolution of Aeronautical Engineering at Parks from its inception to the present.

Introduction

In the fall of 1925, Oliver Lafayette Parks, a Chevrolet salesman, came to Lambert Field, Saint Louis to take flying lessons from a pilot of the Robertson Aircraft Corporation. At that time, no federal regulations governed commercial aviation. Aircraft companies regularly sold courses of flight lessons for $100, with the instructor receiving $25. Parks received his first pilot rating in January 1926. The certificate, numbered 6373, bore the signature of Orville Wright. Six months later, Parks received his transport rating. By July of 1926, he owned two planes, a Standard, and an Eagle Rock. A native Midwesterner, born in Minonk, Illinois, Parks finished high school and served in the Marines in World War I. He arrived in Saint Louis at the same time it was to become a flying center. He enjoyed taking venturesome visitors for rides over Lambert Field, averaging about $300 in an afternoon. The Standard that Parks flew was less than reliable, and he encountered several incidents that brought him to the realization that his flight training had been too short, too hurried, and too narrow. In response, he determined to start a flight-training program for others.[1]

Parks Air College opened on August 1, 1927, in a rented hangar at Lambert Field. Mr. Parks was the only instructor. He had two planes, his old Standard and a Laird Swallow. Harvey Glass, a filling station operator, enrolled and became the first student. Parks often gave rides to others. He happily obliged to give a young man and his girl friend a ride in his Swallow, and the outcome of this flight would change the course of his college from that day forward. The plane went into a spin from which Parks could not recover, and crashed northwest of the airport near St. Stanislaus

Seminary. The passengers escaped with no injuries, but Parks was severely injured with cuts, bruises, broken bones and a damaged left eye. During his four and one-half months in the hospital, his savings dwindled and conditions at the flight school deteriorated. The six enrolled students wanted their money back. In response, Mr. Parks outlined his plans to move the school the following year to its own 113-acre campus across the Mississippi River and increase the pilot training time to 50 hours.

In the spring of 1928, Parks found the future site of his school. He chose a section of Illinois bottomland a mile and a half from the Mississippi River with a clear view of downtown St. Louis. Whether he realized it or not, Mr. Parks had chosen a section of ground that was historic for being the first permanent settlement of Europeans in the central valley, and he was positioned to add a new chapter of history to this region. Even though the college was located in Cahokia, in the initial years he identified the locale of the College as East St. Louis, Illinois.

The first students lived in a farmhouse, and later in a frame dormitory. An early staffer commented, "It was a mud hole. The land was swampy. The levees hadn't yet been built to keep the river out of the bottomland." Despite these conditions, the students kept coming. Mr. Parks also continued his earlier career as a salesman, peddling Travel Air planes as well as using them for instruction at the school. In 1928 alone, he sold 128 of these aircraft.

In 1929 Mr. Parks sent Joseph Wecker to tour the country to inspect other aviation schools and to purchase as many OX-5 engines as he could. Parks intended to follow the example of others in the St. Louis area who had started airplane factories. The Parks Aircraft Manufacturing Company was formed and a factory building was erected at the airport. While the factory produced 65 aircraft with the designations P-1, P-2 and P-3, Parks lost more than a quarter of a million dollars.

The Detroit Aircraft Corporation bought 80 percent of the outstanding stock, but Parks was not content to let others control his operation. Within a year, Parks mortgaged his house for $68,000, sold his two automobiles, and invited investors and creditors to give him a hand. Through this effort, Parks once again controlled his school's destiny. Parks turned the airplane factory building over to the College as a mechanics shop to house the ever-increasing enrollment of mechanics students.[4]

Spring 1933 saw the first edition of the *Parks Air News*. Fred Parks, older brother of the founder, and vice president & registrar of the College, edited the publication. In addition to College news and advertisements, national and international aviation news was reported. Stories of the day included (1) purchase of 900 airplanes for the US Army and 290 new fighting planes for the Navy, (2) flight of 24 Savoia-Marchetti seaplanes in a southerly arc from Italy through West Africa, Brazil, and Central America to the Century of Progress Exhibition in Chicago, and (3) the fine work of Parks graduates in tuning up the Curtiss-Condoor that Adm. Richard Byrd took over the South Pole. The Parks campus was the site of impressive crowds and important visitors. Twenty thousand people came to the campus in August 1933 for the model plane races. Distinguished visitors included Lindbergh, Hawks, and Doolittle.

Mr. Parks started a new educational effort in 1935. He started Parks Air College Airline to train students in all aspects of airline operations. Using the Parks Airport as its base, students managed the airline that flew regular flights to Indianapolis, Memphis and Kansas City.

The 1930's saw the growth of instability in the world. Leaders in Japan, Germany, and Italy became more aggressive in dealing with their own people and neighboring nations. In response to growing unrest, Gen. Harold H. "Hap" Arnold asked civilian air schools to expand their operations and train military flyers. Mr. Parks and others borrowed heavily to expand their schools, knowing that if Gen. Arnold was not able to persuade Congress to fund this effort they would be bankrupt. Forty-two cadets arrived in Cahokia in June 1939. They took flight instruction from seven to eleven o'clock in the morning in the 27 Stearman biplanes, and they attended classes in the afternoon. Two new buildings, Cadet (later Cahokia) Hall was built to house servicemen and Aviation Hall was built to house civilians.

In May 1940, Gen. Arnold asked the flight training operations to triple their enrollments. Mr. Parks responded by adding sites in Jackson, MS, and Sikeston, MO, to existing operations in Cahokia, IL and Tuscaloosa, AL in preparation for training 4000 Army personnel in 1941. Congress and President Roosevelt finally approved funding for the program, and Mr. Parks' $600,000 expansion was placed on sound financial footing. A feature story about Parks College appeared in the national magazine, *Coronet*, in February 1941. That feature story entitled "Harvard of the Air" reported that Parks College "seems the perfect counterpart of the strict little classical college where no compromise with the standards of scholarship is to be considered for a moment."

Mr. Parks trained his students in aviation, but he also considered his school to be a place to mold character in a man. "A man must be clean," Parks insisted. "He must live clean and think clean thoughts and he must be absolutely accurate in everything he does. In other words, he must be a man who can assume a real responsibility--not just part of the time, but all of the time." Mr. Parks tolerated no mischievous behavior from his students. If a student missed a curfew, he was assigned to $100 worth of campus work. If he was caught with liquor in a tavern, it was automatic dismissal. He was equally strict with flight training. The first time a flight student failed to land in the first third of the field or violated any of the flight rules, he was out of the program.[5]

Parks Air News in 1941 reported on the uncertainty in the world and the nation's preparation for war. President Roosevelt set an annual production goal of 50,000 fighter planes, and he also signed a bill that authorized the training of mechanics for the new aircraft. Parks College Dean Fred Roever presented testimony to Congress in support of the mechanic training program.

Shortly after his testimony, the chair called for a recess and the members of the committee went to legislative chambers to hear President Roosevelt declare war on the Empire of Japan.

The Early days of Aeronautical Engineering at Parks

The earliest catalogs and/or course schedules from Parks College are from the fall of 1928. These catalogs list three "courses" of study: Practical Flying Course; Aircraft Industrial Course; Pilots' Ground Course. Although the description of the

The first class of Aeronautical Engineering Students at Parks Air College, in 1934, with instructor Frederick H. Roever at the far end, standing. [2]

Pilots' Ground Course does not contain the term "engineering," it seems that this course contained the instructional elements of what would soon become the aeronautical engineering course of study. The earliest description of the Pilots' Ground Course from the 1929 Parks catalog, Skyward Ho! reads:

Parks College holds the unique stature in the country, with Air Agency Certificate No. 1 from the FAA

"For real success as a pilot in commercial aviation you must know more than merely how to fly, you must know the airplane and aeronautic engine thoroughly—construction, operation, maintenance, and repair. You must know field management, aircraft production methods, aerial navigation, meteorology (the sky and weather), aerodynamics and a general history of aviation."

It was in the November 1933 Outline of Courses for Parks College that the term "aeronautical engineering" first appears. In the general instructions to prospective students regarding how to choose a course of study, one reads:

"Activities in the aviation industry logically divide into four major divisions—executive, engineering, mechanics, and piloting...The aeronautical engineer shares with the pilot, this pride of achievement. By training and experience he becomes intimately acquainted with the laws of nature and allies mechanical systems with them for the constant improvement of aircraft. The designer of airplanes has even greater opportunities for the thrills of achievement than do pilots, who demonstrate what the finished ships can do."

The aeronautical engineering program was designed to take eight terms over two calendar years. The program appears to have been rather structured and intense, with the student taking anywhere from five to nine classes each term, although most of the classes did not meet for every week of the term. Practical work, mathematics, engineering drawing, business subjects, flying and airplane design were at the heart of the curriculum. Each aeronautical engineering student was required to complete 20 hours of practical flying instruction as part of the degree requirements. A typical eight-term program from the 1934 catalog is shown below. It is interesting to note that during the thirties, the course on Airplane Design had 180 contact hours and required the design, construction and flight test of the airplane. Upon graduation, the student received a Bachelor of Science degree in Aeronautical Engineering. The tuition for each term was $100. It was also in this 1933 Outline that the first "Professor of Engineering" is listed among the faculty. This first professor was Mr. Fred Roever, who had earlier been listed as an instructor in the Pilots' Ground course and by 1941, was the dean.

In the 1940's, Oliver Parks association with education brought him close to the then president of Saint Louis University, Patrick J. Holloran, S.J in various fund raising activities [1]. A series of events, including the Second World War, Parks' belief that *"future aviation leaders needed a broader, more academic education"*, his desire for enhancing the educational aspects of the college combined with his gratitude towards the Jesuits who had nursed him back to health after the serious airplane accident in 1928, culminated in Parks Air College being donated to Saint Louis University

The final assembly of the first wind tunnel at Parks College, 36" diameter open jet test section, designed by N.A.C.A, 1936

in 1946. The merging of a down to earth practical engineering and aviation school was not devoid of reverberations from certain segments of faculty within Saint Louis University who opposed all technology efforts as beyond the boundaries of Jesuit tradition. Nonetheless, then president Patrick J. Holloran, S.J. held firm to his course and decision to merge Parks Air College with Saint Louis University. In 1969, Parks College came very close to being separated from Saint Louis University. Father Paul Reinert S.J, then president of the university, was under pressure to relieve the financial crunch of the university[1]. One of the options considered was to "divest" Parks College to Southern Illinois University-Edwardsville. The drama was played out amongst the members of the board of trustees of both Saint Louis University and Southern Illinois University-Edwardsville. In the end, by 1971, this confusing episode ended, with Parks College still under the umbrella of Saint Louis University. The Institute of Technology on the Frost campus of Saint Louis University, which housed several engineering and science programs, was not so lucky. It was shut down in 1971. In 1977, the bachelor's program in Aerospace engineering got its premier accreditation from ABET under the chairmanship of Dr. John A. George and has maintained accreditation to this day.

The Parks administrators of the early fifties (Dean Beck and Father John Choppesky) quickly recognized the importance of space flight and the relevant course work in engineering. They invited Werner von Braun in 1954 to speak to the students. von Braun gave a lecture about possible flights to the moon. In 1958, he was invited back to Parks to receive an honorary doctorate in science. As a special gift to the college, von Braun donated one of the V-2 rocket engines developed in Germany. (This V-2 rocket engine is now on display in the lobby of Earhart Hall on the Frost campus of Saint Louis University). Another space enthusiast of the era, Willi Ley was also invited to speak. This era sparked many students such as Eugene Kranz (mission controller at NASA Houston for the Apollo 13 mission) and Frank Hoban (executive assistant to Werner von Braun, later director of public policy studies related to space at George Mason University) to pursue life - time careers in space. [1]It is noteworthy here to mention that Dr. John A. George taught the first course related to space flight, Orbital Mechanics, in the fall of 1959. By 1965, the department had changed its degree offering from aeronautical engineering to aerospace engineering, in keeping with the rapid innovative advances occurring in space flight. During this period, the American Rocket Society (A.R.S) was born and very active on campus. Eventually, A.R.S joined with Institute of Aeronautical Sciences (I.A.S) to form the American Institute of Aeronautics and Astronautics (AIAA).

Winds of Change

Until 1989, Parks College continued its tradition of exclusively providing undergraduate education on a trimester system, enabling a student to earn a bachelor's degree in Aerospace Engineering in about three years. Simultaneously, the department was recruiting faculty who had a penchant for research, besides teaching. There was additional pressure from the university rank and tenure system for faculty to publish. However, the trimester system and the related 11 – month teaching appointments left little room for young faculty to get involved in scholarly activity. Additionally, there was administrative momentum in 1989 to bring Parks College in line with the "main campus" of Saint Louis University, which had followed the semester system for many years. Thus, Parks College transitioned from a trimester to a semester system beginning fall 1989.

A typical eight-term program from the 1934

AERONAUTICAL ENGINEERING COURSE

FIRST TERM

I **AERODYNAMICS I** (Lecture Room, 5 hrs. per week for 3 weeks)
History of aviation. Theory of flight. Types of airplanes. Stability and balance. Control.

II **ENGINEERING DRAWING I** (Laboratory, 6 hrs. per week for 8 weeks)
Use of drawing instruments. Orthographic projection. Lettering and dimensioning. Drawing of structures.

III **AIRCRAFT MATERIALS** (Lecture Room, 4 hrs. per week for 12 weeks)
Physical properties of all steels. Aluminum, duralinum, magnesium, etc. Properties of woods, fabrics, finishes, and all other materials used in aircraft construction.

IV **METALS** (Shop, 24 hrs. per week for 6 weeks)
Use of shop blueprints. Rolling elbows, cones and spherical shapes. Construction of metal fittings, cowlings, speed rings, wheel pants, tanks, etc. Riveting. Soldering. Metal fitting and sheet metal repairs.

V **METALS** (Lecture Room, 3 hrs. per week for 6 weeks)
Blueprint reading. Proper materials. Bending allowances. Parallel line development. Radius of curvature. Design of sheet metal parts, fittings, etc.

VI **WELDING** (Shop, 24 hrs. per week for 4 weeks)
Setting up welding equipment. Welding of sheet metal, metal fittings, steel tubing, aluminum and alloys. Welding of cluster, fish mouth, butt and other joints. Brazing. Welding complete fuselages. Fuselage alignment.

VII **WELDING** (Lecture Room, 4 hrs. per week for 3 weeks)
Action of different metals under heat. Oxidizing, carbonizing, neutral and hydrogen flames. Inspection of welds.

VIII **INSTRUMENTS** (Shop, 22 hrs. per week for 2 weeks)
Disassembly and adjustment of altimeters, compasses, tachometers, air speed indicators, turn and bank indicator, etc.

IX **INSTRUMENTS** (Lecture Room, 5 hrs. per week for 2 weeks)
Theory of operation, testing and adjustment. Use of instruments to navigate. All flight and engine instruments are covered.

SECOND TERM

I **WOODWORKING** (Shop, 30 hrs. per week for 4 weeks)
Building of wing ribs, wing spars, instrument boards, floor boards, etc. Steaming, bending, laminating. Splicing of spars. Building complete wings, tail groups, etc.

II **WOODWORKING** (Lecture Room, 5 hrs. per week for 4 weeks)
Laying out wing ribs from tables of ordinates. Ribs and beams. Kinds of glues. Kinds of woods used. Bending. Monocoque and plywood construction.

III **PARACHUTES** (Lecture Room, 5 hrs. per week for 1 week)
Types. Materials. How parachutes are packed and inspected. Use in emergency. (No jumps are made.)

IV **RADIO** (Lecture, 5 hrs. per week for 1 week)
Use of radio today. Future developments. Limitations and possibilities. Dept. of Commerce facilities.

V **FABRIC AND FINISHING** (Shop, 30 hrs. per week for 3 weeks)
Covering of wings, tail groups, ailerons and fuselages. Doping, lacquering and varnishing. Use of dope brush and spray gun. Repairs. Sewing.

VI **FABRIC AND FINISHING** (Lecture Room, 5 hrs. per week for 3 weeks)
Fabrics and covering materials. Stitches, knots and seams. Army, Navy and Department specifications. Inspection and tests. Dopes. Sags, runs, tears, pinholes, bubbles, blisters and peeling. Lacquers and varnishes.

VII **AIR LAW** (Lecture Room, 5 hrs. per week for 5 weeks)
Licensing airplanes, pilots and mechanics. Air worthiness requirements. International air law. Dept. of Commerce Rules.

VIII **ASSEMBLY AND RIGGING** (Shop and Lecture Room, 35 hrs. per week for 5 weeks)
Installation. Splices. Attaching and lining up wings, control surfaces, undercarriage, etc. Checking dihedral, sweep back, stagger, angle of incidence, wash in and wash out. Wires, etc. Installation of fuel tanks and lines. Final checking.

THIRD TERM

I **PRIMARY ENGINES** (Shop, 30 hrs. per week for 5 weeks)
Pistons, piston rings, valves, cams, cam shafts and crankshafts. Feed lines. Practical disassembly, study and assembly of aircraft engines, such as Whirlwind, Challenger, Martin, Lambert, Continental, Wasp, etc.

II **PRIMARY AND ADVANCED ENGINES** (Lecture Room, 5 hrs. per week for 6 weeks)
Theory. Ignition, carburetion, superchargers. Lubrication. Fuels and oils. Valves, timing and ignition. Trouble shooting.

III **ADVANCED ENGINES** (Shop, 30 hrs. per week for 4 weeks)
Use of precision instruments. Practical repair operations, such as valve grinding, fitting bearings, magneto and carburetor rebuilding, top overhauls. Installation. Trouble shooting.

IV **PROPELLERS** (Shop, 30 hrs. per week for 1 week)
Adjustment, balancing, tracking, inspection and etching of metal propellers. Use of propeller equipment. Propeller repair, maintenance and installation.

V **PROPELLERS** (Lecture Room, 5 hrs. per week for 1 week)
Static and dynamic balance. Fatigue. Pitch. Track. Blade shapes. Hub mountings. Three and four blade propellers. General inspections. Etching.

VI **ELECTRICAL EQUIPMENT** (Shop, 30 hrs. per week for 1 week)
Electric starters. Lighting circuits. Instrument lighting. Landing lights. Batteries. Generators.

VII **ELECTRICAL EQUIPMENT** (Lecture Room, 5 hrs. per week for 1 week)
Principles of electricity and magnetism. Generators. Storage batteries. Lighting systems. Navigation and landing lights. Thermocouples. Meters. Ohm's Law.

FOURTH TERM

I **MATHEMATICS I** (Lecture Room, 5 hrs. per week for 12 weeks)
Algebra. Positive and negative numbers. Imaginary and complex numbers. Simple equations. Equations with two unknowns. Simultaneous equations. Binomial theorem. Quadratic equations.

II **MATHEMATICS II** (Lecture Room, 5 hrs. per week for 12 weeks)
Trigonometry. Angles and angular measurement. Trigonometric functions. Logarithms. Right triangle. Formulas and identities. Oblique triangle. Graphical representations of trigonometric functions. Analytical trigonometry.

III **ENGINEERING DRAWING II** (Laboratory, 6 hrs. per week for 12 weeks)
Projection drawing. Pictorial projection. Orthographic projection. Developed surfaces and projections. Working drawings. Technical sketching and perspective.

IV **PHYSICS I** (Lecture Room, 5 hrs. per week for 12 weeks)
Displacement, velocity and acceleration. Composition and resolution of velocities and acceleration. Newton's Laws of Motion. Equilibrium. Couples. Work, power, energy. Impact. Hydrostatics. Temperature. Expansion. Colorimetry. Nature of heat. Conduction. Convection and radiation. Properties of gases.

V **AIR TRANSPORT OPERATION** (Lecture Room, 5 hrs. per week for 12 weeks)
Airways and airlines. Federal aids. Meteorology and radio. Airport design. Storage and maintenance facilities. Selection of equipment. Costs of operation. Schedules. Traffic. Personnel.

{ 11 }

Continued on Next Page

AERONAUTICAL ENGINEERING COURSE

FIFTH TERM

I MATHEMATICS III (Lecture Room, 3 hrs. per week for 12 weeks)

Analytical geometry. Cartesian coordinates. Graphs of algebrian functions. Change of coordinate axes. Straight lines and curves. Parametric representations. Polar coordinates.

II ENGINEERING DRAWING III (Laboratory, 6 hrs. per week for 12 weeks)

Descriptive geometry. The plane and its representations. Coordinate planes. Traces of planes on coordinate planes. Traces of lines on planes. Construction of geometric figures in space. Generation of surfaces. Their representation and intersections.

III PHYSICS II (Lecture Room, 5 hrs. per week for 12 weeks)

Light, sound, electricity and magnetism. Illumination. Reflection. Refraction. Prisms and colors. Sound motion. Harmonies. Frequency. Interference. Resonance. Magnetization. Magnetic fields. Electric currents. Volts. Ohms. Galvonometers. Generators and motors.

IV ELEMENTS OF MECHANISM (Lecture Room, 5 hrs. per week for 12 weeks)

Revolving and oscillating bodies. Ropes, belts and chains. Transmission of motion. Gears and gear teeth. Inclined planes. Screw and worm. Cams. Levers. Four bar linkage. Link work.

V MECHANICS I (Lecture Room, 5 hrs. per week for 12 weeks)

Statics. Forces. Parallelogram, triangle and polygon of forces. Parallel forces. Moments. Resultants. Center of gravity. Couples. Link polygon. Stresses. Strength and deflection of beams. Columns and shafts. Euler's and Rankine's formulas.

SIXTH TERM

I MATHEMATICS IV (Lecture Room, 5 hrs. per week for 12 weeks)

Differential calculus. Slopes and areas. Limits. Increments. Derivatives. Sign of a derivative. Differential of area. Theorems on derivatives. Higher derivatives. Differentation of implicit functions. Tangents and normals. Rectilinear motion. Motion in a curve.

II MECHANICS II (Lecture Room, 3 hrs. per week for 12 weeks)

Dynamics and hydraulics. Work, energy and power. Machines. Friction. Velocity and acceleration. Inertia. Kinetic energy. Momentum. Gyrostatic action. Pendulems. Flywheels. Hydraulic pressure. Flow of fluids. Bernoulli's Law. Jets. Turbines and centrifuses.

III MACHINE DESIGN (Lecture and Laboratory, 3 hrs. per week for 12 weeks)

Strength and failure of materials. Riveting and welding. Bolts and screws. Force and shrink fits. Cranks and crankshafts. Spur, helical and bevel gears. Belts and pulleys. Friction drives. Plain, ball and roller bearings. Shafts and shafting. Clutches. Couplings.

IV BUSINESS ENGLISH (Lecture Room, 5 hrs. per week for 12 weeks)

Fundamentals of business English, with special emphasis on preparation of business letters and statements. Preparation and delivery of oral reports. Public speaking.

V COMMERCIAL LAW (Lecture Room, 5 hrs. per week for 12 weeks)

Laws of partnerships and corporations. Liabilities, powers and restrictions of each. Contracts. Agencies Bailments and carriers.

SEVENTH TERM

I MATHEMATICS V (Lecture Room, 3 hrs. per week for 12 weeks)

Integral Calculus. The definite integral. Fundamental formulas. Integrals of complex functions. Areas. Volume of a solid of revolution. Length of a plane curve. Work. Pressure. Center of gravity. Multiple integrals.

II INDUSTRIAL ENGINEERING (Lecture Room, 3 hrs. per week for 12 weeks)

Forms of industrial ownership. Location, arrangement and construction of plants. Organization. Coordination and executive control. Purchasing. Production. Compensation of labor. Cost finding.

III AERODYNAMICS II (Lecture Room, 3 hrs. per week for 12 weeks)

Wing section data. Wing theory. Airplane model tests in Wind Tunnel. Engine and propeller consideration. Wing flaps and slots. Stream function. Two dimensional flow. Circulation. Vortices. Induced drag. Pitching, rolling and yawing.

IV AIRPLANE DESIGN I (Lecture and Laboratory, 5 hrs. per week for 12 weeks)

General design procedure. Critical loading conditions. Reactions. Shears. Moments. Continuous and restrained beams. Simple beams. Torsion. Nose dive to terminal velocities up to 500 miles per hour. Angles of attack. Inverted flight. Wing and tail flutter.

V AIRPLANE DESIGN II (Lecture and Laboratory, 5 hrs. per week for 12 weeks)

Laying out the aerodynamic features of the airplane. Stability. Controllability. Controls. Performance calculations. Top, cruising and landing speed. Rate of climb. Landing and take off run. Effects of high altitude on performance. Streamlining.

VI 10 hours of dual and solo flying instruction.

EIGHTH TERM

I PROPELLERS II (Lecture and Laboratory, 3 hrs. per week for 12 weeks)

Momentum theory. The airfoil. Blade element theory. Aerodynamic tests in wind tunnel. Effect of blade shape. Body interference. Variable and controllable pitches. Materials and construction. Design procedure.

II THERMODYNAMICS (Lecture Room, 3 hrs. per week for 12 weeks)

Heat. Conservation of energy. Ideal mechanisms. Laws of thermodynamics. Expansions and compressions of gases. Gas cycles. Indicator diagrams. Efficiencies. Performance. Fuels. Combustion. Superchargers. Cooling.

III STRESS ANALYSIS (Lecture Room, 5 hrs. per week for 12 weeks)

Applied mechanics. Wing analysis. Load factors. Beam loading. Moments, shears and reactions. Design of struts. Loads in the drag truss. Chassis analysis. Shock absorbers. Fuselage analysis. Tail surfaces.

IV DYNAMICS OF AIRPLANES (Lecture Room, 5 hrs. per week for 12 weeks)

Static equilibrium. Six equations of motion. Equation of impulse and momentum. Differentiation. Integration. Infinite series. Introduction to complex variables. Displacement in space. Gyrostatic vibration of airplane propellers. Stability.

V AIRPLANE DESIGN III (Lecture Room, 5 hrs. per week for 12 weeks)

Completion of a complete airplane design, with exact sizes of parts and materials used. Balance diagrams. Weight analysis. How to get approved type certificate from Department of Commerce for completed design.

VI 10 hours dual and solo flying instruction.

Table1: A comparison of Aeronautical Engineering Curriculum from 1934 and Aerospace Engineering curriculum from 2003.

1934 COURSE OUTLINE	2002 COURSE OUTLINE
FIRST TERM	SEMESTER I
Aerodynamics I	Engineering Chemistry
Engineering Drawing I	Advanced Writing for Professionals
Aircraft Materials	Freshmen Engineering I
Metals Lecture /Lab	Engineering Calculus I
Welding Lecture / Lab	Theological Foundations
Instruments Lecture / Lab	
SECOND TERM	SEMESTER 2
Woodworking Lecture / Lab	Intro to Computer Science
Parachutes	Intro to Computer Aided Design
Radio	Engineering Calculus II
Fabric & Finishing Lecture / Lab	Engineering Physics I / Lab
Air Law	Humanities/Social Sciences Elective
Assembly and Rigging	
THIRD TERM	SEMESTER 3
Primary Engines	Engineering Shop Practice
Primary & Advanced Engines	Small Group Presentation
Advanced Engines	Statics
Propellers / Lab	Engineering Physics II / Lab
Electrical Equipment / Lab	Engineering Calculus III
FOURTH TERM	SEMESTER 4
Mathematics I	Introduction to Aero & Astro
Mathematics II	Electrical Engineering / Lab
Engineering Drawing II	Dynamics
Physics I	Fluid Dynamics / Lab
Air Transport Operation	Differential Equations
FIFTH TERM	SEMESTER 5
Mathematics III	Performance
Engineering Drawing III	Mechanics of Solids / Lab
Physics II	Machine Design
Elements of Mechanism	Linear Vibrations
Mechanics I	Advanced Mathematics for Engineers
	Probability and Statistics
SIXTH TERM	SEMESTER 6
Mathematics IV	Gas Dynamics
Mechanics II	Aerodynamics
Machine Design	Astrodynamics
Business English	Aerospace Structures I
Commercial Law	Linear Systems

	Ethics
SEVENTH TERM	SEMESTER 7
Mathematics V	Propulsion
Industrial Engineering	Aerospace Lab
Aerodynamics II	Stability & Control
Airplane Design I	Aerospace & Structures II
Airplane Design II	Flight Vehicle Analysis & Design I
10 hours of dual & solo flying instructions	Engineering Ethics
EIGHTH TERM	SEMESTER 8
Propellers II	Heat Transfer
Thermodynamics	Flight Vehicle Analysis & Design II
Stress Analysis	Cultural Diversity
Dynamics of Airplane	Technical Elective
Airplane Design III	Technical Elective
10 hours dual & solo flying instructions	

This change provided new opportunities for the aerospace engineering department to explore a master's degree program. An advisory committee was formed, consisting of Ralph Pelican (Parks BSAE '65, currently Branch Chief, Aerodynamics at the Boeing Company), Larry Niedling (Parks BSAE '64 Chief Technology Engineer, Integration, now retired), Robert C. Brown (IT '62 Chief Program Manager, Missile Systems, now retired), Bill Moran (Parks BSAE '65 Sr. Tech Specialist, Aerodynamics), Ray Garrett (Parks BSAE '51 Chief technology Engineer, Strength, now retired), and Victor Meznarsic (Parks BSAE '65, Senior principal Engineer, Aerodynamics). The faculty in the department met with the advisory committee several times over a period of one year to synthesize a master's degree program that was application oriented. After proper approvals from various committees such as the academic affairs committee, Faculty Assembly, the Board of Graduate Studies and the Board of Trustees, the master's degree program started with a modest enrollment of 10 students in the fall of 1991. While it was hoped that the A-12 program then under active development by McDonnell Douglas Corporation would result in many young engineers from McDonnell Douglas Corporation enrolling in the master's program, the turn of events would prove otherwise. McDonnell Douglas Corporation would not win the contract from the Navy for the A-12. Thus, the master's degree program did not get the young applicants it had hoped to recruit. Nonetheless, the master's degree program continued.

The first three graduates were Mr. Karl Nelson (BSAE 1991, MSAE 1993, currently working at NASA, Marshall Space Flight Center as senior engine systems engineer), John Dempsey (BSAE 1991, MSAE 1993, currently working at Raytheon as Principal Systems Engineer) and Mark Johnson (BSAE 1990, MSAE 1993, currently working at The Boeing Company as the manager of Advanced Structures in Phantom Works).The master's degree program has so far produced about 60 graduates, as of December 2002. To bolster the research efforts, the university established an Oliver L. Parks Endowed Chair, named after the pioneer who started the college. The department hired Prof. Paul Czysz (BSAE 1956) as the first Oliver L. Parks Chair during fall 1992. In addition, the department hired Prof. Marty A. Ferman, also from McDonnell Douglas Corporation in support of the master's program. Both of them were at McDonnell Douglas Corporation (now The Boeing Company) for over 40 years prior to their academic appointments. They brought to the department, the "industry flavor". Prof. Czysz retired in May 2002 as Professor Emeritus of Aerospace Engineering.

The fall of the Berlin wall and the Soviet empire during the late eighties clearly had an impact on the aerospace job market in the United States, and perhaps, the rest of the world. This in turn, reflected in a decline in enrollment in undergraduate aerospace engineering across the country. Parks College was not immune to this decline. The central administration at Saint Louis University, under new president, Fr. Lawrence Biondi, S.J, saw an opportunity to merge the duplication of many services such as housing, registrar's office, financial aid office, bursar's office, athletic programs and delivery of courses in humanities and social sciences. The decreased enrollment at the Cahokia campus combined with duplication of services contributed to the decision by the administration in 1995-96 to move the Parks College campus from Cahokia to the Frost campus in Saint Louis. Many meetings of faculty and staff with the administration took place to foster a smooth transition of the Parks campus to the Frost campus. The aerospace laboratories housed in the Avionics Building at Parks College was to be relocated to the power plant building, located on the far east corner of the Frost campus. This power plant building was appropriately named Oliver Hall, for the visionary who started it all.

An aerial view (from east on Lindell Blvd) of McDonnell Douglas Hall grounds and Oliver Hall, previously the power plant building. (left)

The McDonnell Douglas Foundation provided a generous gift of $ 4 million towards the construction of a new building east of Fitzgerald hall, along Lindell Boulevard. The groundbreaking ceremony for what would be called McDonnell Douglas Hall, took place in April 1995. The new building with all the amenities of high speed internet and multimedia classrooms was built in a record time of 1.5 years. Meanwhile, a grand celebration took place on the Cahokia campus in May 1997 marking the end of an era and legacy set by "Lafe" Parks. This event, marking 70 years of existence was appropriately named "Flying Through Time." The relocation of the laboratories from Parks College, Cahokia to Parks College of Engineering & Aviation, Saint Louis, began soon thereafter, in June, July and August of 1997. The low speed wind tunnel had to be carefully disassembled, the huge air tank for the supersonic wind tunnel had to be carefully lifted and the heavy MTS machine had to be moved. By the end of August 1997, all the laboratory equipment had been moved and commissioned for the laboratory classes beginning fall 1997. This was quite an achievement. The faculty and staff moved to the new McDonnell Douglas building during the second week of August 1997, settling in time to start classes for fall 1997. The

An aerial view of the dedication ceremony in front of McDonnell Douglas hall, September 27, 1977

building was formally dedicated on September 27, 1997. Among the invited guests was James McDonnell of McDonnell Douglas Corporation.

Seeing the decline in enrollment in Aerospace Engineering during the early 1990's, the faculty in the department discussed a plan to offer a new bachelor's program in mechanical

engineering. The proposed program had the first two years identical to the aerospace engineering program. After formal approvals from various committees, the department started offering the bachelor's degree program in mechanical engineering in fall 1995. The BSME program received the initial ABET accreditation in 1997, effective retroactively to 1995. Dr. S. Karunamoorthy (currently associate dean of engineering) was instrumental in the establishment of BSME program. In 1996, a student section of ASME was started.

The relocation of the campus from Cahokia, as somber as it was for many

looking east

An aerial view of McDonnell Douglas Hall (left) and Oliver Hall, previously the power plant building. (right), view from Lindell Blvd.

faculty and staff, provided a new opportunity for the students to interact with other students on the main campus. In addition, it provided the students new opportunities to choose from a diverse set of courses in the humanities, social sciences and business. Slowly, beginning 1997, the enrollment in aerospace engineering also increased, partly due to the booming economy and partly due to the brand new campus location and the visibility the campus provided to prospective students.

The low speed wind tunnel motor and diffuser section are being hoisted on the truck in for its journey to its new location on the Frost campus, June 1997

831

**Charles Lindbergh with
Oliver L. Parks, 1927 (1)**

Some notable visitors to campus

Amelia Earhart Putnam, a Parks visitor.

Amelia Earhart. 1935

Harry Stonecipher, CEO of The Boeing
Company visits Parks College, 2000. From left
to right: Dr. C.C. Kirkpatrick, Harry
Stonecipher, Brian Wood (BSAE '00) and
Jeremy Minter (BSAE '00) in Oliver Hall, 2000

The Department Chairs

1978-1996: John A. George
1966-1978: Benjamin Ulrich
1965: Leon Z. Seltzer, acting chair.
1956-1965:Leon B. Trefny

1996-present: K. Ravindra 1955: Mr. Edward J. Flanagan
1953-1954: C. Donald Lundergan
1948-1952: Douglas H. Webber
1945-1947: Edward H. Barker
1934-1945: Frederick H. Roever

Conclusions

The department of Aerospace Engineering at Parks College has produced many engineers who have contributed to the growth and development of aerospace industry and aerospace education in this country and abroad since 1927. The well-rounded, hands-on education in the Jesuit tradition is unique. The college continues the tradition of "learning by doing" education set forth by Oliver Parks nearly 75 years ago.

Acknowledgements

The authors gratefully acknowledge the assistance provided by Dr. John A. George, Dr. Charles.C. Kirkpatrick, Ms. Debbie Farmer and Ms. Dianne Stagg.

References

1. *Parks College, Legacy of an Aviation Pioneer*, published by William Barnaby Faherty, S.J., 1990.
2. Parks Air News, 1943, p3.
3. Skyward Ho, 1934.
4. Parks Today, Alumni Newsletter, Winter 2002.
5. Parks Today, Alumni Newsletter, Spring 2003.

Chapter 65

EVOLUTION OF THE AEROSPACE ENGINEERING PROGRAM AT EMBRY-RIDDLE AERONAUTICAL UNIVERSITY

Charles N. Eastlake, Professor ; Howard D. Curtis, Professor; John R. Novy, Reda R. Mankbadi, Dean, College of Engineering, Embry-Riddle Aeronautical University, Daytona Beach, FL; Richard F. Felton, Associate Dean, College of Engineering, Embry-Riddle Aeronautical University, Prescott, AZ

Abstract

The historical evolution of the Aeronautical/Aerospace Engineering program at Embry-Riddle Aeronautical University is presented. The origin of the University is described first in order to put the nature and focus of the engineering programs in the proper perspective. The engineering curriculum was introduced in Miami, Florida, and later moved to its present home in Daytona Beach, Florida. The engineering program at the second residential campus in Prescott, Arizona, is also described because beyond sharing the established Daytona Beach curriculum it was started and is accredited separately. And the discussion concludes with an evaluation of the programs and a vision for the future by the Dean of the College of Engineering at Daytona Beach.

The origin of Embry-Riddle

J. Paul Riddle, far left, and T. Higbee Embry, third from left, with early Embry-Riddle Waco biplane

On December 17, 1925, the twenty-second anniversary of the Wright brothers' first flight, the Embry-Riddle Company was founded by barnstormer John Paul Riddle and entrepreneur/pilot T. Higbee Embry at Lunken airport in Cincinnati, Ohio. It quickly became one of the most successful dealers for Fairchild Aircraft. In the spring of 1926 the Embry-Riddle School of Aviation began providing pilot and mechanic training in company facilities at Lunken, which at that time was the busiest airport in the country. This choice of location would be comparable to establishing a flight training facility today at Los Angeles, New York, or Atlanta, an indication of Embry and Riddle's determination to become a major player in the emerging airplane industry.

In 1925 Congress passed the Airmail Act, which gave contracts to private companies to carry U.S. mail by airplane. This was the first widespread opportunity for flying businesses to make a reliable and predictable profit. Embry-Riddle was awarded the contract for the Cincinnati-Chicago route in 1927 and needed additional capital to finance these flights. Curtiss Aircraft offered to supply the funding but Fairchild Aircraft was not willing to stand by while a competitor took away their best airplane dealer. So the Fairchild board of directors formed a new subsidiary to loan the money to Embry-Riddle [1]. It was called The Aviation Corporation, more recently known as Avco Corporation, one of the stalwarts of the aerospace industry. By the end of 1929 The Aviation Corporation held a controlling interest in 81companies involved in carrying passengers and airmail and in manufacturing aircraft. When the Great Depression hit in late 1929 businesses throughout the country struggled to stay alive but The Aviation Corporation had mushroomed rapidly and thus had the resources to weather the storm. In fact, it had grown so much that it created another larger holding company called American Airways, which we know today as American Airlines.

In 1930 Embry and Riddle agreed to merge into American Airways. Embry went to California and lived there until 1946. Riddle moved to Miami, FL, and founded several aeronautical businesses. He took on John McKay as a new partner and re-opened the school in 1939 as Embry-Riddle International School of Aviation located in a converted hotel that became something of a landmark known as the Aviation Building. It was just a few blocks east of Miami International Airport, which was the largest and second busiest airport in the country.

Prior to and during World War II the school, still headquartered in Miami but operating out of four airports in two Florida cities, trained thousands of pilots and mechanics for the military services of the U.S. and Great Britain. These dispersed training sites fostered a distributed learning perspective which in 1970 was formalized by opening of the College of Career Education. It is now known as the Extended Campus and it operates instructional units at approximately 130 locations in 36 states and 4 European countries, many of them at U.S. military bases.

The training of British pilots led to an unplanned but highly emotional event in Daytona Beach in 1985. A weekend open house was under way to celebrate the 20th anniversary of the move to Daytona Beach. One of the items being showcased was the Aeronautical Engineering program's new Computer Aided Design (CAD) system. The demonstration involved plotting a detailed and very artistic drawing of two Supermarine Spitfires flying in formation, a file that had been provided with the Prime Medusa software package. By coincidence, seven or eight of the tour participants were Spitfire pilots who had been taught to fly at Embry-Riddle and who later flew in the historic Battle of Britain. When they saw the picture the Spitfire pilots enthusiastically offered to conduct an impromptu seminar for the fascinated ERAU students and faculty while the CAD system, slow by today's standards, plotted a copy of the drawing to be presented to each pilot.

As World War II drew to a close Riddle sold his interest in the school to McKay. When McKay died in 1951 his wife Isabel took over the operation. Riddle continued to live in the Miami area, forming Riddle Airlines and a government sponsored aeronautical school in Brazil. Mr. Riddle maintained contact with Embry-Riddle and made regular visits until his death in 1989. He loved to walk around the campus and engage students in conversation, and was such a low key personality that some of the students were not even aware who he was.

Isabel McKay changed the company to a non-profit school in 1961 and changed its name to Embry-Riddle Aeronautical Institute. In 1963 she hired former Naval aviator Jack Hunt to be President of ERAI. Hunt then arranged and oversaw the moving of the entire operation to Daytona Beach, FL, in 1965. Initial quarters were in "temporary" World War II Navy barracks buildings at Daytona Beach Regional Airport, immediately adjacent to the banked turns at the east end of Daytona International Speedway. Noise during race weeks was every bit as bad as one might imagine!

Paul Riddle, kneeling, 2nd from left, visiting the wind tunnel lab in 1981

Construction of a new campus was begun in 1968 on a 178-acre tract comprising most of the north east corner of the airport property. ERAI was renamed Embry-Riddle Aeronautical University and in 1968 was granted accreditation by the Southern Association of Colleges and Schools.

Daytona Beach campus seen from the East, Daytona Beach International Airport in the background

In 1979 the facilities of Prescott College in Prescott, AZ, were purchased and transformed into the second residential campus. The evolution of the Aerospace Engineering programs at Daytona Beach and Prescott are described separately in subsequent sections.

The presidents of ERAU have been:
Mr. Jack R. Hunt (1963-1984), former Naval aviator and passionate educator. As pilot of the ZPG-2 dirigible Snowbird, he was onetime holder of the world record for the longest non-stop flight, a 9500 mile, 364 hour trip from Massachusetts to Portugal and back to Florida, for which he was awarded the Harmon Trophy by President Eisenhower in 1957.
Dr. Jeffrey H. Ledewitz (1984-1986), long time ERAU administrator who served as Acting President following Hunt's death.
Mr. Kenneth L. Tallman (1986-1990), retired Air Force Lieutenant General and fighter pilot, former Superintendent of the Air Force Academy.
Dr. Steven M. Sliwa (1990-1998), former NASA aerospace engineer, avid pilot, and the administrator responsible for much of the dramatic expansion of physical facilities at ERAU.
Dr. George C. Ebbs (1998-present), founder of a well-known aviation consulting firm before taking over at ERAU.

Full scale stainless steel sculpture of 1903 Wright flyer in front of Daytona Beach campus library.

The curriculum expanded from pilots and mechanics into the engineering field in Miami in the late 1950's. The initial engineering program consisted of associate degrees in both Aeronautical Engineering and Aeronautical Engineering Technology. In 1958 the Associate in Science in Aeronautical Engineering Technology was accredited by the Engineer's Council for Professional Development (ECPD), the predecessor of the Accrediting Board for Engineering and Technology (ABET). In addition to the strong national push to recover the lead in space from Russia, the initial engineering focus at Embry-Riddle was on the needs of the airline industry. Reflecting that focus, both associate and bachelors degrees were created in Aircraft Maintenance Engineering Technology. These technology programs required completion of the FAA Airframe and Powerplant Mechanic's license in addition to the engineering course work because most airlines at that time insisted that their engineers have the A&P license, and a preference for it still exists today.

The Bachelor of Science in Aeronautical Engineering grew out of the associate degree program in 1960, and Embry-Riddle became an affiliate member of the American Society for Engineering Education. Through most of the 1960s a staff of 11 faculty offered both associate and bachelors degree programs in Aeronautical Engineering, Aeronautical Engineering Technology, Aircraft Maintenance Engineering, and Aircraft Maintenance Engineering Technology, though only the associate degree programs were ECPD accredited.

Upon moving to Daytona Beach in 1965, campus life initiated in several World War II Navy buildings at Daytona Beach Regional Airport. This airport had been built as a wartime military training facility, as had most of the airports in Florida, and the "temporary" buildings were still standing. About a third of the 239 students who moved from Miami were in these engineering programs.

Construction of a new campus began in 1968 with an academic complex centered on A-building, the first of a hexagonal group of hexagonal buildings. The Aeronautical Engineering department and several other academic departments occupied the second floor while the library occupied the first floor. Laboratories and classrooms were situated in the remaining four single story buildings, only four because the sixth building in the planned hexagonal pattern was never built. The pie-shaped classrooms were a unique and not entirely popular teaching experience. A closed runway running through the center of the campus was converted into the parking lot. The University Center, now called the John Paul Riddle Student Center, containing the cafeteria and other student services was built directly across the parking lot from the academic complex.

Another complex of hexagonal buildings was built immediately adjacent to the airport taxiway to serve as the Flight Department facilities.

In 1970 the BS in Aircraft Maintenance Engineering Technology was ECPD accredited and while that program is still offered, the word Maintenance was dropped from its name in 1975. In 1976 the BS in Aeronautical Engineering was accredited. Soon thereafter the associate degree programs were phased out.

The AE program grew steadily and a new building to house it was opened in Fall 1978, at which time the enrollment was 503. The first floor housed all AE laboratories (aircraft design, structures and instrumentation, materials, wind tunnel, electrical/electronics) and department faculty and administrative offices were located on the second floor. And it finally included a machine shop dedicated to lab development and maintenance. The southeast quarter of the floor plan did not have a second floor in order to accommodate a closed circuit subsonic wind tunnel. It was designed to fit that space by the new department chairman, Donald Ritchie, and was built by a team of faculty and students. One of the distinctive features of the metal building, the one and only window in the front of the second floor foyer, was an unpopular feature but is a reflection of the attention given to the energy crisis of the mid-70's. Lack of windows was an energy conservation compromise, given the long air conditioning season in Florida. It seemed impossible to prevent people from referring to it as "the tin barn" rather than Canaveral Hall. Despite the somewhat austere exterior appearance the program flourished and everyone was quite pleased when the July 2, 1979 issue of Time magazine ran a feature article on the University, calling it "the Harvard of the skies".

The evolution of the undergraduate AE curriculum into today's 134 semester credit hour curriculum was straightforward in most respects. Probably the most significant early change was the introduction of a strong capstone aircraft design sequence in 1970 by Dr. Ritchie. He wrote two textbooks, one for aircraft preliminary design and one for aircraft detail design. These texts were printed on campus but were also used by a number of other universities. Another issue was the strong laboratory content of the curriculum, which was and is very popular with students and employers alike [4]. It led to some philosophical challenges for the department. There were required separate lab courses taught by the AE department in materials, structures and instrumentation, electrical engineering, wind tunnel testing, aircraft conceptual design, and aircraft detail design. There were also technical elective lab courses in composite materials, flight testing, and acoustics. With so much lab content, the classic distinction between theoretical engineering and hands-on engineering technology became less and less apparent in the ERAU curricula. This caused significant confusion in some ABET accreditation visits. So in 1988 the decision was made to make Aircraft Engineering Technology (ACET) a separate department. Four years later it was combined with Avionics Technology into a broader Department of Technology.

One curriculum matter that deserves separate mention is the initiation of a Master of Science in Aeronautical Engineering in 1985. This program has grown slowly but surely and currently has about 40 students. Research interests are primarily in the areas of computational fluid dynamics, composite materials, acoustics, design optimization, and control of chaotic processes.

Various curricular changes were made in the name of keeping the curriculum up to date. The use of computer aided design (CAD) software was introduced into the capstone design courses in 1983 [5]. The capability of software and hardware evolved so rapidly that six different CAD packages were used in these courses. The latest change was a switch to CATIA, the nearly universal standard of the aerospace industry, in 2001. A composite materials course was created as an elective in 1984 [6]. Two laboratory improvement grants from the National Science Foundation permitted acquisition of rapid prototyping and computer aided manufacturing (CAM) capability in the early 1990s. Dr. James Ladesic's grant provided a stereolithography system and Prof. Charles Eastlake's grant provided a large, 4 x 8 foot bed, computer numerical control (CNC) milling machine [7]. Both were made an integral part of the course deliverables requirement for the capstone design courses. In the context of externally visible changes, the most significant was the change of the program name from Aeronautical Engineering to Aerospace Engineering in 1988. This necessitated only a minor change in the course offerings. The principal focus of the curriculum remained aircraft, but a significant number of Aeronautical Engineering graduates had always gone into the space side of the industry, so the name change simply reflected the reality of what the curriculum prepared graduates to undertake.

In 1995 spacecraft preliminary and detail design courses were initiated by Dr. Yechiel Crispin and in 2002 propulsion preliminary and detail design courses were created. So, as of Fall 2002, students in the AE program can choose a formal specialization track in aeronautics, astronautics, or propulsion.

Lehman Engineering and Technology Center with Space Shuttle launch visible in the background

The dramatically larger Lehman Engineering and Technology Center opened in 1995. The three story building is named after 1943 alumnus and congressman William M. Lehman. It houses engineering, engineering technology, math, physics, and computer departments. AE has 15 spacious, well-lighted laboratories and shops, predominantly on the first floor, including:

Design Lab: 16 Dell 330 1.4 GHz computers, CATIA solid modeling CAD software, NASTRAN finite element analysis, Digital Wind Tunnel (NASA PMARC) panel method aerodynamics code.
Structures Lab: 2 10,000lb MTS universal testing machines, Vibration Test Systems 500 lb force electrodynamic shaker, 20x10x10 foot heavy duty test article holding fixture, Physical Acoustics acoustic emission crack detection system.
Manufacturing Lab: CATIA and Varimetrix CAD and tool path generation software, Komo VR408P three-axis CNC milling machine with 4x8 foot working table.
Stereolithography Lab: The original 3D Systems laser-based stereolithography machine has been replaced by a thermojet "printer" which forms 3D models out of wax.
Composites Lab: Develco oven with vacuum system, overhead vacuum system on 2 4x8 foot layup tables, prepreg freezers, roll fabric storage rack.
Materials Lab: Tinius-Olsen universal tension/compression testing machine with 33,000 lb load capacity, rotational fatigue tester, Tinius-Olsen impact tester, Tinius-Olsen torque tester, 2 Rockwell hardness testers, PME Olympus metallograph with video camera.
Thermal Sciences Lab: 20-700 HP fluid power engine dynamometer, infrared camera, 36 lb thrust turbojet engine and test stand, propeller noise measurement facility.

Wind Tunnel Lab: Open circuit 180 ft/sec wind tunnel with 30x40-inch test section, 2x36-inch 2-D smoke tunnel, 18x24-inch 3-D smoke tunnel, 20-seat classroom.

Electrical / Electronics Lab: Tektronix oscilloscopes, Fluke frequency counters and digital multimeters, Hewlett-Packard oscillators, Heath/Zenith tri-power supplies, Macintosh computers.

Student projects lab: 550 square foot room with swipe-card access system, work tables for 4 work groups, 6 storage cabinets, band saw, belt sander, disk sander, drill press, material storage rack.

In 2001 Dr. Vladimir Golubev created a CFD lab and in 2002 Dr. Yi Zhao received NSF funding for a composites impact testing machine which has just been installed in the materials lab. All of these laboratories are popular stops on campus tours and an unusual sidelight is that virtually all of them are used for undergraduate instruction.

The sequence of Chairs of the Aeronautical/Aerospace Engineering Department has been: James F. Saunders (1965-1966), Dean who created a separate department for AE; Yuri Lawizki (1967-68), the first chair of the department; Ming Hsien (Ken) Wang (1968-1970), long time faculty member [2] who was most senior professor in the University at time of his retirement; Donald J. Ritchie (1970-1978), designed Canaveral Hall and the department's first major wind tunnel and oversaw first ABET accreditation of the AE program; Howard D. Curtis (1978-1986, 1989-93), popular leader who initiated the AE master's degree and is now the senior member of the department; Walter P. Schimmel (1986-1989), thermo-fluids engineer from Sandia Labs who broadened the department's consideration of propulsion topics; David C. Hazen (1993-1996), retired Princeton professor and prominent figure in the development of Lehman center who also introduced the spacecraft capstone design sequence; Allen I. Ormsbee (1996-1999), retired University of Illinois professor and avid sailplane pilot who expanded department participation in ABET; and Reda R. Mankbadi (1999-2002), former NASA senior scientist and technical leader.

The following is a list of the 18 current full time AE faculty, with when they joined the department in parentheses: Yechiel Crispin (1992), Howard D. Curtis (1972), Charles N. Eastlake (1979), Habib Eslami (1987), Vladimir V. Golubev (2001), Tej R. Gupta (1979), James G. Ladesic (1981), Eric v.K. Hill (1986), Lakshmanan L. Narayanaswami (1986), John R. Novy (1979), Eric R. Perrell (2002), Paul L. Quinn (1993), Frank J. Radosta (1982), R. Luther Reisbig (1989), David J. Sypeck (2002), Elaine S. Weavil (1986), John W. Weavil (1983), and Yi Zhao (2001). The stability of the faculty is noteworthy and is a significant factor in the success of the program. More than half have been in the department for over 15 years and seven have been in the department for 20 years or more.

The AE enrollment trend for the last 33 years is illustrated in Figure 1. It shows consistent growth over the majority of that time period. The drop of about 1/3 during the nationwide aerospace slump of the early and mid 1990s was problematic but was not as pronounced as that experienced by most AE programs. In fact, it was during this downturn that ERAU became the largest AE program in the U.S. Growth has been fairly rapid since 1995, reaching 1084 in Fall 2002. The number of BS degrees awarded per academic year is plotted in Figure 2. It has been relatively even at about 100 per year for over 20 years and hit a new peak in 2001-02 at 145.

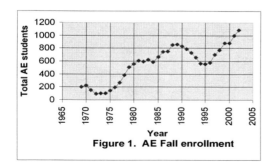

Figure 1. AE Fall enrollment

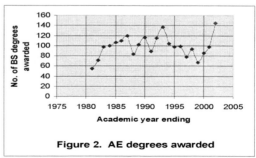

Figure 2. AE degrees awarded

Notable student achievements include the following.

NASA/FAA National General Aviation Design Competition: 2nd Place in 1995, 1st Place in 1996, 1st Place in 1999, 1st Place in 2001.

AIAA Graduate Student Aircraft Design Competition: 1st Place in 1998.

AIAA Outstanding Student Branch, 1996, 97

SAE Aero Design, Radio Controlled Model Cargo Aircraft Competition: 2nd Place overall, 1993; 3rd Place overall, 1995; Best design, Eastern Division, Standard Class, 1997; 1st Place overall, Eastern Division, Standard Class, 1998; Best Design, Eastern Division, Open Class, 1999; 2nd Place overall, Western Division, Standard Class, 2001 SAE Outstanding Student Branch, 1995, 97

NASA/USRA Advanced Design Program, 1990-93 [8]

Society for Unmanned Vehicle Systems International, Intelligent Ground Vehicle Competition, 2nd Place, 2002.

Prototype diesel-powered Cessna 182 in flight, designed and built on campus. AE classes assisted with the design and this effort won the 2001 NASA/FAA National General Aviation Design Competition

Department faculty are active contributors to a spectrum of professional organizations, including American Institute of Aeronautics and Astronautics, American Society for Engineering Education, Accreditation Board for Engineering and Technology, American Society of Mechanical Engineers, American Society for Nondestructive Testing, and Society of Automotive Engineers. And since the university made a conscious decision to cooperate with media requests in 1993, AE department faculty have done well over 300 interviews, including for most major newspapers and virtually all major TV networks. Several faculty have participated with Prof. Eastlake topping the list at 322 interviews. Boeing Welliver Fellowships have been awarded to Drs. Vicki Johnson and Lakshmanan Narayanaswami. The team of Eastlake and Ladesic was a finalist in the 1999 national Boeing Outstanding Educator Award. Finalists for the campus Outstanding Teacher Award include AE faculty Gupta, Narayanaswami, Eastlake, and Radosta, and Dr. Ladesic won the award in 2001. Finalists for the campus Outstanding Researcher Award include Drs. Hill and Patrick, the latter having won in 2001.

To ERAU

AE graduate Susan (Still) Kilrain, the second woman to pilot the Space Shuttle

Among the noteworthy graduates of the AE program are two current space shuttle astronauts, Susan (Still) Kilrain and Nicole (Schaeffer) Stott.

Verification of practical acceptance of the program by industry is that 97.1% of last year's graduates are employed or in graduate school. Principal employers from the past three years are: Lockheed Martin, 18%; Cessna, 15%; Boeing, 14%; United Space Alliance, 13%; U.S. Air Force and U.S. Army (officers, primarily through ROTC), 10-15%; U. S. Navy (civilian), 4%; U. S. Air Force (civilian), 3%; Gulfstream Aerospace, 3%.

The Aerospace Engineering program at Prescott, Arizona

The Aerospace Engineering program at the Prescott Campus started in Fall 1980 under the leadership of Prof. Tracy Doryland (Chair 1980-1984, 1999-2001). .

Initially it was decided that during the first year, only freshmen would be admitted to the program, and each succeeding year another class would be added, but shortly thereafter transfer students were accepted into the program. After lengthy discussions, it was decided that the curriculum would be taken verbatim from the AE program at the Daytona Beach campus.

The next item of business was to establish the laboratories. The first lab was the Graphics Lab, and thirty new drafting tables with the standard mechanical drafting arms were installed in a building of roughly 1500 sq ft. It looked exactly like the rooms in which the faculty had taken their mechanical drawing courses during the 1950-70 period.

Aerial view of the Prescott, AZ, campus.

A major activity during the second year was to purchase the equipment for labs required of juniors (materials, wind tunnel, electrical). This later resulted in a new 5000 sq. ft. building for the wind tunnel laboratory.

The first class consisted of 42 freshmen, and the first graduating class of nine students received their degrees in August 1984. Of the original 42 students, 15 eventually graduated, and including the first graduates, over 700 aerospace engineers have graduated from the Prescott Campus. As would be expected, these graduates have contributed significantly to the betterment of society. One graduate of the first class, Peg Billson, should be particularly noted, as she became the youngest woman Vice President at McDonnell Douglas and currently is Vice President & General Manager for Aircraft Landing Systems at Honeywell in South Bend, IN.

In 1984, Dr. Blaine Butler joined the department (Chair 1984-88), and under his leadership, the AE program received initial ABET accreditation in 1986. During this time, Aerojet Corporation donated a closed circuit wind tunnel with a 32 x 45 x 48 in. test section that added significant capability to the Wind Tunnel Lab. An electron beam microscope also was added to the Materials Lab.

In 1988, Dr. Richard Felton joined the department (Chair 1988-1999), and under his leadership the AE program was reaccredited in 1992 and 1998. Several major changes occurred during this time period. First, Prof. Richard Newcomb started the evolution to computer-aided drafting in 1989 with a couple of computers. By 1994-95 all mechanical drafting equipment had gone the way of the slide rule. Initially CADKEY was the software, then Silverscreen, then Bentley Microstation, and finally CATIA became the software package of choice. Fortunately the campus has been very supportive in always providing the Graphics/CAD/Design lab with current state-of-the-art computers. The lab will move into a new design lab in the new academic complex to be ready Fall 2004.

Second, a broader emphasis was placed on design content in the AE program. In 1990 the experimental aerodynamics course (Wind Tunnel) was offered for the first time during the summer, and 34 students signed on for the course! Prior to that summer, the course had been more or less a series of "canned" experiments. That summer a major design project was incorporated into the course. The projects ranged the entire gamut from the expected aircraft-related ones to unique ones, such as an aerodynamic door for a medical specimen cabinet, and snow fences. The response of the students has been overwhelmingly positive on the design projects.

In the mid 90's a major design project was incorporated into the structures and instrumentation lab course. Since then numerous design projects have been completed. One example illustrates the type of project: Six teams of 4-5 individuals each had to reply to an RFP for a new rocket test stand. The teams had to design, fabricate a model, and test their test stands using Estes rockets. A plot of thrust vs. time was required for whatever Estes rocket their test stand used. Six teams resulted in six different designs. The teams had been told that each should consider themselves as part of a company that was competing for the contract and should protect their design.

Third, in 1992 Dr. David McMaster established a requirement that each design team in the aircraft design courses fabricate a model of their aircraft. The teams tested their models in the wind tunnel to compare their theoretical calculations with experimental data. Shown below are a sample of the nearly 100 models fabricated to date.

Finally, Dr. Govinder Giare introduced several items into the materials and structures courses. A few are mentioned to illustrate the hands-on experience for the students: (1) a large structural fixture which allows testing to failure of various items, such as beams fabricated by students in the second aircraft structures course; (2) casting system that allows students to form the molds of various items and to cast these items using aluminum; (3) composite materials system to expose the students to the fabrication of small

Student design project in the Aerojet wind

parts from composite materials. An example is carbon-carbon aircraft wings.

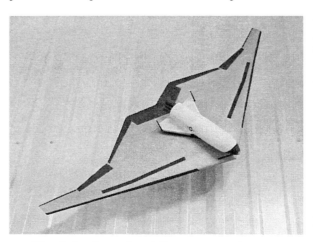

Model of project from capstone design course

Until 1995 the Aerospace Engineering program was mainly focused on aircraft with a little astronautics content. During spring semester 1996 an experimental tunnel spacecraft design was offered. Then with increasing demand by the students, it was obvious that more space courses had to be offered. In 2002 the space option was formalized into the catalog, and currently there is about a 50/50 split between aircraft and spacecraft students. This curriculum addition was the result of the leadership provided by Dr. Ronald Madler (Chair 2001-present).

Since the Aerospace Engineering program at Prescott is strictly undergraduate, there has been limited research conducted. Yet, a number of small research programs have been conducted in the wind tunnel for external clients, including one dealing with the Apache Helicopter. However, there have been numerous research projects for the undergraduate students. Most of these projects were funded by the Arizona-NASA Space Grant Program. One was research on the National Aero-Space Plane, X-30, which received national recognition from the X-30 Program Director.

While the list of faculty and student accomplishments is too long to list, there are a few recognitions that should be mentioned. Dr. Thomas Gally was selected for a Boeing Welliver Fellowship (Summer 2002). Drs. Thomas Gally, Ronald Madler, and John Reis were selected in the recent past by SAE (Society of Automotive Engineers) for the Ralph Teetor Educational Award. A student team, working on a design project of a device to measure the elevator deflection of a Cessna 172 aircraft, submitted their final report to an annual design contest of the Pacific Southwest section of ASEE and won first place.

During the 2002-2003 academic year, two major additions to the AE labs have been acquired. These are a small turbojet engine and a larger student project wind tunnel.

Basic turbojet engine suitable for student operation

New student project wind tunnel with smaller one in background

A College of Engineering was created at Prescott in January, 1999, with Dr. Richard Felton serving as the first Dean. In January, 2002, Dr. Don Rabern succeeded him.

The enrollment in the AE program at Prescott has grown to 408 in Fall 2002. The success of the program is due to the total dedication of the faculty to teaching undergraduates. The following are all previous and current (asterisks indicate current) full-time AE faculty: Jeff Ashworth*, Don Broadhurst, Blaine Butler, Jack Convey, Tracy Doryland*, Hany El Kadi, Richard Felton*, Gabriel Georgiades, Govinder Giare, Xavier Gonzalez, Garry Harrison*, Jim Helbling*, William Keppel, David Lanning*, Ronald Madler*, Steve McIntyre, David McMaster, Donald Moses, Richard Newcomb, Yoel Oved, Curtis Potterveld, Don Rabern*, John Reis*, Mark Sensmeier*, Rachel Shinn*, Karl Siebold*, R.M. Sundar.

The Daytona Beach campus College of Engineering Dean's vision for the future

The Aerospace Engineering Departments are unique in that students and the faculty are truly dedicated to the program. Therefore, the program has had considerable success over the past years. Enrollment in Daytona at the undergraduate level has reached 1100 students, more than twice the next largest AE program in the country. Enrollment in Prescott has been over 400 for the last three years, making it fourth largest. These two together make a strong unit, with curriculum development done jointly under the ERAU "one University" concept. For the last three years (all the years that ratings of individual programs were made) U.S. News & World Report has rated the program as the #1 Aerospace Engineering program among non-PhD-granting institutions.

Many of our faculty are nationally recognized experts in their field, called upon by industry for consulting and as expert witnesses, and by the news media for comments and explanations.

The recent ABET accreditation renewal under EC2000 for our program was successfully completed with highly favorable evaluation comments.

The number of faculty receiving external research grants and active in scholarly activity has considerably increased.

Our students are involved in the application of the latest technologies in their design projects, such as the Sonic Cruiser, diesel powered aircraft, and unmanned aerial vehicles.

The success of the Aerospace Engineering department has favorably impacted the development of all other engineering programs at ERAU. After years without a traditional college/dean structure, the College of Engineering was formed at Daytona on August 15, 2003 with the Chair of Aerospace Engineering, Dr. Mankbadi, serving as the College Dean. Subsequent to a review of current programs within the College, only programs with potential for high quality will remain. In addition to Aerospace Engineering, this includes Software Engineering, Computer Engineering, and Civil Engineering. Other new ABET/EAC accreditable programs are being explored.

Despite significant increases in tuition over the past several years, the total engineering enrollment continues to increase. This has helped us solidify a vision of developing a comprehensive College of Engineering with a focus on aerospace applications and employing the most modern technology. We anticipate our College to be small, prestigious, and highly competitive. We will enable our graduate education to be of top quality by ensuring that we have qualified and dedicated faculty, current in their fields. We are committed to small class size and to providing those classes with modern labs and the latest technologies. We expect the College of Engineering to be active in scholarly and professional activities and to be a center for expertise in the aerospace field. We envision that the productive and successful experiences of our Aerospace Engineering faculty and their students will lead to a top ranking College of Engineering.

References
1. Anonymous, "Avco Corporation - The First 50 Years", Avco Corporation, 1979.

2. Ming Hsien Wang, "Looking Back On My 80 Years", self-published, 2000.

3. John McCollister and Diann Davis, "The Sky is Home, The Story of Embry-Riddle: the World's Leading Aviation/Aerospace University", Jonathon David Publishers, 1996.

4. Eastlake, C.N., "The Tangible Role of Laboratories in the Undergraduate Curriculum," presented at American Society for Engineering Education Conference, Cincinnati, OH, June 1986.

5. Eastlake, C.N., "Incorporating a Computer-Aided-Drafting (CAD) System into an Existing Aircraft Design Course," presented at American Society for Engineering Education Conference, Reno, Nevada, June 1987.

6. Eastlake, C.N., "Initiating an Undergraduate Composite Materials Laboratory in an AE Curriculum," presented at American Society for Engineering Education Conference, Atlanta, GA, June 1985.
7. Eastlake, C.N., and Stanley, A.L., "Establishing a Computer Aided Manufacturing System to Extend the Capability of Traditional Aircraft and Spacecraft Design Courses," Proceedings of the American Society of Engineering Education Conference, Milwaukee, Wisconsin, June, 1997.

8. Ladesic, J., Eastlake C., Kietzmann, N., "The Design of a Primary Flight Trainer Using Concurrent Engineering Concepts," Proceedings of the 9th Annual NASA/USRA Advanced Design Program Summer Conference," NASA/Johnson Space Center, Houston, TX, June 1993.

CHAPTER 66

Boeing School of Aeronautics
Michael Lombardi., Boeing Corporate Historian

The Beginnings

The original articles of incorporation of the Boeing Airplane Company drawn up in 1916 stated that the purpose of the company was to "manufacture aeroplanes and vehicles of aviation," but the visionary founder of the company, William Edward Boeing, also included this forward looking provision: "To maintain, carry on and operate schools of aviation, and for the teaching of all branches of knowledge and of the arts and sciences in any way connected with or useful for the operation of aeroplanes and other vehicles of aviation."

From the start William Boeing was determined to have top engineers trained in the science of aeronautics. It was difficult for Boeing to find good engineers as the only program at that time was at the Massachusetts Institute of Technology. Boeing turned to the University of Washington and encouraged the university to start its own aeronautical engineering program. In return for the University of Washington's assistance in training engineers, William Boeing funded the construction of a wind tunnel at the university which, when it was finished in 1918, was the first tunnel west of the Mississippi and only the third tunnel in the United States.

In 1927 most of the provisions spelled out in the articles of incorporation would start to come to fruition when William Boeing, agreeing with advice from his pilot Eddie Hubbard, bid on the Chicago to San Francisco air mail route. Boeing won the bid and Boeing Air Transport (BAT), the airline subsidiary of the Boeing Airplane Company, was launched. And soon Boeing was once again faced with a shortage of trained personnel.

Formation of the Boeing School of Aeronautics

When BAT began business there were ample former Air Mail Service pilots available to fill all the pilot slots. As the business expanded the leadership of BAT could see a future need for more pilots, mechanics and other personnel trained specifically for the business of commercial aviation. A survey of schools with aviation related programs around the country showed that there were no programs available for training these needed personnel, only the military had anything close to what BAT needed, but there was nothing along commercial lines.

In 1928 Theophilus Lee Jr., an educator and Boeing airmail pilot, suggested that BAT open its own school to train personnel and pilots to support the airline's operations. The Board of directors of BAT, and more importantly, William Boeing, agreed. A site at the Oakland Municipal Airport, at the time the largest municipal airport in the United States, was selected for the school. The school opened on September 16, 1929 under the direction of the same airmail pilot that suggested it: Theophilus Lee Jr. Along with Lee the other officers of the school were Herbert Marsh,

Superintendent of Technical Instruction and George I. Myers Superintendent of Flight Instruction. Marsh left the school after the first year and was replaced by Allen Bonnalie. On November 27, 1929, the Board of Directors of Boeing Air Transport authorized the name, "Boeing School of Aeronautics" to denote the flying school division of BAT.

The First Curriculum

At the time that BAT was developing the School of Aeronautics, The Aeronautics Branch of the Department of Commerce was also concerned with the state of aviation schools in the United States and set out to develop a set of standards to combat the growing concern of substandard schools and training related to the expanding aviation industry. On February 28, 1929 the Air Commerce Act of 1926 was amended and the School Supplement was added to the Air Commerce Regulations. In creating a curriculum for the Boeing School, the Department of Commerce standards were used as a basis but were seen as inadequate for the school since they focused primarily on the training of pilots.

Aerial View of the School at Oakland Municipal Airport

Top Left: Allan F. Bonnalie, Director of Technical Instruction
Top Right: Theophilus Lee, Jr., General Manager of the Boeing School
Lower Left: Hilton F. Lusk, Superintendent of Technical Information
Lower Right: George I, Myers, Superintendent of Flight Instruction

A series of courses were laid out with short, part time, courses that would satisfy the Department of Commerce standards. Master Courses for pilots and mechanics, giving generalized training specifically related to commercial aviation operations, were also developed requiring full time attendance for up to nine months.

Some of the classes that made up the Master Courses included:

Aerodynamics
Aircraft Fabrication
Materials
Engines
Communications
Instruments
Aerial Surveying
Business Methods
Engineering Drawing
Elements of Design
Design Drafting
Meteorology
Law
Field Operations.

The courses were developed to run annually for about 35 students. When the school opened and 100 students immediately applied it was suddenly apparent that the demand for this training was greatly underestimated by the school administration. To accommodate for the demand, the Boeing School quickly expanded its equipment, personnel and facilities and deemphasized the short courses in favor of the more popular Master Courses, which were considered the foundation of the School, and classes were started every three months instead of annually.

BOEING SCHOOL OF AERONAUTICS
OAKLAND, CALIFORNIA - MUNICIPAL AIRPORT

Operations under United Aircraft and Transport Corporation (UATC)
1929-1934

During this period William Boeing and Fredrick Renschler of Pratt and Whitney created an aviation holding company called United Aircraft and Transport Corporation (UATC). Along with Pratt & Whitney, Boeing's airline interests and the Boeing Airplane Company, the corporation would eventually also bring together Chance Vought, Sikorsky, Stearman, Northrop, Hamilton Aero Manufacturing and Standard Steel, as well as a number of other small airlines including Stout, Pacific Air Transport and Varney.

As part of this mega-corporation, The Boeing School of Aeronautics remained a subsidiary of BAT until 1934.

The School benefited greatly from being a part of UATC as the Engine Lab always had an ample supply of Pratt and Whitney engines and the pilot training operations had an impressive fleet of aircraft including 5 model 203 trainers built by the Boeing Airplane Company for the School (In 1935 and 1936 2 more model 203s were built by students at the School). The fleet also included Boeing Model 40 mail planes, a Boeing Model 80 Transport, a Stearman trainer and a Hamilton Monoplane.

Students introduced to one of the school's model 203 trainers

---

Later the school would acquire a Ford Trimotor and a 4-place Stinson Junior. The planes and flight training were under the direction of George Myers who would eventually become the head of pilot employment and training at United Air Lines.

Annually William Boeing would offer scholarships for the school. Applicants for the scholarship had to be ranked in the top third of their class for their entire period of college or university enrollment. The candidates were then required to complete a 2,000 word essay based on one of five different subjects including "Trends of Development in Air Transportation, Progress of Safety in Aviation, Trends of Airport Design, Radio as an Aid to Aviation, and Importance of Proper Co-ordination of Federal and State Laws Governing Air Transportation." William Boeing influenced the direction of the entire corporation with his insistence on experimentation and practical application. At the Boeing School this resulted in the students and instructors designing and manufacturing aircraft.

The school welcomed women in all programs – especially in flight training

Under the instruction of Wellwood Beall, who served as the instructor for aeronautical engineering at the Boeing School and went on to design the Boeing 314 "Clipper" and become Senior Vice President in charge of Engineering for the Boeing Company, students designed and built gliders.

John Willard Thorp, instructor for Engineering Drawing, who went on to design the Lockheed P2V Neptune, is known for designing a series of private aircraft which he began designing and building while instructing at the school.

An interesting experiment occurred at the school on April 12, 1933, when William Besler, flew a modified Travel Air 2000 as the first and only steam powered aircraft. The Boeing School supported the experiment by modifying and balancing the aircraft as well as testing the engine and supporting the historic flight from its facilities at the Oakland Airport.

Operations after the Breakup of UATC: 1934 – 1939

As a result of President Franklin Roosevelt's New Deal legislation, UATC was forced to break up and in order for United Airlines to hold onto its airmail contracts it had to hire new leadership from candidates that were not involved in any airmail negotiations prior to 1934.

The unwarranted government scrutiny and breakup of his corporation convinced William Edward Boeing to retire from the aircraft business. The breakup of UATC resulted in three independent companies; The Boeing Airplane Company, United Technologies, and United Air Lines. As part of the reorganization, The Boeing School of Aeronautics was acquired by United Air Lines on December 28, 1934. During this period the school continued its major courses in pilot and mechanic training, offering a "Boeing Pilot Certificate" and/or "Boeing Airline Operations Certificate." The school also began to attract graduates of accredited engineering colleges. The school recommended its Airline Pilot and Airline Technician's courses to give practical experience that was essential to the field of design engineering.

Beginning with the Fall quarter of 1937, The Boeing School introduced a new program called the Practical Aeronautical Engineer's Course. The course included thorough technical training to fully meet all engineering requirements but also included practical training to familiarize the student with the construction of an airplane and its parts. The pre-requisite for enrollment in the course was two years of pre-engineering at an accredited college or university.

Minimum credits were also required in a number of subjects that included: Plane Trigonometry, College Algebra, Analytical Geometry, Differential and Integral Calculus, Descriptive Geometry, College Physics Mechanical Drawings and Mechanical Shop. Tuition was $1440 for the intensive 24 month course that covered the equivalent of three years of college. Along with the two years pre-requisite the students would have five years of total credits upon completion of the course and would earn an Air Transportation Engineer's Certificate or a Practical Aeronautical Engineer's Certificate. The School claimed in its 1937 brochure: "these are today the only courses of recognized engineering caliber in which technical and practical training have been coordinated to achieve a significant balance between theory and its application to industry."

The curriculum does exhibit a mix of academic, laboratory, and practical shop courses that suggest a very broad scope of instruction with the intent of creating opportunity for employment in a variety of aviation related fields:

First Quarter

	Hours
Drawing- Applied Descriptive Geometry	80
Communications – code	24
Physical Laboratory	80
Mechanics Laboratory – Graphics	36
Apprehensive Work	30
Fabrications – Construction and Repair Shop	80
Mechanics – Graphics	24
Airplanes – history and Introduction	24
Aircraft Power Plant Theory	24
Electricity AC-DC	36
Air Law and Regulations	12
	450

Second Quarter

Drawing – Mechanism	80
Power Plant Laboratory – Thermodynamics	72

Physical Laboratory	80
Fabrications – Construction and repair Shop	80
Mechanics – Analytical	36
Aircraft materials	24
Airplanes – Fundamentals of Flight	36
Aircraft Power Plants – Thermodynamics	<u>36</u>
	444

Third Quarter

Instruments Laboratory	36
Fabrication – Welding – Metals Shop	80
Apprentice Work	48
Design – Applied	80
Power Plants – Engine Shop	80
Aircraft Materials – Mechanics	24
Aircraft Instruments	24
Aerodynamics – Principles	36
Aircraft Power Plants – Engine Dynamics	24
Electricity – Electronics	<u>24</u>
	456

Forth Quarter

Fabrication – Metal Shop	80
Apprentice Work	30
Design – Applied	80
Power Plant – Engine Shop	80
Meteorology – Principles	36
Aerodynamics – Applied	24
Aircraft Power Plant – Installation	24
Metallurgy – Aircraft Alloys	24
Air Law – transport and Commercial	36
Communications – Airways	<u>24</u>
	438

Fifth Quarter (At this point either the Practical Aeronautical Engineering or the Air Transport Engineering course is followed)

	PAE Hours	ATE Hours
Meteorology Laboratory	36	
Power Plant Laboratory - Engine Testing	48	48
Physical Laboratory – Testing	80	80
Fabrications – Metal Shop	E	
Maintenance Shop	E	80
Design – Detail	80	
Communications – Radio Shop	E	80
Aircraft Materials – Structures	48	
Aircraft Instruments – Instrumentation	24	24
Synoptic and Dynamic Meteorology	24	

Aerodynamics – Performance Appreciation	24	24
Aircraft Power Plants – Engine Dynamics	48	
Radio License Seminar	36	
Special Lecture	<u>12</u>	<u>12</u>
	444	444

Sixth Quarter

Meteorology Laboratory	48	
Drawing – vector Analysis Laboratory	36	
Power Plant Laboratory – testing	E	80
Physical Laboratory – testing	80	
Mechanics Laboratory – Control Mechanics	36	
Fabrications – Metal Shop	E	80
Maintenance Shop	E	
Fabrications – Construction and Repair Shop	E	80
Design – Stress Analysis	80	
Dynamic and Synoptic Meteorology	48	
Performance Appreciation	36	
Aircraft Power Plant – Altitude Operation	36	36
Air Transportation – Principles	24	24
Special Lectures	12	12
Shop Practices	<u>24</u>	
	444	444

Seventh Quarter

Airplane - Dead Reckoning Laboratory	36	
Meteorology Laboratory	180	
Drawing – Detail Design Laboratory	36	
Fabrications – Metal Shop	E	
Maintenance Shop	E	
Mould Loft Process	80	
Design – Fittings	80	
Aircraft Power Plant – Engine Shop	E	
Aircraft Materials		
– D of C Transport Design requirements	60	60
Avigation – dead Reckoning Lecture	36	
Dynamic and Synoptic Meteorology	60	
Air Transportation – Airline Economy	36	36
Salesmanship	24	24
Lofting Principles	12	
Special Lecture	12	12
Seminar	<u>24</u>	
	444	444

Eighth Quarter

Avigation – Celestial Laboratory	36	
Drawing – Vector Analysis	36	
Design – Stress Analysis	240	

Design – Airport	60	
Aircraft Materials		
A & N Military Design requirements	36	
Aircraft Materials – Structures	60	
Avigation – Celestial	36	
Business – Operating Procedure	36	
Business – Principles of Accounting	36	
Safety	24	24
Air Transportation – Dispatching Laboratory	240	
	432	432

When these new engineering courses were inaugurated, the country was deep into the Depression, but the Boeing School was able to attract engineering students to its new program with promises of opportunities in the aviation industry. The school claimed that "qualified graduates should experience no difficulty in capitalizing their training." The school was able to make this claim based on the demand made by major aircraft manufacturers for graduates from the school and that those same companies were paying premium wages to Boeing graduates. The school went farther to claim that "not a single qualified graduate has failed to make a good connection."

A person of note who completed this program was Peter M. Bowers, Boeing engineer, designer of the "Fly Baby," and arguably the most prolific and best known author of aviation history.

The job opportunities that the school suggested that its graduates would have excellent opportunities in included: Aeronautical Design Engineering, Automotive Engineering, general Engineering, Production Engineering, Air Transport Engineering, a new field at the time and one that was expected to have more demand than aeronautical engineering due to the rapidly expanding air transport infrastructure and the need for engineers trained in design of airports and airport facilities. The school also saw great opportunities for flight engineering, which at that time was a new field that expected to grow as the size of airplanes grew; it required a pilot's license, mechanical training and an engineering background. Another new and interesting

Typical classroom

field was Sales Engineering, a field that not only included the sale of aircraft, but also the sale of equipment to factories and airports.

In less than ten years the school had grown three fold with three major courses that included 3,712 hours of study in 49 separate subjects compared to 924 hours and 22 subjects in the first year of operation. When the doors opened in 1929 there were 100 students managed by a staff of 16, in 1937 there were nearly 500 students and a staff of 41.

The War Years and the End of the School: 1939-1945

In 1939 the school, took its first steps to becoming a military training center and leaving behind civilian training. During 1939 and 1940 the school took up a contract with the Civil Aeronautics Authority, the school to train pilots for the federal civilian pilot training program. In 1940 the school received a contract from the U.S. Army Air Corp to train airplane mechanics. The Army mechanics training program was large enough to require an expansion of the school and by August 1942 fourteen new buildings were added to the school at the expenditure $550,000, the expansion enabled the school to provide capacity for 1,095 trainees. During the period of the Air Corp mechanic training contract that ended in August 1943, the school trained approximately 5000 mechanics.

With the increase of military use of the Boeing School, wartime restrictions on flying, and shortage of commercial students, the decision was made in August 1942 to suspended all commercial training activities at the school. Any civilian training required by United Air Lines was continued at its training center in Cheyenne, Wyoming. On January 8, 1943 the Board of Directors adopted a resolution discontinuing the name of the school and consolidated its books and accounts with other parts of the company; The Boeing School of Aeronautics was officially closed.

In recognition of the schools facilities having become completely devoted to military training, the name of the former Boeing School of Aeronautics was changed to United Air Lines Training Center.

With the ending of the Army mechanic s contract in August 1943, the school took on a Navy contract to train mechanics and technicians for the Naval Air Transport Service. The Navy program began on September 6, 1943 and ran through November 10, 1945 graduating 1,395 trainees.

With the conclusion of the Navy training contract in 1945 the training center was closed. Today one can still visit the site of the former Boeing School at the Western Air Museum which is housed in one of the School's hangers built during the 1940 expansion.

Chapter 67

From Biplanes to Reusable Launch Vehicles: 75 Years of Aircraft Design at Cal Poly

Russell M. Cummings, Professor, Aerospace Engineering Department
California Polytechnic State University, San Luis Obispo, CA 93407

Abstract

While many universities can boast of making contributions to the development of aeronautical science, few universities have maintained their focus on aircraft design as the key feature of their curricula. Cal Poly's aeronautics education has a constant thread running from its inception in 1927 to the present: teaching students with a "learn by doing" philosophy, applying knowledge through hands-on laboratory instruction and real-world design projects. Aeronautics students graduating from Cal Poly have already experienced the rigors and excitement of designing, building, and often flying their concepts long before they begin their careers in the aerospace industry. While this approach to education is not for everyone, it has stood the test of time and has produced notable leaders in the aerospace industry.

Introduction

Figure 1. Myron Angel, the force behind Cal Poly's founding and the "Learn by Doing" educational philosophy (photo courtesy of University Archives, California Polytechnic State University).

Before discussing the history of the Aerospace Engineering Department, it will be beneficial to familiarize some readers with Cal Poly. The Cal Poly educational philosophy has always been "learn by doing," whether students are studying agriculture, engineering, architecture, or music. While many universities started with a similar educational philosophy, Cal Poly has retained this emphasis throughout its history. Without understanding this unique aspect of Cal Poly, it would be difficult to appreciate its current Aerospace Engineering Department.

The driving force behind the founding of Cal Poly was Myron Angel, a young West Point drop-out who sailed to San Francisco during the gold-rush year of 1849 (see Fig. 1). Having been unsuccessful at making his fortune in the gold fields, he eventually moved to San Luis Obispo by the 1890s and became a leader in a campaign that originally aimed to establish a state "normal"

Figure 2. California Polytechnic School campus, c. 1906 (photo courtesy of University Archives, California Polytechnic State

senator, Sylvester C. Smith of Bakersfield.[1]

school (a teacher's training school) at San Luis Obispo. Eventually he shifted his support to the idea of a polytechnic institute, an idea suggested by the local state

Looking back to his arrival in San Francisco, when he felt completely unprepared for his new life in California, Angel made a case for a technical school and articulated the institution's future: he envisioned a school that would "teach the hand as well as the head, so that no young man or young woman will be sent off in the world to earn their living as poorly equipped for the task as I when I landed in San Francisco in 1849."[1]

On March 8, 1901, legislation founding the California Polytechnic School was signed into law after six years of debate. The mandate was clear:

> *To furnish to young people of both sexes mental and manual training in the arts and sciences, including agriculture, mechanics, engineering, business methods, domestic economy, and such other branches as will fit the students for non-professional walks of life.[1]*

Figure 3. Julian A. McPhee, Cal Poly's President from 1933 through 1966 (photo courtesy of University Archives, California Polytechnic State

We may all laugh at what was considered "non-professional" at the time, but no one can deny that certain aspects of Myron Angel's original desires have permeated the university and the Aerospace Engineering Department ever since.

During its first three decades, Cal Poly evolved into the equivalent of a junior college, and governance was transferred from a local board of trustees to the state Board of Education. The original campus buildings are shown in Fig. 2, including the Recitation and Administration, Household Arts, and Boys' Dormitory buildings.

During the early years of the Depression, the Legislature considered abolishing Cal Poly, but in 1933 Cal Poly received an infusion of energy and leadership when Julian A. McPhee, chief of the California Bureau of Agricultural Education, agreed to become the school's president (see Fig. 3). McPhee assumed leadership of what had been reorganized as a two-year technical college offering instruction in agriculture and industrial fields. During the next 33 years, McPhee guided Cal Poly's transformation into a polytechnic university, culminating in the first baccalaureate exercises being held in 1942. One of McPhee's major contributions to the educational philosophy of Cal Poly was the idea of "the upside down curriculum." He strongly believed that students learned best when they learned within the context of their chosen major or profession, a concept that has strongly influenced the Aerospace Engineering Department at Cal Poly. He insisted that students receive coursework in their chosen field of study during their freshman and sophomore years, not just their junior and senior years.

In addition to the changes at the school that McPhee initiated, Cal Poly's name changed as well: in 1937 the name changed from California Polytechnic School to California State Polytechnic School, and in 1947 became California State Polytechnic College. In the postwar years the first graduate-level programs were added to the curriculum, and in 1956, coeds returned to the campus (after having been asked to leave in 1929 due to a lack of on-campus facilities for women). It was in 1961 that the college became part of the newly formed California State Colleges system (now The California State University). In 1966, Cal Poly's Kellogg-Voorhis campus at Pomona, founded in 1938 as a branch of the San Luis Obispo school, was made a separate state college by the Legislature. The current name of the school, California Polytechnic State University, was formerly adopted in 1972.

The Early Years of the Department

Many early aeronautical programs were started with the support of The Daniel Guggenheim Fund for the Promotion of Aeronautics, including programs at M.I.T., CalTech, New York University, Michigan, Stanford, Georgia Tech, and the University of Washington.[2] These new aeronautics programs, as well as the newly formed National Advisory Committee for Aeronautics (N.A.C.A., the predecessor to NASA), concentrated on the development of aeronautical science, including aerodynamic theory and performance analysis (a more complete history of these early programs may be found in Ref. 3). Cal Poly's aeronautics program, established in 1927, had a very different beginning.

Figure 4. Early airplane flying over downtown San Luis Obispo in 1910 (photo courtesy of University Archives, California Polytechnic State University).

Cal Poly has the good fortune of being located in an area that always was interested in aviation—Fig. 4 shows an early airplane flying over downtown San Luis Obispo in 1910. Airfields were operated in various locations in town, including Clark Field, and in 1939 the current San Luis Obispo airport was opened, in part due to the efforts of Cal Poly Aeronautics students Chris and David Hoover.

Since Cal Poly had started as a vocational school, it seemed natural that programs in engineering would be initiated. In 1927, Cal Poly added programs in Engineering Mechanics and Aeronautics to enhance their already existing programs in agriculture, business, mechanics, home economics, and printing. The fledgling Aeronautics Department, started in the wake of the excitement over Charles Lindbergh's trans-Atlantic flight, followed in the Cal Poly tradition of offering practical, hands-on education for

Figure 5. The Glenmont, the earliest known student-constructed airplane, named for instructors Glen Warren and Monty Montijo, both pictured here with the students who built the airplane in 1928 (photo courtesy of University Archives, California Polytechnic State University).

Courses in engines, welding, and drafting were offered, in addition to courses already offered by other departments (such as electrical and mechanical courses). The department maintained a laboratory where:

Students learn all groundwork connected with aviation. Motors are torn down, overhauled and built up according to precise aeronautical specifications. The shop is equipped with seven aircraft motors of representative types including rotary, vertical and vee-type, also propeller balancing stands and test stands. Motors are given actual running tests.[5]

Although the primary goal was to train students as aircraft mechanics, the student's enthusiasm and creativity resulted in the construction of increasingly impressive aircraft, beginning with The GlenMont, shown in Fig. 5. The GlenMont, named after the department's two instructors, was a six passenger replica of Lindbergh's *Spirit of St. Louis*, a fitting start to the department's design history. It is believed that the GlenMont was the first airplane in the United States built by students (see newspaper article in Fig. 6). Approximately twenty students participated in all aspects of the construction, including structures, electrical systems, and engines. The article states that, "experienced aviators who have inspected the craft declare that it is as fine a piece of workmanship as has been seen in flying circles . . . the excellency of workmanship and the interesting new features of

students. The original instructors, H. Glen Warren and J.G. "Monty" Montijo, decided that the best way for students to learn about aeronautics was to design and build airplanes:

It is the aim of the aeronautics department to prepare the student to take care of an airplane and do necessary repairs to plane and the engine. One or more planes are constructed in the school shop every year and experience is given to all students in overhauling and repairing different types of engines.[4]

Figure 6. Newspaper article describing The Glenmont in 1928 (courtesy of University Archives, California Polytechnic State University).

safety and convenience caused much comment." We are all left to wonder what some of these features may have been.

Figure 7. The Aeronautics Department Building, c. 1935 (photo courtesy of University Archives, California Polytechnic State University).

The students continued designing and constructing airplanes, with the goal of completing a new airplane each year. In 1930 a two-place open seat biplane powered by a Kinner engine was designed and built. In 1931 a three-place high wing monoplane was designed and built, known as "Marty's Cabin Monoplane," powered with a Comet radial engine; this aircraft was similar in appearance to a Curtiss Robin.[6] The students were not "officially" allowed to flight test their aircraft, but the following account shows that there were "unofficial" flight tests going on at Cal Poly in those early years:

We adjourned to the next hangar where the (Curtiss) Junior sat. The gas tank and motor mount were still aboard, with the tach drive, gas line, mag wire, etc., hanging loose. We wondered about the center of gravity, but since the engine was only a little aft of the C.G. we figured that at worst it would tend to be a bit nose heavy. Clyde got in the cockpit and we took turns lifting the tail and estimating that it would be all right. We found two ropes, one 50 ft. long and the other 100 ft. Tied one end of the short rope to a landing gear strut, pulled it out to the front and back to tie it to the other gear strut. We then tied one end of the long rope to the base of the V and pulled the line out to the end of the runway [and tied it to our car]. Clyde advised that he would level off about three feet high and asked me to release the rope when he waved and for Gib to accelerate out of his way for landing. I knelt on the rumble seat facing aft, and held the rope with both hands. It was agreed that in order to avoid a jerk, Gib would start, and stay, in second gear. We got going slowly in second, but soon accelerated. Clyde got off nicely and leveled off while we proceeded down the field till the end approached. He waved, I dropped the rope and pounded on the side of the car; Gibby pulled ahead and off to one side while Clyde landed. 'Flies great,' Clyde said.[7]

There is a growing amount of evidence that this sort of flight testing was common at Cal Poly (including flight testing with

Figure 8. Students working on aircraft in the fabrication laboratory, C. 1930s (photo courtesy of University Archives, California Polytechnic State University

landings at local beaches), and many believe that the spirit of these early students still pervades the department today.

Figure 9. "Learn by Doing": Students taking apart radial engines in the propulsion laboratory (photo courtesy of University Archives, California Polytechnic State University).

The quality of work done by the early students was acknowledged by the government when the Department of Commerce (having created The Aeronautics Branch in 1926[2]) awarded the Cal Poly Aeronautics Department a license as "Approved Repair Station #84" in the early 1930s. The three year program gave students the necessary training and education to become federally licensed Aircraft and Engine Mechanics; examinations and tests were given by the Department of Commerce:[5]

Since the aeronautics shop of the California Polytechnic school is a government-approved repair station and the instructors are licensed mechanics, the student may obtain . . . a government license as an airplane mechanic or an airplane engine mechanic on completion of the course.[8]

Most students chose to receive both licenses, and were highly qualified to repair a wide variety of aircraft.

The department quickly grew and expanded its curriculum to include a variety of courses in what would eventually become a pre-engineering program. Facilities were constructed to house the department (see Fig. 7), including fabrication, propulsion, and aerodynamics laboratories (see Figs. 8-10). New faculty were added to teach courses, including Martin C. Martinson in 1930. In 1932 Roy L. Jones joined the faculty as aeronautical engines and drafting instructor.[9] Many students remember learning to fly from their instructors, an unofficial by-product of their education at Cal Poly. Indeed, a large number of students from this period went on to long careers as airline pilots, where they transitioned over the years from propeller aircraft to Boeing 747s by the time they retired.

By the mid 1930s, Julian McPhee had become president of Cal Poly, and the department's curriculum had expanded to include a variety of aeronautics subjects, as well as a fairly well-balanced general education. The stated goal of the department was to prepare graduates for work in the aeronautical industry, with the ability to, "be intelligent, courteous, reliable, a fast and logical thinker, and full of enthusiasm for . . . work."[10] The 1935 curriculum is reprinted here to show the variety of courses students took in that time period (the number in parentheses represents the numbers of hours the student spent in each discipline during the academic year):[10]

First Year
- Aircraft engines primary class and shop (180)
- Machine shop (180)
- Welding (180)
- Aero-drafting (96)
- Safety precautions (12)
- Physics (189)
- English (135)
- Mathematics (135)
- Physical education (108)

Second Year
- Aircraft engines advanced class and shop (300)
- Woodworking, aircraft (100)
- Aero construction (272)
- Machine shop (58)
- Aero drafting (189)
- Instruments (27)
- Department of Commerce regulations (13)
- Elements of aerodynamics (27)
- Fabric and sewing (27)
- Doping and painting (27)
- Airplane mechanics and rigging (81)
- Sand blasting (13.5)
- Physical education (108)

Third Year
- Parachutes (18)
- Testing materials (18)
- Engines (198)
- Aero drafting (171)
- Heat treating (18)
- Mechanics of the airplane (300)
- Strength of materials (27)
- Aerodynamics (27)
- Sheet metal for aircraft (87)
- Aircraft and shop maintenance (36)
- Cost estimating and repairs (27)
- Meteorology (27)
- Navigation (27)
- Radio installation and service (36)
- Instruments, testing and repairs (27)
- Electroplating (22)
- First Aid (5)
- Fire fighting (9)
- Department of Commerce regulations (27)
- Machine shop (54)
- Welding (54)

Figure 10. Early wind tunnel in the Aeronautics Building (photo courtesy of University Archives, California Polytechnic State University).

Figure 11. Amelia Earhart bringing her Lockheed Electra to Cal Poly for repairs in 1936, being greeted by the department head, M.C. Martinsen (photo courtesy of University Archives, California Polytechnic State University).

The curriculum of the time period shows that the students were being prepared for work as aircraft mechanics, but the special nature of their education often allowed the graduates to work as engineers in the aeronautical industry.

As an example of the quality and fame of the program, in 1936 famed aviatrix Amelia Earhart flew her Lockheed Electra to San Luis Obispo for repairs to be performed by Cal Poly students. Amelia arrived at Clark Field (landing short and clipping a farmer's fence), and was greeted by Aeronautics Department Head Martin C. Martinsen, as shown in Fig. 11. Earhart toured the Cal Poly Aeronautics facilities, and was photographed outside the Aero Lab with her flight instructor, Paul Mantz, and a number of faculty and students (Fig. 12).

Figure 12. Amelia Earhart visiting the Aeronautics Laboratory. From left to right: Martin C. Martinsen (Aeronautics Dept. Head), Paul Mantz ("King of the Hollywood Pilots" and Earhart's flight instructor), Phil Jensen, Amelia Earhart, and Harley Smith (photo courtesy of University Archives, California Polytechnic State University).

In 1937 the faculty was supplemented with the addition of Roy F. Metz, who came to Cal Poly with experience from his military service in World War I and work with both United and Pan American airlines.[11] Students continued constructing and repairing aircraft throughout this time period, with the types and complexity of aircraft expanding continuously. New methods of construction and new advances in aeronautics had greatly improved the capabilities of aircraft. Engine cowlings, stressed-skin construction, high performance engines, retractable landing gear, and high lift systems all began appearing regularly on airplanes, and Cal Poly students learned about all of these

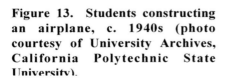

Figure 13. Students constructing an airplane, c. 1940s (photo courtesy of University Archives, California Polytechnic State University).

Figure 14. Ben Snow's sketch of John McKeller's Flying Wing, including pusher/tractor propeller configuration, retractable gear, and two seats.

advances with first-hand experience (see Fig. 13).[2]

One of the most interesting of all faculty and student designed aircraft of this period was John McKeller's flying wing. Mr. McKeller had designed and patented a control system for a flying wing aircraft, and in 1937-38 a low wing, twin engine aircraft in a pusher/tractor configuration was built (see Figs. 14 and 15). Unfortunately the flying wing was destroyed in a high-speed taxi test and never flew.[6]

Prior to World War II, the Cal Poly Aeronautics program cooperated with the Civil Aeronautics Authority in the nationwide Civil Pilot Training Program. Cal Poly aeronautics faculty taught 72 hours of ground school and organized 35 hours of flight training by local flight instructors at the San Luis Obispo airport. According to an article by M. Eugene Smith, campus historian of the time, by early 1942 Cal Poly had graduated 118 pilots and, "32 of these trainees had enlisted in the Army Air Corps, 14 in the Naval Air Corps, and one each in the Royal Air Force and Canadian Royal Air Force."[1]

Early in 1943, as the U.S. involvement in World War II began gearing up, the U.S. Navy designated the campus as a site for a Naval Flight Preparatory School (one of only twenty in the country). The U.S. Army Corps of Engineers built an airstrip on campus (see Fig. 16), and over 3,600 naval aviation cadets completed training during World War II. The Navy also designated Cal Poly as a "Fleet School" in Nov. 1943, which enabled an additional 840 Navy and Marine Corps enlisted personnel to come to Cal Poly for aviation training. In July 1944, Cal Poly was chosen as one of eight colleges to conduct the Naval Academic Refresher Unit, a new naval aviation training program. This program continued until 1946, training over 1,100 pilots.

These programs supplemented the Aeronautics Department faculty with fifteen additional instructors who had diverse experience in aviation and flying. Some of those instructors, like Leo

Figure 15. John McKeller's Flying Wing after construction.

Figure 16. Construction of the airstrip by the Army Corps of Engineers, c. 1940s (photo courtesy of University Archives, California Polytechnic State University).

F. Philbin, would stay at Cal Poly after the war and help to expand and improve the department.[11] After the war, a wave of practical-minded veterans, including faculty and former students, returned to Cal Poly using the G.I. Bill. This helped inject fresh vigor into the department's programs, with the curriculum, facilities, and enrollment expanding rapidly due to the existence of the flight school. Again, the practical side of aviation was at the forefront of Cal Poly's aeronautics education.

The Transition to an Accredited Engineering Program

By the 1950s the Aeronautics Department had grown significantly, both in size, faculty, facilities, and in changes to the curriculum, and changed its name to Aeronautical Engineering. After World War II, it had become apparent that Cal Poly needed to upgrade from primarily a "junior college" into a full-fledged university. For the Aeronautical Engineering Department, and the other engineering departments on campus, this meant transitioning into accredited engineering programs. Faculty were hired and facilities were upgraded to accomplish this transition. In addition to M.C. Martinsen and Roy F. Metz, the department had a total of eight faculty, with Lester W. Gustafson, who had worked at Lockheed and Hughes Aircraft Company, serving as department head. The primary concern of the faculty at the time, however, was how to retain the "hands on" flavor of the Cal Poly education while adding the more rigorous aspects of a science-based curriculum. A hangar was constructed adjacent to the airstrip (with an annex added later) that would accommodate most of the department's activities (see Fig. 17).

Figure 17. Cal Poly hangar and airstrip complex, 1970 (photo courtesy of University Archives, California Polytechnic State University).

During the 1950s the curriculum changed considerably as well. While the department had maintained many of the courses that enabled students to become aircraft mechanics (including drafting, machine shop, welding, woodwork, and hydraulic systems), a fourth year of study

Figure 18. Students working on an experimental glider in the Hangar, c. 1970 (photo courtesy of University Archives, California Polytechnic State University).

had been added with a variety of new coursework to enable the students to understand the science of aeronautics: calculus, physics, and a full year of aerodynamics.[12] By 1960 courses had also been added in fluid mechanics, stress analysis, chemistry, electrical engineering, gas dynamics, and rocket propulsion. Another crucial addition to the curriculum was Senior Thesis (later called Senior Project), which required students to perform a design and analysis of some aeronautical system, resulting in high quality work often worthy of a graduate thesis. Similar changes were also made at the Pomona campus that would enable their aeronautical engineering department to become a top undergraduate program as well.

These changes resulted in the new aeronautical engineering program being accredited by the Engineers' Council for Professional Development (the predecessor of the current Accreditation Board for Engineering and Technology) in 1969, along with the electrical, industrial, and mechanical engineering programs. In spite of these massive changes in the curriculum, however, the department maintained its earlier emphasis on construction and design of aircraft, something that would make Cal Poly fairly unique for the following decades, as other universities transitioned to purely science-based engineering programs. The benefits of the unique education at Cal Poly is evidenced by the unusually large number of graduates from this period that went on to outstanding careers in aerospace, including Burt Rutan (Scaled Composites), Robert "Hoot" Gibson (former NASA Chief Astronaut), Dean Borgman (President of Sikorsky), Jim Phillips (VP for Boeing's Long Beach Division), Bob Wulf (retired VP for Engineering and Technology at Northrop Grumman), Paul Martin (former VP at Lockheed's Skunk Works and now VP of Engineering at Sikorsky), and James Crowder (Boeing Senior Technical Fellow). Cal Poly also began its long-running success in AIAA design competitions during this period when Burt Rutan won the AIAA National Student Paper Competition for his aircraft design concepts in the mid 1960s. A great deal of the success from this time period can be attributed to the aircraft design professors of the time, including Clifford Price and Al Andreoli, who tried to instill in their students the importance of thinking "outside the box" and trying new, innovative concepts.

By the early 1970s the transition to an engineering program was complete, and the addition of faculty with more advanced degrees (M.S. and Ph.D degrees), coupled with industry experience, enabled Cal Poly to maintain the unusual balance that a "learn by doing" tradition required. Students continued designing and building airplanes, as shown in Fig. 18, and the aircraft design course, coupled with the Senior Project, enabled students to pursue "hands on" projects while also completing the engineering science coursework that goes along with an accredited engineering degree. Another vital aspect of the student's education was Cal Poly's encouragement of both co-operative education (alternating school with work in industry) and extra-curricular projects associated with student clubs. Both of these aspects continue today to help make Cal Poly's education full and enriching.

Evidence of Julian McPhee's famous "upside down curriculum" was also obvious during the transition phase of the department. McPhee had been a strong advocate for giving students coursework in their first year that related to their chosen field of study. Cal Poly's Aeronautical Engineering Department had always accomplished this by having students take introductory courses in drafting and basic aircraft engine concepts, but many of those courses had been replaced by a freshman series of courses in aerospace fundamentals that taught students about design, basic aeronautics, and laboratory experience. Aircraft design was now an integral part of the senior year, with students taking two quarters of coursework in the subject, as well as Senior Project, which almost always involved design aspects.

In the mid 1970s, John Nicolaides came to Cal Poly from Notre Dame University as department head. Nicolaides had been deeply involved in early missile and rocket development for the Navy, and had a lasting impact on the aerospace program at Notre Dame (see Ref. 14 for more details on his Notre Dame years). His years at Cal Poly were to see him spend a great deal of time involving students in experimentation and flight tests on his novel parafoil concepts. Nicolaides had developed the parafoil while at Notre Dame, but he was trying to find practical uses for the device other than just as a replacement for a parachute. Many of his ideas, while seeming quite strange at the time, have become commonplace today, including the use of parafoils on ultralights and for the recovery of RPVs and UAVs. Fig. 19 shows Nicolaides flying his early "ultralight" in the 1970s (a project that the author was involved in while an undergraduate at Cal Poly).

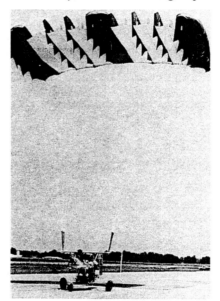

Figure 19. John Nicolaides landing his parafoil airplane—a predecessor to modern ultralights (from Ref. 15).

In the late 1970s the department expanded by adding a graduate program consisting originally of a Master of Engineering degree; eventually an M.S. degree in Aeronautical Engineering would be offered as well. To be consistent with Cal Poly's history and tradition, the graduate degree program required that students conduct research and publish a master's thesis, rather than having a course-only degree as had become popular at many other aeronautical programs. The goal of the program was to enhance both the graduate student's experience (by giving them higher-level tools to use in their professional experience) and to bring real-world research into the undergraduate classroom. In order to do this, Cal Poly began close ties with NASA's Dryden Flight Research Center and Ames Research Center for flight-related research projects, including free-flight vortex tracking, remotely-piloted vehicles, and free-wing/free-trimmer aircraft. Eventually, Cal Poly graduate students would perform cutting-edge research using flight tests, wind tunnel tests, and computational simulation for new and innovative design concepts. Many graduate programs began by making the claim of research aiding instruction, but Cal Poly continues to retain these goals as serious elements of their M.S. program.

Further evidence of the design capabilities of Cal Poly students was seen during the 1980s and 1990s when a large number of students worked on the DaVinci project over an eight year period of time. DaVinci was Cal Poly's attempt to win the Sikorsky Prize being offered by the American Helicopter Society to encourage the construction of human-powered helicopters. Under the guidance of faculty member William Patterson, students designed and constructed a total of three DaVinci concepts. All of these projects required a great deal of teamwork and coordination, especially since students were coming and going from the team over the years. Novel concepts in aerodynamics, composite construction, mechanics, and controls were required to build the successful helicopter. On December 10, 1989, the students attained the first hover for a human-powered helicopter (a total of 6.8 seconds off the ground), and became the only group of students in the world to hold a certified aviation record (see Fig. 20).[1,16]

Figure 20. Students flying DaVinci III in the Cal Poly Gym—in 1989 this aircraft set the National Aeronautics record for the first hover of a human-powered helicopter.

The Present and The Future

The Aeronautical Engineering Department was renamed the Aerospace Engineering Department in the late 1990s in response to a growing need for engineers with experience in space-related subjects. Many programs had renamed themselves "aerospace" in the 1960s and 1970s, but the change was often "in name" only. Cal Poly resisted such a change until the facilities and faculty were in place to offer students a true astronautics option for their education. In the short period of time since that change was made, Cal Poly has continued to apply its expertise in aircraft design to this new realm of endeavor.

The most important observation that can be made about the current department is that it embodies the modern day continuance of the "learn by doing" philosophy that was practiced when the department began in 1927. Yes, the drafting, welding, and hydraulics classes are gone, but Cal Poly students still learn about CAD, modern welding techniques, and state-of-the-art controls systems. Where a group of students once stood in a field, proud of their recently completed aircraft (see Fig. 5), modern-day students conduct launches of candidate designs for reusable launch vehicles (see Fig. 21). Where students once learned to fly airplanes, current students now develop design concepts at our Educational Flight Range, an on-campus airstrip for radio-controlled aircraft, as well as flight testing aircraft in our flight test course (one of only a handful of such courses in the country—see Ref. 17 for more details about university flight test programs).

Since the late 1980s, Cal Poly has brought its aircraft design curriculum farther than it has ever been; under the leadership of department head Doral R. Sandlin, Cal Poly greatly strengthened the aircraft design sequence. Participation in NASA's University Space Research Association (USRA) brought valuable insight into how to give senior-level students a capstone design experience while retaining the unique nature of our "hands on" approach to education.[18,19] The current aircraft design sequence lasts for the entire senior year, and requires students to perform a team-centered, detailed design of an aircraft, including most of the systems. In the mid 1990s, Cal Poly also participated in NASA's Aeronautics Multidisciplinary Design and Fellowship (AMDAF) program that aided in the development of computer-based tools that allow design students to perform analysis beyond the scope of their analytic background, including making detailed control systems for their aircraft and analyzing the handling qualities of their designs.[20]

Participation in these programs has proven a great help in making our aircraft design program one of the best in the country. Cal Poly student aircraft design teams have been participating in the AIAA Team Aircraft Design Competition for many years, and over the past twelve years, with multiple teams submitting proposals to satisfy industry-generated RFPs, we have been fortunate to win First Place five times, Second Place seven times, Third Place three times, and have two Honorable Mentions. In addition, students have participated and succeeded in other AIAA design competitions, including the Individual Aircraft, Team Engine, Design/Build/Fly, and Micro Aerial Vehicle (MAV) competitions. Design faculty with significant industrial experience, such as recent faculty members Robert van't Riet and Dave Hall, are keys to this level of design education.

In recent years Cal Poly has taken its role as an aircraft design center even farther by pioneering the teaching of aircraft design to sophomores.[21] We noticed that a great dilemma has long existed in engineering curricula: capstone design courses have been seen as just that—capstones. Students are required to complete large numbers of mathematics, science, engineering science, and program-dependent analysis courses before they "complete" their education with design. In aeronautics, this progression has included background instruction in aerodynamics, stability & control, propulsion systems, and structures, with a supporting cast of coursework in mechanics, strength of materials, electronics, materials science, and social sciences & humanities. But the skills and abilities that make students successful in their analysis courses often do not serve them well when they take design. In addition, many of the students at Cal Poly came to our program because of our history in aircraft design. These students were frustrated as they "waited" to get to the design sequence, often forgetting or neglecting the important material that was presented in their other courses. By offering aircraft design to sophomores using the "upside down" curriculum concept of Julian McPhee, we have put our student's education into context—the impact of this on future groups of students is still to be determined.

Another aspect in the continuation of the "hands on" philosophy is the continued use of laboratory courses to enhance education. While all engineering programs require students to conduct laboratory experiments, Cal Poly has always believed that laboratory experiences are extremely important. The current Aerospace Engineering Department has nearly ten laboratory courses (course made up almost entirely of laboratory experiences for the students), including freshman and sophomore design labs, as well as aerothermodynamics, wind tunnel testing, controls, structures, propulsion, and design labs. In addition, students take a number of other laboratory courses in their engineering science and science curricula.

Figure 21. Cal Poly launches StarBooster, a fully reusable first stage booster.

As the Aerospace Engineering Department celebrates its 75th year in existence, a celebration that corresponds with the university celebrating its Centennial, it is clear to see that the faculty and students, as well as the alumni, have a sense of who and what they are. Cal Poly does not struggle with questions of educational goals and outcomes, since our philosophy has always included a well-rounded, systems approach to solving aircraft design and analysis problems. The future is bright for our department as we continue to supply the aerospace industry with bright, young, aircraft and spacecraft design students capable of meeting the modern challenges of flight, both in the atmosphere and in space.

Conclusions

Cal Poly's Aerospace Engineering program cannot boast of faculty developing earth-shattering theories or conducting cutting-edge experiments, as many of the early aeronautics programs eventually were able to do. Instead, Cal Poly has kept in touch with its early roots and given students an excellent well-rounded, hands-on education with an emphasis on design that prepares them well for work in the aerospace industry. Our success is evidenced by the current aircraft projects that Cal Poly alumni have had high-level influence on: F-117, B-2, JSF, Boeing 717, Boeing 777, all current Boeing and Sikorsky helicopters, etc. I would hope that the students of 1927 could look at the current students and be proud—I equally hope that the current students can look back to the students of 1927 with respect and awe, and all can appreciate the benefits of the unique education that Julian McPhee instituted with his "learn by doing" philosophy, and which Myron Angel envisioned over one hundred years ago.

Acknowledgements

Whenever a large amount of historical information is gathered, a great debt is owed to the people at archives and libraries for helping to find the dates and pictures and quotes. I am indebted to Catherine Trujillo of the University Archives, and Dena Ross and Bobbi Binder of the Aerospace Engineering Department at Cal Poly for their valuable help. In addition, I want to thank the many alumni, faculty, and staff members of the Aerospace Engineering Department who have written accounts and descriptions of the department throughout the decades—this paper would not have been possible without their recollections.

References

1. *Cal Poly: The First Hundred Years*, California Polytechnic State University, San Luis Obispo, CA, 2001.
2. R. Bilstein, *Flight in America: From the Wrights to the Astronauts*, Johns Hopkins University Press, Rev. Ed., 1994.
3. B. W. McCormick, "The Growth of Aerospace Education Following its Beginning," AIAA Paper 2002-0560, Jan. 2002.
4. *The California Polytechnic Catalogue, 1928-1929*, California Polytechnic, San Luis Obispo, CA, 1928, p. 22.
5. *The California Polytechnic Catalogue, 1927-1928*, California Polytechnic, San Luis Obispo, CA, 1927, p. 7.
6. Dave Hoover, "More Alumni Memories," *Cal Poly Aeronautical Engineering Newsletter*, Spring 1996.
7. Roy Moungovan, "More Alumni Memories," *Cal Poly Aeronautical Engineering Newsletter*, Spring 1996.
8. *Bulletin of the California Polytechnic School, 1933-1934*, California Polytechnic, San Luis Obispo, CA, 1933, p. 15.

9. *Bulletin of the California Polytechnic School, 1936-1937*, California Polytechnic, San Luis Obispo, CA, 1936, p. 10.

10. *Bulletin of the California Polytechnic School, 1933-1934; Aeronautics Department Supplememt, 1935*, California Polytechnic, San Luis Obispo, CA, 1935.

11. *Circular of Information and Announcement of Courses, 1944-1945*, California State Polytechnic, San Luis Obispo, 1944, p. 16.

12. *California State Polytechnic College Bulletin, 1950-1951*, California State Polytechnic College, San Luis Obispo, CA, 1950, p. 129.

13. *California State Polytechnic College Bulletin, 1970-1971*, California State Polytechnic College, San Luis Obispo, CA, 1970, p. 119.

14. T. J. Mueller and R. C. Nelson, "Aeronautics to Aerospace at the University of Notre Dame," AIAA Paper 2002-0566, Jan. 2002.

15. J.D. Nicolaides, "Flight Test Results of a Powered Parafoil System," AFFDL TR-76-15, 1976, p. 31.

16. *World and United States Aviation and Space Records*, National Aeronautics Association of the USA, Arlington, VA, 1992.

17. J. Wolf and A. Sansone, "The U.S. Air Force Academy's Flight Test Course—Preparing Tomorrow's Flight Testers," AIAA Paper 2002-1048, Jan. 2002.

18. D. R. Sandlin and R. van't Riet, "The Cal Poly Aircraft Design Program," AIAA Paper 93-1111, Feb. 1993.

19. D. Soban, "Aircraft Design Education—A Student's Perspective," AIAA Paper 93-3993, Aug. 1993.

20. R. M. Cummings and H. J. A. Freeman, "Integrating Multidisciplinary Design in an Undergraduate Engineering Curriculum," *SAE Transactions*, Vol. 106, No.1, 1997, pp. 1665-1670.

21. R. M. Cummings and D. Hall, "Aircraft Design for Sophomores," AIAA Paper 2002-0958, Jan. 2002.

Chapter 68

Aeronautical Engineering at the Aero Industries Technical Institute, Inc.

Conrad F. Newberry, D. Env.

Introduction

Aero Industries Technical Institute (AITI) was located at 5245 (5261) West San Fernando Road, Los Angeles, California [1,2,3]. The Institute consisted of several buildings on a five-acre campus. The buildings essentially constituted a thoroughly modern aircraft production facility. Although the AITI promotional literature indicates a Los Angeles address for the School, the physical location of the Institute is in Glendale, California, just a few blocks from the Grand Central Air Terminal (GCAT), formerly the Glendale Airport. Some sources indicate that AITI was sponsored by three of the largest aircraft manufacturers: Lockheed, Northrop, and Consolidated Aircraft. Sources also indicate that the Institute was needed because IN THE 1930S there was an insufficient national supply of riveters, sheetmetal workers and aircraft mechanics for the aircraft industry [4].

The exigencies of the late 1930s promoted the expansion of the aircraft industry and a corresponding expansion of the U. S. Army Air Corps. The expanding Army Air Corps resulted in a corresponding need for more military aircraft mechanics. Military mechanics in sufficient number were needed to keep the growing military aircraft inventory available for combat operations.

AITI Programs

The initial AITI program started in Glendale (Los Angeles), California. It appears that at some time either AITI opened a branch campus in Oakland, California or moved the entire operation to Oakland.

Glendale

According to their 1937 Institute Brochure, AITI had only one Course - for Aircraft Mechanics. The Course had both a theoretical and a practical component. However, they had three ways to complete the Course: (1) a complete day Course, (2) a complete night Course for prospective students who had to work during the day, and (3) a combination Course whereby the student completed the theory portion of the Course at home and the practice portion in residence at the Institute. The Course consisted of some 800 contact hours of instruction, roughly five and one-half months of instruction, with a tuition fee of $ 410 (remember this was 1937). Instruction was highly efficient and effective. The curriculum was developed by a group of aviation industry leaders, engineers, and production personnel. Students were taught exactly what they would need to know to be effective in industry, on equipment identical to what they would be using in industry. The student could start the Course at any time (on any day). However, the Institute administrators preferred that a new student start on a Monday morning, if possible [1].

Course information could be obtained only from a bonded registrar of Vocational Services, Inc. The student age varied from roughly eighteen to forty years of age. Only American citizens of sound physical condition and mental alertness were considered for admission to the Institute. Prospective students were expected to be of good moral character. Also, prospective students were urged to get their applications in early because class sizes were limited. Each student received a copyrighted set of AITI course notes and lessons, the cost of which was covered by his tuition [1].

Classes started at 0830 hours. The lunch period was from 1200 to 1300 hours, and classes resumed from 1300 to 1630 hours. Examinations were given periodically. New lessons were initiated every week, approximately [1]. The instruction was considered to be quite rigorous [5].

Other Courses had been added to Institute offerings by 1940. By that time, AITI provided vocational training for craftsmen and mechanics as well as professional education for aeronautical engineers. In 1940, AITI provided prospective students with a choice of five Courses in Aircraft Mechanics and Aeronautical Engineering. One of the five Courses was a twelve-month Course in Aircraft Mechanics approved by the U. S. Civil Aeronautics Administration (CAA) [2,3].

Oakland.

At some point in time, AITI either opened a new campus in Oakland, California or the Institute moved to Oakland. In the September 1947 issue of Flying magazine AITI was advertised as being based at the Oakland Municipal Airport. Fall was said to be the time of heavy enrollment. Complete courses were given in Aeronautical Engineering and in Master Aviation Mechanics. At that time classes were scheduled from 0700 hours to 1300 hours. This left the afternoon open for those students who needed part-time employment [6,]. The Oakland campus of AITI had nine World War II surplus aircraft, including a P-47, P-51 and a B-24 [7].

Mr. O. D. McKenzie was the President of the Oakland AITI operation [6]. It seems likely that this was the same O. D. McKenzie who earlier had been the Registrar at the Curtis-Wright Technical Institute of Aeronautics [8].

AITI Graduates

Some 750 aircraft mechanics and aeronautical engineers graduated from AITI in 1939. They joined hundreds, if not thousands, of other AITI graduates in the pre-war aviation workforce [2]

Aero Industries Technical Institute Graduates	
Name	**AITI Course**
Heber Bradford	Aircraft Mechanic
Joseph F. Cozzens	Aircraft Mechanic
John Haldeman	Aircraft Mechanic
George Kozelisky	Aircraft Mechanic
Robert L. Masters	Aircraft Mechanic
Richard H. Miller	Aircraft Mechanic
Philip Shoemaker	Aeronautical Engineering
Earl Turner	Aircraft Mechanic

Table 1

Aero Industries Technical Institute Graduates

A number of AITI graduates have been identified during the preparation of this paper. Some of these graduates (Glendale operation) and the Course from which they graduated are listed in Table 1.

Advisory Council

The Aero Industries Technical Institute Advisory Council (Glendale) was a Who's Who of the aviation industry in 1937. There were eleven members of the Advisory Council: Messrs. D. L. Brown (President, United Aircraft Corporation), Donald W. Douglas (President and Chairman of the Board, Douglas Aircraft Company), Major Reuben H. Fleet (President, The Consolidation Aircraft Corporation), Jack Frye (President, Transcontinental & Western Air), Robert E. Gross (President, Lockheed Aircraft Corporation), J. L. Kindelberger (President, North American Aviation), John K. Northrop (President, The Northrop Corporation), J. E. Schaefer (President, The Stearman Aircraft Company), I. I. Sikorsky (President, Sikorsky Aircraft), C. A. Van Dusen (Vice-President and Works Manager, The Consolidated Aircraft Corporation), and E. E. Wilson (President, Chance Vought Aircraft). The Executive Board consisted of Messrs. Robert E. Gross, John K. Northrop, and C. A. Van Dusen.

AITI (Glendale) management, policies, and operations were directly supervised by the Executive Board, with the assistance of the Advisory Council members. The membership of both the Advisory Council and the Executive Board would lend support to the contention that AITI was sponsored by the aircraft industry in an attempt to provide a sufficient number of trained and educated mechanics and engineers for both the aircraft industry and the U. S. military.

Closure

The author does not know the exact starting and closing dates of the Aero Industries Technical Institute training and education operation. However, the Institute is believed to have at least graduated students in each of the years 1937 through 1947.

Acknowledgements

The author hereby acknowledges the assistance of Mr. William F. Chana and the San Diego Aerospace Museum personnel (Ms. Pamela S. Gay, Librarian and Mr. Allen Renga, Archivist) who were able to locate a number of documents related to the Aero Industries Technical Institute. Mr. Charles Wike and the staff at the Glendale Public Library have been generous in their assistance to the author in the preparation of this paper. Also, Mr. William T. Larkins provided significant assistance in locating information about AITI.

References.

1.   _____, Aero Industries Technical Institute, Inc. Brochure, 1937.

2.   _____, Aero Industries Technical Institute, Inc. Advertisement, Aviation, volume 39, number 2, February 1940, p. 123.

3.   _____, Aero Industries Technical Institute, Inc. Advertisement, Aviation, volume 40, number 9, September 1941, p. 169.

4.   _____, http://newdeal.feri.org./library/n98.htm, August 26, 2003.

5.   Masters, Robert L, Personal (telephone) Communication, September 8, 2003.

6.   _____, Aero Industries Technical Institute, Inc. Advertisement, Flying, volume 41, number 3, September 1947, p. 91.

7.   Larkins, William T., Personal (e-mail) Communication, September 11, 2003.

8.   _____, Curtiss-Wright Technical Institute of Aeronautics, Prospectus 1933-34, Los Angeles (Glendale), California, 1933, p. 2.

Chapter 69

Aeronautical Engineering at the Polytechnic College of Engineering

Conrad F. Newberry, D. Env.

Introduction

The Polytechnic College of Engineering was located in Oakland, California at the corner of Madison and Thirteenth Street and at the Oakland Airport. The Polytechnic College of Engineering (PCE) was founded in 1898 and became known for its practical, well-trained graduates [1]. It offered Bachelor of Science degrees in several fields of engineering, e.g., Electrical Engineering [2] and Aeronautical Engineering. Their graduates met the on-the-job requirements of industry.

Thirteenth and Madison is almost in the center of downtown Oakland, California. Oakland Airport is in a southwesterly direction from the center of the City. It is known that on October 17, 1941 the PCE School of Aeronautics was occupying Hanger 7 at the Oakland Airport [3]; the start and termination dates of the occupancy are not known to the author.

Programs

PCE had been graduating aeronautical engineers for some time prior to the fall of 1941. However, in the fall of 1941, PCE formally established their School of Aeronautics. At that time, the PCE School of Aeronautics offered seven aeronautical programs to prospective students: aeronautical engineering, aircraft drafting and design, aircraft and engine (A & E) mechanics, aircraft welding, aircraft instruments, production mechanics, and aircraft engines. In 1941, the PCE School of Aeronautics was one of only a few institutions in the nation to grant the Bachelor of Science degree in Aeronautical Engineering [1].

Paul R. Hill is said to have been a Professor of Aeronautics in the PCE School of Aeronautics [4].

Closure

Sometime after World War II, the Polytechnic College of Engineering is believed to have been absorbed into Oakland (California) City College, then part of the Oakland Public School System, now part of the Peralta Community College District. In September of 1947 Flying magazine published a list of some 280 aviation ground schools (including schools that had complete aeronautical engineering programs); the Polytechnic College of Engineering was not on the list [5,6].

Acknowledgments

The author acknowledges the grateful assistance of the Western Air Museum and Mr. William T. Larkins in obtaining information on the Polytechnic College of Engineering.

References

1. _____, Polytechnic College of Engineering Advertisement, Aviation, volume 40, number 9, September 1941, p. 147.

2. _____, http://www.energyplan.com/Resume.html, September 7, 2003.

3. Larkins, William T., Personal (e-mail) Communication, September 11, 2003.

4. _____, http://www.ufomig.com/unconventional_flying_objects.htm, September 9, 2003.

5. Larkins T., William, Personal (e-mail) Communication, September 11, 2003.

6. _____, "Here's Your Ground School," Flying, volume 41, number 3, September 1947, pp. 51-53.

Chapter 70

The Evolution of Aeronautical Engineering Education at the Curtiss-Wright Technical Institute of Aeronautics/Cal-Aero Technical Institute

Conrad F. Newberry, D. Env.

Introduction

The Curtiss-Wright Technical Institute of Aeronautics (CWTIA) was founded as the training division of the Curtiss-Wright Corporation [1]. The school was located at the $ 3,000,000 Curtiss-Wright Grand Central Air Terminal (the Glendale Municipal Airport until it became known as the Grand Central Air Terminal circa 1928), an international airport, located at 1226 Airway in Glendale, California. The Grand Central Air Terminal was dedicated in 1923 (when it was known as the Glendale Municipal Airport) but was not officially opened until February 22, 1929. At this time, the Grand Central Air Terminal was, perhaps, the most modern airport in southern California. Grand Central was also the general location of the genesis of several aircraft manufacturers such as the Airplane Development Company (Cord/Vultee) and the Hughes Aircraft Company. The Hughes H-1 (a media designation) was designed and built at or near

Figure 1
Grand Central Air Terminal, Glendale, California
(Courtesy of The Boeing Company)

Grand Central [2,3]. The Grand Central Air Terminal control tower, administration, and engineering building is shown in Fig. 1; the DC-1 prototype is shown in front of the building.

CWTIA classes were held on the second floor of the building. The drafting room has been described as being scientifically lighted with both skylights and large windows; one of the finest drafting rooms in the world. The décor has been described as pleasing; the room was completely equipped [1].

Major C. C. Moseley established the Institute in 1929 with the guidance and direction of the Curtiss-Wright Corporation. CWTIA students were able to observe the daily operations of two transcontinental airlines: TWA (The Lindbergh Line) and American Airlines (featuring sleeper plane service). In addition, Pan American Airlines operated flights from Grand Central to Mexico as well as Central and South America. CWTIA claimed that within a twenty-five mile radius of its facilities, more people were employed in aviation than in any other such community in the world [1,3]. Figure 2 shows an aerial view of the Grand Central Air Terminal.

Figure 2
An Aerial View of the Grand Central Air Terminal
(Courtesy of John Underwood, Ralph Johnson Collection)

Figure 2 is an aerial view of the Grand Central Air Terminal looking in a southwesterly direction. One can see the "turf" runway (the grassy area) to the right and pretty much at right angles to the paved runway. One can also see the bend of the Los Angeles river in the background; Griffith Park is seen beyond the river.

C. C. Moseley

Corliss C. Moseley was a well-known aviator in the 1920s and 1930s. Moseley had attended the University of Southern California. During the spring of his junior year, in 1917, he enlisted in the U. S. Army Air Corps. He served in France, spending two years as a pilot in the 27th Pursuit Squadron, 1st Pursuit Group. Other officers in the 1st Pursuit Group included Edward V. Rickenbacker (twenty-seven aerial victories), Frank Luke (eighteen aerial victories), Joseph Wehner, Raoul Lufbery ("Lufbery Circle," seventeen official aerial victories), and Quentin Roosevelt. In 1919 Moseley was made Commandant of Cadets at the Army Air Corps Primary Flying School [1,3,4].

The decade following World War I was the era of the "barnstormers." Military and civilian pilots vied for altitude, range, endurance, and speed records. In November of 1920, Lt. Moseley won the First International Pulitzer Race Trophy with a speed of 156.54 mph (a new American speed record) [4]. Subsequently, he was, for several years, a special Air Corps Aviation lecturer at the United States Military Academy. He served as a test pilot at the Army Air Corps Engineering Division at Dayton, Ohio. He was asked to organize the first Air Corps National Guard squadron in California. In 1925, Moseley resigned from the Army to organize (as Director and Vice-President of Operations) Western Air Express (Western Airlines), one of the early commercial carriers of airmail. In 1929, Moseley was selected to represent the Curtiss-Wright Corporation's interests on the Pacific Coast; in addition to his duties as President of the Curtiss-Wright Technical Institute of Aeronautics. Moseley served in this dual role until December 31, 1933 [1,3,4,5].

On January 1, 1934 Moseley obtained complete ownership and control of the Curtiss-Wright Technical Institute of Aeronautics, and concurrently leased the Grand Central Air Terminal from the Curtiss-Wright Corporation. He thereby controlled both the Institute and the Air Terminal. He purchased the Grand Central Air Terminal from the Curtiss-Wright Corporation in 1944. Hereafter, Institute will refer to either or both the Curtiss-Wright Technical Institute of Aeronautics and the Cal-Aero Technical Institute [1,3,5].

In June of 1944, with complete ownership of both the Institute and the Air Terminal, Moseley changed the name of the Institute from the Curtiss-Wright Technical Institute of Aeronautics to the Cal-Aero Technical Institute. However, the name change was largely administrative [5]. The Institute logo did not change. The Institute colors of Travelair Blue and Orange did not change. For some months Cal-Aero Institute students found the name Curtiss-Wright Technical Institute on pages of their Cal-Aero Institute course notes and manuals. The mission of the Institute did not change; however, by June of 1944 Moseley was totally in charge.

C. C. Moseley was endowed with the entrepreneurial spirit. Furthermore, the 1925-1945 era was one of great expansion and fluidity for the aviation industry. Aside from his varied and distinguished military career, Moseley, at one time or another, was Director and Vice President of Western Air Express Corporation (eventually TWA), Director of American Airlines, President of Aircraft Industries, President of Curtiss-Wright Technical Institute of Aeronautics, and President of Cal-Aero Technical Institute. During World War II he organized several contract flying schools (Cal-Aero Academy) for the U. S. Army Air Corps. In 1955 he organized the Grand Central Rocket Company (eventually the Lockheed Propulsion Company). Often he held several positions concurrently. He certainly cast a large shadow on the aviation community during the early part of its Golden Era.

CWTIA/CATI Mission

The mission of first the Curtiss-Wright Technical Institute of Aeronautics (CWTIA) and then the Cal-Aero Technical Institute (CATI) was to provide an individual student with the maximum practical training in aeronautics in the minimum time. This was consistent with the Institute's stated intent to maximize the student's return on his investment (his time and tuition). The intensive aeronautical engineering training course was math, physics, and engineering - void of all non-essential topics - no frills - no humanities or social sciences [1,3,6,7,8]. CWTIA promotional material suggested that graduating from their Master Mechanics course could increase a student's lifetime earnings (circa 1938) by some $46,000 (at that time the average lifetime earnings for a high school graduate was estimated to be $40, 000) [1,3].

Student Body

The CWTIA Registrar was, for many years (during the 1930s and early 1940s), O. D. McKenzie [1,3,9]. Typically, the students ranged from eighteen to thirty years of age, although men as young as sixteen were eligible (age wise) to enroll in the courses. The Institute was non-sectarian, however, students were expected to be of "good moral character and serious purpose;" they were expected to abide by Institute regulations. There were no rules governing student conduct; they were expected to do the "right thing" [1].

It seems that in the early days of the Institute, students had to secure their own room and board. Some type of dormitory arrangement seems to have been available during the Cal-Aero Era.

It does not appear that the Institute was initially co-educational. However, Ms. Dorothy M. King was accepted by CATI for the aeronautical engineering program in 1947. She was told that she would have to find her own living accommodations; marriage altered her plans and she did not attend CATI. Furthermore, there were a number of women in a group of Israeli Air Force students taking the Master Mechanic Course during the early 1950s [10].

The student body was diverse in the sense that it contained individuals from many states and foreign countries. One snapshot of the student body indicated that there were students from twenty-six states (there were forty-eight at the time) and six foreign countries; roughly half of the U. S. students were from the eastern half of the country. By 1935, it appears that the numerical size of the student body was in the neighborhood of 100 individuals. The faculty, staff, and student body circa 1935 are shown in Fig. 3.

Figure 3
1935 CWTIA Faculty, Staff, and Student Body
(Courtesy of John Underwood, Ralph Johnson Collection)

The exigencies of the late 1930 pre-World War II years conspired to increase the size of the CWTIA student body. For example, in 1940 CWTIA was the only West Coast school (there were only seven nationwide) to contract for the training of U. S. Army personnel as mechanics. These personnel started to arrive at CWTIA in August of 1939, approximately 35 students every two weeks. These Army personnel were trained side-by-side with civilian students. In 1940, it is estimated that CWTIA had a civilian enrollment in excess of 1000 students [11].

Throughout its lifetime, the Institute had a peak enrollment of some 1500 students. Figure 4 shows the CWTIA student body and faculty circa 1939.

A comparison of Figs. 3 and 4 suggests a marked increase in CWTIA enrollment from 1935 to 1939.

Figure 4
1939 CWTIA Faculty, Staff, and Student Body
(Courtesy of John Underwood, Ralph Johnson Collection)

Schedule

Classes were held five days a week, Monday through Friday. Class work started at 0800 hours and ended at 1630 hours. There were short breaks at 1030 and 1420 hours, with an hour for lunch. Time cards (and time clocks) were used to monitor student arrival and departure times. Tardiness and unacceptable absences had an adverse effect on a student's grades [1,3].

Instruction

The teaching paradigm was *Learn by Practice* - not by theory. This was a variation of the *Learn by Doing* paradigm employed in a number of other programs. The CWTIA instruction was individualized and personalized - the reason students could enter the Courses at any time. The drafting room was the "home room" for CWTIA students. Classes in mathematics, aerodynamics, and structures, for example, were taught at certain times and in specific cycles. The incoming student would start in the drafting room and hone his drafting skills while waiting for, say, the next mathematics class to be given [12].

Morning and afternoon breaks allowed students to relax a bit between learning sessions. Although the Instituted had no athletic teams, volleyball courts were available for use during breaks and after school hours. The Institute did not support organized sports. The eight-hour school day was intended to simulate the environment the student would encounter when he entered the aviation industry [1,3].

Classes were developed in such a way as to enable students to complete their studies in the classroom [1,4]. However, many, if not most, students completed many of their assignments in the evening or on weekends (homework!) [12,13].

886

None of the Institute Courses required student involvement in flying activities. However, if they were so inclined, Institute students in the late 1930s could avail themselves of the Grand Central Flying School (Plosserville!) operated by Mr. J. B. Plosser. The Grand Central Flying School (GCFS) was the only flying school permitted to operate from the Grand Central Air Terminal. The GCFS was approved by the Department of Commerce and was equipped with modern aircraft [1,3,5].

CWTIA/CATI Programs

In July of 1941, the Curtiss-Wright Technical Institute of Aeronautics was offering eight courses of study: (1) Aeronautical Engineering, (2) Post Graduate Aeronautical Engineering, (3) Master Aviation Mechanic, (4) Specialized Engine, (5) Specialized Airplane, (6) Specialized Aircraft Sheet Metal, (7) Aeronautical Drafting (home study), and (8) Aircraft Blue Print Reading (home study). It should be noted that even in the early 1950s, a number of aviation factory workers could not read aeronautical blueprints proficiently. Only the Aeronautical Engineering course will be discussed in detail herein. According to promotional material, CWTIA was aviation's most distinguished school of aeronautics. Anthony ("Tony") H. G. Fokker, Gerard (Jerry) F. Vultee, Walter L. Seiler, and Frank Hawks wrote testimonials for CWTIA [1,3,7,8]].

The Aeronautical Engineering and Master Aviation Mechanic Courses were considered to be the major career Courses. The Aircraft Sheet Metal and Post Graduate Aeronautical Engineering Courses were considered to be supplementary courses. The Aeronautical Drafting and the Aircraft Blueprint Reading Courses were strictly home study courses [1]. The CWTIA was a U. S. Army Air Corps contractor in 1939 [11].

With respect to competing educational or training programs, the Institute considered that its Courses had two major advantages: (1) students could enter the Courses at any time, and (2) graduation was distributed throughout the year. By being able to enter courses at any time, a student could graduate sooner and thereby enter the aviation work force sooner than would be the case otherwise. Distributed graduation resulted in a smaller number of individuals entering the aviation industry at any given time; as opposed, for example, to having all the graduates in one year enter the industry at one time. As a corollary, distributed graduation was seen to promote higher earnings for the aviation work force by avoiding a surplus of unemployed engineers [1]. The CWTIA Aeronautical Engineering Course tuition is estimated to have been approximately $1,500 (50 weeks) [12].

The Post Graduate Aeronautical Engineering Course was developed for individuals who had a degree in some field of engineering other than aeronautics. This Course consisted of studies in airplane production drafting, layout and assemblies, structural design, aerodynamic design, military and transport design, and shop work. The Course took the student through the layout and design of one complete airplane. The student completed major production drawings, stress analysis, and aerodynamic analysis of one complete airplane [3].

Institute students were involved in the design and construction of real airplanes. Several classes of CWTIA students were involved with the design and construction of the CWTIA Bunting; the aircraft flew for the first time circa 1938 [14]. Curtiss-Wright Technical Institute students designed and built the Curtiss-Wright COUPE; the COUPE was a two or three-place all-metal, low-wing, cantilever monoplane with a fixed landing gear (with pants!). The COUPE was produced by the Curtiss-Wright Airplane Company plant located in Robertson, Missouri [3]. Having won a race or two himself, Moseley felt that his students would be at the leading edge of aeronautical engineering, if they were involved with the design and construction of racing airplanes (remember this was the 1930s). Accordingly, CWTIA students were involved in the design and/or construction of the Burrows R-6, the Rider R-6 (*Eight Ball*), and the Crosby C6R3 and CR-4 [5].

Grading System

Examinations were thorough and frequent. As in any natural progression, the CWTIA program evolved with time. The 1938 and 1943 grading systems are contrasted in Table 1. Over time, grading became more explicit.

CWTIA Grading Systems			
Circa 1938		**Circa 1943**	
Above Average	Recommended	Excellent	96-100
Average	Passing	Very good	90-95
Below Average	Conditional	Average	80-89
Failure		Pass	70-79
An average grade was representative of average work in industry. CWTIA issued a special recommendation for above average students.		Failure	Below 70
		CWTIA did not recommend a student for employment, if his average grade was below 70	

Table 1
CWTIA Grading Systems

Aeronautical Engineering Course

The CWTIA aeronautical engineering syllabus was designed to produce a competent, effective aeronautical engineer in fifty forty-hour weeks (the basic course). The instructional format was essentially a trimester system. Each class was a trimester, some sixteen-plus weeks, in length.. The slow and/or somewhat unprepared might take up to two years to complete the course of study. Any topic or subject not directly relevant to aeronautical engineering was eliminated from the Course syllabus. Mathematics and mechanical drawing were considered to be the foundation of any program or course in engineering. Instruction in the Institute's Engineering Division was under the supervision of the Chief Engineer [1,3].

The well prepared student would have taken (in high school, junior college, or university) algebra, geometry (plane, solid, and descriptive), trigonometry, logarithms, vectors, statics, kinetics, mechanics, and properties of materials prior to entry in the Aeronautical Engineering Course. Otherwise, he would have had to complete this work at the Institute, and, thereby, taken longer than fifty weeks to graduate. A slow learner might take longer than fifty weeks to complete his studies, whereas a faster, more ambitious student might complete the syllabus in less than fifty weeks [1].

Subjects were taught concurrently with sufficient practice and review. The Institute faculty believed that their concept of interdisciplinary teaching kept the interest of the student at a peak level, thus, the intensity of learning was similar to that found in the normal work environment of the aviation industry. Each student conducted his studies with the interdisciplinary application of mathematics, drafting, design, and analysis of a typical airplane piece of work, which he later constructed as a part of his **practical** training [1].

Engineering Mathematics. The Institute syllabus in engineering mathematics included such topics as algebra, geometry (plane, solid, and descriptive), rules, equations, vectors, logarithms, properties of materials, mechanics, statics, and kinetics. Students entering with a good prerequisite knowledge in these subjects could proceed at a more rapid pace through the Aeronautical Engineering Course than could those without [1,3].

Airplane Drafting. Some of the topics included in the airplane drafting syllabus included orthographic projection, intersections, developments, aircraft standards, fits and tolerances, shop sketching of components, layout and detail drafting of a typical airplane, castings, assembly drawings, and material specifications. The student also learned to use special drafting and/or drawing instruments such as ship curves, the planimeter, and the beam compass [1,3].

Drafting room practice, shop limits, shop notes and shop terms were an integral part of the Course. The principles and conventions of mechanical drawing as applied to aeronautical practice served as an overlay of these topics [1,3].

Applied Aerodynamics. This segment of the Course considered aerodynamics as applied to aircraft design problems. Topics included properties of the standard atmosphere, aerodynamic coefficients, mean aerodynamic chord (mac), parasite drag of components, motion of the airplane, mechanics of flight, engine effects, dynamic loads, supercharging, and the comparison of model and full-scale data [1,3].

Pitch, roll, and yawing motions about their respective axes were part of the student's introduction to stability and control. Also, the important topics of downwash, tail-setting angle, the variation of lift and drag with angle of attack, and other stability and control parameters and topics were addressed [1,3].

Aircraft performance parameters such as thrust, torque, propellers, power required, power available, weight effects, altitude effects, propulsive efficiency, were considered in this portion of the syllabus. Other topics included variable camber flaps, spinning, flutter, and buffeting phenomena [1,3].

Manufacturing Economics. This part of the syllabus was concerned with industrial engineering principles and their impact upon aircraft production. Topics included factory organization, management, factory layout, general shop systems, planning and progress of manufacturing operations, tooling systems, inspection methods, and final assembly [1,3].

The student was introduced to the importance of production costs and sales upon aircraft design. Students visited local aircraft factories and laboratories to see the practical application of manufacturing economics. Vultee (ADC) had a manufacturing facility located at Grand Central in close proximity to the Institute [1,3].

Aircraft Structures. This part of the curriculum addressed the application of applied mechanics and strength of materials to the stress analysis of a typical aircraft. Critical load conditions, load factors, margins of safety, and factors of safety were discussed. Other topics that were considered included elements of beam theory; load, shear, moment, and deflection diagrams; allowable stresses; form factors; shape effects; structural weight data; and stress analysis [1,3].

The importance of framed (truss) and shell (monocoque) structures to aircraft design were part of the CWTIA syllabus. Bending, lateral buckling, column (short and long) buckling, and combined loads were included in this segment of the Course [1,3].

Airplane Design. This part of the student's syllabus included the principles of machine design and mechanisms applied to aircraft practice. Aircraft design topics included types of aircraft, types of construction, design specifications and functions, weight estimates, dimensions, areas, wheel loads and sizes, powerplant selection and arrangement, wing configuration, aspect ratio and planform. Parameters and topics related to stability and control included static balance (center of gravity, center of pressure, and center of resistance); control surface design, areas, and balance ratios; and dihedral angles [1,3].

The student was taught to make analytical and graphical performance estimates from both model and full-scale data. Design layouts, assemblies, and details of wings, fuselages, landing gear, control surfaces, control systems, and engine mounts were included in the design syllabus. Other topics included joints and connections utilizing bolts, pins, rivets, screws, nails, welding, brazing soldering, clamping, splicing, gluing, sewing (e.g., fabric surface coverings), doping, and covering. Also, aluminum alloy sheet metal design was addressed [1,3].

Practical Shop Work. This portion of the curriculum considered the fabrication of wood and metal parts from blueprints of an original design sponsored by the Institute. Woodworking activities involved wing assemblies, Pratt and Warren truss ribs and assembly jigs, solid and box-type spars, splicing, wingtip bends, covering, and doping. Metal working activities included plate fittings, cutting, bending, forming, strut and wire connections, splicing, wrapping, soldering, and sheet metal construction [1,3].

Epitome. An abridgement of the CWTIA aeronautical engineering course, equivalent semester units of the separate elements, and required/recommended textbooks is shown in Table 2 [1]. There were at least one or more specific courses in each of these curriculum areas.

Aeronautical Engineering Course Curriculum Areas		
Curriculum Area	**Equivalent Semester Units**	**Author/Textbook/Publisher (Estimated Cost)**
Engineering Mathematics	4	Palmer, *Practical Mathematics for Home Study* W. G. Smith, *Practical Descriptive Geometry*, McGraw-Hill Poorman, *Applied Mechanics*
Airplane Drafting	3	T. E. French, *Manual of Engineering Drawing*, McGraw-Hill [15] ($ 3.00) Anderson, *Airplane Drafting*
Applied Aerodynamics	4	Carter, *Simple Aerodynamics* ($ 4.50) W. S. Diehl, *Engineering Aerodynamics*, Ronald Press [16] ($7.00) K. D. Wood, *Technical Aerodynamics*
Manufacturing Economics	1	Kimball, *Principles of Industrial Organization*
Airplane Structures	8	Klemin, *Airplane Stress Analysis* ($ 7.00) Niles and Newell, *Airplane Structures* [17]
Airplane Design	15	Warner and Johnson, *Aviation Handbook* E. P. Warner, *Airplane Design Aerodynamics*, McGraw-Hill [18]
Practical Shop Work	8	Younger, *Airplane Construction and Repair*
Total	43	

Table 2
Elements of the Aeronautical Engineering Course

Students were required to purchase only those textbooks for which a price is indicated. The other textbooks were recommended and available to students in the classroom upon request. As an aside, the author paid $4.25 for his copy of *A Manual of Engineering Drawing for Students and Draftsmen* (Seventh Edition) by Thomas E. French in 1948; the first edition was published in 1911 [15]. The author's copy of *Engineering Aerodynamics* by Walter S. Diehl, Revised Edition [16], was purchased for the sum of $7.00 in 1956.

In terms of equivalent semester units, it will be noted that airplane design (fifteen units) is the largest element in the CWTIA Aeronautical Engineering Course, while aircraft structures (eight units) and shop work (eight units) are in a not too far distant tie for second. The practical character of the CWTIA Aeronautical Engineering Course is quite apparent.

Sometime after the name change from CWTIA to CATI in 1944, the Aeronautical Engineering Course expanded from 50 weeks (three trimesters) to two years (six trimesters). This expansion is believed to have occurred between 1944 and 1946. Table 3 compares the two Engineering Courses.

Table 3 indicates that, compared to the mature CWTIA program, the mature CATI program has roughly twice the mathematics content; more science is evident by the inclusion of the physics class. The aerodynamics content increased by almost a factor of six. Communication skills were enhanced by the addition of the report writing class. The stress analysis content was almost doubled, and the design content was more clearly defined, as was the shop content.

Comparison of Curtiss-Wright and Cal-Aero Curricula Overtime			
CWTIA (circa 1940)		**CATI (circa 1953)**	
Course Title	**Contact Hours**	**Course Title**	**Contact Hours**
First Year		**First Year**	
Drafting I	125	Elementary Drafting, Sheet Metal Parts	160
Drafting II	100	Aircraft Drafting, Machine Parts	160
Drafting III	100	Aircraft Drafting, Assemblies I	160
Drafting IV	100	Aircraft Drafting, Assemblies II	80
Drafting V	100	Aircraft Drafting, Descriptive Geometry	80
Descriptive Geometry	100	Design & Drafting, Machine Parts Layout	160
Algebra	65	Design & Drafting, Sheet Metal Layout	160
Trigonometry	65	Engineering Math I, Algebra	120
Calculus	65	Engineering Math II, Trigonometry	120
Engineering Procedures	65	Engineering Math III, Calculus	120
Mechanics I	65	Elementary Physics	40
Mechanics II	130	Mechanics I, Statics	160
Mechanics III	65	Elementary Stress Analysis	80
Materials & Processes	65	Basic Aerodynamics	120
Shop	240	Aerodynamics Laboratory I	40
Structures I	130	Engineering Shop I, Sheet Metal	80
Structures II	130	Engineering Shop II, Machines - Foundry	80
Structures III	65	Technical Report Writing	40
Lofting	100	Design Fundamentals	40
Aerodynamics	65		
Design	225		
		Second Year	
		Design & Drafting, Design Sketching	80
		Design & Drafting, Parts & Fittings I	80
		Design & Drafting, Parts & Fittings II	80
		Design & Drafting, Tool Design	80
		Design & Drafting, Stressed Skin Parts	160
		Design & Drafting, Mechanisms	160
		Design & Drafting, Airplane Design I	160
		Design & Drafting, Airplane Design II	160
		Mechanics II, Dynamics	160
		Stress Analysis II, Structural Parts	120
		Stress Analysis II, Laboratory	40
		Stress Analysis III, Primary Structures	160
		Stress Analysis IV, Primary Structures	160
		Advanced Aerodynamics	160
		Aerodynamics Laboratory II	80
		Performance Estimation	80
		Power Plants, Propulsion Methods	80
Total	2165	Total	4000

Table 3
A Comparison of the Mature CWTIA and CATI Aeronautical Engineering Programs

Master Aviation Mechanic Course

While there are many individuals who can fly airplanes, and many who can design and build them, it seems that there are never enough mechanics to keep existing aircraft available for flying. This appears to have been true generally since the first powered flight. In fact, one figure-of-merit associated with any new aircraft design is MMH/FH - maintenance man-hours per flight hour.

Table 4 illustrates the CWTIA coursework required to complete the Master Aviation Mechanic Course. The CWTIA Mechanics Division was under the supervision of the Chief Instructor. The Master Aviation Mechanic Course required forty weeks, five-days a week, eight-hours a day for completion [1,3].

Master Aviation Mechanics Course		
Component	Contact Hours	Component Length (Weeks)
Airplane Mechanics	440	11
Engine Mechanics	400	10
Metal Fabrication	440	11
Related Technical	80	2
Line Maintenance	80	2
Traffic and Weather	80	2
Navigation	20	0.5
Meteorology	20	0.5
Aircraft Instrumentation	20	0.5
U. S. Air Regulations	20	0.5
Total	1600	40

Table 4
CWTIA Master Aviation Mechanic Course Epitome

The Mechanics Course, as suggested in Table 4, included airplane mechanics, engine mechanics, propellers, blueprint reading, metal fabrication, welding, navigation, aircraft instrumentation, meteorology, airport traffic control, and Department of Commerce rules and regulations. The navigation, traffic and weather, meteorology, and U. S. air regulation segments of the Course were elective courses. The student could substitute additional time in one or more of the major Course components in lieu of one or more of these elective courses [1]. These 43 units represent almost a year and a half of regular college work.

College Equivalence

The CWTIA Aeronautical Engineering Course was considered to be the equivalent of forty-three semester units at a conventional college or university. These forty-three units were distributed over seven subject areas: engineering mathematics, applied aerodynamics, airplane structures (stress analysis), airplane drafting, manufacturing economics, airplane design, and practical shop work (wood, metal, assembly, and installation). These forty-three units were approved by the California Board of Education. These units were fully accepted by the Glendale and Pasadena Junior Colleges, and by the Santa Barbara State Teachers' College [1].

CWTIA diplomas were awarded to students who were regular in attendance, had displayed the proper attitude, completed the necessary work, passed the examinations in a satisfactory manner with a grade average of better than 70 (%), and were not under any financial obligation to the Institute. In addition, nearby Junior Colleges gave credit to their students who completed aviation electives taken at the CWTIA. The California State Department of Education, Division of Secondary Education, approved the CWTIA curricula in both the Aeronautical Engineering and Master Aviation Mechanic Courses. Aviation students living in the geographical areas served by Junior Colleges, e.g., Glendale and Pasadena Junior Colleges, could attend CWTIA classes and complete the requirements for an Associate of Arts degree. The Associate of Arts degree was awarded upon completion of the requirements summarized in Table 5 [1].

In the early 1940s, Mr. Williard I. Staples was the instructor in charge of Teacher Training. Also, he was the Machine and Auto Shop Instructor in Charge of Vocational Teachers' Examination in Aeronautics for the California State Board Education. CWTIA was approved by the California State Board of Education to give instruction in aeronautical courses to teachers [1].

Requirements for the Associate of Arts Degree, with Emphasis in Aeronautics	
I. Graduation from a standard high school curriculum	
II. The satisfactory completion of the following:	
Subject	**Semester Units**
English	6
Health	2
Physical Education	2
Political Science	2
Social Science	6
Science or Mathematics	6
Aeronautical Electives (CWTIA)	43
Total	67

Table 5
Requirements for the Associate of Arts Degree

CWTIA provided instruction for the aeronautical elective coursework. Table 5 indicates that sixty-seven semester units were required for the Associate of Arts degree [1]. A minimum of sixty semester units was required generally for the Associate of Arts degree; pre-engineering students often completed more than 60 units. As an aside, the author completed seventy-nine semester units (four semesters) in obtaining his Associate of Arts Degree; not uncommon for pre-engineering majors.

CWTIA Faculty

The first CWTIA Chief Engineer (Dean/Director of Engineering, circa 1929) was Mr. Gerard (Jerry) F. Vultee, former Chief Engineer at Lockheed Aircraft Company (and later head of Vultee Aircraft Company). Circa 1932, Mr. Vultee became the Chief Engineer of the Aircraft Development Company (ADC) at the request of Mr. Errett Lobban Cord. Subsequently, a number of CWTIA students became involved with the development of the Cord Transport, the V-1[5]. When Mr. Vultee left CWTIA for his new position at ADC, Mr. Stanley Harold Evans was made CWTIA Dean of Engineering [14]. In the late 1930s Mr. F. R. Shanley became CWTIA Chief Engineer (he later became Professor of Engineering at UCLA) [3]. The Chief Engineer in 1940 was Mr. Walter C. Clayton [12].

At one point in time (circa 1943), the CWTIA seems to have had a faculty of eleven members and an administrative staff of seven members. In this snapshot, the Engineering Division had, what appears to be a senior faculty of three under the direction of John George Miller, Director of Training [1]. Individual class instructors were often imported from industry and, in some particular cases, from university programs [12]. During the early 1940s, Donald W. Douglas, Jr. was a CWTIA instructor [19]. As with any school, the size of the faculty and staff depended, to some extent, upon the size of the student body.

Mr. Carleton E. Stryker became the Chief Engineer circa 1943. Mr. Herbert Rawdon and Mr. Howard Poyas assisted Mr. Stryker. All three had extensive experience in the aviation industry [1].

Before coming to the CWTIA, Mr. Stryker had studied engineering at the University of Michigan. He had worker for the Curtiss Airplane and Engine Company, Aircraft Engine Company, Northrop Aircraft Corporation, and the Airplane Development Company, a subsidiary of the Cord Corporation. Also, Mr. Stryker had served as the Chief Engineer of the Western College of Aeronautics. During World War I, he served with the U. S. Navy and as an instructor at Columbia University [1].

Mr. Herbert Rawdon supervised CWTIA instruction in stress analysis. He had studied engineering at Ohio Northern University, with graduate work at the University of Wichita and the University of Washington. Mr. Rawdon had worked for the Travel Air Manufacturing Company, Curtiss-Wright Airplane Company, Boeing Airplane Company, and the Lockheed Aircraft Corporation [1].

Mr. Howard Poyas supervised CWTIA instruction in drafting and elementary design. He had studied aeronautical engineering at the University of Alabama. He had held several positions at the Earl Aviation Corporation [1].

Following the expansion of the CATI Aeronautical Engineering Course from 50 weeks to two years, additional faculty members were required. Mr. John T. Lane taught in the Master Mechanic Course curriculum at CATI from 1949 to 1952 [20]. By 1954 Mr. A. J. Victor was the CATI Chief Engineer [13].

Institute Students

During the preparation of this paper a number of Institute graduates were located. The names of these individuals, and their Institute Course are shown in Table 6.

CWTIA/CATI Graduates		
Student Name	**Institute Name**	**Institute Course**
Donald E. Ayres	CATI	Master Aviation Mechanic
Robert W. Bemer	CWTIA	Aeronautical Engineering
Frederick "Crash" Blechman	CATI	Aeronautical Engineering
Donald W. Douglas, Jr.	CWTIA	Aeronautical Engineering
James M. Hunt	CWTIA	Aeronautical Engineering
B. Ralph Johnson	CWTIA	Aeronautical Engineering
Lawrence Korodi	CATI	Master Aircraft Mechanic
Harry A. Palmer	CATI	Master Aviation Mechanic
Ernest Prete, Jr.	CATI	Aeronautical Engineering
Herbert E. Torberg	CWTIA	Aeronautical Engineering

Table 6
Institute Graduates

Advisory Council

The CWTIA Advisory Council looked like a Who's Who in Aviation during the 1920s and 1930s. The sixteen-odd members of the Council included such individuals as Donald W. Douglas, E. L. Cord, John K. Northrop, Gerard ("Jerry") F. Vultee, Robert E. Gross, Albert Menasco, R H. Fleet, T. P. Wright, and Robert Porter [1].

Closure

One might say that the Institute suffered at least three traumatic changes or episodes during its existence. The first would have been the expansion of its program to accommodate the influx of military students prior to and during World War II. Some 1,200 Army Air Corps men were trained at CWTIA. The second major event was the name change from the Curtiss-Wright Technical Institute of Aeronautics to the Cal-Aero Technical Institute. The third event was the expansion of the Aeronautical Engineering Course from three trimesters to six trimesters in length.

The last of the military students graduated in August of 1952. Subsequently, the fortunes of the Institute declined. The student body eventually dropped from a maximum of some 1,500 students to less than 200. After losing money for three years running, the Institute closed its doors after its last trimester in 1954 [5].

During its operation, the Institute provided the American aviation industry, not just the component in southern California, with hundreds, if not thousands, of capable aeronautical engineers. It was a valuable national asset during World War II. The Institute, together with other technical schools, provided a significant portion of the non-degree aeronautical engineers who, in conjunction with the many degreed engineers supplied by numerous universities, fueled the growth of the aerospace industry during the second half of the twentieth century. In 1975, for example, a study of the composition of the American aerospace engineering work force indicated that, nation-wide, between 16.0 % and 42.5 % of the corporate designated aerospace engineers did not possess college or university degree [21]. It seems likely that a good portion of these non-degreed engineers were technical school graduates.

Acknowledgements

The author hereby acknowledges the assistance of Mr. William F. Chana and the San Diego Aerospace Museum personnel (Ms. Pamela S. Gay, Librarian and Mr. Allen Renga, Archivist) who were able to locate a number of documents related to the programs offered by the Curtiss-Wright Technical Institute of Aeronautics. Messrs. John W. Underwood, Donald W. Douglas, Jr., Herbert E. Torberg, Robert W. Bemer, Donald E. Ayres, Frederick Blechman, and Dan MacPherson, provided the author with much information and insight into the institution that was the Curtiss-Wright Technical Institute of Aeronautics/Cal-Aero Technical Institute. Mr. Charles Wike and the staff at the Glendale Public Library have been generous in their assistance to the author in the preparation of this paper.

References

1. _____, Curtiss-Wright Technical Institute of Aeronautics Brochure, (no date, believed to be) circa 1943.

2. Schoneberger, William A., *California Wings: A History of Aviation in the Golden State,* Windsor Publications, Inc., 1984, pp. 134, 170.

3. _____, Curtiss-Wright Technical Institute of Aeronautics Brochure, circa 1938.

4. Sunderman, James F., Major USAF, Editor, *Early Air Pioneers, 1862-1935,* Franklin Watts, Inc., New York, New York, 1961, pp. 109-151.

5. Underwood, John W., *Madcaps, Millionaires, and "Mose,"* Heritage Press, Glendale, California, 1984, pp. 40, 60, 68,76,79,108,132.

6. _____, CWTIA Advertisement, Aviation, volume 38, number 12, December, 1939.

7. _____, CWTIA Advertisement, Aviation, volume 40, number 7, July, 1941, p. 9.

8. _____, CWTIA Advertisement, Aviation, volume 40, number 8, August, 1941, p. 15.

9. _____, Curtiss-Wright Technical Institute of Aeronautics Prospectus 1933-34, Los Angeles (Glendale), California, 1933, p. 2.

10. Ayres, Donald E., Personal (e-mail) Communication, September 10, 2003.

11 _____, "Training Military Personnel at Curtiss-Wright Tech," Aero Digest, volume 37, number 1, July, 1940, pp.97, 98.

12. Torberg, Herbert E., Personal (telephone call, documents) Communications, September 4, 2003.

13.   Blechman, Frederick (Crash), Personal (telephone call, documents) Communications, September 4, 2003.

14.   Underwood, John W., Personal (letter) Communication, September 9, 2003.

15.   French, Thomas E., Revised by Charles J. Vierck, *A Manual of Engineering Drawing for Students and Draftsmen,* Seventh Edition, Mc-Graw Hill Book Company, Inc., New York, New York, 1911, 1918, 1924, 1929, 1935, 1941, 1947.

16.   Diehl, Walter S., *Engineering Aerodynamics,* Revised Edition, The Ronald Press Company, New York, New York, 1928, 1936.

17.   Niles, Alfred S. and Joseph S. Newell, *Airplane Structures,* Volumes I and II, Third Edition, John Wiley & Sons, Inc., New York, New York, 1929, 1938, 1943.

18.   Warner, Edward P., *Airplane Design Aerodynamics,* Mc-Graw Hill Book Company, Inc., New York, New York, 1927.

19.   Douglas, Donald W. Jr., Personal (telephone) Communication, August 28, 2003.

20.   Lane, John T., Personal (telephone) Communication, September 11, 2003.

21.   Cornish, Joseph J. III, "AE: Man in the Middle," Astronautics & Aeronautics, volume 13, number 6, June 1975, pp. 45-49.

Chapter 71

AEROSPACE ENGINEERING AT NORTHROP UNIVERSITY

Paul A. Lord

The following is a brief chronological history of the Aerospace Engineering program at Northrop University. This program was somewhat unique in that it began as a department within an aircraft company, Northrop Aircraft, and later became a fully accredited engineering program.

The Aerospace engineering program at Northrop University (formerly Northrop Institute of Technology) was designed to prepare graduates for immediate productivity in industry. In order to accomplish this objective the curriculum was rich in engineering application and design content. Even though the primary objective was to prepare people for employment in industry it also graduated students who were well prepared to pursue graduate degrees. Many of the alumni attended highly recognized universities such as the Universities of California and the California Institute of Technology where they earned graduate degrees. Many interesting senior projects were carried out. One of the most notable was the design and construction of a man-powered vehicle named "White Lightning" which was the first vehicle of its type to attain a speed in excess of 55 miles per hour.

Because the graduates were well prepared for employment the department enjoyed a very fine reputation with the aerospace firms that hired the graduates. Because of this, the department had very strong ties with the industrial firms that supported the program.

There was considerable turmoil at the university starting in the mid to late 1970s ending with the closure of the BS programs in 1991. Since the college-level programs no longer exist, and no archives are available, it has been difficult to obtain reliable information on enrollment and the dates of some events.

I, and those who have kindly assisted in obtaining the necessary information (especially Mrs. K. L. Strite, widow of Dean Strite, and Robert E. Poteet, a former faculty member), believe the following information to be correct but if former faculty members or students would like to make corrections or additions please feel free to contact me by e-mail at palord@csupomona.edu.

Chronological History

1942 – The school started as a technical training program during WW II to train people for Northrop Aircraft. It was identified as Department 95. Initially the program was open to only military personnel. John K. Northrop and James L. McKinley established the department. The program originally was named "Northrop Aeronautical Institute."

1945 – New buildings were built for the program at Northrop Field, which later was renamed Hawthorne Airfield. At that time the programs were a two-year curriculum in Aeronautical Engineering Technology and a 50-week program in Airframe and Powerplant Maintenance.

1946 – The program was opened to civilian students. The graduates of the training program were much sought after by Northrop and other aircraft companies. The program was opened to its first class of tuition-paying students on March 4.

1947 – The Aeronautical Engineering Technology program was accredited by the Engineers' Council for Professional Development, ECPD, which later became the Accreditation Board for Engineering and Technology, ABET.

1953 – The school separated from the aircraft company and combined with the California Flyers School of Aeronautics and became a proprietary school to train people for engineering positions. It was a two-year program that lacked the humanities and social science content for a BS program. It was founded by John K. Northrop and James L. McKinley. The school moved from the Northrop Company facilities to a campus located on Arbor Vitae St. in Inglewood, CA.

1957 – Kenneth L. Strite became the Dean of Engineering replacing Harry R. Filson. Dean Strite served until retirement in 1973. He was a member of the faculty for 27 years.

1958 –The first BS degree in Aeronautical Engineering was established and the school became a non-profit college. The Superintendent of Public Instruction of the State of California authorized the school to confer the Bachelor of Science degree.

1959 – The name of the college was changed to Northrop Institute of Technology.

1960 –NIT was accredited by the Western Association of Schools and Colleges as a specialized institution to grant the Bachelor of Science degree. The U.S. department of Education noted that NIT was the largest private undergraduate aerospace engineering school west of the Mississippi.

1961 – Herbert W. Hartley became president of the college. James L. McKinley, founding president, became Chairman of the Board of Trustees and Chairman of the Executive Committee.

1962 – The school became non-profit and started a four-year BS program in Aeronautical and Astronautical Engineering. The school allowed students to enter at the beginning of any of the four quarters and graduated students at the end of each quarter; thus it was possible for a student to complete the BS degree in three years.

1960s – The Aerospace Engineering Department established a very active student branch of the American Institute of Aeronautics and Astronautics and several times was recognized as the outstanding student

Figure 1: McKinley residence hall.

branch by the Los Angeles Section of AIAA. The J. L. McKinley residence hall, shown in Figure 1, was constructed.

1963 – Paul A. Lord joined the faculty. He later became Chairman of the Aerospace Engineering Department and Associate Dean of Engineering.

1969 – Established a Masters of Science degree in Aeronautical Engineering.

1970 – Opened the new engineering and library buildings. These buildings are shown if Figures 2 and 3. The engineering building contained classrooms, offices and laboratories for fluid mechanics, energy conversion, design & fabrication, thermodynamics, simulation & control, metallurgy, electronics and computers. In addition, an aerodynamics laboratory containing an open-return subsonic wind tunnel with a 3X3 foot test section and a supersonic tunnel with a 3X3 inch test section was developed in a adjacent building. This laboratory was supported by a National Science Foundation grant. Figure 4 shows the students who designed and helped to build the subsonic wind tunnel and Professor Paul A. Lord, the faculty advisor.

Figure 2: Northrop engineering building.

Figure 3: Northrop Alumni Library

Figure 4: Wind tunnel
design team with two proposals

1970s – The school developed programs for students who had received the two-year degree so they could obtain the necessary humanities and social science courses and thus be awarded the BS degree. The Sigma Gamma Tau honor society for Aerospace Engineering was chartered in the department.

1972- ECPD (ABET) accreditation of the BS in Aeronautical Engineering was received. B. J. Shell became the university president.

1975 – The name was changed to Northrop University to acknowledge the addition of business and law schools. The school continued to receive strong support from alumni and industry.

1976 – Eighty-five aerospace engineering students were enrolled.

1978 – This was the year the White lightning won the man-powered vehicle competition with a speed in excess of 55 mph. See Figure 5.

1979 – Aerospace Engineering faculty members were P. A. Lord (chairman), J. M. Allman, K. Kao, A. J. Kuprenas, R. E. Poteet and T. Sugimura.

1975 to 1991 – During this time period there were so many deans of the engineering program that it is difficult to reconstruct who was in charge of the engineering programs as well as the aerospace program at any particular time after 1980.

The Los Angeles Times reported on September 12, 1989 that the school was in danger of loosing its accreditation by the Western Association of Schools and Colleges and that the university President, B. J. Shell had abruptly resigned two weeks previously.

The Los Angeles Times, September 23, 1989, reported that Northrop University "was struggling against allegations of financial and ethical improprieties." The problems were not related to the engineering programs, but were related to an international business program created for foreign students. The article also states that the Western Association of Schools and Colleges was considering revoking the university's accreditation.

1980 –Paul A. Lord resigned after 17 years on the faculty to accept a faculty position at California State Polytechnic University, Pomona.

1991 – The May 10th edition of the Los Angeles Times reported that due to financial trouble and low enrollment the school would discontinue the degree programs in June.

This concludes this short, and sometimes turbulent, history of Aerospace Engineering education at Northrop University. The author of this report wishes to take this opportunity to say hello to all of the Northrop University graduates whom he came to know and admire during his tenure and to wish them continued good fortune.

1980 Aerospace Engineering Curriculum

AEROSPACE ENGINEERING

P. A. Lord, Chairman
J. M. Allman
K. Kao
A. J. Kuprenas
R. E. Poteet

The field of aerospace engineering is developing into a broader field than ever before. It has always been directed toward the analysis and design of vehicles, but usually restricted to airborne or space vehicles. Because of changes in national and regional priorities, the aerospace engineer of today is moving into the more general area of transportation vehicles. In addition to aircraft and spacecraft this includes high speed ground vehicles such as trains, automobiles and ground effect machines and water vehicles, both the submerged and surface types.

Interest in wind as an energy source is a natural area for aerospace engineers to apply their knowledge of aerodynamics and propeller design.

The Aerospace Engineering department of Northrop University is aware of these trends and has modified its courses and curriculum to better prepare its graduates to participate in the activities necessary to develop new and improved transportation and energy systems. The Master of Science degree program in Aerospace Engineering is described in the graduate section of this bulletin.

A minimum of 197 quarter units of credit are required for completion of the Bachelor of Science degree program in Aerospace Engineering.

BACHELOR OF SCIENCE IN AEROSPACE ENGINEERING

An ECPD Accredited Engineering Curriculum

Recommended Undergraduate Program for a Bachelor of Science in Aerospace Engineering — A sample undergraduate program, including all of the current requirements for a Bachelor of Science in

Aerospace Engineering, is printed below. This program has been designed as a logical sequence of course work to give students an idea of what to expect in their academic careers and to guide them in scheduling their programs. It is for information and advisory purposes only. Class schedules are published prior to the beginning of each quarter. Students are advised to see their advisor to schedule course and programs.

FRESHMAN YEAR

1st Quarter

		Qtr. Hrs. Credit
GM 120	Calculus and Analytic geometry I	4
GS 112	General Chemistry I	4
GE 106	English Composition and Literature	3
E 105	Introduction to Engineering	3
	Humanities Elective	3
		17

2nd Quarter

GM 121	Calculus and Analytic Geometry II	4
GS 113	General Chemistry II	4
GS 119	Physics For Scientists and Engineering I	4
GH 112	United States History	4
		16

3rd Quarter

GM 122	Calculus and Analytic Geometry III	4
GS 220	Physics For Scientists and Engineers II	4
GE 107	English Composition and Literature II	3
GE 202	Oral Communication	3
E 110	Engineering Drafting and Design	3
		17

SOPHOMORE YEAR

1st Quarter

		Qtr. Hrs. Credit
GM 200	Calculus and Analytic Geometry IV	4
GS 221	Physics For Scientists and Engineers III	4
ME 215	Statics	4
	Humanities Elective	3
E 225	Engineering Materials	2
		17

2nd Quarter

GM 260	Ordinary Differential Equations	4
ME 216	Dynamics	4
GS 222	Physics For Scientists and Engineers IV	4
CS 120	Fortran	3
E 235	Processes of Materials	2
		17

		Qtr. Hrs. Credit
GM 261	Advanced Engineering Mathematics I	4
ME 340	Fluid Mechanics I	4
ME 301	Thermodynamics I	4
	Humanities Elective	3
CE 200	Writing for Industry and Business	2
		17

JUNIOR YEAR

1st Quarter

		Qtr. Hrs. Credit
AE 323	Aerodynamics I	4
	Humanities Elective	3
GM 262	Advanced Engineering Mathematics II	4
ME 356	Mechanics of Solids I	4
ME 341	Fluid Mechanics Laboratory	1
		16

2nd Quarter

AE 424	Aerodynamics II	4
ME 302	Heat Transfer I	4
ME 311	Engineering Metallurgy	4
EE 300A	Introduction to Electronic Engineering	4
		16

3rd Quarter

ME 303	Thermodynamics II	4
AE 360	Aircraft Structures I	4
AE 463	Control Systems Analysis	4
AE 493	Flight Dynamics	4
ME 370	Mechanics of Solids Laboratory	1
		17

SENIOR YEAR

1st Quarter

		Qtr. Hrs. Credit
	Technical Elective	3
AE 425	Gas Dynamics	4
AE 428	Aerodynamics Laboratory I	1
AE 461	Aircraft Structures II	4
	Humanities Elective	3
E 400	Design Procedures	1
		16

2nd Quarter

E 483	Engineering Economy	4
	Technical Elective	3
	Humanities Elective	3
AE 457	Powerplants	4
AE 444	Vehicle Design I	3
		17

3rd Quarter

AE 454	Vehicle Design II	3
	Technical Electives	7
	Humanities Elective	3
AE 429	Aerodynamics Laboratory II	1
		14